Advanced Analytics in Mining Engineering

Ali Soofastaei
Editor

Advanced Analytics in Mining Engineering

Leverage Advanced Analytics in Mining Industry to Make Better Business Decisions

Springer

Editor
Ali Soofastaei
Artificial Intelligence Center
Vale, WA, Australia

ISBN 978-3-030-91588-9 ISBN 978-3-030-91589-6 (eBook)
https://doi.org/10.1007/978-3-030-91589-6

This Springer imprint is published by the registered company Springer Nature Switzerland AG
The registered company address is: Gewerbestrasse 11, 6330 Cham, Switzerland

Preface

As an author on business and advanced analytics management, I have many ideas in my mind. I think all of them are great, of course, but it is always challenging to know which theoretical concepts can reach a practical result in advance.

After writing some papers and chapter books about advanced analytics and the mining industry's machine learning application, I found much demand for speaking and consulting on the subject. In that work, I talked with hundreds of managers and analytical professionals in countries worldwide. I also worked with many professors in different universities in America, Europe, Asia, and Australia to extend my knowledge of applied analytics in the mining industry.

Moreover, I have worked as an AI program leader with technology developers such as IBM, Accenture, Deloitte, and Oracle to develop suitable products for prestigious mining companies based on AI. These applications now play a critical role in predicting, optimizing, and making decisions for operation and maintenance in mining companies such as BHP, Rio Tinto, Vale, Angelo America, and Peabody Energy. After many years working in this area, I decided to write a comprehensive book to guide the researchers and industrial managers to find the analytical opportunities better and making the best decision to deploy the new science in their work.

In front of the research and development group in mining companies, there are some barriers to use practical advanced analytical approaches to solve their business problems.

The first barrier is the lack of bright and trained people who need to design innovative analytical solutions for the problem. There are two different groups of graduates who are looking for job positions in mining companies. The first group is mining engineers who do not have any data analysis experience. The second group is the IT and computer engineers who do not have any mining background. Therefore, each mentioned group cannot provide the mining companies' requirements individually. The digital mines need people familiar with the mining operations and have enough knowledge and experience to use the data analytical approaches.

The second barrier is the lack of valuable collected data to develop advanced analytical solutions. In the last decades, many new companies and start-ups have been established to make and use different tools for data collections in mine sites. However, there is no validated guideline to help the mine managers collect the necessary and correct data from equipment and process. As a result, a massive amount of noisy data is collected from mine site equipment, and the main part is not useable.

The third barrier is developed specific analytical applications to solve the unique business challenge. The mining operations are linked together, and any change in any particular process can dramatically affect the upstream and downstream activities. The main part of developed analytical tools for the mine sites focused on a specific operation. However, we need the use the integrated approached to minimize the harmful side effects overall.

The fourth barrier is the maturity level of analytics in the mining industry. The traditional mine managers' mindsets need to be changed. In the digital mine era, we should predict and optimize instead of scene and response. AI and machine learning models can help us predict failures and avoid them, and the optimization models will support the management decisions.

The advanced analytics for mining engineering book has been designed to tackle the barriers mentioned above. The book can be used as a reference book to teach at universities, and students can use it as a reference in their research. The book covers the students and research requirements to get familiar with the analytical approaches in mining engineering. This book also can help the technology developers and companies to identify the essential parameters in the mine sites and providing suitable tools to collect valuable data for the mining operations. The book chapters have been designed based on the mining value chain operations, and there is a logical connection between the chapters to help the readers make integrated solutions. Many practical examples are designed for the chapters that help mine managers get familiar with the benefits and limitations of advanced analytics in future digital mines. The prediction, optimization, and decision-making tools introduced in the book can give a clear vision of the future of mining to managers and researchers in the mining industry.

I believe that we are at the beginning of an exciting journey to apply advanced analytics, AI, and machine learning approaches to solve the mining companies' challenges. Digital mines will be developed, and we need to support the young generation who will be the future digital revolution leaders in the mining industry. This book aims to share the knowledge and experience of authors who have worked in the analytics field in mining as executives, managers, specialists, and researchers.

I hope this book can help the people who dream of making future mining safer, more creative, and more productive.

Brisbane, Australia
July 2021

Dr. Ali Soofastaei

Acknowledgements

I am sincerely grateful for all the universities, research centers, technology developers, and companies that shared their analytical successes and frustrations while completing this book—some of the professors and industrial managers in mining who were working with me during this research journey.

Several analytical vendors have also helped this practical research around the book topic. The managers from Microsoft, IBM, Accenture, Oracle, Deloitte, and SAS kindly supported the project from beginning to end.

I also need to thank all professors and researchers who helped me enhance the quality of the presented materials in this book. Special thanks to the University of Queensland, Pennsylvania State University, University of California Berkeley, John Hopkins University, Arizona State University, University of Colorado, and the Queensland University of Technology.

This project could not be completed without the help of large mining companies and equipment providers. I acknowledge the support of my friends from Rio Tinto, BHP, Vale, Angelo American, Coal India, Caterpillar, Komatsu, Hitachi, and Liebherr.

In terms of individual thanks, I would like to thank all authors who have worked with me closely to complete a comprehensive book for the mining industry. In this way, I am grateful for the help of Milad Fooladgar, a buddy that I could not finish the project successfully without his technical and management support.

About This Book

Most mining companies have a massive amount of data at their disposal. However, they cannot use the stored data in any meaningful way. The powerful new business tool-advanced analytics enables many mining companies to aggressively leverage their data in key business decisions and processes with impressive results.

In this book, Dr. Soofastaei and his colleagues reveal how all mining managers can effectively deploy advanced analytics in their day-to-day operations—one business decision at a time.

From statistical analysis to machine learning and artificial intelligence, the authors show how many analytical tools can improve decisions about everything in the mine value chain, from exploration to marketing.

Combining the science of advanced analytics with the mining industrial business solutions, introduce the "Advanced Analytics in Mining Book" as a practical road map and tools for unleashing the potential buried in your company's data.

Contents

About the Editor

Dr. Ali Soofastaei is an artificial intelligence (AI) scientist and an industrial global project leader. He leads innovative industrial projects in AI applications to improve safety, productivity, and energy efficiency and reduce maintenance costs.

He holds Bachelor of Engineering in Mechanical Engineering and has an in-depth understanding of energy management (EM) and equipment maintenance solutions (EMS). In addition, the extensive research he conducted on AI and value engineering methods while completing his Master of Engineering also provided him with expertise in applying advanced analytics in EM and EMS.

He completed his Ph.D. at The University of Queensland in AI applications in mining engineering. He led a revolution in using deep learning and AI methods to increase energy efficiency, reduce operation and maintenance costs, and reduce greenhouse gas emissions in surface mines. As Postdoctoral Research Fellow, he has provided practical guidance to undergraduate and postgraduate students in mechanical and mining engineering and information technology.

For more information, visit www.soofastaei-publications.com

Chapter 1
Advanced Analytics for Mining Industry

Ali Soofastaei

Abstract Digital transformation (DT) quickly and fundamentally modifies corporate entities or organizations because of digitalization development. This development requires a progressive assessment of the technology used for strategy modification, value chain, management, and business models with important repercussions for customers, business, employees, and the public. As a result, companies launch DT initiatives to examine customer needs and create operational models that exploit new competitive opportunities. In this context, customer value proposals and the operation model's reconfiguration from the main solutions to manage changes in the digital age. In this regard, companies have been launching their DT initiatives to upgrade their operations in industries in which the product is primarily a commodity, such as the mining industry. In addition, they modify their operating model to reflect users' preferences and expectations in every activity within the value chain. This approach requires integrating business activities and optimizing the management and monitoring of data associated with each value chain's essential activity. Although there are numerous possibilities for future development, the present level of digital mining transformation is low. Then, a question arises: How should DT initiatives to enhance mining companies' operational models be launched and implemented effectively? This chapter discusses the mining industry's importance, focusing on the sector's main issues to answer this question. Then, through research, the DT initiative's key aspects are presented to improve the operational model of the large, diversified mining company; challenges and success factors in a given context are identified and classified. Moreover, the research efforts and obtained results that focus on the role and importance of the DT phenomenon in mining to use digital technologies more widely and efficiently are discussed. Mining companies shift their strategy to include new technologies and adopt new business and operating models; they have done so more quickly and globally than ever before. The combination of market volatility, changing global demand, radically different input economies, the expansion of mining operations for locating additional reserves, and commitment to operational excellence contribute to a seismic shift in the industry. Decades

A. Soofastaei (✉)
Vale, Australia
e-mail: ali@soofastaei.net
URL: https://www.soofastaei.net

A. Soofastaei (ed.), *Advanced Analytics in Mining Engineering*,
https://doi.org/10.1007/978-3-030-91589-6_1

of reducing costs and an aging workforce have led to reduced resource adaptation within mining companies. DT, a rapidly changing set of new technologies, opens new opportunities to improve business efficiency, build accurate and agile planning, increase provider awareness, and cooperate with business partners throughout the value chain. DT can lead to significant differentiation and competitive advantage in the mining industry. This industry has been disrupted by mining automation, new analysis capabilities, a digital workforce, and remote and autonomous operations. All of these factors must be closely examined to boost growth and efficiency. DT and its associated opportunities and risks are crucial to mining businesses. DT has been succeeded by a more interconnected and information-based operation of human interactions. The next wave of industry differentiation is created by the possibilities of new operating models and new optimization levels. This chapter explains the varying levels of acceptance of DT in the mining industry.

Keywords Digital transformation · Advanced analytics · Mining · Prediction · Optimization · Decision-making

Introduction

By using advanced analytics, businesses can improve their decision-making and take informed actions, which, in turn, would boost their performance. Mine managers have relied for too long on their intuition or "golden gut." That is, essential decisions were established not based on correct data but on the decision-makers' experience and unassisted assessment. More than 40% of mining firms' significant decision-making is on fact and managers' gut feelings [1]. Intuitive and experiential decisions sometimes work well but often end in disaster [2]. It is no coincidence that companies cited as excellent analytical competitors often perform well. Analytics are a viable path in most industries and complement other methods that can lead to organizational success [3].

Benefits of an Analytical Approach

An analytical approach can

- help manage and steer the business in turbulent times—analytics give managers tools to understand their business dynamics, including how economic and marketplace shifts may influence business performance and
- help managers understand what is working.

Rigorous analytical testing can establish whether an intervention is triggering desired changes in the business or whether changes are simply the result of random statistical fluctuations; companies can use the analysis results to

- leverage previous information technology (IT) and information investments to obtain greater insight, faster execution, and more business value in many business processes and
- cut costs and improve efficiency.

Optimization techniques can minimize asset requirements, and predictive models can anticipate market shifts and enable companies to move quickly to reduce costs and eliminate waste to

- manage risk—greater regulatory oversight will require more precise metrics and risk-management models;
- anticipate changes in market conditions—companies can detect patterns in the vast amount of customer and market data to which they have access; and
- have a basis upon which to improve decisions over time.

If individuals use clear logic and detailed supporting data to inform decisions, they, or someone else, can examine and improve the decision process more efficiently.

What Does "Advanced Analytics" Mean?

Advanced analytics mean enhanced analysis, sophisticated data, and systematic reasoning to make the best possible decisions for the company. Its use involves answering the following questions:

- What kind of analysis?
- What types of data?
- What kind of reasoning?

There are no hard and fast answers; we contend that almost any advanced analytics process can be effective based on systematic approaches. Many methods of analysis are available, from the latest intelligence optimization techniques to tried-and-true root-cause analysis types. Perhaps the most common method is statistical analysis, in which data are used to make inferences about a population from a sample. Variations of statistical analysis can be used for a massive variety of decisions—from identifying whether a past event resulted from intervention to predicting likely events. Statistical analysis can be robust, but it is often complex and sometimes employs untenable assumptions about the data and business environment.

When the tasks are completed correctly, statistical analysis can be both simple and valuable. Specialists can examine a series of points on a two-dimensional graph, for example, and notice a pattern or relationship. Are there outliers in a pattern that require explanation? Are some values out of range? Visual analysis helps us stay close to data using exploratory data analysis. The key is always to become more analytical and fact-based in specialists' decision-making approaches and use the appropriate analysis level to make decisions.

Like environmental sustainability and energy efficiency, some mining business areas have not historically applied data or analysis. The specialists could become more analytical by creating simple metrics of critical activities, reporting them regularly, and acting on the emerging patterns. This initial step would accomplish a great deal; however, convincing an organization to agree upon metrics in a new area is no simple task. In other areas, such as maintenance, there are much detailed data available from SAP, workshops, and procurement, which will enable the operators to make effective decisions about providing equipment components (parts) to reduce maintenance costs.

Where Do Analytics Apply?

Analytics can help transform just about any part of the business or organization. Many organizations begin where they make their money—they aim to transform customer relationships. They use analytics to segment their customers and identify their best buyers. They analyze data to understand customer behavior, predict customer wants and needs, and offer suitable products and promotions. They ensure that the money and resources devoted to marketing focus on the most effective campaigns and channels. They price products for maximum profitability at levels that they know their customers will pay. In addition, they identify the customers that are most likely to switch to a competitor and intervene to retain them.

Supply chain and operations is also an area in which analytics are commonly employed. For example, the most influential supply chain organizations optimize inventory levels and delivery routes. They also segment their inventory items by cost and velocity, build their facilities in the best locations, and ensure that the right products are available in the right quantities. If they are in service businesses, they measure and fine-tune service operations. Further, human resources, a traditionally intuitive domain, increasingly use analytics to hire and retain employees. Just as sport teams analyze data to select and retain the best players, firms use analytical criteria to select valuable employees and identify those most likely to depart. They also determine which operational employees should work at which time to maximize sales and profits.

Not surprisingly, analytics can also be applied to the most numerical business areas: finance and accounting. Instead of simply placing financial and non-financial metrics on scorecards, leading firms use analytics to determine the genuinely drive financial performance factors. Moreover, in this era of instability, financial organizations and other firms use analytics to monitor and reduce risk. Thus, despite the recent problems in the investment industry, the serious business of investment cannot proceed without analytics.

Indeed, the use of analytics will continue to grow businesses, nonprofits, and governments, in part because the volume and variety of available data will continue to grow. As more processes are automated and instrumented with sensors, the only way to control them efficiently is to analyze the vast volumes of data they produce.

For example, at present, intelligent grids use analytics to optimize and reduce energy usage for sustainability. In the future, we may live on a "smart planet," with the ability to monitor and analyze all aspects of the environment. It is already an analytical world, and it will only become more so.

What Kinds of Questions Can Analytics Answer?

Every organization needs to answer some fundamental questions about its business. Taking an analytical approach begins with anticipating how the information will address common questions (see Table 1.1).

These questions are organized across two dimensions:

1. Time frame—are we examining the past, present, or future?
2. Innovation—are we working with the known formation or gaining new insight?

The presented matrix in Table 1.1 identifies the six critical questions that data and analytics can address in organizations. The first set of challenges is using information more effectively. For example, the "past" information cell is the realm of traditional business reporting rather than analytics. By applying the rules of thumb, the alerts can be generated about the present what is happening right now (e.g., whenever an activity is outside its typical performance pattern). In addition, using simple extrapolation of past patterns creates information about the future, such as forecasts. These questions are helpful to answer, but they do not clarify why something happens or how likely it is to recur.

The second set of questions requires different tools to dig deeper and produce new insights. Insight into the past is gained by statistical modeling activities that explain how and why things occur. Insight into the present takes the form of recommendations about what to do right now—for example, which additional product offering might interest a customer. Last, insight into the future is derived from prediction, optimization, and simulation techniques to create the best possible future results.

Together, these questions encompass much of what an organization needs to know about itself. The matrix can also be used to challenge existing uses of information. It may be found, for example, that many of the business intelligence activities are in the top row. Moving from purely information-oriented questions to those involving

Table 1.1 Key questions addressed by analytics

	Past	Present	Future
Information	What happened? (Reporting)	What is happening now? (Alerts)	What will happen? (Extrapolation)
Insight	How and why did it happen? (Modeling, experimental design)	What is the next best action? (Recommendation)	What is the best/worst that can happen? (Prediction, optimization, and simulation)

insights is likely to give a much deeper understanding of the dynamics of the business operations.

Digital technologies, such as cellular, cloud, big data, advanced analytics, sensors, Internet of Things (IoT), and artificial intelligence (AI), have an exponential impact on business and society, together with opportunities to combine on an innovating platform [4]. As a result of digitalization, DT can be defined as a "business model driven by changes in the application of digital technology in every aspect of human society" through the process of digitalization [5].

DT includes integrating digital technologies in all business fields, transforming how an organization operates, and providing efficient and thorough customer value. Companies have implemented the DT initiative to change their working, learning, communication, and working relationship to

- more easily identify the customer-related values; and
- restructure business models based on new possibilities—how these models are delivered to improve their competitiveness [6–9].

Although DT can be achieved in various scenarios with the help of different digital technologies, it affects corporate transformation rather than the technology infrastructure itself, changing business processes and business models, which significantly affects the corporate culture and employees [10–13].

Companies radically reconsider using digital technology to improve customer experience, operating processes, models, and business strategies to respond successfully to market changes [9, 12, 14–19]. Concentrating on one of the above areas and implementing special initiatives, specific industries find ways to implement the DT initiative quicker than others [20]. According to its strategic objectives, the industrial environment, competitive pressures, and customer expectations, every business must find the best way to implement DT individually. Further, individual industries and their contributions to their national economies are unique in different countries. Therefore, DT has other properties, and digital strategies have unique objectives in different parts of the world and individual industries [6, 21].

DT generally begins with initiatives to enhance operations: internal processes in primary industries that provide physical products, semi-products, or raw materials, such as mining. The operating model adapts to customer preferences and needs to affect every activity in the value chain. Therefore, all business activities must be integrated into the value chain, and these activities should be monitored and managed optimally.

The mining industry is expected to adopt large-scale digitalization in the future. Here, the application and connection of devices and items containing electronics, sensors, and software is expected to have particular importance for mutual communication and data exchange via the Internet or using alternative methods [22]. Further, digital processing opportunities to improve quality control, maximize productivity, and improve working conditions and product quality are highlighted among the various potential benefits of digital mining transformation [23]. However, all this requires avoiding traps, such as resource shortages, lack of knowledge, cooperation, awareness, and the underestimation of digitalization [24]. Conversely, the non-selective application of new technologies, which led to increased complexity and decreased productivity, is a paradox in productivity [25].

The extant literature indicates that the current state of DT in the mining industry is potentially significant. Therefore, the question arises as to how DT initiatives can be effectively implemented in mining companies. This section seeks to understand better the implementation of the mining DT initiative and identify challenges and factors of success within the given contexts, considering the observed lack of DT research in the mining industry.

Discussions on digital operations occur in most industries, including analytical issues, cloud, mobility, and the IoT. The mining industry is no exception, and mining executors wonder whether the so-called DT can positively affect their organizations from top to bottom. In the present industry context, such discussions are given special attention to a combination of market volatility, changing global demand, radically different input economies, the expansion of mining operations for locating additional reserves, an emphasis on a longer asset life cycle, and a commitment to operational excellence, and policy changes worldwide all contribute to a seismic shift.

Decades of cost reduction and an aging labor force have led to limited resources for mining companies. However, a fast-changing set of new technologies now appears to offer new opportunities to improve operational effectiveness, develop more precise and flexible planning, increase vendor awareness, and collaborate in the entire value chain with business partners. However, many uncertainties regarding DT effectiveness and ways to achieve DT have accompanied the expectations about DT. More importantly, few mining businesses have been able to enhance efficiency using DT.

In this context, managers have three questions that appear reasonable and are hence worth examining in this study:

- What is the meaning of DT in mining?
- How can DT affect mining companies' operations through the mine value chain?
- How can mining enterprises obtain the total value of DT?

Digital Transformation in Mining

DT is firmly associated with the increased speed at which individuals and companies receive more information and improve business performance. A straightforward relationship may be established between digital and new methods of generating and consuming information. IoT will enhance the quantity of information accessible. Data distribution and consumption will be facilitated by mobility. Social media and online collaboration will enhance the exchange of knowledge. The use of high-volume real-time data in analytics facilitates decision-making on complex issues. The cloud will make the management and storage variability required by making this information more agile and scalable. DT is occurring in the mining industry, in which highly complex digital solutions have been implemented.

DT is occurring in the mining industry, in which highly complex digital solutions have been implemented, but the rate of DT adoption is insufficient.

Big Data Solution for Digital Transformation in Mining

In the mining industry, big data in advanced analytics has resulted in a considerable revolution. For example, the International Mining Journal recently surveyed 10 of the 20 top mining companies globally. It revealed that big data analytics would stimulate the next wave of efficiency gains in ore extraction, analysis, transport, and processing, allowing for faster and more informed decision levels [26].

Every effort is required in a competitive market to enhance margins through operational intelligence. Thus, analysis has a significant role in increasing the use of assets, boosting productivity, and addressing delays in material flow. Sensors embedded in mining operations can help achieve these goals. These sensors generate vast quantities of geoscience, asset, and operating data in real time. Wi-Fi and 3G/4G-LTE speed improvements make it possible to collect data in real time, from the point of extraction to the ore's final transfer to plants. These data can be analyzed through massive parallel processing, and analysis results can be distributed rapidly to stakeholders. This is possible because modern big data platforms can assimilate vast amounts of heterogeneous inputs from multiple sources in real time. In turn, these predictive and prescriptive analytics, extracted in real time, drive business excellence.

Data sources may be classified as direct or indirect (independent) in the mining industry. For example, instruments such as conventional geodetic surveys and global positioning satellites are direct sources that provide measurements on natural systems. Indirect sources refer to systems that collect data, such as fleet management systems, SCADA, and DCS data, as a by-product of processes or transactions.

An ore-body modeling technique is used to improve ore recovery. Geological patterns are provided for the model to determine drill holes. Therefore, accurate data from multiple systems combined with real time (or near real time) analysis are essential to make the right mining decisions (see Fig. 1.1). These decisions can be used for exploitation, manufacturing, and operations in mining. Data can also be used for monitoring and reporting measurements and KPIs. Additional reasons for operational bottlenecks encountered, such as unplanned delays in the maintenance of trucks, long lorries, LHDs, laboratory samples undergoing quality management, and batch processing, are also identified.

In addition to providing decision-making insights, the big data analytics platform can offer prescriptive decisions. For instance, the correlation of causal variable patterns, such as loading time, transport time, dumping time, idling period, and time to the average total cycle time, would affect the output rate. In addition, the use of historical and real-time data analysis to verify arrangements can help forecast output (see Fig. 1.2).

The flow of materials plays a significant role in the mining value chain. Therefore, the impact of uncharged events caused by mechanical LHDs, trucks, critical transport

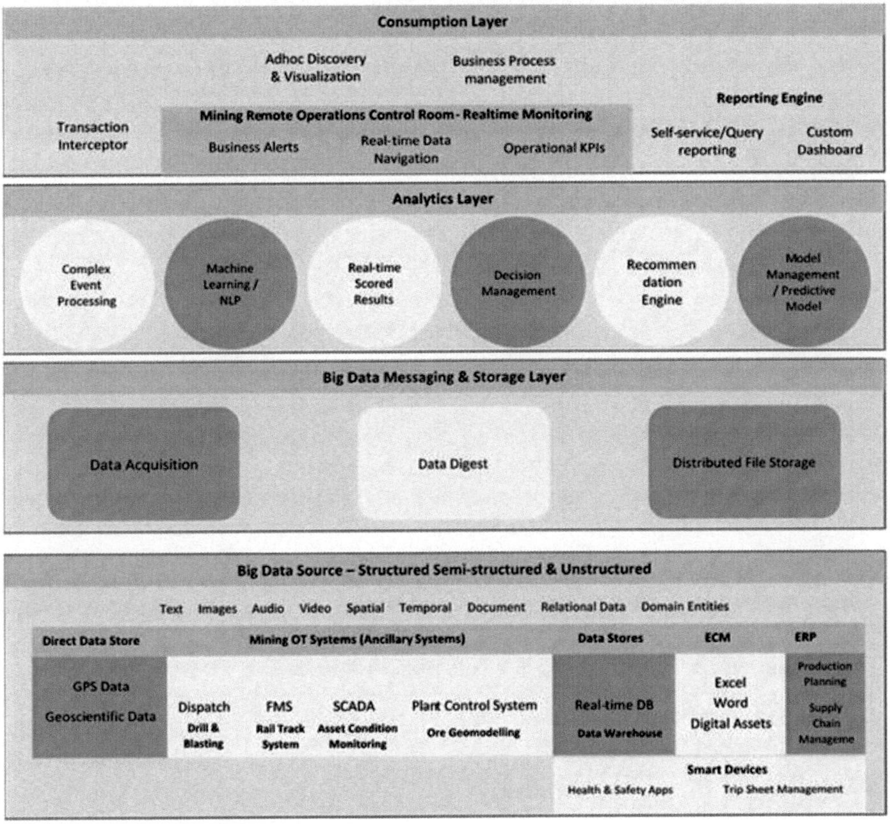

Fig. 1.1 Big data advanced analytics resolution structure

Fig. 1.2 Big data advanced analytics on mine material movement

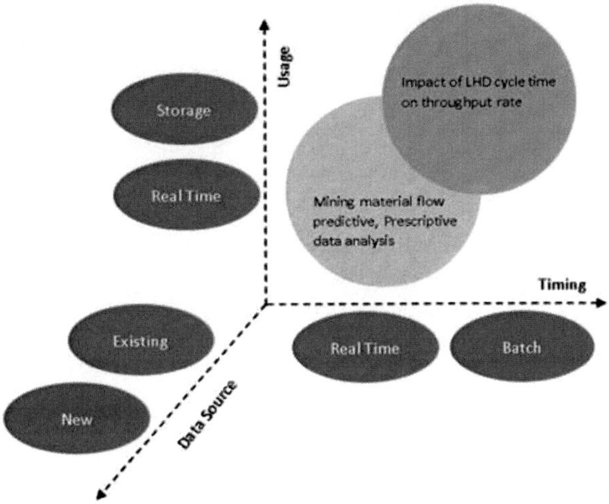

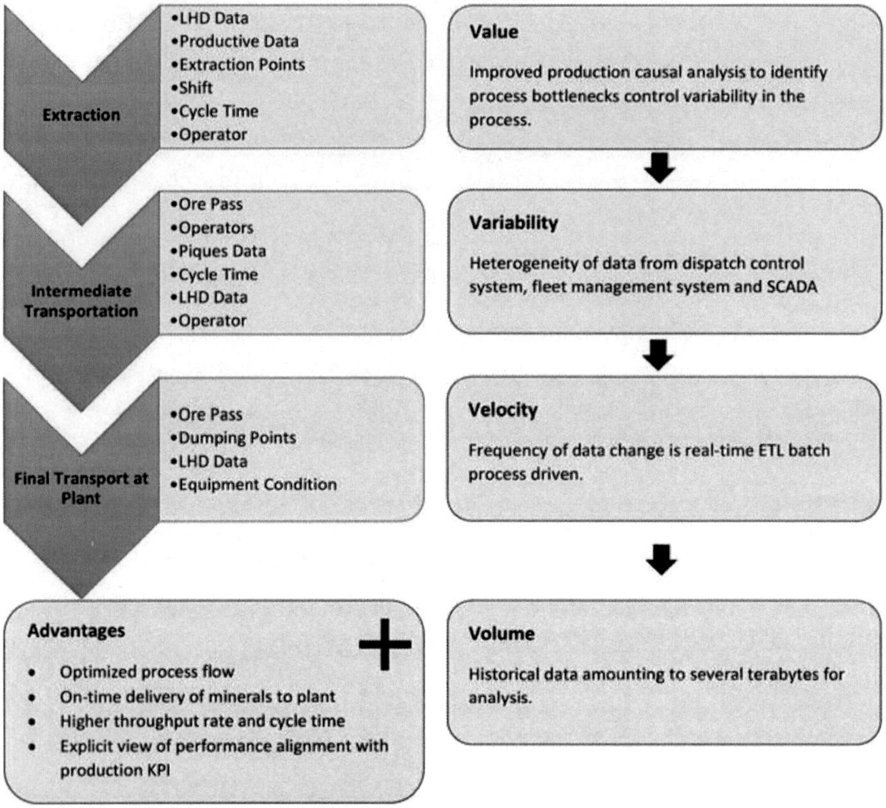

Fig. 1.3 Causal and correlation advanced analysis using business intelligence

means, queues, and overheads is analyzed. This includes an analysis of the results. In addition, several other causal variables that affect production throughput can be analyzed daily/monthly through methods such as machine learning, continuously matching patterns, and statistical predictive models. The big data analytics platform equipped with these models offers multi-value levels for extraction, midway transport, and final plant transportation by leveraging information value, volume, speed, and variability. The causal data used in each process step to improve operational efficiency and increase ore yield are shown in Fig. 1.3.

The Value of Digital Transformation

DT's inherent value derives from how companies can do things differently. Simply from the concepts already presented, we can develop

- efficient use of fast-growing data to make decisions in real time and for insight and continuous optimization;
- an unprecedented autonomous operation, remote asset management, equipment, and human tracking possibilities;
- increased cooperation and knowledge-sharing among the organization's operational teams and beyond; and
- solutions to new business demands that are much more flexible and responsive.

In turn, this new approach can have practical effects on a range of processes, such as those for

- expected and situation-based asset management;
- improved mine fleet management;
- decreased operational uncertainty;
- management techniques; and
- environmental sustainability, energy efficiency, and safety controls.

Improving management practices can yield specific results in terms of business value, including

- greater throughput and productivity;
- enhanced quality of results;
- decreased operational expenses;
- increased safety;
- decreased environment risks; and
- enhanced working investment.

How to Obtain the Entire Digital Transformation Benefit

The approach through which mining enterprises manage their DT programs can be understood from the technology they use, the individuals who drive this process, and the methods they use for this purpose. Companies that seek to implement digital solutions must have a suitable technological environment and the willingness to use the required technologies. The new requirements of DT have challenged both the technology environment and the structure of organizations. Among these complexities, there appear to be three cornerstones that companies must address to turn digital goals into outcomes:

1. **Integration**—DT's purpose is to capture the value of increasing business information only if it can be easily accessed by different systems, sites, processes, and business units.
2. **Multi-paced**—DT offers an innovative approach to gain the desired results and an opportunity to engage in rapid experiments and increase understanding while accelerating outcomes and identifying potential failure. It also results in innovation and agile development. This novel strategy is not a substitute for the traditional approach, which uses longer development cycles to develop essential

and corporate solutions. Instead, it will coexist with the traditional methods and technologies.

3. **Resilience**—Currently, supporting operations' technological environments is increasingly integrated with IT environments, and new well-defined boundaries are being established using mobile solutions and the cloud. This integration results in security threats to which operational teams are not accustomed. Companies will not consistently develop digital solutions without solutions to these threats.

It is essential to admit that mining firms, in general, are dependent on several technology environments and organizations that meet different business needs, almost independently of each other, and lack the elements of integration, multi-speed, and resilience. One of these many organizations is the IT group, which already operates most corporate systems and infrastructure and is closer to the business sector.

The various operating technology (OT) organizations responsible for automation and business systems are also installed at different units, operations, and locations (e.g., supervisory systems and manufacturing systems). Thus, these OT organizations can achieve increasing third-party participation in the technology ecosystems of companies, including externalizers, cloud suppliers, and suppliers of "smart" industrial products.

Digital Transformation Challenges

DT is a complicated enterprise with considerable challenges, as many writers have stated in various industries and organizations. From this perspective, the vision and the initiation of the business model's change are the keys to successful DT. For example, Westerman et al. [14] argue that irrespective of whether old or new tools are used, there are significant challenges related to management and employees, and not only to technology. They classify problems by the implementation phase into challenges relating to initiation (lack of impetus, regulation, reputation, and unclear business situations) and the organization (lack of capacities, problems of culture, and lack of IT—incremental vision and coordination issues). According to Berman [6], whether DT focuses on customer value reconfiguration or customer modeling, the crucial challenges are to monetize new solutions to offer customer value or set business needs for better activities.

Earley [25] states that it is challenging to recognize and apply innovative technology in an existing business model while considering increases in the additional cost of the new instruments and infrastructure and avoiding the so-called paradox in this context of increased complexity and decreased productivity. In this respect, Earley views making decisions on adopting new technologies as a significant challenge, given the complexity of DT.

In addition, De Carolis et al. [27] emphasize the significance and value of information in Industry 4.0 to raise awareness of the added value that DT provides and state that managing large quantities of data is the main challenge for traditional production enterprises. In connection with the growing significance of data, Fehér et al. [11] mention that the main challenge in implementing DT relates to accepting novel technologies. As a further challenge, they identify that the current IT infrastructure lacks flexibility, especially in aggregating and managing unstructured data from different sources and linking existing systems with a new front-end system.

Conversely, Nahrkhalaji et al. [17] explore nonprofit organizations and state that the change in the organization's culture and organizational, strategic, and managerial challenges are crucial challenges because IT personnel do not ensure success. Wolf et al. [24] consider individual thought, lack of knowledge of digitalization, and lack of awareness of digitalization's importance as the main challenges or impediments to successful DT.

Challenges of Digital Transformation in the Mining Industry

Since its emergence, the mining industry has faced generic challenges, such as an increasingly aging labor force, high operating costs, reduced ore deposit quality, and a challenging work environment. DT is a crucial initiative but requires hard work and substantial investment to overcome these challenges. Gao et al. [9] observe that many businesses are not agile enough to respond to the changes fast-tracked by DT. Simultaneously, those who do not recognize this opportunity face various challenges to implementing complementary improvements in different parts of the company competitively. The challenges include the absence of competencies (lack of radical change, insufficient staff, commitment, and investments), goal ambiguities (passive skills and the governance model and uncertain expectations), and the possibility for companies in the metal industry to exploit the opportunities to improve business operations through digital technologies (complex operating environment, safety, legislation, and social responsibility).

Moreover, Deloitte [28], which investigated global trends in the mining industry, states that the lack of digitally knowledgeable personnel in the mining industry undermines the digitization process. Ernst & Young [29] highlight a gap between DT's potential and the absence of data in the mining industry and that there is a lack of knowledge about the importance of digital disconnection and models successfully implemented to increase productivity. The further challenges it highlights are the lack of access to capital, the need for increased transparency, license retention (concessions) issues, and cybersecurity risks.

Success Factors of Digital Transformation

A few researchers have examined how DT can be successful from both the organizational and customer perspectives [19]. Berman [6] maintains that companies can successfully transform their business model through a consistent plan to integrate digital and physical operational components, whereas Westerman et al. [20] note that governance abilities are successful in the critical execution of DT. Earley [25] supports this, stating that leadership involves viable digital expertise for internal and external users. In this way, the digital experience should influence the content and functionality of the users. Data play an essential role in the entire process, making it a strategic resource that must be managed and stored.

Leadership is viewed as an essential factor of DT. There are three critical components for a successful DT:

1. A digital approach that creates a digital technology value proposal;
2. An operational backbone that facilitates business excellence; and
3. A digital service policy that allows rapid innovation and reaction to new market opportunities.

The awareness and training of managers and employees must be increased. This can help them access and monitor data and statements in real time. For Erjavec et al. [11], the main path to a successful DT requires cultural transformation. Pflaum and Gölzer [13] agree, emphasizing the importance of creating a digital business vision, aligning its operational goals, and viewing DT as a timed race in which speed depends on a digital department.

According to Nahrkhalaji et al. [17], organizations must review existing business models in the context of DT to raise their maturity levels, improve decision-making processes, identify the right leader, resolve the complexities and insecurity imposed by new competition and collaboration models, and involve clients. DT combines old and new technologies, a revolution, and a change. Thus, Wolf et al. [24] state that the critical success factors for DT are creating new areas, connecting across the organization's boundaries, applying agile methods, motivating new things to be tried, and ensuring active management. Osmundsen et al. [18] state that apart from the initiator, the objectives, and the consequences of DT, the following are the factors that influence the success of DT.

- An agile and supportive organizational culture;
- Excellent management of transformation actions;
- Information management;
- The contribution of employees and managers;
- Improvement of knowledge system abilities;
- Enhancement of dynamic capabilities;
- Advancement of a digital business approach;
- Business agreement; and
- Information structure.

Liere-Netheler et al. [12] have been working on a successful DT framework, and diverse investigators have addressed success factors in different technologies, including users' experience [19].

Accomplishment Factors of Digital Transformation in the Mining Industry

Deloitte [28] mentions that a company's use of information to improve production, decrease costs, enhance efficiency, and increase security changes in mining is observed in how a firm extracts resources and how it utilizes the information. In line with this view, Deloitte [28] claims that DT's success depends not on capitalizing on the latest technologies and applications but rather on integrating digital thought into the very center of corporate strategies and business practices to change corporate decision-making. This requires the evolution of leadership skills and a clear vision of how a future digital mine can transform development, information flow, and supporting processes.

Because productivity growth is an important mining goal, DT can be successfully achieved; Ernst & Young [29] identify four main components:

1. complementing the DT and throughput plan;
2. creating a complete business approach (from the mines to market);
3. managerial culture—ensuring a supportive organizational culture, particularly in terms of leaders' influence; and
4. focusing on waste disposal.

Although numerous writers have identified and investigated DT's challenges and success factors, research on the mining industry is generally lacking. Therefore, it is evident that further research is necessary for the mining industry to overcome the digital interruption lag, validate existing knowledge, and identify new information in different situations.

Where Lies the Value?

Big data analytics can benefit the mining industry in several critical ways. It can

- ensure a continuous flow to the processing plant from the ore extraction point;
- maximize ore output by reducing production bottlenecks;
- reduce non-productive time, such as unplanned maintenance, delays, wastage, and wait time;
- assist management to make informed decisions on the "as-is" production process, covering the value chain from extraction to delivery at plants and beyond; and
- support on-the-fly assay results and interpretation analysis to help geoscientists make informed decisions.

Organizations should be able to scale horizontally, considering that such a platform ensures that low total ownership costs are achieved without the seller's lock-in.

Lessons Learned

The findings presented in the previous sections can be consolidated into three main issues:

1. mining is an essential activity but also presents significant challenges to mining companies;
2. DT can be the key factor to product development; and
3. effective DT programs can only be prepared for companies that can coordinate at the corporate level.

These issues can take various forms in terms of governance, organizational arrangements, and combined skills. The major challenge is that the business should efficiently formulate and consistently implement a DT strategy, as growing, larger companies have demonstrated. However, the adoption of this corporate management approach should not result in centralization. Instead, innovative and flexible technology solutions are locally required because a firm's nature differs according to business units, geography, and business type.

This means that enterprise-wide and local solutions must coexist. Therefore, corporate solutions must focus on significant cross-operational initiatives and the foundations to facilitate and direct local solutions without simply killing them. Thus, it has been recommended to companies interested in obtaining the total value of DT.

- Ask the business where it is going.
- Ensure that they establish a company-level digital agenda.
- Create the foundations and capacity to become agile.
- Orchestrate the innovation.

Digital Mine Framework

While large companies have invested in automation and technology innovation for several years, the broader mining industry has also adopted a digital agenda. Rapid advances in technology and decreasing costs overall have meant that harnessing the power of digital has become more practical and achievable to the extent that it is now becoming an imperative. However, in discussing this topic, many mid-tier and junior miners ask: "How do I know where to start with digital, and what technologies should I invest in to obtain the best bang for my buck for my business, when I do not have the kind of budget or team that the majors have?" To help answer this question, we need to introduce the "digital mine framework" (see Fig. 1.4).

This framework explains the future state of digital mining organizations and how this might transform the core mining processes, the flow of information, and supporting back-office operations.

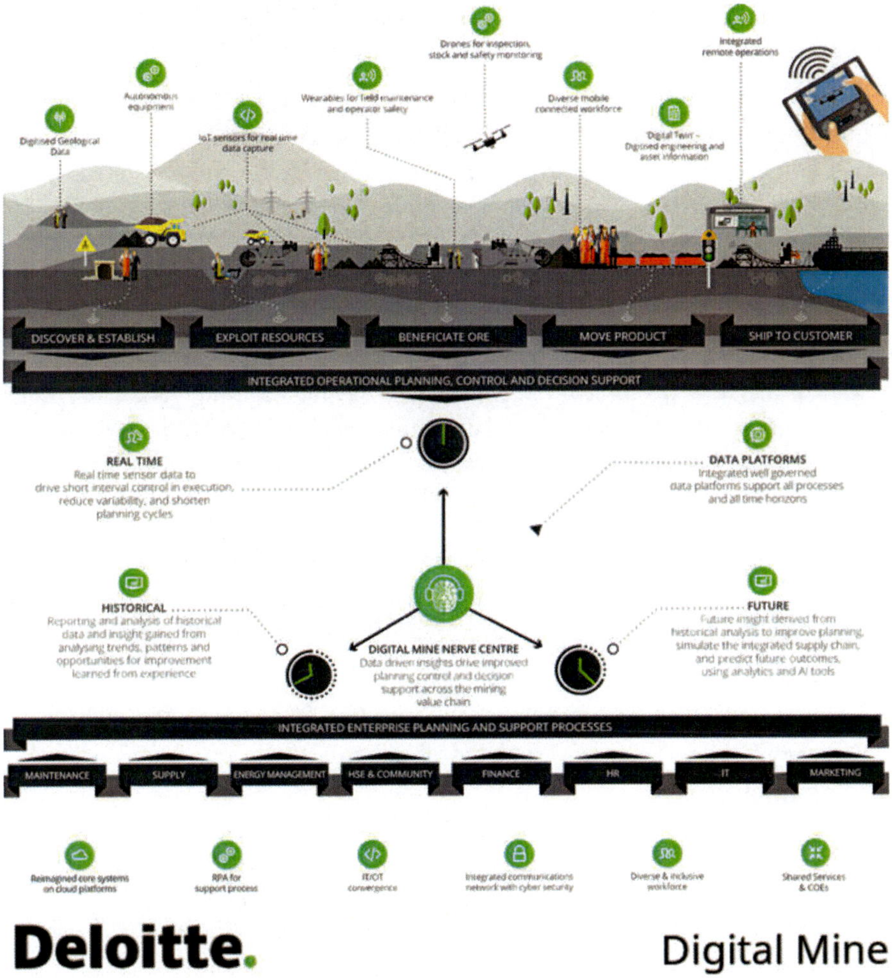

Fig. 1.4 Deloitte Digital Mine Framework (https://www.deloitte.com/)

Core Mining Processes Automating Physical Operations and Digitizing Assets

There are five essential features in the core operational processes of the “digital mine”:

1. **Automation and remote operation**

The majors have shown how autonomous mining equipment (including drills, trucks, and trains) and remote operations can improve safety and productivity and reduce costs in large-scale operations. Although automation may not be feasible for smaller players' existing processes, they should consider it an option for new mines and leverage the capability and investment made by many equipment manufacturers and service providers.

2. **Real-time data capture**

Advances in IoT technology enable a connected network of low-cost, highly capable sensors to capture data in real time to facilitate integrated planning, control, and decision support. Moreover, these capabilities do not require significant investment and are increasingly bundled by equipment manufacturers and service providers into their offerings.

3. **Digital twins digitized geological, engineering, and asset data**

A digital model of the physical environment, constructed using geological, engineering, and asset information, can be continuously updated with data collected from sensors and location-aware mobile devices. This approach would enable better planning, prediction, and simulation of future outcomes. Furthermore, this model does not have to cover all operations and assets; it can focus on areas with the most significant potential value and should be considered in planning and designing new functions.

4. **Drones**

Given that the capability of unmanned drones is improving and that their cost is reducing steadily, they can be used for data collection, inspection, stock control, and safety monitoring.

5. **Wearables**

Similar advances in wearable technologies make them worthy of consideration for providing field maintenance and real-time machine inspection instructions, which would improve operator-based care and safety.

The Digital Mine Nerve Center Data-Driven Planning, Control, and Decision-Making

The "nerve center" of a digital mine will bring together data across the mining value chain in multiple time-horizons; improve planning, control, and decision-making to optimize volume, cost, and capital expenditure; and improve safety. Mid-tier and junior miners can gain significant returns by investing in the following:

1. **Improved visualization, reporting, and analysis of historical data**

Improved business intelligence and visualization tools can now be implemented more quickly and cheaply to integrate data from multiple sources and reduce the reliance on legacy enterprise resource planning (ERP) systems and Excel spreadsheets.

2. **Short interval control and operational improvement**

Real-time data captured from processing equipment and machinery sensors during operation (as described previously) will enable short interval control to identify key drivers of process variability and drive rapid and focused operational improvements. More timely data will also allow ore-body models, mine plans, and financial models to be updated more frequently and shorten the planning cycle.

3. **Future modeling, prediction, and simulation**

Rapid advances in analytics and AI tools enable insights to improve planning, simulate the integrated supply chain, and predict outcomes.

4. **Integrated data platforms**

An integrated and well-governed data platform supported by specialist data scientists and analysts can enable each of the three types of analysis described above. This does not have to be a significant investment; it can begin with a small, focused team or be purchased as a service.

Re-imagined ERP and Automated Support Processes

The effects of digitization will extend beyond the core operations to the supporting processes and systems of functions such as supply, HR, and finance. Smaller mining organizations have an opportunity to go further than the majors by avoiding legacy on-premises solutions and achieving a more innovative back-office via the following:

1. **"Re-imagined" cloud-based ERP**

Core systems, such as ERP, can be migrated to cloud-based solutions with a low cost of ownership and can be purchased "as a service." Smaller players who have previously not considered integrated ERP solutions because of the cost of licenses and on-premises implementation may now be able to afford them.

2. **Robotic process automation**

There has been rapid growth in robotic process automation in the past two years to automate repetitive human activities and reduce costs and errors in back-office support processes, especially in shared service centers. However, the benefits of such automation are also achievable on a smaller scale and should be considered for targeted pilots on approach "pain points."

3. **Network connectivity and bandwidth**

A historical constraint to leveraging the benefits of technology has been the limitations of communication networks, especially in remote locations. However, using the LTE spectrum and capabilities such as software-defined networks has reduced the cost of supporting a mobile workforce across all platforms.

4. **IT-OT convergence**

More integrated IT and OT management will enable the automation and digitization of core and support processes.

5. **Cybersecurity**

Cybersecurity solutions will mitigate the risks of more excellent connectivity (as more equipment and devices are connected to the Internet).

Impact on People

The people's implications of the digital mine should not be underestimated. As the digital mine becomes more of a reality, our interpretation of what work is and how it is delivered is changing rapidly. The digital revolution will affect work, employees, and the workplace in a multitude of ways. Various media have put forth various viewpoints about the impact of automation on jobs, but history shows that the prior industrial and technological revolutions have positively impacted jobs over time. Further, although some roles decline, new positions are created.

Technologies will enable work to be moved to locations that can support a more diverse and inclusive workforce, increased human–machine interaction, and new and different skills. In addition, the future mobile and connected workforce will increasingly expect an enhanced user experience consistent with the home consumer experience.

Conclusion

This chapter tried to introduce digital transformation in mining. This introduction included the opportunities and challenges of transferring from traditional mining to digital one. Some success factors of DT were discussed, and the learned lessons of digital approaches to solve the business problems in mining were reviewed. A comprehensive digital mine framework was introduced, and all its components were explained in detail. At the end of this chapter, the impact of DT on people working in the mining industry was discussed.

References

1. Soofastaei, A. 2020. *Data analytics applied to the mining industry.* CRC Press.
2. Martins, P., and A. Soofastaei. 2020. Making decisions based on analytics. In *Data analytics applied to the mining industry*, 193–221. CRC Press.

3. Soofastaei, A., et al. 2018. Energy-efficient loading and hauling operations. In *Energy efficiency in the minerals industry*, 121–146. Springer.
4. World Economic Forum & Accenture. 2017. Digital transformation initiative 2017. Cited 2021 Mar 20. Available from: https://www.accenture.com/t20170411t115809z__w__/us-en/_acnmedia/accenture/conversion-assets/wef/pdf/accenture-telecommunications-industry.pdf.
5. Stolterman, E., and A.C. Fors. 2004. Information technology and the good life. In *Information systems research*, 687–692. Springer.
6. Berman, S.J. 2012. Digital transformation: Opportunities to create new business models. *Strategy & Leadership.*
7. Konde, P., and P. Tundalwar. 2018. Agile approach-digital transformation. *International Journal of Research in Engineering, Technology and Science* 8: 1–8.
8. Zimmermann, A., et al. 2018. Evolution of enterprise architecture for digital transformation. In *2018 IEEE 22nd International Enterprise Distributed Object Computing Workshop (EDOCW).* IEEE.
9. Gao, S., et al. 2019. Digital transformation in asset-intensive businesses: Lessons learned from the metals and mining industry. In *Proceedings of the 52nd Hawaii International Conference on System Sciences.*
10. Raab, M., and B. Griffin-Cryan. 2011. *Digital transformation of supply chains: Creating value—When digital meets physical.* Capgemini Consulting.
11. Kő, A., P. Fehér, and Z. Szabó. 2019. Digital transformation—A Hungarian overview. *Economic and Business Review for Central and South-Eastern Europe* 21 (3): 371–495.
12. Liere-Netheler, K., et al. 2018. Towards a framework for digital transformation success in manufacturing.
13. Pflaum, A.A., and P. Gölzer. 2018. The IoT and digital transformation: Toward the data-driven enterprise. *IEEE Pervasive Computing* 17 (1): 87–91.
14. Westerman, G., et al. 2011. Digital transformation: A roadmap for billion-dollar organizations. *MIT Center for Digital Business and Capgemini Consulting* 1: 1–68.
15. Fitzgerald, M., et al. 2014. Embracing digital technology: A new strategic imperative. *MIT Sloan Management Review* 55 (2): 1.
16. Henriette, E., M. Feki, and I. Boughzala. 2015. The shape of digital transformation: A systematic literature review. *MCIS 2015 Proceedings* 10: 431–443.
17. Nahrkhalaji, S.S., et al. 2018. Challenges of digital transformation: The case of the non-profit sector. In *2018 IEEE International Conference on Industrial Engineering and Engineering Management (IEEM).* IEEE.
18. Osmundsen, K., J. Iden, and B. Bygstad. 2018. Digital transformation: Drivers, success factors, and implications. In *MCIS.*
19. Sahu, N., H. Deng, and A. Mollah. 2018. Investigating the critical success factors of digital transformation for improving customer experience. In *International Conference on Information Resources Management (CONF-IRM).* Association for Information Systems.
20. Westerman, G., D. Bonnet, and A. McAfee. 2014. *Leading digital: Turning technology into business transformation.* Harvard Business Press.
21. Lammers, T., L. Tomidei, and A. Regattieri. 2018. What causes companies to transform digitally? An overview of drivers for Australian key industries. In *2018 Portland International Conference on Management of Engineering and Technology (PICMET).* IEEE.
22. Singh, A., U.K. Singh, and D. Kumar. 2018. IoT in mining for sensing, monitoring, and prediction of underground mines roof support. In *2018 4th International Conference on Recent Advances in Information Technology (RAIT).* IEEE.
23. de Sousa, J.S., I.O. Rocha, and R.M. de Castro. 2018. Digital transformation applied to bauxite and alumina business system—BABS 4.0. In *Proceedings of 37th International ICSOBA Conference.*
24. Wolf, M., A. Semm, and C. Erfurth. 2018. Digital transformation in companies—Challenges and success factors. In *International Conference on Innovations for Community Services.* Springer.
25. Earley, S. 2014. The digital transformation: Staying competitive. *IT Professional* 16 (2): 58–60.

26. Soofastaei, A. 2019. Energy-efficiency improvement in mine-railway operation using AI. *Journal of Energy and Power Engineering* 13: 333–348.
27. De Carolis, A., et al. 2017. Guiding manufacturing companies towards digitalization a methodology for supporting manufacturing companies in defining their digitalization roadmap. In *2017 International Conference on Engineering, Technology, and Innovation (ICE/ITMC)*. IEEE.
28. Deloitte. 2018. The top 10 issues shaping mining in the year ahead. Tracking the trends 2018, July 3. Available from: https://www2.deloitte.com/content/dam/Deloitte/us/Documents/energy-resources/us-er-ttt-report-2018.pdf.
29. Ernst & Young. 2017. The digital disconnect: Problem or pathway? Cited 2021 Mar 20. Available from: https://assets.ey.com/content/dam/ey-sites/ey-com/en_gl/topics/digital/EY-digital-disconnect-in-mining-and-metals.pdf?download.

Chapter 2
Advanced Analytics for Modern Mining

Diego Galar and Uday Kumar

Abstract Technology is growing very fast, and we are facing the Fourth Revolution in industry. Digital transformation has found its way in to many different traditional and modern industries. Digitalization and automation are two common words in mining these days. However, there are many challenges in the mining industry to reach the appropriate maturity level for Industry 4.0. Data collection systems, cloud-based storage, intelligent data architecture, creating online data hubs are some examples of digitalization challenges. Moreover, it is essential to use advanced analytics models applying artificial intelligence and machine learning models for automation, prediction, and optimization. Mine managers need an intelligent indigent virtual assistant to make better decisions and develop smart mines. This chapter some key components of intelligent mining and helps researchers understand Industry 4.0 in the mining context.

Keywords Mining · Industrial Internet of Things · Machine-to-machine communication · Advanced analytics · Artificial intelligence

Introduction

The modern-day mining industry is mechanized, automated, and capital intensive. Systems deployed in modern mines are to be robust, reliable, and to perform safely with designed performance levels most of the time or even round the clock. However, due to unforeseen events and processes, design deficiencies, or operational and environmental stresses, these systems are often unable to meet production requirements

D. Galar (✉) · U. Kumar
Luleå University of Technology, Luleå, Sweden
e-mail: diego.galar@ltu.se; diego.galar@tecnalia.com

U. Kumar
e-mail: uday.kumar@ltu.se

D. Galar
TECNALIA, Basque Research and Technology Alliance (BRTA), 48170 Derio, Spain

A. Soofastaei (ed.), *Advanced Analytics in Mining Engineering*,
https://doi.org/10.1007/978-3-030-91589-6_2

Fig. 2.1 Mining 4.0

in volume and other critical system performance indicators (such as energy consumption). In addition, unplanned stoppages in highly mechanized and automated mines generally lead to wastage of resources and quality problems.

To ensure a high level of performance from the system and reduce the wastage of resources and efforts, one needs to predict the system behavior as correctly as possible and initiate measures to eliminate disturbances, thereby removing uncertainty from production processes. This prediction has to be in real time for automatic systems, focusing on estimating parameters like the remaining useful life of components and the degree of deviation from regular operation, and the status of various machines and systems in the production chain. Risk scenarios must be visualized for various alternative operational scenarios in real time considering the total risks. Most disturbances are due to the poor reliability of operating systems.

The chapter gives an overview of mature, new, and emerging technologies described by the generic term Industry 4.0. The application of Industry 4.0 concepts to mining is generally referred to as Mining 4.0 (Fig. 2.1). It has the potential to more effectively manage capital-intensive mining production systems and production processes.

The term Industry 4.0 refers to the Fourth Industrial Revolution. This revolution involves the integration of systems and the creation of networks connecting people with machines. The real world is united with the world of machines and the virtual world of the Internet. We now have access to any information at any time, from anywhere. Thanks to this revolution, it will be possible to continue developing automation, optimizing products and processes, collecting and processing many data in real time, performing preventive maintenance of machinery and equipment, and rapidly adapting to changes in the market. Enterprises implementing this concept gain maximum flexibility and a significant advantage over the competition [1].

Technological solutions driving Industry 4.0 include the following: autonomous robots, simulation and forecasting techniques, vertical and horizontal integration, Industrial Internet of Things (IIoT) or direct communication between machines, innovative methods of collecting and processing vast amounts of data (big data), including the use of the cloud, additive technologies, technologies of augmented reality (AR) and virtual reality (VR), cyber-physical systems (CPS), such as digital twins using artificial intelligence (AI), and neural networks. These are often referred to as the pillars of the Fourth Industrial Revolution [2].

The mining sector lags behind other sectors in implementing digitalization and automation is moving in the right direction [3]. Mining 4.0 means automation in all mining processes, including extraction, transportation, and processing. As a result, mines will have a reduced negative impact on the environment, see increased productivity and efficiency, including mine management, and enjoy improved employee health and safety [4–7]. In addition, Mining 4.0 allows sustainable production; for example, technologies associated with Industry 4.0 make it possible to save on transportation costs, treatment, and disposal of non-useful minerals [8].

Another aspect of Mining 4.0 is the "low-impact mine" in densely populated areas. This is a mine with a minimal negative impact on the environment and urban areas. Even so, high productivity, high security, and efficient resource consumption are expected [4, 9].

The chapter focuses on new and emerging technologies incorporating big data analytics, IIoT, AR/VR technologies deployed, or potentially deployed in mines. It covers both technological and management issues of implementing Mining 4.0 solutions.

Why Mining 4.0?

Mining 4.0 applies the concept of Industry 4.0 to mining, fostering research on the development of activities related to the concept of intelligent mining [4]. In other words, it is a general term for the digitalization and automation

of mining operations, both above ground and underground. It is inspired by various factors, such as cost reduction, health, safety, higher availability in day-to-day operations of mining equipment, longer equipment lifetime, reduced uncertainty of mining operations, increased productivity, and real-time data availability and monitoring. Furthermore, it is based on the understanding that the mining industry can only maintain its competitive advantage if it can combine reliable technologies with innovative business models to create new products [10–12].

Many factors are driving the implementation of Mining 4.0. An obvious factor is the availability of Industry 4.0 technologies and software which can be adapted to mining. In addition, the latest advances in artificial AI, big data analytics, and sensor technology enhance productivity and equipment lifespan by offering accurate and precise predictive maintenance and diagnostic tools in all industries, including mining.

Outstanding mining performers use IIoT solutions to minimize downtime and reduce the costs associated with scheduled and emergency maintenance in the mines. Sensor data from vehicles and other assets are collected on various brake parameters such as system pressure, brake pad wear, position of the brake and its piston, brake fluid levels, and temperature. Data are sent to the cloud, where advanced AI-driven data analytics is used to extract meaningful information on the status of the brakes and their components and provide crucial predictions on expected equipment failure. The results are usually accessible remotely, from anywhere. As a result, operators have a clear and comprehensive real-time overview which allows them to identify anomalies and determine the best time to conduct scheduled maintenance activities. Regular inspections can be halved with this type of architecture, resulting in substantial decreases in downtime and maintenance costs without affecting the equipment's service life. Many mining operators and equipment manufacturers have realized that it helps them to extend the lifespan of critical components. Moreover, as inspection activities often require traveling to remote areas, the mining business could reduce its visits to these sites and maximize uptime in the event of equipment failure.

Another factor driving the implementation of Mining 4.0 is the ongoing steady decline of productivity in mining operations. In addition, and as a consequence, sources of finance for mining operations, especially in the early stages, are declining. Thus, despite a steadily increasing demand for the products of the world's mines, today's operating conditions face enormous challenges.

Mining needs to re-invent itself to reverse the long-term decreasing productivity and regain financial security. A step-by-step marginal optimization process will not be able to achieve this. Mining 4.0 may offer a solution to this problem [3].

Industry 4.0 Versus Mining 4.0

Mining 4.0 does not mean a one-to-one transfer of the Industry 4.0 objectives to industrial raw material production. Instead, Mining 4.0 means the advance of automation during extraction, transportation, and processing. Figure 2.2 shows the road map of the Swedish mining sector to reach the vision of fully autonomous mining operations without human presence in production areas. The darker gray box illustrates the current status. The lighter gray box illustrates the first steps that need to be taken to meet the following objectives:

- production;
- construction and disassembly of infrastructure;
- monitoring of equipment and the mine environment; and

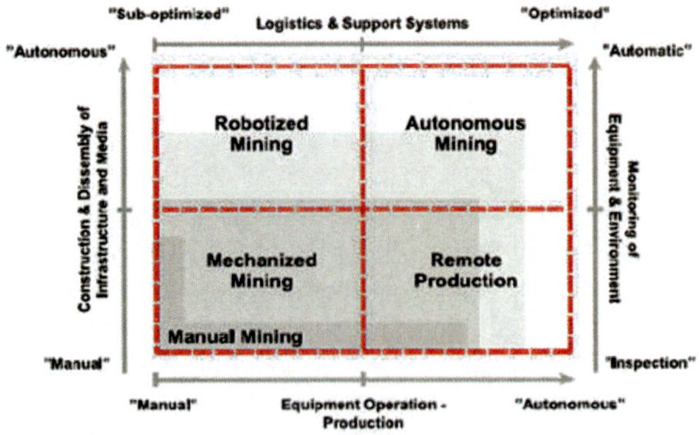

Fig. 2.2 Road map to autonomous mining operations

- logistics and support systems.

It is only possible to optimize raw material production in Western high-wage countries by automating highly productive mechanized mining operations. In addition, resource-sparing and sustainable raw material production are only possible with a high degree of selective mining, which saves transportation, treatment, and disposal of economically unusable minerals (waste).

The objectives of Mining 4.0 include the following:

- selective production of raw materials;
- autonomous production, transport, and processing;
- minimal impact on man and the environment.

The development of Mining 4.0 requires great efforts in the fields of sensor technology and M2M communication. M2M stands for the automated exchange of information between devices, including machines, vehicles, and containers, or with a central control unit, increasingly using the Internet and the various access networks such as the mobile network.

In contrast to automation, autonomous devices operate with a set objective but within their own decision-making space to achieve this objective. Autonomous devices, therefore, need a variety of sensors and AI. The AI evaluates all available sensor data (sensor model) and, within the limits of the permitted degrees of freedom, independently generates the decision as to which actuators are set (actor model) [4].

As indicated in Fig. 2.3, the following technologies are required to meet the main objectives of Mining 4.0 [4]:

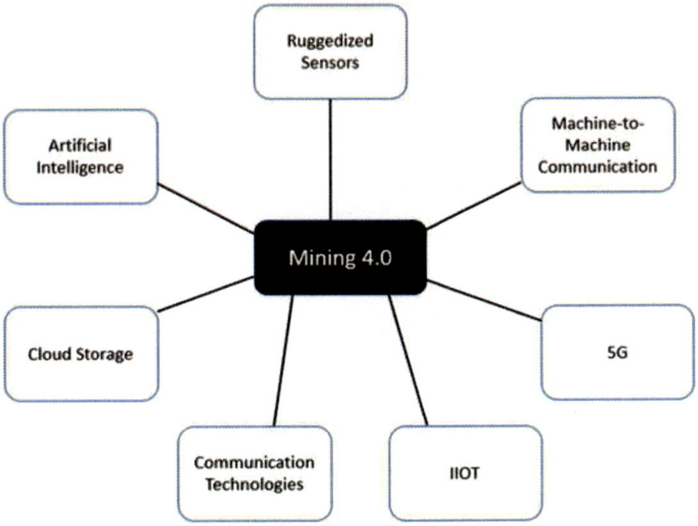

Fig. 2.3 Core technologies for Mining 4.0 [4]

- ruggedized sensors;
- M2M communication; and
- AI.

There are many other requirements for future mining projects, including security, environmental impact, cost-effectiveness, and social acceptance [4].

Robust Sensor/Military Standard

A reliable sensor system is a prerequisite for the automation of machines and processes. Here, the mining industry is highly demanding. For example, embedded sensors installed on a machine/system must deal with vibration and temperature; environmental sensors must be resistant to dust, water, light, mechanical stress, fluctuating temperature ranges, and should also be explosion-proof, if necessary (Fig. 2.4) [4].

Embedded sensors have long been used as input variables for control and monitoring applications in machines and systems. Embedded sensors are also increasingly used in the condition diagnostics of machine and system components.

Environment sensors or sensor networks are used in the mining industry to record the state or properties of the environment within the mine. The environmental sensor system includes such tasks as

- position determination, e.g., positioning, collision avoidance → mine model;
- material determination, e.g., material identification, boundary layer detection, material quality → mineral deposit model;
- ventilation determination, e.g., temperature, quality, flow rate → ventilation model; and
- mine water determination, e.g., water level, water quality, flow rate → mine water model.

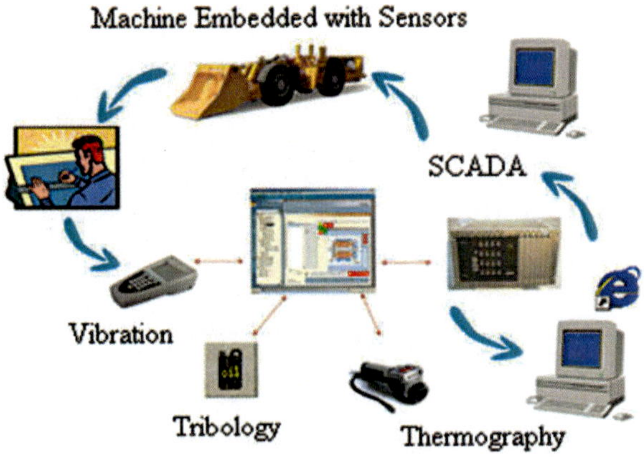

Fig. 2.4 Conventional and embedded sensor measurements

In general, the mining industry requires the following sensor technology:

- small sensors that can be applied quickly and in as many places as possible (small, embedded);
- wireless radio-based sensors that can be applied to each other without any significant installation effort and risk of cable damage (sensor network); and
- energy self-sufficient sensors, extraordinarily energy-efficient and equipped with low-energy, energy-harvesting power packs.

Networks of environmental sensors are not yet widely used in mining. On the one hand, there is a lack of reliable sensors designed for the environmental conditions in the mining industry to carry out the required tasks. On the other hand, the lack of a communication infrastructure does not allow allow sensor networks to be integrated into the existing mine communication. Here, M2M communication is of particular importance [4].

M2M Communication

The autonomous production processes in Industry 4.0 require secure and standardized M2M communication. Today, sensors and actuators are integrated into vast mining facilities via a Fieldbus, such as Modbus, Profibus/ProfiNet, or EtherNet. However, it is always challenging to integrate Fieldbuses. Therefore, the clear definition of engineering interfaces that are easy to use for all parties involved and integrating different sensor systems into higher-level planning systems are decisive criteria, especially in a complex and heterogeneous system, such as a mine.

The Open Platform Communications Unified Architecture (OPC UA), first published in 2011 under the standard IEC 62541, provides a communication stack for Internet-based communication between machines and systems from different manufacturers. The OPC UA relies on the IoT and defines the "language" in which machines exchange information using Internet technology.

OPC UA consists of a server stack and a client stack (Fig. 2.5). The server provides information on a machine/system via the defined services: alarm, data, and history. The client accesses the server via the Internet with a protocol defined in OPC UA (OPC.tcp://<server_server_server_name: 4840) [4].

Although OPC UA and OPC sound alike, the former does not develop the latter but is an entirely new technological development, even if it uses some of OPC's conceptual approaches. OPC, now commonly called OPC "Classic," relies on a proprietary protocol from Microsoft, and users are limited to the MS Windows world. In addition, the protocol is not quite routable, which makes worldwide data exchange over the Internet more challenging. Figure 2.5 compares OPC UA and OPC [4].

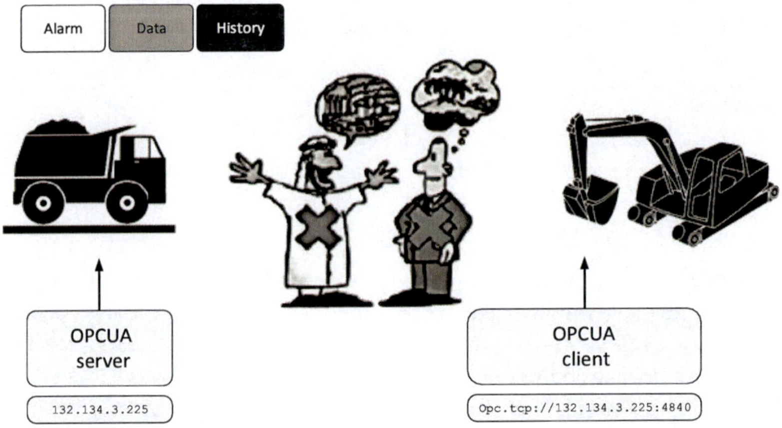

Fig. 2.5 M2M communication standard OPC UA [4]

The Role of the Operator in Mining 4.0

In Mining 4.0, miners are experts who oversee production. However, they do not do this from a control room. Instead, they have access to data on processes and machine health as they move around the mine. In addition, they interact with other stakeholders in real time: operators, experts, suppliers, and customers. Digital twins can even be used to control production, and, for that, the miner does not even need to be in the mine.

The miner's senses and memory are enhanced by technology, including sensors embedded in clothes. This means operations are continuously monitored, reducing the possibility of human error caused, for example, by working conditions or fatigue [6].

A typology for Industry 4.0 has been modified for mining operators [6]. Miner 4.0 includes the following items:

- Bio-mechanical support: The miner has super strength, i.e., biomechanical support, resulting in increased strength and endurance.
- AR: The miner uses AR to integrate information from the digital and physical worlds. For example, if there is a problem with a particular piece of equipment, the maintainer can communicate directly with the equipment manufacturer. The manufacturer can see the problem and resolve it through special glasses that send and receive live video.
- VR: The miner uses VR for training. VR can simulate high-risk situations (e.g., fires) or the use of new equipment and is already widely applied in mining. In theory, the control room and production control could be set into a VR environment.

- Wearable sensors: The miner uses wearable sensors to monitor location (GPS) and health-related metrics. Many mines are already using advanced positioning systems, and research has examined the use of sensors to monitor health.
- Intelligent personal assistants: The miner uses technology to interface with machines, computers, databases, and other information systems. For example, radio-frequency identification (RFID) tag systems can be combined with a smartphone app to report failed equipment quickly.
- Robots: The miner uses robots to perform tasks that are incredibly repetitive and strenuous. Semi-autonomous machines, such as loaders or haul trucks, are commonly found in mines. The machines take over the repetitive and strenuous task of driving to and from the muck pile.
- Social networks: The miner uses the organization's social networking services to interact with other workers and the Internet of Things (IoT). The introduction of underground Wi-Fi and 5G has made this possible. With Wi-Fi, for example, operators can use group chats to exchange information.
- Big data: The miner uses big data analytics.

These will affect how the work is done. New jobs will be created, and old ones will evolve. As a result, tomorrow's miner will be different from today's miner [6].

Mining 4.0 in Developing Countries

As we have noted, the mining sector has been slower than other sectors to implement Industry 4.0. However, Mining 4.0 has been implemented in "rich" mining-dependent countries, such as Australia and Canada (e.g., driverless trucks and automated trains). It is now entering mining-dependent countries in the developing world, notably in Africa. Some experts forecast disruptive negative consequences in developing countries, resulting from job losses, and reduced local procurement. Table 2.1 shows some insights into current and future developments [3].

Mining 4.0 constitutes a structural (also called "disruptive") innovation. It is characterized by technologies and management systems that enable a new way of mine operation which significantly differs from current and past practices [3]:

- underground and open pit machines communicate directly with the control room;

Table 2.1 Global expenditures for IoT in mining [3]

Item	Share (%)
Subsystems	4
Sensors (oil and gas)	5
Robots	12
Vehicles	22
Field devices	57

- state and location of all equipment are known and documented at any time;
- smart sensors in machines allow prediction of failure of components and communication between machines;
- all data and information are directly transmitted;
- mine operators are remote from the locations of the machines, operating and monitoring machines in control rooms, sending instructions to machines from there;
- mine operators can configure cutting cycles according to the varying conditions of the mine faces or other geological mine structures;
- machine instructions can be simulated and tested before being sent to machines;
- real-time data are available, and operations are run in real time;
- real-time data allow analysis tools to improve decision-making on scheduling and processing.

Figure 2.6 shows how the Swedish agenda for innovation in the mining industry comprises integrating remote-control rooms, autonomous vehicles, and processes to meet the expectations of performance and sustainability aiming for zero waste and zero accidents.

Results and outcomes expected from the adoption of Mining 4.0 are summarized in Table 2.2.

Advantages of Mining 4.0 as a structural innovation essentially center on economic benefits through better use of available time, costs savings from reduced

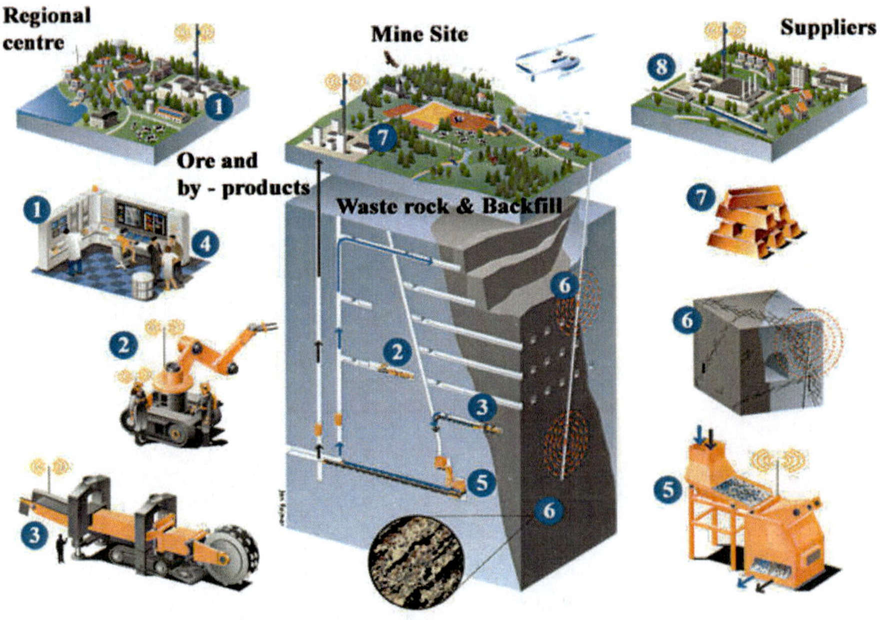

Fig. 2.6 Features of sustainable mines for the future (RTC 2012)

Table 2.2 Results and outcomes from the adoption of Mining 4.0 [3]

Real-time dispatch Real-time production reporting Production optimization Reduced staff numbers in mines Production continues over shift change	Reduction of overhead costs Better fuel efficiency Reduced maintenance efforts Reduced procurement at mine sites
Proximity awareness Traffic management Electronic tagging Automatic ventilation control	Improved safety conditions Changes in the qualification of staff
Better predictability Better and faster adjustments	Better environmental performance through better monitoring and control

employment, energy savings, and less equipment maintenance efforts—all of which must be balanced against the higher capital outlay [3].

Intelligent Mining: Digital Transformation

The mining industry has made several technological changes from the early twentieth century to the twenty-first century by implementing new mining systems and technologies. As a result, it has advanced from the "modern mine" to the "real-time mine," and the industry has initiated a new term, the "intelligent mine."

To take part in the digital revolution and use it to their advantage, mining companies must change. However, many continue to focus on how they have always done things, perhaps not understanding the impact digital transformation (DT) can have on how work is done, the people doing the work, and the workplace itself.

In all parts of the world, mining companies have two interlinked challenges: what to do in the face of declining ore grades and how to maintain operating efficiency. Nor is this all: expensive infrastructure, long-term planning horizons, and lengthy implementation timelines exacerbate the fundamental problems. Last but by no means least, operating environments are dangerous, making worker safety a critical concern.

Mines sticking to traditional approaches face diminishing returns—overworking their remaining assets and working harder for smaller gains. However, the cost of digital technology is decreasing, and digital solutions are a viable option. Intelligent mining through digital transformation may be a solution to all the concerns listed above.

Being digital is about the ability to see differently, think differently, and act differently. Faster and better decisions will reduce performance variability, improve equipment use, and enhance safety. Instead of a reactive management approach, organizations will take a proactive one. Value realization will increase as organizations make this transition:

- Evidence-based knowledge: Through digital technology, organizations will gain evidence-based insights, allowing them to do the same work but to do it better;
- Integrated information: Organizations will take advantage of digital technology to automate processes and make system-wide decisions; and
- Integrated systems: Organizations will integrate and optimize all technologies across the mining value chain, allowing them to work differently.

Many mining companies have already started their digital transformation [12, 13]. However, there are some remaining challenges:

- mines may opt for solutions that do not suit their goals;
- technology providers may not be able to implement their solutions;
- mines may not understand the impact DT will have on the workforce or the changes to be made in the work processes;
- mines may not understand the technologies or how to implement them;
- governance models may not support the more agile digital principles;
- mines may not know who "owns" the digital transformation; and
- instead of focusing on project outputs, mines need to refocus and think about business outcomes.

Be that as it may, mining production techniques are quickly evolving toward intelligent and fully automated production, ranging from mechanization tools to automated single-equipment and independent systems [14].

Intelligent mining technology integrates information technology with mining technology, as shown in Fig. 2.7. Automation and information technology have advanced to such an extent that fully automated and centralized mining equipment will be applied practically very shortly [15].

As Fig. 2.8 shows, the intelligent mine uses information technologies, integrated communication, sensors, VR, AI, computer technologies, and fully automated control equipment. As a result, precise, accurate, and reliable decision-making and production process management are enabled by real-time monitoring; mine production is

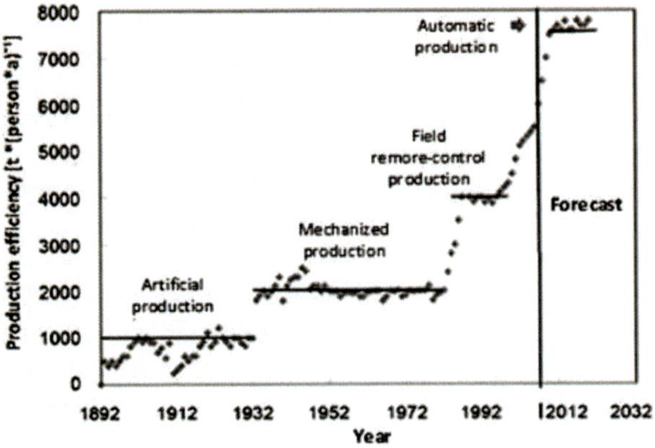

Fig. 2.7 Comparison of production efficiency and mining technology development [16]

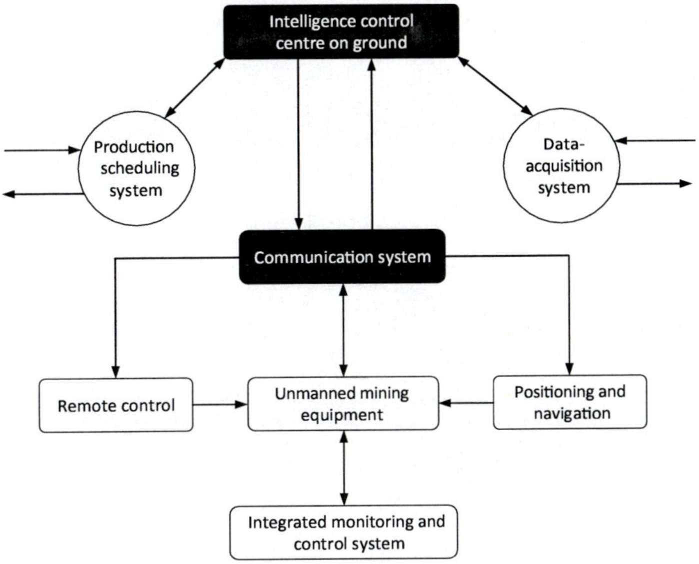

Fig. 2.8 Fundamentals of intelligent mining [16]

maintained at an optimum level, mining efficiency improves, and economic benefits accrue. Put another way, and the mine is green, safe, and efficient [16].

Intelligent mining technology comprises three layers: executive, transport, and control [17].

- Executive layer: includes mining equipment such as drills, trucks, and charging vehicles;
- Transport layer: includes wireless communication systems, information-acquisition systems, and positioning and intelligent navigation systems; and
- Control layer: a system-level platform responsible for process scheduling and control. All intelligent mining functions are implemented through the control layer.

The intelligent process is shown in Fig. 2.9. In the initial step, a mining plan is designed by analyzing the reserves and geological conditions and developing a production schedule. In the next step, an intelligent scheduling and control platform is created to carry out the plan. Finally, in the practical enactment of the plan, control instructions are sent through the transport layer to a piece of equipment to perform a particular mining task in a specific position and time. The control layer collects real-time information on the location of the equipment so that it can be adjusted as needed [16].

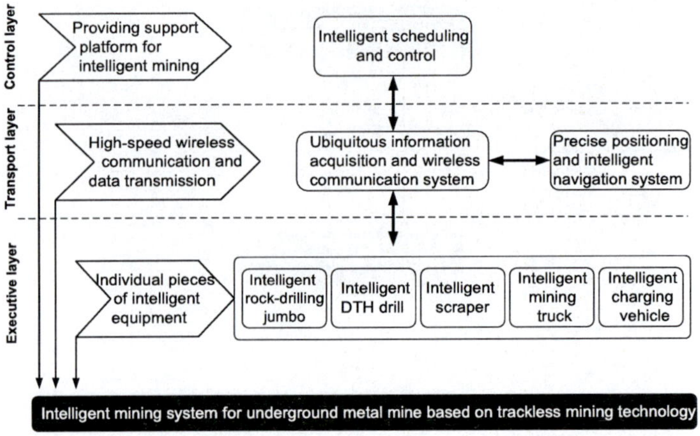

Fig. 2.9 Diagram of intelligent mining technology [16]

The Future Is Intelligent Mining

Intelligent mining is about broader organizational transformation, not just the creation of a "digital mine." It will impact how decisions are made, what skills are required, how to engage with the workforce and communities, and how to best use resources, such as energy. In addition, it will yield insights into where waste occurs, with types of waste ranging from inherent variability in the orebody to management-induced variability. Digital solutions can reduce or even eliminate this waste. Figure 2.10 shows the key performance indicators for the Swedish mining sector for 2030, where a total alignment of automation and waste reduction configures the industry's vision (see Fig. 2.10).

Moving to digital solutions and intelligent mining results in the following [18]:

- operational performance enhanced and enabled by augmented decision-making;
- optimal management of resources;
- deployment of the right skills in the right place at the right time; and
- creation of alliances to support the new approach to work.

Digital transformation goes beyond the use of technology, however. As the following sections explain, the journey to intelligent mining requires more changes than technology.

Making Decisions

Complete, timely, and accurate data enable better decision-making. The decision-making processes of intelligent mines include the following features [18]:

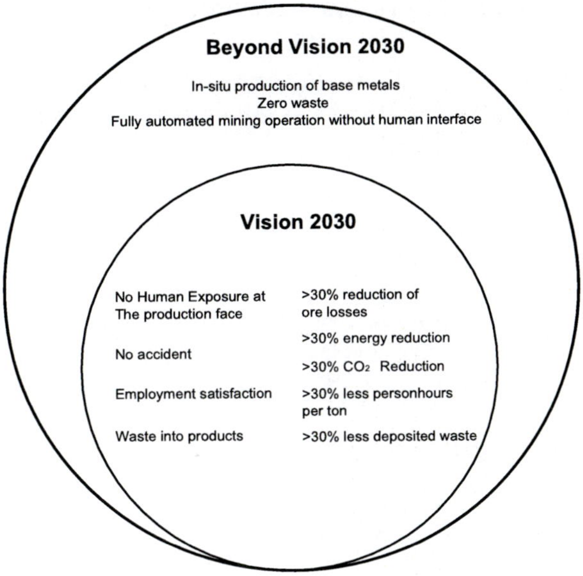

Fig. 2.10 Vision 2030 and key performance indicators for the Swedish mining industry

- decisions are input-driven and forward-looking: with proactive intervention, problems can be solved even before they occur;
- exceptions drive decisions: a performance deviation triggers an action;
- decisions are integrated and holistic: a decision related to an unplanned event becomes collaborative, fact-based, and transparent;
- decisions are prescriptive: decision-making based on predicted outcomes is driven by analytics and AI. Pattern recognition and historical analysis underlie a full range of decisions from safety to commercial decisions; and
- decisions are augmented: operators' decisions are augmented by digital decision support; they are given options and technical guidance through, e.g., AR engagement.

Resource Management

By integrating these resources across the organization, intelligent mines will better manage variable resources of all types, from labor to equipment, energy, and infrastructure. For example, conveyor systems can be integrated with energy monitoring across the value chain to optimize system utilization. Thus, energy becomes subject to resource allocation in much the same way as employees and materials. When this variable resource is optimized across the value chain, sustainability will improve, and utility consumption will drop [18].

Required Skills and Resource Deployment

As we have already made clear, intelligent mining will change how people work, how jobs are designed, and how the working environment is set up. To ensure success, intelligent mines need to understand the following:

- jobs will be re-invented;
- people will need to acquire new skills, so proactive upskilling and reskilling are imperative;
- systemic change will lead to workplace innovation, including the creation of digital cultures and experiences; and
- organizations need leaders with the vision required to lead a digital strategy.

Sharing Value

Adopting digital technology reduces human input, and a common assumption is that it will lead to job losses, with a resulting negative socioeconomic effect on surrounding communities. However, if they adopt a shared value approach, mines can positively impact the community. The concept of shared value posits that an organization's competitiveness is interwoven with the community's health in which it operates.

A shared value perspective will be useful in the following areas [18]:

- Strengthen digital infrastructure: Technology infrastructure investments can improve the local digital infrastructure. This leads to improved quality of life and community development.
- Improve access to education: Online learning platforms provide valuable skill development. In addition, community-oriented education projects become affordable through digital platforms, allowing the mine to impact a crucial area.
- Develop new skills: The digital transformation has created a need for new skills. This allows mines to invest in local development to cultivate digital skills in the community. These include intangible skills (e.g., curiosity, emotional intelligence (EQ), critical thinking, adaptability, creativity, and resilience) and technical skills (i.e., based on data and algorithms).

Beyond the community, the concept of shared value can extend to links between the mining organization and its suppliers. Digital transformation provides technological advances that can create new business opportunities for suppliers, including shared tendering platforms, partnerships with original equipment manufacturers (OEMs), and renewable energy partnerships.

Organizational Transformation

Organizational transformation involves shifting from traditional and traditional ways of thinking to flexible and agile models and new (digital) ways of thinking. Every organization will have a unique transformation. That said, all transformations encompass the following areas [18]:

- automated operations/digitized assets;
- data-driven planning/decision-making; and
- automated, integrated support processes.

An organization's particular goals will determine how (and how much) it opts to change. However, change is the one thing all organizations have in common, whether it constitutes organization-wide transformation or a one-time use case. Regardless of its scale, the transformation will change technology, processes, and ways of work.

To successfully transform, organizations must integrate their core and support processes through a robust technology architecture. Their workforce must be augmented by digital decision support and real-time performance management. Moreover, their leaders must be resilient, driven, and flexible [18].

Organizational transformation comprises five steps [19]:

1. Develop a digital strategy and manage transformation: The digital strategy and initiatives must be clearly defined at the organizational level. Digital transformation should begin with understanding the desired future state and the value created—a key challenge delivering short-term results while aiming for long-term transformation. New approaches can be tested in a pilot and applied in phases. Another challenge is the plethora of products and platforms offering digital solutions. No single vendor has a solution for the entire digital mine organization to manage several providers, and the number and type will change over time.
2. Automate operations and digital assets: Digitally transformed companies will operate in an ecosystem of stakeholders, including providers and partners. Hence, they will need good program management and robust integration capabilities. Unfortunately, many mining organizations have not taken advantage of the possibilities offered by remote operations centers to change their operations and must remain focused on a single product in a single country. In addition, asset-intensive organizations (e.g., mines) face challenges (e.g., data integrity) managing asset information throughout the life cycle. Thus, there are many challenges. The first step in automating operations is defining the most significant value and impact areas and focusing digitization on those areas.
3. Create a digital mine nerve center: Mining companies must use data to resolve a range of problems, and intelligent business decisions require timely access to relevant information. However, most organizations use a fraction of the data they are already collecting, never mind the volume of data that could be captured via

IoT. Problematically, many are still struggling with the limited business intelligence capability inherent to historical enterprise resource planning (ERP) environments and non-integrated operational systems. Moreover, even if they want to change, it can be hard to find people with the necessary skills and experience in data science and analytics. Establishing the capability for an insight-driven organization requires finding those employees and developing and embedding data science and analytic skills into the organization to create a working nerve center.

4. Implement supporting platforms and enablers: Digital transformation can be slowed down or even blocked by core systems such as ERP systems—these systems are tough to remove, expensive to run and maintain, and inflexible in the face of change. Upgrading out-of-date core systems to cloud-based applications and platforms can lead to total cost of ownership (TCO) benefits and a better user experience. For example, robotic process animation (RPA) can replace some tasks currently performed by humans, thus offering the opportunity to reduce the costs of support processes and shared services. Mining companies could test the waters to gain valuable experience by taking advantage of support systems.
5. Lead a diverse, connected workforce: Define what work is and how to do it quickly and constantly. Human–machine interactions will impact work, workers, and the workplace. Importantly, those in the future workforce will demand an improved user experience based on their experience as consumers.

Simply stated, the digital mine requires organizational transformation: a new approach to leadership, a new culture comprising data-led decision-making and diversity, a new way of operating, and new skills. This is not easy to accomplish, especially given the scarcity of digital talent. Companies must bring in new people and reeducate their current workers if they are to make the transition.

IIoT in the Mining Industry

The IoT is "a global network infrastructure composed of numerous connected devices that rely on sensory, communication, networking, and information processing technologies" [20].

The IIoT is the application of these technologies in industry [21]. IIoT emphasizes the idea of consistent digitalization and the connectivity of all productive units [21], combining the strengths of traditional industry with Internet technologies [22]. When IoT is applied in an industrial context (i.e., IIoT), it is possible to immediately benefit from the analytics obtained, contributing to process optimization, machine health, the safety of workers, and asset management. IIoT can assist real-time platforms in remotely monitoring and operating a complex production system with minimal human intervention. Hence, it can be beneficial for hazardous industries, such as mining, by increasing the safety of personnel and equipment while reducing operating costs.

IIoT can manipulate vast volumes of data for insights, running algorithms and information models that can cross multiple plants and correlate data from corporate information technology (IT) solutions or smart sensors not connected to PLCs.

Many different sensors are currently used in mine-related activities, such as geophones in exploration and blast control, piezometers in dewatering, and toxic gas detectors in working frontlines. However, a fully integrated automated system is challenging in practice due to infrastructural limitations in communication, data management, and storage. Moreover, mining companies have a tendency to continue with traditional methods instead of relying on untested novel techniques [23].

The mining industry collects and manipulates both legacy and new information, thus requiring advanced analytics. In addition, IIoT demands a wide range of new technologies, and many of them are not yet mature and require the development of new skills. One essential skill is managing the inherent immaturity of solutions in this space and designing architectures that can evolve to accommodate future technologies. Given the urgent need to choose one or more alternatives in the context of technological diversity and varied options in the market, the mining industry faces the tough choice of waiting for technology maturity or developing a definition that can be extendable [24].

Neubert [25] argues that IIoT can bring operational and economic benefits by discovering inefficiency in processes, increasing supply chain efficiency, allowing faster production, and reducing costs. In addition, IIoT includes the application of M2M communication and intelligent machines capable of delivering data that can be used for future advanced analysis.

When faced with potentially innovative technologies, industries are often cautious in their adoption because they almost invariably involve investment and organizational changes [26]. A common strategy to address these challenges is to use cloud computing to create IIoT platforms while reducing the need for high initial investments. Cloud platforms offer services that cover a large part of the infrastructure needed to implement IIoT. The merging of cloud computing and IIoT brings powerful synergies to industry. Cloud computing makes it possible to leverage on-demand computing power and remotely operate software, data storage, and application processing capabilities. However, the market offers several technology options with different service delivery choices, often difficult to compare in applied technology and cost drivers [24].

The mining industry is considered a key industry for the application of IIoT-related technologies [20]. Specific areas that benefit from these technologies are safety, maintenance, and production processes [24].

In 2003–2015, fueled by growing demand combined with supply constraints, commodity prices increased, and mining industry profits soared. During this favorable phase, the mining industry did not prioritize productivity and cost reduction [27]. After the "commodity super-cycle," companies found themselves in a reduced price scenario, often producing below cost. This required the reinvention of operations and a drastic reduction of costs [28], and the mining industry increasingly accepted a digital revolution as a form of cost reduction and operational efficiency [24].

Challenges the Mining Industry Should Understand

There is growing interest in the use of IIoT technologies in various industries [29]. However, one of the biggest challenges is to choose the proper foundation [24]. Many IoT solutions are making their way into the industry marketplace, but according to Velosa and colleagues [30], "There is widespread confusion on the scope and composition of enabling technology for IoTs business solutions." Choosing the wrong technology can jeopardize the expected return to the business and the process of adopting IIoT.

The mining industry has specific technological features that need to be taken into account before choosing a platform. Industrial plants are usually in remote locations and feature harsh working environments. Devices must be robust and redundant since remote locations typically do not have the proper infrastructure. There are several challenges mines should understand before beginning the IIoT technological platform selection [24].

Technological Challenge

The mining industry must understand different technological challenges when changing into the Mining 4.0 domain—improving the current infrastructure by deploying the new technologies. For instance, five technologies are widely used in the development of IIoT-based products and services: radio-frequency identification (RFID), wireless sensor networks (WSNs), middleware, cloud computing, and IIoT application software [31]. Three main components enable IIoT: hardware, middleware, and presentation [32]. Others include field communication, social networks, and communication protocols such as Zigbee and 3G/4G [20].

Data Islands and Legacy

Data or information islands occur when data are distributed across incompatible platforms, making it challenging to combine the information. For example, mining industries typically have several industrial plants, and many automation technology solutions were implemented over decades. These solutions were built on a mix of open and proprietary communication protocols that were the industry standard during their development.

These systems often do not easily interoperate, resulting in communication issues that can block IIoT capabilities in the mining industry. When data are scattered throughout the plants and the enterprise, integrating and analyzing them manually becomes resource-intensive and time-consuming [24].

Amount and Variety of Data

Another challenge is the amount and variety of data generated at each plant. Large volumes of data come in many formats: from structured numeric data in relational databases to unstructured text documents and sensor data to video and audio data [24].

Data Ownership

The issue of data ownership is a significant challenge. In IIoT, industries typically want to retain the data and the intelligence to analyze the data and make the best use of all data available. The owner is usually the data generator or proprietor of the mechanism generating the data. For example, a manufacturer may collect data from a wide fleet across many companies and thus identify trends. The information generated from the analysis of one customer's assets can benefit other customers experiencing similar situations. In these cases, the vendor or entity does not assume ownership of the data but may ask for rights to use the customer's data for a specific purpose. Commercial relationships must be analyzed carefully and reviewed to ensure that the industrial data owner retains all data generated and all the knowledge related to the intelligent use of the data [24].

Cybersecurity

The ability to share data instantly with everyone and everything creates a significant cybersecurity threat. This challenge increases as more sensitive information is captured. Consequently, the data and communication paths need to be secured. IIoT devices are "conspicuous infection points," and specific security planning is needed for their defense, along with rules for the data governance of the collected information [24].

Data Management

Data management refers to the architectures, practices, and procedures for managing the data lifecycle. In the IIoT context, data management is a layer between the devices that provide data and analytic services. Data management systems summarize data online while providing storage, logging, and auditing facilities for analysis, considering areas of querying, indexing, process modeling, transaction handling, and integration of heterogeneous systems [33, 34]. Abu-Elkheir and colleagues [34] state that "this expands the concept of data management from offline storage, query processing, and transaction management operations into online–offline communication/storage dual operations."

The mining industry has many data sources, with different data types distributed in systems, such as enterprise resource planning (ERP) systems, manufacturing execution systems (MES), PIMS, and many legacy IT systems. In this environment, data management becomes complex, requiring data lakes, master data, migration, and replication. In addition, data management must overcome constraints like network latency. These challenges are compounded when different infrastructures are involved, such as cloud providers and on-premise services [24].

Analytical Applications

Operators need increased visibility and better insights on the health of processes and machines (using multi-parameter analysis and multi-parameter views), so they can detect anomalies and prevent issues before they occur. Asset performance management can provide operators with answers to critical questions, including how often equipment fails so that it can be prioritized, how equipment should be maintained, and how unexpected failures and downtime can be avoided. Condition-based maintenance (CBM) tools can also trigger (open) maintenance work orders automatically [24].

Business Challenge

When transforming the mining business into Mining 4.0, four areas should undergo transformation. These are processes, projects, organizational culture, and technology necessary for upgrades and transformations. These, in turn, incur costs, and require knowledge and human resources. Companies must continually update their analysis of production data. The challenge is meeting the requirements of technology and computerization. The new technology and training the workforce to use it is a challenge for the mining business. The organizational culture will be changed with the introduction of new technologies and processes, and the workforce must be retrained to adapt to the new situation.

Corporate and Regulatory Challenge

There are a few corporate and regulatory challenges to implementing Mining 4.0. Governments worry about the disappearance of jobs and want to develop opportunities and a new skilled workforce. Yet there is a need for digital standardization and regulation. This includes improvements in the digital infrastructure and the development of suitable framework conditions to satisfy industrial needs. Further, there is a need to consider economic, sustainable, environmental, and social goals.

High-Level IIoT Architecture for Mining Industry

The layered architecture of IIoT is divided into five domains: control, operation, information, analytics, and business and application domains. The information and analytics domains can be deployed on-site in a single mine site at the edge or off-site at the cloud. The off-site information domain at the cloud is connected to all mine sites and domains. The analytics and the business and application domains follow the information domain. The flow of information, commands/requests, and decisions are shown in Fig. 2.11 with arrows [35].

Edge Control Domain

The control domain in the mine site consists of various components, such as OT software, IT software, sensors, actuators, devices (mobiles), and other physical systems. This domain is responsible for functions carried out by industrial control and automation systems involving reading data from sensors, applying rules and logic, and sending commands to actuators to control the physical systems. For example, the control domain in the mining site usually requires high timing accuracy. The components or devices that perform such tasks in the control domain are typically located close to the physical systems they control, but they can also be dispersed geographically. As a result, the maintenance staff may not be able to easily access these devices and physical systems after deploying them in the mine site [35].

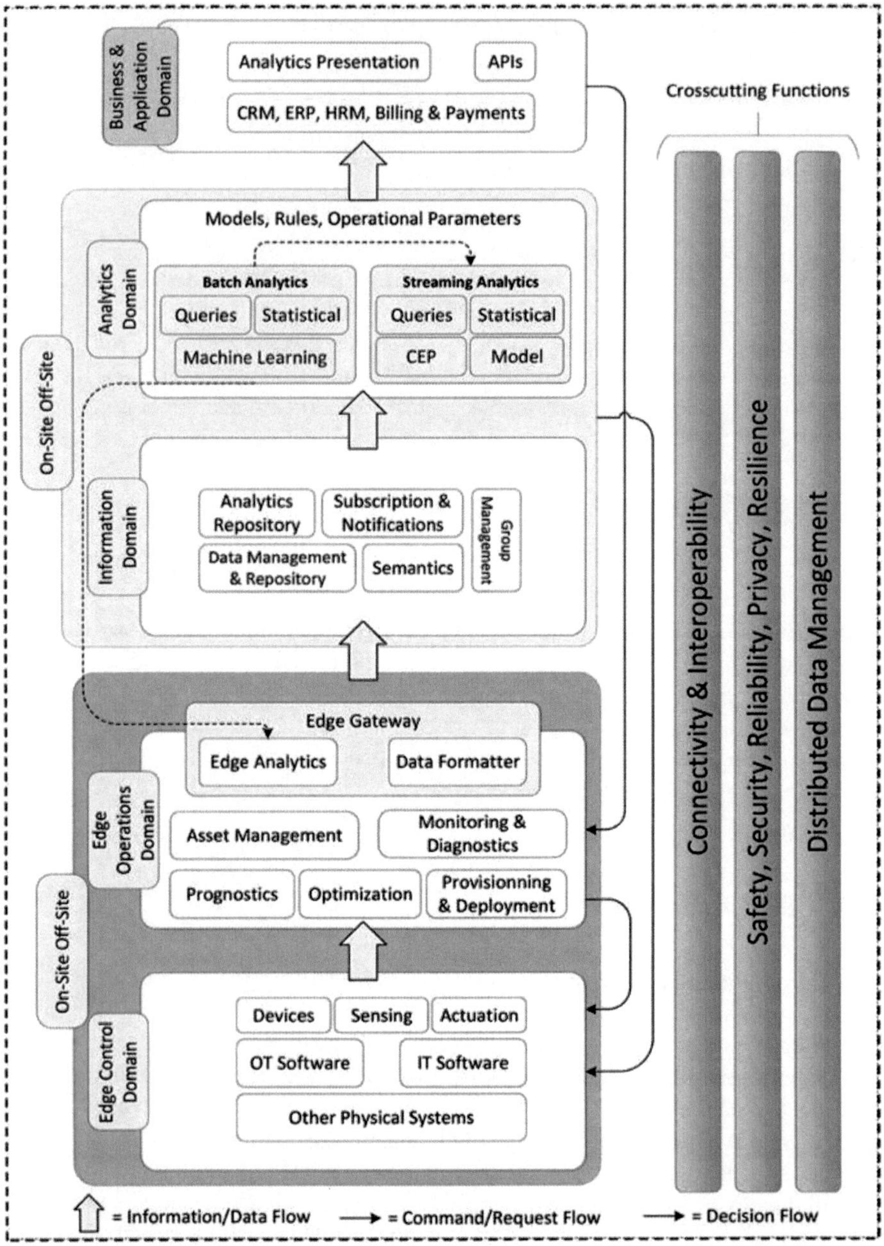

Fig. 2.11 Synthesized high-level IIoT architecture for the mining industry [35]

Operation Domain

There are different types of mines, such as surface and underground mines. Therefore, the operation domain can be inside the mine or on the surface. The operation domain is responsible for the management and operation of the control domain. The functions in this domain are in charge of provisioning and deployment, asset management, monitoring and diagnosis, optimization, and prognostics [35].

Information Domain

The information domain is responsible for managing and processing data. This domain acts as an information hub with the functions for gathering data from various domains, most notably from control and operation domains, and can reside on-site (locally in the mine) or off-site (in the cloud). This domain is the centralized data repository in the mining industry; it stores all the data from various departmental systems, including exploration, geologists, crushers and mills, waste management, sales, and business [35].

Analytics Domain

The analytics domain is responsible for obtaining data from the information domain, mainly from the analytics repository, and performing analytics. Like the information domain, it can reside on-site (locally in the mine) or off-site (in the cloud). However, analytics in the mining industry has challenges as events in the physical world can alter operation and safety. These events can be unintended or dangerous and may impact people's health or damage property and the atmosphere.

The analytics domain in the mining industry can use both batch and stream analytics processing models. The analysis of batch data is an incredibly effective method of storing vast data gathered over time. Stream analytics is the process of analyzing data that stream from one device to another almost instantaneously [35].

Business and Application Domain

The business and application area is a technical environment for the execution of practical enterprise and application logic. It includes business functions that facilitate end-to-end business operations. Business functions include ERP, CRM, HRM, billing, and payments. In addition, data from the information and analytics domains are used for report generation and analysis [35].

Cross-Cutting Functions

The various domains described above focus on primary system functions required to support IIoT implementation in the mining industry. However, additional functionality is also required. These supporting features, the so-called cross-cutting functions, are as follows [35].

- Connectivity and interoperability: Connectivity and interoperability are essential for IIoT systems. However, they are difficult to achieve in the mining industry because of vertical silos and legacy systems. Figure 2.12 shows a synthesized architecture offering a connectivity and interoperability cross-cutting function to ensure heterogeneous systems can communicate.
- Distributed data management (DDM): Data are distributed between various mine sites and edges. DMM is used to manage data produced by various devices and applications in different mine sites. DMM involves three steps as follows:
  - data collection
  - data aggregation
  - data storage.

  Connectivity between them is a prerequisite.
- Safety, security, reliability, privacy, and resilience: To ensure security in system-wide components, a collection of security functions, as cross-cutting functions, must be incorporated in each operational component and its communications, such as encryption and authentication. The overall security of an IIoT system is based on how these systems have implemented security and how securely they are integrated.

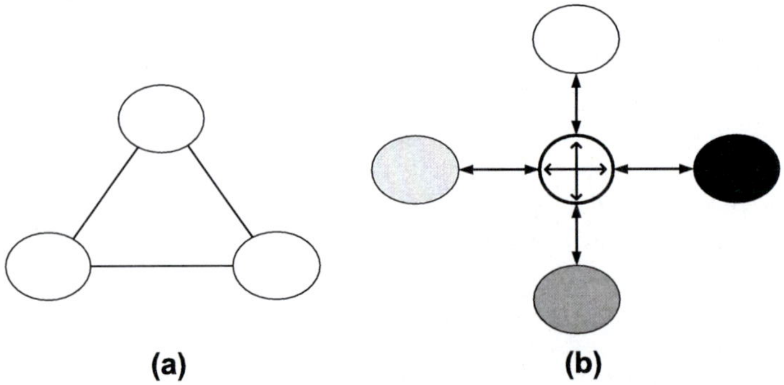

Fig. 2.12 Interoperability models: **a** common metamodel, **b** broker-based model [35]

Digital Themes and Initiatives in Mining

As discussed in the previous sections, digitalization is poised to affect the mining industry in several key areas. In addition, specific digital initiatives (see Fig. 2.13) are expected to significantly impact mines and their stakeholders, including the workforce, other industries, the environment, and society at large [36].

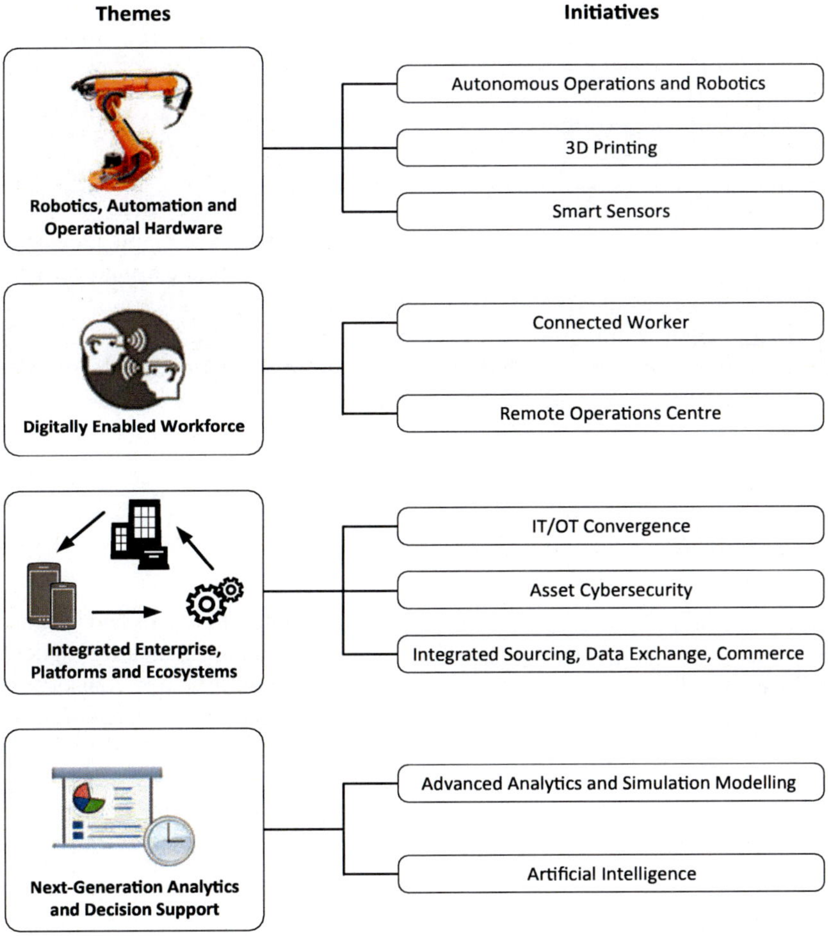

Fig. 2.13 Digital themes and initiatives in mining [36]

Automation, Robotics, and Operational Hardware

Automation and robotics are growing in popularity in the mining industry, and digitally enabled hardware tools are taking over activities traditionally carried out by human-controlled machinery. Some of the most important new technologies are robotic trucks, trains, diggers, 3D printing, automated exploration drones, autonomous stockpile management, autonomous robots to recover recycling material, and pit drones. Over time, their capabilities will improve, and costs will drop, leading to their greater incorporation into the industry [36].

Digitally Enabled Workers

The widespread adoption of digital technologies that empower field workers will revolutionize mining operations. With the rapid proliferation of mobile devices with advanced capabilities, many companies have introduced digitally enabled working methods. The most important technologies are connected worker/mobile devices, digital simulation training/learning, remote operations centers (ROCs), logistics control towers, and virtual collaboration. In addition, improvements in connectivity and other technologies (e.g., wearables, IoT, AR, and VR) have opened up opportunities [36].

Integration of Enterprise, Platforms, and Ecosystems

The mining industry could generate significant value for itself and society by connecting IT to operational technology (OT) and exchanging data throughout the supply chain and beyond. Some of the most critical technologies to be considered here are the convergence of information technology (IT) and operational technology (OT); integrated sales and operations planning; asset cybersecurity; plugged-in, cloud-enabled backbones; intelligent sensors; integrated, agile supply chains; digital monitoring, tracking and analysis of environmental/health and safety indicators; advanced track-and-trace technology [36].

Analytics and Decision Support

Mines can leverage algorithms and AI to process data, thus providing real-time decision support and aiding in future projections. Essential technologies include: self-aware mine/cognitive networks; exploration; field development planning; production forecasting; predictive asset management/maintenance; advanced analytics for demand forecasting; digital twin; digitally enabled fraud detection and anti-corruption traceability; advanced analytics for production optimization and maintenance; integrated dynamic constraint optimization across the supply chain; AI for operations support; ore valuation and simulation modeling [36].

Recommendations for a Successful Digital Transformation

Successful digitalization will require collaboration between industry leaders, communities, and policymakers. The following sections include considerations for relevant stakeholders [36].

Industry Leaders

For successful digital transformation, industry leaders must do the following:

Align strategy and operations with innovation: Build a focused strategy that incorporates digital transformation, and align it with the business model, the business processes, and the organization;
Look beyond the organization: Looking outside the organization's current portion of the value chain and connecting or engaging in new areas with buyers, suppliers, and customers can add value;
Work to improve data access and data relevance: Focus on applicable insights and share them with the appropriate organizational levels;
Engage and train a digital workforce: Tomorrow's digital worker must be prepared today;
Find ways to work with and compensate local stakeholders for the responsible use of their resources; and
Make new partnerships and strengthen existing ones: Optimal digitalization requires integration with local stakeholders and creating new models for the operation and ownership of mining assets [36].

Communities, Policymakers, and Governments

For successful digital transformation, communities, policymakers, and governments must do the following:

- Define digital standards and regulations: Interoperability and information sharing should be encouraged across the digital ecosystem. However, the requisite level of security must be maintained;

- Prioritize key performance indicators (KPIs): Several KPIs have been used to tie mining organizations to local communities. However, none gives a causal view, and no standard suggests how to best invest in both industry and society; and
- Work on transparency and traceability: Digital platforms can be developed to trace and share information on production, the environment, and community engagement [36].

Conclusion

In the digital era, the mining industry needs to be updated, just like other traditional industries. Everybody knows digitalization can be very beneficial. However, the old industries, like mining, need to know the basics before going through the Industry 4.0 process. This chapter has tried to open a new window to researchers and mine managers interested in helping the mining industry improve its analytical maturity to tackle the challenges of digital transformation. It has clarified some critical subjects in Mining 4.0 and explained the role of operators and managers in the generation of smart mine sites. It has explained future of intelligent mining in detail, and has described organizational transformation. Finally, it has discussed advanced analytics as the main component of digital mines, and clarified the role of automation, robotics, digital platforms, optimization models, and decision-making applications.

References

1. Ślusarczyk, B. 2018. Industry 4.0: Are we ready? *Polish Journal of Management Studies* 17.
2. Jurdziak, L., R. Błażej, and M. Bajda. 2018. Cyfrowa rewolucja w transporcie przenośnikowym–taśma przenośnikowa 4.0. *Transport* 2: 40.
3. Bongaerts, J.C. 2020. Mining 4.0 in the context of developing countries. *African Journal of Mining, Entrepreneurship and Natural Resource Management (AJMENRM)* 1 (2): 36–43.
4. Bartnitzki, T. 2017. Mining 4.0—Importance of Industry 4.0 for the raw materials sector. *Artificial Intelligence* 2: M2M.
5. Rylnikova, M., D. Radchenko, and D. Klebanov. 2017. Intelligent mining engineering systems in the structure of Industry 4.0. *E3S Web of Conferences*. EDP Sciences.
6. Lööw, J., L. Abrahamsson, and J. Johansson. 2019. Mining 4.0—The impact of new technology from a workplace perspective. *Mining, Metallurgy & Exploration* 36 (4): 701–707.
7. Roldán, J.J., et al. 2019. A training system for Industry 4.0 operators in complex assemblies based on virtual reality and process mining. *Robotics and Computer-Integrated Manufacturing* 59: 305–316.
8. Faz-Mendoza, A., et al. 2020. Intelligent processes in the context of Mining 4.0: Trends, research challenges, and opportunities. In *2020 International Conference on Decision Aid Sciences and Application (DASA)*. IEEE.
9. Soofastaei, A. 2020. Data analytics for energy efficiency and gas emission reduction. In *Data analytics applied to the mining industry*, 169–192. CRC Press.
10. Sishi, M., and A. Telukdarie. 2020. Implementation of Industry 4.0 technologies in the mining industry—A case study. *International Journal of Mining and Mineral Engineering* 11 (1): 1–22.

11. Nanda, N.K. 2019. Intelligent enterprise with Industry 4.0 for the mining industry. In *International Symposium on Mine Planning & Equipment Selection.* Springer.
12. Soofastaei, A. 2020. Digital transformation of mining. In *Data analytics applied to the mining industry*, 1–29. CRC Press.
13. Deloitte. 2018. The top 10 issues shaping mining in the year ahead. Tracking the trends 2018, July 3. Available from: https://www2.deloitte.com/content/dam/Deloitte/us/Documents/energy-resources/us-er-ttt-report-2018.pdf.
14. Wang, L.G. 2015. Digital mining technology and upgrading of mine technology in China. *World Non-Ferrous Metal* 7–13.
15. Carter, R.A. 2014. Equipment selection is key for productivity in underground loading and haulage. *Engineering and Mining Journal* 215 (6): 46.
16. Li, J.-G., and K. Zhan. 2018. Intelligent mining technology for an underground metal mine based on unmanned equipment. *Engineering* 4 (3): 381–391.
17. Gustafson, A., et al. 2014. Development of a Markov model for production performance optimization. Application for semi-automatic and manual LHD machines in underground mines. *International Journal of Mining, Reclamation, and Environment* 28 (5): 342–355.
18. Deloitte. 2018. Intelligent mining. Delivering real value, March 2018. Available from: https://www2.deloitte.com/content/dam/Deloitte/global/Documents/Energy-and-Resources/gx-intelligent-mining-mar-2018.pdf.
19. Deloitte. 2019. Intelligent mining. Delivering real value. For private circulation only, October 2019. Available from: https://www2.deloitte.com/content/dam/Deloitte/in/Documents/about-deloitte/in-mm-ntelligent-mining-brochure-noexp.pdf.
20. Da Xu, L., W. He, and S. Li. 2014. Internet of Things in industries: A survey. *IEEE Transactions on Industrial Informatics* 10 (4): 2233–2243.
21. Blanchet, M., et al. 2014. *Industry 4.0: The new industrial revolution—How Europe will succeed*, vol. 11, 2014. München: Hg. v. Roland Berger Strategy Consultants GmbH.
22. Schmidt, R., et al. 2015. Industry 4.0—Potentials for creating smart products: Empirical research results. In *International Conference on Business Information Systems.* Springer.
23. Anastasova, Y. Internet of Things in the mining industry—Security technologies in their application.
24. de Moura, R.L., L.D.L.F. Ceotto, and A. Gonzalez. 2017. Industrial IoT and advanced analytics framework: An approach for the mining industry. In *2017 International Conference on Computational Science and Computational Intelligence (CSCI).* IEEE.
25. Neubert, R. 2016. IIoT offers economical, operational benefits. *Plant Engineering* 4: 9–10. In Focus.
26. Clegg, S., M. Kornberger, and T. Pitsis. 2016. *Administração e organizações: uma introdução à teoria e à prática.* Bookman Editora.
27. Lala, A., et al. 2016. *Productivity at the mine face: Pointing the way forward*, 2–12. New York: McKinsey and Company.
28. Ng, E. 2013. *Commodities super-cycle is 'taking a break': Runaway prices in commodities markets have ended, but long-term demand for commodities on the mainland is strong.* South China Morning Press.
29. Li, Y., et al. 2012. Towards a theoretical framework of strategic decision, supporting capability and information sharing under the Internet of Things. *Information Technology and Management* 13 (4): 205–216.
30. Velosa, A., Y. Natis, and B. Lheureux. 2016. Use the IoT platform reference model to plan your IoT business solutions, September 17. Available from: https://www.gartner.com/en/documents/3447218-use-the-iot-platform-reference-model-to-plan-your-iot-bu.
31. Lee, I., and K. Lee. 2015. The Internet of Things (IoT): Applications, investments, and challenges for enterprises. *Business Horizons* 58 (4): 431–440.
32. Gubbi, J., et al. 2013. Internet of Things (IoT): A vision, architectural elements, and future directions. *Future Generation Computer Systems* 29 (7): 1645–1660.
33. Cooper, J., and A. James. 2009. Challenges for database management in the Internet of Things. *IETE Technical Review* 26 (5): 320–329.

34. Abu-Elkheir, M., M. Hayajneh, and N.A. Ali. 2013. Data management for the Internet of Things: Design primitives and solution. *Sensors* 13 (11): 15582–15612.
35. Aziz, A., O. Schelén, and U. Bodin. 2020. A study on industrial IoT for the mining industry: Synthesized architecture and open research directions. *IoT* 1 (2): 529–550.
36. World Economic Forum & Accenture. 2017. Digital transformation initiative 2017. Cited 2021 Mar 20. Available from: https://www.accenture.com/t20170411t115809z__w__/us-en/_acnmedia/accenture/conversion-assets/wef/pdf/accenture-telecommunications-industry.pdf.

Chapter 3
Advanced Analytics for Ethical Considerations in Mining Industry

Abhishek Kaul and Ali Soofastaei

Abstract Advanced analytics and AI technologies are driving digital transformation in the mining industry. Although these technologies are delivering results, their recommendations for people-based decisions are subject to ethical considerations. Unethical AI models can expose a mining company to reputational, regulatory, and legal risks. This chapter introduces the concept of ethics and discusses AI's ethical principles in the mining industry using case studies on energy, workforce, automation, maintenance, and safety. Further, practical guidelines and recommendations are provided on how to use AI ethically across the project lifecycle. Lastly, fairness metrics and bias mitigation algorithms are described to evaluate and remove bias in AI.

Keywords Advanced analytics · Artificial intelligence · Mining · Ethical consideration

Introduction

Advanced analytics holds excellent power to solve some of the most complicated mining industry problems.

Statistical methods, machine learning (ML), deep learning (DL), artificial intelligence (AI) are the main techniques used in advanced data analytics. Mining companies are increasingly using these techniques to improve revenue and reduce cost by making decisions about people, processes, and technologies, accessing worker productivity, exploring the next mine site, or predicting equipment maintenance schedules.

This chapter will use AI as the umbrella term to reference advanced analytics technologies, including algorithms, ML, neural networks, DL, natural language

A. Kaul (✉)
IBM, Singapore, Singapore

A. Soofastaei
Vale, Brisbane, Australia
URL: http://www.soofastaei.net

A. Soofastaei (ed.), *Advanced Analytics in Mining Engineering*,
https://doi.org/10.1007/978-3-030-91589-6_3

processing, computer vision, and more. As defined by the European Commission High-Level Expert Group on AI [1], "AI refers to systems that display intelligent behavior by analyzing their environment and taking actions—with some degree of autonomy—to achieve specific goals. AI-based systems can be purely software-based, acting in the virtual world (e.g., voice assistants, image analysis software, search engines, speech and face recognition systems), or AI can be embedded in hardware devices (e.g., advanced robots, autonomous cars, drones or Internet of Things applications)." Although AI is delivering results, its recommendations for people-based decisions are subject to ethical considerations. Issues arise if advanced analytics models are biased based on gender, age, or ethnicity and are not ethical in providing recommendations.

If AI is not ethical, companies are exposed to reputational, legal, and regulatory risks as seen in the industry. For example, Goldman Sachs is being investigated by regulators for discriminatory credit card limits for Apple cards based on gender [2]. Regulators are also investigating the UnitedHealth Group algorithm that a study found prioritized care for healthier white patients over sicker black patients [3], and after many years of development, Amazon scrapped the AI recruiting tool that showed bias against women [4]. These risks arise if there is a lack of understanding in training data, model results, or fairness metrics.

Ethics in AI has been an active field of research for decades. If we look back in history, in 1950, Asimov has introduced the three laws of robotics [5]:

- First Law: A robot may not injure a human being or, through inaction, allow a human being to come to harm.
- Second Law: A robot must obey the orders given by human beings except where such orders would conflict with the First Law.
- Third Law: A robot must protect its existence as long as such protection does not conflict with the First Law or Second Law.

Although his laws are not adequate and Asimov's stories are on science fiction, they define how an ethical robot would behave in society. In early 2000, as AI models gained mainstream adoption, many governments, private enterprises, civil society, the multi-stakeholder organization invested in developing ethical frameworks. Technology giants like IBM, Microsoft, Amazon, Google, Facebook, and more have already published ethical guideless to prevent risk [6].

The mining industry has always been focused on ethics and has developed many regulations to protect the people and environment. However, with the adoption of AI models at large, ethics in AI has become a significant topic and requires companies to come together and define ethical guidelines for AI in mining. Unethical AI models can expose mining companies to reputational, regulatory, and legal risks.

For example, an AI model that predicts fuel consumption for haul trucks shows operator behavior is a significant predictor, and older operators have a lower fuel consumption compared to younger male operators; what should the manager do? Should he question the efficacy of the AI model, request to re-train the model without demographic variables, change the hiring policies to favor older operators, or invest in training programs for the younger operator to improve driving behavior or set up devices in trucks that suggest ideal speed on different gradients, weather conditions, or completely do away with operators by investing in automated trucks? These questions form a part of the ethical dilemma essential to discuss and have guidelines for miners.

This chapter provides case studies and guidelines for mining companies to understand the ethical considerations for AI models. We will first define what ethics is, then dive into the guidelines published on ethics in AI. Next, we will discuss the ethical principles for AI leveraging one of the prominent frameworks. For each principle, a mining case study is explained in the area of energy, workforce, automation, maintenance, and safety with practical guidance on how to use AI ethically. Next, we will deep dive into the AI methodology and define ethical risks at each stage. Lastly, we will detail the AI fairness metrics that mining companies should consider and briefly discuss the bias mitigation algorithms deployed to ensure AI model fairness.

It is essential to understand that AI is not about replacing people, but reskilling people and deploying AI solutions will improve the quality of work in mine sites and reduce the human failures and hazards in the mine site. However, to be successful, we have to work with AI and ensure its ethical use.

Defining Ethics

What is ethics? As per the Oxford dictionary, ethics means "the moral principles that govern a person's behavior or the conducting of an activity." Britannica dictionary has a similar definition "Ethics, also called moral philosophy, is the discipline concerned with what is morally good and bad and morally right and wrong." Merriam-Webster defines ethics as "the discipline dealing with what is good and bad and with moral duty and obligation." In summary, ethics is based on well-founded standards of right and wrong that prescribe what humans ought to do, usually in terms of one's rights, obligations, benefits to society, fairness, or specific virtues [7]. The three main ethical theories are deontology, consequentialism, and virtue ethics. They help to identify, describe, and eventually overcome ethical issues.

In the mining industry, ethics has been an important topic in terms of the environment and society. It impacts climate, deforestation, dams, relocation of people, the safety of miners, conflict mining, and more [8]. As a result, many codes, agreements have been published in the mining industry to ensure ethical, sustainable, and safe mining practices like Global Reporting Initiative, International Cyanide Management Code, Equator Principles, International Finance Corporation's Performance Standards, Extractive Industries Transparency Initiative, Natural Resource Charter, United Nations' "Ruggie Principles," Task Force on Climate-related Financial Disclosures, Dodd-Frank Act: Section 1502, Environmental Social Corporate Governance (ESG) reporting initiative, and more. These provide mining companies a way to act ethically and responsibly toward the environment and society. Further, industry bodies like the International Council on Mining and Metals publish Mining Principles, which serve as best practice frameworks on sustainable development.

However, there is limited guidelines on ethics for the use of AI in the mining industry. Many times, the answer to ethical questions is tied to the organization's values. The organization should unpack values with detailed guidance on when and how to use AI for the data science and digital transformation teams. Detailed guidance includes compliance requirements and regulations and advises teams on what metrics the models should use. For example, when data scientists are building a resume screening application for truck operators, should they use:

- Demographic parity, which states that the proportion of each segment of a group (like gender or ethnicity) should be selected in equal numbers; or
- Equal opportunity parity, which states that the proportion of each segment of a group (like gender or ethnicity) should be selected in proportion to the population of that group.

Guidance on these questions is essential to ensure the ethical use of AI.

Guidelines on Ethics for AI

AI is a broad term encompassing technologies that can mimic, automate, and outperform human intelligence capabilities. When we try to apply ethical principles to AI, there are limited common and agreed guidelines globally. Many countries, organizations, and associations have published AI policy guidelines. However, it is voluntary to adhere. In this section, we discuss and summarize the guidelines available today.

Country-specific guidelines provide a broad level objective for the use of AI to ensure human-centric, safe, and trustworthy AI. Most of these guidelines make the organization use AI responsible and accountable for their decisions and ask for the same ethical standards in AI-driven decisions as human-driven decisions. Examples include the Obama administration's "Preparing for future of AI" [9], China's White Paper on AI Standardization [10], European commissions "Ethics guidelines for trustworthy AI" [11], and OECD "Recommendation of the Council on AI" [12]. In addition, organizations developing AI solutions like IBM [13], Google [14, 15], Microsoft [16], Deepmind [17], and others have also published guidelines on the ethical use of AI systems. Further, there have been contributions to guidelines from multi-stakeholder associations like IEEE [18], Partnership in AI [19], and also from civil society like the "Toronto Declaration: Protecting the rights to equality and non-discrimination in ML systems" [20] and more.

Some of the recently published papers provide comparisons and themes of AI Ethics guidelines published. Notably, the paper by Fjeld et al. for Harvard University Berkman Klein Center for Internet and Society on "Principled AI: Mapping Consensus in Ethical and Rights-based Approaches to Principles for AI" [21] has analyzed 32 of the initiatives by governments, private sector, civil society, and multistakeholder organization from 2016 to 2019. The study clusters guidelines into key themes around privacy, accountability, safety, and security, transparency and explain ability, fairness and non-discrimination, human control of technology, professional responsibility, and promotion of human values.

Another paper published by Jobin et al. in Nature magazine on the global landscape of AI ethics guidelines suggests globally five emerging ethical principles that are deemed necessary—transparency; justice, fairness and equity; non-maleficence; responsibility and accountability, and privacy [22]. However, the paper also concludes that there is significant divergence in how these principles are interpreted and implemented.

The paper by Hagendorff et al. for Springer Minds and Machine on "The Ethics of AI Ethics: An Evaluation of Guidelines" [23] discusses the guidelines from an impact perspective and suggests that unless more detailed, action-oriented guidelines are not released, doing the right thing, self-governed, recommendation guidelines may not be enough for the ethical deployment of advanced analytics and AI applications.

A recent study by the Panel for the Future of Science and Technology (STOA) of the European Parliament discusses significantly 'How can we move from AI ethics to specific policy and legislation for governing AI?' and recommends policies to govern ethical AI. The key part of their policy option is to implement controls in terms of "ethical technology assessment," "data hygiene certificate," and more [24]. Data quality plays a key role in AI model efficacy. Reviewing the sourcing, acquisition, diversity, data labeling, and data hygiene certificate would ensure that good quality data have been used to train the AI model.

When looking at the major players in the mining industry, there are limited guidelines published on ethics in AI. Most of the publication focuses on sustainability and ethical code of conduct. As more and more AI applications get deployed in the mining industry, there needs to be a focus on defining the ethical guidelines for AI.

Global Mining Group (GMG) is an industry collaboration body between OEMs/OTMs, regulators, research organizations, universities, and others in the mining industry. It has one active project on providing guidelines for "Implementation of AI" [25]. In addition, ethics and education are topics being developed as part of the "Implementation of AI" guideline.

In subsequent sections, we will cover practical examples of ethical considerations for the mining industry based on my experience working with mining, industrial companies.

Mining Industry Ethical Principles for AI

Mining companies adopting AI solutions should have the same standards and guidelines on AI ethics similar to their business conduct guidelines. As published in a paper by Jobin et al. on the global landscape of AI ethics guidelines [22], globally, five emerging ethical principles are deemed necessary—transparency; justice, fairness and equity, non-maleficence, responsibility and accountability, and privacy.

This section attempts to interpret these principles with a view on mining industry use cases and what considerations AI project managers, developers, and users of these

Table 3.1 Ethical principles and mining use cases

Ethical principle	Mining industry-use case	Description of use case
Transparency	Energy	AI model for reducing fuel consumption leveraging operator demographics
Justice, fairness and equity	Workforce	AI model for automated resume screening application
Non-maleficence	Automation	AI model for autonomous mining equipment
Responsibility and accountability	Maintenance	AI model for predictive maintenance application using operator logs
Privacy	Safety	AI model for worker safety using surveillance video (CCTV)

applications understand when developing, interacting, or buying these applications (Table 3.1).

The next section discusses each ethical principle and describes how it needs to be considered for mining industry use cases.

Transparency

AI transparency refers to explaining the ability, interpretability, disclosure of the algorithmic models, including their training data, accuracy, performance, bias, and other metrics.

When AI models are used in the mining industry, it is essential to understand why a particular recommendation is made. Explaining model results helps get business and data science teams on the same platform when discussing model outputs and transparency in training data and model metrics. In addition, by explaining how an AI model makes the recommendations, the business can identify and ultimately correct biases in the models. This helps to ensure that the model is ethically compliant [26, 27].

Depending on the level of influence a mining company has while adopting the solution, i.e., developed by in-house data science teams or purchased commercially from external vendors (COTS), details on training data, bias, and explain ability charts can be reviewed.

Let us discuss this in detail using the AI model for reducing fuel consumption in haul trucks.

Fuel is an important cost contributor for haul trucks in surface mining. Multiple parameters affect fuel consumption like the type of truck, payload, distance, hours, weather, and operator behavior (includes speed, maneuvering, acceleration, and braking) (Fig. 3.1).

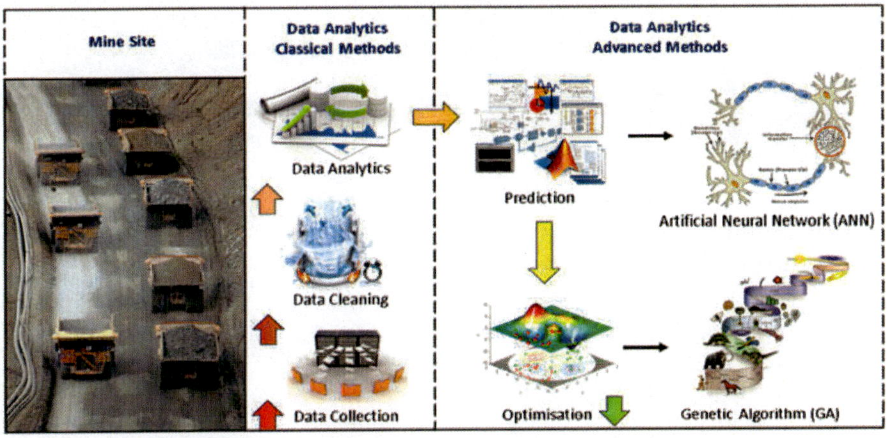

Fig. 3.1 AI solution for fuel consumption

Input Data Acquisition and Suitability

When building the model, data scientists will choose the data elements, i.e., features that go into the model. For example, let us assume that when the model was built, based on the training data, the team concluded that operator (driver) behavior is an essential predictor of haul truck fuel consumption. Therefore, the team further included the demographic data points of operators (like age, race, gender, education, etc.) to identify patterns in data that influence operator behavior.

Depending on the training data set, for example, country, mine site, number of operators, this analysis may have bias and may not hold for the general case. For example, maybe the model can get biased to predict low fuel consumption for older male workers based on one mine site operation.

It is recommended not to include demographic data in the analysis and rely solely on the unique mine equipment operator identifier in such applications. If the use of demographic data is needed, then depending on the level of influence a mining company has, de-biasing techniques can be applied on training data, model, or post-processing the results. The section on fairness metrics and de-biasing will detail more on these techniques.

Explaining Model Results

Artificial neural networks (ANNs) are generally applied to data to understand the top factors influencing fuel consumption and recommend changes to controllable factors, reducing fuel consumption per ton of ore mined using a genetic algorithm (GA) in the optimization phase of the project.

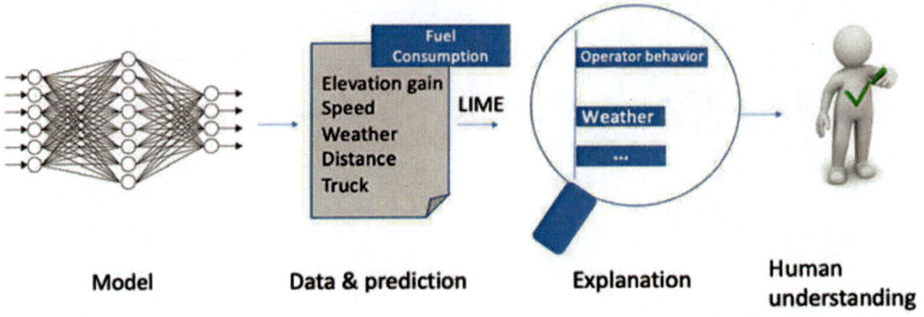

Fig. 3.2 LIME explanation for AI model

If we treat this as a black-box model, we should know why the model recommends a particular fuel burn rate. This visualization can be explained using a [28] Shapley Additive Explanations (SHAP) or [29] Local Interpretable Model-Agnostic Explanations (LIME), or activation atlases [30] (Fig. 3.2).

This visualization explains the model but does not represent causality.

Ideally, in this use case, the business objective of the AI model should be fine-tuned to provide guidelines to an operator to influence their behavior like speed on the inclined wet road, maximum acceleration, etc., and provide a mechanism to monitor for deviations in operator actions to AI recommendations.

In summary, transparency in terms of the model explains that ability is important to understand from an ethics standpoint. However, in some cases, if the solutions are acquired from external vendors, and they do not require any person-related data (e.g., predictive maintenance application taking in thousands of equipment tags), it may be acceptable to have opaque models, so long as the business teams using the model ascertain results accuracy post-deployment.

Justice, Fairness, and Equity

Justice means that AI algorithms are fair and do not discriminate against particular groups intentionally or unintentionally. There have been numerous publications on fairness and how to identify, mitigate bias in algorithms.

When advanced analytics and AI models are used in the mining industry, it is important to understand that the model's recommendations are just, fair, and equal.

Data drive AI models to detect patterns and make predictions and recommendations. Hence, human historical biases influence predictions. There are many fairness metrics like statistical parity difference, disparate impact, equal opportunity difference, Theil index, and more that can be used to calculate the algorithm fairness. Further, once the fairness metrics are calculated, bias mitigation algorithms can be deployed to bring the metrics to an ideal condition like reweighing, optimized preprocessing, adversarial de-biasing. We will discuss details of these techniques in the section on fairness metrics.

For this principle, we will use the case on AI model for automated resume screening applications.

Many AI-powered automated resume screening solutions are available commercially or can easily be developed by in-house data science teams. The primary task for these models is first to extract the data from candidates' resumes like names, experience, education, and more into a table. The second step is to match this as per the company's requirement, in terms of both of job profile specification and the historical success criteria of employees (Fig. 3.3).

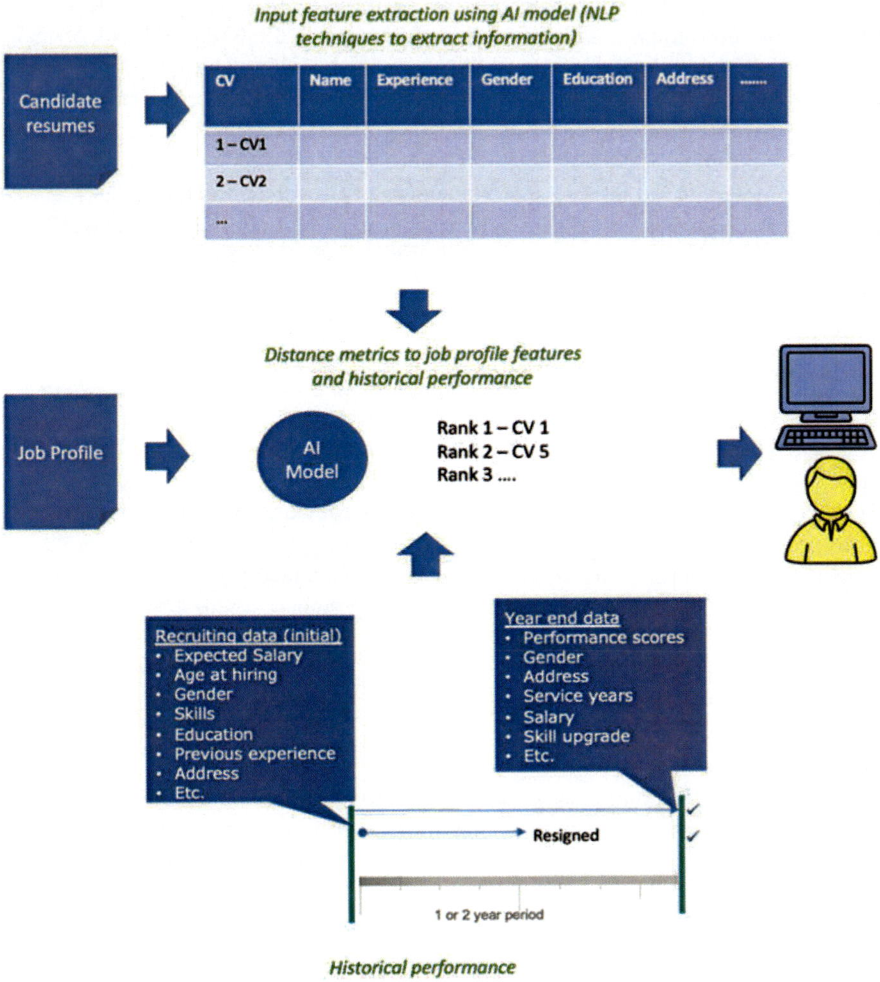

Fig. 3.3 Automated resume screening application

Choosing Features

When developing these models, one of the tasks is to choose the data inputs (features) for training the model. Removing gender features from the model will not make it fair. Many historical human biases need to be considered.

Data science generally works with business experts to define what features to consider and test them out iterative. Fewer instances of particular groups—minority groups, unrepresented data, or proxy variables may make the algorithms unfair. For example, considering the length of service in a prior role may not be fair to the spouse of transferable job candidates. Similarly, applicant address may become the proxy variable for income or ethnicity data.

Further, the model performance should be measured for accuracy and fairness metrics like demographic parity for gender (we select the equal number of males and females) or equal opportunity parity (we select males and females in the ratio of males the number of applications). The metrics depend on organization values; however, there will be some tradeoff inaccuracy when we force the model to meet the fairness criteria.

Mining organization leaders should encourage conversations on model features, guide their teams on which fairness metrics to use, and understand their impact on accuracy.

Continuous Monitoring

Data science teams would check the model accuracy, fairness metrics, and extensive testing model during the model development. Similar exercises should be done at regular intervals to monitor the model and check for deviations from accuracy metrics or fairness metrics.

Mining organizations should encourage regular evaluation of model metrics and fine-tuning if the models are rolled out to globally disparate regions.

Non-maleficence

This term is used to define users' safety, security, and the commitment that the AI model will not cause harm, such as discrimination, violation of privacy, malicious hacking, or bodily harm.

In the mining industry, when advanced analytics and AI models are used for equipment automation, then the considerations for non-maleficence arise for both individuals and society in terms of harm to privacy, discrimination, loss of skills, adverse economic impacts, risks to the security of critical infrastructure, and possible negative long-term effects on societal well-being.

Let us discuss this in detail using the example of automation for haul trucks. In haul trucks, autonomous haulage systems (AHS) technology has been deployed in many mine sites and provides significant business benefits. For example, in Rio Tinto's autonomous fleet, each AHS-operated truck was estimated to have operated for 700 h more than conventional trucks and 15% lower load and haul unit costs [31]. Moreover, although these trucks are safe to operate, they have had few incidents in the past [32, 33].

With current technology, the AHS-operated truck follows a preprogrammed path from one point to the next point and is monitored through a remote-control room with remote operator manual fallback. As automation technology evolves, the trucks will become truly autonomous to adjust their paths to accommodate obstruction, other variations which will lead to moral dilemmas as described for automatic cars [34, 35] (Fig. 3.4).

IEEE has done significant work in ethically aligned design [37], which provides the framework to guide the non-technical implications of these technologies, especially related to ethical aspects. Eight principles of ethically aligned design have been published—human rights, well-being, data agency, effectiveness, transparency, accountability, awareness of misuse, and competence. In addition, further IEEE P7000 series of standards has been developed specifically to addresses specific issues at the intersection of technological and ethical considerations.

ICMM White Paper and Guiding Principles for Functional Safety for Earth-moving Machinery [37] also details non-deterministic ways to evaluate the safety of autonomous systems [37].

Eventually, automation has to be balanced with social implications. Operators will work on computer screens and not on the physical equipment, which will decrease lower-skilled labor requirements. Mining companies should look at reskilling their workforce along with deploying autonomous technology [38, 39].

Responsibility and Accountability

Responsibility and accountability refer to the AI acting with integrity, clarifying responsibility, data ownership, and legal liability. There has been much debate on who is ultimately responsible—is it the AI model or the humans who built it.

In the mining industry, when AI models are used across the value chain from exploration and development, operations and supply chain, marketing and sales for decision making, responsibility, and accountability arise.

For example, in the area of energy optimization, if the AI model recommends the optimized set points for equipment from a cost perspective, then two questions arise. First, is the AI responsible for automatically updating the set points, or is the operator responsible? Secondly, is this set point going to adversely impact other company KPIs like timely delivery or environmental compliance, and who decides?

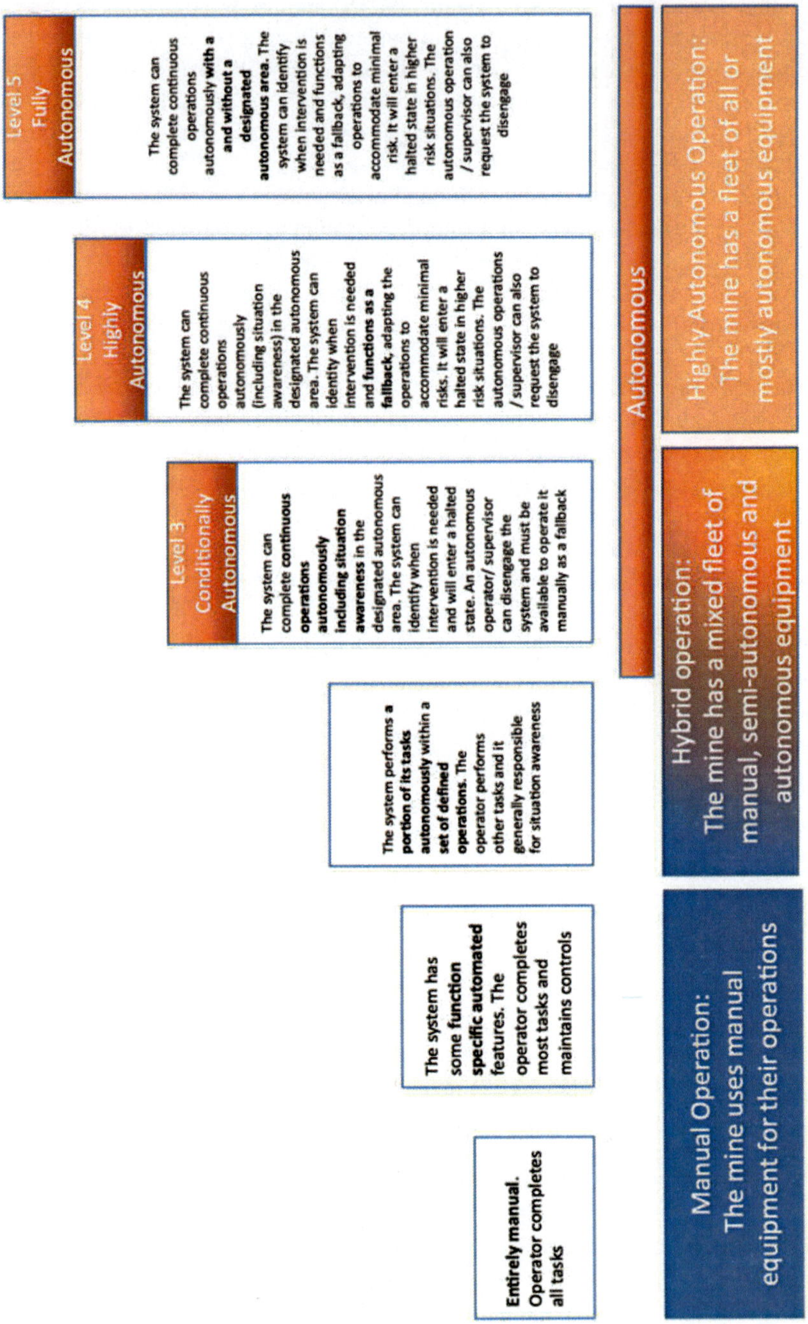

Fig. 3.4 GMG implementation of autonomous systems report [36]

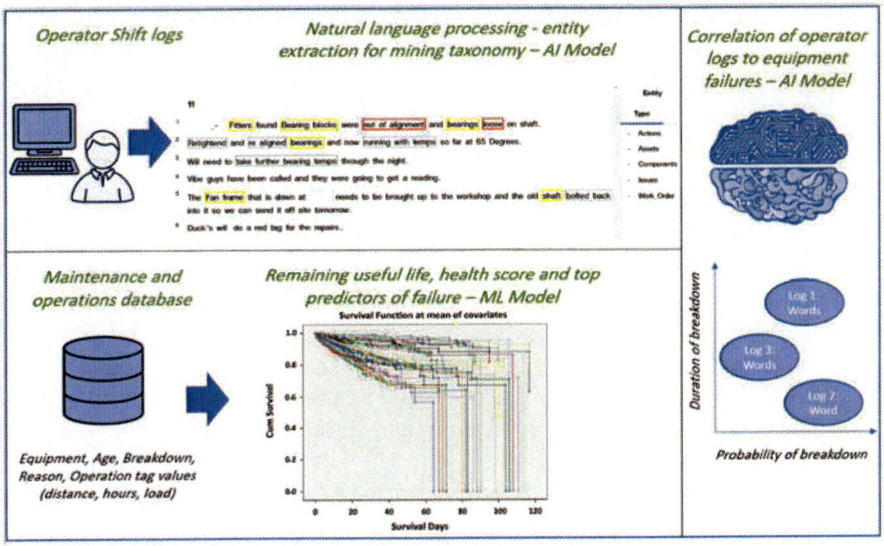

Fig. 3.5 Predictive maintenance application using operator logs

Equipment downtime significantly affects the productivity and safety of mining operations. The main goal of predictive maintenance is to shift the unplanned breakdowns to planned maintenance activities, increase the equipment lifetime, optimize maintenance schedules, and ensure safe operations. ML algorithms like Cox regression, logistic regression, gradient boosting, neural nets are applied to predict the remaining useful life (RUL) and predict the health score of equipment using historian, maintenance data. Further, natural language processing (NLP) techniques like Word2Vec, BERT are applied on operator shift logs to gain a deeper understanding of operational events like faults, trips, overriding, noise, resetting observed, actioned, and documented by the operator, which are then co-related to maintenance failures to provide deeper insights (Fig. 3.5).

One of the data points used in the analysis is the operator shift logs. Operator privacy is one of the considerations for analyzing logs, however beyond privacy, when analyzing operator logs, if language linguistics analysis is used for deciphering personal traits, personal attributes, like modeling, and then used for decision making that affects the operators professionally or personally, then this becomes a relevant topic for ethical consideration.

In such applications, it is recommended to use the co-relation of events (nouns, verbs) with maintenance failures only for the predictive maintenance application without diverting into other areas of linguistics modeling. Further, it is recommended to de-bias the NLP models, which may cause some drop-in accuracy points but helps to keep the recommendations fair and unbiased.

Similar use cases can be discussed for ethical consideration in other parts of the value chain, like recommending mine sites for exploration or optimizing routes for logistics or timing for chartering vessels.

Privacy

Privacy means that your personal information is kept confidential and only shared with consent. Many countries have passed laws and regulations to protect the privacy of their citizens, like general data protection regulation.

Digital surveillance technologies have seen exponential growth during the coronavirus pandemic with governments, organizations tapping into mobile phones, video footage for contact tracing, face mask compliance, safe distance control, and more [40].

Furthermore, data are the gold mine that is used to build AI models. Data science teams with the best intentions always try to add more and more datasets, cross-correlate them to improve performance, and sometimes fail to consider how they may be doing something improper for workers and customers.

When advanced analytics and AI models are used to say HSE non-compliance, worker safety, health, and customer data, privacy considerations arise in the mining industry.

Let us discuss this in detail using the AI model for increasing worker safety by using video surveillance data (CCTV).

Video surveillance data (CCTV cameras) are used in many work areas to ensure the security and safety of the site. Typically, hundreds of cameras feed data to the site security/safety office. Since it is impossible to review the human operator's feed from all cameras in real-time, the use case for video surveillance tends toward post-facto video retrieval and analysis for historical incidents and disputes. With advancements in computer vision technologies, AI models are trained with the surveillance video feed to perform automated analysis for object detection, object classification, object tracking and raise proactive alerts in real-time to detect and mitigate any safety or security violations (Fig. 3.6).

Surveillance is itself an ethically neutral concept. What determines the ethical nature of a particular instance of surveillance will be the considerations which follow, such as justified cause, the means employed, and questions of proportionality. While it can be argued that monitoring remotely via a camera is no different from historical times when security personnel were physically present at the worksite, there are local privacy laws and regulations which must be complied with while using surveillance, face recognition technologies.

AI technologies in computer vision are leapfrogging every few months, like understanding the spatiotemporal relationships of objects, which can monitor people's behavior based on the change of posture, time of day, relationship with equipment, movement between areas, and more.

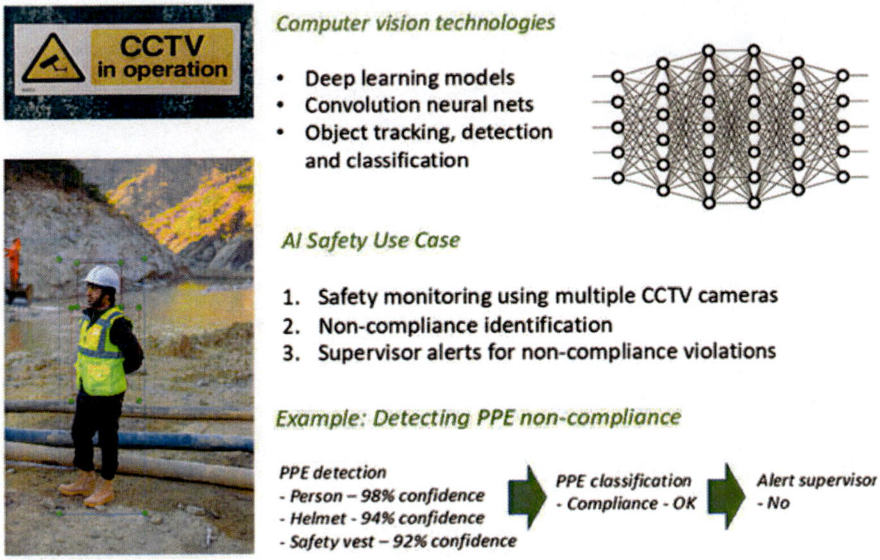

Fig. 3.6 Worker safety using surveillance video

When employing AI for surveillance monitoring, ethical considerations should balance workers' privacy–trust–autonomy and workers' safety–security–behavior. In addition, mining companies should have policy guidelines beyond regulatory compliance to ensure ethical use of data. They should also have defined identity and access management standards in place and protect the privacy of their workers and customers.

In the next section, we deep dive into the AI methodologies and mitigate risk while developing AI models.

Deep Dive into the AI Methodology and Ethical Risks

As AI grows in every area of the mining industry, the ethical risks associated with people and the planet continue to grow (Fig. 3.7).

In order to understand the ethical risks and their drivers, it is important to deep dive into the technical aspects and understands how advanced analytics and AI models are built.

ML, DL, and advanced analytics techniques are used to build AI models. Generally speaking, these algorithms learn from data (training data) and build a decision rule based on historical instances of data. This decision rule is then used to predict the future.

AI model learning is about understanding the statistical patterns in historical data. If historical data reflect an existing social bias against a particular input variable or

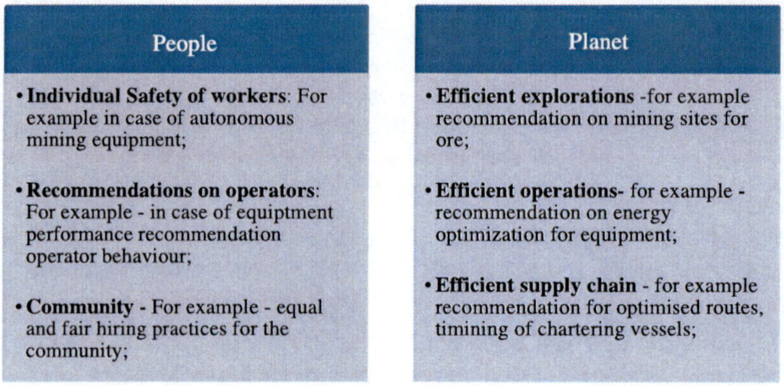

Fig. 3.7 People and planet risks from AI models

feature like gender, then the model will develop that historical bias. This may lead to the model outcome being unfair to applications from the female gender in the above example. Even if we remove the gender feature from Table 3.2, the salary range or college degree could still be a proxy to gender [41].

However, this is not the only reason for the models being unfair. Another reason is the amount of historical data. Models learn from the historical instance of data and become more accurate with more and more data. Even if we had unbiased data, if the number of data points for a minority group is less compared to the general population, the accuracy of the minority group would still suffer [42].

One widely used data science methodology to build advanced analytics and AI models is the cross-industry standard process for data mining (CRISP-DM) [43]. The methodology has six phases: business understanding, data understanding, data preparation, modeling, evaluation, and deployment (Fig. 3.8).

We will examine each phase of this methodology and discuss the potential ethical risks and drivers (Table 3.3).

During AI model development, mining companies can use risk mitigation strategies for each risk and systematically address them to ensure that the AI models are ethical [44].

Let us take an example of each ethical risk and discuss mitigation strategies;

1. Incorrectly defining the use case of advanced analytics

Table 3.2 Training data (historical inputs and outputs) for AI model

Input data variable or features			Outcome hiring result
College degree	Salary range	Gender	
Yes	High	M	Yes
Yes	Low	F	No
No	N/A	F	No
…	…	….	….

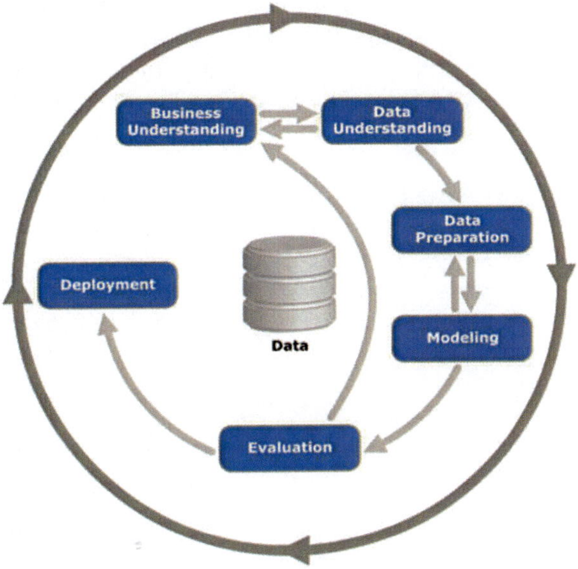

Fig. 3.8 Process diagram showing the relationship between the different phases of CRISP-DM

Table 3.3 CRISM-DM phases and potential ethical risk in each phase [43]

CRISP-DM	Ethical risks arise from …
Business understanding and data understanding	1. Incorrectly defining the use case of advanced analytics 2. Misinterpreting or mislabeled data field in the input data
Data preparation and modelling	3. Using a protected field (gender, race) or proxy of that field in the model 4. Using non-representative data set for the problem 5. Using a complex model that is difficult to understand
Evaluation	6. Poorly published metrics for fairness, explain ability of the model
Deployment	7. Wrong choice of technology for model deployment; and 8. Lack of training for the people using the results of the model

- For example, if the business objective is defined as maximum throughput without consideration for maintenance or safety aspects;
- Mitigation Strategy: The mining company should upfront discuss the ethical implication of the use case and that should form part of the selection criteria.

2. Misinterpreting or mislabeled data field in the input data

- For example, if the input data column 0 and 1, which has been coded as male (0) and female (1), is misinterpreted in the opposite direction;
- Mitigation Strategy: Data governance and cataloging should be done before embarking on model development.

3. Using a protected field (gender, race) or proxy of that field in the model
   - For example, if operator ethnicity has been included in building the AI model as a data point for mobile mine equipment operator behavior.
   - Mitigation Strategy: Multiple iterations can be done with and without a protected variable to visualize if the results are the same or different. If the results are the same, then there is a possibility of the proxy variable.
4. Using non-representative data set for the problem
   - For example, if the AI model training data has been selected from a mine site where the demographic of the population is from the older age group; and
   - Mitigation Strategy: Data quality checks and bias mitigation algorithms can be employed.
5. Using a complex model that is difficult to understand
   - For example, if the team has deployed a DL model which has difficulty explaining the results and their basis, instead of a simple model, where both are giving similar accuracy
   - Mitigation Strategy: Business teams should ask for model explain ability and why a particular recommendation is made
6. Poorly published metrics for fairness explain the ability of the model
   - Only accuracy metrics of AUC or MAPE are published, and no fairness metrics for protected variables like gender or race are not published
   - Mitigation Strategy: The mining company should define the fairness metrics guidelines for people-specific use cases.
7. Wrong choice of technology for model deployment
   - For example, for real-time processing, older hardware is used which cannot process the model results in a reasonable timeframe, say for CCTV safety issue detection
   - Mitigation Strategy: Sizing should be done before the final choice of model deployment has been finalized.
8. Lack of training for the people using the results of the model
   - For example, if the team using or maintaining the model does not know how to monitor the model metrics and check for deviations.
   - Mitigation Strategy: Proper model documentation and regular model metrics should be a part of the model development phase.

In the mining industry, there is always the debate between buying (COTS) versus build applications. Most of the examples discussed above are from a build perspective. However, evaluating the explain ability and fairness metrics model is essential for organizations to mitigate ethical risks, even from a buying perspective.

Discussion on AI Fairness Metrics

This chapter would not be complete without a discussion on model fairness metrics. The objective of an ethical AI model is to maximize accuracy with fairness constraints.

First, let us separate fairness from statistical bias. Statistical bias is defined as the difference between an estimator's expected value and the actual value. For example, a fuel consumption index forecasting model that always overpredicts fuel consumption is statistically biased. Thus, even if we remove the statistical bias, the model may not be fair.

Today, many fairness metrics and definitions available today of group fairness, individual fairness, blindness, representational harm, and more. Arvind Narayanan has summarized fairness definitions brilliantly in his talk on "21 fairness definition and their politics" at ACM Fairness, Accountability and Transparency Conference 2018 [45]. Key fairness metrics are.

- **Group fairness**—do outcomes systematically differ between demographic groups or other population groups.
- **Individual fairness**—is defined that similar individual should be treated similarly.
- **Blindness**—using a protected variable like gender in the training data, but not while predicting.
- **Representational harm**—is caused when the AI model learns societal stereotypes and mirrors the same bias.

In order to understand group and individual fairness metrics, let us look at a simple example of a binary classifier that recommends hiring candidates who applied to a mining company. If we hypothetically compare this recommendation to the actual success of candidates, from saying alumni database, then we have four outcomes (Table 3.4).

Table 3.4 Confusion matrix for hiring decision

	Succeed in mining industry jobs (from alumni database)	Did not succeed in mining industry jobs (from alumni database)
AI model recommended for hiring	True positive (TP)	False positive (FP)
AI model did not recommend hiring	False negative (FN)	True negative (TN)

As a mining industry hiring manager, the model will typically evaluate the number of people recommended for hiring and succeeded in mining over the percentage of the total applicants. They were recommended for hire, i.e., the metric positive predicted value or precision defined as TP/(TP + FP). Simply put, if the model recommends many people, but they all leave and join another industry, then the model is not good.

However, as a candidate who applied for the job, he will be worried whether the model incorrectly classified him for not being hired even if similar candidates had succeeded in the industry, i.e., false-negative rate or miss rate defined as FN/(FN + TP). Simply put, how many times did not model recommend a successful candidate.

Moreover, as a community, the question may be asked if the selected candidates are demographically balanced. In this case, the model recommendation for hiring candidates would be evaluated against the demographic variable (gender, race, etc.) and should be selected in equal numbers, i.e., demographic parity.

Lastly, as the data scientist, model accuracy is of interest, accuracy defined as (TP + TN)/(TP + FP + TN + FN).

Suppose we look at scientific studies, more than 14 metrics on which a binary classifier like the one used for hiring recommendations may be evaluated [46]. All these metrics are related to fairness but from a different perspective. In reality, it is impossible to satisfy all the fairness metrics. In Chouldechova's [47] paper which discusses the fairness of recidivism prediction instruments, i.e., AI models that provide decision-makers with an assessment of the likelihood that a criminal defendant will re-offend at a future point in time, she suggests that not even three criteria of false-positive rate balance, false-negative rate balance, and predictive parity can be satisfied together.

To guide data science teams through these metrics, the organization needs to define how to interpret the organization's values.

For example, in hiring decisions:

- If the organization defines group fairness as "we are all equal," then the assumption is that all groups have similar abilities with respect to a given task (even if it cannot be appropriately observed). In this case, college grades may not be evaluated since they contain structural bias, and the distribution among different groups should not be mistaken for the group's ability. Hence, the demographic parity metrics like disparate impact and statistical parity difference should be used to evaluate the model, and
- If the organization defines group fairness as "what you see is what you get." The assumption is that the observation reflects the ability for a given task. In this case, good college grades represent candidates' ability to be successful at the job. Hence, equality of odds metrics like average odds difference and average abs odds difference should be used to evaluate the model.

Other metrics lie between these two views of "we are all equal" and "what you see is what you get" [48]. In order to guide which metrics to use, a simple fairness tree is presented Fig. 3.9.

Table 3.5 describes the definition of fairness metrics and guidance on what value is considered fair.

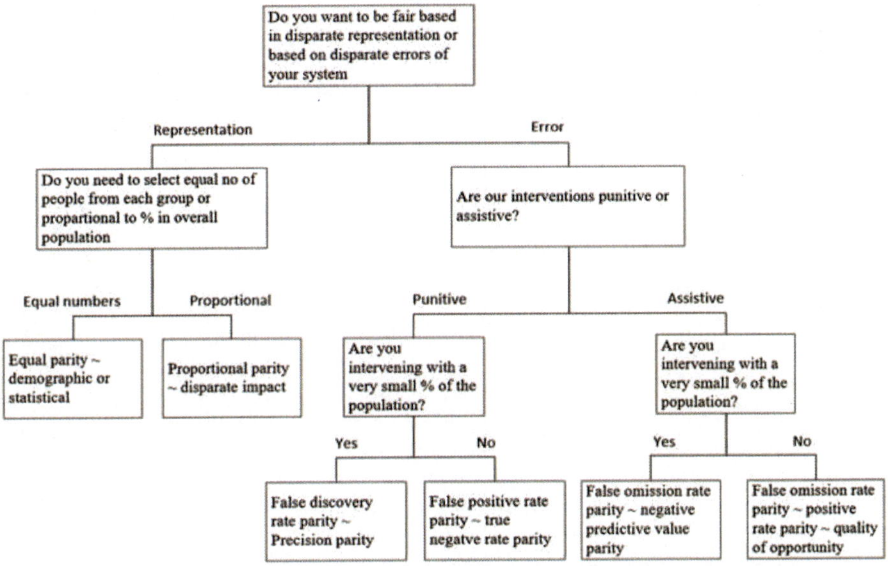

Fig. 3.9 Fairness tree, bias, and fairness audit toolkit [49]

Once the fairness metrics are calculated, multiple algorithms can be deployed to bring the metrics to ideal condition. The algorithms can be deployed on training data (preprocessing), changing the learning model (in-processing), and after the model results (postprocessing) [50]. Table 3.6 provides some of these algorithms and explanations.

Lastly, for other metrics like blindness, as published in the paper by Hardtl et al., "equal opportunity in supervised learning" [51], the model can still learn proxies by not using a protected attribute in prediction, and hence, this technique is ineffective.

In the case of representational harm, which is caused when the model learns from societal stereotypes and mirrors the same bias, the question is—does the model bias represent the real existing societal imbalance. A paper by Kay et al. on "Unequal Representation and Gender Stereotypes in Image Search Results for Occupations" [52].

The search results mirror societal stereotypes. For example—if you search for CEO, only male images show up. The question that is yet to be answered is: Can shifting the representation of gender in image search results shift people's perceptions about real-world distributions?

Table 3.5 Fairness metrics definition and ideal value

Metric	Definition	Idea value
Statistical parity difference	It is computed as the difference in the favorable outcomes received by the unprivileged group to the privileged group	The ideal value of this metric is 0
Disparate impact	It is computed as the ratio of a favorable outcome for the unprivileged group to that of the privileged group	The ideal value of this metric is 1 A value < 1 implies a higher benefit for the privileged group, and a value > 1 implies a higher benefit for the unprivileged group
Average odds difference	It is computed as the average difference of false-positive rate (false positives/negatives) and true positive rate (true positives/positives) between unprivileged and privileged groups	The ideal value of this metric is 0. A value of < 0 implies a higher benefit for the privileged group, and a value > 0 implies a higher benefit for the unprivileged group
Equal opportunity difference	This metric is computed as the difference of true positive rates between the unprivileged and the privileged groups. Thus, the true positive rate is the ratio of true positives to the total number of actual positives for a given group	The ideal value is 0. A value of < 0 implies a higher benefit for the privileged group, and a value > 0 implies a higher benefit for the unprivileged group
Theil index	It is computed as the generalized entropy of benefit for all individuals in the dataset, with alpha = 1. Thus, it measures the inequality in benefit allocation for individuals	A value of 0 implies perfect fairness

Note Each of these metrics should be evaluated against the protected group. For example, in hiring decisions, we should calculate these metrics for gender and try to intervene to get closest to fair value

Table 3.6 Fairness algorithms to mitigate bias

Algorithm	Deployed on	Definition
Reweighing	Training data [preprocessing]	Weights the examples in each (group, label) combination differently to ensure fairness before classification
Optimized preprocessing	Training data [pre-processing]	Learns a probabilistic transformation that can modify the features and the labels in the training data
Adversarial de-biasing	Model/classifier [in-processing]	Learns a classifier that maximizes prediction accuracy and simultaneously reduces an adversary's ability to determine the protected attribute from the predictions. In addition, this approach leads to a fair classifier as the predictions cannot carry any group discrimination information that the adversary can exploit
Reject option-based classifier	Results/prediction [postprocessing]	Changes predictions from a classifier to make them fairer. Provides favorable outcomes to unprivileged groups and unfavorable outcomes to privileged groups in a confidence band around the decision boundary with the highest uncertainty

Conclusion

The examples and use cases described in this chapter provide insights into ethical issues which arise while adopting AI technologies in the mining industry. It is essential to understand that unethical AI models expose mining companies to reputational, regulatory, and legal risks.

In order to mitigate risks first, leaders should detail company values with examples that can help data science teams to make decisions about ethical dilemmas.

Second, the data science teams should be transparent on the training data, model results explanation, and model metrics (fairness and accuracy). Many techniques are described in this chapter to help the data science teams to understand, evaluate, and mitigate bias using the appropriate fairness metric.

Lastly, when the business teams or IT teams are procuring tools with AI technologies (COTS), they should leverage European Parliament STOA study recommendations to request for Data Hygiene Certificate (DHC). Data quality plays a key role in AI model efficacy. By reviewing the sourcing, acquisition, diversity, labeling of data, DHC would ensure that good quality data have been used to train the AI model.

Optionally, the team could also request an ethical technology assessment, a written document detailing the AI solution's dialogue between ethicists and technologists.

References

1. Ethics guidelines for trustworthy AI. 2021 [cited 2021 Jul 29]. Available from https://digital-strategy.ec.europa.eu/en/library/ethics-guidelines-trustworthy-ai.
2. Apple Card algorithm sparks gender bias allegations against Goldman Sachs [cited 2021 Jul 30]. Available from https://www.washingtonpost.com/business/2019/11/11/apple-card-algorithm-sparks-gender-bias-allegations-against-goldman-sachs/.
3. Mathews, M.E.A.W. 2019. New York Regulator Probes United Health Algorithm for Racial Bias. 2019 [cited 2021 Jul 29]. Available from https://www.wsj.com/articles/new-york-regulator-probes-unitedhealth-algorithm-for-racial-bias-11572087601.
4. Dastin, J. 2018. Amazon scraps secret AI recruiting tool that showed bias against women. 2018 [cited 2021 Jul 30]. Available from https://www.reuters.com/article/us-amazon-com-jobs-automation-insight/amazon-scraps-secret-ai-recruiting-tool-that-showed-bias-against-women-idUSKCN1MK08G.
5. Asimov, I. 1981. The Three Laws. Compute! The Journal of Progressive Computing 1981 [cited 2021 Jul 30]. Available from https://archive.org/stream/1981-11-compute-magazine/Compute_Issue_018_1981_Nov#page/n19/mode/2up.
6. Blackman, R. 2020. A Practical Guide to Building Ethical AI. 2020 [cited 2021 Jul 30]. Available from https://hbr.org/2020/10/a-practical-guide-to-building-ethical-ai.
7. Shanks, T., et al. 2010. What is ethics?
8. Rieu, M.C.J.M.L. 2020. Can Mining Ever Be Ethical? 2020 [cited 2021 Jul 30]; Available from https://www.man.com/maninstitute/mining-ever-ethical.
9. White House, O. 2016. Preparing for the future of artificial intelligence. Tech. Rep. Executive Office of the President-National Science and Technology.
10. Original CSET Translation of "Artificial Intelligence Standardization White Paper". 2018 [cited 2021 Jul 30]. Available from https://cset.georgetown.edu/research/artificial-intelligence-standardization-white-paper/.
11. HLEG, A. 2019. *Ethics Guidelines for Trustworthy AI*. European Commission.
12. Instruments, O.L. 2019. *Recommendation of the Council on Artificial Intelligence*. Organization for Economic Cooperation and Development.
13. Cutler, A., M. Pribić, and L. Humphrey. 2019. *Everyday Ethics for Artificial Intelligence*. PDF, IBM Corporation.
14. Google. 2021. People AI research [cited 2021 Jul 30]. Available from. https://pair.withgoogle.com.
15. Google. 2021. Responsible AI practices [cited 2021 Jul 30]. Available from https://ai.google/responsibilities/responsible-ai-practices/.
16. Microsoft. 2021. Responsible AI [cited 2021 Jul 30]. Available from https://www.microsoft.com/en-us/ai/responsible-ai.
17. Deepmind. 2021. Safety and ethics [cited 2021 Jul 30]. Available from https://deepmind.com/safety-and-ethics.
18. IEEE. 2021. Ethically aligned design [cited 2021 Jul 30]. Available from https://ethicsinaction.ieee.org/.
19. Partnership in AI [cited 2021 Jul 30]. Available from https://www.partnershiponai.org/about/.
20. Toronto Declaration. 2021. Protecting the rights to equality and non-discrimination in machine learning systems. 2018 [cited 2021 Jul 30]. Available from https://www.accessnow.org/the-toronto-declaration-protecting-the-rights-to-equality-and-non-discrimination-in-machine-learning-systems.

21. Fjeld, J., et al. 2020. *Principled Artificial Intelligence: Mapping Consensus in Ethical and Rights-Based Approaches to Principles for AI*. Berkman Klein Center Research Publication.
22. Jobin, A., M. Ienca, and E. Vayena. 2019. The global landscape of AI ethics guidelines. *Nature Machine Intelligence* 1 (9): 389–399.
23. Hagendorff, T. 2020. The ethics of AI ethics: An evaluation of guidelines. *Minds and Machines* 30 (1): 99–120.
24. Wynsberghe, A. 2020. *Artificial Intelligence: From Ethics to Policy*. Panel for the Future of Science and Technology.
25. GMG. 2021. Implementation of AI in mining [cited 2021 Jul 30]. Available from https://gmggroup.org/projects/implementation-of-ai-in-mining/.
26. Burkhardt, R., N. Hohn, and C. Wigley. 2019. *Leading Your Organization to Responsible AI*. McKinsey Analytics.
27. Soofastaei, A. 2020. *Data Analytics Applied to the Mining Industry*. CRC Press.
28. Explain Your Model with the SHAP Values. 2019 [cited 2021 Jul 30]. Available from https://towardsdatascience.com/explain-your-model-with-the-shap-values-bc36aac4de3d.
29. Hulstaert, L. 2018. Understanding model predictions with LIME. 2018 [cited 2021 Jul 30]. Available from https://towardsdatascience.com/understanding-model-predictions-with-lime-a582fdff3a3b.
30. Schubert, C.O.L. 2019. Introducing Activation Atlases. 2019 [cited 2021 Jul 30]. Available from https://openai.com/blog/introducing-activation-atlases/.
31. Rio Tinto to expand autonomous truck operations to the fifth Pilbara mine site. 2018 [cited 2021 Jun 1]. Available from https://www.riotinto.com/en/news/releases/Automated-truck-expansion-Pilbara.
32. Latimer, C. Autonomous truck and water cart collide on-site UPDATE. 2015 [cited 2021 Jun 1]; Available from: https://www.australianmining.com.au/news/autonomous-truck-and-water-cart-collide-on-site-update/.
33. Soofastaei, A. 2016. Development of an advanced data analytics model to improve the energy efficiency of haul trucks in surface mines.
34. Moral Machine [cited 2021 Jun 1]. Available from https://www.moralmachine.net/.
35. Soofastaei, A. et al. 2016. Reducing fuel consumption of haul trucks in surface mines using artificial intelligence models.
36. GMG. 2019. Guideline for the implementation of autonomous systems in mining. 2019 [cited 2021 Jun 1]. Available from https://gmggroup.org/wpcontent/uploads/2019/06/20181008_Implementation_of_Autonomous_Systems-GMG-AM-v01-r01.pdf.
37. White Paper and Guiding Principles Functional Safety for Earthmoving Machinery 2020 [cited 2021 Jul 30]. Available from https://www.cmeig.com.au/wp-content/uploads/CMEIG-EMESRT-ICMM-White-Paper-and-Guiding-Principles-for-Functional-Safety-on-Earthmoving-Machinery-Ver.-0.5-1.pdf.
38. Bellamy, D., and L. Pravica. 2011. Assessing the impact of driverless haul trucks in Australian surface mining. *Resources Policy* 36 (2): 149–158.
39. Soofastaei, A., et al. 2016. A discrete-event model to simulate the effect of payload variance on truck bunching, cycle time, and hauled mine materials. *International Journal of Mining Technology* 2 (3): 167–181.
40. Janiszewska-Kiewra, E., J. Podlesny, and H. Soller. 2020. Ethical data usage in an era of digital technology and regulation. 2020 [cited 2021 Jul 30]. Available from https://www.mckinsey.com/business-functions/mckinsey-digital/our-insights/tech-forward/ethical-data-usage-in-an-era-of-digital-technology-and-regulation.
41. Barocas, S., and A.D. Selbst. 2016. Big data's disparate impact. *California Law Review* 104: 671.
42. Hardt, M. 2014. How big data is unfair, Understanding unintended sources of unfairness in data-driven decision making. 2014 [cited 2021 Jul 30]. Available from https://medium.com/@mrtz/how-big-data-is-unfair-9aa544d739de.
43. Chapman, P., et al. 1999. The CRISP-DM user guide. In *4th CRISP-DM SIG Workshop in Brussels*, Mar 1999.

44. Cheatham, B., K. Javanmardian, and H. Samandari. 2019. Confronting the risks of artificial intelligence. *McKinsey Quarterly*: 1–9.
45. Narayanan, A. 2019. Fairness definition and their politics. 2019 [cited 2021 Jul 30]. Available from https://shubhamjain0594.github.io/post/tlds-arvind-fairness-definitions/, https://www.youtube.com/watch?v=jIXIuYdnyyk.
46. Powers, D.M. 2020. Evaluation: from precision, recall and F-measure to ROC, informedness, markedness, and correlation. arXiv:2010.16061.
47. Chouldechova, A. 2017. Fair prediction with disparate impact: A study of bias in recidivism prediction instruments. *Big Data* 5 (2): 153–163.
48. d'Alessandro, B., C. O'Neil, and T. LaGatta. 2017. Conscientious classification: A data scientist's guide to discrimination-aware classification. *Big Data* 5 (2): 120–134.
49. Fairness Tree [cited 2021 Jul 30]. Available from http://aequitas.dssg.io/static/images/metrictree.png.
50. Friedler, S.A., et al. 2019. A comparative study of fairness-enhancing interventions in machine learning. In *Proceedings of the Conference on Fairness, Accountability, And Transparency*.
51. Hardt, M., E. Price, and N. Srebro. 2016. Equality of opportunity in supervised learning. *Advances in Neural Information Processing Systems* 29: 3315–3323.
52. Kay, M., C. Matuszek, and S.A. Munson. 2015. Unequal representation and gender stereotypes in image search results for occupations. In *Proceedings of the 33rd Annual ACM Conference on Human Factors in Computing Systems*.

Chapter 4
Advanced Analytics for Mining Method Selection

Fatemeh Molaei

Abstract There is no single acceptable deposit mining process. Two or more practicable approaches are generally feasible. Each approach includes some fundamental problems. The optimal approach is, therefore, one that offers the least problems. Each orebody is special, with its properties, and in such flexible mining work, engineering judgment greatly affects the decision. Therefore, it seems evident that only an experienced engineer who has improved his knowledge by working in many mines and acquiring expertise in various methods can rationally select mining methods. In recent years, artificial intelligence (AI) has been recognized as an effective tool for solving practical challenges in mining and geoengineering.

Keywords Mining method selection · Advanced analytics method · Underground mining · Surface mining

Introduction to Mining Method Selection

Mining methods are the methods used to extract mineral resources from the earth. No single mining process can extract all mineral resources because of the complications of mineral resources' geometric and geological features. Given the uniqueness of each mineral resource, only the required mining method must be implemented for extracting a specific resource so that the applied method can be as technically and operationally compatible as possible with the geometrical and geological conditions for the mineral resource [1]. This is not an easy issue, given the multiplicity of influential factors in selecting an acceptable mining process.

Significant underground mining indicators such as net productivity output, excavation expense, ore losses, ore dilution, and final financial effects depend on selected and applied mining methods. Indeed, the most important aim of the mining method is to achieve lower excavation costs and thus more significant financial benefit. Mining

F. Molaei (✉)
University of Arizona, Tucson, AZ, USA
e-mail: fatemehmolaei@email.arizona.edu

Stantec, Chandler, AZ, USA

A. Soofastaei (ed.), *Advanced Analytics in Mining Engineering*,
https://doi.org/10.1007/978-3-030-91589-6_4

method selection cannot, however, be based solely on these criteria. Safe working conditions, effective mineral excavation, and dilution of the mineral are also essential features of the mining method chosen and directly impact the financial effects of the mining method.

Many considerations show that the selection of the underground mining method depends on many factors. These factors can be divided into three major categories as follows:

- Geological mining factors include ground conditions, hanging wall and footwall, the thickness of the ore, general form, dip, plunge, depth below the surface, quality distribution, resource quality, etc.
- Mining-technical factors including annual productivity, applied machinery, environmental considerations, mine recovery, methodological versatility, mechanical and mining rate, and
- Economic factors include capital cost, operation costs, mineable ore tones, orebody grades, and mineral volume [2].

These criteria can become the primary determinant for selecting the process, but the apparent predominance of one factor does not prevent a careful assessment of the parameters.

While experience and engineering judgment still provide significant input in choosing a mining process, subtle variations in the characteristics of each deposit can typically only be interpreted through a thorough review of available data. Therefore, it becomes the geologist and engineer's duty to work together to ensure that all considerations are considered in selecting the mining process.

The selection of mining methods underlies each mining activity and is essential to optimize economic return to estimate the capital and operating costs of the alternatives. Due to its operating costs and being an important part of mining and design, this selection is a significant challenge for mine management. Above all, the effective mining method improves employee safety and ensures production [3, 4].

Some mining companies now strive to make more of a science than art with machine learning (ML). For example, they work to examine all available geological information using artificial intelligence (AI) to find better ear boiling sites worldwide. When ML is used to find new fields, these efforts become more precise and help the mining industry be rentable [5, 6]. Moreover, AI has been investigated and introduced as a beneficial tool in decision-making and mining technique selection to overcome the challenges mentioned.

AI is a computer science field that deals with machinery that can perform a specific task without any particular instructions. The ultimate objective in the AI field is to make machines intelligent to the human level. The subfields or particular AI applications are deeper learning and ML, designed to educate machines through recognizing patterns and extracting knowledge from previous cases. From prospecting to production, closing, and reclamation, ML and AI can be applied from the outset of mining to the end of the mine life cycle.

Guray et al. [7] therefore suggested an AI model for the selection of mining methods based on 13 separate expert systems. This system can support engineers in selecting the most efficient appropriate mining method. The study aimed to design a hybrid system, which can be employed as a tool for mining method selection by systematically incorporating the intuition of expert engineers into the selection criteria through their abilities to learn. An ore age agent is the interface and decision media at the center of the system.

Evaluation of Mining Methods and Systems

The selection of mining methods is one of mining engineering's most important and controversial practices. The ultimate goals of selecting mining methods are to yield maximum benefit, maximize the recovery of mineral resources, and provide miners with a healthy environment by choosing the least problem method from the feasible alternatives. Selecting a relevant mining method is a complex task requiring several technological, economic, political, social, and historical factors. If an orebody has been formed by deliberation and sufficient information has been obtained to support further study, the next important step is to choose the most suitable method or method of excavation for the deposit. Underground methods are used when the deposit depth or the overburden-to-ore (or coal or rock) stripping ratio or both become costly for surface mining extraction.

Mining system selection is not a well-defined procedure since it includes several subjective variables or parameters interacting. Therefore, in this method, several controllable and uncontrollable parameters should be considered [8]. According to scientific and technical research, these criteria should be reached for each deposit.

There are several approaches to select optimum mining methods which were studied previously [9–14]. Ooriad et al. [15] determine how these variables can improve the dynamic and demanding role of MMS in the decision-making procedure. In addition, because the structure of an orebody is unique, using a specific mining method without meeting the needs of a particular orebody would not be a practical solution. Finally, Kant [16] pointed out that MMS is a critical process because, as implemented, it will be irreversible. As a result, the mining method chosen will govern all decisions made during the mine's life cycle, and this decision may impact a project's economic potential.

An appropriate mining method needs to be chosen to maximize profit and the extraction of the deposit and provide a safe working environment for the workers.

Across all commodities, there have been previously proven methods that have been used. Some were successful in putting their plans into action. Others, on the other hand, needed to be more developed to be successful. In the 1960s, techno-economic methods were adopted. However, the method was insufficient since it was solely focused on the financial prediction of the financial effect of a particular mining method [2]. In the early 70s, Boshkov and Wright introduced a qualitative underground mining method. In such geological formations, only geological and geotechnical considerations were taken into account. Mining methods that have been used in similar geological conditions were chosen as alternatives ones [17].

The approaches used for the selection of mining methods by researchers, academicians, practicing mining engineers include

- AHP;
- MCDA;
- Fuzzy logic including the application of fuzzy sets;
- Fuzzy numbers;
- Fuzzy AHP;
- Decision theory and operations research applications;
- Nicholas theory; and
- Combination of two or more approaches.

Table 4.1 illustrates a summary of studies which were used soft computing technologies and MCDM methods.

Table 4.1 Summary of representative studies of MMS with SC technologies and MCDM methods [18]

Author	Soft computing technologies					MCDM methods		
	EXS	FUA	ANN	YAM	MC	FUE	AHP	TOPSIS
Yun and Huang [19]		*						
Bandopadhyay and Venkata Subramanian [20]	*							
Gershon et al. [21]	*							
Xioohua et al. [22]	*		*					
Guray et al. [7]	*							
Bitarafan and Ataei [23]		*		*				
Ataei et al. [10, 24]		*			*		*	
Alpay and Yavuz [25]				*			*	
Mahadevi et al. [12]								
Azadeh et al. [1]		*					*	
Namin, et al. [17]		*				*	*	*
Gupta and Kumar [26]							*	
Yavuz [27]				*			*	
Javanshirgiv and Safari [28]		*						*
Bakhtavar et al. [8]		*						*
Balusa et al. [29]							*	

EXS expert system, *FUA* fuzzy algorithm, *ANN* artificial neural network, *YAM* year's method, *FUE* fuzzy entropy, *MCDM* multiple criteria decision making, *AHP* analytic hierarchy process, *MC* *Monte Carlo*, *TOPSIS* technique for order performance by similarity to an ideal solution

Mining Methods Classification System

The approaches to mining method selection can be divided into three categories: methods of profile and checklist, methods of numerical ranking (scoring), and decision-making models [30].

Samimi Namin et al. [30] presented a novel fuzzy multi-criteria decision-making model to select the mining method. For this reason, they introduced fuzzy decision-making (FDM) software tool based on the Fuzzy TOPSIS.

A large number of scholars and engineers have developed many innovative mining methods and theories. The mining industry has significantly benefited from advanced high-tech computing technology combined with enhanced machinery. In reality, modern mining represents a sophisticated synthesis of the fundamental sciences.

The mining manager often encounters numerous complex decision-making problems in actual mining activities without sufficient information or information to overcome these problems. An inappropriate decision could endanger people's lives and irreversibly harm the mining economy, given the enormous size of mining capital. The main causes of the problems in mining decision making can be classified as follows:

- Uncertainties in commodity markets;
- Geological and geotechnical uncertainties of rock mass;
- Lack of clarity of qualitative and linguistic expressions of mining-associated factors;
- The subjectivity of individual decision-makers;
- Uncertain effect of weights of single, multiple, and mutual relationships of mining-related factors; and
- Possibility of undefined mechanisms of rock mass behaviors under particular conditions [18].

As mentioned before, the selection of mining methods is a crucial issue in a mining planning process, and the goal of MMS is to choose between alternatives the most appropriate mining process for a given mineral deposit. Therefore, MMS influences the economy, safety, and productivity of mine products significantly. Moreover, the MMS is recognized as a MADM issue that requires numerous factors, such as technical and industrial issues, financial issues, mining-related policies, and environmental and social issues. The conceptual framework work of MMS is shown in Fig. 4.1.

As illustrated in Fig. 4.1, the MMS processes require not only the consideration of many key criteria but also their sub-criteria. The factors included in the MMS process, i.e., 3D depositary features, geological and geotechnical environment, environmental and economic considerations, and other industrial factors. Moreover, policy and social constraints, machinery, and supply conditions for employees are also important factors. Furthermore, the range of criteria that significantly impact the selection process is rather challenging to define. In addition, a few mines can be mined using a single mining method, but most need two or more viable methods to be used.

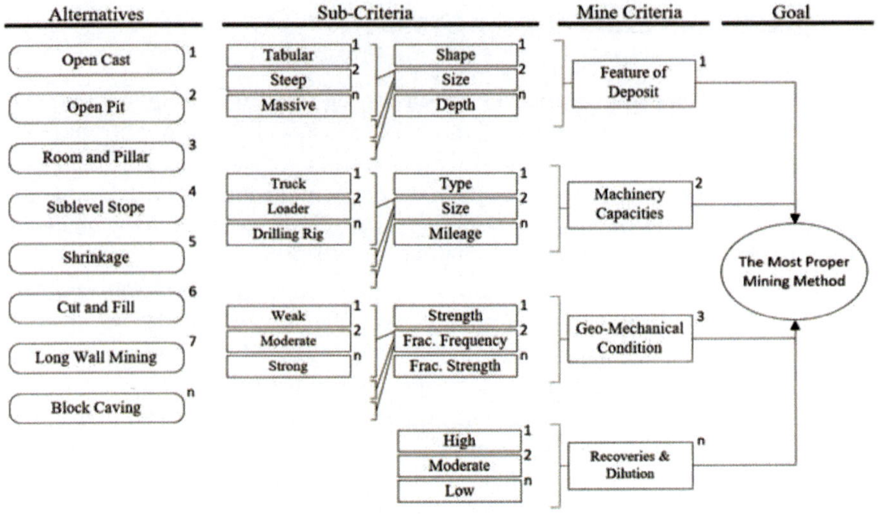

Fig. 4.1 Conceptual framework of mining method selection

In open-pit mines, engineers and scientists are always very concerned about the hazards of blasting activities. The challenges of engineers and scientists in reducing explosion-inducing issues always include explosive soil vibration, air overpressure, fly rock, back break, dust, and toxicity [31–33]. Due to the adverse effects of blasting, many mines were forced to close. In addition, natural hazards are also seen as an engineer and manager rival. Research on the stability of banks and slopes has been carried out continuously because of many uncertainties and the influence of explosions and earthquakes. Similar problems have also been mentioned for underground mines. Furthermore, mine pressure can lead to the collapse of the underground mines due to the heterogeneous rock environment, explosion-led soil vibration, and earthquakes. The potential dangers of fire and suffocation in underground mines are toxic gases such as methane, CO, and SO_2 [34–36].

Comparison of Underground Mining Methods

Underground mining methods are primarily chosen based on the deposit's geological and spatial context. As a result, candidate methods can be selected and rated based on projected operational/capital costs, production rates, labor availability, materials/equipment availability, and environmental concerns. After that, the method that provides the most realistic and optimized combination of protection, economics, and mining recovery is chosen.

In addition to various external factors such as labor supply and the financial state of the operating company, the following are the key factors that affect the mining method to be used:

- The size and shape of the deposit;
- The character of the ore, whether high or low in grade;
- Whether the values are regularly or irregularly distributed;
- Physical character, whether hard or soft, tough or brittle;
- The character of the country-rock;
- Immediate and future demands for ore;
- Amount of development work done or that may be necessary;
- Amount of water to be handled;
- Cost of power, timber, and supplies;
- Ventilation; and
- Whether drilling is to be done by hand or with machine drills.

Underground mining methods are classified into three categories based on the amount of support required: unsupported, supported, and caving, reflecting the importance of ground support (see Fig. 4.2 and Table 4.2).

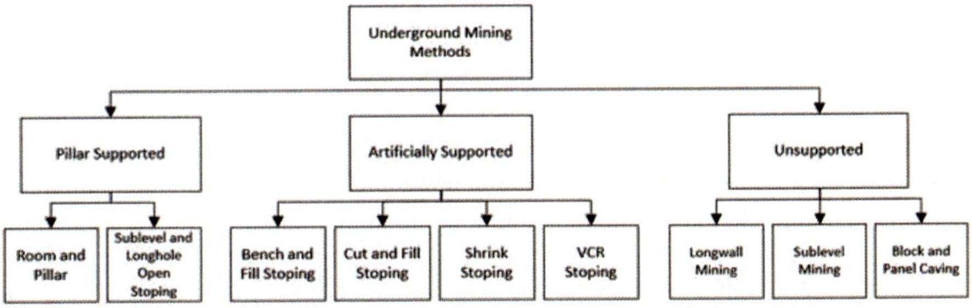

Fig. 4.2 Underground mining methods classifications

Table 4.2 Mining methods capacity

Method	T/Manshift	Avg. T/day	Relative operating (cost per ton)
Cut and fill	12–48	500–1500	20–70
Shrinkage	20–28	200–800	20–50
Room and pillar	15–150	1500–10,000	7–20
Open stopping	20–115	1500–25,000	7–25
Sub-level caving	65–180	1500–50,000	7–17
Block caving	300–2000	10,000–100,000	1–2.5

Comparison of Surface Mining Methods

Mechanical and aqueous surface mining methods are historically classified into two categories. Mechanical methods use mechanical means to break down the rock, while aqueous methods use water or another solvent, such as an acid, to break down the ore and remove it.

Zha et al. [37] performed a comparative study on three mining methods, including the local steep slope mining method (LSSMM), underground mining method (UMM), and highwall mining system (HMS), in the mining of ultra-thick coal seams of an end slope in surface mine. Six indicators were used to develop the TOPSIS model, including recovery ratio, technological complexity and adaptability, impact on surface mining production methods, the safety of production, and economic benefits.

Selection Process for Hard Rock Mining

In underground hard rock mines, the choice of the correct mining method is significant, as it directly affects the recovery and dilution of ore and the amount of development required. In addition, mining techniques often assess the level of technology and, in turn, influence the type of equipment needed, cycle time and chronological succession of operations, yield (tons per year), and the risk (financial and operational) involved [16].

All known parameters related to the problem should be evaluated to make a correct decision to select the underground mining method on hard rock. Although an increase in the number of relevant criteria makes the problem more complex and more challenging to reach a solution, this may also increase the correctness of the decision. Many traditional approaches can accept minimal parameters due to the increasing complexity of the decision process and may be generally defective. Therefore, it is clearly seen that it is essential to evaluate all the established parameters associated with selecting the mining method by integrating the decision-making process.

With its properties, each orebody is only one of its kind, and decisions to select a mining method must fit the properties of the orebody under consideration. Thus, for a deposit, there is no single effective mining process. Two or more feasible strategies are usually available as choices for a mine planning engineer. Some benefits, as well as inherent problems, are involved with each process. Therefore, one that offers the minor problems and the best operational effectiveness is the optimal form. While knowledge of mining conditions and engineering decisions still provide considerable input into the choice of a mining process, it is important to analyze subtle variations in the characteristics of each ore deposit solely through a thorough review of the existing information. In order to ensure that all elements are seen in the excavation method selection process, it becomes the responsibility of the practicing engineers to work together [16].

Because of the high demand for raw materials from other industries, innovative mining technologies must be used to increase mining efficiency while reducing negative environmental impacts. AI is regarded as a reliable method for solving mining difficulties [18].

Therefore, in situations where one or more criteria are unknown, techniques fail to provide a solution. In this case, techniques can be used to select mining methods optimally. Furthermore, the existing techniques have been limited because new findings from technological developments and science/sector studies have changed many selection criteria and the related mining methods.

The problems mentioned above may be overcome by artificial neural networks (ANN). ANNs are based on a biological neural network and are well recognized as one of the most adaptable methodologies used in the last two decades. ANNs are computer programs that can provide similar or different situations with solutions (regardless of information lack) by learning from relationships between cause and effect in samples. In addition, the system can update its knowledge base with new data from technological developments and science/sector studies via dynamic learning processes in ANNs.

By definition, a computation system designed to simulate the way the human brain analyzes and processes information is an artificial neural network (ANN). It is the basis of AI and resolves problems that would prove impossible or difficult through human or statistical standards. In addition, ANNs have the capacity to self-learn, so that better results can be achieved when there are more data.

Özyurt and Karadogan [38] used a model based on ANN and game theory to select underground mining methods. This model can predict if the information lacks science/sectoral studies by following technological evolutions and new findings if learning is continuous. They applied the back-propagation algorithms (Fig. 4.3) to predict the outputs. Weight and bias are dynamically updated in these networks so that a mean error is minimized.

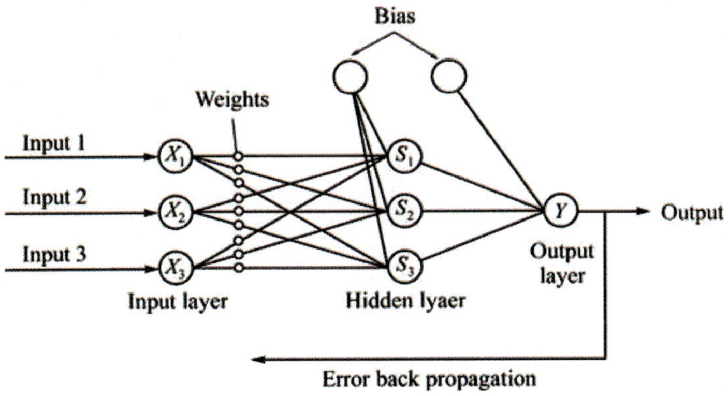

Fig. 4.3 Algorithm of the developed model for underground mining method selection [38]

In a network, a hidden neuron j is connected to a number of inputs $x_i = (x_1, x_2, x_3, \ldots. x_n)$

$$\text{Net}_j = \sum_{i=1}^{n} x_i w_{ij} + \theta_j \tag{4.1}$$

where x_i shows the inputs; w_{ij} denotes the weights; θ_j is the bias neuron (optional); n is the number of inputs. Finally, the net output Y from the hidden layer is calculated using a logarithmic sigmoid function:

$$Y_j = \sum_{i=1}^{n} x_i w_{ij} + \theta_j j = f(\text{Net}_j) = 1/[1 + e^{-(\text{net}_j + \theta_j)}] \tag{4.2}$$

The goal of artificial neural network training is to reduce error by altering the weights of the network. For example, if a vector of N predictions is generated from a sample of n data points on all variables, and Y is the vector of the observed variable being predicted, then the within-sample MSE of the predictor is computed as (Fig. 4.4).

$$\text{MSE} = 1/N \sum_{1}^{n} (Y_i - Y_i')^2 \tag{4.3}$$

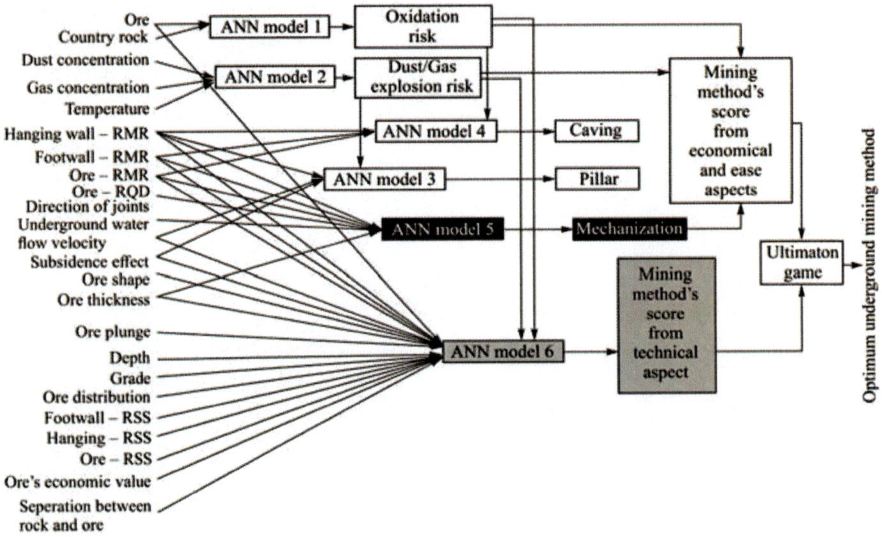

Fig. 4.4 Algorithm of the developed model for underground mining method selection

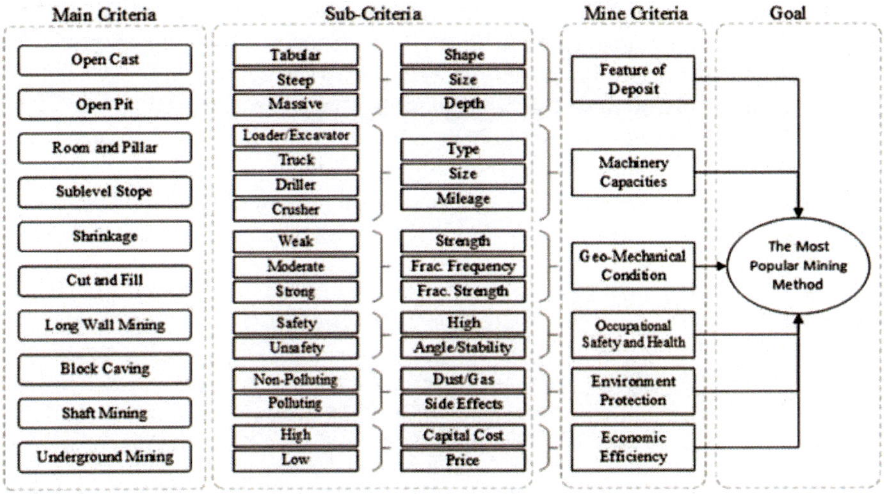

Fig. 4.5 Framework of mining method selection based on criteria

where Y_i is the target, and Y_i' is the output [38].

The two main pillars of modern automation technology are AI and ML. Almost every industry has been influenced by these two fields, with billions of dollars invested in advancing frontiers. Unfortunately, the mining industry was late to acknowledge this advanced sector's importance. However, even if it is at the slowest rate, research in mining, focused on technology development through master learning and AI, accelerates [38].

One of the most important aspects of mining is selecting a mining method. It determines the efficiency of the economy, technology, and the environment. As shown in Fig. 4.5, mining technique selection is seen as an issue with multiple decision-makers. In reality, some mines exclusively use one type of mining. However, some mines with more challenging conditions necessitate the employment of a mixture of mining techniques. As a result, engineers and researchers face difficulty selecting the best mining method.

Selection Process for Underground Soft Rock Mining

Soft rock mining underground is a very challenging operation. The substances extracted using this process, e.g., coal and certain minerals, are usually less than hard rock mining. For this reason, most soft rock mining operations in the underground are expected to produce vast quantities of material daily and are very profitable to be economically sustainable.

The financial success of industrial mineral ventures, such as potash, relies on low-cost recovery of high tonnages. Sometimes these deposits are tabular, and the overweight is relatively low, which causes underground high extraction issues. This makes the deposition relatively low. Much of the miners' and mechanics engineers' tools and techniques are based on practitioners' experiences of hard rock mining. These encounters do not consider the remarkable physical features of soft rock leading to a conservative nature, although this is a good starting point [39].

References

1. Azadeh, A., M. Osanloo, and M. Ataei. 2010. A new approach to mining method selection based on modifying the Nicholas technique. *Applied Soft Computing* 10 (4): 1040–1061.
2. Bogdanovic, D., D. Nikolic, and I. Ilic. 2012. Mining method selection by integrated AHP and PROMETHEE method. *Anais da Academia Brasileira de Ciências* 84 (1): 219–233.
3. Peskens, T. 2013. Underground mining method selection and preliminary techno-economic mine design for the Wombat orebody, Kylylahti deposit, Finland. 2013, Delft University of Technology.
4. Soofastaei, A. 2020. *Digital Transformation of Mining Data Analytics Applied to the Mining Industry*, 1–29. CRC Press.
5. Molaei, F., et al. 2020. A comprehensive review on internet of things (IoT) and its implications in the mining industry. *American Journal of Engineering and Applied Sciences* 13 (3): 499–515.
6. Soofastaei, A. 2020. *Data Analytics Applied to the Mining Industry*. CRC Press.
7. Guray, C., et al. 2003. Ore-age: A hybrid system for assisting and teaching mining method selection. *Expert Systems with Applications* 24 (3): 261–271.
8. Bakhtavar, E. 2013. Transition from open-pit to underground in the case of Chah-Gaz iron ore combined mining. *Journal of Mining Science* 49 (6): 955–966.
9. Wang, C., et al. 2019. Monte Carlo analytic hierarchy process for selection of the longwall mining method in thin coal seams. *Journal of the Southern African Institute of Mining and Metallurgy* 119 (12): 1005–1012.
10. Ataei, M., H. Shahsavany, and R. Mikaeil. 2013. Monte Carlo analytic hierarchy process (MAHP) approach to selection of optimum mining method. *International Journal of Mining Science and Technology* 23 (4): 573–578.
11. Sitorus, F., J.J. Cilliers, and P.R. Brito-Parada. 2019. Multi-criteria decision making for the choice problem in mining and mineral processing: Applications and trends. *Expert Systems with Applications* 121: 393–417.
12. Naghadehi, M.Z., R. Mikaeil, and M. Ataei. 2009. The application of fuzzy analytic hierarchy process (FAHP) approach to selection of optimum underground mining method for Jajarm Bauxite mine, Iran. *Expert Systems with Applications* 36 (4): 8218–8226.
13. Soofastaei, A., et al. 2016. Development of a multi-layer perceptron artificial neural network model to determine haul trucks energy consumption. *International Journal of Mining Science and Technology* 26 (2): 285–293.
14. Soofastaei, A. 2016. Development of an advanced data analytics model to improve the energy efficiency of haul trucks in surface mines.
15. Ooriad, F.A., et al. 2018. The development of a novel model for mining method selection in a fuzzy environment; case study: Tazareh Coal Mine, Semnan Province, Iran. *Rudarsko-geološko-naftni zbornik* 33 (1): 45–53.
16. Kant, R., et al. 2016. A review of approaches used for the selection of optimum stoping method in hard rock underground mine. *International Journal of Applied Engineering Research* 11 (11): 7483–7490.

17. Namin, F.S., et al. 2012. FMMSIC: A hybrid fuzzy based decision support system for MMS (in order to estimate interrelationships between criteria). *Journal of the Operational Research Society* 63 (2): 218–231.
18. Jang, H., and E. Topal. 2014. A review of soft computing technology applications in several mining problems. *Applied Soft Computing* 22: 638–651.
19. Yun, Q., and G. Huang. 1987. A fuzzy set approach to the selection of mining method. *Mining Science and Technology* 6 (1): 9–16.
20. Bandopadhyay, S., and P. Venkatasubramanian. 1988. Rule-based expert system for mining method selection. *CIM (Canadian Institute Mining and Metallurgy) Bulletin (Canada)* 81 (919).
21. Gershon, M., S. Bandopadhyay, and V. Panchanadam. 1993. Mining method selection: a decision support system integrating multi-attribute utility theory and expert systems. In *Proceedings of the 24th International Symposium on the Application of Computers in Mine Planning (APCOM).*
22. Xiaohua, W.Y.T.G.C. 1995. A study on the neural network based expert system for mining method selection. *Computer Applications and Software* 5.
23. Bitarafan, M., and M. Ataei. 2004. Mining method selection by multiple criteria decision making tools. *Journal of the Southern African Institute of Mining and Metallurgy* 104 (9): 493–498.
24. Ataei, M., et al. 2008. Mining method selection by AHP approach. *Journal of the Southern African Institute of Mining and Metallurgy* 108 (12): 741–749.
25. Alpay, S., and M. Yavuz. 2009. Underground mining method selection by decision making tools. *Tunnelling and Underground Space Technology* 24 (2): 173–184.
26. Gupta, S., and U. Kumar. 2012. An analytical hierarchy process (AHP)-guided decision model for underground mining method selection. *International Journal of Mining, Reclamation and Environment* 26 (4): 324–336.
27. Yavuz, M. 2015. The application of the analytic hierarchy process (AHP) and Yager's method in underground mining method selection problem. *International Journal of Mining, Reclamation and Environment* 29 (6): 453–475.
28. Javanshirgiv, M., and M. Safari. 2017. The selection of an underground mining method using the fuzzy topsis method: A case study in the Kamar Mahdi II fluorine mine. *Mining Science* 24.
29. Balusa, B.C., and J. Singam. 2018. Underground mining method selection using WPM and PROMETHEE. *Journal of the Institution of Engineers (India): Series D* 99(1): 165–171.
30. Samimi Namin, F., et al. 2008. A new model for mining method selection of mineral deposit based on fuzzy decision making. *Journal of the Southern African Institute of Mining and Metallurgy* 108 (7): 385–395.
31. Nguyen, H., and X.-N. Bui. 2019. Predicting blast-induced air overpressure: A robust artificial intelligence system based on artificial neural networks and random forest. *Natural Resources Research* 28 (3): 893–907.
32. Nguyen, H., et al. 2020. A comparative study of different artificial intelligence techniques in predicting blast-induced air over-pressure 1 (2): 187.
33. Guo, H., et al. 2021. Deep neural network and whale optimization algorithm to assess flyrock induced by blasting. *Engineering with Computers* 37 (1): 173–186.
34. Irsyam, M., et al. 2018. Geotechnical approach for occupational safety risk analysis of critical slope in open pit mining as implication for earthquake hazard. In *IOP Conference Series: Materials Science and Engineering*. IOP Publishing.
35. Hustrulid, W.A., M.K. McCarter, and D.J. Van Zyl. 2000. Slope stability in surface mining.
36. Jiang, N., et al. 2017. Propagation and prediction of blasting vibration on slope in an open pit during underground mining. *Tunnelling and Underground Space Technology* 70: 409–421.
37. Zha, Z., et al. 2017. Comparative study of mining methods for reserves beneath end slope in flat surface mines with ultra-thick coal seams. *International Journal of Mining Science and Technology* 27 (6): 1065–1071.

38. Özyurt, M.C., and A. Karadogan. 2020. A new model based on artificial neural networks and game theory for the selection of underground mining method. *Journal of Mining Science* 56 (1): 66–78.
39. Coleman, T. 2021. Economic Success in Underground Soft Rock Mining. SRK News Issue #43 [cited 2021 Jul 2]. Available from https://www.srk.com/en/publications/economic-success-in-underground-soft-rock-mining.

Chapter 5
Advanced Analytics for Valuation of Mine Prospects and Mining Projects

José Charango Munizaga-Rosas and Kevin Flores

Abstract This chapter investigates the opportunities for using advanced analytics techniques to establish the value of a mining proposition. The value of mining projects is usually summarized into the financial technical model, which combines technical and financial aspects. The usual main inputs to such models are discussed and the potential sources for the information required in building such models. After the basic framework is established, sources of uncertainty are considered, and ways of characterizing some of them are illustrated. Finally, the chapter concludes with the presentation of the value at risk metric as one possibility that can be used to inform the impact of the underlying financial uncertainty on the value of the project. A case study to illustrate some of the concepts discussed is presented.

Keywords Valuation · Mining · Uncertainty · Risk · Advanced analytics

Introduction

Mineral economics has been defined recently as the "academic discipline that investigates and promotes understanding of economic and policy issues associated with the production and use of mineral commodities" [1]. As such, the area is intended on the one hand to provide the mining professionals, who already can assess the technical feasibility of a project, with tools to judge the economic potential of a mineral deposit in order to perform further exploration activities or to develop the project into an operating mine, and on the other hand acknowledging that the mineral economists must recognize the significance of the geologic, technical, and political influences on mining and mineral processing activities when evaluating the economic prospect of a project [2].

J. C. Munizaga-Rosas (✉)
Facultad de Ingeniería de Minas, Universidad Nacional del Altiplano, Puno, Perú
e-mail: jmunizaga@unap.edu.pe

K. Flores
Departamento de Ingeniería de Minas, Universidad de Chile, Santiago, Chile
e-mail: kfloresc@ing.uchile.cl

A. Soofastaei (ed.), *Advanced Analytics in Mining Engineering*,
https://doi.org/10.1007/978-3-030-91589-6_5

To produce an economic valuation of mineral commodity projects, it is necessary to borrow several tools from other more traditional disciplines such as [1]:

- economics;
- finance;
- management;
- statistics;
- econometrics;
- geology;
- mining and petroleum engineering;
- mineral processing;
- fuel science and technology; and
- metallurgy.

It is important to note that some of those disciplines were not originally developed specifically for mining. However, they have found good use in this otherwise daunting task. The multidisciplinary nature of the discipline is maybe one of its most important characteristics, and because of this, it requires a certain degree of specialization. It is not uncommon for mining professionals to learn the underpinning of economics and financial techniques, or on the contrary, to have economists that learn several of the technical aspects of mining to provide appropriate assessments of mineral projects.

As seen from an investment point of view, the resource industry possesses three important characteristics [3]:

- Mining investment is partially or entirely irreversible, and the initial cost is at least partially sunk.
- Investment in mining projects involves uncertainty over future rewards.
- Mining project investment has some leeway regarding timing, with the mining investor can delay investment to obtain more information about future investment conditions.

Those characteristics transform the associated decision problem into a difficult one. However, once the investment has been decided, it is usual to stick to a mining project even if the cash flows are not performing as expected (within a particular performance band). To the previous list, we add the following general characteristics of mineral deposits [4]:

- They are initially unknown, which means that they need to be discovered, and the success rate of these discoveries is low,
- They are fixed in size, which means that the life of the mineral asset is limited and finite,
- They are variable in quality, which means that the value of a mineral deposit, and consequently, the financial viability of a mining project is affected by the size and grade aspects,
- They are fixed in location, which means that there is a geographical distribution of deposits and that location for those deposits cannot be chosen, or equivalently, the mine cannot be moved to a more convenient location.

All the aspects that have been previously mentioned point out the necessity of having appropriate tools to measure the worth of a project and its financial viability. It is important to note that mining projects are intense in the capital, usually requiring hundreds of millions of investment dollars to proceed. Consequently, there is a high component of capital recovery present in the production cost of mineral commodities, companies are financially exposed for an extended period due to the long pre-production periods associated with mining ventures, the mineral resources are non-renewable, which is the most distinguishing characteristic of the industry, the project is usually indestructible which has secondary, and recycling markets as a consequence and the risks are not only financial, but they also encompass geological, engineering, political, social, environmental, etc.

Careful planning and design are usually required before the decision point of going or not with the project into a producing mine usually require investments in the order of several million dollars to assess the technical and financial viability of the project. Thus, it is essential to use a valuation framework that is particularly adapted to the nature of the mining projects.

This chapter focuses on one topic within the mineral economics discipline but observed through the lens of data requirements, uses that data and outputs created with the assistance of models for valuation of mine prospects and mining projects. Other topics such as mineral commodity market analysis, regulation, and policy can be found in some specialized textbooks or articles that the reader is encouraged to consult [4].

FTMs for Valuation in Mining

To provide value for a mining project, we may ask which approaches to valuation exist out there. A general list is provided [5], which is discussed with mining in mind:

- **Cost approach**: How much could it cost to reproduce an asset of identical quality? It needs to be noted that every project in mining is unique in terms of location, quality, size, etc. This approach is complicated to use in practical terms. In some cases, particularly when new projects are assessed in established mining districts, it becomes “easier” to use this approach. However, there are always going to be discrepancies because of the unique nature of mineral deposits,
- **Market approach**: Using comparable sales, which is believed could reflect the balance between supply and demand. This approach is also difficult because there are not too many transactions available that can be used for similar deposits. To further exacerbate the problem, some mining investment decisions are heavily affected by global market conditions, and as such, there is a lack of consistency between transactions. For example, gold is generally accepted as a financial commodity, which means that in times of global uncertainty (e.g., the current pandemic), the price is heavily affected by financial considerations rather than normal supply–demand relationships.

Table 5.1 General template to determine cash flows

+	Income
−	Tax credits
−	Expenditures
−	*Depreciation*
−	Taxable expenditures
=	**Profit before tax**
−	Tax
=	**Profit after tax**
+	Depreciation
−	Non-taxable expenditures
+	Other Benefits and Credits
=	**Cash flow**

- **Income approach**: Using future earnings as the yardstick of the worth of a mining project. This option is also tricky as we never really get to know the full extent of mineralization. El Teniente mine in Chile is a typical example of a deposit that has extended much more than its initial projected life, and it is not alone in this feat.

A basic template for the determination of cash flows for any project (not necessarily mining) is given in Table 5.1, where (+) means that the component adds to the cash flow and (−) means that the components subtract from the cash flow.

The template outlined in Table 5.1 depends on the country in which it is applied, and the differences are mainly due to differing legislation concerning taxes, allowances, royalties, etc. However, the basic idea remains the same: We account for all income, subtract all allowed expenditures, credits, and depreciation to obtain the profit before tax. After the tax is applied, depreciation and other allowances are put back into the cash flow, and any other non-taxable expenditure is subtracted to determine the cash flow. The valuation of any project is based on the consideration of the resulting cash flows.

The peculiar nature of mining requires careful consideration of the technical aspects that influence the income and expenditures of a project. Concepts such as mineral recovery, dilution, and others play a role when building cash flows. A financial technical model (FTM) is usually a spreadsheet that carefully considers a project's financial and mining aspects in a consistent and integrated valuation framework.

Typical Items Included for FTMs in Mining

As mentioned before, an FTM summarizes both the financial and technical aspects of a project consistently. The level of detail will depend on the effort of the modeler, and it could range from very simplistic to overly complex models.

Recalling the unique characteristics of mining projects will lead to the main components that require careful consideration of at least the following factors when building an FTM:

- capital investment expenditures;
- operational expenditures;
- cost of capital;
- price projections;
- projected exchange rates;
- interest rates projections;
- cost escalation factors;
- resource model;
- mine plan;
- mineral extraction rates;
- life of mine;
- recovery;
- dilution;
- cut-off grade;
- depreciation policies; and
- closure/rehabilitation costs.

We will discuss some of these in more detail, given the possible opportunities they present from an analytics point of view. Others are so important that they will be discussed separately in other sections of this chapter. After the discussion, a general template will be offered to the reader. The template is not intended to be a one-shoe-fits-all solution for project valuation but just an illustration of the minimum level of detail those models have.

Exchange Rate

The exchange rate plays an essential role in the valuation of mining projects. Most of the time, the commodities and mineral products are traded in US dollars in the international markets. To further exacerbate this problem, projects usually pay salaries and contractors in local currency depending on the project's location, and it may be the case that several currencies are used in a single project.

It may be thought that the forecasting of exchange rates would be an excellent application of advanced analytics techniques. However, there is evidence [6] that the forecasting power of economic models is not better than the naïve prediction of the random walk model. As disheartening as this news may be, still do not eliminate the requirement to at least guess the type of future exchange rate used in the project evaluation. Other approaches have proved co-integration for specific pairs of exchange rates and proposed models intended to perform better than the naïve prediction of the random walk model. Hann and Steurer [7] compare a neural network and classical error-correcting model to see if the use of neural networks outperforms linear models, and conclude that this is the case, and postulate that hybrid approaches combining nonlinear models such as neural networks with econometric methods should go a long way toward a more accurate prediction. Recent work uses many other techniques to improve what has been termed the Meese and Rogoff puzzle and present a more optimistic view for the problem [8–10].

Interest Rate

The interest rate is the amount a lender charges for using assets (usually money) expressed as a percentage of the amount lent. In its essential form, it represents a form of rent paid to the owner of the money lent to compensate him for the inconvenience of not having those funds to take advantage of other (potentially more profitable) opportunities, rent that potentially helps the lender to cover debt not paid by some borrowers, and finally bearing in mind that due to inflation the purchasing power of a dollar today will usually be more significant than the same dollar in future times.

Projects are usually evaluated in terms of net present value (NPV), which, as we will see in more detail, uses the interest rate (or discount) as one key parameter needed by the calculation. However, despite its importance in the process, discount rate selection remains elusive even for seasoned financial analysts [11]. One of the complications is that the interest/discount rate should represent the risk of the asset in question by replicating the return required by investors from assets of similar risk. In mining, this is a difficult question to answer as every project possesses unique characteristics. Hence, estimating the risk for a particular project is difficult, not to mention how difficult it becomes to compare risks between projects.

Traditional valuation regularly employs the weighted average cost of capital (WACC) to estimate discount rates [11]:

$$\text{WACC} = \left(\frac{E}{A}\right)R + \left(\frac{D}{A}\right)R_{\text{d}}(1 - t_{\text{c}}) \tag{5.1}$$

where

- E is the market value of equity;
- D denotes debt;
- A is defined as debt plus equity;
- R is the cost of capital;
- R_{d} denotes the interest rate on debt; and
- t_{c} represents the corporate tax rate.

From the previous list, the only parameter that is difficult to estimate is R. Usually, the well-known capital asset pricing model (CAPM) is used:

$$R = R_f + \beta(R_m - R_f) \tag{5.2}$$

where

- R is the required rate of return to equity holders;
- R_f represents the risk-free interest rate;
- β is the relative risk of the stock; and
- R_m measures the average market return.

The more challenging estimate for the previous parameters is β, where most advanced analytics potential disruptions reside. All parameters in Eq. (5.2) can be estimated from market data. As can be seen from the previous formula, the rate of return R is a combination of the risk-free rate and the average market excess return (relative to this risk-free rate), adjusted by the relative risk of the stock. The real risk-free rate can be calculated by subtracting the current inflation rate from the treasury bond yield, matching the investment duration. Hence, it is a complex estimation problem but still doable. Newer methodologies that allow measuring factors that influence the determination of interest rates could be implemented. For example, [12] attempt to use sentiment analysis techniques to assess the impact of public sentiment in the financial decision of monetary policy.

For the estimation of β, the classical methodology involves running a regression model of the form $r_s = \alpha + \beta r_b$, with r_s represents the stock variable percent return and r_b is the market benchmark variable percent return. The model uses historical data when running the regression and thus is affected by the time horizon used in its construction. If structural changes are present in the data used in this form of supervised learning model, then the estimate of the interest rate R could be misleading. At least, one recent study compares machine learning (ML) algorithms and the CAPM model [13]. Using ML algorithms to forecast future returns, this is done and concluded that the proposed approach outperforms the CAPM on out-of-sample test data. Another approach attempted just for calculating β is based on a method using maximal overlap discrete wavelet transform, allowing to investigate the β behavior at different time horizons [14].

Cost Escalation

Cost escalation can be defined as changes in price levels driven by underlying economic conditions [15]. Cost escalation attempts to reflect the cost of specific goods or services in given periods. It is very similar to the concept of inflation and deflation, but the big difference with those is that cost escalation is not general in nature but applies to an item or set of items. As such, cost escalation is not primarily driven by changes in money supply (at the core of inflationary/deflationary processes) but is very likely driven by technological changes, practices, and supply–demand imbalances. For example, advances in technology can introduce downward trends in the cost of producing goods or services by making them cheaper. In addition, new regulations that substantially affect the way services are provided could increase their price over time. Finally, if there is a localized shortage of some products or services, their price will substantially increase. This last point is particularly prevalent in mining, where the overall price cycles for commodities are not heavily linked to short-term fluctuations. However, it behaves cyclically over periods spanning several years at a time, which cause boom and boost periods which affect the costs of products and services needed in the extraction and processing of mineral products, a typical example being the price of mining tires during the 2008 mining boom [16], where it was reported that some mineral producers paid as much as $56,000 a tire for its larger vehicles, while some sold for as much as $280,000 in Internet auctions over the same period.

As macroeconomic conditions heavily influence cost escalation, the study of this phenomenon belongs to the "core" skill and knowledge area of economists [15]. Estimating cost escalation factors represents an excellent opportunity for advanced analytics in the mining business, but not too much is done practically. As the escalation process is driven by influences other than the money supply, it requires many other sources of information to understand technological change evolution, projected practice changes, and, more importantly, supply–demand imbalances. From a methodological point of view, Touran and Lopez [17] provide a review of different approaches to cost escalation estimation in the construction industry, which include simple average and exponential smoothing, Box–Jenkins's approach (ARIMA), causal methods, regression, neural networks, qualitative methods, and surveys. They highlight three essential conditions that the data must satisfy to work with quantitative methods:

- historical information is available;
- this information can be quantified in the form of historical data; and
- there is an assumption of continuity.

Of the three data conditions presented in [17], the most troublesome in the mining industry is the third condition. The assumption of continuity does not match the usual boom and bust cycles where it is common to see essential changes in commodity prices in short periods. Mineral products are heavily linked to human activity. Consequently, any significant changes such as wars, pandemics, market instabilities, and structural changes of some markets will introduce substantial changes in the fundamentals of the market structure for the commodity, which will be reflected in booms and busts cycles for the escalation processes each one possessing different severity.

Life of Mine

The life of mine (LOM) can be defined as the number of periods the mining project will operate. The definition appears simple enough to be understood from the name itself. However, its determination is heavily dependent on a series of factors of a technical nature. First, we need to realize that to know how much time a given asset will be operated, a detailed design, mine plans, schedules, etc., need to be known. Usually, this is not trivial at the outset as in the early stages of the process (scoping, pre-feasibility, etc.), limited to few pieces of information are available to make an informed decision. Hence, the uncertainty levels associated with this input are usually high in the early stages and diminishing as more information is gathered and design/planning decisions are fixed.

If no detailed design is available, it is still possible to get a notional LOM estimate using Taylor's rule [18–20], which in its original form was formulated as:

$$\text{Tonnes per day} = 0.014 \times (\text{Expected Tonnes})^{0.75} \tag{5.3}$$

This rule can be rewritten in a more general form as:

$$C = bT^a \tag{5.4}$$

where C is the metric tons per day, T is the reserve tonnage in metric tons, and a and b coefficients are estimated. Estimation of Eq. (5.4) can be done using linear regression techniques for the simple case the coefficients a and b are constant. However, newer nonlinear techniques could lift the implicit assumption of constant elasticities that model (5.4) enforces. The usual regression model is a particular case of supervised learning algorithm, so nothing prevents practitioners from using other algorithms such as regression trees, and neural networks that in theory could produce a better fit for actual data. However, practitioners must be aware that the ML algorithms could lead to overfitted models and unreliable predictions in out-of-sample cases if improperly used.

Recovery and Dilution

Mining recovery is a value that measures an estimate of in-situ ore obtained after mining has taken place. It is frequently expressed as a percentage, conveying the notion of mining lost during the mining activity and processing. Mining dilution is somehow like recovery; however, it has a different nature.

According to AusIMM, JORC code, 2012 edition [21] requires that for a reserve estimate, at least a prefeasibility level study is presented where reasonable assumptions support the mining recovery and mining dilution modifying factors. Based on the information contained in the code, it is not difficult to identify the following technical elements as having an impact on the quantity of material that is eventually left either in-situ or that we are unable to process appropriately:

- mining method;
- mine design;
- mine schedule;

- overburden;
- geotechnical considerations;
- drilling decisions and techniques;
- grade control strategies;
- mining and processing assumptions; and
- metallurgical models.

To illustrate the idea behind recovery, let us consider a small example. If a given copper deposit with an average grade of 2.5% Cu presents a recovery of 90%, this implies that the conversion from resource to reserve due to mining and processing has an efficiency of 90%; i.e., we lose 10% of in-situ ore, and this cannot be extracted and processed. On the other hand, if the in-situ deposit has 300 Mt (million tons) of ore, then this means that we will lose 30 Mt of ore at 2.5% Cu; i.e., we will "lose" in the order of 1.65 million pounds of the final product, which at a value of US$52/lb is equivalent to close to US$86 million that is left in the mine and cannot be recovered at all.

The recovery factor can be seen alternatively as the loss that the mining and processing impose on the extraction of the ore, as a 100% recovery is hard if not impossible to achieve. On the other hand, and for the purposes of illustration only, if a 5% dilution is present in the same hypothetical example, this means that 5% of the material sent to the plant is waste which, among other things, affects the grade and as mentioned before only increases the cost with no income attached to its processing. Following the illustration used before, in this case, the grade can be recalculated ("diluted") as:

$$\frac{(270 \times 0.95 \times 0.025 + 270 \times 0.95 \times 0.0)}{270} = 0.02375 = 2.375\%$$

Which in the present example has an equivalent impact of increasing the production cost by the extraction and treatment of 13.5 Mt of waste material, translated into representative cash costs per pound of US$1.5/lb gives in the order of MUS$44,600 of cost attached to material that does not produce income.

It has been traditional in mining to use a global estimate for both recovery and dilution, independently for each one. However, more recently, the research has focused on creating resource models that provide appropriate estimates of metallurgical characteristics of the material, such as hardness and metallurgical recovery [22, 23]. One of the important characteristics that make this problem complex is that metallurgical recovery is non-additive; i.e., we cannot calculate recovery for a given area by taking averages or some other usual operations, which implies that the assessment of its spatial variability is not direct. Thus, geometallurgy is currently a very active research area and a multidisciplinary technical challenge.

On the other hand, selecting the support for the resource model produces an impact related to the achievable selectivity of an extraction method given the block model at particular support. The general problem is named the ore selection problem [24, 25] and has been given some consideration in mining to try to understand the nature of it, and hopefully attempt assessing the impact of some decisions made in the resource model, the interested reader is encouraged to review those works.

These two modifying factors (dilution and recovery) present significant opportunities for using advanced analytics techniques in mining. Indeed, there are some models, but there are still plenty of opportunities to produce an impact. Dominy et al. [23] present a list of challenges from a metallurgical point of view but also from a more global point of view for the industry which we reproduce here:

- declining ore grades;
- orebodies geometrically and/or internally more involuted;
- processing of more challenging ores with refractory and/or textural complexities;
- deep-seated deposits with potentially high in-situ stress regimes;
- increasing quantities of mine waste that need to be managed appropriately;
- higher energy, water, and chemical costs;
- stricter environmental/permitting and social conditions (the social license to mine);
- increasing demand for complicated to process specialist/critical metals (e.g., rare earth elements and lithium);
- commodity market volatility; and
- difficult funding environment.

Cut-Off Grade

Any mining company needs to decide and control the output rate of ore and mineral products by determining the quantity and quality of ore it extracts from the mine on each period [26]. Economic models of extraction and ore quality have usually assumed that the ore quality is homogeneous within each deposit, in the best of cases adopting an extraction policy that considers a variable ore quality between periods but homogeneous within each period until depletion of the deposit is achieved. It is not a surprise that a prevalent extraction policy consists of mining the ore sequentially from high to low quality as this policy, in principle, increases the value of the earlier cash flows, thus providing a higher NPV for the project. After extraction, it is prevalent to have a processing stage that concentrates the ore to such grades that allow the refining of final metallic products. There are always exceptions to this typical workflow, and it depends on the type of mineral product. For example, coal is extracted from either open-cut or underground mines with no further processing, and definitively no refining included.

The cut-off grade (COG) is the grade value that enables the decision-maker to discriminate between ore and waste within a given mineral deposit. If the material grade in the mineral deposit is above COG, it is classified as ore, and if the material grade is below COG, it is classified as waste [27]. The deposit portion classified as ore is usually sent to the processing plant for further processing to transform it into mineral products (concentrate, metal, etc.). However, sometimes the material can be temporarily stored in stockpiles to delay the processing until future periods.

Miners call the problem of choosing the lowest quality of extracted ore the cut-off grade problem [26]. The optimum cut-off grades, which change every period and are likely to be declining due to the impact they have on NPV, indeed depend on the metal price, but also the operating costs of mining, processing, and refining stages, but also consider the limiting capacities of these interrelated stages [27].

From the point of view of applying advanced analytics in determining a representative cut-off grade or COG policy, it needs to be acknowledged that COG determination is one of those very complex problems in mining due to definitional interrelationships. This is a case of a chicken or egg type of problem or a circular logical argument with difficult to break loops. To illustrate the point, think about the idea behind COG. It requires knowing the potential profit (or loss) to be made at every block in the resource model. However, the profit depends on the costs, and the costs are determined by the extraction method used, which in turn depends on the desired selectivity, which in turn depends on being able to differentiate ore from waste. The granularity of the problem also plays a role in the process as the project's economics is ultimately defined based on cash flows. However, cash flows can be defined at aggregate levels such as days, weeks, months, and years. Therefore, each level of granularity will be based on financial performance derived from cash flows that are materialized as coming from a proposed extraction sequence, which out of necessity, at some point, depends on being able to differentiate between ore and waste, or the selectivity of the mining method.

One possible way of breaking this otherwise pervasive loop is to use Hall's "Hill of Value" methodology [28, 29]. The method, in a nutshell, consists of "flexing" parameters that impact the value of the project, usually by taking a partition into a discrete number of points within a specific range for a set of variables. This discretization for each variable is then used to create all possible (feasible) combinations of factors, and for each unique combination, the economic value is calculated. The name "Hill of Value" comes from the fact that when two factors are used, the resulting surface for the NPV looks like a Hill. The maximum value observed is the peak of the Hill, but also that combination of factors (e.g., production rate and cut-off) that maximizes the NPV, thus helping break the otherwise logical loop presented before. The approach is simple, but the logic is crisp. For example, if we try a set of possible combinations where a production rate and a COG are assumed, we can observe the resulting NPV. We only need to try this for a set of possible combinations to identify where the highest value is achieved; i.e., we avoid calculating a COG from the economics of the project but assessing the financial performance of a particular COG, and the method allows us to analyze the profitability of the project for those parameters chosen. Despite its simplicity, the computational problem can get huge very quickly. The number of cases to study depends on the number of flexed cases and the number of points used on each flexing.

Depreciation

How different assets belonging to a project can be depreciated are dependent on what the tax system allows. Hence, depreciation is location-dependent, and it is influenced by the location where the project declares its address.

The basic idea is that economic depreciation consists of a stream of deductions that ideally replicate the decline in value of the assets over time, and it is acknowledged that risk affects depreciable assets [30]. For example, it is uncertain how long an asset will operate, and some assets will wear out and fail earlier than others. This last is particularly true in mining as it imposes extreme operating conditions in some cases, thus changing the way the asset should be operated.

Usually, a tax depreciation schedule is set in advance when the asset is put into service. What differentiates depreciation rules is the way an asset can be depreciated. One possibility is to use a straight-line depreciation, where the asset life is estimated, and a series of equally sized deductions can completely write off the asset's value. Another possibility is to use a percentage of the residual value of the asset. Also, some tax regimes allow accelerated depreciation of assets to stimulate investment which means that the asset can be depreciated faster than usual.

As mining is a capital-intensive endeavor and requires copious capital investments, appropriate depreciation should be considered. In financial, technical models, this is usually reflected in the cash flows, but its treatment does not do justice to the complexity of the topic.

Producing appropriate estimates of the asset's age-price profile is a complex problem, particularly in mining, as the equipment operates in a location-dependent manner. It is believed that this aspect is the most promising one for advanced analytics opportunities in mining.

Closure/Rehabilitation Costs

Almost every mining legislation in the world considers the requirement of presenting closure plans and remediation alternatives to the competent authorities. These closure plans are utilized to determine the potential liabilities that the mining companies must cover after the extraction of material from the earth's crust has been performed. Legislation requires mining companies to estimate beforehand the impact of their operations in the environment and the local communities and to present plans to remediate the impact or rehabilitate the damage they have produced if this is possible. After the estimate has been produced, funds need to be provisioned and eventually invested in specific instruments for potentially funding rehabilitation requirements in case the company cannot comply with its obligations.

If the company can demonstrate that it has produced all the actions approved in the different remediation/rehabilitation plans presented to the authorities and authorized by them, then it can get back the funds it has provisioned back. Otherwise, the government will use part or all the funds to act on behalf of the company to remediate the externalities produced in the extraction process that has not been yet addressed.

This problem is another interesting one, and it should be considered with caution. The main problem we identify in mining relates to technological development. Usually, the closure will happen when the operations have ceased, which can happen several decades later. However, remediation/closure costs need to be estimated today to get the plan approved, and little or nothing is known about future treatment options for the potential externalities introduced into the environment and society. Therefore, there is a big gap between the plan built, the impact assessed, and the actual closure time. So, there is a significant source of uncertainty regarding this modifying factor. As a related version of this decision problem, we have the possibility of mining companies investing in research to develop new technologies. Hence, the problem is identified as the problem of estimating the future performance of a technology portfolio and deciding where to put the research funds. It is believed that as further models are developed to characterize technological developments, and roadmap techniques become more widely used in the industry. Then it will be possible to estimate better and plan the closure activities and the provision of funds for those.

Deterministic Project Valuation

Once the cash flows are determined and all the components, both technical and financial are included in the technical, financial model, it is then possible to estimate the financial result of the projected mining exercise to try to assess its value.

This is usually done by considering the *discounted, or net present, value* (NPV) of a stream of cash flows by using the well-known formula:

$$\mathrm{NPV} = \mathrm{CF}_0 + \frac{\mathrm{CF}_1}{(1+\rho)} + \frac{\mathrm{CF}_1}{(1+\rho)^2} + \cdots + \frac{\mathrm{CF}_{n-1}}{(1+\rho)^{n-1}} + \frac{\mathrm{CF}_n}{(1+\rho)^n} \tag{5.5}$$

With CF_i the cash flow for periods i and ρ being the interest rate. As a rule, if a project has $\mathrm{NPV} > 0$, it means that discounted income is higher than discounted costs, and thus, the project is profitable. However, in the case of mining, there are usually other considerations that take place.

As a variant for the NPV method just outlined before, the *profitability index* (PI) is another one used in project evaluations. The profitability index is sometimes called the benefit–cost ratio or present value index, and it is calculated by taking the present value of cash inflows divided by the present value of cash investment outflows (not considering operational costs). The formula is given by:

$$\mathrm{PI} = \frac{\sum_{i=0}^{n} CI_i/(1+\rho)^n}{\sum_{i=0}^{p} CO_i/(1+\rho)^n} = \frac{\text{Present Value of Cash Inflows}}{\text{Present Value of CAPEX (initial outflows)}} \tag{5.6}$$

where p is the number of periods of investment (within n), CI_i is the cash inflows for each period $i \in \{1, \ldots, n\}$, CO_i is the cash outflows for each period

$i \in \{1, \ldots, p\}$, $p \leq n$, and ρ is the discount rate. The decision criteria to use with this indicator is to accept a project with a PI value greater than one; i.e., a project for which the discounted cash flows can recover the initial outflows.

Another related indicator is the *present value ratio* (*PVR*) which is the ratio of net present value (NPV) to present worth cost (PWC), the PVR is calculated by dividing the NPV of a project by the net present value of the capital expenditure outflows, and it essentially measures the NPV contribution of the project per unit of investment. The formula is given by:

$$\mathrm{PI} = \frac{\mathrm{NPV}}{\sum_{i=0}^{p} \mathrm{CO}_i/(1+\rho)^n} = \mathrm{PI} - 1 \tag{5.7}$$

The last commonly used discounted cash flow indicator is the *internal rate of return* (*IRR*). The IRR is the annualized effective compounded return rate that the invested capital can earn, and essentially, the IRR represents the yield on the investment. The way to determine the IRR is by finding the root of the equation $\mathrm{NPV}(\rho) = 0$; i.e., the rate ρ^* that solves the previous equation satisfies:

$$\mathrm{NPV}(\rho^*) = \mathrm{CF}_0 + \frac{\mathrm{CF}_1}{(1+\rho^*)} + \frac{\mathrm{CF}_1}{(1+\rho^*)^2} + \cdots + \frac{\mathrm{CF}_{n-1}}{(1+\rho^*)^{n-1}} + \frac{\mathrm{CF}_n}{(1+\rho^*)^n} = 0 \tag{5.8}$$

Which can be equivalently written as the following polynomial equation in ρ:

$$CF_0(1+\rho)^n + CF_1(1+\rho)^{n-1} + \cdots + CF_{n-1}(1+\rho) + CF_n = 0 \tag{5.9}$$

The fundamental theorem of Algebra needs to be recalled here, which is one version states that a polynomial of degree n can have at most n distinct roots, and if a root is a complex, then the conjugate is also a root. This result implies that the IRR may not be unique. It could be any one of n possible roots if it were the case that the resulting polynomial has n different real roots. Without going into too much detail, we can state that if the sign of a project cash flow changes more than once, there will be multiple IRR. In that case, NPV is preferred. Typically, in mining projects, there is a direct investments stage where the CAPEX cash flows materialize (negative cash flows); after the investment stage, usually, cash flows remain positive during the life of mine.

As the IRR represents the yield on the investment, the higher the IRR is, the more attractive the project. Therefore, from this point of view, a simple decision rule based on this indicator is to rank projects by descending IRR and consider only those whose IRR is greater than the discount rate determined for the project, i.e., consider a project whose yield is attractive enough to compensate for the lost opportunities due to the investment and financial risks which should be included in the discount rate.

Other indicators not using discounting are the *Payback Period, Return on Investment, and Total Profit*. The payback period (PP) is also known as the payout period, and it is one of the most used evaluation criteria by mining companies. The calculation involves determining how long it takes to fully recover the initial investment in a project, i.e., the number of required periods so that cash income can cover the initial investment. It is elementary to calculate, but we need to understand that the PP will remain unchanged no matter how long the project continues after the payback. The return on investment (ROI) measures the performance of an investment. It is the ratio of money gained or lost on an investment relative to the amount of money invested. Clearly, if the ROI is negative, then we lose money, and it will be acceptable only when it is more significant than one. Finally, the total profit is the cumulative net cash flow over the life of the project.

Example of Items Considered in a Project's FTM

A particular case of a mineral processing plant that treats copper tailings, coming from both a mining operation and inactive tailings dams, and obtaining copper fines as a product and molybdenum as a by-product will be used for illustration purposes. This case has been chosen because it is a real-life example with sufficient data availability to allow the model to be sufficiently accurate. Despite being a particular example for a given operation, this case shows the level of information included in FTMs. Applying it to any other project of different characteristics will undoubtedly introduce some changes related to the technical nature of the project. However, the structure will remain similar. The project to be used for illustration purposes is already in operation, so capital expenditures are considered a sunk cost. In addition, since there is no capital investment, there is no depreciation. Costs of capital and interest are also not included, nor are the exchange rates. Also, being a tailing processing and not a mining extraction, elements such as waste to ore, dilution, and cut-off grade are not considered. This demonstrates the versatility of the FTM technique: It can be easily adapted to the case at hand. In any case, Table 5.2 summarizes the items considered in constructing the FTM for this project.[1]

It needs to be stated that the project has two inactive tailings dams, one of them is called *Tailings* in this table, and it is currently in operation, and another one but not currently being exploited. Therefore, to simplify the model presented herein, the rows relating to the other tailing dam have not been included as they are analogous to those of *Tailings*.

[1] Confidentiality prevents from fully revealing the company and its location.

Table 5.2 FTM items

FTM item	Unit	Example
Metal prices		
Copper price	US$/lb	3.3
Molybdenum price	US$/lb	8.88
Production		
Fresh tailings		
Fresh processed tailings	t	3,750,000
Fresh copper grade	%	0.113%
Fresh copper recovery	%	22%
Fresh molybdenum grade	%	0.008%
Fresh molybdenum recovery	%	6%
Fresh produced copper	lb	2,055,238
Tailings A		
processed tailings	t	1,841,146
Copper grade	%	0.252%
Copper recovery	%	49%
Molybdenum grade	%	0.021%
Molybdenum recovery	%	20%
Produced copper		5,012,041
Total copper produced	lb	7,067,279
Molybdenum produced	lb	210,160
Income		
Fresh copper sold	US$	$ 6,782,287
Tailings copper sold	US$	$ 16,539,735
Molybdenum sold	US$	$ 1,866,224
Total income	US$	$ 25,188,246
Operational costs		
Fresh Costs	US$	$ 3,678,877
Tailings Costs	US$	$ 8,971,553
Total Costs	US$	$ 12,650,430
Income—costs	US$	$ 12,537,816
Fixed costs	US$	$ 6,000,000
Royalty		
Fresh DET royalty	US$	$ 650,164
Royalty DET tailings	US$	$ 2,081,250
Royalty DET moly	US$	$ 267,803
Royalty company	US$	$ 114,492
Total royalty	US$	$ 3,113,710

(continued)

Table 5.2 (continued)

FTM item	Unit	Example
Profit before tax	US$	$ 9,424,106
Tax	US$	$ 3,385,210
Profit after tax	US$	$ 6,038,896
Cash Flow	US$	$ 38,896

Example of Data Requirements for a Mining Project Valuation

To obtain reliable results from the model, it is necessary to have key information to perform the economic evaluation. In the FTM used as an example before, information was acquired from a technical report shared by the head office of the mining company in question, from which all data used in the FTM was obtained.

Some of the elements that this case includes but is not detailed in the previous FTM are:

- *Mine plan*: Although there is no mine plan as such to carry out the tailings dam operation, there are different units to be exploited (fresh tailings from the current mining operation and tailings from two different inactive tailings dams) and the order of their exploitation, according to the plant's treatment capacity.
- *Life of mine*: Making an analogy with a mining operation, the operation contemplates a time horizon at the end of which all the material is processed, and the operation ends. It is essential to consider this limit because it delimits the economic evaluation.
- *Mineral extraction rates*: We have the yearly tonnages to be processed for each extractive unit, which gives the total yearly tonnage.
- *Copper and molybdenum grades*: Each unit has an estimated copper and molybdenum grade for every processing year.
- *Recovery*: Although there is no detailed recovery for each stage of the process (flotation and metallurgy), there is an overall recovery rate for the process that will tell us how much fine metal is produced from the entire content of the tailings.

With all the above, it is possible to calculate the amount of fine metal available for commercialization.

- *Price projections*: Considering a single price for metals over the entire evaluation horizon is deemed not precise enough to assess the risk to the project derived from price variations, particularly for this marginal project. Therefore, models to predict/simulate future commodity prices over the entire evaluation horizon are used to estimate the revenues received from the sale of all the metal produced.
- *Operational expenditures*: From the number of copper fines produced, the operational costs of the process are calculated. These range from energy and labor costs to refining, transportation, and administration costs. With this, the total expenditures can be calculated.

- Furthermore, there is information on additional costs associated with royalties arising from the nature of the operation (a plant that treats tailings that do not belong to it).

It is worth mentioning that a tailing reprocessing project is very marginal, so it is very susceptible to sources of risk, which can take this project from being modestly profitable to unprofitable. Therefore, it is essential to address all quantifiable uncertainty sources carefully.

With all these elements, the evaluation was carried out using the following steps:

- Two price trajectories were generated, one for copper and one for molybdenum. These trajectories were generated using a geometric Brownian motion model. Since commodities tend to behave similarly, the correlation between the historical price series for both metals was calculated and, using Gaussian copulas, and trajectories were generated that respected this correlation.
- From tonnage, grade, and recovery, pounds of copper and molybdenum to be marketed are obtained:

$$F_i = T_i \times G_i \times R_i \times 2204.6 \tag{5.10}$$

- Once the quantities of copper and molybdenum to be sold are obtained, the total income can be obtained with the price of both metals:

$$\text{Income}_i = F_i \times P_i \tag{5.11}$$

- To obtain the operating expenditure, the refined copper produced is multiplied by the sum of the operating costs:

$$\text{Expenditure} = F_c \times \left(c_e + c_g + c_l + c_r + c_t + c_a + c_o\right) \tag{5.12}$$

  These are energy, grinding media, labor, refining, administration, transportation, and other direct costs.
- With the income and expenditure, profits before tax are obtained. This is because there is no depreciation or credits to add.
- Tax is calculated taking profits before tax and multiplying it by the tax percentage, given by the characteristics of the company to be evaluated:

$$\text{Tax} = \text{Profit} \times \%_{\text{Tax}} \tag{5.13}$$

- In addition, other royalties are considered, one corresponding to payment to the mining company that owns the tailings and another of administrative nature within the company.
- Finally, subtracting these expenses from the profit before taxes gives the profit after taxes, and since there is no depreciation or interest, the cash flow is obtained. A discount rate of 7% per annum was used, which was converted to a monthly rate.

Besides the above, several assumptions were established to develop the model further. First, to have a more realistic case of economic evaluation, a month-by-month evaluation was carried out instead of year-by-year. However, in the report provided by the company, all the data was shown every year, so assumptions had to be made regarding that:

- It was assumed that the annual tonnage for all units was spread evenly over the months of the year. This may not be true, but it is a good option without a more comprehensive mine plan. Also, it was assumed that grade and recovery remain uniform month to a month each year.
- An assumption was used to determine the percentage royalty paid to the mining company that owns the tailings. In their technical report, the plant owner states that a sliding scale royalty assigns the percentage to be charged, but it is not specified therein. Instead, they only detail the minimum and maximum value, but no more. Thus, the assumption is made that for all copper and molybdenum prices within the minimum and maximum ranges, the average of the minimum value percentage and the maximum value percentage is applied. This assumption reduces variability in the process, but it is a better option than not considering the tax at all in the absence of information. Another option to simulate a sliding scale royalty is to make a linear relationship between the price and the percentage tax using the starting and end points as values; however, this method incorporates a large amount of variability into the model and is still an unreliable representation reality.

Comments Regarding Costing Inputs

Costing is a problematic exercise. To produce an appropriate costing, much detail is required, and even in that case, nothing can prepare the professional responsible for doing the costing to face the unknown. Nevertheless, independent of the method used to obtain them, costs are a fundamental component in determining profit. Therefore, it is classical from an economic point of view to consider the total cost function defined as:

$$\text{Total Cost} = \text{Total Fixed Cost} + \text{Total Variable Cost}$$

The fixed costs include all those costs that need to be paid irrespective of the production level. The variable costs are incurred when production happens and are proportional to the production level. The costs of mining are usually divided into two main components. One component is the capital expenditure or CAPEX, and the other component is the operating expenditure or OPEX. CAPEX relates to all the investments needed to start the mining project and is usually spent on the initial years of the project. CAPEX is the funds needed to sustain production and include the direct costs of production plus all other expenses such as management salaries and head office overheads.

It needs to be noted here that there are usually two approaches to estimating costs: capital or operating. The first approach is to perform top-down costing, usually done using formulas that depend on some global factors such as ore throughput. Those models are based on experience and/or statistical collection of costs in different operations. Usually, a linear regression (on a double-log model specification) is used to fit the costing model. The top-down option is the method of choice in the early stages of any project as it allows quickly to get an order of magnitude estimate of the costs for the project. The other option is to consider a bottom-up approach that starts from a very detailed design and schedule of the mining project operations; each cost component can be estimated based on this design and schedule. This option usually requires much work, and it is mainly used on detailed valuations.

CAPEX

Mular [31] presented a rule of thumb for a capital estimation known as the six-tenths rule. The rule uses the following relationship to estimate CAPEX estimation:

$$\frac{\text{Cost 1}}{\text{Cost 2}} = \left(\frac{\text{Capacity 1}}{\text{Capacity 2}}\right)^{0.6} \tag{5.14}$$

where Capacity 2 and Cost 2 relate to a known similar operation in a similar environment, Capacity 1 relates to the operation under consideration. From this relationship, Cost 1 can then be estimated as:

$$\text{Cost 1} = \text{Cost 2}\left(\frac{\text{Capacity 1}}{\text{Capacity 2}}\right)^{0.6} \tag{5.15}$$

Another rule that can be used is the so-called annualized cost per ton, which allows estimating CAPEX for another similar operation, defined as:

$$\text{Cost} = \frac{\text{Total Capital Cost}}{\text{Tonnes per Year}} \tag{5.16}$$

Other indicative relationships can be found in [19] for a list including but not limited to:

- site clearing;
- shaft sinking;
- hoisting equipment and headframe;
- development and stope prep;
- equipment;
- ventilation, drainage, and water supply systems; and
- etc.

For higher-level capital estimates, the practitioner can use [32], generally suited to:

- preliminary capital cost estimation;
- preliminary and control are operating cost estimations.

Above this level, generic cost estimation methods are not suitable and detailed design is required as the basis for estimation.

OPEX

There is no single reliable source for estimating operating costs. Estimates are essentially obtained from actual mine statistics, and there are commercial services that curate this information and provide it for a subscription fee. Some potential sources can be used with different levels of reliability:

- operating mines with similar infrastructure;
- manufacturers;
- consultants, industry cost services;
- government and industry authorities;
- company's operations;
- contractor quotations; and
- rules and formulas.

Indicative OPEX can be obtained using a formula like the one used by Mular with a slight modification [33]:

$$\frac{\text{Cost 1}}{\text{Cost 2}} = \left(\frac{\text{Capacity 1}}{\text{Capacity 2}}\right)^{\alpha} \tag{5.17}$$

where a list of different values for α is presented depending on the cost to be estimated (see Table 5.3).

Comments Regarding Some Income Inputs

The income depends, among other things, on commodity price and the quality of the material that can be extracted from the earth's crust. In this section, some essential definitions regarding the difference between resource and reserve are provided.

The discussion is based on the JORC Code [21]. The name JORC stands for "Joint Ore Reserves Committee," set up in 1971 and has been working on defining reporting standards. It is currently incorporated into the Australian and New Zealand

Table 5.3 Exponents for calculating operating costs [33]

Operation	Estimated cost/capacity	Exponent
Open-cut labor costs	Tons per annum mined	0.5
Open-cut mine supplies	Tons per annum mined	0.5
Underground mine labor	Tons per annum mined	0.7
Underground mine supplies	Tons per annum mined	0.9
Treatment plant labor	Tons per annum mined	0.5
Treatment plant supplies	Tons per annum mined	0.7
Open-cut mine and mill electric power	Tons per annum mined	0.5
Underground mine and mill electric power	Tons per annum mined	0.7
Average cost	Annual capacity	0.6

stock exchanges' listing rules and applies to all solid minerals, including diamonds, other gemstones, stone, aggregate, and coal.

The code sets out minimum standards, recommendations, and guidelines for public reporting exploration results, mineral resources, and ore reserves. The guiding principles of the JORC Code are:

- *Transparency*: requires that the reader of a public report be provided with sufficient information, the presentation of which is unambiguous, to understand the report and not be misled by this information or by the omission of material information that is known to the Competent Person;
- *Materiality*: requires that a public report contains all the relevant information that investors and their professional advisers would reasonably require and reasonably expect to find in the report to make a reasoned and balanced judgment regarding the exploration results, mineral resources, or ore reserves being reported. Where relevant information is not supplied, an explanation must be provided to justify its exclusion,
- *Competence*: requires that the public report be based on work that is the responsibility of suitably qualified and experienced persons subject to an enforceable professional code of ethics (the competent person).

The categorization that the code defines is usually summarized in Fig. 5.1.

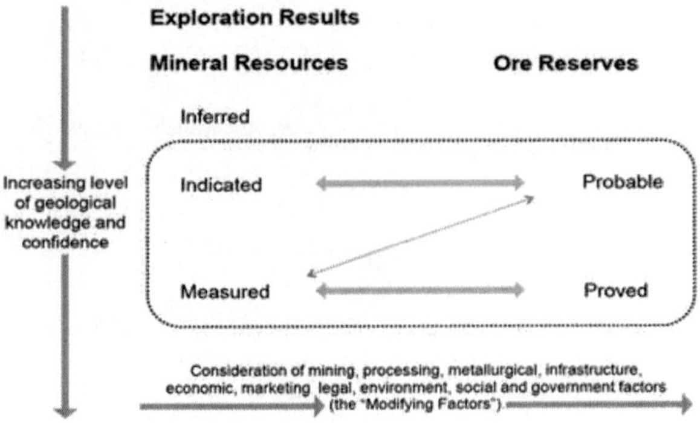

Fig. 5.1 Relationship between exploration results, mineral resources, and ore reserves [21]

Resource Models

According to the JORC Code, a Mineral Resource is defined as:

> A 'Mineral Resource' is a concentration or occurrence of solid material of economic interest in or on the Earth's crust in such form, grade (or quality), and quantity that there are reasonable prospects for eventual economic extraction. The location, quantity, grade (or quality), continuity, and other geological characteristics of a mineral resource are known, estimated, or interpreted from specific geological evidence and knowledge, including sampling. Mineral resources are subdivided, in order of increasing geological confidence, into Inferred, Indicated and Measured categories.

The difference between "inferred," "indicated," and "measured" mineral resource categories is based on the confidence that the modeler has in the estimate. In the case of "inferred" resources, the quantity and grade (quality) are estimated based on limited geological evidence, and the geological and grade continuity can be implied from geological evidence but cannot be verified. In the case of "indicated" resources, the quantity, grade, densities, shape, and physical characteristics are estimated with sufficient confidence to allow the application of modifying factors in sufficient detail to support mine planning and evaluation of the economic viability of the deposit. For a "measured" mineral resource, the quantity, grade, densities, shape, and physical characteristics are estimated with confidence sufficient to allow the application of modifying factors to support detailed mine planning and final evaluation of the economic viability of the deposit.

Reserves

According to the JORC Code, a Mineral Reserve is defined as:

> An 'Ore Reserve' is the economically mineable part of a measured and/or indicated mineral resource. It includes diluting materials and allowances for losses, which may occur when the material is mined or extracted and is defined by studies at pre-feasibility or feasibility

level as appropriate that include application of modifying factors. Such studies demonstrate that, at the time of reporting, extraction could reasonably be justified.

The reference point at which reserves are defined, usually, the point where the ore is delivered to the processing plant, must be stated. It is essential that, in all situations where the reference point is different, such as for a saleable product, a clarifying statement is included to ensure that the reader is fully informed as to what is being reported.

For reserves, the difference between "probable" and "proved" relates to the confidence in the modifying factors applied to the resource model. A "probable" ore reserve is the economically mineable part of an indicated, and in some circumstances, measured mineral resource with a confidence lower than that applying to a proved ore reserve. Conversely, a "proved" ore reserve is the economically mineable part of a measured mineral resource for which there is a high degree of confidence in the modifying factors.

Sources of Uncertainty in Mining Projects

We should start defining uncertainty as to general enough that a broad audience can understand. We start with this because there is a common misconception, and uncertainty is treated equivalent to risk, while they are different concepts. To make things simple, we quote a definition from the National Research Council in the USA [34]:

The term uncertainty is normally used to describe a lack of sureness about something or someone, ranging from just short of complete sureness to an almost complete lack of conviction about an outcome. Doubt, dubiety, skepticism, suspicion, and mistrust are common synonyms. Each synonym expresses an aspect of uncertainty that comes to play in risk analysis. Uncertainty with respect to natural phenomena means that an outcome is unknown or not established and is therefore in question. Uncertainty concerning a belief means that a conclusion is not proved or is supported by questionable information. Uncertainty concerning a course of action means that a plan is not determined or is undecided.

André Journel [35] provides some guidelines for modeling uncertainty that are reproduced here:

- First and foremost is honesty: *accept that all uncertainty assessments are but models, and there is no "best" model.* The randomization process from which the uncertainty assessment is derived should be stated clearly, i.e., in layman terms.
- The second guideline relates to the definition of the randomization process: It should be as to *freeze* (i.e., reproduce) *any piece of information deemed intrinsic to the estimation at hand* and make the rest variable.
- The third guideline is to *perform systematic calibration of any data used.*

Journal's vision relates to using specific techniques for assessing uncertainty by means of a randomization process. Models of uncertainty are based on previous decisions regarding the randomization process used to assess the unknowns. Therefore, choices of a given randomization model determine what can be observed from an uncertain situation. Despite the focus on the randomization process, what is conveyed by Journel's guidelines is that our ability to model uncertainty depends on the appropriateness of both the technique used to create realizations of the uncertain process and the data used, which is a very general message.

Uncertainty by itself is an elusive concept that is difficult to systematize. For example, how can we model a lack of sureness or lack of conviction regarding a particular situation? The most common approach is to consider uncertain parameters as modeled by using random variables. However, this approach requires adopting some specific assumptions regarding the random variables in use, which, if improperly chosen, could be misleading. Furthermore, in some domains, such as earth sciences phenomena, there is no mathematically feasible way to have a proper mathematical description that can model the inherent complexity of the natural world. Consequently, assumptions made in the name of mathematical modeling can be potentially unrealistic.

When dealing with uncertainty, it needs to be recognized our inherent inability to predict the future. Despite the apparent limitations of human beings, if the process under study exhibits some regularity, then gathering information by collecting samples will contribute to characterize it. Collecting information on the phenomenon under study helps reduce the inherent uncertainty, i.e., giving more certainty about the otherwise uncertain outcomes of the phenomenon. Probability measures (distributions) are a generally accepted tool to quantify uncertainty. Probabilistic tools enable the modeler to systematically incorporate information and sometimes beliefs about the phenomenon under study.

In mining, there are several sources of uncertainty. The following list is very likely an incomplete one but illustrates the nature of the phenomenon:

- geological;
- operational;
- market;
- government;
- environmental;
- social;
- occupational health and safety;
- processing;
- geomechanics;
- planning;
- human resources; and
- etc.

In the following subsections, we will deepen the discussion for a limited selection of uncertainty sources. The idea is to show the specialized tools that appear in each area with a view toward modeling uncertainty, which can later be used to assess risk. The selection is heavily biased toward the authors' previous experience, but this implies that the chosen areas are more relevant than others. The discussion is presented herein mainly for illustration purposes.

To illustrate, some examples of the specific situation with the presence of uncertainty are detailed below:

- delays due to blasting misfires and delay in blasting sequence;
- failure in machinery, such as belt conveyors, shovels, and crushers;
- failure in supply, such as water, oil, and gear oil;
- emergencies, such as fires and rockfalls;
- unplanned maintenance;
- unplanned truck queues.

Example 1: Geological Uncertainty

From a geological point of view, the word *uncertainty* recognizes that the measurements and observations deviate from the objective reality. It refers to the imperfect and inexact knowledge of the world [36]. The geologic conditions under which a mine will operate are defined as the "state of nature." This state cannot be changed but is poorly known due to the limited information. It has been acknowledged that getting more or better information can reduce geologic uncertainty [37].

The main types of uncertainty in geology are [36]:

- *Inherent natural variability* of geologic objects and processes;
- *Sampling error*;
- *Observation error* of geological features in the field;
- *Measurement error*;
- *Errors of the mathematical evaluation of geological data*;
- *Propagation of errors*; and
- *Conceptual and model uncertainty*, the inadequacy of identification and classification of geological structures, and other specific features.

Information is collected and interpreted to develop a geological model used as input for the mine planning process, thus adding another level of uncertainty. Model creation requires a high level of geological, mathematical, and statistical expertise. Deposit modeling requires modeling discrete features such as a fault or geological zones and continuous variables such as grades or metallurgical properties. Uncertainty in the grade distribution can be quantified by geostatistical simulation [37].

Martin and Sparrow [38] developed a model relating exploration and exploitation by linking them within the same framework. Embedded within a computer simulation model of the exploration process, they propose a multi-stage stochastic optimization process model to determine the deposit's optimal exploitation pattern, stressing the interdependence between planning and exploration.

Geological Uncertainty Management in Mine Planning

The influence of geological uncertainty in mining project valuation is mainly about the influence grade uncertainty has on the definition of mining plans. The mine plan determines the material to be extracted from the mine and its destination, which after processing in the plant, it can be sold as some form of product with commercial value, which is transformed into cash flows of the mining project. Hence, it is essential to assess this factor to see how robust the mine plans are. Almost all the recent literature is heavily concentrated on open-pit mining. Nevertheless, a notable exception from Grieco and Dimitrakopoulos [39] demonstrates an application of conditional simulation-based risk analysis techniques for an underground stope design problem.

The team of R. Dimitrakopoulos has led almost all treatment of uncertainty in mine planning and scheduling. In [40], the following uncertainty modeling and risk assessment methodology were proposed:

- Development of possible deposit descriptions. This is achieved through geostatistical conditional simulations;
- Development of transfer functions that describe the mining process;
- Uncertainty modeling and risk assessment (for the response of the transfer functions).

In reference to [41], the influence of deposit uncertainty on mine production scheduling was investigated. The paper shows the application of the methodology already described in [40] and performs a comparison with traditional scheduling (based on a kriging block model). In [42], conditional simulations are still in use, but a new and faster technique is utilized. The main conclusion of this work was that having an accurate assessment of uncertainty arising from grade variability can enhance the decision-making process, showing quantitative supporting evidence of it. The paper also provides a case for using new geostatistical techniques in the context of mine planning and optimization with uncertainty. Along different lines of work but in a related vein, Richmond [24, 43] studies the so-called *ore selection problem*, consisting of deciding the classification of rock into ore or waste. He uses geostatistical conditional simulations to develop risk models for the problem.

Dimitrakopoulos et al. present in [44] a maximum upside minimum downside approach to the problem of determining a robust extraction strategy considering grade uncertainty. The key idea was to slightly change the risk analysis procedure described in [40]. In this case, for each simulated orebody, an optimal plan/design was obtained (using *mixed integer programming* (MIP) models). Then, each plan/design is evaluated against all the other scenarios to approximate the robustness of each plan/design. The process allows to determine the *upside* and *downside* for each plan/design option. The chosen design will ideally be the one that maximizes the upside and minimizes the downside. This approach is easy to implement and provides reasonable insight for the decision-making process, unfortunately is a heuristic for which optimality cannot be proven.

Basic Geostatistical Concepts

There are several good textbooks in geostatistics that the reader is encouraged to consult to deepen its knowledge if he/she so desires. This subsection will present a summary of the main geostatistical techniques used in the mining industry. One of the classical references is the textbook from Journel and Huijbregts [45]. Goovaerts [46] treats the geostatistical concepts intuitively and shows how the techniques are used outside mining, specifically in environmental applications. On the other hand, Chilès and Delfiner [47] offer a comprehensive treatment of the field in a mathematically oriented fashion recommended to both graduates and practitioners. Finally, Olea [48] provides a more mathematical treatment of the subject mainly oriented to engineers and earth scientists.

The whole idea behind geostatistics is that natural processes can be represented through the concept of a *regionalized variable* (RV) [49, 50]. RV is a function $f(\vec{x})$ of the point $\vec{x}$. An RV is presented through two contradictory aspects:

- a *random* aspect which takes care of the potential high irregularity and unexpected variations from one point to another; and
- a *structured* aspect which means that the RV should reflect on the structural characteristics of a regionalized phenomenon.

A good conceptualization of RV is obtained using random functions (RF). An RF can be defined as a mapping that assigns to each point of the space $\vec{x}_0$ a random variable $Z(\vec{x}_0)$. Note that this concept accounts for the random aspect and, under some reasonable hypotheses, also accounts for the structured aspect of RV.

To perform statistical inference for the RF, it is usual to adopt the following hypotheses[2]:

- stationarity of the random function; This implies a constant mean $m(Z(\vec{x})) = m\left(Z\left(\vec{x} + \vec{h}\right)\right)\forall\vec{x}, \vec{h}$ And second-order stationarity which implies the existence of a covariance function $C\left(\vec{h}\right) = \mathbb{E}\left\{Z(\vec{x})Z\left(\vec{x} + \vec{h}\right)\right\} - m^2\forall\vec{x}, \vec{h}$, where m the mean is constant because of stationarity and $\mathbb{E}(\cdot)$ represents the expectation; and
- Ergodicity of the RF.

The whole purpose of geostatistics is to provide a value of the attribute of interest at an unknown location $Z(\vec{x}_k)$. Two different problems in geostatistics arise estimation and simulatio*n*. In estimation, the problem can be described as giving the "best"[3] value of the attribute of interest in an unknown location. Simulation solves

[2] There are basically two lines of thought in geostatistics. Those who assume a set of stricter hypotheses and those who do not. Stricter hypotheses mean more constraints in the set of admissible random field models; however, the imposition of more restrictive hypotheses produces better mathematical tools in the sense of simplicity of application. The methods differ slightly between those two schools, but the building blocks remain the same. We will adopt here the more restrictive hypotheses.

[3] The use of "best" involves the concept of optimality of a given performance measure.

the problem of obtaining the "true"[4] probability distribution at unknown locations. There are plenty of techniques available to solve those two problems. It is not the purpose of this subsection to review them as it would take an entire separate textbook to do so (hence the suggestion is to look at the classical references). Among all the possible choices, we will review the two most representative: ordinary kriging and sequential Gaussian simulation.

One of the essential concepts in geostatistics is the description of spatial continuity relations for spatial phenomena. With a model of spatial continuity, points located in space can be related (correlated via the covariance function). For both estimation and simulation, it is necessary to know how points correlate in space. With this knowledge, values can be assigned to an unsampled location considering the existing relationships between an estimated or simulated point and the existing conditioning data (measured). The quintessential tool used in geostatistics for spatial continuity description is the semivariogram. The semivariogram measures the average dissimilarity that exists between samples located at specific distances for certain directions.

For two locations of an RF $Z(\vec{x})$, the similarity can be calculated employing the difference between two samples $Z(\vec{x}_1)$ and $Z(\vec{x}_2)$ of a spatial domain [51]:

$$\gamma^*(\vec{x}_1, \vec{x}_2) = \frac{(z(\vec{x}_1) - z(\vec{x}_2))^2}{2} \tag{5.18}$$

To make this dissimilarity depend on the distance and orientation of a couple of points we define, denoting $\vec{h} = \vec{x}_2 - \vec{x}_1$:

$$\gamma^*(\vec{h}) = \frac{\left(z\left(\vec{x}_1 + \vec{h}\right) - z(\vec{x}_1)\right)^2}{2} \tag{5.19}$$

Calculating the average dissimilarity, we get what is called in geostatistics the experimental semivariogram:

$$\gamma^*(\vec{h}) = \frac{1}{2n_{\vec{h}}} \sum_{\alpha=1}^{n_{\vec{h}}} \left(Z\left(\vec{x}_\alpha + \vec{h}\right) - Z(\vec{x}_\alpha)\right)^2 \tag{5.20}$$

where $n_{\vec{h}}$ is the number of observations.

What is traditionally done in geostatistics is to fit a theoretical model for the semivariogram. There are some properties that a variogram function must satisfy. The main one will be discussed later in more depth. The theoretical semivariogram function is usually denoted by $\gamma\left(\vec{h}\right)$ to differentiate it from the experimental semivariogram. Under the previous hypotheses, the following relationship holds:

[4] The use of "true" could seem abusive at first glance, and it is clearly not possible to know the true distribution of the random variable $Z(\vec{x})$ (recall the definition of random function). By true we mean a softer condition in which we reproduce the statistics for the stationary RF.

$$\gamma\left(\vec{h}\right) = C\left(\vec{0}\right) - C\left(\vec{h}\right) \tag{5.21}$$

The semivariogram is characterized by the range that specifies the points in space that are correlated up to which distance. The nugget effect indicates the microscale variability of the deposit, the sill value usually normalized and equal to 1, and behavior in a neighborhood of the origin. One of the main conditions that a function must satisfy to qualify as a reasonable semivariogram model is that of being positive definite [47, 52]; otherwise, the variance of the linear combination of regionalized variables $Z(\overrightarrow{x}_i)$ could be negative.

Kriging

Following [53], we first need to note that the covariance allows us to calculate the mean and variance of any linear combination,

$$Z(\vec{x}) = \sum\nolimits_k \lambda_k Z(\vec{x}_k) \tag{5.22}$$

for any location $\overrightarrow{x}$:

$$\mathbb{E}\{Z(\vec{x})\} = \sum\nolimits_k \lambda_k \tag{5.23}$$

$$\text{Var}\{Z(\vec{x})\} = \sum\nolimits_i \sum\nolimits_j \lambda_i \lambda_j C\left(\vec{x}_i - \vec{x}_j\right) \tag{5.24}$$

The idea behind kriging is to estimate the value at an unknown location $\vec{x}_k$ by using the information of a set of samples $\{Z(\vec{x}_k)\}$ by using a linear combination of those values:

$$Z^*(\vec{x}) = \sum\nolimits_k \lambda_k^* Z(\vec{x}_k) \tag{5.25}$$

For doing this, we need to determine the weights λ_k^* in the previous summation. They are determined to minimize the error variance. Also, note that the unbiasedness condition $\sum\limits_k \lambda_k = 1$ is required to preserve the mean (constant due to stationarity assumptions). Thus, the problem to solve is:

$$\text{Min Var}\left\{Z^*(\vec{x}) - Z(\vec{x})\right\} \tag{5.26}$$

$$\text{s.t.} \sum\nolimits_k \lambda_k = 1 \tag{5.27}$$

which gives us the following system (first-order conditions of optimality):

$$\sum_j \lambda_j C(\vec{x}_i - \vec{x}_j) - m = C(\vec{x}_i - \vec{x}) \forall i \tag{5.28}$$

$$\sum_j \lambda_j = 1 \tag{5.29}$$

This kriging variant is called *ordinary kriging*. Note that it is assumed that the mean is constant over the entire study area, but it is not determined before doing the calculations; it is assumed to be a characteristic of the random function. If it is calculated in advance, the system varies slightly, and the resulting kriging system is known as *simple kriging*.

Finally, keep in mind that estimation by kriging is "best" in the least-squares sense because the local error $\mathrm{Var}\{Z^*(\vec{x}) - Z(\vec{x})\}$ is minimum. A shortcoming of the least-square criterion, however, is that the local variation of the Z values is smoothed. Another drawback is that the smoothing depends on the local data configuration; it is small, close to the data locations, and increases as the estimated location gets farther away from sampled locations. This uneven smoothing yields kriged maps that artificially appear more variable in densely sampled areas than in sparsely sampled areas. For all these reasons, interpolated maps should not be used for applications sensitive to the presence of extreme values and their patterns of continuity, typically soil pollution data and physical properties (permeability, porosity) that control solute transport in soil. A better alternative is to use simulated maps which reproduce the spatial variability modeled from the data [46].

Sequential Simulation

Sequential simulation algorithms are based on the following decomposition of any given probability distribution function (PDF):

$$f\left(x_{\sigma(1)}, x_{\sigma(2)}, \ldots, x_{\sigma(n)}\right) = f\left(x_{\sigma(n)} | x_{\sigma(n-1)}, \ldots, x_{\sigma(2)}, x_{\sigma(1)}\right) \cdots f\left(x_{\sigma(2)} | x_{\sigma(1)}\right) . f\left(x_{\sigma(1)}\right) \tag{5.30}$$

For every possible permutation function $\sigma(\cdot)$ and $|$ indicating conditioning (probabilistic conditioning).

The previous decomposition, applied to geostatistics, implies that if a realization of an RF wants to be produced, the conditioned decomposition can be used instead of the joint PDF and generate an instance by visiting the nodes we want to be simulated one by one for a given permutation.

Consider the simulation of the continuous attribute Z at N grid nodes $\vec{x}_j$ conditional to the data $Z(\vec{x}_i) i \in \{1, \ldots, n\}$. Sequential simulation amounts to model the *conditional cumulative distribution function* (CCDF) $F\left(\vec{x}_j; z|(n)\right) = \mathrm{Prob}\left(Z\left(\vec{x}_j\right) \leq z|(n)\right)$, then sampling it at each of the N nodes visited along a random sequence. To ensure the reproduction of the semivariogram model, each CCDF is made conditional not only to the original n data but also to all values simulated at previously visited locations. The value to be simulated at each node is drawn from a distribution with the PDF conditioned to the original data plus previously simulated locations. Each permutation provides a different conditioning path and hence a different realization.

Two major classes of sequential simulation algorithms can be distinguished, depending on whether the series of CCDF's is modeled using the multi-Gaussian or the indicator formalism. The probability field approach also requires the sampling of N successive CCDF's. However, unlike in the sequential approach, all CCDFs are conditioned only to the original n data. Reproduction of the semivariogram model is here approximated by imposing an autocorrelation pattern on the probability values used for sampling these CCDF's.

Unlike previous simulation algorithms, creating a stochastic image is formulated as an optimization problem without reference to a random function model in simulated annealing. Instead, the basic idea is to gradually perturb an initial seed image to match target constraints, such as the reproduction of the semivariogram model.

Some Comments

Both methods (kriging and simulation) are exact methods; i.e., they honor the conditioning data. This can be translated mathematically as $Z_{\mathrm{S}}^{*}(\vec{x}_{\mathrm{d}}) = Z_{\mathrm{E}}^{*}(\vec{x}_{\mathrm{d}}) = Z(\vec{x}_{\mathrm{d}})\forall\vec{x}_{\mathrm{d}}$, where $Z_{\mathrm{S}}^{*}(\vec{x}_{\mathrm{d}})$ is the simulated value, $Z_{\mathrm{E}}^{*}(\vec{x}_{\mathrm{d}})$ is the estimated value, $Z(\vec{x}_{\mathrm{d}})$ is the actual value (known and measured), and $\vec{x}_{\mathrm{d}}$ is the locations for the experimental data.

Journel and Huijbregts [45] show that the variance of a simulated location is twice the variance of the estimation for the same location:

$$\sigma_{\mathrm{S}}^{2}(\vec{x}) = 2\sigma_{\mathrm{E}}^{2}(\vec{x}) \tag{5.31}$$

where $\sigma_{\mathrm{S}}^{2}(\vec{x})$ is the simulation variance and $\sigma_{\mathrm{E}}^{2}(\vec{x})$ is the estimation variance. A guideline is provided in [54] concerning the criteria that must be used to accept conditional simulations for their use. The minimum criteria mentioned are:

- data reproduction;
- histogram reproduction;
- reproduction of summary statistics; and
- variogram reproduction.

Example 2: Commodity Price Uncertainty

Traditional mining investments are in general capital intensive and, in practical terms, irreversible projects, meaning that once the investment takes place, it cannot be recouped without substantial economic loss [55], often with long but limited economic lives. Moreover, the economic viability of these investments depends typically on the uncertain materialization of world market prices of metal commodities and on how certain project-specific risks materialize [56]. Thus, forecasting future prices for any commodity is a significant problem that needs to be solved to provide some level of approximation to the actual value of a mining project. Furthermore, mining projects are very diverse, with some projects providing better certainty of producing profitable results than others.

To address the uncertainty of commodity prices, their future values can be modeled with stochastic differential equations (SDEs) [56]. One of the most used SDE is the geometric Brownian motion (GBM), which assumes a customarily distributed return over any time interval. In addition, there are other SDE models, such as the mean-reverting model (MR). MR models are built based on the market tendency of prices to revert toward the average production costs in the long term.

It is standard industry practice to "choose" a future long-term average price and use this constant price as the basis of evaluation. Indeed, this approach provides a first approximation to the value of the project. However, if the project is marginal and this could mislead the decision-maker. A second idea that comes to mind when trying to incorporate an unknown future price into the evaluation model is to consider that the price is fixed over the valuation horizon. However, as it is uncertain, then it is drawn from a probability distribution, probably a normal distribution or some similar well-known model. This second option is better than the first because it provides a set of possible NPVs for the project, and its distribution can be observed and analyzed. Unfortunately, this approach also fails because the evolution of prices possesses a much richer structure, there is an autocorrelation structure and the most straightforward model that can be thought that relates a price of one period with that of the previous of has the form:

$$\hat{p}_{t+1} = p_t + \varepsilon_t \tag{5.32}$$

where $\hat{p}_{t+1}$ is the price we want to forecast in the next period, p_t is the price (known) for the current period, and ε_t is noise or perturbation. Rearranging terms, we can see that essentially the expression reads as $\hat{p}_{t+1} - p_t = \varepsilon_t$, i.e., the difference between two consecutive days (or return) characterizes the evolution of the process. The statistical problem associated with this type of modeling essentially consists of appropriately characterizing the random variable ε_t. There are several possibilities for choosing appropriate candidates for ε_t, the most ordinary and straightforward being that $\{\varepsilon_t\}_{t \in T}$ is a set of independent and identically distributed (i.i.d.) variables, usually following a normal distribution with zero mean and heteroskedastic variance. There are critics of this approach from within the financial and economic communities, pointing out the unrealistic nature of this assumption.

The main object of study when modeling financial time series (or commodity prices, for instance) is to characterize the return defined as $p_t - p_{t-1}$. For reasons of a practical nature that will become clearer later in the chapter, the following two versions of the return are frequently used in applications:

$$\text{Relative return: } R_t = \frac{p_t - p_{t-1}}{p_{t-1}} \tag{5.33}$$

$$\text{Logarithmic return: } r_t = \ln\left(\frac{p_t}{p_{t-1}}\right) \tag{5.34}$$

The reader should note that the logarithmic return allows easy calculation of accumulated returns within a finite time horizon because it can be observed that:

$$\begin{aligned}\sum_{i=0}^{k-1} r_{t-1} = \sum_{i=0}^{k-1} \ln\left(\frac{p_{t-1}}{p_{t-(i-1)}}\right) &= \sum_{i=0}^{k-1} \left(\ln(p_{t-1}) - \ln\left(p_{t-(i+1)}\right)\right) \\ = \ln(p_t) - \ln(p_{t-k}) &= \ln\left(\frac{p_t}{p_{t-k}}\right) = r_t(k)\end{aligned} \tag{5.35}$$

This means that the cumulative logarithmic return in each period composed of several days can be calculated simply as the sum of the daily returns. This also means that operating with logarithmic returns is simple and probably one reason they are preferred over the relative return.

Based on this concept of logarithmic return, a model can be defined which is based on the following reasoning:

$$\frac{\hat{p}_{t+1} - p_t}{p_t} \approx \ln\left(\frac{\hat{p}_{t+1}}{p_t}\right) = \varepsilon_t \tag{5.36}$$

Con $\varepsilon_t \sim \mathcal{N}\left(\mu, \sigma^2\right)$. With this, we can write the following:

$$\ln\left(\frac{\hat{p}_{t+1}}{p_t}\right) = \varepsilon_t \Leftrightarrow \frac{\hat{p}_{t+1}}{p_t} = e^{\varepsilon_t} \Leftrightarrow \hat{p}_{t+1} = p_t e^{\varepsilon_t} \tag{5.37}$$

which is a multiplicative model as opposed to that in Eq. (5.36), which is additive. Note that Eq. (5.37) can be written in an additive way as:

$$\ln\left(\hat{p}_{t+1}\right) = \ln(p_t) + \varepsilon_t \Rightarrow \ln(p_{t+1}) = \ln(p_t) + \varepsilon_t \tag{5.38}$$

Starting from Eq. (5.7) and applying it repeatedly, we obtain the following:

$$\begin{aligned}&\ln(p_k) = \ln(p_{k-1}) + \varepsilon_{k-1} \Leftrightarrow \ln(p_k) = \ln(p_{k-2}) + \varepsilon_{k-2} + \varepsilon_{k-1} \\ &\vdots \\ &\ln(p_k) = \ln(p_0) + \sum_{i=0}^{k-1} \varepsilon_i \Leftrightarrow \ln(p_k) - \ln(p_0) = \sum_{i=0}^{k-1} \varepsilon_i\end{aligned} \tag{5.39}$$

As $\varepsilon_t \sim \mathcal{N}\left(\mu, \sigma^2\right) \forall t$ (i.i.d), then, the last expression implies that starting at p_0, after k steps, the logarithmic price deviation concerning the initial price $\ln(p_0)$ is a random variable distributed following a $\mathcal{N}(k\mu, k\sigma^2)$. This means that the standard deviation of this difference is $\sqrt{k}\sigma$; i.e., the process evolves as $\sqrt{\Delta t}$, with Δt the time elapsed.

All this previous discussion is the basis for the continuous version of this process. Usually employing stochastic differential equations, the generalized Wiener process can be used to model price evolution [11]:

$$\mathrm{d}p = \mu \mathrm{d}t + \sigma \mathrm{d}z \tag{5.40}$$

With dz representing the Wiener process $\mathrm{d}z = \varepsilon\sqrt{\mathrm{d}t}$, with $\varepsilon \sim \mathcal{N}(0, 1)$. Essentially, this is the Brownian motion process from physics. The first component of the process ($\mu \mathrm{d}t$) represents the trend that the series exhibits, and the second component ($\sigma \mathrm{d}z$) model the volatility of the stochastic process. Equation (5.8) represents what is known as the arithmetic Brownian motion process. In financial applications, it is common to use what is called GBM:

$$\frac{\mathrm{d}p}{p} = \mu \mathrm{d}t + \sigma \mathrm{d}z \tag{5.41}$$

Finally, another model with some level of popularity in applications is called mean reverting process, described with the following SDE:

$$\frac{\mathrm{d}p}{p} = \lambda(\mu - p)\mathrm{d}t + \sigma \mathrm{d}z \tag{5.42}$$

Which is the case of no drift ($\mu = 0$) is known as the Ornstein–Uhlenbeck process.

Value at Risk and Alternative Risk Metrics for Mining Projects

Risk Analysis in Investment Appraisal

Risk analysis (or "probabilistic simulation" based on the Monte Carlo simulation technique) is a methodology by which the uncertainty encompassing the primary variables is processed to estimate the impact of risk on the projected results. It is a technique by which a mathematical model is subjected to several simulation runs, usually with the aid of a computer. Successive scenarios are built using input values for the project's key uncertain variables selected from probability distributions during the simulation process. The simulation is controlled so that the random selection of values from the specified probability distributions does not violate known or suspected correlations among the project variables. Finally, the results are collected and analyzed statistically to arrive at probability distributions of the potential outcomes of the projects and to estimate various measures of project risk.

The stages of a risk analysis process can be broken down to [57]:

- *Forecasting model preparation*: This stage requires creating a forecasting model that defines mathematical relationships between numerical variables related to future forecasts.
- *Risk variables selection*: A risk variable is defined as one which is critical to the viability of the project in the sense that a small deviation from its projected value is both probable and potentially damaging to the project's worth. To select risk variables, sensitivity analysis and uncertainty analysis are usually used. Sensitivity analysis measures responsiveness by changing the value of a given variable. Uncertainty analysis is the attainment of some understanding of the type and magnitude of uncertainty encompassing the variables to be tested and using it to select risk variables.
- *Probability distributions definition*: Probability distributions are used to express the beliefs and expectations of experts regarding the outcome of a particular future event quantitatively. They are usually calibrated using collected data.
- *Correlation conditions setting*: Identifying and attaching appropriate probability distributions to risk variables are fundamental in a risk analysis application. However, using the computer to generate scenarios based on the probability distributions is correct only if no significant correlations exist among the selected risk variables. Therefore, we need to respect the correlation structure of the risk variables and generate them; otherwise, inconsistent results are expected.
- *Simulation runs*: This is the stage in which the computer takes over. Once all the assumptions, including correlation conditions, have been set, it only remains to process the model repeatedly until enough results are gathered to make up a representative sample of the nearly infinite number of combinations possible. During a simulation, the values of the risk variables are selected randomly according to the probability distributions and respecting the correlation structure. The results of the model are then computed and stored for posterior analysis.
- *Analysis of results*: This is the final stage. It consists of the analysis and interpretation of the results collected during the simulation runs stage. Statistical analysis is the usual tool to use for performing this.

Methodologies for Assessing Risk

The first thing that needs to be noted is that the inability to predict the future must be acknowledged when dealing with uncertainty. However, if the process under study exhibits some form of regularity, information gathering for some key variables could effectively contribute to characterize the uncertain nature of the phenomenon. Gathering information relating to the uncertain phenomenon under study gives some knowledge about the possible outcomes that the phenomenon exhibits. Probability measures are traditionally used as a modeling tool in quantifying/modeling uncertainty. The systematic incorporation of information and beliefs about the phenomenon under study enables forecasting of future behavior based on current conditions. This area is where most of the advanced analytics opportunities reside in mining, at the characterization level. Data collection, curation, and use in decision-making is the spine of most mining activities, and to some extent, most investment decisions are made based on mathematical models fitted using data.

Risk assessment lies at the frontier of technical competencies in mining. If more mining professionals can work with some of the tools that enable risk assessment, it could have a beneficial impact on the industry. Nowadays, the typical miners' attitude toward risk is to ignore it simply, mainly because the industry lacks proper tooling for dealing with uncertainties, even though the whole industry is plagued with uncertain phenomena. A second important point to note in this discussion is that decision-makers worry about uncertainty because they need to make decisions involving an uncertain future. If it were not the case, and the future somehow could be predicted with precision, then every decision that could be taken would have no associated risk, and then life would be much simpler. However, because uncertainty is pervasive, it is imperative to adopt some definition/measure of the risk involved when deciding in the presence of uncertainty to make good decisions in mining. Some examples of risk measures are provided in [58]:

- *Expected Regret*: It is defined as the expected value of the (loss) distribution beyond a threshold α, i.e.,

$$G_\alpha(\vec{x}) = \int_{\vec{y}\in\mathbb{R}^m} \left[f(\vec{x},\vec{y}) - \alpha\right]^+ p(\vec{y})d\vec{y} \tag{5.43}$$

  With $[u]^+ = \max\{0, u\}$, $\vec{y}$ a random vector is representing scenarios and $p(\cdot)$ a probability density function.
- *Conditional value at risk*: It is the expected value of the losses exceeding VaR_k, i.e.,

$$C\mathrm{VaR}_k = \phi_k(\vec{x}) = (1-k)^{-1} \int_{f(\vec{x}.\vec{y})\geq\alpha_k(\vec{x})} f(\vec{x}.\vec{y})\ p(\vec{y})\mathrm{d}\vec{y} \tag{5.44}$$

  Or equivalently can also be defined as

$$C\mathrm{VaR}_k = \mathrm{VaR}_k + \mathbb{E}\{f(\vec{x}.\vec{y}) - \mathrm{VaR}_k | f(\vec{x}.\vec{y}) > \mathrm{VaR}_k\} \tag{5.45}$$

where VaR_k is defined as $\mathrm{VaR}_k = -F_X^{-1}(k)$ with F_X^{-1} denoting the inverse of the distribution F_X. Risk measures must satisfy some properties to be acceptable and coherent. For more details, consult [58–60].

Financial Risk

This subsection is based on [61]. The financial risk associated with a financial project can be defined as the probability of not meeting a particular target profit (maximization) or cost (minimization) level referred to as Ω, i.e.,

$$\text{Risk}(x, \Omega) = \mathbb{P}(\text{Profit}(x) < \Omega) \tag{5.46}$$

where $\text{Profit}(x)$ denotes the actual profit, resulting after the uncertainty has been unveiled and a scenario realized. When independent and mutually exclusive scenarios are used in uncertainty assessment, the above probability calculation can be expressed as:

$$\text{Risk}(x, \Omega) = \sum\nolimits_{s \in S} \mathbb{P}(\text{Profit}_s(x) < \Omega) \tag{5.47}$$

where $\text{Profit}_s(x)$ represents the profit of x under scenario s. The extension of this definition to the case where a continuous probabilistic distribution represents the uncertainty is:

$$\text{Risk}(x, \Omega) = \int_{-\infty}^{\Omega} f(x, \xi) \mathrm{d}\xi \tag{5.48}$$

There exists a connection between risk and expected profit. For example, for the continuous probability case, the expected value of the profit can be written as:

$$\mathbb{E}(\text{Profit}(x)) = \int_{-\infty}^{\infty} \xi f(x, \xi) \mathrm{d}\xi \tag{5.49}$$

By using the definition of financial risk, this leads us to the following:

$$\mathbb{E}(\text{Profit}(x)) = \int_{0}^{1} \xi d\, \text{Rist}(x, \xi) \tag{5.50}$$

Which, after integration by parts, becomes:

$$\mathbb{E}(\text{Profit}(x)) = \overline{\xi} - \int_{\underline{\xi}}^{\overline{\xi}} \text{Risk}(x, \xi) \mathrm{d}\xi \tag{5.51}$$

Which is valid for the case $\text{Profit}(x)$ is bounded (as it is usually the case) and $\underline{\xi} < \text{Profit}(x) < \overline{\xi}$. This last relation shows us that when $\text{Risk}(x, \xi)$ is minimized, the expected profit is maximized. Similar relationships can be developed between financial risk and other risk measures, but we believe the previous discussion illustrates the types of arguments used, so we encourage the interested reader to consult [59] for further information.

Operational Definition of Value at Risk

As we can see from the discussion of the previous subsubsection, the mathematics involved in defining risk is heavy and not too amenable, which could explain why people, in general, try to run away from this topic. To economically evaluate a project, we must make inferences and projections of geological, technical, and economic variables that will allow us to build the project's future cash flows and then determine their expected (NPV). This estimated value is usually madebased on incomplete information and random variables whose value is uncertain; we cannot know with certainty in advance. This introduces uncertainty to the project, which means that the project's actual NPV (or cash) may be higher or lower than the estimate obtained from the assessment (expected NPV). In this context, the financial risk is understood as the possibility that the actual value is less than expected.

Typical essential tools used in preliminary risk assessment are *sensitivity* and *scenario* analysis. Sensitivity analysis seeks to quantify and visualize a project's sensitivity to variations of the uncertain variables, or equivalently, against errors in the estimation. The method assumes a base or expected situation; identifies the most significant uncertain variables: prices, production volumes, costs, investments, etc.; generates optimistic and pessimistic values of uncertain variables; and finally estimates the NPV of the project modifying one variable at a time, keeping the other variables constant. It is easy to apply and easy to understand. However, it just permits analyzing the impact of one parameter at a time. The interpretations of concepts such as optimistic and pessimistic are subjective, it does not consider possible interrelationships between uncertain variables, and finally, it does not deliver a single measure of risk. Scenario analysis tries to address some of the deficiencies present in a sensitivity analysis by considering the interrelationships between uncertain variables. For this purpose, coherent sets of variables are generated, called scenarios, often pessimistic, expected, and optimistic. The NPV for each scenario is calculated, and the impact of the set of variables is assessed.

A more advanced method consists of using a Monte Carlo simulation methodology. For this methodology:

- Statistical distributions of each uncertain variable are modeled together with their correlations,
- Computationally generated values are repeatedly generated for the uncertain variables, using the corresponding probability distribution,
- With each set of simulated values (realization), a cash flow value is computed,
- Distribution of the NPV is generated, in which the project value is the mean, and the risk is quantified by the dispersion of the "empirical" distribution.

The whole process can be summarized in Fig. 5.2.

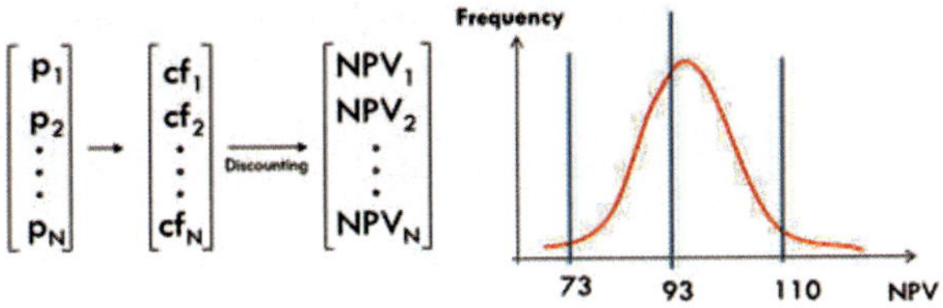

Fig. 5.2 Monte Carlo process for risk assessment

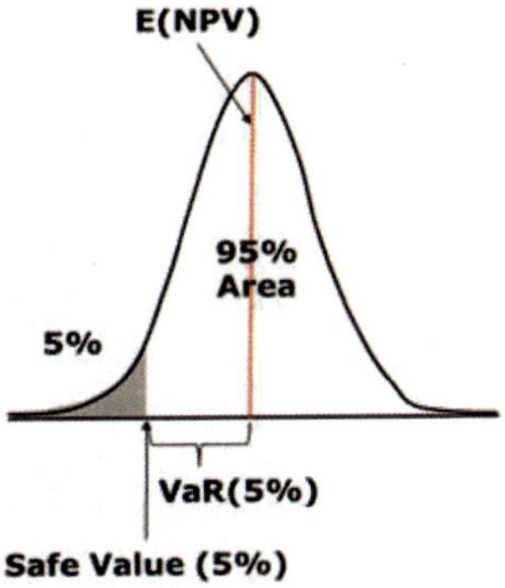

Fig. 5.3 Operational definition of value at risk

With the NPV distribution characterized in this way, the value at risk (VaR) can be defined as the maximum loss of value concerning its expected NPV that could occur with a certain confidence level, or with a formula:

$$\text{VaR}(x\%) = \mathbb{E}(\text{NVP}) - \text{Safe Value }(x\%) \tag{5.52}$$

where safe value is the percentile $x\%$ for the distribution (usually 5%). Graphically (see Fig. 5.3).

There is more about VaR that could be discussed, but for the purposes of the current presentation is not deemed necessary. Instead, the reader interested in deepening their conceptual and practical understanding of this important concept is referred to [62].

Case Study: Valuing a Marginal Mining Deposit Facing Substantial Price Uncertainty

The case study has already been introduced in some of the previous examples in this chapter. The case study focuses on a mineral processing plant treating tailings coming from both a running mining operation and two inactive tailing dams in a region in Chile.[5] This setting involves two main characteristics: one, the minerals contained in the tailings make the process economically feasible, but in a very marginal way;

[5] Due to confidentiality not much can be said about this operation.

Table 5.4 Company economic analysis, molybdenum: 8.80 US$/lb, base case: 3.30 US$/lb

Copper price [US$/lb]	2.50	3.00	3.30	3.50	4.00
NPV 7% [US$ millions]	$207	$341	$403	$430	$548

and two, it operates in a highly volatile price scenario, which is an essential source of uncertainty, being capable of making a marginal project start losing money. So, it is important to address this risk to know how much money is on the firing line. This will be done by calculating the value at risk for this project based on the commodity price uncertainty.

To incorporate commodity price uncertainty into the financial, technical model, different price trajectories will be made using a model that uses Brownian motion to predict the next period's price variation. In turn, the price trajectories for copper and molybdenum are generated according to the correlation between these metals to replicate the behavior seen in real life of commodities, which evolve together.[6] The correlation was determined using historical price data for both metals. These price trajectories were obtained using *R* statistical software, which was also used to combine these trajectories with the technical financial model generated in Excel software.

Since the company has successfully revalued tailings, it was chosen as a case study to build the technical, financial model (FTM) that led to the evaluation using the price trajectories generated with the tool, the items, and assumptions regarding this FTM have been already discussed in previous subsections where it was used to illustrate other aspects of the valuation, so it is not going to be repeated here. The operational data was obtained from a technical report dated March 2019, which the company gently provided.

To create and use the company's FTM, the exploitation of all the plant's feeds is considered: the Primary *Tailings* Dam, the Secondary *Tailings* Dam, and the fresh tailings obtained from a neighboring mining operation (ETD). The exploitation horizon for the project is projected to have a useful life lasting until 2037.

An economic analysis section within the technical report presents several NVPs for different copper prices, assuming fixed prices and a molybdenum price along with all copper price cases. These NPVs have been created using a deterministic FTM, which in a sense it has been presented is an attempt at producing some sensitivity analysis regarding price. One of the things that are not clear from the information presented, but it is believed to be the case, is that the economic valuation considered fixed commodities prices over the project's life. To see how different the valuation can be, an experiment was performed where price trajectories were generated using a GBM model (see Table 5.4).

Although these values are positive and consistent, they do not give a good idea of how much money is susceptible to loss due to variations in copper prices. As

[6] This technique uses the Cholesky decomposition of the covariance matrix, and it is an advanced topic, to learn more about this and other connected topics the reader is directed to the excellent reference [63].

mentioned previously, using a different modeling tool, this impact will be better assessed.

The model was implemented in the statistical language *R*. The most relevant code in this implementation is the price generator. For an evaluation horizon, n and several prices m, perform r repetitions of price trajectories. These generations are "market independent," which means that no market agent influence is considered. The only consideration included in the code is that a negative price remains zero due to the impossibility of a price commodity being null. Besides this, every price can be generated, no matter how low it is.

It needs to be recalled that a realization of a price path is governed by Eq. (5.41) which can be discretized (after integration) as:

$$\ln(P_{t+1}) = \ln(P_t) + \mu'\Delta t + \sigma.N(0, 1).\sqrt{\Delta t} \tag{5.53}$$

With $\sigma = \text{DevSt}\left\{\ln \frac{P_t}{P_{t-1}}\right\}$ and $\mu' = \mathbb{E}\left\{\ln \frac{P_t}{P_{t-1}}\right\}$. If $\Delta t = 1$ as it is usually the case, the expression is further simplified to $\ln(P_{t+1}) = \ln(P_t) + \mu' + \sigma \cdot \mathcal{N}(0, 1)$. The important loop in the *R* code that implements this model is (in boldface we highlight the use of the GBM discretization):

```
for (j in c(1:r)){
    z <- mvrnorm(n,mu=rep(0, m),Sigma=sigma)
    for (i in c(2:n)){
        result[i,,j] <- result[i-1,,j] + mu_0 + sigma_0*z[i,]
        if (result[i,1,j]< 0){result[i,1,j]<-0}
        if (result[i,2,j]< 0){result[i,2,j]<-0}
}}
```

Figures 5.4 and 5.5 show the typical outputs of the process.

After the prices are generated, they are transferred into the Excel spreadsheet in which the technical, financial model is hosted once for each repetition. This allows using EXCEL as a black-box model, which is used to calculate the project's NVP.

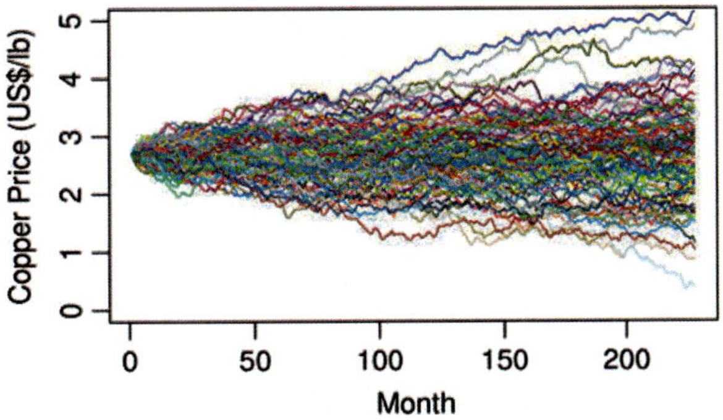

Fig. 5.4 Realizations of copper price trajectories

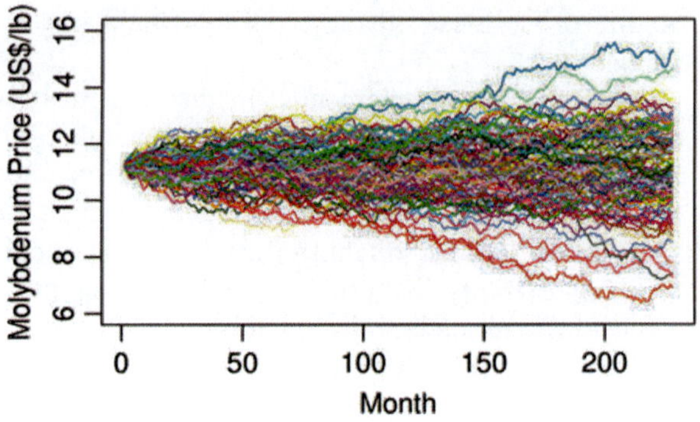

Fig. 5.5 Realizations of molybdenum price trajectories

Table 5.5 Expected value and value at risk [VaR(5%)] for the tailings processing project

Mean [US$ millions]	$419
Value at risk [US$ millions]	$209

After the repetitions have been processed, it is possible to analyze the histogram, calculate the mean and the 5% percentile, and determine the project's VaR (5%) (see Table 5.5).

From the VaR (5%) calculation, we understand that the project's maximum loss concerning the expected value is $210 million (see Fig. 5.6).

Also, it can be observed from the table that the average NVP yielded by the proposal is $419 million, which is very close to the base case determined in the report provided by the company. From the histogram of all the values of NPV created by

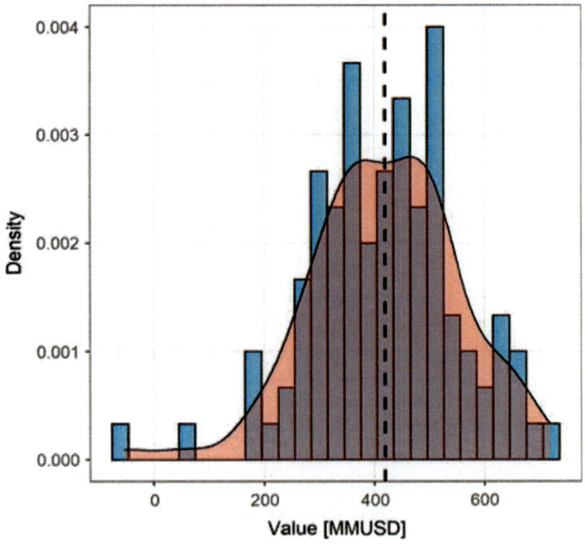

Fig. 5.6 NVPs histogram obtained from the model

the Monte Carlo process, it is easy to see that the Company's previous sensitivity analysis exercise produces values that are strictly contained within the distribution obtained by the simulation process used. However, it also provides a value at risk [VaR(5%)] for the project that is not as clearly informed from the analysis performed by the company.

The histogram obtained by the simulation exercise even shows negative values, which indicates that in regrettable and improbable cases, the price evolution of commodities could be such that the company could incur losses in this project. However, this last aspect is no quantifiable from the sensitivity analysis used by the company, which shows that the conclusions that the company could arrive at are potentially unreliable and misguiding of what could happen in the future; for according to the sensitivity analysis, it does not matter which future price is projected, the company will always have a positive NPV. In contrast, the simulation analysis shows that very low or even negative NPV values could be produced. The latter information is of value to stakeholders, who need this type of data to determine whether to incur a particular investment or take more conservative actions with their capital. Thus, illustrating the usefulness of the method used and its superiority when compared with more traditional techniques.

Advanced Analytics Opportunities in Valuation of Mining Projects

The valuation of mining prospects and mining projects heavily rests on financial data to produce results. As such, it shares some similarities in some respects with other areas of human endeavor, more particularly with Fintech initiatives, and it is believed that by looking at recent developments in that area, it is possible to envision somehow what the future looks like, in terms of opportunities, for valuation of mining projects. As such, the collection of possibilities presented herein is not an extensive one but attempts to identify opportunities not currently in use in mining but, given their current predominance in Fintech developments, they could soon become a trend.

One key trend in Fintech these days is the use of finance on-demand models for personalized portfolio management. Portfolio management tools/models are becoming more and more refined at the time of writing and are currently able to suggest investment opportunities based on projected income, current investment habits, risk appetite, and more. It is evident that the investment timeframe differs between the normal investor and mining investments, as the investment and pre-production periods are usually in the order of years for the mining case. Nevertheless, the mathematical tools used to manage financial portfolios can be aptly adapted to decide on a portfolio of mining investments. Questions such as: How to allocate budget for exploration every year throughout the portfolio? How to financially structure the portfolio? etc., are the first ones that come to mind. In addition, learning from past investment decisions can approximate risk appetite from the investors of a particular mining project and thus help in better focusing the development of financial assessment of mining prospects (recall that this process takes time and money to arrive at appropriate certainty levels).

Another trend observed in Fintech that may find its way into the mining industry is optimizing processes. As mentioned before, the financial assessment requires a long time and technical and financial information generation. The process is currently an extremely manual one. However, it is believed that as algorithms can learn the way certain decisions are taken, then it is possible to automate some of these valuation workflows, to produce better information that will allow the decision-maker to decide with better confidence than the current practice. The main idea here is to reduce the valuation process's work footprint and increase the overall efficiency, thus allowing a better characterization of the underlying uncertainty, enabling better risk quantification in the decision process. A crucial role in process automation is improving and refining big data analytics to provide more significant insights into the business. Big data contains great variety, arriving in increasing volumes and with more velocity, known as the three Vs. The mining case consists more and more of a big data scenario for analyzing and using information. Mining is becoming flooded with data coming from many operational and non-operational sources that require careful curation and storage, avoiding becoming trapped in the traditional functional partition mentality that has characterized the mining industry. The use of big data in the mining industry will undoubtedly become an innovation driver. It enables a deeper understanding of interdependencies between entities within the mining system, whether they are physical or abstract, but that for sure has an impact on the profitability of the mining venture.

Along the lines of big data analytics, and as a complement to it, the mining industry has started to consistently talk about "Mining 4.0," which is just a name that encompasses the opportunity behind the connected nature of devices and processes into systems that can be effectively exploited to produce better information, both in quality and quantity. The traditional process of cost estimation, an essential component of cash flows determination, is performed most of the time by using either a top-down approach where formulas are used to produce indicative values which are inherently imprecise, or otherwise much work needs to be introduced to model a system in full detail and produce a bottom-up estimation. With the ready availability of massive information coming from operations, appropriate reconciliation of production figures, costing, etc., it is possible to produce significantly better estimates that can be used to forecast costs (CAPEX or OPEX), thus eliminating manual processing and dubious assumptions.

Indeed, the computational infrastructure required to support this level of information is not the usual one that exists in typical mining operations. Furthermore, recent developments in the use of cloud computing allow the virtualization and scalability of computational systems and the implementation of remote-control centers that enable sharing of information between different operations at the corporate level and between operations, which enable obtaining better estimates for a series of other KPIs relating to production, equipment availability, processing results, etc. There is a strong presence of functional silos in mining, each with its preferred software, file formats, and processing. This characteristic of the industry presents an integration and interoperability challenge that has not yet been solved. The increased availability of data is yet to be introduced into the decision models used in the different functional areas of the industry.

As outlined in previous sections of the present chapter, there are specific opportunities in the valuation process. Fundamentally, estimates for the different values used in the valuation models could be improved using models that take advantage of ML or artificial intelligence methods. The foremost opportunity for the industry is aligned with the appropriate use of more data to improve the estimates and produce simulations of uncertain values to assess the risk associated with the decisions taken. This requires technology and a change of mentality in the industry that has the excellent opportunity to embrace uncertainty in the decision-making process instead of avoiding uncertainty. One of such recent trends is the use of digital twin models, which are becoming increasingly popular for characterizing aspects of the operations and having short-term forecasting capabilities of the response of a particular mining system when confronted with specific circumstances.

Conclusions

This chapter has provided an overview of how mine project valuations can be produced. It has focused on the main aspects affecting a project's value and illustrating one of those aspects with a simplified case study. As the reader can perceive, the topic is extensive, and a chapter like this one cannot go into too much detail. Otherwise, a whole book would be needed. Mineral economics is a multidisciplinary endeavor connecting several specific technical and economic/financial areas under a single umbrella. As such, the area presents plenty of opportunities for the use of advanced analytics techniques.

The most significant opportunity and probably the most urgent to tackle appropriately relates to the excellent characterization of all sources of uncertainty within the industry. It is a well-known fact that mining is severely affected by uncertainty. However, little or no use of uncertainty information is incorporated into the valuation process. In some cases, this is due to technical challenges. For example, mine planning models are not currently capable of producing robust mine plans based on information coming from scenarios. However, most of the time, the poor treatment of uncertainty within the mining industry starts with the lack of familiarity that the mining professionals have with the techniques that could be used. Instead of embracing uncertainty and risk as a business opportunity, they usually look to the other side and ignore these factors because of their inability to deal with them. We know that this last statement could sound harsh; however, it is a well-reflected statement that has matured over 20 years of professional practice where one of the authors has consistently observed this phenomenon.

Mining uses heaps of information. There is a tendency to collect much more information than needed, thinking this could somehow counterbalance the aforementioned technical deficiencies. Nevertheless, data needs to be appropriately curated and stored. This is the second challenge in the mining industry where the presence of functional silos does not allow the proper interoperability of technical solutions within mining. It is not surprising that the most used file format in the industry is a comma-separated value file (CSV). If this is not worrying, then it should start being. It indicates the industry's problems that those different areas do not work concurrently, and whichever information is shared, it does so via a CSV file which is very likely to be obsolete quickly as the different areas continue their work. It sounds even ironic, but a mine plan, costing, and cash flows are determined for a mine that will never exist most of the time. We believe that this is a must soon. However, the roadmap is not yet that clear.

Very recent initiatives in the mining industry have put the use of *digital twins* in the spotlight. It cannot be denied that the concept is enticing. However, it is not a mature approach yet and has not solved the underlying issues mentioned before regarding lack of interoperability between functional areas and poor uncertainty characterization. The newer developments are certainly taking advantage of the increased power of computers that enable running models such as a digital twin for a mining operation. We believe this trend will continue. However, as mentioned before, it does not tackle the fundamental issues, so it may fade away if it does not provide the industry's value.

The case study that was presented is a simplified version of an actual operating project. Due to confidentiality commitments, it was not possible to provide more detail about it. However, the inclusion of this case study within the chapter was done to illustrate the nature of the analysis type. However, it needs to be noted that we are still far from the most advanced possibilities out there with this case study. Some self-criticisms for the case study shown include:

- Brownian Motion-type models are just too perfect to be accurate. They assume that the returns are independent and normally distributed;
- However, returns tend to have more giant tails than what normality assumption produces. The statistical evidence is gathered using the Hurst exponent, which measures, among other things, how much long-term memory a time series could exhibit; and
- A whole school of thought (initiated by B. Mandelbrot) lifts some of these assumptions, which give rise to a different way to approach the problem. Mainly, returns follow a power law, which, married with the long-term memory of the commodities time series, give rise to a different mathematical tool. One possibility is to simulate using fractional Brownian motion, which is definitively more mathematically involved.

Despite the natural attitude that arises indicating that the case study always could have been better, we still hope that what we have presented herein serves as an inspiration to the practitioner, who can look for further information to try his models when valuating mining projects/investments. As the industry adopts advanced techniques to produce valuations, better and more complete communication among different areas will occur, and better financial performance will be assessed before commencing the operations.

References

1. Gordon, R.L., and J.E. Tilton. 2008. Mineral economics: Overview of a discipline. *Resources Policy* 33 (1): 4–11.
2. Gocht, W.R., H. Zantop, and R.G. Eggert. 1988. International mineral economics, 1 ed. In *Mineral Exploration, Mine Valuation, Mineral Markets, International Mineral Policies*, 271. Berlin Heidelberg: Springer-Verlag.
3. Foo, N., H. Bloch, and R. Salim. 2018. The optimization rule for investment in mining projects. *Resources Policy* 55: 123–132.
4. Gentry, D.W., and T.J. O'neil. 1984. *Mine Investment Analysis.*
5. Wirtz, H. 2012. Valuation of intellectual property: A review of approaches and methods. *International Journal of Business and Management* 7 (9): 40.
6. Meese, R.A., and K. Rogoff. 1983. Empirical exchange rate models of the seventies: Do they fit out of sample? *Journal of International Economics* 14 (1–2): 3–24.
7. Hann, T.H., and E. Steurer. 1996. Much ado about nothing? Exchange rate forecasting: Neural networks vs. linear models using monthly and weekly data. *Neurocomputing* 10 (4): 323–339.
8. Ca'Zorzi, M., M. Kolasa, and M. Rubaszek. 2017. Exchange rate forecasting with DSGE models. *Journal of International Economics* 107: 127–146.
9. Colombo, E., and M. Pelagatti. 2020. Statistical learning, and exchange rate forecasting. *International Journal of Forecasting* 36 (4): 1260–1289.
10. Fu, S., et al. 2019. Evolutionary support vector machine for RMB exchange rate forecasting. *Physica A: Statistical Mechanics and its Applications* 521: 692–704.
11. Garvin, M.J., and C.Y. Cheah. 2004. Valuation techniques for infrastructure investment decisions. *Construction Management and Economics* 22 (4): 373–383.
12. Yasir, M., et al. 2020. An efficient deep learning-based model to predict interest rate using Twitter sentiment. *Sustainability* 12 (4): 1660.
13. Ndikum, P. 2020. *Machine Learning Algorithms for Financial Asset Price Forecasting.* arXiv: 2004.01504.
14. Xiong, X. et al. 2005. Wavelet-based beta estimation of China stock market. In *2005 International Conference on Machine Learning and Cybernetics.* IEEE.
15. Hollmann, J., and L.R. Dysert. 2007. *EST.01-Escalation Estimation Working with Economics Consultants.*
16. Mining trucks' tires are now more pricey than a Merc. 2011 [cited 2021 Jul 30]. Available from https://www.smh.com.au/business/mining-trucks-tyres-now-more-pricey-than-a-merc-20110630-1grrb.html.
17. Touran, A., and R. Lopez. 2006. Modeling cost escalation in large infrastructure projects. *Journal of Construction Engineering and Management* 132 (8): 853–860.
18. Long, K.R. 2009. A test and re-estimation of Taylor's empirical capacity-reserve relationship. *Natural Resources Research* 18 (1): 57–63.
19. O'Hara, T.A., and S.C. Suboleski. 1992. Costs and cost estimation. *SME Mining Engineering Handbook* 1: 405–424.
20. Singer, D., W. Menzie, and K.R. Long. 2000. *A Simplified Economic Filter for Massive Sulfide Deposits Underground Mining.* US Department of the Interior, US Geological Survey.
21. Code, J. 2012. Australasian code for reporting exploration results, mineral resources, and ore reserves. *AusIMM, Melbourne* 44: 320.
22. Carrasco, P., J.-P. Chilès, and S.A. Séguret. 2008. Additivity, metallurgical recovery, and grade. In *8th International Geostatistics Congress.*

23. Dominy, S.C., et al. 2018. Geometallurgy—A route to more resilient mine operations. *Minerals* 8 (12): 560.
24. Richmond, A. 2002. Applying four different risk models in local ore selection. *Natural Resources Research* 11 (4): 299–314.
25. Richmond, A., and J.E. Beasley. 2004. An iterative construction heuristic for the ore selection problem. *Journal of Heuristics* 10 (2): 153–167.
26. Krautkraemer, J.A. 1988. The cut-off grade and the theory of extraction. *Canadian Journal of Economics* 1988: 146–160.
27. Asad, M., and E. Topal. 2011. Net present value maximization model for optimum cut-off grade policy of open-pit mining operations. *Journal of the Southern African Institute of Mining and Metallurgy* 111 (11): 741–750.
28. Hills, A.J., B.E. Hall, and J. Poniewierski. 2014. Mining scenario analysis and mine variable value. In *12th AusIMM Underground Operators Conference*. Adelaide.
29. Hall, B. 2014. *Cut-Off Grades and Optimizing the Strategic Mine Plan*. Australasian Institute of Mining and Metallurgy.
30. Strnad, J. 1999. Tax depreciation and risk. *SMUL Review* 52: 547.
31. Mular, A. 1978. *The Estimation of Initial Capital Costs*, 52–70. Montreal: Canadian Institute of Mining and Metallurgy.
32. Noakes, M., and T. Lanz. 1993. *Cost Estimation Handbook for the Australian Mining Industry*.
33. The Australian Institute of Mining and Metallurgy. 2012. *Cost Estimation Handbook—Monograph 27*, 2nd ed. The Australian Institute of Mining and Metallurgy.
34. Committee on Risk-Based Analysis for Flood Damage Reduction et al. 2000. *Risk Analysis and Uncertainty in Flood Damage Reduction Studies*. National Academy Press.
35. Journel, A.G. 1994. Modeling uncertainty: Some conceptual thoughts. In *Geostatistics for the Next Century*, 30–43. Springer.
36. Bárdossy, G., and J. Fodor. 2001. Traditional and new ways to handle uncertainty in geology. *Natural Resources Research* 10 (3): 179–187.
37. Rendu, J.-M. 2002. Geostatistical simulations for risk assessment and decision making: The mining industry perspective. *International Journal of Surface Mining, Reclamation, and Environment* 16 (2): 122–133.
38. Martin, D., and F. Sparrow. 1984. The treatment of uncertainty in mineral exploration and exploitation. *Annals of Operations Research* 2 (1): 271–284.
39. Grieco, N., and R. Dimitrakopoulos. 2018. Grade uncertainty in stope design—Improving the optimization process. In *Advances in Applied Strategic Mine Planning*, 573–589. Springer.
40. Dimitrakopoulos, R. 1998. Conditional simulation algorithms for modeling orebody uncertainty in open pit optimization. *International Journal of Surface Mining, Reclamation, and Environment* 12 (4): 173–179.
41. Smith, M., and R. Dimitrakopoulos. 1999. The influence of deposit uncertainty on mine production scheduling. *International Journal of Surface Mining, Reclamation, and Environment* 13 (4): 173–178.
42. Dimitrakopoulos, R., C. Farrelly, and M. Godoy. 2002. Moving forward from traditional optimization: Grade uncertainty and risk effects in open-pit design. *Mining Technology* 111 (1): 82–88.
43. Richmond, A. 2003. Financially efficient ore selections were incorporating grade uncertainty. *Mathematical Geology* 35 (2): 195–215.
44. Dimitrakopoulos, R., L. Martinez, and S. Ramazan. 2007. A maximum upside/minimum downside approach to the traditional open-pit mine design optimization. *Journal of Mining Science* 43 (1): 73–82.
45. Journel, A., and C.J. Huijbregts. 1978. *Mining Geostatistics*. New York: Academic Press.
46. Goovaerts, P. 1999. Geostatistics in soil science: State-of-the-art and perspectives. *Geoderma* 89 (1–2): 1–45.
47. Chiles, J.-P., and P. Delfiner. 2009. *Geostatistics: Modeling Spatial Uncertainty*, vol. 497. John Wiley & Sons.

48. Olea, R.A. 2012. *Geostatistics for Engineers and Earth Scientists*. Springer Science & Business Media.
49. Matheron, G. 1969. Cours de géostatistique. l'Ecole national supérieure des mines de Paris.
50. Mathereon, G. 1970. La théorie des variables régionalisées et ses applications. *Cah Cent Morphol Math* 5: 1–212.
51. Wackernagel, H. 1993. Cours de géostatistique multivariable. ENSMP, Fontainebleau, France.
52. Armstrong, M., and R. Jabin. 1981. Variogram models must be positive-definite. *Journal of the International Association for Mathematical Geology* 13 (5): 455–459.
53. Rivoirard, J. 2003. Cours de géostatistique multivariable. Cours C-172, CG, Ecole des Mines de Paris.
54. Leuangthong, O., J.A. McLennan, and C.V. Deutsch. 2004. Minimum acceptance criteria for geostatistical realizations. *Natural Resources Research* 13 (3): 131–141.
55. Gligoric, Z., et al. 2014, Evaluation of underground zinc mine investment based on fuzzy-interval grey system theory and geometric Brownian motion. *Journal of Applied Mathematics* 2014.
56. Savolainen, J. 2016. Real options in metal mining project valuation: Review of the literature. *Resources Policy* 50: 49–65.
57. Savvides, S. 1994. Risk analysis in investment appraisal. *Project Appraisal* 9 (1): 3–18.
58. Szegö, G. 2002. Measures of risk. *Journal of Banking & Finance* 26 (7): 1253–1272.
59. Frittelli, M., and E.R. Gianni. 2002. Putting order in risk measures. *Journal of Banking & Finance* 26 (7): 1473–1486.
60. Rockafellar, R.T., and S. Uryasev. 2002. Conditional value-at-risk for general loss distributions. *Journal of Banking & Finance* 26 (7): 1443–1471.
61. Barbaro, A., and M.J. Bagajewicz. 2004. Managing financial risk in planning under uncertainty. *AIChE Journal* 50 (5): 963–989.
62. Philippe, J. 2001. *Value at Risk: The New Benchmark for Managing Financial Risk*. NY: McGraw-Hill Professional.
63. Asmussen S., and P.W. Glynn. 2007. *Stochastic Simulation: Algorithms and Analysis,* vol. 57, Springer Science & Business Media.

Chapter 6
Advanced Analytics for Mine Exploration

Sara Mehrali and Ali Soofastaei

Abstract Due to declining commodity prices in a constantly dynamic environment, there has always been a desire to maximize profits and achieve value from limited resources. Traditional experimental and numerical simulation techniques have failed to provide comprehensive and optimized solutions in a bit of time. With the enormous volume of data produced daily, a solution to meet the industry's challenges was imminent. The various opinions of the expert are fraught with additional challenges to achieving timely and cost-effective solutions. Data analysis has contributed significantly to several areas. This chapter overviews the various applications of advanced data analysis and machine learning in mine exploration. After an introduction to exploring the geological features and genetic models of mineral deposits will be discussed. Later in this chapter, the role of advanced analytics in minerals prospecting and exploration will be explained. Finally, at the end of this section of advanced analytics in the mining engineering book, the details of mining geophysical and geochemical aspects when the analytics approaches are used will be described.

Keywords Advanced analytics · Mining · Exploration · Mineral deposits · Prediction · Optimization

Introduction

Exploration may include various activities to help determine whether or not there are minerals under the earth. Mining can be possible in the future if the exploration finds minerals that can be commercially extracted [1]. Exploration is the method by that geologic info is collected and analyzed to identify mineral deposits and deciding the economic practicability of their extraction [2].

S. Mehrali (✉)
Simulation and Data Processing Laboratory, School of Mine Engineering, University of Tehran, Tehran, Iran
e-mail: sara.mehrali@ut.ac.ir

A. Soofastaei
Vale, Australia
URL: https://www.soofastaei.net

A. Soofastaei (ed.), *Advanced Analytics in Mining Engineering*,
https://doi.org/10.1007/978-3-030-91589-6_6

Exploration is an essential step in locating minerals that would most likely be extracted. In addition, mineral demand is rising; for example, we need a lot of copper to build renewable energy assets like wind turbines. This demand aids in the expansion of exploration [1].

As mining's overall prospects brighten, however, two chronic issues remain: the declining rate of latest mineral discoveries and, therefore, the rising cost for every new ounce or weight unit of metal or mineral additional to reserves.

The typical prospector relied entirely on direct observation of mineralization in outcrops, sediments, and soil when looking for valuable minerals. While direct observation is still common, today's prospectors use a combination of geological, geophysical, and geochemical instruments to provide indirect indicators and narrow the search range. Anomalies, or variations between what is found at a particular location and what would usually be anticipated, are the goal of modern techniques.

The modern era of mineral exploration began after World War II. After seven decades of geological prospecting globally, almost all the "easy" deposits at the surface have been discovered and extracted.

As today's mineral explorers search out difficult deposits, they become reliant on costly deep drilling and data-heavy surveying.

Advanced analytics, artificial intelligence (AI), and machine learning (ML) are some of the new terms that today's mineral explorers use in the vocabulary of their work because they seek new ways to achieve exploration, discovery, and production through improved accuracy in deposit modeling capacity, resulting in earlier access to information which is possible.

Advanced analytics techniques are being applied in different industries to improve strategic and operational processes by optimizing decisions. This chapter examines how to use these analytical techniques for the fundamental challenges of the mining industry.

Advanced analytics can block the loop between analyzing data and acting. Using analytics to make better decisions does not require specialist skills or much time, and analytics is included in the business and regularly to improve decision-making at all levels of the organization, from high-level management to operations and maintenance personnel. Embedding analytics in the business processes brings better decisions backed up by data integrated from several resources in real time. Finally, predictive tools suggest or, in some cases, automate practical courses.

The primary goal of advanced exploration analytics services is to reduce the (extremely high) risks associated with exploration decisions taken by company executives in border areas, marginal exploration areas, and deep horizons in mature basins. In discovery, there are no solutions that are 100% error free. According to advanced exploration analytics, the best way to handle risks is to use all available data on hydrocarbon development, migration, and accumulation to forecast the exploration potential of a target formation. Customers can save millions of dollars in unnecessary costs by using patented computer technology services, reducing the chance of making lousy exploration decisions [3, 4].

Advanced exploration analytics offers metallogenic research services to mining companies and mineral exploration investors. A preliminary review should be conducted to reduce investment risks before allocating large budgets for mining, minerals discovery, and other related operations. The metallogenic analysis can be used to find possible mineralization sites for various commodities, including gold, platinum, silver, copper, zinc, and rare earth metals [3].

This chapter looks at how advanced analytics techniques can be used in the mining industry to solve the aforementioned issues. Statistical analysis, predictive modeling, optimization, risk modeling, advanced visualization, expert analysis, and automated decision-making are among the techniques evaluated. These capabilities assist business decision-makers with limited time and statistical knowledge in making better decisions from large amounts of data.

Advanced analytics tools aim to collect all of an organization's data and historical experience to make better decisions. Advanced analytics goes beyond merely delivering reports containing historical data explaining "what happened" to achieve this aim. Advanced analytics seeks to improve business processes by automating routine decisions ("embedded analytics") and assisting decision-making by automating data collection and validation and distributing decisions to the required levels within an enterprise.

Figure 6.1 shows how the organization can make gradual progress to achieve such a goal through several stages of analytical maturity.

Advanced analytical capabilities are involved in advancing the maturity of the value chain of mining organizations in partnership. The main distinction is the focus of the business process on analysis, not the data-focused reporting currently provided for extraction operations. Advanced analysis of integrated collaboration by:

- Ability to optimize for different organizational roles;
- Embedded analysis of business processes;
- Facilitates streamlined and optimized business processes;

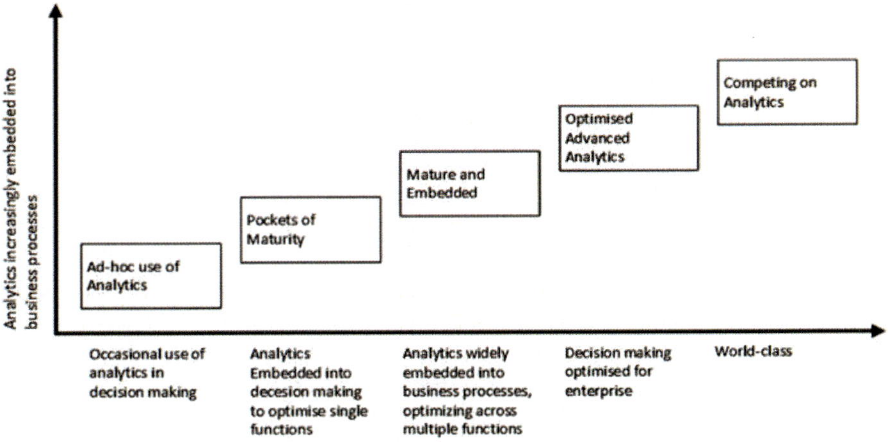

Fig. 6.1 Analytics maturity model [5]

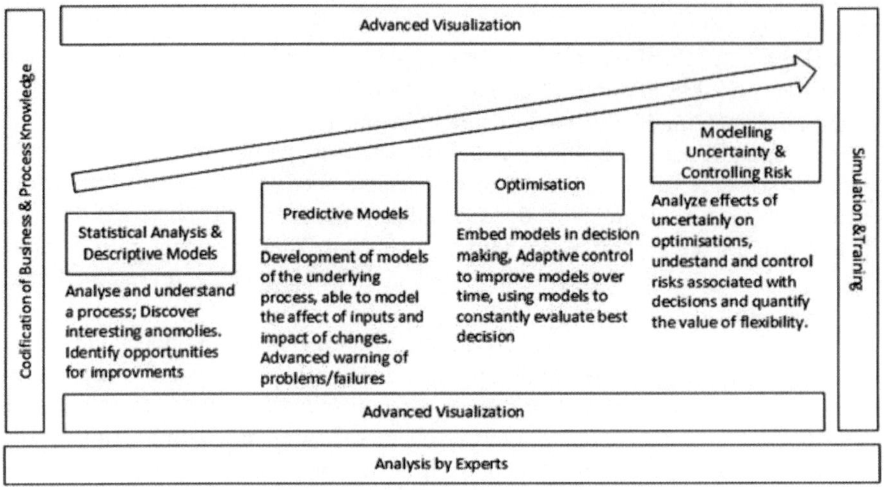

Fig. 6.2 Advanced analytics model for the mining industry [5]

- Typical usage of real-time information across multiple decision-making functions; and
- Information exchange and visibility of priorities and limitations outside the company for stakeholders.

Figure 6.2 illustrates the development of advanced analytical techniques that begin with the statistical analysis of data and culminate in modeling uncertainty and risk control.

A mining organization at the top of the advanced analytics maturity curve will employ a wide range of analytical techniques across the organization and integrate analytical efforts into multiple business functions.

Advanced Analytics Opportunities for Mine Planning

Changes in reserve figures, product rates, or costs are generally factored into life-of-mine (LOM) plans every year. A block model of the orebody is used to start the mine planning process. The predicted benefit for each block in the model is determined by subtracting the estimated costs of mining, mineral processing, and transportation from the estimated amount of recoverable mineral content for each block. A sequence of "nested" pit shells will be produced based on the variable, the expected price of the mineral. The pit's potential "push-backs" are described by successive pits. The optimum cutoff grade sequence for each pit is calculated using the grade tonnage distribution of each push back. Cutoff grades decide where resources go, whether it is to a processing facility, a marginal ore stockpile, or a waste dump. The best cutoff grade series is usually decreasing since this maximizes cash flow during the project's early stages, raising the project's net present value.

Advanced analytics can be used for long-term and strategic mine planning in the following fields.

- First, advanced analytical capabilities would be essential in the mining industry's transformation to network-centric operations. Second, increased levels of business process convergence and greater automation across the value chain would characterize network-centric operations. This pattern can be seen in the early deployments of remote operation centers (ROC) by some mining companies.
- Second, advanced analytical capabilities directly support network-centric operations by embedding analytics in automated operational decisions and enabling decisions tailored to national business purposes.

Data privacy is critical in the modern world of network-centric operations. However, until investing in advanced analytics, the mining industry must be prepared to invest in data quality assurance. This could be an important first step toward advanced analytics [5, 6].

Machine Learning

Innovative mining technology has recently advanced, allowing for the production, collection, and sharing of large amounts of data in real time. As a result, the mining industry is actively conducting ML studies that use these results. The findings indicate that ML research, which has been in the mining industry since 2018, has been chiefly used for mineral exploration. The support vector machine is the highly used ML model, followed by deep learning models. The root means square error and coefficient of determination were used to assess the ML models. Because of their ability to think, humans learn from their everyday lives. They will, for example, benefit from their training as well as their thoughts and memories. Unlike humans, computers, on the other hand, learn by algorithms. This is referred to as ML.

ML simulates human understanding by encouraging computers to gain information from the real world, allowing them to perform better on specific tasks based on newly acquired knowledge. "A computer program is said to learn from experience E of a particular class of T tasks and assess P success if its performance in T tasks, as measured by P, improves with E experience," according to the definition of ML [7].

Mining has a significant impact on the global economy. The use of ML to transform the industry's technical, economic, and environmental landscape is one of the factors that can fuel its growth and make it more profitable for investors.

ML, which heavily relies on manipulating and analyzing big data, can transform the technology paradigm of the future. ML is based on computational algorithms that simulate human intelligence and can imitate the environment with the help of big data.

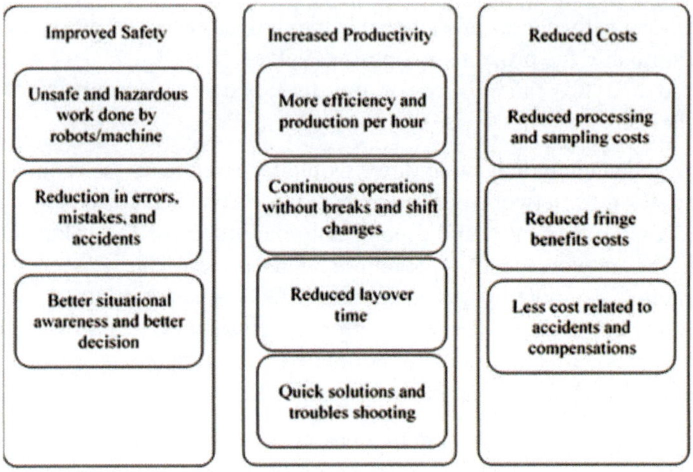

Fig. 6.3 Some significant advantages of these technologies

ML can usher in a new age of mining by lowering prices, increasing productivity and operational performance, and lowering the industry's environmental footprint through the use of intelligent systems [8].

Some of the most intriguing elements of ML are listed here:

- Elf learning systems or computers that can evolve;
- They are educating robots to function as intelligently as humans;
- A device that runs on its own after initial implementation and does not need much human interaction.

ML is a self-learning algorithm that learns over time to improve decision-making. ML refers to the process of teaching a robot to perform specific tasks. First, it must be trained until it understands how to navigate and function (Figs. 6.3 and 6.4).

Mineral Deposit Geological Characteristics and Genetic Models

Mineral deposits, often known as ore deposits, are natural concentrations of one or more mineral commodities that reach a level where economic extraction is feasible. They are formed as a result of many geological processes but are included in specific geological settings. Similar genetic methods may form deposits with similar characteristics and define a "class" of deposits. Geological processes that may concentrate minerals include fractional crystallization of magmas, the release of volatiles from a crystallizing magma, magmatic interactions with host rocks, metamorphic reactions of fluid production, organic interactions, and chemical changes circulation of

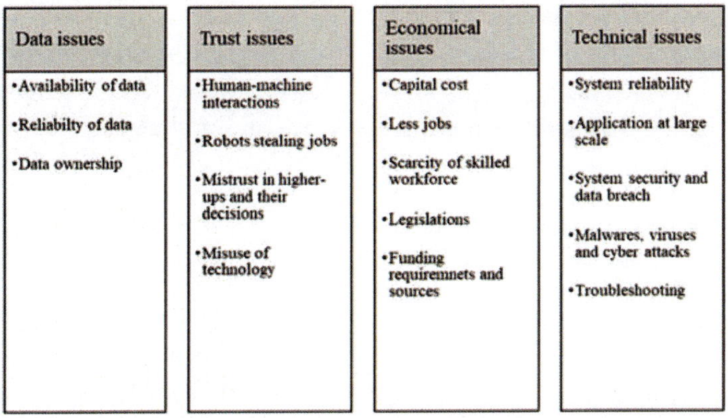

Fig. 6.4 Some of the most significant obstacles to ML in mining

heated groundwater, and weathering is another case. In summary, "types of mineral deposits" are mineral deposits that have a set of geological features and contain a specific commodity (or commodity) that generally distinguishes them from other types [9, 10].

Ideally, mineral deposit types should reflect how the mineral deposit was actually formed, and in this case, one might be able to use the term genetic model. However, there is considerable debate among geologists about forming a particular deposit in many cases, so classifications based solely on a specific genetic model are problematic. Another option is to classify deposits based on experimental properties (minerals, host rock association, etc.), resulting in a unique fingerprint of a particular deposit (i.e., a descriptive model). Even if no two mineral deposits are the same, experimental descriptions of deposits show that natural grouping can be freely defined in a small number of categories or types. In turn, these categories are usually matched with genetically derived models. Thus, even with purely descriptive physical classifications, a close connection between this and models is often defined using genetic criteria [9].

Before the twentieth century, models of mineral deposit formation were often based on the polarity of Plutonism theories (all deep igneous roots) or Neptuneism (all sedimentary roots) on the origin of rocks, and it is only in the twentieth century that modern views are relevant. It emerged in the formation of mineral deposits. Since the late 1960s, plate tectonic theory has revolutionized geological thinking, and in the mid-1980s, Guilbert and Park [11] developed a scheme to classify mineral deposits based on plate tectonics [9].

Ore Genesis (Formation Processes)

The preferred classification approach is to connect deposit types to ore formation processes directly. This offers more precise criteria for understanding the deposit regarding related features (such as association with associated igneous assemblages, change patterns). As a result, better exploration models for their discovery and evaluation will emerge. First, however, descriptive models are needed to assist engineers in evaluating absolute reserves (selection of exploration tools, elements for analysis in geochemical exploration, etc.).

Internal processes, hydrothermal processes, metamorphic processes, and surface processes are the four major groups that Evans [12] divides ore formation processes into. The first three processes are concerned with subsurface phenomena, while the last three processes are concerned with events that occurred on the earth's surface. Therefore, the three main sections of Orth magmatic, hydrothermal, and surface deposits are preferred in this chapter. To better understand the essence of the hydrothermal process, it should be subdivided into magmatic hydrothermal, metamorphic hydrothermal, diagenetic hydrothermal, and surface hydrothermal.

More qualified terms such as syngenetic and epigenetic help discuss the formation of rock deposits. Syngenetic is used for mineral deposits that are interpreted to be formed simultaneously with enclosed rocks. Syngenetic deposits such as host rocks such as sedimentary rocks from magma crystallization result from sedimentation within the seafloor sediments of the hydrothermal vents, or perhaps by the physical deposition of heavy mineral sand. Conversely, epigenetics is used for mineral deposits later formed from enclosed rocks, resulting in deposits found in the form of masses, layers, or disseminations that are seen to cut or reprint the original rock. An example of this might be a gold quartz vein crossing through deformed greenstone.

Hypogene is another helpful term used to describe mineralization created by deep earth processes, while supergene refers to techniques located on the main mineralization. These terms are often synonymous with primary and secondary terms, respectively. In the context of a sulfide deposit formed, for instance, as a magmatic segregation of chalcopyrite that is then partly oxidized by weathering to, for example, chalcocite, the chalcopyrite is considered to be hypogene (or primary), while the weathering product chalcocite would be termed supergene (or secondary).

Key Deposit-Forming Processes

Orth Magmatic Deposits

Orth magmatic deposits consist of primary magmatic processes (i.e., cooling and crystallizing magmas). They are hosted inside the rocks that form from there. Therefore, they are syngenetic by definition. Deposits may form as a result of the following:

- Solid phases crystallize as differentiation as magma cools;
- Crystallized minerals are enriched from residual liquids that are formed by cooling and crystallization of magma;
- Formation of sulfide melt, which is caused by the inflexibility resulting from the simultaneous silicate melt; or
- A magma carries the xenobiotic or xenocryst phases that it picks up after passing through the earth's crust.

Hydrothermal Deposits

Hydrothermal deposits are made up of hot water (hydrothermal fluids) circulating in the earth's crust. The direct evidence for the presence of these liquids in the earth's crust is superficial manifestations such as hot springs and fumes. Still, indirect evidence is provided for the existence of hydrated minerals even in crystalline rocks. Hydrothermal fluids may originate in igneous masses or from deep-crust metamorphic reactions. Still, they may also result from heated meteoric or seawater flows or the release of trapped water from sedimentary basins diagenetic change.

Tracing the primary source of this spectrum of liquids is now largely possible by using blockage studies of fluids trapped in minerals and using a wide range of isotopic markers. Some liquids are directly related to or out of the cooling igneous intrusions at one end of the hydrothermal deposits range. At the other end, surface water is heated because it is dispersed in the earth's crust's more profound and warmer parts. Technically, the term hydrothermal should only be used for predominant liquids in water, whereas gaseous liquids should be known as pneumatolytic. In addition, hydrothermal deposits can be syngenetic or epigenetic, depending on the location and time of deposit formation.

Magmatic Hydrothermal

As the magma body cools and crystallizes, magmatic hydrothermal fluids form. In some instances, the magmatic system may serve as a passive source of heat, directing the circulation of liquids out of the magma through the adjacent broken crust. However, magmas, especially felsic magmas like those that form granite rocks, contain large quantities of miscible water carried in the magma itself in many other cases. The concentration of the magma increases as it cools and crystallizes, forming an immiscible fluid phase that absorbs other components that tend to detach from the silicate melt into a hydrous-fluid phase. At higher temperatures and pressures, gases can also be decomposed by melting silicate. However, the gases are not miscible and can break into a liquid phase or be released entirely as vapor when the magma cools and penetrates shallow sections of the crust (pneumatolytic liquids).

Chloride ions, which form high-temperature brine and are effective solvents for many metals such as iron, copper, lead, and zinc, are concentrated components in the fluid phase that can dissolve from magma. These metal-rich fluid phases may move away from the magma and interact with minerals and fluids in previously crystallized magma or foreign rocks, causing chemical reactions that change the minerals and fluids, resulting in new mineral phases, including ore minerals. These fluids may also concentrate accumulative gases like carbon dioxide (CO_2), sulfur dioxide (SO_2), and hydrogen sulfide (H_2S), as well as transport dissolved components with vapor-containing fluids with magmas' aid.

All mineral deposits formed by replacement or fracture around the cooled magma chamber are epigenetic because they overprint the host rocks. However, they are associated with the magmatic activity. Other rare metals may accumulate longer in silicate magmas but may be concentrated in the final hydrous-fluid phase, trapped inside the cooling magma chamber at the last moment. In this case, the deposits are defined as syngenetic because the deposit crystallizes in the magma by crystallizing like Orth magmatic sediments. Many deposits may be formed from these processes depending on the path of magma crystallization and its interaction with external components.

Metamorphic Hydrothermal

Metamorphic hydrothermal fluids are formed as a result of mineral chemical processes that can release volatiles. They are typically water dominated but may contain gases such as CO_2.

Heat or external strain, or (usually) a mixture of both, causes metamorphism in rocks. For example, the deep burial of a rock mass over time or its replacement by the intrusion of a nearby magma body can both provide heat. In addition, metamorphic pressure may reappear during deep burial or as a result of structural processes. Both of these deposits are epigenetic by definition, and their effects cause preformed rocks to be overprinted.

Diagenetic Hydrothermal

Diagenetic hydrothermal fluids are formed from trapped pore waters that bind freely to clays and other minerals before being released during compaction and lithification. This is related to hydrocarbon processing on a large scale in a sedimentary basin under burial and lithification. The released water will dissolve soluble salts (convert to brine), allowing several cations and ligands to be transported to a deposition point and form an ore deposit. Evaporative substrates can be a specific source of salt that the base water can dissolve in these sedimentary basins.

The basins undergoing diagenesis are heated, so the primary brine can become a very effective solvent for dissolving large amounts of metals. This major saltwater may migrate to depositional environments through crustal faults and permeable horizons. If these fluids are discharged in a submarine or lacustrine environment, where chemical deposition leads to layered deposition of minerals, epigenetic deposits can form. Epigenetic deposits can also include where the fluid interacts with previously lithified rocks, causing mineral changes and deposition.

Surface or Seafloor Hydrothermal

When water from the penetration depth of air or seawater descends, it is heated deeper in the crust, resulting in surface or seafloor hydrothermal fluids. This process is particularly visible in areas with a high crust heat flow, standard when the earth's crust is diluting. This is a natural occurrence on the seafloor, where a new ocean is propagated by the seafloor and created by the eruption of submarine volcanoes. In addition, such hydrothermal fluids may develop on land in areas of crustal depletion, which are frequently linked to underwater volcanism. Although these phenomena may not be specifically related to igneous activity, fluids in the deep crust may easily interfere with residual heat and convection.

Superficial manifestations of this process may be the presence of hot springs in-ground hydrothermal vents or the seafloor. Deposits formed by these processes can be epigenetic. They replace the formed rocks or maybe syngenetic, for example, in the case of seafloor hydrothermal vents where expiratory fluids may lead to layer formation and masses of chemical ore minerals. On the other hand, syngenetic deposits can also form when hot springs vent to the surface and minerals are deposited in layered groups.

Surface Upgrade (Low Temperature)

Surface processes at low temperatures (less than 50 °C) may also result in the formation of economically valuable ore deposits. Certain minerals are redistributed and accumulated due to physical processes such as physical erosion, transportation, and deposition. The minerals' different physical and chemical activity that make up the primary rock causes these deposits to form. Hydraulic (water) or aeolian (air) processes are examples of physical processes (wind).

Weathering is another significant ore formation process that involves mobile solutions causing chemical changes and redistribution of components in surface rocks. Differential chemical properties of minerals at the surface and the interface with the surface crust may result in residual upgrades, chemical dissolution mechanisms, and chemical precipitation to concentrate the metal/mineral of interest. In such instances, meteorites at the surface drive the formation of ore by water movement, while similar analogous processes may occur at the seafloor. The components can be dissolved and re-established in the desired mineral locations or surface interfaces by these groundwaters. The evaporation process, which is a surface phenomenon, is another critical process for ore formation. Soluble salts are lost when water is lost, such as in an evaporation basin or when water evaporates from the earth's surface due to the sun's heat energy.

The Role of Genetic Models in the Evaluation of Deposits and Estimation of Mineral Resources

Genetic models are rarely discussed in mineral resource estimation, even though they are critical to getting things right. Early in the exploration process, the ability to conceive/understand and apply the correct genetic model is essential for deposit discovery and the quality of the final deliverable product, which includes calculated, known, and inferred resources. It is important to note that mineral deposit tonnage and grade estimates are just as good as the genetic/geological model they are based on.

Any genetic model can include a wide variety of deposits, such as the Lac des Îsles deposits in Canada, the Bushveld Complex in South Africa, the Stillwater Complex in the USA, and the Great Dyke in the UK (Zimbabwe). It has several characteristics in general. However, on the other hand, each deposit is particular and must be viewed to comprehend, analyze, identify, and improve it fully.

Mineral deposits may be homogeneous or heterogeneous, depending on the genetic model. Complex genetic models, in general, result in heterogeneous mineralization. Since many estimation elements are uncertain, estimating and classifying mineral resources become more complex as deposits become more heterogeneous. Figure 6.5 illustrates this.

A deposit's regularity and predictability are dependent on and intimately linked to the geological environment in which it formed. The more we understand how deposits shape and the factors that influence their production and eventual reform, the more we will be able to analyze, identify, and plan for their future growth. When classifying tools, qualified people (QP) often consider drill hole distance and areography. This is acceptable, but the fact is that radiation output continuity is a property of geological processes, not a drill hole or areography. Therefore, not only should "genetic/geological reliability" be high enough to ensure a sufficient basis for estimating and classifying tonnage and grade, but it should also provide a foundation for assessing the validity and application of geostatistical studies. Because:

- Changes in the genetic model can dramatically alter tonnage calculation; errors resulting from inaccurate or inadequate genetic/geological interpretation are more severe than errors resulting from grade estimation; and
- Rather than the other way around, the grade of interpolation should be governed by genetic/geological characteristics.

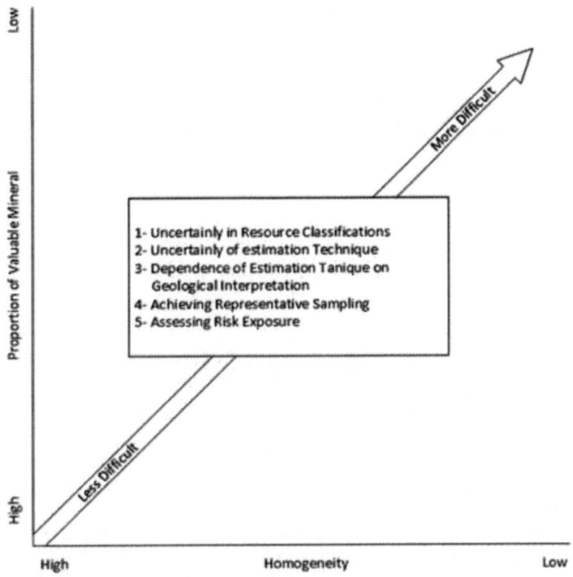

Fig. 6.5 General trend of increasing difficulty in estimating and classifying mineral resources

When the genetic model is well understood, and the geological interpretation/domain is accurate, there may be a small overall grade gap resulting from various interpolation methods.

As a consequence, for the resource group, the significance of genetic models in estimating mineral resources has been calculated as follows:

Mineralization or other economically preferred natural resources may be categorized as a measured mineral resource when the existence, quality, quantity, and distribution of data allow them to estimate the tonnage, grade, or quality of minerals near the range. Furthermore, the shift in estimation has no material impact on the deposit's future economic viability. This category necessitates a high degree of self-assurance and a thorough understanding of geology and mineral deposit controls. Since degree continuity is not always consistent with the geological/lithological continuity direction, a high level of certainty in mineralization controls, i.e., the genetic model, is needed.

ML has advanced and grown in recent years to include data set analysis in the quest for clues to understand better and interpret mineral systems. However, the genetic models created in this manner fall far short of the high degree of reliability needed for the evaluated resource category [13].

Minerals Prospecting and Exploration

"Mineral" means any substance that exists naturally on the earth and has a specific chemical formula or rocks with economic importance formed and under a geological

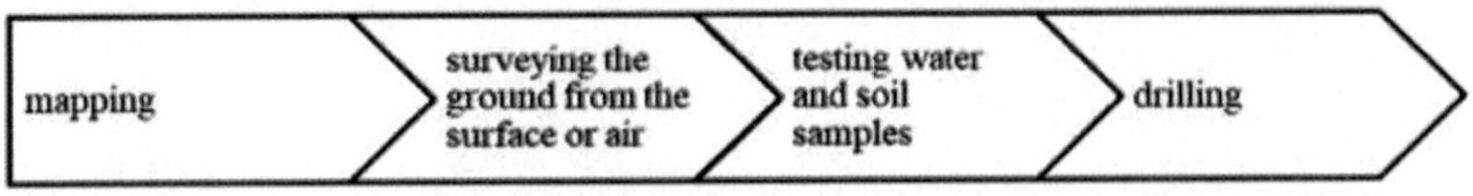

Fig. 6.6 Exploration for minerals may involve [14]

process and can be obtained from the earth by digging, drilling, dredging, quarrying, hydraulicking, sluicing, or other mining methods.

"Mineral exploration" means exploring geological resources such as minerals, rocks, fuels, gemstones, alluvial, glacial, colluvial, and residual deposits for various uses in construction, industry, metal extraction, tourism, decoration, and energy. In addition, it is the process of grading and reserve identified potential geological resources for economic extraction.

"Mineral prospecting or prospecting" means the geological study of searching and locating prospective minerals, fossils, precious metals, or mineral specimens.

The grade of a valuable mineral within ore refers to how concentrated it is. In total deposits, the grade can vary significantly. Furthermore, even though a mineral is still present in the ore, it is not economical to extract it below a particular grade. The mine cutoff grade is the term for this. If the substance has already been processed, there is a point at which production becomes unprofitable. The mill cutoff grade is this.

The break-even grade is the point at which the costs of mining and processing minerals exceed the revenues. Materials of a higher grade are considered, and everything below it is considered waste. As a result, estimating the production costs and the estimated price for the product is needed to decide which part of a mineral can be considered a reserve of exploitable rock. The type of mining method, the degree of mechanization, the mine life, and various other factors all influence extraction costs. As a result, deciding on the best scheme for a specific deposit is a difficult task.

Primary exploration activities, such as mapping, begin in a wide area and are gradually narrow in scope. The aim is to determine if the minerals are large enough to be processed commercially [14] (Fig. 6.6).

Prospecting and Exploration

Prospecting is the first step in finding an economically viable mineral deposit and analyzing it in light of current economic and business conditions to determine whether or not further investment is feasible for the prospect. Identifying the area of interest and collecting geophysical, geological, and financial data are part of prospecting. In addition, sampling, laboratory work, borehole logging, and further investigation of opportunities are part of the exploration process. These measures necessitate the collection and extensive use of data. However, using conventional methods such as human resources, site visits in remote areas, manual sampling and assaying, and

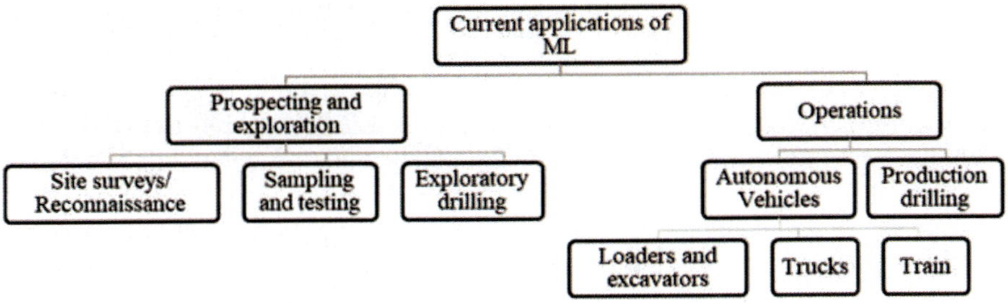

Fig. 6.7 Areas of the mining industry where ML is in progress

rudimentary methods can take anywhere from two to twenty years until the deposit's actual value is determined.

Using maps, aerial photos, satellite imagery, and other available data, prospectors look for anomalies in the structure, lithology, geological features, and planting and plant growth patterns. Geological, topographic, mineralogical, and mapping data can be fed into ML systems and data analysis tools, which can then be used to find anomalies and variances in the data and possible areas of interest [8] (Fig. 6.7).

Geophysics Prospecting

"Geophysics" refers to the recording or mapping of the physical properties of rocks and minerals through the application of physics.

Geophysical prospecting due to its applications in the exploration, evaluation, and extraction of earth resources publishes the best results in primary research in geophysics.

Rock physics is an essential aspect of earth science, with applications ranging from earthquake research to geomechanics, oil and gas storage, carbon sequestration, and geotechnical engineering. We examine how the physical properties of the rock are influenced by the different components of the rock, ranging from the minerals that form a solid structure to the pore network and the various liquids and gases that saturate the pores. These are the physical properties that we test in laboratories and the field: mechanical power, velocity, and attenuation of the elastic wave, resistance, and permeability. With new theoretical developments in models, advances in laboratory equipment and tests, and the acquisition and modern analysis of field data, the study of rock physics is constantly developing.

In geophysical prospecting, rock properties such as density, magnetic resistance, natural magnetism, electrical conductivity, dielectric permittivity, magnetic permeability, seismic wave velocity, and radioactive decay are detected using gravity, magnetic, electrical, seismic, and radiometric methods.

To locate ore deposits, geophysical prospecting incorporates the sciences of physics and geology. The use of Geiger counters to locate radioactive uranium deposits and magnetic surveys to find iron deposits are two famous examples of geophysical prospecting.

Mineral exploration has successfully used five main geophysical methods: magnetic, gravitational, electrical, radiometric, and seismic. Unfortunately, some of these techniques necessitate the use of sophisticated and costly equipment and highly skilled operators. Others, on the other hand, are straightforward and inexpensive. The magnetic and radiometric methods are listed below, while seismic and gravity instruments can be used for specific purposes. For example, a gravity meter tests the density of material underground, while seismic equipment measures the speed of sound in the soil. Unfortunately, with these two instruments, reviews take a long time.

Magnetic Methods

The magnetometer was one of the first instruments used in archeological geophysics. A magnetometer detects minute variations in the magnetic field of the earth. A sensor is placed on a vertical column and connected to a small console where the measurements displayed a widely used instrument. A magnetic map of the target region is generated by taking point-to-point measurements at a distance of 1–2 m apart. Magnetometers help find iron objects and fired earth from bricks, roof tiles, kilns, and furnaces. Magnetometer surveys usually are quicker than other geophysical instruments, and they can be done in brushy, wooded, or rock environments. However, urban areas and areas with igneous rocks are less suitable for them [15].

Magnetic prospecting relies on the idea that certain minerals, including magnetite, are natural magnets. Since the local magnetic field skews the earth's magnetic field, the needle in the compass placed near the magnetite-rich rock acts erratically. Responsive magnetic instruments record minerals like ilmenite (iron titanium oxide), hematite (iron oxide), and pyrrhotite (iron sulfide). Gamma is a standard unit of measurement for magnetic field resistance.

Vast irregularities on magnetic iron ore objects are specifically detected using instruments such as magnetometers and dipping needles. Magnetic readings over poor magnetic host rocks may deviate from local average or "background" values by 10–500 gamma. Still, readings over the formation of magnetic iron may deviate by 100–100,000 gamma. Magnetometers and dip needles may also locate hidden rock formations with magnetic properties that vary from nearby formations. The operator can, for example, be aware that copper is found in igneous rocks such as quartz monzonite. Thus, a magnetometer may be used to detect quartz monzonite under the dirt, talus, or other covers whether it has more magnetism than the surrounding rocks. Similarly, magnetite or ilmenite grains in "black sand" placer deposits can affect the magnetometer and dipping needle. As a result, these methods may be used in the quest for gold or "heavy minerals" in "black sand."

The dip needle is an axial magnetic needle enclosed in a case that can be held vertically and rotated in a vertical plane. The needle is horizontally calibrated for use by changing the balance weight attached to a needle arm. The needle stays horizontal if there is no interfering magnetic mass. Still, if the needle is "pulled" horizontally, it thus dips at various angles depending on the magnetic force, disturbing mass, and needle orientation due to magnetic field. The steeper the angle of dip of the needle, the more magnetic the rock mass is. Dip needles are primarily used in the quest for iron deposits.

Magnetometers are more complicated than dip needles and come in a variety of shapes and sizes. Schmidt-type magnetometers, often known as magnetic balances, are constructed-dipping needles or compasses with the angle of dip (to measure the vertical portion of the magnetic field) or declination (to measure the horizontal component) calibrated to obtain gamma readings. Rather than determining the total field, such a magnetometer is used to measure the difference in the magnetic field. They need a tripod, careful leveling, and careful handling due to their delicate structure. Nevertheless, they should be able to be read with a few gammas of accuracy.

Fluxgate, torsional, and nuclear precession magnetometers are examples of other types of magnetometers that are usually easier to operate than Schmidt-type magnetic balances.

Magnetic surveys may be carried out in a grid or a sequence of sections. The distance between the stations is determined by the size of the anticipated deposit and the form of deposit. For example, minor magnetic anomalies exposed to weak or moderate magnetic rocks can require stations spaced 10–20 ft. apart. Still, if it is suspected of magnetic rocks over a wide area, stations spaced up to 100 ft. apart may be required. In any form of magnetometer survey, power lines, rails, vehicles, and other large metal objects should be avoided because they produce strong local magnetic fields that obscure the rocks' inherent anomalies.

Radiometric Methods

Subatomic particles emit from naturally radioactive elements like uranium and thorium, breaking down or decomposing into other elements or isotopes. The most popular particles produced during this process are gamma rays (similar to X-rays but with higher frequencies), alpha particles (helium atom nuclei), and beta particles (electrons).

Portable counters that detect variations in radioactivity strength, such as Geiger and Scintillation, have been commonly and successfully used for prospecting for uranium and thorium deposits in recent years. These instruments can detect minute variations in the number of radioactive elements in rocks, but they cannot explain which element causes radioactivity. Chemical analysis of a sample of radioactive rock may be used to make these distinctions. A Geiger counter is a tube filled with helium, argon, or krypton used to detect radiation. A heavy pressure wire runs through the tube's base. As gamma or beta particles are emitted through the tube from a radioactive source, some of the rays interfere with gas molecules, resulting in electrically charged particles drawn to the central wire and cause electrical pulses. Counts per minute can be calculated from electric pulses.

Scintillometers use crystals of certain compounds, such as sodium iodide, that emit light flashes when hit by radiation. A photoelectric cell detects a flash of light or scintillation and records the number of flashes per unit of time using electronics. Counting the minutes can be moved to the dial pad. Geiger counters are less reliable than sensitivity meters.

Crossing the terrain when listening to the earphones count or watching the counter dial is the most common way to use radiation counters in the field. Radioactive deposits can produce readings that are ten to one hundred times higher than the "background." On the other hand, the radiation is undetectable if even a few tens of centimeters shield the deposit. Therefore, when using a portable clock, be cautious about interpreting the results until they have been checked by proper sampling and chemical analysis [16].

It is possible to predict the position of sites or site features using an excellent geophysical survey. This should help reduce drilling costs while also allowing for the exploration of a wide variety of features.

Although remote sensing cannot detect features deeper than about 1 m, geophysical explorations can detect features at any depth, though the maximum dimension should generally be greater than that depth.

Geochemical Prospecting

The chemical composition of rocks, minerals, and soils is referred to as "geochemical." The quest for anomalies in geochemical prospecting is focused on systematic measurements of trace elements or chemically affected properties. To detect unusual quantities, samples of soil, lake and river sediments, glacial deposits, rocks, plants and humus, animal tissues, microorganisms, gases, air, and particles are collected and analyzed.

Many mineral deposits are not found on the earth's surface. They could be concealed under rock layers or covered by a dense soil cover. These deposits can be beneficial in more advanced geochemistry, geophysics, and geobotany techniques. The majority of these methods necessitate advanced training and, in some cases, costly equipment.

Geochemical prospecting focuses on measuring one or more chemical properties of rock, soil, glacier debris, stream sediments, water, or plants systematically. The content of a main "trace" element is usually the chemical property that is measured. The aim is to find areas in soils or rocks with high concentrations of specific factors that will lead a prospector to a secret deposit. Geochemical "anomalies" are high concentrations of indicator elements in rocks or soils. The amount of a critical component of a sample can be minimal, but if it is enormous compared to the surrounding region, it is called an anomaly. For example, a few "big" samples would be geochemical anomalies if most soil samples contained around 0.00001% (0.1 part per million) silver, but a few numbers had 0.0001% (1 part per million). On the map, graphs of analytical results can indicate areas that need to be explored further.

Primary and secondary geochemical variations are distinguished—the dispersion of elements in mineral-forming solutions to the outside causes primary anomalies. The scattering of metals laterally or vertically along the fractures can result in a "halo" leak hundreds of feet away from the deposit, resulting in "high" concentrations surrounding deposit metals. This form of the halo is beneficial in prospecting because it can be hundreds of times larger than the deposit surrounding it, making it easier to find.

Weathering disperses the elements, resulting in secondary anomalies. Some primary minerals, including cassiterite, are chemically resistant and are borne by streams as crushed stone. Other minerals may dissolve, and metals in groundwater and surface water may be re-deposited locally or transported in solution. Cassiterite, for example, is a primary mineral. Plants and trees can absorb some of the metal in the solution, which can then accumulate in living tissue. Numerous studies on sulfide deposits' residual soil metal content have shown that the distribution of anomalous metal levels in the soil typically correlates to the highest metal concentrations in the underlying rock.

All weathering products flow out of the region into a drainage basin funnel into streams and rivers. Half of the weathering substance is in the form of chemicals in solution, while the rest is in sediments. Both can be sampled and checked, and their chemical makeup represents the chemical makeup of the drainage basin's rocks. The study of metals extracted by weathering, alluvial flow, or carried in the solution can be associated with an ore deposit in a drainage basin. By sampling water and sediment from each successive branch and counting the number of metals above average, the deposit can be located within the basin. This approach focuses the quest on the most promising places.

In geochemical surveys, surface pollution is a constant threat. Materials derived from mining are the most important sources of emissions. The minerals in these compounds are oxidized and dissolved, causing soil contamination, sediments, water currents, and masking natural anomalies. In addition, smelter gases, wind chimney dust, and metallic artifacts that humans release into the natural world can contaminate soils and rocks.

Geochemical prospecting analytical methods must be sensitive enough to measure small quantities of the main elements, precise sufficient to display minor variations in concentration, quick enough to generate a large number of samples in one day to be analyzed, and inexpensive. Wet chemical techniques are typically restricted to rapid colorimetric methods that necessitate specific equipment and reagents. Instrument techniques (such as emission spectroscopy and X-ray fluorescence methods) necessitate costly equipment and specialized personnel, but they are typically less expensive to determine when thousands of samples are examined.

Wet Chemical Methods

For geochemical exploration, a 27-element analysis method has been developed. These methods vary from fundamental techniques that can be performed in more complicated procedures involving a well-equipped laboratory.

Instrument Techniques

Emission spectroscopy, X-ray spectroscopy, and radio activation units are the most widely used instruments for large-scale prospecting. This is because they make it possible to detect more chemical elements more quickly.

Emission spectrographic methods are standard because they produce 20–50 elements per sample. First, a sample is vaporized for analysis by the heat of an electric arc or spark, which causes the atoms in the sample to emit light. This light enters a spectrograph, dispersed by a prism or diffraction grid, resulting in a spectrum with clear lines of a particular wavelength for each element. Then, the abundance of the different features in the sample can be calculated based on the intensity of these lines as recorded by photographs or electronics.

Atomic absorption is the inverse of diffusion spectrograph, in which unstimulated atomic vapors absorb light emitted directly in the excited state. Mercury detectors make use of this phenomenon. Mercury is a precious metal in and of itself, but it is also found in trace quantities in ores like silver, gold, lead, zinc, and copper. As a result, mercury may suggest the existence of these other metals. Furthermore, mercury is a volatile element easily transported as a gas through small cracks and porous rocks. As a result, a mercury halo serves the prospector better than halos created by many other elements.

X-ray spectrometers (also known as X-ray fluorescence spectrograph) use X-ray bombardment to excite atoms in solid samples and transform their energy into a characteristic of each element's radiation spectrum. An X-ray spectrometer is a device that uses this property to calculate the element concentration in a sample. To protect the operator, high voltage and a suitable protector are needed. The theory behind radio activation is that when some elements are bombarded with high-energy gamma rays, they emit neutrons. For example, beryllium is bombarded with gamma rays; for example, beryllium emits neutrons captured by the counter. As a result, when an instrument runs over rocks, beryllium can be detected [16].

Conclusion

Exploration plays a critical role in improving the mine value chain, and this role has been highlighted based on the limited earth resources for mining. The traditional methods used to find the new resources of mine materials do not work now, and it is needed to use innovative approaches based on advanced analytics to have more benefits in digital mining transformation. Applied AI and ML approaches are valuable tools in this field. This chapter overviewed the various applications of advanced data analysis and ML in mine exploration. After a comprehensive introduction to exploring the geological features and genetic models of mineral deposits, this chapter opened a new window in front of researchers to get familiar with the advantages and challenges of using advanced analytics in mine exploration. Finally, the details of mining geophysical and geochemical aspects when the analytics approaches are used were discussed. The chapter concluded that there are plenty of benefits of using AI and ML methods in mine exploration. However, there is a long way to reach sufficient maturity in data, system, technology, and human to use the maximum capacity of advanced analytics approaches in mine exploration.

References

1. Government, V.S. 2020. Understanding minerals exploration. Cited 17 June 2021. Available from: https://earthresources.vic.gov.au/community-and-land-use/understanding-exploration#:~:text=Exploration%20is%20an%20important%20step,in%20exploration%20across%20the%20state.
2. Australia, M.C. n.d. Exploration. Cited 17 June 2021. Available from: https://minerals.org.au/exploration#:~:text=Exploration%20is%20the%20process%20by,economic%20feasibility%20of%20their%20extraction.
3. Analytics, A.E. 2018. Cited 17 June 2021. Available from: http://www.advexanalytics.com/.
4. Soofastaei, A. 2020. *Data analytics applied to the mining industry*. CRC Press.
5. Bassan, J. 2008. The application of advanced analytics in mining—Safer, smarter, sustainable operations. In *Australian Mining Technology Conference*.
6. Soofastaei, A. 2020. Advanced data analytics. In *Data analytics applied to the mining industry*, 31–50. CRC Press.
7. Jung, D., and Y. Choi. 2021. Systematic review of machine learning applications in mining: Exploration, exploitation, and reclamation. *Minerals* 11 (2): 148.
8. Hyder, Z. 2019. Artificial intelligence, machine learning, and autonomous technologies in mining industry. *Journal of Database Management*.
9. Herrington, R. 2011. Geological features and genetic models of mineral deposits. In *SME mining engineering handbook*. SME.
10. Martins, P., and A. Soofastaei. 2020. Process analytics. In *Data analytics applied to the mining industry*, 131–148. CRC Press.
11. Guilbert, J.M., and C.F. Park. 1986. *The geology of ore deposits*, 715–720. New York: W.H. Freeman.
12. Evans, A.M. 2009. *Ore geology and industrial minerals: an introduction*. John Wiley & Sons.
13. Murahwi, C. 2018. The role of genetic models in the evaluation of deposits and estimation of mineral resources.
14. Government, V.S. 2020. Understanding minerals exploration.
15. Bevan, B. 1995. Geophysical-prospecting. *American Journal of Archaeology* 99 (1): 88–90.
16. U.S. Geological Survey. *Suggestions for prospecting*. Washington: Department of the Interior.

Chapter 7
Advanced Analytics for Surface Mining

Danish Ali

Abstract Surface extracting is the most widely used mining method globally. Surface mining can be divided into open-pit and open-cut methods of metal and coal mining. The surface mining will be more complicated when the depth of the pit increases, and the miners need to go deeper to reach the ore. Advanced analytics and using the new technology can potentially help mining companies find a cost-effective way to extract material. However, using advanced analytics needs fundamental requirements, such as a modern data collection system. In addition, applying advanced analytics is useful in different aspects such as prediction and optimization. This chapter tries to clarify the role of advanced analytics to improve surface mining operations, including design, plan, load, haul, crush, and equipment maintenance.

Keywords Mining · Advanced analytics · Surface mining · Optimization · Mine planning · Mechanical extraction

Introduction

Surface mining is a form of mining that deals with extracting minerals near the earth's surface. Nearly two-thirds of the world's annual mineral production is carried out through surface mining operations. The majority of surface mining takes place in the USA, Canada, and Australia. Surface mining yields high productivity with low operating costs. However, it requires a significant capital investment. It is safer compared to underground mining and involves fewer complications about resource allocation. Figure 7.1 provides the types of surface mining methods. Based on the excavation mode, the surface mining is broadly divided into two categories, i.e., mechanical extraction methods—which further includes open pit, open cast, strip mining, terrace, auger mining, and quarrying—and aqueous methods—which includes placer mining and solution mining. For every type of surface mining operation, rock/soil is removed above the ore body.

D. Ali (✉)
Mining Engineering and Management, South Dakota School of Mines and Technology, Rapid City, SD, USA
e-mail: Danish.Ali@sdsmt.edu

A. Soofastaei (ed.), *Advanced Analytics in Mining Engineering*,
https://doi.org/10.1007/978-3-030-91589-6_7

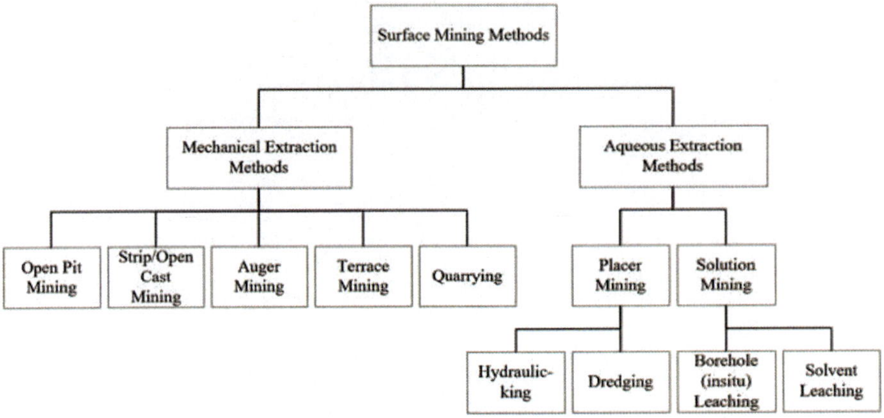

Fig. 7.1 Classification of surface mining methods

Open-Pit Planning and Design

Open-pit planning involves capital resource allocation, reserve estimation, equipment selection, short-term planning, long-term strategic plans, and risk assessment. Advanced master data management, workflow management, and predictive analytics capabilities may be utilized to examine multiple data types in real time and optimize short and long-term planning, scheduling, and transportation with advanced data analytics.

Ore reserve estimation involves the determination of tonnage and quality of the ore reserve. It is one of the primary tasks upon which any open-pit mine is designed. Intelligence-based advanced data analytic techniques can be used for estimating the ore reserves using the drill-hole data set. This application has already been demonstrated by Dutta et al. [1], where neural network and support vector machine, two of the highly advanced data analytic techniques, were used to estimate the reserves, and algorithms performed better than the traditional kriging method.

The selection of equipment is by far the most crucial decision made during mine planning. It impacts the overall mining plan, production plan, and mine design, as well as crucial economic aspects. Therefore, any mistake made during this process could prove catastrophic regarding the profitability and viability of the mining project, especially considering the involvement of substantial capital resources [2, 3]. With the usage of intelligent data analytic-based algorithms, mining engineers and managers could get much-needed assistance during such a complex multi-variant decision-making process. The fuzzy set theory introduced by Zadeh [4] can be ideal for such data-based complex decision making, which involves multi-range widely distributed variables [5]. Fuzzy logic uses the concept of approximate reasoning to arrive at the final response output, based on the degree of membership of the input variables and the fuzzy if–then rules. Başçetin et al. [6] introduced a computer program based on fuzzy set logic to automate the equipment selection process. Open-pit mining equipment requires enormous capital investment, and the decision is often made with having incomplete or non-obtainable information. Furthermore, fuzzy logic performs exceptionally well in such cases, thus making it ideal for the equipment selection process, which is also exhibited by Bazzazi et al. [7], where fuzzy set logic usage was demonstrated for open-pit mining equipment selection for any particular operation.

Generally, geostatistical methods have been used for resource classification within an orebody. However, fuzzy logic can provide an improved framework to encounter uncertainties and a gradual transition between adjacent classes. Ozkan et al. [8] have successfully displayed the application of the fuzzy logic-based model for resource classification for the lignite basin reserves, thus solving the shortcomings of conventional methods.

Reclamation is an essential aspect of the mining process that ensures long-term viability. The procedure includes determining a suitable post-mining use for the affected land so that future generations can efficiently exploit it. Multiple factors, such as cost, pit geometry, ore mineralization, area features, mining activity, available resources, local community needs, and so on, go into deciding on appropriate post-mining land use from the available options of agriculture, forest, lake, storage unit, park, backfilling, and so on. This makes it a complex multi-variate decision-making process that can be handled efficiently through fuzzy logic. Bandopadhyay and Chattopadhyay [9] have demonstrated the application of fuzzy logic for optimizing the post-mining land use following the mining activity. The efficiency and effectiveness of the fuzzy logic approach for ensuring sustainable reclamation for the mining sector were proven by Kommadath et al. [10]. The successful application of fuzzy logic for optimizing the post-mining land use for an open-pit copper mine was exhibited by Bangian et al. [11]. Furthermore, Anis et al. [12] have successfully applied a GIS-based fuzzy logic approach for an efficient post-mining land utilization for an open-pit coal mine.

Ensuring a safe working environment should be the topmost priority for any mining company, which requires a careful assessment of any risky situations during the mining operations [13]. The underlying uncertainties during such an assessment can be tackled using an intelligent data analytics-based system. The effectiveness and superiority of fuzzy logic over the conventional decision matrix methods were exhibited by Iphar and Cukurluoz [14] for a detailed risk assessment and safety evaluation for a coal mine.

Advanced data analytics-based system, especially based on fuzzy logic approach, has proven its worth and efficacy when dealing with almost every mine planning decision. Therefore, any mining company can use the existing systems or built their own based on the fuzzy set logic to assist their team of engineers for every complex multi-variant decision-making process involved during mine planning using their own set of input variables. Moreover, recently introduced advanced neural network-based adaptive control systems [15–16] built with robust event-triggering mechanisms can also be used for developing systems capable of handling the uncertainties during open-pit mine planning and operational execution.

Mechanical Extraction, Loading, and Hauling

In surface mine operations, mechanical or hydraulic shovels matched with large capacity dump trucks almost all of the time make up the excavating, loading, and haulage equipment. These equipment, primarily haul trucks, contributes significantly to injuries and accidents. According to MSHA, out of 250 fatalities from 1990 to 2001 in surface mining, 40% were attributed to powered haulage. An average of 675 accidents and 21 powered haulage fatalities occur each year in surface mining, and 20% of these injuries and fatalities involve dump trucks [17]. Based on 2000 and 2007 data, Ruff et al. [18] showed that the third most severe equipment accidents in all mines are from dump trucks, and 44% of the fatal accidents in surface mines among mobile machines are caused by dump trucks. Based on the findings by Aldinger et al. [19], 46.3% of accidents involving haul trucks occur during operations, and 37.7% of these accidents are caused due to jarring, which causes operator back injuries. According to a 2019 MSHA report, 39.5% of powered haulage accidents involved dump trucks resulting in severe injuries to operators [20]. MSHA and mining companies can process this injury and accident data using advanced intelligence-based data analytics techniques to understand the problem and develop solution strategies to control these injuries. Moreover, machine automation and its smart implementation should focus on all the relevant stakeholders to ensure safe mining operations.

In economics, truck haulage has the highest operational cost, accounting for 50–60% of total surface mining costs [21]. So in order to ensure profitability, the availability and utilization of dump trucks need to be increased. Moreover, the efficiency and the effectiveness of truck haulage require enhancement as well.

Various manufacturers and mining businesses have put autonomous haul truck technology to the test. Each of the three major haul truck manufacturers (Komatsu, Caterpillar, and Hitachi) has worked with a variety of mining corporations (including Rio Tinto, BHP, and Fortescue) to test their technologies in real-world mine conditions [22, 23]. Positive results have already been reported with implementing this haul truck semi-automation technology; however, it is still far from being an actual autonomous system, which should be the ultimate goal. This technology can only flourish after implementing a real-time data-based advanced analytic system along with intelligent computational methodology. The system needs to use all the information from nearby machines and the environment, thus virtually providing the framework for managing the interaction between the equipment (autonomous or operated). Otherwise, it would require that the equipment interaction be either eliminated or very closely managed, putting significant limitations, thus outweighing the automation benefits. Autonomous haul truck manufacturers need to understand that removing the driver means replacing them with a system that automatically performs all those functions and intelligence. One of the simplistic examples is the ability of the machine to not just detect the objects in its path and the nearby environment. However, it should have the ability to recognize those objects and intelligently decide upon the consequent action based on the received percept rather than hard coding the actions or having a general action for any object. Haul truck manufacturers can achieve this by developing a new perception technology using advanced data analytic techniques with artificial intelligence and machine learning. This would allow these machines to determine the actual state of the environment around them, have a broader sense of awareness during machine operation, and intelligently carry out all the operations under all possible environmental conditions.

When a truck is loaded using a high-impact shovel, the operator is subjected to whole-body solid vibrations, resulting in long-term disabilities and injuries if they persist for a long time [24, 25]. The vibrations are attenuated by the hydro-pneumatic suspension struts installed within the large dump trucks. As the truck ages, the suspension system loses its capability of effective vibration attenuation during the shovel dumping process [26]. Contamination of hydraulic oil, mixing the oil and nitrogen, and rupturing the gas accumulator diaphragm are the main causes of suspension system deterioration [26]. Output vibration levels at various components can be used to evaluate the effectiveness of any suspension system. For a haul truck, the vibration levels at the operator's seat require a determination as to the truck ages to evaluate the effectiveness of the hydro-pneumatic suspension system. The performance of the suspension system can be modeled using advanced data analytics-based algorithms, which was also demonstrated by Ali and Frimpong [27]. The study mentioned above exhibited a successful application of advanced AI and machine learning algorithms, including artificial neural networks (ANN), fuzzy logic (FL), and hybrid neural fuzzy interference system (HyFIS), for modeling the hydro-pneumatic suspension system performance, which is given by the output RMS accelerations at the operator's seat of the large capacity dump truck. With the implementation of these advanced self-learning models within the truck control system, the performance of the suspension system can be evaluated in real time, autonomously, throughout the shovel dumping operation. The truck operator or the site maintenance crew can easily use this system to monitor the condition of the truck's suspension system, which will allow them to make precise decisions about the suspension system maintenance. Once the data-based system is functional, on-time suspension system maintenance can be ensured. The suspension system will perform up to its potential, effectively attenuating the vibrations produced during the shovel dumping operation.

Since the high-impact shovel loading operations (HISLO) result in sizeable dynamic impact force at truck bed surface, it is imperative that real-time monitoring of these forces is ensured using advanced data analytic methodology. Ali and Frimpong [28] demonstrated applying a deep learning algorithm to develop a framework for achieving the above objective. Implementing such an intelligence-based data-driven monitoring system would allow the operation optimization to ensure reduced impact forces at the truck bed surface. This will reduce the vibration levels, thus improving the health and safety of the operator and ensuring a safer working environment. Therefore, haul truck manufacturers and mining companies can implement these advanced data analytics-based systems to improve operation safety.

Makers of autonomous technology should use advanced data-based AI and machine learning algorithms for mining operations. Existing models can be modified/trained to create a specifically designed framework to instill intelligence and smartness in machine activities. With the application of such algorithms and techniques, mining machinery, especially the autonomous trucks, can be developed which can have the ability to determine the different states within its path, communicate with closely present machines, and should be able to adjust operating conditions to ensure optimization of the use of the machine. Implementing intelligently advanced haulage technology would lay the foundation for an enhanced mining setup with improved operations in efficiency, safety, effectiveness, and productivity.

Earthmoving and material excavation is another critical machine operation in surface mining, typically done by shovels and draglines. The connection between an excavator and the ground is complex and powerful in this activity. The essential factors that characterize these interactions involve the earth material's type, properties, and physical characteristics. Furthermore, the random occurrences of tree roots, boulders, and any such obstructions make the process even more complicated. Therefore, all this information must be recorded and processed in real time for any excavating machine to operate effectively and efficiently. Moreover, in operation autonomy, the controller needs to be designed to detect changes in the operating conditions, adjust the digging configuration, and respond in real time with utmost precision.

With regard to material excavator autonomy, partial success has been reported. For example, a sensor-based framework was developed by Bullock and Oppenheim [29], which automated the backhoe excavation operation by using forces sensed during digging. The system, however, requires significant improvement as it worked upon encountering the rock obstruction, thus cannot be used for any real-world excavation operation. On the other hand, Komatsu, Ltd. and Nippon Hodo Co have collaborated to develop an automated wheel loader capable of carrying out operations, i.e., scooping, hauling, and loading, as reported by Gocho [30]. Neural network and fuzzy logic are the two main data-based learning algorithms used by Shi et al. [31] to automate the excavation operation in general. However, its finite state model restricted its applicability beyond specific environments.

Image recognition techniques are one of the most effective methods that can be used to gather information about the surroundings of excavators like draglines. Then vision-based techniques can be applied with the image data as those advanced data analytic methods can enhance the efficiency, effectiveness, and safety of any excavation operation. Hainsworth [32] reported a vision-based model for detecting the dragline bucket pose with partial success. A highly advanced vision-based method, HOG, was used by Rezazadeh Azar et al. [33] for developing a pattern recognition-based system capable of detecting the dump truck and its loading cycles during operation. Memarzadeh et al. [34] demonstrated the application of HOG method for developing a framework for recognizing various equipment like the excavator and the dump truck. Furthermore, Golparvar-Fard et al. [35] have used the same method for identifying the various operational phases of an excavator operating at any excavation site.

Dragline is one of the most efficient and low-cost excavation tools in surface mining operations [36], and it is imperative to increase its efficacy, availability, and productivity by using advanced data modeling techniques. Somua-Gyimah et al. [37] demonstrated the application of a convolutional neural network (CNN) which is a state-of-the-art deep learning approach for terrain recognition and object detection during any mining excavation operations. This approach can overcome the disadvantages of the HOG method, including high computational cost and low efficiency.

As it can be seen, during these recent years, the frontiers for surface mining machinery have significantly advanced in the right direction. As a result, mining companies and equipment manufacturers can collaborate to design and implement such advanced intelligence-based data analytic systems to improve the efficacy, effectiveness, and overall productivity of these excavating, loading, and hauling mine machinery and the corresponding operations.

In addition to improving these heavy mining machinery's operational efficiency and effectiveness, intelligence-based advanced data analytic techniques can be developed and used for intelligent maintenance monitoring. Such automated systems can be designed which, with the help of real-time operational data, monitor the health of every piece of equipment. They should predict when a particular part of equipment would fail and thus would let the concerned person know when the preventive maintenance or replacement is due. Using such advanced technologies, downtime of heavy mining equipment like excavators, wheel loaders, and haul trucks can be effectively reduced or even eliminated, thus improving equipment availability, reducing operational costs, and increasing project profitability.

In-Pit Crushing and Conveying (IPCC)

In-pit crushing and conveying is a type of haulage system through which the blasted rock is broken down into smaller chunks using a crusher, and then it is transported using the conveyor system [38]. Alternatively, the crushed rock can be hauled using dump trucks. However, the goal of IPCC is to provide an efficient alternative to the conventional haul truck system [39].

The process works with the excavation being done using the conventional wheel loaders or excavators. Then the crusher receives the excavated material. Crusher can be fixed, semi-mobile, or fully mobile, depending on how often it has to be moved. Once the feed has been crushed to an appropriate size, it is hauled to the stockpile or dump area for further processing, generally outside the pit limits. This system allows for a more continuous mining setup, as the material excavation and haulage can occur simultaneously.

Advanced data analytic techniques can assist the mining companies in deciding on an appropriate crusher for the mine site. The selection of crushers depends on several rock properties (density, compressive strength, material abrasiveness, porosity, and moisture content), application-related parameters (feed size, production required, final product size), and mine depth. Intelligence-based systems can be developed with which the whole process data (including the excavation rate, capacity, and bucket parameters) and the material data can be gathered, analyzed, and further used in real-time processing and analysis for effectively controlling the IPCC system and thus improving the overall operational efficiency.

Design, Construction, and Maintenance of Haul Roads

Haul roads are considered to be the lifeline of surface mining operations. The dump trucks would not move the waste/ore from the bench to the processing plant or dumping site without proper haul roads. However, due to periodic maintenance or poor road construction, haul roads in poor conditions significantly impact workers' safety, equipment health, and overall mine productivity. On the other hand, a well-maintained and well-constructed haul road ensures operational safety and keeps the equipment maintenance to a minimum, thus reducing the operating cost and maximizing the overall profitability [40].

The haul road design depends upon the largest and the heaviest vehicle to traverse the road, the turning radius, braking condition, operating speeds of the haul truck, haul distance, geometric restrictions, and climatic conditions [41]. The haul road's geometric design involves the road's layout and alignment (both horizontal and vertical), stopping distances, crossfalls, and berm wall design [42]. Stopping distance is the total minimum distance required by any vehicle to stop without any collision. It depends on the driver reaction time, the vehicle's retardation action time delay, and the time taken by the vehicle to come to a complete stop [43]. Therefore, the design would depend on the vehicle with the weakest braking capacity at a maximum allowable speed [44]. Crossfall is the elevation difference between the crest of the road and its outer edge. It is widely used to ensure proper drainage. Berm wall is designed at the outer edge of the road with an appropriate height and width to prevent the haul truck from tripping off. The fundamental objective of geometric design is to produce a safe and efficient haul road considering the operating constraints. The structural design of the road ensures that the road has the required strength to support the load without requiring excessive maintenance throughout the life of the haul road.

Haul road needs to be efficiently designed because the operating cost of truck haulage largely depends upon the resulted road conditions. Therefore, the road design should be such that it ensures an efficient and safe truck haulage operation during the life span of the haul road. Baek and Choi [45] have demonstrated the use of raster-based least-cost path analysis for designing the haul roads and then simplifying the zigzag road sections with the application of the Douglas–Peucker algorithm. Mine operational data and the existing road design guidelines were used for an algorithmic-based haul road design. Such haul road design methods are greatly improved through advanced analytics-based techniques. A reinforcement learning algorithm and an intelligence-based customized heuristic can optimize the haul road design using the mine operational data set while satisfying the given structural and geometric constraints.

Haul road maintenance is one of the critical elements to ensure. Data can be complied encompassing the suspension strut pressures of haul trucks during the everyday surface mining operation. A sharp fluctuation in strut pressure of any operational haul truck would indicate a damaged haul road spot requiring an immediate repair. Advanced data analytics tools can develop a machine learning-based classification model to identify such spots at haul road using the suspension system pressure data. Real-time automated identification of damaged haul road sections through data-driven intelligence-based systems would allow for immediate on-time maintenance resulting in an improved haul road quality and structural integrity. This would result in safe and efficient operation with enhanced equipment life cycle and reduced operating cost, thus improving the overall working environment and mine profitability.

Conclusion

This chapter aimed to cover an important subject in surface mining. This subject was using advanced analytics to improve the efficiency and productivity in surface mines. New technology has equipped mining companies to access the huge collected data from different mine sections and equipment. The collected data are potential power, but the real power is the extracted knowledge from the data. Advanced analytics play a critical role in converting raw data to valuable information. This information can help mine managers to optimize the mining processes. This chapter focused on different surface mining areas such as mine planning, loading, hauling, crushing, construction, and maintenance to show the opportunities of using advanced analytics in mining. The presented study showed that advanced analytics would be applied more and more in the near future, and mining companies will benefit from this new approach.

References

1. Dutta, S., et al. 2010. Machine learning algorithms and their application to ore reserve estimation of sparse and imprecise data. *Journal of Intelligent Learning Systems and Applications* 2 (02): 86.
2. Ali, D., and S. Frimpong. 2020. Artificial intelligence, machine learning, and process automation: Existing knowledge frontier and way forward for the mining sector. *Artificial Intelligence Review* 53 (8): 6025–6042.
3. Martins, P., and A. Soofastaei. 2020. Making decisions based on analytics. In *Data analytics applied to the mining industry*, 193–221. CRC Press.
4. Zadeh, L. 1965. Fuzzy collection. *Information and Control* 8: 338–356.
5. Ali, D., et al. 2018. An evaluation of machine learning and artificial intelligence models for predicting the flotation behavior of fine high-ash coal. *Advanced Powder Technology* 29 (12): 3493–3506.

6. Öztaş, O., A. Başçetin, and A. Kanli. 2006. EQS: A computer software using fuzzy logic for equipment selection in mining engineering. *Journal of the Southern African Institute of Mining and Metallurgy* 106 (1): 63–70.
7. Bazzazi, A.A., M. Osanloo, and B. Karimi. 2011. A new fuzzy multi-criteria decision-making model for open-pit mines equipment selection. *Asia-Pacific Journal of Operational Research* 28 (03): 279–300.
8. Ozkan, E., M. Iphar, and A. Konuk. 2019. Fuzzy logic approach in resource classification. *International Journal of Mining, Reclamation, and Environment* 33 (3): 183–205.
9. Bandopadhyay, S., and A. Chattopadhyay. 1986. Selection of post-mining uses of land via fuzzy algorithm. In *Proceedings of the 19th International Symposium on the Application of Computers in Mine Planning (APCOM), SME/AIME.*
10. Kommadath, B., R. Sarkar, and B. Rath. 2012. A fuzzy logic-based approach to assess sustainable development of the mining and minerals sector. *Sustainable Development* 20 (6): 386–399.
11. Bangian, A., et al. 2011. Fuzzy analytical hierarchy processing to define optimum post-mining land use for pit area to clarify reclamation costs. *Gospodarka Surowcami Mineralnymi* 27: 145–168.
12. Anis, M., et al. 2017. Fuzzy logic approach for post-mining land use planning: A case study on coal mine of Pt. Adaro Indonesia-South Kalimantan. *Indonesian Mining Journal* 20 (2): 81–91.
13. Rehman, A.U., et al. 2020. Effect of text message alerts on miners evacuation decisions. *Safety Science* 130: 104875.
14. Iphar, M., and A.K. Cukurluoz. 2020. Fuzzy risk assessment for mechanized underground coal mines in Turkey. *International Journal of Occupational Safety and Ergonomics* 26 (2): 256–271.
15. Ghafoor, A., et al. 2019. ETNAC design enabling formation flight at liberation points. In *The 2019 American Control Conference (ACC).* IEEE.
16. Ghafoor, A., S. Balakrishnan, and T. Yucelen. 2018. Modified state observer-based decentralized neuroadaptive controller for large-scale interconnected uncertain systems. In *The 2018 Annual American Control Conference (ACC).* IEEE.
17. Ruff, T. 2002. Hazard detection and warning devices: safety enhancement for off-highway dump trucks. In *Compendium of NIOSH research.*
18. Ruff, T., P. Coleman, and L. Martini. 2011. Machine-related injuries in the US mining industry and priorities for safety research. *International Journal of Injury Control and Safety Promotion* 18 (1): 11–20.
19. Aldinger, J., J. Kenney, and C. Keran. 1995. *Mobile equipment accidents in surface coal mines. Information circular/1995.* Twin Cities, MN: Bureau of Mines, Twin Cities Research Center.
20. MSHA. 2019. *Mine injury and worktime.*
21. Mukhopadhjay, A. 1989. Selection, maintenance, and relations of various parameters for off-highway hauling tires. In *Off-highway haulage in surface mines*, ed. T.S. Golosinski, and V. Srajer, 153–159. Balkema.
22. Parreira, J. 2013. *An interactive simulation model to compare an autonomous haulage truck system with a manually-operated system.* University of British Columbia.
23. Soofastaei, A. 2020. Digital transformation of mining. In *Data analytics applied to the mining industry*, 1–29. CRC Press.
24. Ali, D. 2016. *Mechanics of impulse force reduction for mitigating dump truck vibrations under HISLO conditions.* Missouri University of Science and Technology.
25. Soofastaei, A., et al. 2016. A comprehensive investigation of loading variance influence fuel consumption and gas emissions in mine haulage operation. *International Journal of Mining Science and Technology* 26 (6): 995–1001.
26. Ali, D., and S. Frimpong. 2018. Artificial intelligence models for predicting the performance of hydro-pneumatic suspension struts in large capacity dump trucks. *International Journal of Industrial Ergonomics.*

27. Dindarloo, S.R. 2016. *Dynamic impact of aging dump truck suspension systems on whole-body vibrations in high-impact shovel loading operations.* Missouri University of Science and Technology.
28. Ali, D., and S. Frimpong. 2021. DeepImpact: A deep learning model for whole-body vibration control using impact force monitoring. *Neural Computing and Applications* 33 (8): 3521–3544.
29. Bullock, D.M., and I.J. Oppenheim. 1989. A laboratory study of force-cognitive excavation. In *Proceedings of Sixth International Symposium on Automation and Robotics in Construction.*
30. Gocho, T. 1992. Automatic wheel-loader in the asphalt plant. In *Proceedings of the 9th International Symposium on Automation and Robotics in Construction.*
31. Shi, X., Lever, P.J., Wang, F-Y. 1996. Experimental robotic excavation with fuzzy logic and neural networks. In: *Robotics and Automation, 1996. Proceedings., 1996 IEEE International Conference.* IEEE
32. Hainsworth, D. 1996. *Dragline automation.* Australian coal association research program project report (C3007).
33. Rezazadeh Azar, E., S. Dickinson, and B. McCabe. 2013. Server-customer interaction tracker: Computer vision-based system to estimate dirt-loading cycles. *Journal of Construction Engineering and Management* 139 (7): 785–794.
34. Memarzadeh, M., M. Golparvar-Fard, and J.C. Niebles. 2013. Automated 2D detection of construction equipment and workers from site video streams using histograms of oriented gradients and colors. *Automation in Construction* 32: 24–37.
35. Golparvar-Fard, M., A. Heydarian, and J.C. Niebles. 2013. Vision-based action recognition of earthmoving equipment using spatio-temporal features and support vector machine classifiers. *Advanced Engineering Informatics* 27 (4): 652–663.
36. Wei, Z., Q.-X. Cai, and S.-Z. Chen. 2007. Study on dragline-bulldozer operation with variations in coal seam thickness. *Journal of China University of Mining and Technology* 17 (4): 464–466.
37. Somua-Gyimah, G., et al. 2019. A computer vision system for terrain recognition and object detection tasks in mining and construction environments. In *SME Annual Conference.*
38. Pekol, A. 2019. Evaluation and risk analysis of open-pit mining operations. *BHM Berg- und Hüttenmännische Monatshefte* 164 (6): 232–236.
39. Nehring, M., et al. 2018. A comparison of strategic mine planning approaches for in-pit crushing and conveying, and truck/shovel systems. *International Journal of Mining Science and Technology* 28 (2): 205–214.
40. Thompson, R. 2010. Mine haul road design and management best practices for safe and cost-efficient truck haulage. In *Society for Mining, Metallurgy and Exploration 2010 Conference Proceedings.* Pre-print. Society for Mining, Metallurgy, and Exploration.
41. Tennant, D., and B. Regensburg. 2001. *Guidelines for mine haul road design.*
42. Thompson, R. 2011. 10.6 Design, construction, and maintenance of haul roads. In *SME mining engineering handbook*, vol. 1, 957–977. Society for Mining, Metallurgy, and Exploration.
43. Elam, R., E. Teaster, and M. Lawless. 1999. *Haul road inspection handbook.* MSHA handbook series. Handbook number PH99-I-4. Arlington, VA: US Department of Labor.
44. Hustrulid, W.A., M. Kuchta, and R.K. Martin. 2013. *Open pit mine planning and design, two-volume set & CD-ROM pack.* CRC Press.
45. Baek, J., and Y. Choi. 2017. A new method for haul road design in open-pit mines to support efficient truck haulage operations. *Applied Sciences* 7 (7): 747.
46. Ghafoor, A., et al. 2018. Event-triggered neuro-adaptive controller (ETNAC) design for uncertain linear systems. In *2018 IEEE Conference on Decision and Control (CDC).* IEEE.

Chapter 8
Advanced Analytics for Surface Extraction

Ali Moradi-Afrapoli and Hooman Askari-Nasab

Abstract Nowadays, approximately 90% of the minerals are extracted using surface mining methods. Surface mining is the process of extracting minerals located at the surface or near the surface. Although at least nine different surface mining methods have been introduced, thus far, open-pit, and strip mining have the highest contribution in raw material extraction from Earth. However, deposits that are being mined using these two methods are substantially expensive, both in the capital and operational costs requiring several managerial decisions to be made for the sake of lowering proportions of the total costs. Analytics has had an inevitable role in this matter. It has been involved in the decision-making procedures from method selection to finding the best location for in-pit crushers in surface mines. This chapter elaborates on how analytics contributes to different steps of extracting material using surface mining methods.

Keywords Surface mining · Open-pit mine · Mine planning · Decision-making · In-pit crushing · Equipment selection

Introduction

Surface mining is defined as exploiting ore from the ground's surface with no exposure of operation crew to underground spaces. The surface mining operation is performed using nine different methods categorized under two main classes of mechanical methods and aqueous methods (Table 8.1). The mechanical methods class responsible for 90% of total surface mining production is defined as a class of ore extraction that applies mechanical processes to mine material from the Earth in a dry environment [1, 2].

A. Moradi-Afrapoli (✉)
IntelMine Laboratory, Laval University, Quebec, Canada
e-mail: ali.moradi-afrapoli@gmn.ulaval.ca

H. Askari-Nasab
Mining Optimization Laboratory (MOL), University of Alberta, Alberta, Canada
e-mail: hooman@ualberta.ca

A. Soofastaei (ed.), *Advanced Analytics in Mining Engineering*,
https://doi.org/10.1007/978-3-030-91589-6_8

Table 8.1 Surface mining methods

Class	Method
Mechanical	Open pit
	Strip
	Quarry
	Auger (high wall)
Aqueous	Dredging
	In situ
	Hydraulic
	Surface techniques
	Evaporate

From all nine surface mining methods listed in Table 8.1, 90% of tonnages mined using surface mining methods are mined by open-pit and strip mining [3]. Strip mining is more common in tabular near the surface flat-lying deposits like coals, whereas open-pit mining is more common in vertically aliened deposits. A literature survey shows that more than 52% of total industrial-scale mining operations worldwide are open-pit metal mines [4]. Based on the importance of open-pit mining across the different surface mining methods, this chapter will focus on elaborating on the application of advanced analytics in different planning and design stages of the open-pit mining method.

As defined in [4], open-pit mining is the process of mining near-surface deposits using horizontal benches. It has two main differences comparing to strip mining and quarrying. First, in open-pit mining, the overburden must be moved out of the pit rim and been disposed of in an external disposal area, whereas in strip mining, the overburden can be disposed of inside the mined area. Second, compared to the quarrying, the open-pit mining method selectively mines the ore, whereas, in quarrying, an aggregate or a dimensional stone is produced.

All open-pit mines have at least three main infrastructures: benches, haul road networks, and dumps. Material is excavated on a series of layers with a uniform thickness called a bench [5, 6]. As depicted in Fig. 8.1, an open pit consists of three types of benches. Active (working) benches are benches where shovels are mining material. Inactive benches where no production activity is taking place at the moment but has the potential to be activated in the future and catch benches where material falling from the top benches is caught and stopped from falling onto the active areas.

Haul road networks are a series of haul roads connecting different loading points to different dumping points and connecting to other service areas inside and outside the open pit. Haul roads in open-pit mines are constructed by either cutting or filling the floor of mining benches in deep hard rock mines and paving the pit floor in shallow soft rock mines. A typical haul road in an open-pit mines consists of three major components: travel lane, safety berm, and drainage ditch. As a rule of thumb, road width for two-way traffic, which is the most common road in open-pit mines, must be greater than four times the largest truck width [5, 7].

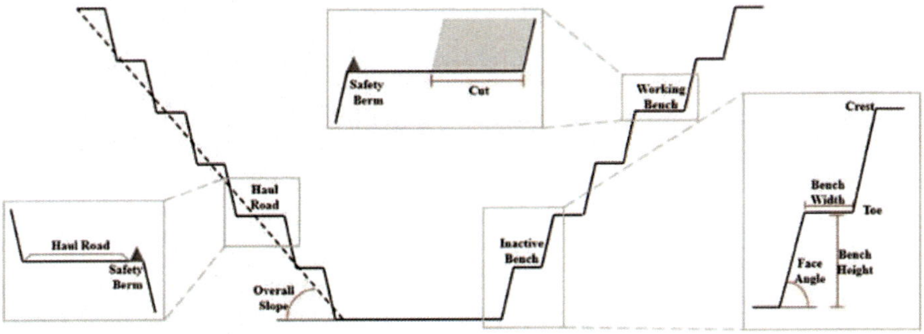

Fig. 8.1 Open-pit geometry [8]

The third major infrastructure of an open-pit mine is its waste dump. The waste dump refers to the dump of mined material with no to little economic value at the time of its placement [8].

The mining production fleet works within the three above-mentioned infrastructures to produce economic value for shareholders. This brief introduction being said, the following sections in this chapter present the application of advanced analytics in each stage of planning, design, and production from surface mines and, more specifically, open-pit mining methods.

Open-Pit Planning and Design

Before starting the extraction of raw material from the ground, the business needs a step-by-step actionable plan that provides a detailed guideline for the production to maximize the net present value of the deposit. Thus, the open-pit planning and design are a decision-making process that provides a detailed, actionable plan for profitably producing the raw material from the area of interest [9].

Open-pit planning and design are generally outlined for three different time frames. Short-range plans are providing information for a time frame between a shift to a week. Medium-range plans range between a week to a year of operation, and the long-range plans provide production schedule and design for a year of operation to the life mine. In this chapter, our focus is on long-range open-pit planning and design.

There are seven main steps in generating the long-range mine plan for an open-pit mine. Initially, a resource model is needed to be prepared for the area of interest.

Then, in the second step of the open-pit planning and design, we apply a pit optimization algorithm to generate a set of pit shells. After laying out the set of possible pit shells, the planner needs to choose the final pit. The final pit is the pit in the set of possible pit shells that maximizes the mine's net present value. When the final pit is selected, we need to establish the desired production rate and generate a plan for material production within the final pit. In the last two steps of the long-range open-pit planning and design, a number of desired pushbacks and their basic designs are chosen, and finally, the production schedule is optimized. A detailed step-by-step open-pit mine planning and design can be found in [5].

Optimization is the branch of advanced analytics of the highest demand in the open-pit mining planning and design stage. Determination of ultimate pit limit and production scheduling are two main steps in the open-pit mine planning and design where optimization algorithms are required to make optimal decisions.

The final pit limit is defined as the extent of the mineable reserve and the waste material that is economically and technically viable and safe to be mined for extracting the ore [10]. The optimum final pit limit is determined by maximizing net present value (NPV) from mining a set of blocks concerning slope constraints. In Eqs. (8.1) and (8.2), we present the formula for calculating the block economic value for each block in the block model and the basic formula for NPV calculation.

$$\begin{aligned}\mathrm{BEV}_{ijk} = {} & \mathrm{Ore}_{ijk} \times g_{ijk} \times \mathrm{Recovery} \\ & \times (\mathrm{Price} - \mathrm{Cost\ of\ Selling}) \\ & - \mathrm{Ore}_{ijk} \times \mathrm{Cost\ of\ Processing} \\ & - \mathrm{Total\ Rocks}_{ijk} \times \mathrm{Cost\ of\ Mining} \end{aligned} \tag{8.1}$$

$$\mathrm{Max}\,P = \sum_{ijk} \mathrm{BEV}_{ijk} \tag{8.2}$$

where BEV_{ijk} is the block economic value of the block located in the orebody with index i, j, and k. Ore_{ijk} is the content of ore in the same block with an average grade of g_{ijk}.

Two significant steps must be taken. At first, the block economic value (BEV) must be calculated using stochastic parameters in nature, such as tonnage of ore, grade, recovery, price, and cost. Then in the second step, the pit limit must be determined using a set of uncertain BEVs inputs. Again, analytics is to the rescue here. It all started with Dowd's risk-based algorithm [11]. In the risked-based mine planning algorithm developed by Dowd (Fig. 8.2), all the input parameters with the stochastic behavior are imported as the distributions fitted on the data. Each step of the algorithm randomly samples from the distributions instead of using a deterministic value for each input parameter.

After the risked-based algorithm developed by Dowd, several other algorithms have been developed by researchers to schedule open-pit production incorporating uncertainties in the procedure. For example, Godoy and Dimitrakopoulos [13] proposed a five-step algorithm that provides an open-pit mine production schedule under uncertain conditions. Osanloo et al. [12] brief the proposed five steps as follows:

1. Generate a series of simulated ore body;
2. Design final pit and push backs for the simulated ore body;
3. Find the optimal production rate for the life of mine;

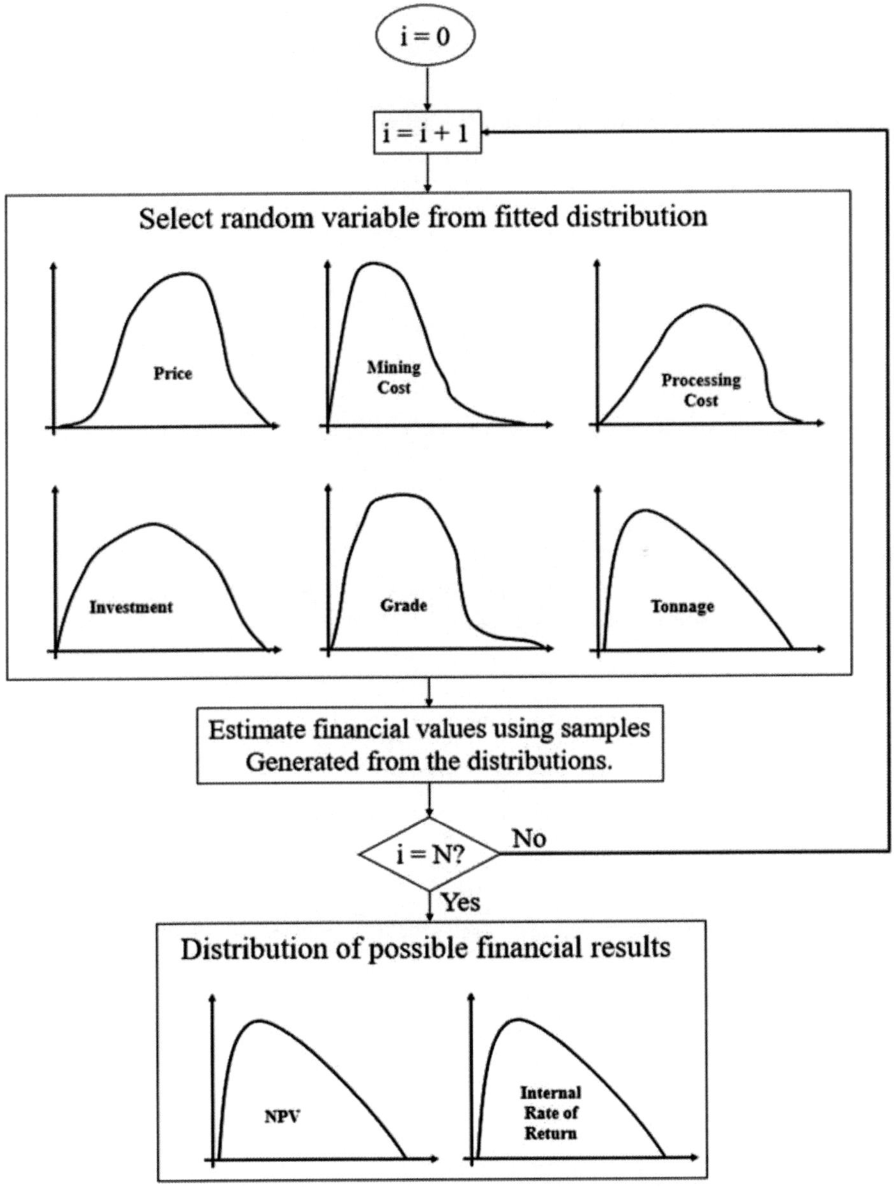

Fig. 8.2 Risked-based mine planning [11, 12]

4. Generate a production plan for each simulated ore body; and
5. Combine obtained mining sequences to a single production schedule.

To accommodate grade uncertainty in open-pit mine production planning, Ramazan and Dimitrakopoulos [14, 15] proposed another algorithm. Here, after receiving the simulated orebody, the MIP generates a production scheduling pattern. This pattern is used as a guideline to calculate the probability of mining each block in a given period. Then any block with a chance between zero and one is considered in a new optimization model with the objective function presented in Eq. (8.3).

$$\text{Maximize} \sum_{t=1}^{T} \sum_{n=1}^{N} v_n^t \times x_n^t \tag{8.3}$$

where T and N are the total number of periods and blocks in the model. v_n^t is the NPV that will be generated by mining block n in period t. x_n^t is one of the block n is scheduled to be mined in period t and zero otherwise.

The stochastic open-pit production schedule model is, however, limited with the following constraints:

The material sent to the plant must meet the upper and lower bounds for the average grades, as presented in Eqs. (8.4) and (8.5).

$$\sum^{N} (g_n - G_{\min}) \times O_n \times x_n^t \geq 0, \quad n = 1 \tag{8.4}$$

$$\sum_{n=1}^{N} (g_n - G_{\max}) \times O_n \times x_n^t \leq 0 \tag{8.5}$$

where g_n is the grade of material in block n, $G_{\min}$ and $G_{\max}$ are lower and upper bounds of the average feed grade to the processing plant, and O_n is the tonnage of ore to be mined from block n.

The plant also has a capacity constraint that must be followed by the quantity of material being sent to it in each period. Equations (8.6) and (8.7) present an example of how the plant capacity is constrained in stochastic production planning.

$$\sum_{n=1}^{N} O_n \times x_{nt} \geq \text{PC}_{\min} \tag{8.6}$$

$$\sum_{n=1}^{N} O_n \times x_{nt} \geq \text{PC}_{\max} \tag{8.7}$$

where $\text{PC}_{\min}$ and $\text{PC}_{\max}$ are the minimum production requirement and maximum production capacity of the plant.

The mining operation has its limitations. For example, we cannot produce more than the available equipment capacity, and we need to maintain waste material production throughout the periods. These are met using Eqs. (8.8) and (8.9).

$$\sum_{n=1}^{N} (O_n + W_n) \times x_n^t \geq \mathrm{MC}_{\min} \tag{8.8}$$

$$\sum_{n=1}^{N} (O_n + W_n) \times x_n^t \leq \mathrm{MC}_{\max} \tag{8.9}$$

where W_n is the waste tonnage scheduled to be mined from block n.

We first need to mine a series of blocks located above the block based on slope requirements. Then, the constraint provided in Eq. (8.10) is used to ensure that all blocks situated above the current block are mined ahead of the current block or at least at the same period.

$$Yx_k^t - \sum_{y=1}^{Y} \sum_{r=1}^{t} x_y^r \leq 0, \quad t = 1, 2, 3, \ldots, T \tag{8.10}$$

where Y is the total number of blocks to be mined before mining a given block. k is the index of the block to be mined in period t. y is the index for Y blocks to be excavated before the given block.

The optimization model is also a constraint to mine all the blocks in reserve only once. Therefore, Eq. (8.11) makes sure that all the blocks in reserve are mined only once.

$$\sum_{t=1}^{T} x_n^t = 1 \tag{8.11}$$

We use the scheduling model presented above on all the ore body realizations. By doing that, we can calculate the probability of each block being mined in a given period. The probability calculation generates three categories of blocks: blocks with zero chance to be mined in a given period meaning that they are not mined in the period of concern. Blocks with probability equal to one indicating that the entire block must be mined in the given period. Moreover, the third category is the category of blocks with the probability of being mined in a given period between zero and one. Blocks in the third category are considered in a second optimization model with an objective function presented in Eq. (8.12).

$$\text{Maximize} \sum_{t=1}^{T} \left[\sum_{n=1}^{N} c_n^t \times x_n^t - \sum_{m=1}^{M} w \times d_m^t \right] \tag{8.12}$$

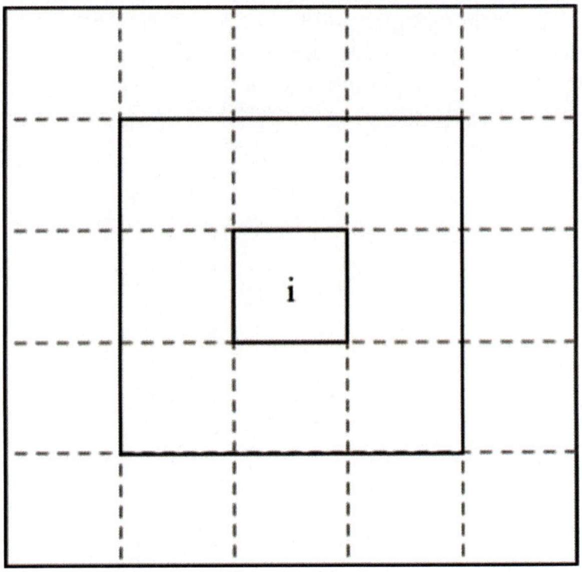

Fig. 8.3 Inner and outer windows for block i

where M is the number of blocks in the smoothness constraints, as shown in Fig. 8.3 c_n^t is calculated as $(v_n^t \times p_n)$, where p_n^t is the probability of mining block n in period t. w is the cost of unit deviation caused by smoothing the schedule, and d_m^t is the deviation from the smooth pattern when mining block m.

The model is constrained by the probability of blocks having the desired grade, equipment access, and mobility listed in Eqs. (8.13) to (8.15).

$$\sum_{n=1}^{N} (P_n - 100) \times O_n \times x_n^t + Y_1^t \times \mathrm{TO} = 0 \tag{8.13}$$

$$-\sum_{j=1}^{\mathrm{nb1}} K1_j \times O_j \times x_j^t + K2_i \times \mathrm{OT}_i^t - Y_{2i}^t \leq 0 \tag{8.14}$$

$$-\sum_{j=1}^{\mathrm{nb2}} K1_j \times O_j \times x_j^t + K2_i \times \mathrm{OT}_i^t - Y_{3i}^t \leq 0 \tag{8.15}$$

where T and N are the total number of periods and blocks in the model. P_i is the probability of block i having a grade within the desired interval. OT_i^t is the tonnage of ore scheduled to be mined from block i in period t. TO is the total tonnage of ore to be mined in the period of concern. Y_1^t is a deviation from desired probability in period t. $K1_j = 1/\mathrm{TO}_j$, $K2_i = \mathrm{nb1}/\mathrm{TO}_i$ for the inner window, and $K2_i = \mathrm{nb2}/\mathrm{TO}_i$ for the outer window is the coefficient to convert ore tonnage to percentage. TO_j is total ore available in mining block j. nb1 and nb2 are the total number of blocks within the inner and outer windows presented in Fig. 8.3. Y_{2i}^t and Y_{3i}^t are deviations from the smoothness of the inner and outer windows, respectively.

Based on the same concept, the model can be solved using a genetic algorithm (GA) [16]. The GA assumes p_n^t in above model as chromosome for block n to be mined in period t then estimates the p_n^t randomly in each stage of evolution using its mutation or crossover operation.

Mechanical Extraction, Loading, and Hauling

As presented in Table 8.1, the five most common mechanical excavation methods are open pit, stripping, quarry, and auger. These five mechanical methods account for 90% of the total surface mining productions all around the world. Therefore, after deciding on using the mechanical surface mining method to excavate material from the deposit, step one is to choose between those mentioned above four mechanical methods. Based on Hartman and Mutmansky [1], selecting the proper method depends on four major factors, including the shape of the deposit, its depth, thickness, and size, as presented in Table 8.2.

Other mining method selection strategies exist in the literature. The mining method selection strategy developed by Nicholas [17] and its modified version known as UBC mining method selection [18] are two commonly used strategies. These two strategies also use qualitative ranking systems to select the best mining method for the deposit in hand.

The mining mentioned above method selection techniques are manual methods. However, to make more precise decisions on the mining methods, one can combine different analytics techniques. Two of the main analytics techniques to select between several alternatives are the analytic hierarchy process (AHP) and the analytic network process (ANP). Saaty developed the AHP and ANP techniques at the University of Pittsburgh. Hierarchies of feedback networks are generated by AHP or ANP model developers considering criteria to derive ratio scales. These scales are then used to select the best alternative among the available choices. In a nutshell, AHP and ANP work based on developing priorities for available options and criteria used to judge the available choices [19]. Thus, an AHP model consists of three primary levels: goal, criteria, and alternative (Fig. 8.4).

Table 8.2 Mechanical extraction method selection factors

Mining method	Shape of deposit	Depth of deposit	Thickness of deposit	Size of deposit
Open pit	Any shape	Any depth	Thick	Large
Stripping	Tabular	Low depth	Thin	Large
Quarrying	Tabular or massive	Any depth	Thick	Moderate size
Augering	Tabular	Flat	Thin	Remnant

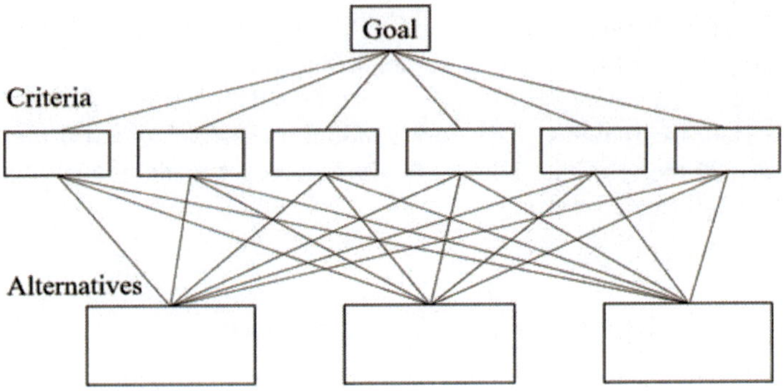

Fig. 8.4 AHP diagram [19]

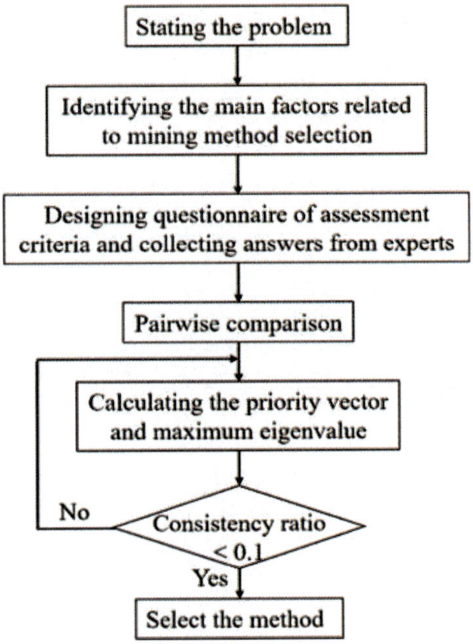

Fig. 8.5 Flowchart for selecting mining method using AHP technique [20]

As depicted by Fig. 8.4, the first concept of the AHP is to structure the problem at hand as a hierarchy of goals, criteria, and alternatives. The second concept of the AHP is to perform a pair-wise comparison of elements at each hierarchy step. The third concept covering by AHP is to synthesize judgment on each piece over different levels of hierarchy. To solve an AHP structured problem, we first need to determine each criterion's importance compared to other criteria. Then we need to calculate the relative importance of each alternative to each criterion. And then, calculate the overall priority weight of each of the choices.

Table 8.3 Simple example of a questionary for assessing the importance of each factor

Criterion	Importance				
	None: 1	Minor: 2	Substantial: 3	Fundamental: 4	Highest: 5
Shape of deposit					
Depth of deposit					
Thickness of deposit					
Size of deposit					

Following, we provide (Fig. 8.5) an example of selecting a mining method using the AHP technique based on [20].

A sample questionary to send to the experts to quantify critical criteria is provided in Table 8.3. Based on the deposit you have in hand to select a mining method, you prepare a list of important factors affecting the method selection procedure. Then, prepare a questionary like the one presented in Table 8.3. Distribute the questionary among the area experts and ask them to fill it based on your deposit information. It is worth noting that Table 8.3 is a simple example, and the questionary list can vary from one deposit to the other one.

If you have listed many factors using the questionary, you decide which essential factors must be included in the method selection procedure as a criterion. After finding the most critical factors in the method selection, we construct the hierarchy presented in Fig. 8.6.

Now we need to determine the relative importance of criteria with respect to the goal. For that, we use the pair-wise comparison matrix. To create the pair-wise comparison matrix, we use the so-called fundamental scale provided in Table 8.4.

The pair-wise comparison matrix is an n-by-n matrix where n is the number of criteria associated with our problem. For our specific example, the pair-wise comparison matrix is constructed as presented in Table 8.5.

The value in each cell of the pair-wise matrix presented in Table 8.5 depends on the decision-maker. However, one criterion compared to itself has equal importance. Thus, the diagonal values of the pair-wise comparison matrix are always one. Another

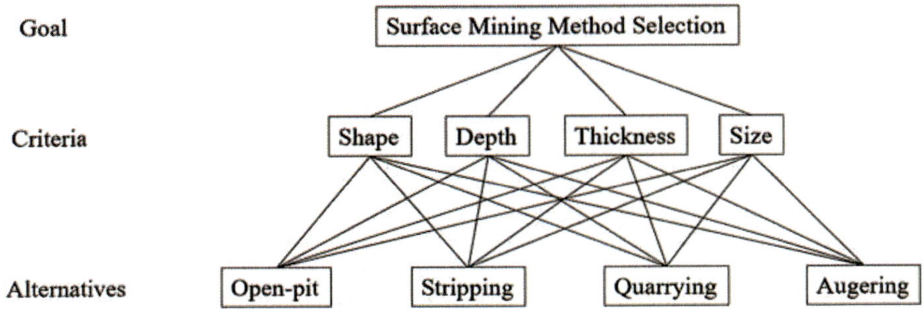

Fig. 8.6 AHP model structure for surface mining method selection

Table 8.4 Fundamental scale ore the scale of relative importance table to be used for filling up the pair-wise comparison table [19]

Definition	Intensity of importance
Equal	1
Weak	2
Moderate	3
Moderate plus	4
Strong	5
Strong plus	6
Very strong	7
Very, very strong	8
Extreme	9
Inverse comparison	1/2, 1/3, 1/4, …, 1/9

Table 8.5 Pair-wise comparison matrix of the surface mining method selection problem

	Shape	Depth	Thickness	Size
Shape	1	5	4	7
Depth	1/5	1	1/2	3
Thickness	1/4	2	1	3
Size	1/7	1/3	1/3	1
Sum	1.59	8.33	5.83	14

important note regarding the pair-wise comparison matrix is that if the importance of criterion *A* to criterion *B* is equal to *x*, then the importance of criterion *B* to criterion *A* is $1/x$. We then calculate each column's sum and insert it in the last row of the pair-wise matrix.

In the next step, we create the normalized pair-wise comparison matrix by dividing each value in a column to the same column's cumulative value, as depicted in Table 8.6.

After normalizing the importance, we calculate each criterion's weight by averaging values in each row of the normalized pair-wise comparison table (Table 8.6). Each criterion's weight is now calculated by averaging each row's values (Table 8.7).

Table 8.6 Normalized pair-wise comparison matrix

	Shape	Depth	Thickness	Size
Shape	1/1.59 = 0.6289	5/8.33 = 0.6002	4/7 = 0.6861	7/14 = 0.5000
Depth	0.2/1.59 = 0.1258	1/8.33 = 0.1200	0.5/7 = 0.0858	3/14 = 0.2143
Thickness	0.25/1.59 = 0.1572	2/8.33 = 0.2401	1/7 = 0.1715	3/14 = 0.2143
Size	0.14/1.59 = 0.0898	0.33/8.33 = 0.0400	0.33/7 = 0.0572	1/14 = 0.0714
Sum	1.59	8.33	5.83	14

Table 8.7 Weight of criteria

	Shape	Depth	Thickness	Size
Shape	0.6289	0.6002	0.6861	0.5000
Depth	0.1258	0.1200	0.0858	0.2143
Thickness	0.1572	0.2401	0.1715	0.2143
Size	0.0898	0.0400	0.0572	0.0714
Sum	0.6289	0.6002	0.6861	0.5000

Table 8.8 Preparation for calculating λ_{max}

	Shape	Depth	Thickness	Size	Weighted sum value	Weighted sum/criteria weight
Shape	1 × 0.60	5 × 0.14	4 × 0.20	7 × 0.06	2.52	4.18
Depth	0.2 × 0.60	1 × 0.14	0.5 × 0.20	3 × 0.06	0.55	4.02
Thickness	0.25 × 0.60	2 × 0.14	1 × 0.20	3 × 0.06	0.81	4.16
Size	0.14 × 0.60	0.33 × 0.14	0.33 × 0.20	1 × 0.06	0.26	4.05

Now we need to calculate the consistency of the criteria to check if the calculated values are correct. To calculate the model's consistency, we use the pair-wise comparison matrix values in Table 8.8, which are not normalized. Then, we multiply each value in a column with the criterion weight of that column. In the next step, we calculate the weighted sum value for each criterion by summing all the values in each row. By dividing the weighted sum value by the criteria weight and averaging over all the criteria, we find λ_{max}.

$$\lambda_{max} = \frac{4.1762 + 4.0225 + 4.1553 + 4.0488}{4} = 4.1007$$

Now that we have λ_{max} value, we can calculate the consistency index using Eq. (8.16).

$$\text{CI} = \frac{\lambda_{max} - n}{n - 1} \tag{8.16}$$

where n is the number of criteria and CI is the consistency index. In our example, CI = (4.1007 − 4)/(4 − 1) = 0.03358. We now can calculate the consistency ratio using Eq. (8.17).

$$\text{CR} = \frac{\text{CI}}{\text{RI}} \tag{8.17}$$

Table 8.9 RI values for up to ten criteria in AHP [21]

n	1	2	3	4	5	6	7	8	9	10
RI	0	0	0.58	0.90	1.12	1.24	1.32	1.41	1.45	1.49

where CR is the consistency ratio and RI is the random index, the consistency index of the randomly generated pair-wise matrix of the same size. Saaty [21] calculated RI for one to ten criteria that is presented in Table 8.9.

Based on Table 8.9, consistency ratio in our example is

$$\mathrm{CR} = \frac{\mathrm{CI}}{\mathrm{RI}} = \frac{0.03358}{0.90} = 0.0373$$

A pair-wise matrix is consistent if it is CR < 0.10. Based on that, the pair-wise matrix in our example is consistent.

Now that we are sure about the consistency of the pair-wise comparison matrix for the selected criteria, and we follow the same steps but this time for the alternative methods and develop pair-wise comparison matrices for them based on each criterion to calculate the weights. Table 8.10 presents the pair-wise comparison matrices for our example based on each criterion. In Table 8.10, O, S, Q, and A stand for open pit, stripping, quarrying, and augering, respectively.

In the final step of the AHP for surface mining method selection, using Eq. (8.18), we determine the best surface mining method to mine the deposit.

$$\mathrm{Max}\left\{\sum_{i=1}^{N} \mathrm{CW_i} \times \mathrm{MW}_j \,|\, j \in \{1, \ldots, M\}\right\} \tag{8.18}$$

where CW_i is the criterion weight criterion i from N criteria and MW_j is the j alternative weight from M alternative solutions available. Equation (8.18) shows the solution alternative, which obtained the highest rank among all the methods.

Selection and Sizing of Excavating, Loading, and Hauling Equipment

In open-pit mining, equipment selection and sizing are critical problems to be solved by the mining engineers. This problem is divided into two main subproblems of loading equipment and transferring equipment. For the loading equipment, we need to follow three steps: (1) finding the bucket capacity, (2) determining the machine's geometry, and (3) finding the best match from the machines that are currently available in the market [22].

The bucket size is determined using Eq. (8.19) after finding the mine's required production rate using the long-term plan.

Table 8.10 Generating pair-wise matrices for solution alternatives concerning different criteria

Shape	O	S	Q	A	Weight	Weighted sum	Weighted sum/weight	CI and CR
O								
S								
Q								
A								
Depth	O	S	Q	A	Weight	Weighted sum	Weighted sum/weight	CI and CR
O								
S								
Q								
A								
Thickness	O	S	Q	A	Weight	Weighted sum	Weighted sum/weight	CI and CR
O								
S								
Q								
A								
Size	O	S	Q	A	Weight	Weighted sum	Weighted sum/weight	CI and CR
O								
S								
Q								
A								

$$\mathrm{BC} = (\mathrm{Pr} \times \mathrm{Ct})/(\mathrm{Ff} \times \mathrm{Ma} \times \mathrm{Ef} \times 3600) \tag{8.19}$$

In Eq. (8.19), BC is the bucket capacity for the loading equipment required to meet the production plan, Pr is the rate of the production based on the production plan, Ct is the nominal cycle time of the desired equipment, Ff is the bucket fill factor, Ma is the mechanical availability of the equipment, and Ef is the operational efficiency. In the next step, using multiple iterations, the loader geometry is determined based on the benches' geometry in the designed pit. Based on the loader's selected geometry and its bucket capacity, the decision-maker compares available options from different OEMs and makes decisions on the make and model of the loading equipment. Analytics comes to operation in this stage with the AHP method to select the best loader. The steps detailed in the previous section are required to be taken to make the best decision. Then, based on the production rate and stripping ratio size of the loader, fleet is determined.

In the next step of the equipment selection and sizing for open-pit mines, it is required to determine the size and number of trucks needed to transport material

Table 8.11 Number of passes required to load a truck with a loader [22]

Truck size (*t*)	Loader bucket size (m^3)				
	15	20	25	35	45
136	5	4	3		
181	6	5–6	4	3	
217		6	5	3–4	
290			6	4–5	3–4
360				6	4–5

from the pit to the destinations. Mining engineers typically select trucks with enough capacity to be loaded with three to five passes by the mine's available loaders. Thus, based on the size of the loaders in the mine's production fleet, using the AHP method, the decision-maker decides on the mark and model of the required trucks among those that can be filled by the selected loaders in three to five passes. Again, the decision-maker follows the previous section's steps to select the mine's best truck types. A typical example of how different types of trucks match different load types is presented in Table 8.11.

Based on the selected loaders' bucket capacity and a table of truck options, as shown in Table 8.11, three sets of truck fleets are chosen. A fleet of small trucks can be loaded by five to six passes using the selected loaders. A fleet of large trucks can be loaded by three to four passes using the selected loaders and a fleet of trucks, including both small and large trucks. After that, we need to calculate the number of trucks required in the fleet to meet the scheduled production.

The common practice in mining determines the number of trucks in the fleet using a deterministic formula presented by Burt and Caccetta [23] based on the match factor definition. The match factor (MF) is defined as presented in Eq. (8.20) for a homogeneous truck fleet and Eq. (8.21) for a heterogeneous truck fleet:

$$\text{MF} = \frac{\text{trucks}}{\text{loaders}} \times \frac{\text{loading cycle}}{\text{truck cycle}} \tag{8.20}$$

$$\text{MF} = \frac{1}{\left(\sum_j \frac{\text{loaders}_j}{\sum_i \text{trucks}_i \times (\text{loading cycle})_{i,j}}\right) \times \text{truck cycle}} \tag{8.21}$$

where i and j are the index of truck types and loader types, respectively, based on the match factor as presented in Eqs. (8.20) and (8.21), and the mining production system falls into one of the three domains of the under-truck system (MF < 1), balanced system (MF = 1), and over-truck system (MF > 1). The number of trucks of each type in the fleet is determined to keep the match factor of the production system at one.

Although the required number of trucks in the fleet is determined using the MF method, the method has some drawbacks, one of which is that it does not account for the technical uncertainties in its calculations. Analytics comes to action here, and

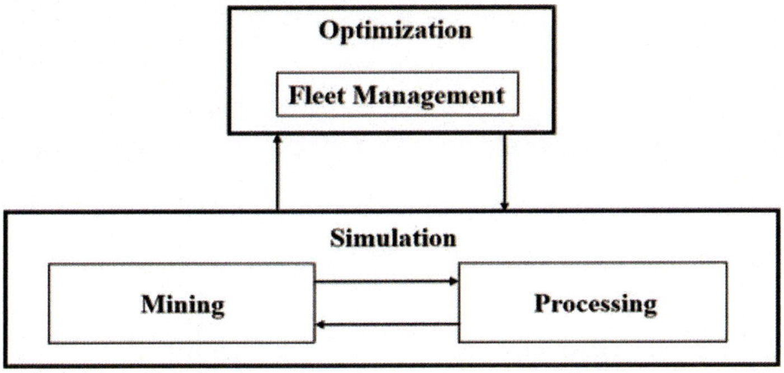

Fig. 8.7 Schematic of the analytics system to solve truck fleet sizing problem in surface mines [26]

using descriptive analytics, simulation, and optimization helps the mining industry to more precisely determine the number of trucks required to meet the production targets. It is worth noting that the analytics approach to determine the size of the truck fleet uses the MF method as its starting point [24, 25].

At the first step of using analytics to determine the optimal truck fleet size, understanding the surface mine's material handling system is essential. The material handling system consists of the mining operation, processing, and decision-making systems, which in most cases are the fleet management systems. The analytics system that needs to be developed to solve the truck fleet sizing problem must include these components (Fig. 8.7).

As presented in Fig. 8.7, a simulation and optimization framework must be developed for the mine where the simulation part of the framework mimics the operations and the optimization part of the framework mimics decision-making tools. To develop the simulation and optimization framework, one requires three major software: discrete event simulation software, optimization software, and Microsoft Excel. These three software are needed to talk with each other, as presented in Fig. 8.8.

After selecting the proper software and setting up the framework in software, we need to prepare the required inputs. Thus, we collect historical data for our selected truck type and loader type from another mine in a new mine. In the case of a currently active mine, we collect historical data from our mining operation. Using data analysis methods, we fit different distributions on the data and find the best distribution using testing methods such as Kolmogorov–Smirnov (KS).

A list of input parameters must be collected, preprocessed, analyzed, and fitted with the proper distribution. After finding the fitted distribution for each input parameter, we store them in an Excel file readable from the simulation software. The input parameters depend on developing our simulation and optimization framework and the detail we consider in the model development. However, for a truck and shovel surface mining operation, the significant inputs are components of shovel loading time and truck cycle time, such as idling in the queue at the shovel and velocity.

After setting up the input Excel file with the required input parameters stored in it, we develop the simulation model and connect it to an external optimization software

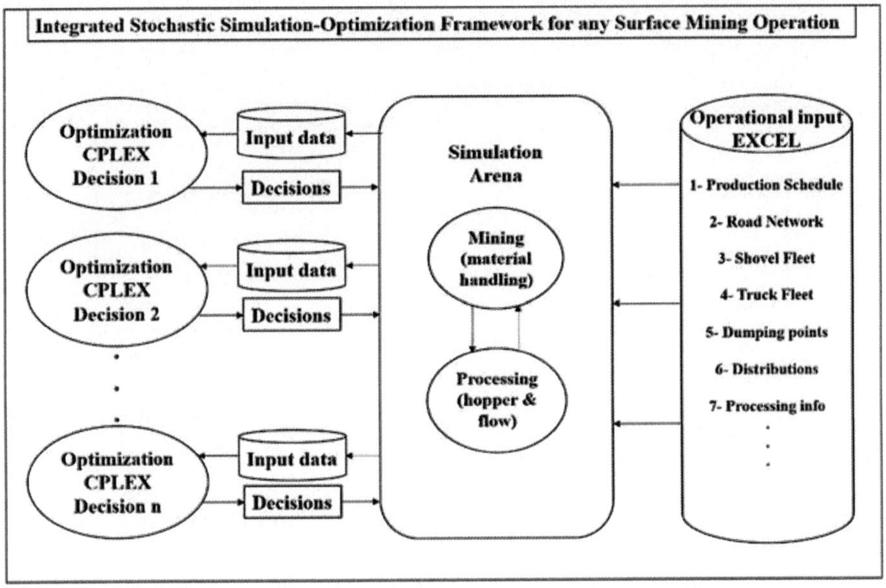

Fig. 8.8 General layout of the simulation and optimization software to be used in the analytics system for determining truck fleet size in surface mines [27]

such as IBM CPLEX [28]. In the optimization software, we need to set up all the decision-making models that the mine is or will be using in its operation. One of the essential decision-making models to be provided here is the fleet management system in use in the operation. Afrapoli, in [27], presents a comprehensive explanation of how to develop the framework.

The framework is ready to determine the optimal number of trucks required to meet the production schedule by connecting the simulation model to the optimization models and the data file. At this stage, one should define production scenarios based on the number of trucks in the fleet to test them and find the best possible combination of trucks and loaders fleet. Three sets of scenarios must be defined: a set of scenarios for the fleet of only small trucks, a set of scenarios for the fleet of large trucks, and scenarios for the fleet of small and large trucks. Here, to develop scenarios, one can start from a random number of trucks in the fleet add or deduct the number of trucks. This will increase the run time for the procedure. Thus, it is promising to start with the number of trucks determined by each category's MF method and then develop ten scenarios in the under-truck system and ten scenarios in the over-truck system as presented in Table 8.12.

In Table 8.12, T_s, T_l, and T_m are the number of trucks required to meet the production in the small, large, and mixed truck fleet, respectively. These numbers are calculated using the MF method. Following the method mentioned above, 63 scenarios are developed. The best fleet is then chosen by comparing their impacts on

Table 8.12 Scenarios to be created for determining the optimal truck fleet size in surface mines

Fleet type	Under truck	Balanced	Over truck
Small trucks	$T_s - 10, \ldots, T_s - 1$	T_s	$T_s + 1, \ldots, T_s + 10$
Large trucks	$T_l - 10, \ldots, T_l - 1$	T_l	$T_l + 1, \ldots, T_l + 10$
Mixed fleet	$T_m - 10, \ldots, T_m - 1$	T_m	$T_m + 1, \ldots, T_m + 10$

the key performance indicators of the operation, such as daily and hourly production, grade quality, and quantity.

In-Pit Crushing

The first in-pit crushing and conveying (IPCC) system was installed in a limestone quarry mine located east of Hanover in Germany [25]. Since then, three different IPCC systems have gained popularity in surface mining sectors: fixed, mobile, and semi-mobile crushers. The main reason for transferring the surface mining operations from the conventional truck and shovel operations to IPCC is that the mine gets deeper and the haul roads get longer, so transportation costs using conventional trucks get higher. This makes the implementation of IPCC economically viable. However, according to Utley [29], the major drawbacks of the IPCC systems are their short-term movement inflexibility, the dependency of their production to their loading process, inflexibility when different blending is required, etc. However, Osanloo and Paricheh [30] believe that the most reliable way to overcome trucking difficulties in surface mines is to implement IPCC. They also provide data on IPCC's economic advantage over conventional truck and shovel systems showing that by implementing IPCC in 25-year mine life, the system will produce a 200% higher NPV than the truck and shovel system. Furthermore, the IPCC is environmentally friendlier than conventional truck and shovel mining systems as its total energy consumption and greenhouse gas emissions are 20 and 5% less than the truck and shovel systems [30].

Accepting all the advantages of IPCC implementation in surface mines has a significant impact on mine planning, which is not handleable using conventional mine planning tools as a significant change occurs by adding IPCC to the problem. Thus, new optimization models are needed to incorporate IPCC impacts into the mine plan. Conventional long-term production plans provide the time of mining each block in the block model to maximize the mine's net present value. However, by adding IPCC to the equation, the long-term plans must provide IPCC location and time for relocating it along with the production plan. Paricheh et al. [31] proposed a two-step optimization model to incorporate IPCC in the long-term production plan by optimizing its location and finding the best relocation time. Their two-step model is presented here.

$$\text{Minimize } Z = \sum_{k=1}^{r}\sum_{j=1}^{p}\sum_{i=1}^{m_k+1} F_{kij}x_{kij} + \sum_{k=b}^{r} C_k y_k \tag{8.22}$$

Subject to

$$y_k = \frac{1}{2}\sum_{j=1}^{p} w \quad \forall k \tag{8.23}$$

$$z_{kj} - z_{(k-1)j} \le w_{kj} \quad \forall j, k \tag{8.24}$$

$$z_{(k-1)j} - z_{kj} \le w_{kj} \quad \forall j, k \tag{8.25}$$

$$\sum_{j=1}^{p} z_{kj} = P \quad \forall k \tag{8.26}$$

$$\sum_{j=1}^{p} x_{kij} = 1 \quad \forall i, k \tag{8.27}$$

$$x_{kij} - z_{kj} \le w_{kj} \quad \forall i, j, k \tag{8.28}$$

$$x_{kij} = \begin{cases} 1 & \text{If face } i \text{ is assigned to point } j \text{ in period } k \\ 0 & \text{Otherwise} \end{cases} \tag{8.29}$$

$$Z_{kj} = \begin{cases} 1 & \text{If crusher } i \text{ is assigned to point } j \text{ in period } k \\ 0 & \text{Otherwise} \end{cases} \tag{8.30}$$

$$y_k = \begin{cases} 1 & \text{If crusher is relocated in period } k \\ 0 & \text{Otherwise} \end{cases} \tag{8.31}$$

where *i*, *j*, and *k* are indices over the number of mining faces to be mined in a certain period, candidate in-pit crusher locations, and extraction periods, respectively. Moreover, r, p, and m_k are the total number of periods, the total number of candidate locations for in-pit crushers, and the total number of mining faces to be mined in period *k*, respectively. F_{kij} is the total cost of transporting material from the mining face *i* to candidate in-pit crusher location *j* in period *k*. Capital and operating cost of in-pit crusher to move material from candidate location *j* to the processing plant is considered at each period by adding an extra transportation cost to the equation as $m_k + 1$. They also added the relocation cost of C_k that consists of all the costs associated with relocating the in-pit crusher in period *k*. x_{kij}, z_{kj}, and y_k are decision variables, and w_{kj} is a logical variable.

The optimization model presented above Eq. (8.22) minimizes total material transportation costs, including truck transport and IPCC transport. Equations (8.23) to

(8.25) incorporate costs of IPCC relocation to the decision-making model. Equation (8.26) confirms that P number of crushers are open in period k of the mine life. The model uses Eq. (8.27) to make sure that in period k, active mining face i has been assigned to a destination. Equation (8.28) constraints the model to assign the active mining faces to the crusher locations that are open in period k. There are three other constraints in the model, where Eqs. (8.29) to (8.31) used for binary variables.

The best locations to install and re-install the in-pit crusher over the mine's life are determined by minimizing the total transportation cost using the decision-making model presented in Eqs. (8.22) to (8.31).

Conclusion

Approximately 90% of the raw material are mined using surface mining methods. Among all the surface mining methods, open-pit mining has the most contribution to this material movement. Thus, in this chapter, we outlined how different analytics methods can help surface mine and, more specifically, open-pit mine managers in their decision-making with and without the incorporation of uncertainties. Four primary analytics approaches that can be extensively used in surface mining activities are descriptive analytics, analytics hierarchy process, optimization modeling, and predictive analytics. Descriptive analytics such as exploratory data analysis methods is useful for operational data preprocessing and analysis for trucks and shovels or other types of material handling systems. AHP is often used as a procedure with which the proper methods are selected. The best practice in making optimal decisions in long-/short-term planning and equipment dispatching and IPCC allocations uses optimization models. Furthermore, finally, predictive models and, more specifically, discrete event simulations can be used to evaluate technology implementations and changes to the mining operation.

References

1. Hartman, H.L., and J.M. Mutmansky. 2002. *Introductory mining engineering*. Wiley.
2. Martins, P., and A. Soofastaei. 2020. Process analytics. In *Data analytics applied to the mining industry*, 131–148. CRC Press.
3. Bohnet, E. 2011. Comparison of surface mining methods. In *SME mining engineering handbook*, 3rd ed, ed. P. Darling, 405–413.
4. Wetherelt, A., and K.P. van der Wielen. 2011. Introduction to open-pit mining. In *SME mining engineering handbook*, vol. 1, 1840.
5. Hustrulid, W.A., M. Kuchta, and R.K. Martin. 2013. *Open pit mine planning and design, two volume set & CD-ROM pack*. CRC Press.
6. Soofastaei, A. 2020. Digital transformation of mining. In *Data analytics applied to the mining industry*, 1–29. CRC Press.

7. Soofastaei, A., et al. 2016. A discrete-event model to simulate the effect of truck bunching due to payload variance on cycle time, hauled mine materials and fuel consumption. *International Journal of Mining Science and Technology* 26 (5): 745–752.
8. Orman, M., R. Peevers, and K. Sample. 2011. Waste piles and dumps. In *SME mining engineering handbook*, vol. 1, 667–680.
9. Whittle, D. 2011. Open-pit planning and design. In *SME mining engineering handbook*, vol. 1, 877–901.
10. Achireko, P.K., and S. Frimpong. 1996. Open pit optimization using artificial neural networks on conditionally simulated blocks. In *Proceeding of 26th International APCOM Symposium.*
11. Dowd, P. 1994. Risk assessment in reserve estimation and open-pit planning. *Transactions of the Institution of Mining and Metallurgy-Section A-Mining Industry* 103: A148.
12. Osanloo, M., J. Gholamnejad, and B. Karimi. 2008. Long-term open pit mine production planning: A review of models and algorithms. *International Journal of Mining, Reclamation and Environment* 22 (1): 3–35.
13. Godoy, M., and R. Dimitrakopoulos. 2004. Managing risk and waste mining in long-term production scheduling of open-pit mines. *SME Transactions* 316 (3).
14. Ramazan, S., and R. Dimitrakopoulos. 2004. Recent applications of operations research and efficient MIP formulations in open pit mining. *SME Transactions* 316.
15. Dimitrakopoulos, R., and S. Ramazan. 2004. Uncertainty based production scheduling in open pit mining. *SME Transactions* 316.
16. Paithankar, A., and S. Chatterjee. 2019. Open pit mine production schedule optimization using a hybrid of maximum-flow and genetic algorithms. *Applied Soft Computing* 81: 105507.
17. Nicholas, D. 1992. Selection procedure. In *SME mining engineering handbook*, vol. 405, 424. Society for Mining Engineering, Metallurgy and Exploration. Inc.
18. Miller-Tait, L., R. Pakalnis, and R. Poulin. 1995. UBC mining method selection. In *Mine planning and equipment selection.* Vancouver, BC: University of British Columbia.
19. Saaty, T.L., and L.G. Vargas. 2012. The seven pillars of the analytic hierarchy process. In *Models, methods, concepts & applications of the analytic hierarchy process*, 23–40. Springer.
20. Jamshidi, J., et al. 2009. The application of AHP approach to selection of optimum underground mining method, case study: Jajarm bauxite mine (Iran). *Archives of Mining Sciences* 54 (1): 103–117.
21. Saaty, T.L. 1990. How to make a decision: The analytic hierarchy process. *European Journal of Operational Research* 48 (1): 9–26.
22. Berkhimer, E.N. 2011. Selection and sizing of excavating, loading, and hauling equipment. In *SME mining engineering handbook*, 3rd ed, ed. P. Darling, 931–939.
23. Burt, C.N., and L. Caccetta. 2007. Match factor for heterogeneous truck and loader fleets. *International Journal of Mining, Reclamation and Environment* 21 (4): 262–270.
24. Afrapoli, A.M., M. Tabesh, and H. Askari-Nasab. 2019. A multiple objective transportation problem approach to dynamic truck dispatching in surface mines. *European Journal of Operational Research* 276 (1): 331–342.
25. Soofastaei, A., et al. 2016. Development of a multi-layer perceptron artificial neural network model to determine haul trucks energy consumption. *International Journal of Mining Science and Technology* 26 (2): 285–293.
26. Afrapoli, A.M., M. Tabesh, and H. Askari-Nasab. 2019. A stochastic hybrid simulation-optimization approach towards haul fleet sizing in surface mines. *Mining Technology* 128 (1): 9–20.
27. Afrapoli, A.M. 2019. A hybrid simulation and optimization approach towards truck dispatching problem in surface mines.
28. CPLEX, I.I. 2016. *Optimization studio CPLEX user's manual.* New York, NY: IBM Corporation.
29. Utley, R.W. 2011. In-pit crushing. In *SME mining engineering handbook*, 941–956.

30. Osanloo, M., and M. Paricheh. 2020. In-pit crushing and conveying technology in open-pit mining operations: A literature review and research agenda. *International Journal of Mining, Reclamation and Environment* 34 (6): 430–457.
31. Paricheh, M., M. Osanloo, and M. Rahmanpour. 2018. A heuristic approach for in-pit crusher and conveyor system's time and location problem in large open-pit mining. *International Journal of Mining, Reclamation and Environment* 32 (1): 35–55.

Chapter 9
Advanced Analytics for Surface Mine Planning

Jorge Luiz Valença Mariz and Ali Soofastaei

Abstract Planning is one of the fundamental tasks for surface mine projects, and it can play a critical role in the success and results obtained after the development and operationalization of the project. This role will be more important when the miners need to go deeper to access the ore, especially when its grade tends to decrease. The quality of the surface mine planning decision depends on the available data, information, and experience of the decision makers, being imperative that robust decision criteria, new technologies and advanced analytics tools are employed in order to make the most of the huge amount of data generated at each instant. This chapter focuses on advanced analytics approaches such as machine learning and artificial intelligence as valuable decision-making tools to improve the surface mine planning process, as well as mathematical optimization techniques already successfully employed in solving ultimate pit and production scheduling problems, using rigorous algorithms, heuristics and metaheuristics. This chapter also covers stochastic mine planning and discusses future mine planning.

Keywords Advanced analytics · Mine planning · Surface mine · Optimization

Introduction

Mining is a high-risk activity, mainly due to the impossibility of having a definite geological knowledge and the impossibility of foreseeing the fluctuating prices of the commodities on stock markets, which widely affects the economic assessment and its adherence to reality. Mine planning is a fundamental stage in every mining project, whose impacts of unwise and inaccurate decisions can be dreadful to the enterprise since the entire operation phase depends on this and the geological modeling, activities that must be re-evaluated every time new data arises. After the supposition that the long-term geological model is reliable and the classification of the resources is

J. L. V. Mariz (✉)
Universidade Federal do Rio Grande do Sul (UFRGS), Porto Alegre, Brazil

A. Soofastaei
Vale, Brisbane, Australia

A. Soofastaei (ed.), *Advanced Analytics in Mining Engineering*,
https://doi.org/10.1007/978-3-030-91589-6_9

approved, the long-term mine planning team has the challenge of considering all the factors that can potentially restrict the economic extraction of the metals of interest, such as technical, social, legal, economic, environmental, and financial constraints to classify the reserves. However, modern mining cannot be based solely on deterministic analyses, whose assumptions are static throughout the lifetime of the projects. At the same time, reality will inexorably present a range of uncertainties that, if ignored, can significantly reduce project profits or even make them unfeasible. While some of the industry's "best practices" date back to techniques developed half a century ago, the progressive evolution of computational capacity has allowed the development of techniques such as advanced analytics, which employ machine learning, forecasting, cluster analysis, multivariate statistics, simulations, neural networks, among other tools for more sophisticated and robust analysis. This chapter aims to reconstruct the progress made in the field of mine planning from the beginning of the use of computers to the present day in addressing its main problems, notably the definition of the ultimate open-pit mine scheduling, encompassing a plethora of techniques employed and publications [1, 2].

Open-Pit Planning and Design

In a simplistic perspective, most professionals would aim the maximization of the net present value (NPV) in long-term scenarios with different schedules, pushbacks, cutoff grades, ultimate pit shells, and stockpiles attached to assumptions related to tonnages and grades of the orebody, mining, and processing methods and their related costs, sales volume, commodity prices, and discount rates. Tactical mine planning provides more particularized information, allowing a project to extract certain areas of the mine with more precision and considering targets of shorter reach in terms of time. On the other hand, operational mine planners must maintain the mine's adherence to the long- and medium-term plan while aiming to meet production, processing, and economic constraints in daily time horizons. Several increments of data are required to provide their models, such as geological models with tight configuration drilling, metallurgical and mineralogical data, blast drilling data, among others [3–5].

Over the years, there was a tendency to subdivide the large problem of mine planning optimization into smaller individual problems due to the computational impossibility of dealing with real models. In 1989, Jeff Whittle stated that the pit outline with the highest value cannot be determined until the block values are known; the block values are not known until the mining sequence is determined, and the mining sequence cannot be determined unless the pit outline is available.

Figure 9.1 presents the mine planning subdivision in smaller steps [6, 7].

Along with the computational capacity evolution, techniques to deal with uncertainty and simultaneously optimize several constraints have been implemented in problems with great block models, like direct block scheduling (DBS) and stochastic mine planning. Meanwhile, there is no ultimate solution, and despite the continuous

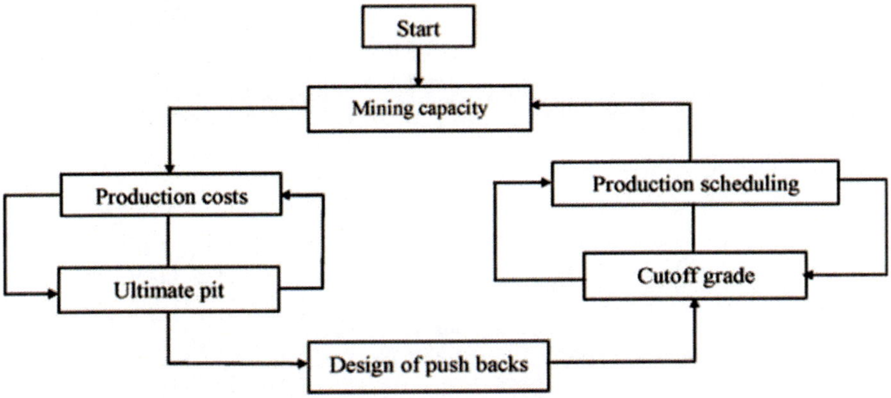

Fig. 9.1 Stages of conventional mine planning [7]

evolution of the algorithms and proposals, this discussion will probably remain for years.

The Ultimate Pit Problem

Since the introduction of block model concepts by Axelson (1964), one of the primary concerns in open-pit mine planning is determining the ultimate pit, which consists of the contour of the mine represented in a block model that aims to maximize the profits of the enterprise. The structural requirement is slope angles of pit walls, defined in geotechnical studies and applied to the economic block model. Converting a geological to an economic block model depends on and relies on the arbitrary definition and combination of economic factors to generate cutoff grades, obtained by solving a benefit function that decides first if the block must be mined, then whether it should be sent to the processing plant or the waste deposit. The benefit function considers the tonnage of products contained in a block and the set of economic parameters (selling price, costs of extraction, processing, and recovery of each block) attached to its extraction [8–10].

In the early days, nevertheless, the manual approach for final pit determination was an ordinary operation. Erikson (1968), Soderberg and Rausch (1968), and Koskiniemi (1979) proposed ways to this problem, where a cross section was chosen, parameters like the depth of the pit bottom and the slope angle of one wall were definite, and then the angle of the other wall and the pit bottom length was iterated. Figure 9.2 presents one example of this approach, currently in disuse [11–13].

On the other hand, for situations where a block model was not a reliable representation of a deposit, as in steeply dipping, relatively narrow vein/strata form deposits, Luke (1972) proposed a computer-assisted pit design based upon sections, but these were particular cases. In the last decades, heterogeneity modeling techniques have

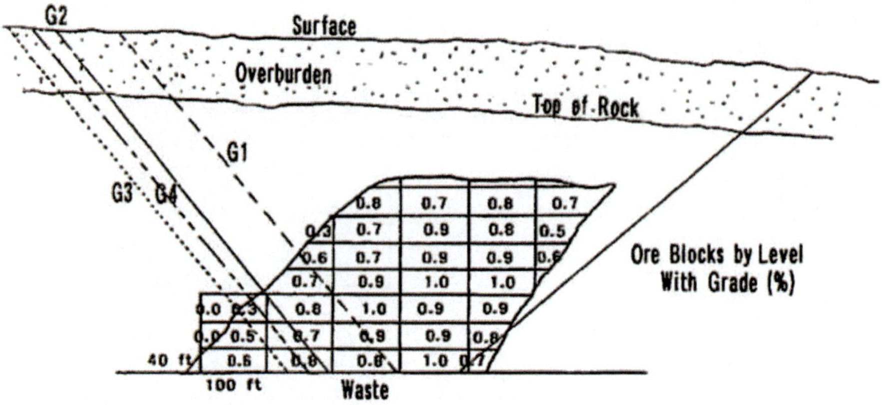

Fig. 9.2 Manual pit limit determination with bottom in waste [13, 14]

evolved, experimented with, and implemented in the petroleum industry, such as those based on variograms, objects, and events, in addition to the most recent multipoint geostatistics, which is based on training images [15, 16].

Considering the block models as the most common and reliable representations of mineral deposits, the relationship among the blocks will be determined by the objective function, the precedence relationship, and the constraints considered in the mathematical optimization model, which can be rigorous or heuristic. Figure 9.3 exemplifies the precedence relationship of blocks, where the number of upper blocks to be extracted influences the slope angles of the mathematical pit of determined methodology, generally from 35° to 55° [4, 17].

Nevertheless, this approach is deterministic, providing just one solution to each set of geological and economic parameters, being necessary to reconsider them to proceed with a sensitivity or a risk analysis. Another issue is that this approach disregards the temporal interaction over the parameters, which precludes changes in the production rate over the life of the project and disregards the effect of inflation on capital. The first computational solutions to the ultimate pit problem date back to the

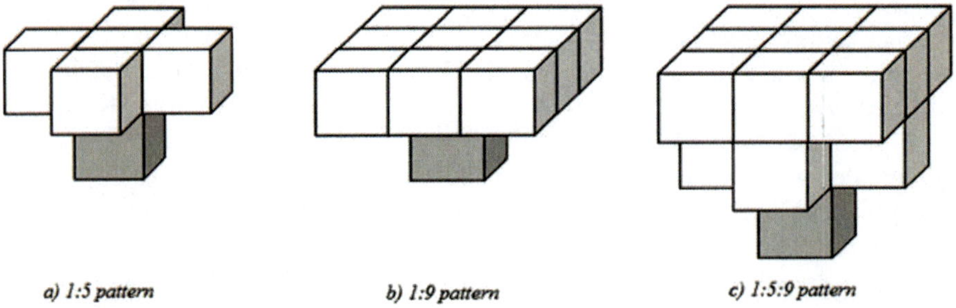

Fig. 9.3 Examples of precedence relationships in block models [17]

1960s, and their bases are still applied in solutions we use today, being rigorous or heuristic. Kim (1978) describes four classes of optimizing algorithms for the ultimate pit problem: dynamic programming, linear programming, network flow, and graph-theoretic algorithms, besides the heuristic algorithms, although many algorithms incorporate approaches and features from different classes simultaneously [4, 18].

Some heuristic approaches became very popular, like the cone-based algorithms, especially when computers did not support the complexity of real mining problems. Proposed by Pana (1965), the trial-and-error moving cone algorithm searches for cones whose total weight of the blocks within it is positive. Therefore, the generated pit is updated considering that each cone in the optimized pit is profitable in itself, when in fact, a pit optimized can be formed by a set of cones that, even though some are individually negative, provide a positive overall solution. However, even with the improvements implemented by authors like Lemieux (1979), Wrigth (1999), Khalokakaie (2007), Galić, Janković, and Mrakovčić (2009), and Elahi Zeyni, Kakaie, and Yousefi (2012), the moving cone proposition never proves itself an optimal solution [19–24].

Another popular heuristic approach was Korobov's (1974) proposition. First, an extraction cone respecting the pit slope angle is built for each positive unit in the economic block model so that the algorithm distributes positive versus negative values within each extraction cone to update the value of the source block, being the cone included in the solution whether the value of the block remains positive after the distribution. Then, if a cone is added to the solution and all blocks within it have been fully exploited, the algorithm proceeds to the next iteration. However, if that added cone has been partially exploited, the algorithm restarts and the blocks that have not yet been extracted have their original values restored. This process is repeated until all positive blocks have been analyzed. Despite the major improvements proposed by Dowd and Onur (1992, 1993), in which the joint testing of cones was implemented to decide whether or not to remove a cone, this method could not guarantee an optimal solution [25–27].

In the last decades, some new cone-based heuristics have arisen. Khalokakaie (1999) proposed the utilization of elliptical interpolation based on four points (positioned in N, S, E, and W) for 3D models or two points (E and W) for 2D ones. Shishvan and Sattarvand (2012) propose implementing a spline interpolation based on an unlimited number of angles in several reference directions, whose intersections between lines and horizontal planes are located in the centroids at each level of the block and generate a set of points for the interpolation process. Xu et al. (2014) incorporated the ecological costs of an open-pit mine in a cone-based algorithm to solve the ultimate pit problem, quantifying the lost value of direct and indirect eco-services, prevention and restoration costs, and cost of carbon emission from energy use. The case study presented demonstrates that it is possible to internalize ecological costs in mine projects and that the pit optimization design and profitability change considerably when these costs are accounted for. Gilani and Sattarvand (2015) propose using nonlinear interpolation (inverse of the power of the distance) based on a finite number of directions and points defined utilizing azimuths. Its formulation allows its application in rotated models and tends to smooth the cone due to the possibility of

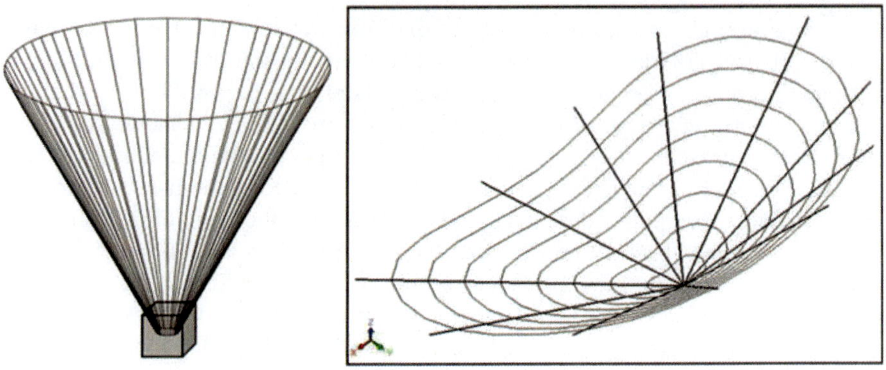

Fig. 9.4 Examples of cone templates for a homogeneous angle and by spline interpolation [17]

using a higher number of directions in the interpolation. Figure 9.4 exhibits examples of cone templates for a homogeneous angle and by spline interpolation. Sadeghi, Dehghani, and Shokri (2021) used the flashlight heuristic, an algorithm capable of examining and improving the value of the ultimate pit limit at each step with an overview, avoiding following defined steps as the floating cone and Korobov's algorithms do. In this algorithm, all positive blocks are identified in the first step and placed in the dark and then from bottom to top and from right to left (or left to right) will be determined whether each addition of positive block (and its consequent upper cone) improves the final solution. A hypothetical flashlight is applied, in which all blocks that were illuminated determine the ultimate pit limit for an iteration, then the positive blocks are evaluated and placed on a blocklist or in a whitelist, the first consisting of blocks that must be forgotten, while the second groups the blocks that can be in the main or replaced flashlights. An application in a cross section of an iron ore deposit showed that the flashlight heuristic reached the same value as a rigorous approach (the dynamic programming algorithm discussed below), obtaining great improvements compared to the floating cone and Korobov's algorithms [17, 28–32].

In parallel, some rigorous algorithms were proposed to solve the ultimate pit problem since the 1960s. In their seminal paper, Lerchs and Grossmann (1965) proposed two rigorous algorithms based on dynamic programming to solve the 2D ultimate pit limit problem and the other based on graph theory to solve the 3D problems. Taking advantage of the processing speed of the first approach, their dynamic proposition begins by the addition of an artificial zero column and a zero row to the left and top, respectively. Then, an overall cumulative sum is executed from top to down, in a way each block sums all values above it in the same column. Hereafter, another overall cumulative sum is performed, this time from left to right, starting at the rear upper left block; for each block, the values of the three immediate neighbor blocks on the left (top diagonal, left, and bottom diagonal) are examined, and the most positive value is selected to be summed to the original block value. Finally, tracing back from right to left, the major value of the first real row must be encountered. Thus, the three immediate neighbors to the left of each block must be analyzed, and the significant value must be identified and incorporated into the pit outline solution, whose blocks in its interior should provide the maximum cumulative economic value. Figure 9.5 presents some steps of Lerchs–Grossmann's dynamic programming algorithm [14, 32, 33].

Nevertheless, this 2D approach is impractical in defining a 3D pit geometry by assembling all optimal cross sections and, as a result, violates the pit slope constraints. Several proposals were made to extend the 2D Lerchs–Grossmann's dynamic programming algorithm to solve the 3D ultimate pit problem. Johnson and Sharp (1971) created a fast algorithm that repeats the 2D optimization on all cross sections, subsequently optimizing a longitudinal section; however, they failed to attend to the slope constraints, and their solution demands inherent smoothing on pit bottom and ends. Barnes (1982), though, pointed that this proposition could be used as a bounding algorithm in generating an outer bound for the true optimizing algorithms. Koenigsberg (1982) solution searches beyond the block value information in its particular cross section, also investigating the relationship between levels in a column, between columns in a section, and between sections. Nevertheless, a generalized pit slope constraint is still unfeasible because the method is recursive and requires a pit slope consistent throughout the model, which is also highly tied to the geometry of the blocks [32–36].

In contrast, Wilke and Wright (1984) proposed using a 3D increment in the form of removal cones over and enclosing every block, while the basic procedure of the 2D dynamic programming proposition was preserved and remained valid. This increment successfully both solves the smoothing problem and overcomes the problem of degeneration characterizing Koenigsberg's algorithm, while at the same time it always guarantees the true optimum design. Over the years, some rigorous and heuristic modifications arise, as proposed by Shenggui and Starfield (1985), Wright (1987), and Yamatomi et al. (1995). The popularity of the Lerchs–Grossmann algorithm evolved in proportion to the computational capacity, which made it possible to obtain optimal ultimate pit limits with savings in time and resources [32, 34, 36–40].

In their graph formulation, Lerchs and Grossmann (1965) recast the 3D block model into a directed graph (LG algorithm) where nodes correspond to block centroids and arcs correspond to dependencies, whose direction of dependency can be illustrated as an arrow. It is an iterative algorithm that converges to the maximum valued closure in a directed graph using a dual approach. It starts with an infeasible solution and iterates through primal infeasible solutions closer bounded until the first primal feasible solution is identified, which is optimal. A solution is determined by a set of strong nodes, which identify that a block must be mined. At each iteration, a tree is built by replacing an arc whose strong node depends on a weak one. Therefore, a primal feasible solution occurs when a set of strong nodes that are not directly dependent on weak nodes provides closure. Figure 9.6 presents a directed graph representation of a block model [32, 41].

Vallet (1976) proposed a variant algorithm, which instead of dividing the nodes into strong and weak classifies the sets considering the average weight obtained from the B/V ratio, in which B stands for the total weight and V the volume of the set, which he refers to as strength. This classification does not affect the status of

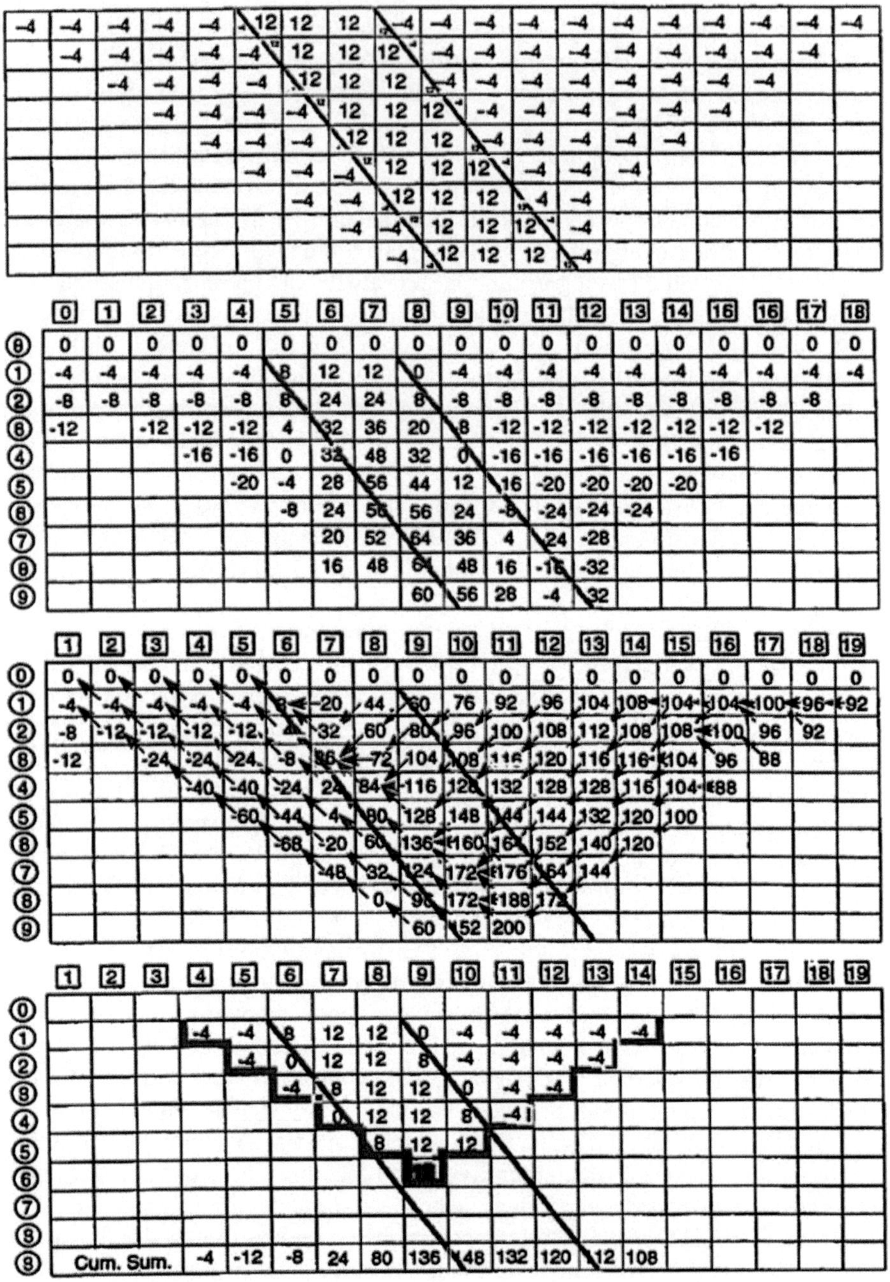

Fig. 9.5 Lerchs–Grossmann's dynamic programming algorithm procedures for ultimate pit limits determination [14, 32]

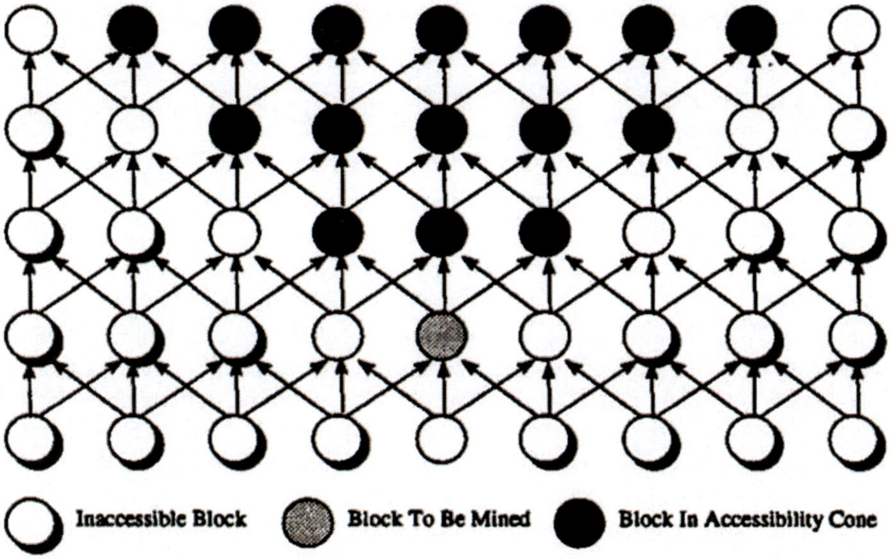

Fig. 9.6 Directed graph representation of block model [41]

subsets, affecting only the order of processing of the merger arcs. In 1986, Whittle Programming Ltd. presented the Whittle 3D open-pit optimization package, thus utilizing the LG algorithm for the first time in a commercial application. Huttagosol (1988) proposed a modified graph tree algorithm, while Zhao (1992) formulated the model as a directed graph composed of many trees that were faster than the first by indirect proof. Zhao and Kim (1992) proposed a solution that differs from the original LG algorithm in that blocks are reassembled when a block in a profitable set is identified below a block in a nonprofit set. Notwithstanding the claim that its algorithm is faster, the authors provided no direct computational comparison [32, 33, 42–45].

The network (or maximum) flow was first applied to the ultimate pit problem by Johnson (1968), who proposed the construction of a network containing a source and terminal nodes, in which each node embodying the ore blocks is attached by arcs to the source node, and those denoting the waste blocks are attached to the terminal node. Moreover, each ore block node is linked to any covering waste block nodes so that the arcs personify the pit slope constraints. Each arc capacity connecting the source node to the ore block nodes is designated with its respective block value, while those that connect the terminal node to waste block nodes are designated with the absolute block value; the arcs connecting the ore block nodes to the covering waste block nodes are assigned with infinity capacity. Once the network is created, the problem is finding the maximum flow from the source to the terminal node. Figure 9.7 presents an example of a network [46].

Picard (1976) proposed a network flow model to solve the maximum closure problem, which finds its main application in mining engineering for determining

$$E = \begin{array}{c} \\ x_1 \\ x_2 \\ x_3 \\ x_4 \end{array} \begin{array}{cccccc} (x_1x_2) & (x_1x_3) & (x_2x_3) & (x_4x_4) & (x_3x_2) & (x_3x_4) \\ \end{array}$$

$$\begin{bmatrix} 1 & 1 & 0 & 0 & 0 & 0 \\ -1 & 0 & 1 & 1 & -1 & 0 \\ 0 & -1 & -1 & 0 & 1 & 1 \\ 0 & 0 & 0 & -1 & 0 & -1 \end{bmatrix}$$

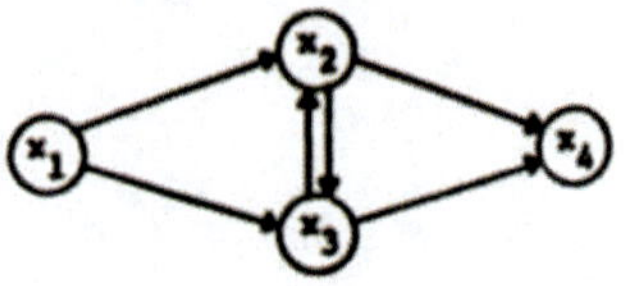

Fig. 9.7 Example of a network [46]

optimum pit mine contours. Several algorithms solve the maximum flow problem, such as the network simplex method of Dantzig, the augmenting path method of Ford–Fulkerson, the blocking flow method of Dinic, and the push–relabel method of Goldberg and Tarjan. Cai (1989) developed an algorithm guided by principles of maximum network flow. However, it adopts the moving cone approach, and the proposition tries to avoid the re-allocation situations, as it was identified as an obstacle that avoids a simulation-oriented pit design algorithm from attaining the optimal solution. The algorithm presented 99.985% of optimality for the case study, a result superior to the moving cone algorithm by 1%, with an execution time 2.08 slower than this. Faaland, Kim, and Schmitt (1990) proposed a flow algorithm that works directly on the mining graph to solve the maximum closure, achieving the fastest results for different small- to medium-sized problems in 86% of cases compared to the other nine algorithms. Giannini et al. (1991) proposed another implementation of the network flow algorithm, which uses Dinic's maximum flow algorithm to compute the optimal pit contour. Yegulalp and Arias (1992) compared the performances of the excess scaling algorithm (a variant of the push–relabel maximum flow algorithm) and the LG algorithm, proving that the second was the fastest. Caccetta et al. (1994) compared other methods to the LG algorithm, which was computationally superior only in dense networks containing many arcs. Underwood and Tolwinski (1996) developed a network flow algorithm based on the dual to find the maximum closure of a graph, closely related to the LG algorithm, although faster. Both algorithms investigate the dependence of strong and weak nodes in the same way, differing in how they update the change in flows when new arcs between the strong and weak nodes are introduced [19, 32, 43, 47–55].

Hochbaum and Chen (2000) compared the performances of the push–relabel and LG algorithms, in which the first achieved better performance for several evaluated instances, although the second consumes less space and memory, thus reducing overhead. Hochbaum (2001) presents a Lerchs–Grossmann-based algorithm to solve the minimum-cut problem on closure graphs without the maintenance of flow, which instead of applying a flow concept employs sets of nodes that have a total positive weight that is closed in relation to some subgraph. Hochbaum (2008) introduces the pseudo-flow algorithm to solve the maximum flow problem, which does not explicitly generate flows, using only pseudo-flows, directly solving a problem equivalent to the maximum blocking-cut problem. The pseudo-flow algorithm can be initialized from any pseudo-flow to reach an optimal solution quickly, whether the initial pseudo-flow is “close” to the optimal solution. Mwangi et al. (2020) formulated the Boykov–Kolmogorov (BK) maximum flow algorithm to solve the ultimate pit problem. They compared the results of two instances of copper and gold/copper deposits to the push–relabel algorithm so that the BK reduces the time consumed to achieve the same solution as the push–relabel algorithm by 12% and 16%, respectively [32, 43, 56–58].

Linear programming (LP) was first implemented in the ultimate pit problem by Meyer (1969). He addressed the problem for an inclined ore body using a linear objective function and the pillar concept, an approach that reduces the number of variables and restrictions compared to blocks. However, it restricts the selectivity and the assumption of local variability of the model. The author solved two single-slice problems by the simplex tableau method, although he claims that multiple slices and ore bodies can be included in unique analysis. Figure 9.8 shows the single- and multi-slice ore claims proposed by the author [59].

Gershon (1983a), whose real focus was the scheduling problem, proposed four different formulations for sequencing production in mines while reducing the problem’s complexity to one period instead of several, which in practice consists of the final pit problem. The first three formulations are based on LP, integer programming (IP), and mixed-integer programming (MIP) for application in block model,

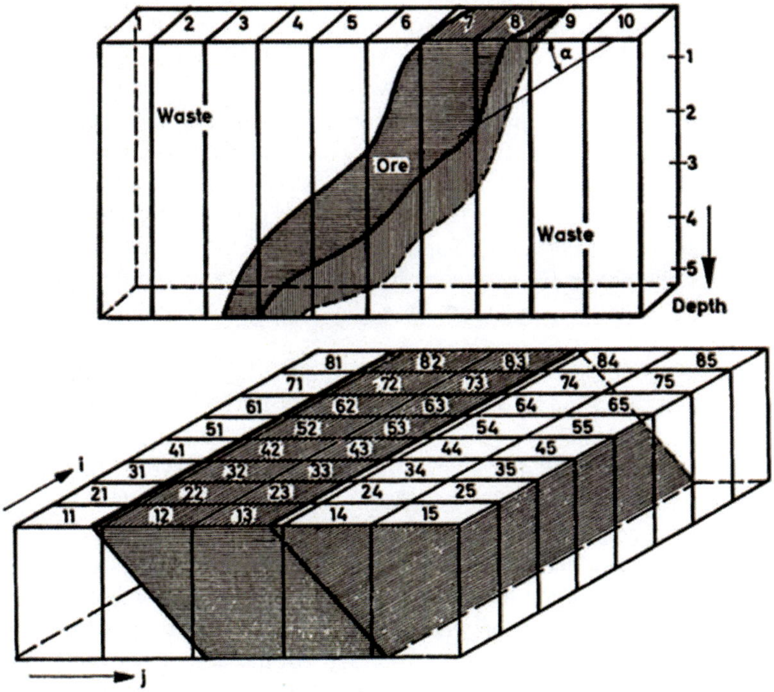

Fig. 9.8 A single slice of an ore claim (α-maximum slope angle) and the multi-slice ore claim proposed by Meyer (1969) [59]

while the fourth is an LP for columns as proposed by Meyer (1969). Huttagosol and Cameron (1992) addressed the ultimate pit problem as a transportation problem, whose objective was to define quantities from suppliers (sources) to consumers (destinations) in order to minimize the total cost and satisfy the demands without exceeding the supplies. The network simplex method was used to solve the problem, and computational experiments attested that the LG algorithm was faster than the network simplex method. Underwood and Tolwinski (1998) proposed a dual simplex approach to solving the ultimate pit problem and compared its performance with the LG algorithm in two instances, finding that their algorithm was slightly faster due to the way of updating the changes in flows when a new arc between the strong and weak nodes is inserted [32, 43, 59–62].

The pseudo-flow algorithm presented by Hochbaum (2008) also has a simplex variant named pseudo-flow simplex, which was the fastest simplex algorithm known to solve the maximum parametric flow problem up to that time. Khodayari (2013) formulated a solution using a network model as an LP problem that can be applied to 2D and 3D block models. In his method, only the positive blocks (those that pay the costs of extracting their negative covering blocks) can remain in the final pit so that these waste block costs can be paid for by a positive block itself or in conjunction with other positive blocks having the same waste blocks over their cones [57, 63].

Another approach to solving this problem is the parameterization method pioneered by Matheron (1975), which expresses the solution (ultimate pit) as a function of a predefined parameter (e.g., commodity price) in a manner that multiple solutions are engendered concomitantly for all possible values inputted. The given example is a single parameterization. However, in theory, it is possible to generate solutions for two (double parameterization) or even multiple input parameters (multiple parameterizations). Instead of using a revenue block model, this approach generally uses a block grade model and proceeds an analysis as a function of cutoff grade, dividing the problem into technical and economic parts. The technical part consists of finding a set of feasible optimal pits that maximize the amount of metal given a selected tonnage. Each pit corresponds to a different cutoff grade; hence, the economic parameters are applied to these pits to define the economically optimum pit [28, 64].

Particularities of the method can be verified in several publications, among which are François-Bongarçon and Marechal (1976), François-Bongarçon and Guibal (1982), and François-Bongarçon (1994). Dagdelen and François-Bongarçon (1982) present a complete double parameterization of reserves using an inexpensive (computer costs) single parameterization algorithm. Whittle (1988) applied the parameterization concepts to address the uncertainty of economic parameters used in the computation of the economic block model, reducing the number of variables employed to generate the revenue block model to two, main (mining cost) and secondary variables (ratio between processing and mining cost). Finally, Coléou (1989) introduced specific algorithms to obtain all the potentially optimal pits and discussed the influence of fluctuation in the parameters. Including existent relations among parameters can generate tight bounds for quickly detecting optimum pits even when using other maximization techniques. Although the method is not rigorously optimal and for criticism due to the simple manner in which the economic optimization is handled, it can potentially solve several mining problems, including the scheduling problem [28, 65–70].

The use of artificial intelligence in mining is also not new. An artificial neural network (ANN) is a computer system based on the functioning of the neural networks that make up human brains, which is based on a collection of interconnected layers of nodes (neurons) that maps sample input vectors to an output target. The map is controlled by the capacity of the neurons' internal connections, called the synaptic weights, which are adjusted as learning proceeds through sample analysis until a generalization to predict the outcomes for other samples is achieved. A single artificial neuron impacts the signal travel through the hidden layers according to the weights of the received signal (a nonlinear function of the sum of its inputs) and if the threshold is surpassed in the selected transfer function. Denby, Schofield, and Bradford (1991) described neural networks in relation to structure, major network paradigms, and training, also presenting potential applications for using this technology in mining engineering in the early 1990s. Figure 9.9 presents an artificial neuron with its attributes (left) and the structure of a neural network (right) with an input layer, a hidden layer, and an output layer [71, 72].

Achireko and Frimpong (1996) developed an algorithm to overcome the limitation of not addressing the random field properties related to the ore grades, reserves, and commodity prices in optimizing ultimate pit limits. A modified conditional simulation was employed to address the random field properties of the reserves and ore grade, and an ANN is used to classify the blocks according to their conditioned values. The error backpropagation algorithm optimized the pit limits to minimize the output error in a multilayer perceptron considering the pit slope constraint. The optimized pit value obtained by this methodology applies a modified conditional simulation and a multilayer feedforward neural network (MCS/MFNN) algorithm, as compared to the LG algorithm, achieving the same results but faster and in random

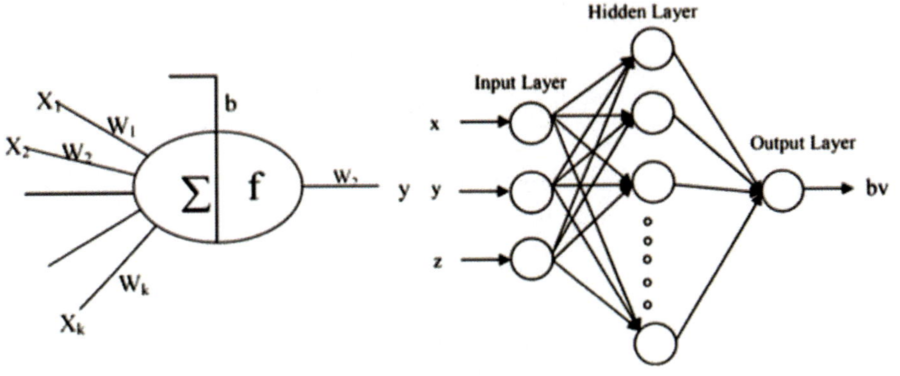

Fig. 9.9 An artificial neuron with transfer function f, where $x_1, \ldots, x_k$ are network inputs and f is the transfer function (left), and the structure of a neural network block classification (right) [71]

multivariable states. Frimpong and Achireko (1997) applied the MCS/MFNN algorithm to optimize a 2D section of a gold deposit in Zimbabwe, comparing the results to the LG algorithm. Frimpong, Whiting, and Szymanski (1998) proposed an intelligent pit optimizer (IPOP), which consists of the MCS/MFNN, a pit search algorithm, and a stochastic optimization annealing process to obtain a set of optimal pits in a random field. The IPOP algorithm jointly addresses simulations of ore reserves and commodity prices to generate target values and economic blocks and has been tested in 2D sections of a gold mine in Zimbabwe, reaching the same results as the LG algorithm, albeit with 50% less time spent. Achireko (1998) used an MCS/MFNN algorithm to obtain the final 2D and 3D pit of an undisclosed gold mine while using a multiple regression model (MRM) and an MFNN to obtain stochastic gold prices based on a sociopolitical–socioeconomic dataset [32, 73–76].

Asa (2002) proposed some neurogenetic mining block predictors based on geostatistics and machine learning, applying the techniques to a gold deposit and a coal deposits, followed by the proposition of a combination of simulated annealing and neurogenetic optimizer (SA-NGO) algorithm to solve the ultimate open-pit problem, achieving financial results slightly lower than the LG algorithm. Frimpong, Szymanski, and Narsing (2002) developed a computational intelligent (CI) algorithm to optimize surface mine layouts with stochastic optimization annealing process (SOAP), which was tested and compared to the LG algorithm in a 2D section of a gold mine in Zimbabwe, achieving the same results with about 50% less CPU time. The SOAP component showed that under market, geological and operating dynamics, the probability of reaching the expected value or higher is about 50%, with a 1% coefficient of variation. Sayadi, Fathianpour, and Mousavi (2011) proposed the three-dimensional pit limit optimization/artificial neural network (3DPO/ANN), which consists of three stages: the generation of an economic block model of the deposit, the classification of the blocks by an ANN-based on their economic values, and the determination of the ultimate pit limit while considering the variable slopes. With the ability to incorporate any bound impurity into the block grades, the methodology was applied to optimize the ultimate pit limit of the Esfordi phosphate mine in Iran, and the results were compared to the LG algorithm, achieving almost the same results when the slope was the only constraint, on the other hand reaching 30% more profit when the main purpose was to control ore impurities [32, 71, 77, 78].

Another approach is the use of metaheuristics to solve the ultimate pit problem, which is a series of algorithmic ideas that upgrade the heuristic methods and make their application possible to a large number of problems, frequently based on nature and biology, such as the genetic algorithm (GA) inspired by Darwin's theory of evolution. Since variables representing operational functions can be combined and encoded as a binary string, the structure includes the random generation of a set of viable solutions (starting population), each representing an individual in a gene pool. Then, a fitness calculation is performed, where each string in the population is evaluated to measure how well the network performs. Nevertheless, using a probability-based approach, where highly suitable strings are more likely to reproduce, the selected strings are copied to the next generation and modified randomly, employing operators such as mutation and crossover. Thus, strings that increase their fitness value (chance

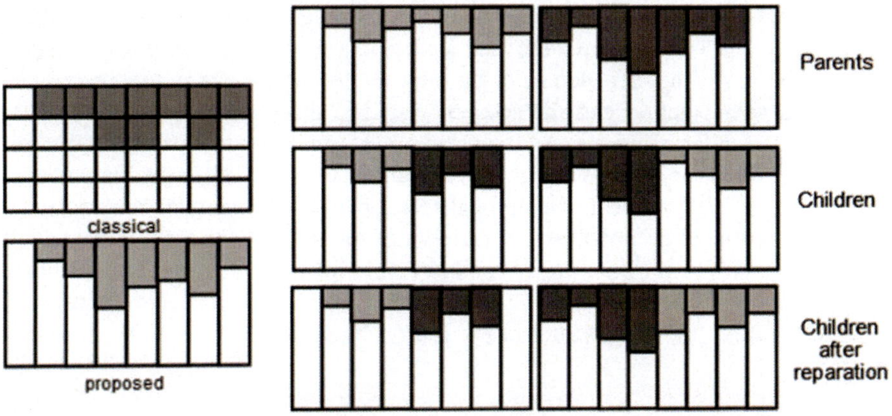

Fig. 9.10 Differences between classical block model representation of an open pit in 2D and a set of a column representing chromosomes with the total percentage of extraction (left) and a reparation mechanism for scheduling crossover (right) [82]

of reproduction) are included in the next generation and iteration, unlike strings that decrease their fitness value. Some common GAs are ant colony optimization (ACO), particle swarm optimization (PSO), bee colony optimization (BCO), among others. Figure 9.10 presents the classical block model representation of an open pit in 2D and a set of a column representing chromosomes with a percentage of extraction in each column between 0 and 1 (left), and a reparation mechanism for scheduling crossover (right) [79–82].

Denby and Schofield (1993) introduced the generally optimized pit (GO-PIT), optimizing the ultimate pit and the production schedule by employing a GA approach. The first step in the self-learning methodology is to define a series of viable pits that know the production of the forthcoming generations (pits). In contrast, a relatively wide series of random schedules (including some highly constrained) was employed as a strategy to examine the significance of the first population. The efficiency parameter for a given schedule is obtained by an individual pit fitness function, which performs an NPV calculation for the simulated pit accounting for functions that link factors such as grade, cutoff grade, extraction timing, recovery factors, discount factors, mineral market values, and mining and processing costs. Furthermore, there is a relationship between the fitness value of chromosomes and the probability of reproduction from one generation to the next, which means that the schedules that remain in the forthcoming generations have high fitness values and do not violate constraints. Denby and Schofield (1994) applied this methodology in 2D examples, while Denby and Schofield (1995) deepen their efforts in order to consider risk evaluation in their scheduling procedure [80, 83–86].

On the other hand, Denby, Schofield, and Hunter (1996) improved the GO-PIT to incorporate 3D analysis, applying it to a gold–copper vein deposit for three different slope angles, comparing the results with other package optimization, obtaining a very similar NPV for the ultimate pit and the production schedule in just one step, and avoiding iterations required by the traditional approach. Finally, Denby Schofield, and Surme (1998) consolidated the later studies in a way that divides the scheduling problem into two distinct elements, the first related to the development of a 3D GA to evolve a solution and the second related to the method assessment of the fitness of particular solutions. As the basic optimization engine is decoupled from the fitness evaluation method, flexible scheduling can occur, while the main genetic operators employed to update the optimal solution include randomization, mutation, crossover, and local optimization. A variety of fitness functions are also presented, whose objectives aim to maximize the NPV, minimize early stripping, or balance ore production of multivariate minerals and stripping, in addition to presenting two cases to highlight the possibilities of the approach [86–88].

The Scheduling Problem

Once the ultimate pit limit has been identified, the problem of determining the extraction sequence must be solved, through optimization criteria for either maximizing the integral cash flow or maximizing the discounted cash flow (NPV). Several authors, inspired by operational research problems, developed LP and IP models to solve this problem, such as the precedence-constrained knapsack problem (PCKP), where multiple periods fill the knapsack, the constraints are production capacity ones, such as mining and processing capacity, and precedence restrictions represent operating block access considerations. Although these constraints individually do not present computational challenges, in combination, they produce an NP-hard problem, whose optimal solution generally cannot be obtained within an acceptable computation time, so that metaheuristics are usually implemented in the search for approximate solutions. Many of the approaches demonstrated for solving the ultimate pit problem have been extended to the scheduling problem, although the solutions based on IP, MIP, and mixed-integer linear and nonlinear programs (MILP and MINLP) will be presented in the following topic, focused on direct block scheduling (DBS), as well as the stochastic approaches will be presented separately [81].

Preferably, it is better to optimize the final pit by maximizing NPV instead of profit, since the discounted ultimate pit is a subset of or equal to the undiscounted ultimate pit, so that a model obtained from the second method is unlikely to be the most profitable when put into practice. Thus, the size of the pit limit reached from both approaches is a relevant indicator for estimating their performance, although the complexity of the intended model can often (and has done so for decades) preclude a direct approach due to computational restrictions. Furthermore, in the traditional undiscounted design methodology, following the ultimate pit determination, production planning must be defined for different time intervals (e.g., short, medium, and long term), whose ranges can be defined as 20–30 years for long-term, 1–5 for medium-term, and 1- to 6-month period for short-term mine scheduling [4, 89, 90].

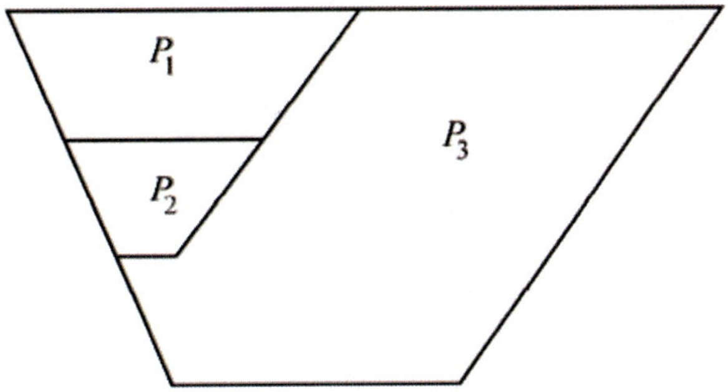

Fig. 9.11 Representation of an open pit containing three pushbacks with very different volumes, resulting in gap problems [91]

Another algorithm described by Lerchs and Grossman (1965) uses the concept of nested pits to solve the long-term extraction sequence problem, generating consecutive pits by reducing the value of all blocks in the pit by a constant amount, thus greedily defining the extraction sequence from the smallest nested pit to the largest. Whittle (1988) used heuristic techniques to incorporate discounting and pushback design to scheduling based on the nested pit methodology. The pushbacks (or phases) are manageable volumes of material or individual pit units containing their respective working front and mining dynamics, allowing the development of short-term planning. Pushbacks are designed so that the nested pits range from the smallest containing mainly high-grade ore and higher added value, for the largest pit encompassing ore with the lowest value per ton, and some heuristics suggested for this task are those proposed by Dagdelen and François-Bongarçon (1982), Gershon (1987), Whittle (1998), Wang and Sevim (1995), and Ramazan and Dagdelen (1998). Meagher, Dimitrakopoulos, and Avis (2014) elaborated an extensive review on optimizing pushbacks and the gap problem. Figure 9.11 represents an open pit containing three pushbacks with very different volumes, resulting in gap problems [4, 32, 41, 69, 91–95].

Seymour (1995) altered the LG scheduling algorithm by incorporating the pit size as a parameterization variable in order to produce pits with maximum value related to this parameter, as the algorithm returns the pit designs found in the upper convex hull of the point set in a graph of economic pit value versus parameter value. Thus, when the upper convex hull is proportional and there are viable pits at or nearby the expected parameter values, pushbacks can be developed that do not suffer from non-uniform sizes based on these pits. Although if the pits that stand on the upper convex hull are distant from each other in terms of size, gap problems will consequently emerge because the algorithm does not provide pits with the desired dimensions. Figure 9.12 shows both cases, a well-defined upper convex hull (left) and a deficient one (right) [32, 91, 96].

The first to use dynamic programming to an open-pit production scheduling was Roman (1971, 1974), who integrated the ultimate pit limit optimization and block sequencing into an algorithm that determines the position of the last block in a 2D pit that must be mined at the beginning. The decision variable is whether or not to exploit a new block, and depending on the slope constraints, all viable forms of sequencing

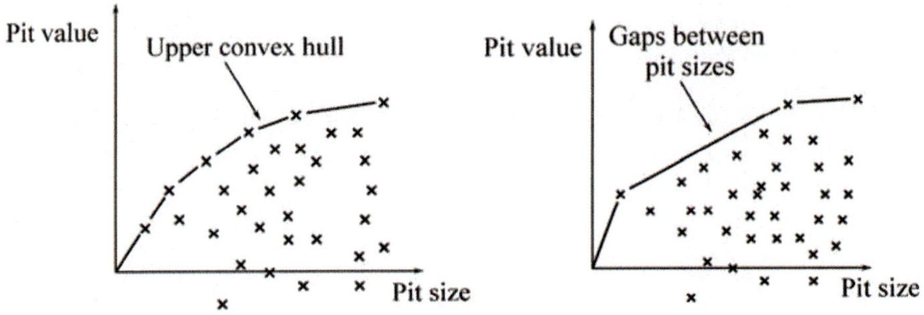

Fig. 9.12 A well-defined upper convex hull (right) and a deficient one with a large gap (left) are plotted on a graph of pit value versus pit size [91]

the blocks above the selected one are verified, and the optimal sequence is determined using an NPV calculation. Hence, blocks close to the pit boundary should be assessed if they contribute or not to generate a positive NPV, those that do not conduct to a positive NPV are excluded, and a new pit sequence and NPV are calculated, a process that remains until it is not necessary removing any blocks from the pit. Wright (1989) states that dynamic programming is efficacious in determining the ultimate pit limits and identifying the incremental pit boundaries, which define the production requirements and equipment capabilities employed to define those incremental pits that meet the related constraint, also presenting a case study. Gershon and Murphy (1989) present a dynamic programming approach that optimizes the cuts for a single drill hole from which the ore qualities are evaluated, and layers are aggregated from the strata and defined as ore or waste, being determined which strata should be mined to maximize NPV. Dowd and Onur (1992, 1993) utilized dynamic programming for production scheduling and ultimate pit design using a variable cone algorithm, and Onur and Dowd (1993) proposed a methodology that considers the ultimate pit limits as provided, defines block scheduling, smooths schedule, and includes haul roads based on various monetary and geological criteria [4, 5, 26, 27, 97–101].

Tolwinski and Underwood (1992) developed a method that combines dynamic programming and stochastic optimization to solve long-term production scheduling problems, recognizing that this problem is tantamount to finding the way with the greatest value in a graph. As in realistic instances containing a large number of nodes, the use of dynamic programming requires the enumeration of all of them (blind search method), which is an NP-hard problem. They introduced the informed search method, relying on artificial intelligence and heuristic rules to learn as the path values are upgrading with each simulation. Tan and Ramani (1992) considered a linear program and a dynamic program to define a production schedule covering multiple periods subject to equipment capacity constraints, in which the linear program evaluates discrepancies in schedules with different discount rates and equipment availability, and if there are no discrete variables, block sequencing decisions do not occur. Elevli (1995) presented a dynamic program that uses local search strategies to maximize the NPV subject to hard sequencing constraints and soft constraints on mining and processing capacity, illustrating the methodology with a small example. Tolwinski and Golosinski (1995) and Tolwinski (1998) proposed a method inspired by the depth-first search strategy of dynamic programming, in which the nested pits obtained previously are evaluated, and pushbacks that meet all operational constraints are defined. Hence, the rock mass within the limits of the ultimate pit is split into elementary units named atoms, the objective function and constraints are determined based on target variables representing an attribute function, and the viable schedules (sequence of atoms) are arranged in a tree data structure. Two kinds of trees are built: The first, whose arcs are correlated to atoms, is created using the depth-first search technique; the second tree is constructed using total accumulated values of the attributes so that the arc lengths represent the unit of time employed as a reference for scheduling. After eliminating non-feasible solutions in this tree, dynamic programming is used to find the optimized schedule [4, 5, 102–106].

Sevim and Lei (1998) circularly described the interactions between the ultimate pit limits, the cutoff grade, the mining sequence, and the production rate. The authors developed a methodology that combines heuristics and dynamic programming to simultaneously obtain the optimal mining sequence, ore and waste extraction, the ultimate pit limits, and the mine lifetime in a three-phase model. The first phase forms a block model from the deposit limits and applies a limiting algorithm to define the largest viable pit. A range of cutoff grades is accounted for in the second phase, in which a set of nested pits is obtained within the largest pit for each cutoff grade, each pit containing the largest quantity of metal of all feasible pits of the same size. All viable pushback sequences are built from the pits and assessed in the third phase, considering their NPV. Erarslan and Çelebi (2001) developed a profit maximization model that determines the optimal pit contour as a function of optimum production scheduling subject to factors such as cutoff grade, mining start location, equipment selection, annual production rates, excavation sequence, and the synchronization of the mining plant operations during the sequencing stage of the mine. They employed dynamic programming to numerate multiple volumes and defined the optimal pit size. Wang and Sun (2001) proposed an approach utilizing a dynamic pit sequencing structure to integrate the cutoff grade identification, the definition of production rates, the ultimate pit contours, and the production schedule, modeling the distinct choices for the next pit phase as a network for each period [5, 107–109].

Gu et al. (2011) developed a method of dynamic sequencing that concomitantly optimizes the ultimate pit and the production schedule of a coal mine, initially establishing a geological seam model that represents the bedded coal deposit and then generating the sequence of geologically optimum ultimate pits, along with a sequence of geologically optimum pushbacks. Thus, the optimized pushbacks are submitted to a dynamic programming structure, and the schedule that presents the highest NPV is achieved for each final pit, chosen as the ideal pit reaching the highest overall NPV, and its best-related schedule is the superior overall production schedule. Wang, Xu, and Gu (2013) presented a model that first generates a sequence of geologically optimal pits and then dynamically optimizes them in relation to the geometry, the number of phases, and the location of each pit, simultaneously maximizing the overall NPV. In addition, the dynamics of the problem concerning time and space is addressed by the model, which can also accommodate distinct wall slopes and bench heights, being tested using a large real-life deposit model containing nearly a million blocks. Figure 9.13 presents a cross section of a hypothetical seam as described by Gu et al. (2011), the moving cone elimination process, and pushbacks in a final pit [110, 111].

Latorre and Golosinski (2011) introduced and Nanjari and Golosinski (2013) later expanded an algorithm based on dynamic programming and heuristic, which defines open-pit limits, the optimum mining sequence, an optimum width of pushbacks. The model aims to maximize the NPV of a deposit and permits incorporating cutoff grades, costs, price vectors, production rates, dilution, mining, and metallurgical recovery, among others. Test runs reached ultimate pits and production schedules, obtaining higher NPV values than those achieved through parametric techniques [112, 113].

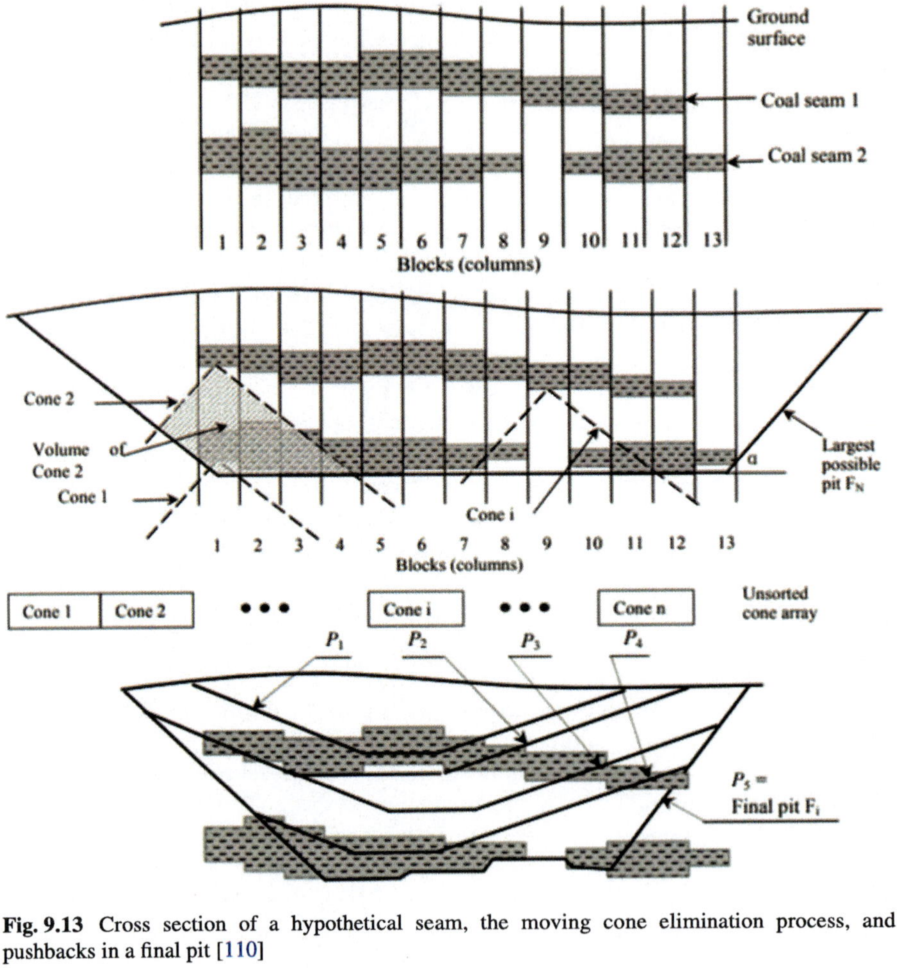

Fig. 9.13 Cross section of a hypothetical seam, the moving cone elimination process, and pushbacks in a final pit [110]

Since the 1960s, LP formulations were used to solve open-pit production scheduling problems, such as the one proposed by Albach (1967), who formulated the long-range production plant for a lignite mine as a chance-constrained programming problem, developing a model that requires the maximization of a linear form subject to linear and nonlinear constraints. In order to simplify the computational requirements for large problems, he reformulated the model to a straightforward LP model, in which an iteration procedure is derived by which the solution to the original nonlinear problem is found. Different acceptable risk levels were computed by introducing uncertainty explicitly into the model, resulting in a risk–profit surface from which management can choose the optimum plan according to its risk preference function. Manula and Gezik (1968) proposed an LP model to solve the optimal production plan for a crushed stone operation that aims to determine an economical means of producing and stockpiling a multi-item stone product produced in three different production modes in order to minimize the production and stockpiling costs under seasonal demand conditions. Johnson (1968, 1969) used an LP model to optimize the production scheduling of a mine, considering the effect of time on the value of money, distinct types of processing, and dynamic cutoff grades, decomposing the monolithic multi-period production scheduling model into a series of subproblems employing the Dantzig–Wolfe decomposition principles.

Nevertheless, each subproblem is addressed as a single-period problem with the same features as the ultimate pit limit problem and is relatively simple to solve the master problem. This proposition, however, while generating optimal individual outcomes for each period, is not able to fully solve the long-term scheduling problem, as there are situations where a part of the block can be extracted while all covering blocks have not yet been removed, as stated by Osanloo et al. (2008). Wilke and Reimer (1977) employed an LP model to solve the short-term production scheduling problem, maximizing shovel utilization (each shovel being weighed to an associated priority) subject to operational constraints, such as blending and equipment capacity [4, 46, 114–118].

Chanda and Wilke (1992) developed a linear goal programming model combined with deterministic mining simulation to define a short-term production schedule in mines containing stratiform orebodies. Shovel regions in the orebody model are the planning elements of the model, not requiring block models. The algorithm evaluates the status of the entire shovel regions when starting a scheduling period concerning precedence and geometrical constraints, then supplies the optimal shovel regions aiming to maximize the metal content in the blended ore, and minimizes deviations in tonnage and grade. Moreover, a post-simulation smooths the LP solution to fit it into an executable plan. Youdi, Qingziang, and Lixin (1992) use a linear goal programming model to produce a medium-term schedule that feeds a manual short-term planning structure using CAD tools. Fytas, Pelley, and Calder (1987) and later Fytas, Hadjigeorgiou, and Collins (1993) proposed a short-range production scheduling model based on LP that calculates the optimal schedule of multiple short periods (e.g., months) that fulfill a series of mining and processing constraints. The simplex model employed can schedule the extraction of partial blocks within multiple user-defined shovel regions, where the blocks consist of candidates to be selected as the best combination to fulfill the restrictions of each short-term period [119–122].

Sundar and Acharya (1995) presented an integrated computer system for short-range planning that uses an LP model to choose the blocks to be blasted and then a stochastic programming model to select the blasted areas to be excavated and decide the quantity of ore and waste to be exploited from the selected blocks. A chance-constrained program considers the randomness of ore's quality and quantity features and changes in the operating time demand. The model also provided a temporary allocation of dumpers to support the excavation plan and was implemented at an iron ore mine in India, accounting for the mill's transport capacity and blending ore blocks to meet demand. Chanda and Dagdelen (1995) proposed a linear goal programming model for addressing blending in the short-term mine sequencing production scheduling, whose defined objectives were maximizing an economic parameter and minimizing the sum of the absolute deviations from the grade target tonnage. Deviation variables were used to calculate penalties (or premiums) incurred for failing to meet contractual quality specifications since goal programming allows violations of certain constraints to meet practical mine production scheduling requirements. An application example of the model for a coal mine was presented. Hu, Wei, and Fang (1995) presented an improved PERT network method with resource constraints combined with a linear goal programming model to define the short-term mine sequencing, considering practical mining sequences in the scheduled area according to a given long-term plan. Based on the shovel's operation, the PERT network computed the earliest (and latest) beginning and finishing times for mining, the general (and local) time difference for mining each block, and the critical path in the PERT network. The multiple objectives were to meet the requirements of ore production (precedence, waste stripping, and blending) and the work hours assignments of shovel and trucks, and then the goal programming is responsible for finding a mining schedule that meets the constraints [5, 123–125].

Vujić and Ćirović (1996) proposed a fuzzy linear programming model (FLP) to solve the mine sequencing problem. Instead of maximizing the economic return, the aim of the approach is formulated on how to find a satisfactory income given by a triangular fuzzy number. The constraints are also transformed into fuzzy numbers, and the uncertainty of the model is directly related to the narrowing of the boundaries of the fuzzy triangular numbers. The authors discussed the use of the methodology in an open-pit bauxite mine. Pendharkar (1997), in turn, presented an FLP model to solve the mine sequencing problem in the coal industry, aiming to find an acceptable profit level for the fuzzy constraints imposed. The author fuzzified the model by maximizing the fuzzy profit and quality membership functions, which are related since the need to guarantee product quality results in lower profits for the company. Once the assumption of k values results in satisfaction index values for the evaluated restrictions, it is possible to calculate the influence these assumed values for k will have on the company's profit. Djilani (1997) proposed a combination of LP and simulation parameterization approach to defining a mine production schedule. Inspired by grade values rather than revenue values, the algorithm first activates a waste stripping module to uncover a minimum tonnage of ore at the start of each period, followed by an LP stage that defines an optimal mine scheduling for three periods from subsets of blocks submitted to the predefined ultimate pit. Finally, a mining simulator detaches extracted blocks from the block model for the first period and adapts the subsets of blocks for presentation to a new round of waste stripping or LP module. The program continues its iteration by scheduling three periods at once, choosing only the blocks whose extraction was planned for the first periods, adjusting the subset, and then scheduling three more periods whose results were not rigorous [126–128].

Halatchev (1993, 1996, 1999) proposed an LP model to define long- and short-term production scheduling of a polymetallic ore deposit with multiple mines, which has the capacity to cope with constantly changing market conditions, as well as for a significant increase in the company's profit. The model provides the annual quantities of a given economic metal recovered from ore treatment. The objective function represents an expression of the project's NPV as a parameter for evaluating the economic success of the mine sequencing using a methodology that determines the net smelter return. Halatchev (2002) states that the mine sequencing optimization is concerned with two features, spatial and temporal, and proposes a formulation that includes the temporal aspect, defining the viable distribution of extraction time for the quantities of ore and waste from each mine bench. An LP model was developed inspired by combinatorial analysis, which supplies a multivariate investigation of mine sequencing to maximize the NPV as a parameter for evaluating the successfulness of the production sequencing. This approach determines a process for defining the optimal time and order of inclusion of all the bench schedules in the mine's production sequencing arrangement over the life of the mine. From a theoretical standpoint, there are a number of alternatives of mine production schedules in which the optimal alternative must be investigated among all the generated alternatives. Applying the methodology in a copper mine in Bulgaria addresses the variation of NPV according to the mining factor, in the range of 15.84%, which can be used as a reserve against price volatility. Figure 9.14 presents the cumulative time graph of the mine from the theoretically possible variants 1–6 when considering the sequences of the first three cutbacks of the studied mine, each representing an alternative in the composition of the open-pit mine working zone, exponentially increasing the number of possibilities when new benches are added [129–132].

Halatchev and Kozan (2003) proposed an LP model to optimize open-pit coal production that addresses the specific features of the exploitation of a single coal seam and the use of truck-shovel and dragline systems, adding to the model foundations of mine sequencing, waste postponement, and coal quality optimization. Halatchev (2004) presented an LP model to optimize the shovel selection in coal mines using the principles proposed in Halatchev (2002), and Halatchev (2005) expanded his work by developing a framework to optimize mine sequencing with stabilization at multistages of the mining rate that incorporates a combinatorial analysis to generate all production scheduling alternatives that are theoretically feasible. The model considers the interaction among market, mine, and mill, and the objective function aims to maximize metal recovery and postpone waste extraction. The sequencing optimization structure uses spatial and temporal cumulative graphs, in

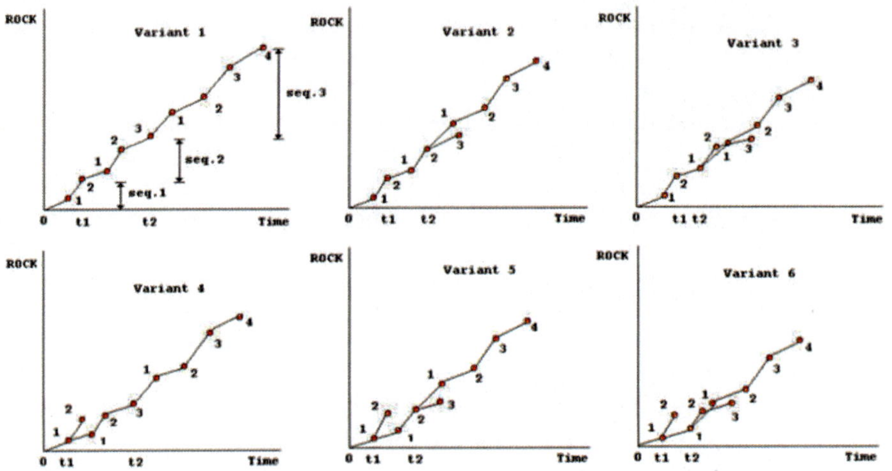

Fig. 9.14 Cumulative time graph of the mine from the theoretically possible variants 1–6 when considering the sequences of the first three cutbacks of the mine studied [132]

which the optimal mine sequence represents a piecewise-linear function (in order to achieve the multistage stabilization of the mining rate) in the viable domain that can be determined as the one that reaches the optimum working width, meeting the requirements of the machinery and the penalties inserted. The application results in a synthetic open-pit gold mine result in a discrete model of the NPV variation, revealing the potential of the proposition to address the geological uncertainty and price volatility in the short- and long-term schedule [132–136].

Halatchev (2015) deals this time with the spatial aspect of optimizing the mine sequence, which is addressed when determining the optimal number of pushbacks in an open-pit mine. The LP formulation utilizes a combinatorial analysis to generate pit shells and group them in an optimal number of pushbacks to maximize the project value, thus providing a multivariate investigation of the optimal production schedule to maximize NPV. A case study in a hypothetical open-pit gold mine is provided, and the results highlighted the complexity of investigating the variation of NPV as a function of pushback geometry. Souza et al. (2018a) developed a system inspired by goal programming capable of achieving blending constraints of grades and tonnages, reducing stockpile fluctuations for short- and medium-term mine scheduling. The target is minimizing operational costs and quality deviation while respecting restrictions such as distance among the feed piles, grade control, equipment, and plant capacity [137, 138].

The formulation of the open-pit mine sequencing as a network flow also received propositions. Picard and Smith (2004) proposed a parametric maximum flow approach to find intermediate contours aiming to maximize the benefit-cost ratio while advancing toward the ultimate pit as proposed in Picard (1976). Based on this pit with the maximum benefit-cost ratio, the successive application of the parametric algorithm can re-optimize the perturbed maximum closure problem and find intermediate contours of decreasing benefit-cost ratios. Topal and Ramazan (2012) developed a network flow model aiming at the optimization of the strategic planning and scheduling of a multi-mine operation, in which sources of ore, processing plants, and circuits, ports, and shipping facilities are denoted by nodes, while conveyor belts, roads, and railway lines represent the arc interrelations between these nodes. The algorithm aims to maximize the NPV while considering operational demands and product requirements to meet customer demand for quantity and quality of products supplied, addressing key issues such as ideal start-up time of the plant and the mining of distinct, optimal plant, mine, and mine train capacities. The model was employed to optimize the 50-year life span schedule for a large mining district in Western Australia, encompassing mines containing more than 13 plants and 100 pits, and reached an improvement in the NPV of around 10–20% compared to other tested software, achieving the solution in approximately 2 s [47, 139, 140].

Amankwah, Larsson, and Textorius (2014) formulated the multi-period production scheduling problem as a maximum flow problem in a time-expanded mine graph, extending Picard's results (1976) and considering only temporal-spatial precedence as Lagrangian dualized constraints. First, a copy of the mining graph is created for each period. Thus, the directed arcs in this expanded graph model the temporal precedence association among nodes in consecutive mine graph copies, similar to the directed arcs that model the spatial precedence interconnection within each mine graph. Next, Liu and Kozan (2016) proposed two graph-based algorithms for large-scale ultimate open-pit and scheduling problem inspired on network flow graph and conjunctive graph theory, achieving faster results than the best-known solutions until that time when applying the formulations to available block models in MineLib, a series of datasets for testing published by Espinoza et al. (2013). Figure 9.15 presents a time-expanded mine graph encompassing three time periods and the minimum cut (i.e., the maximum flow dual) determining the mine schedule in the proposition of Amankwah, Larsson, and Textorius (2014) [47, 141–143].

The mathematical IP, MIP, and MILP models that seek the simultaneous definition of the ultimate pit and production scheduling make up a special type of planning called DBS, which will be covered in a specific topic later. Nevertheless, formulations using these types of programming that do not seek to concomitantly define the ultimate pit and mining sequence, as a definition of long-, medium- and short-range mine scheduling for a final pit already determined, will be mentioned below. As aforementioned, Gershon (1983a) proposed a MIP and an IP formulation for sequencing production in mines, allowing partial block extraction. Smith and You (1995) developed an IP formulation to define a short-term mine sequencing that minimizes the production costs of a given phosphate mine operation and plant design and requirements to provide the minimal variation in the optimal plant feeding characteristics. Smith (1998) presented a MIP model based on goal programming for a single-period production schedule, whose objective was to minimize the sum of the product of the deflection from targets and the associated penalty weight, penalties that provide a way to rank the significance of meeting the distinct goals (ore, waste, gold and silver,

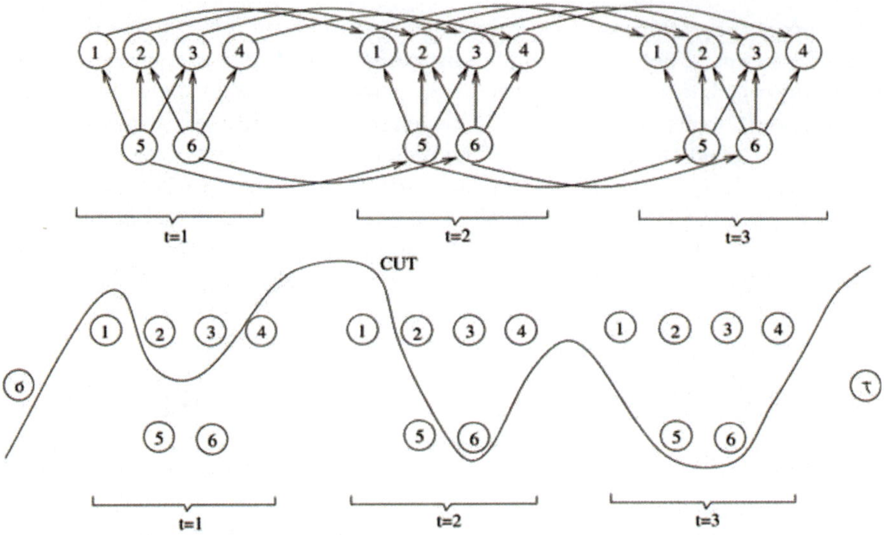

Fig. 9.15 A time-expanded mine graph encompasses three time periods and the minimum cut determining the mine schedule in the proposition of Amankwah, Larsson, and Textorius (2014) [141]

production) and scaling variables (tons or ounces). Smith (1999) proposed a mixed-integer goal programming model for optimizing inventory stockpiles at the mill and the mine and short-term production scheduling of a phosphate mine [60, 144–146].

Rehman and Asad (2010) presented a MILP model to address the short-term scheduling problem with blending optimization of a cement industry instance, whose objective was to minimize the total cost of supplying raw materials and were considered raw materials quantity and quality constraints. The model was applied to a limestone quarry and cement manufacturing operation in Pakistan, and significant cost savings were obtained by comparing the results of schedules produced by the trial-and-error methodology previously applied in this operation. Ben-Awuah and Askari-Nasab (2011) proposed a mixed-integer goal programming to address the mine planning of oil sands encompassing different material types, defining the extraction sequence of dyke material and waste from a predetermined ultimate pit that maximizes the NPV of the operation. The formulation relies on binary integer variables to handle precedence and continuous variables to handle ore and dyke material exploitation, being necessary to set up the deviational goal variables, penalty costs, and priorities. The model is flexible since trade-offs can be made between reaching a goal and maximizing NPV, and the case study provided a schedule consistent with regulatory requirements and the exploitation of the oil sands deposit. Askari-Nasab et al. (2011) formulated a MILP model to solve the short-term mine sequencing, aiming at minimizing the costs incurred by mining and haulage, processing, waste rehabilitation, and rehandling, while they sequenced the block extraction over multiple short-term periods. The model addresses multiple destinations, processes, waste dumps, and stockpiles, and the algorithm decides on the period, destination, and route that ore and waste must follow to minimize costs while meeting all the mining and processing requirements. A case study of monthly short-term planning of an iron ore deposit, with 3089 blocks on 4 benches aggregated into 150 mining cuts, was carried out over a year to validate the model [147–149].

Eivazy and Askari-Nasab (2012) developed a multi-destination MILP model to solve the short-term mine sequencing that minimizes the overall cost in mines, such as mining, processing, haulage, rehabilitation, and rehandling costs. Solved by a branch-and-cut (BC) algorithm, the model controls technical constraints while incorporating blending and buffer stockpiles, mining in a preferred horizontal directional and ramp decisions. Three scenarios with distinct horizontal mining directions in an iron ore mine were accessed considering the monthly production schedule over 12 months. The scenario that encompasses the minimum amount of drop cuts is chosen as the superior practical mine plan. L'Heureux, Gamache, and Soumis (2013) developed four strategies to solve a MIP model addressing the short-term production scheduling problem that incorporates shovel movement and block selection for each blast, with periods ranging from days to three months. The objective function minimizes operating costs, such as shovels, drilling, and blasting, while it imposes many constraints. The strategies consist of (1) setting certain variables to zero and then specifying priorities for the branching step, (2) adding a discount on the objective function to eliminate identical solutions, (3) imposing a minimum number of displacements for each shovel, and (4) adding constraints to define integer values more quickly and reinforce the links between variables. The strategies were applied together or individually, and improvements in the number of nodes, gap, and computational time were discussed. Yavarzadeh, Abodallheisharif, and Neishabouri (2014) proposed a MILP model to minimize the deviations from the average grade at a short-term open-pit production schedule, whose restrictions were related to the grade blending, reserve, processing and mining capacity, geometry, and the number of faces. The algorithm was successfully applied to define the monthly and daily plans of a copper mine in Iran [150–152].

Mousavi, Kozan, and Liu (2015) proposed a MIP model to maximize the profits from an hourly based period production scheduling considering precedence, mining and processing capacities, grade control, and stockpiles. The authors discussed the intractability of problems with many blocks by MIP and constrained programming (CP) formulations, comparing the results achieved by both methodologies in 25 examples. In addition, they stated the inefficiency of the CP model to solve the problem in a reasonable time. Kozan and Liu (2018) formulated a MIP model to solve a multi-resource and multistage sequencing problem to optimize the drilling and blasting simultaneously, and excavate operations in an open-pit mine under equipment capacity constraints to maximize throughput and minimizing total equipment inactive times at each stage. The model approached an iron mine in Australia and considered as mining attributes and constraints the features of equipments, the length between each pair of blocks, the expected windows of blocks and the interdependency between movement times of equipment and operating times of blocks. Flores-Fonseca, Linfati, and Escobar (2021) proposed a MILP approach to solving the operational problem that consists of defining the optimal extraction sequence of blocks considering mining shovels. Five shovels were accessed in three structured instances, and a comparison of the Gurobi and IBM CPLEX solvers was performed, reaching the first smaller gaps within the defined time limit (3600 s) than the second [153–155].

Owing to the great intricacy of the production scheduling problem, many heuristic methods have already been employed to search for feasible solutions. Even if some of the proposals already mentioned are based on rigorous methods, it is common that at some point, a relaxation of restrictions, a simulation stage, or a local search strategy is incorporated into the algorithms; otherwise, many of the problems studied would be unfeasible. Gershon (1985, 1986) introduced a heuristic procedure for production scheduling, which involves solving a series of single-period blending problems (such as LP formulations) by choosing regions for exploitation in each period to optimize a predefined long-range objective. Gershon (1987) proposed another heuristic approach, which calculates a desirability index (positional weight) for each block to be mined, based on the deposit's quality and position, and extracts the most desirable blocks and the preceding ones included in a downward cone, thus recalculating the positional weights and repeating the entire procedure. Wang and Sevim (1992, 1993, 1995) developed a heuristic that eliminates the gap in the pit sequence by generating a sequence closer to the desired increment, and although it does not guarantee the optimal succession of nested pits, it requires less time than demanded by the parameterization method. The heuristic considers that finding the maximum metal pit in a block model is equivalent to removing the minimum metal blocks. A downward cone explores the blocks, and a ranking in ascending order of the cones found is developed considering their average metal grade. Furthermore, a pit is built by unifying the cones with lower metal tonnage, which are eliminated from the block model, and then a new iteration begins in the remaining blocks so that different pit shells are created until the end of the process. Figure 9.16 presents the creation process of maximum metal pits using downward cone models [91, 93, 156–160].

As mentioned earlier, Denby and Schofield (1993, 1994, 1995) introduced GO-PIT, which can optimize the ultimate pit and the mine sequence using a GA approach, applying the methodology to 2D examples and incorporating risk assessment in the process. Latter, Denby, Schofield, and Hunter (1996) improved the GO-PIT to incorporate 3D analysis, and Denby Schofield and Surme (1998) subdivided the scheduling problem into two elements, one related to the development of a 3D GA to evolve a solution and the other related to the assessment of the suitability of specific solutions. Sevim and Lei (1996) proposed a heuristic algorithm capable of simultaneously determining the mine sequence, the ultimate pit limits, the production rates of ore and waste, and the mine lifetime, also allowing the determination of the best milling cutoff grade. The methodology first considers the generation of a block model constrained by the orebody boundaries. Thus, a bounding algorithm determines the largest feasible pit and discards those blocks that are outside. The second stage consists of assuming a set of processing cutoff grades and defining each one using a heuristic succession of nested pits containing the largest amount of metal among the feasible pits having the same size, a sequence named maximum metal pits.

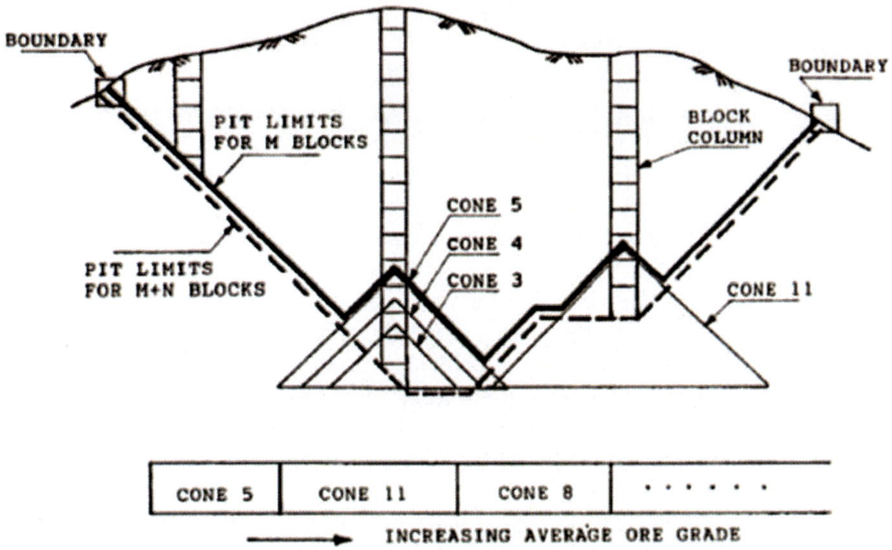

Fig. 9.16 Creation of maximum metal pits using downward cone models [159]

In the third stage, all viable pushbacks from these pits are formed and evaluated using the NPV as a parameter, and finally, a search is done to find the superior sequences of operational slope pits, resulting in a final slope pit when sufficient iterations are performed. In other words, for all assumed cutoff grades, after the creation of a series of operational slope maximum metal pits within a series of broad final slope pits, the algorithm builds networks of all viable sequences and then searches the one with the highest NPV [83–88, 160, 161].

Kumral and Dowd (2002, 2005) introduced a simulated annealing (SA) approach to optimize production scheduling. First, an LP structure is employed to recognize blocks that can be blended without infringing constraints, in addition to defining the ultimate pit limit using the LG algorithm. Hereafter, a two-stage mine sequencing optimization is performed, the first being a Lagrangian parameterization stage, which seeks sub-pits inside the ultimate pit that meet the tonnage constraints, resulting in a suboptimal initial solution. Then, a multi-objective SA is performed, whose objectives are to minimize deviations from the expected tonnage, alterability, opportunity, and penalty costs for each content variable, further improving the suboptimal schedule. The approach was demonstrated in an iron ore deposit in Western Australia, where the SA stage improved the Fe recovery and reduced the silica content; besides, the cost of the first objective has not been decreased, and the costs addressed on second and third objectives were reduced in the ten periods analyzed in comparison with the results of the Lagrangian parameterization. Samanta, Bhattacherjee, and Ganguli (2005) proposed a GA to optimize the grade control planning over 24 months in a 98-layer bauxite deposit to minimize silicon and aluminum oxide deviation targets subject to sequencing constraints. Zhang (2006) employed a GA approach to aggregate blocks aiming to diminish the size of the scheduling problem, comparing the performance of his proposition with that obtained by CPLEX when solving BHP Billiton instances, with or without aggregation, and finding that his procedure was two to four times faster in obtaining solutions with the same quality [5, 32, 162–165].

Ferland, Amaya, and Djuimo (2007) proposed a PSO procedure to solve a variation of the mine sequencing problem aiming to maximize NPV subject to capacity constraints in each assessed period. They introduced a greedy randomized adaptive search procedure (GRASP) stage to produce an initial population (swarm). They thus evolved it according to a particle swarm procedure to search for the viable domain of the solution's genotype representations, which are inspired by priority value encoding. The computational tests indicate that increasing the quality and size of the initial population increases the solution generated, and the results seem to be quite insensitive to the values of the parameter of the particle swarm procedure. Sattarvand (2009) proposed five ACO methodologies to solve the long-term mine sequencing problem. Various programming variables (number of planning periods) were considered for each block as the pheromone trails, which symbolize the attractiveness of a blocked rest in the lowest location of the pit in that column for a designated mining period. The pheromone trails' values are thus initialized based on a mine sequence produced by the LG algorithm and Wang-Sevim's heuristic (1995), receiving relatively high values for blocks that are nearby the lowest location of each pushback, which helps in the construction of a series of random schemes that are relatively close to the initial solution. Several mine schedules are built in each iteration of ACO based on current pheromone trails, which are updated as evaporation occurs and deposition of new pheromone in each block is guided by the quality of the solutions generated so that those with high probabilities of being the deepest blocks in their columns in given period tend to have their pheromone level increased. The analysis of a hypothetical iron ore deposit with 1000 blocks by the ACO methodologies revealed that the initial mine sequence improved up to 34%, demanding a sufficient computational time. Moreover, the max–min ant system (MMAS) algorithm proved to be the most exploratory variant, while the ant colony system (ACS) was the fastest method. Shishvan and Sattarvand (2015) extended this work by accessing the results produced by these two ACO methods in a real size copper–gold deposit, improving the value of the initial mine sequence produced by the conventional algorithms up to 12% demanding an adequate computational time and satisfying at least 90% of the planning targets [32, 92, 160, 166, 167].

Myburgh and Deb (2010) proposed a GA called evolution capable of solving complex multi-objective capacity constraint production scheduling problems for large open-pit iron ore mines, encompassing hundreds of thousands to millions of variables. Encoding each block as unique integers to define the three-dimensional location in the pit, the mining sequence (chromosome) is represented by a permutation of all block numbers from left to right. Ore and waste targets must be specified, and violations achieved with a schedule are calculated by period. The genetic operators used include a standard binary tournament selection, custom crossover, and mutation operators designed to emphasize each best current solution without losing mutations' diversity. Khan and Niemann-Delius (2014) compared the results of implementing continuous PSO and guaranteed convergence PSO (GCPSO) methodologies to solve the long-term mine sequencing problem. Taking the ultimate pit and the production schedule obtained through an IP formulation, the suggested approach converts the problem into a depth search problem instead of scheduling a block model. At the beginning of the PSO approach, to create an initial set of random feasible solutions, a GRASP process is implemented, followed by an encoding scheme that determines, through one of the PSO algorithms mentioned, the lowest block over a given column to be exploited in a given period. Hence, a back transform procedure is employed to define the period assigned to each block after each iteration. Thus, two constraint handling techniques are used, a solution normalization to back transform infeasible solutions into a feasible one considering the required slope angle and a penalty method for capacity constraint violation. In addition, three PSO topology variants were compared in continuous PSO and on GCPSO, as a global best (Gbest—the best particle of its immediate *K* neighbors influences each particle), a Gbest with a multi-start procedure, and a local best (Lbest—each particle is influenced only by the best particle of its immediate *K* neighbors) strategy. Numerical experiments on two different hypothetical copper deposits have shown that Gbest is more likely to be trapped in an optimum local location, and Lbest is relatively slower in its convergence rate. However, the multi-start strategy overcame the early convergence problem of the Gbest alternatives of PSO and GCPSO algorithms and generated better solutions with minor gaps. Khan (2015) extended this work by assessing the implementation of the bat and modified bat algorithm (BA and MBA), both of which required less computational time and iteration to converge, although they presented more inconstant convergence demeanor than the PSO algorithm. A differential evolution (DE) algorithm was also compared to the other metaheuristics, which proved to be a more robust solution, achieving better results in the percentage of the gap, although the time required was generally longer than the other methodologies. Figure 9.17 presents examples of PSO topologies Gbest (a) and Lbest with $k = 2$ (b) and $k = 4$ (c) [168–170].

Asl and Sattarvand (2016) presented an imperialist competitive algorithm (ICA) to solve the ultimate pit and long-term mine scheduling problems. Inspired by the

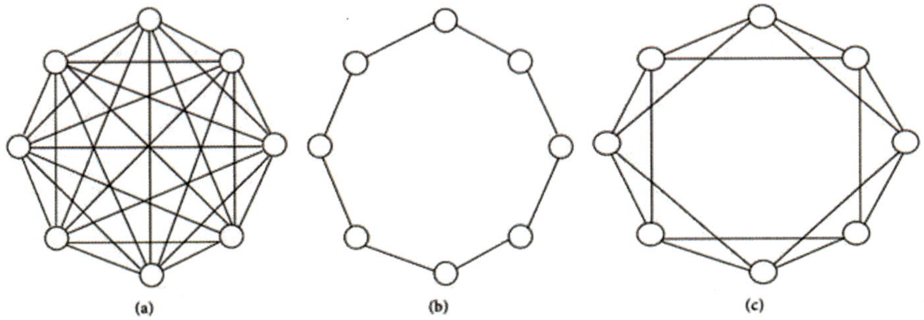

Fig. 9.17 Topologies Gbest (**a**), Lbest with $k = 2$ (**b**) and $k = 4$ (**c**) [169]

sociopolitical process of imperialism, the initial population of solutions (countries) are divided into imperialists and colonies according to the quality of the objective function in each solution. In contrast, the group of an imperialist and his colonies is named empire. As the imperialists attempt to spread their characteristics to the related colonies over time, some revolutions can happen, and colonies can leave their empire for others if there is a greater chance of promotion. Thus, the initial population contains a series of solutions (production schedules) of which the most mighty (in terms of NPV) are chosen as imperialists.

In contrast, the others depict colonies randomly partitioned among the imperialists considering their strength so that the empires with the greatest strength attach more colonies. An array scheme was used in order to encode the feasible solutions, and the ICA assimilation process was altered by randomly choosing a sub-array in the imperialist array and replicating it to the colony array, whose modified array generally does not respect the slope angles and block precedency, demanding a normalization process. A revolution operator precludes the ICA from being trapped in local optima by replacing the weakest colony in each iteration with a new random one, and if a colony reaches a location with a reduced cost function than its imperialist, they change their positions. The ICA's objective is to minimize the sum of the cost function value of all countries or minimize the negative NPV, and generally, the empires attempt to assume control of the colonies from other empires, diminishing the power of the weakest empires until the most powerful one prevails. The algorithm was implemented in a 2D hypothetical 200-block copper deposit, and Fig. 9.18 presents the assimilation operator for the production scheduling problem and the normalization process of the assimilated colony [171].

Alipour et al. (2017) implemented a GA to solve production scheduling problems and compared its performance to a linear formulation in a 2D hypothetical 200-block copper deposit. In order to maximize NPV subject to precedence, mining, and processing capacity, the GA advances to start a population of solutions, thus upgrading them using multiple iterations of operations, such as generation based on a fitness function, mutation, crossover, and normalization. The numerical experiment has shown that the gap between the GA and the rigorous solutions is less than 5%. Alipour et al. (2020) extended their work by proposing a GA approach to solve 3D open-pit production scheduling problems, implementing their solution in the Marvin orebody dataset, encompassing 53,271 blocks. As CPLEX was unable to rigorously solve the programming running for 25 days, the authors used the commercial DBS software SimSched for comparison so that the GA reached an NPV superior by 4%, although the execution time of the DBS algorithm was 27 times faster [81, 172].

Ramezanalizadeh et al. (2020) introduced an artificial bee colony (ABC) formulation to generate a block sequence in a multi-objective MIP model that regards the composition of waste rocks to consider the potential acid formation (PAF) rocks. Therefore, aiming to maximize the NPV and minimize the consequences of disposing harmful waste from sulfide lithologies, as rehandling costs and acid mine drainage, enclosing the PAF rocks by non-acid formation (NAF) lithologies is regarded in waste disposal operation. The main rock dump must be filled in from the beginning so that NAF blocks surround all PAF blocks to avoid the need to deposit NAF blocks

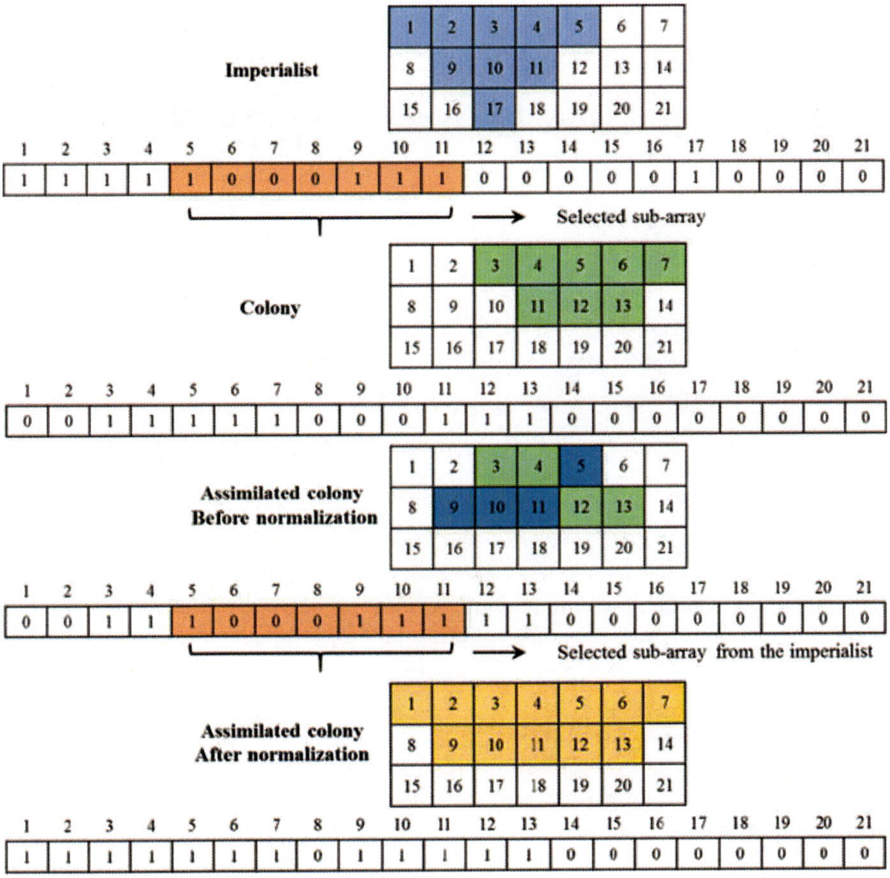

Fig. 9.18 Assimilation operator for production planning problem and the normalization process of the assimilated colony [171]

extracted at an inappropriate time in the temporary stockpile and later re-handle for the main dump. Therefore, a constructive ABC heuristic is employed to produce the first feasible solution, and then, a greedy mechanism is implemented to improve the solution by local movements. The foraging behavior of bees inspires the ABC metaheuristic, and a bee colony consists of food founts, employed and unemployed bees, in which the location of a food fount indicates a viable solution and its value depicts the adequacy of the solution.

In contrast, an employed bee is designated to each food fount. The unemployed bees are separated into observer and explorer bees, and while the former try to encounter a food fount using the information gathered by employed bees, the latter randomly investigate the territory to encounter new food founts. Computational experiments in three instances have shown that this ABC approach achieved solutions with a gap of less than 2% compared to the CPLEX solution, in addition to

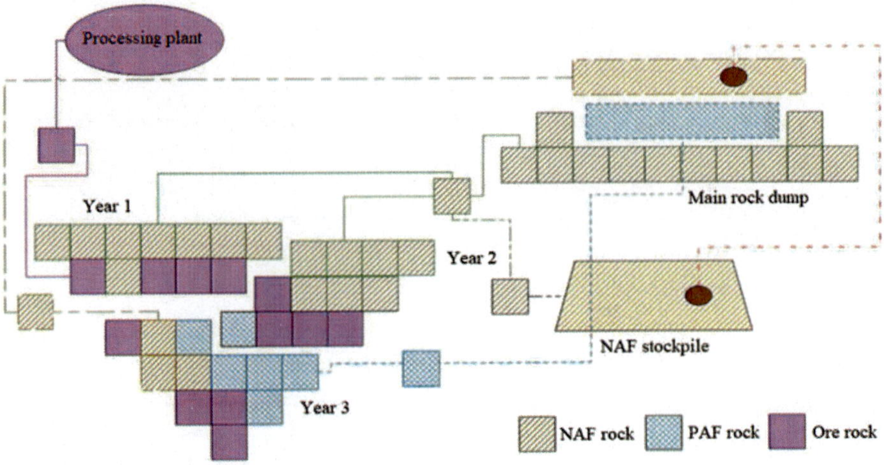

Fig. 9.19 Production scheduling problem considering the management of potential acid formation rocks [173]

the two instances whose solutions could not be found by rigorous methods, while the ABC methodology has found it. Figure 9.19 shows the production scheduling problem considering the management of potential acid formation rocks [173].

Although many of the propositions mentioned above combine different methodologies, such as heuristics, rigorous algorithms, or both, into one solution, they have been grouped into the last topics considering their main optimization technique. Other propositions are classified as hybrids, chiefly in combinations of rigorous and heuristic solutions to solve short-term production scheduling problems. The formulations that combine rigorous and heuristics to solve long-term mine sequencing problems will be presented in the next topic, reserved for the DBS approach, unless the cases are scheduled after obtaining the ultimate pit or any pushback handling. Pendharkar and Rodger (2000) proposed a combination of nonlinear programming (NLP) and genetic search to address problems arising from the coal industry. The main objectives of the model are to determine the schedules of production, transportation, and blending, and furthermore, the selection of markets so that the objective function was defined as maximizing the discounted net profit before taxes subjected to capacity, quality, and demand constraints. Like other GA propositions, the population size was defined, and the chromosomes were subject to crossover and mutations, the approach being tested in a hypothetical coal model. Souza et al. (2010) proposed a hybrid algorithm combining GRASP and variable neighborhood search (VNS) heuristics to optimize the production schedule of multi-pit MIP models based on goal programming. Considering a time span of one hour, the objectives are to minimize deviations from production and quality targets and diminish the number of trucks demanded operation by including shovel and truck allocation constraints. Eight movements representing neighborhoods have been developed, such as movements of the number of trips, load, relocate trip from a truck, relocate trip from a pit,

pit operation, truck operation, swap trips, and swap shovels. An initial solution is built in two stages, the first of which consists of allocating the shovels and distributing trips to the waste pits using a greedy heuristic, while the second consists of defining these items for the ore pits by means of GRASP. Hereafter, the proposed hybrid algorithm evolves the local search for a variable neighborhood descent (VND) procedure until, in a given number of iterations, there is no improvement when a mechanism is employed to "shake" the search for a random neighborhood from the list of movements mentioned. Limiting the execution time to two minutes, the methodology was tested on dataset of an iron ore mine company in Brazil, and a validation was carried out comparing the results to those obtained when using the CPLEX optimizer with the same data. The computational experiments demonstrate that the hybrid algorithm finds near-optimal solutions in most instances with a gap of less than 1%, requiring short computation times [174, 175].

Tabesh and Askari-Nasab (2011) introduced a two-stage clustering algorithm for aggregating blocks in order to diminish the size of MILP formulations for long-term production scheduling problems. An agglomerative hierarchical algorithm is applied in the first phase, aggregating blocks into selective mining cuts established in a similarity index relying on ore grades, rock types, and distances between blocks. The second phase consists of a Tabu search (TS) process that tries to decrease the number of arcs (precedence) between new clusters and those of the bench below, diminishing the binary resolution variables. This methodology is guided by an upward fashion, in which each bench is aggregated individually and improved by TS, so the algorithm moves toward the overhead bench. A case study was developed at an iron ore mine, the results of which showed that the size and shape of the clustered blocks are responsible for 10–15% of the NPV, also inflicting in the practicality of the project. Tabesh and Askari-Nasab (2013) expanded their work by developing an automatic procedure for designing mining polygons based entirely on hierarchical clustering. Their new algorithm addresses mining cut dimensions as required for different periods, projecting homogenous shapes in rock types and grades, making it possible to evaluate the dilution and quality of the material supplied to the plant, also considering the preferential mining direction [176, 177].

Tabesh, Mieth, and Askari-Nasab (2013, 2014) developed a hybrid methodology to solve MILP models, including a controlled optimal open-pit phase design, the characterization of selective mining cuts, and the optimization of long-term mine sequencing from a predetermined ultimate pit. First, a binary IP pushback design algorithm that benefits from the structural properties of the PCKP mathematical formulation and uses Lagrangian relaxation and heuristics to find near-optimal pushback design in large size models is employed, aiming to maximize the mine's profit while controlling the tonnage of ore and waste within each pushback. Subsequently, a clustering approach based on Tabesh and Askari-Nasab (2011, 2013) is employed to generate mining units within these pushbacks with controlled features that, together with the bench phases also generated within the pushbacks, serve as planning units in the MILP formulation, which generates production schedules that maximize the NPV and are practical in terms of operation. The performance of the proposed methodology in two case studies of real size iron ore deposits showed that the pushbacks generated are uniform in terms of tonnage and more practical in the form of the minimum mining width compared to the parametric pushback design approach. Furthermore, the NPV generated in the smaller deposit increased by 2–4% compared to the traditional approach. Tabesh, Askari-Nasab, and Peroni (2015) extended previous works, including stockpiling in the MILP formulation to optimize blending requirements, and proposed a nonlinear and linearized alternative to the original formulation. Figure 9.20 presents the flowchart of the hybrid algorithm suggested by Tabesh, Mieth, and Askari-Nasab (2013, 2014) [176–180].

Blom et al. (2014) proposed an MINLP model to address the short-term production scheduling problem for a network of multiple open-pit mines and ports, proposing a decomposition-based heuristic to solve it. The problem requires a collaborative adjustment of the grade and quality targets of each mine to ensure that the blending constraints at each port are met, thus composing a one-week extraction plan whose objective function is to minimize the deviation of the composition of the port's products in relation to the customer's requirements. Moreover, the MINLP model is decomposed into smaller MIP models representing each mine and port entity in the network. Thus, the decomposition-based heuristic first solves the mines and mine-to-port transport and blending in a sequence of iterations. Therefore, through a sequence of improvements, each mine is led to a solution that maximizes the

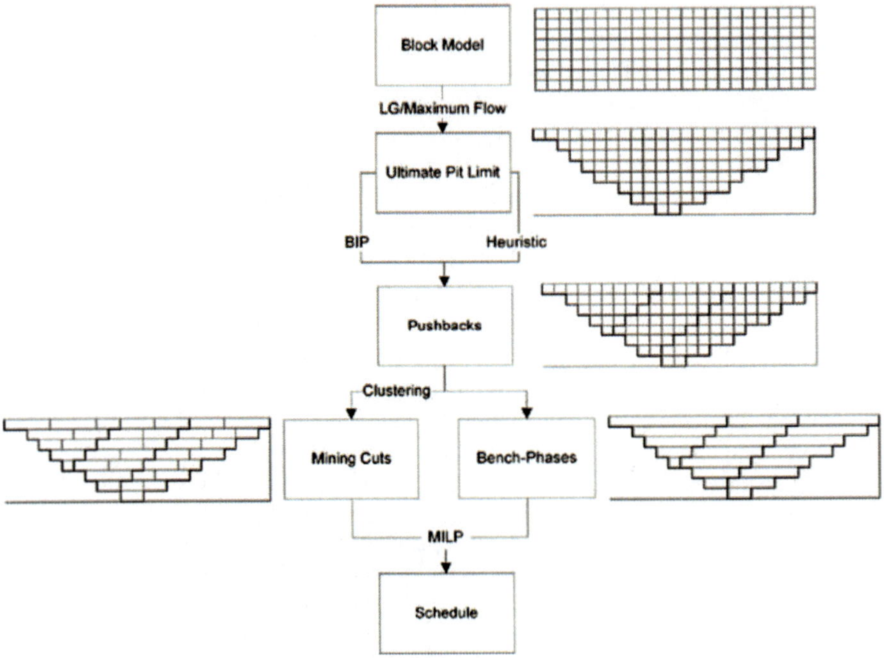

Fig. 9.20 Flowchart of the hybrid algorithm suggested by Tabesh, Mieth, and Askari-Nasab (2013, 2014) [179]

revenue while enabling each port to blend its products correctly. A case of a network with eight mines and two ports was assessed, and the results were compared with those obtained using a piecewise-linear relaxation and the ALT heuristic, in which the proposed framework achieved equal or better solutions in terms of computational time spent for the most tests. Blom, Pearce, and Stuckey (2016) extended the previous work by adding multiple periods to the short-term production scheduling problem for mine and port networks, incorporating additional constraints to the MINLP model, expanding the decomposition-based algorithm, and solving the MIP subproblems using a rolling horizon heuristic [181, 182].

Mousavi, Kozan, and Liu (2016a) proposed two-hybrid BB and SA algorithms (one includes a large neighborhood search, LNS) to solve large instances of the short-term open-pit production schedule problem, considering block sequencing, precedence, mining and processing capacity, grade control, and stockpile management. Unlike other neighborhood searches, LNS employs a destroy-and-repair mechanism to ruin part of the solution and then rebuild a new one, which in the case of mining scheduling it consists of destroying the destinations assigned to blocks and reallocating new destinations by BB method. Computational experiments compared the performance of a MIP formulation of the problem using CPLEX and both heuristics proposed in several instances. In cases where rigorous solutions are obtained, heuristics also find the same solutions or deviate less than 1% on average of CPLEX solutions, especially where large neighborhood searches are employed. On the other hand, there are instances in which rigorous solutions could not be found, though heuristic solutions have been proposed. A case study was proposed in a real iron ore mine considering 12 periods of two weeks, and the results obtained and sensitivity analysis are carried out. Mousavi, Kozan, and Liu (2016b) proposed three hybrid metaheuristic combinations for MIP models to allow solvers to find feasible solutions for large instances of short-term mine sequencing problems combining SA, TS, and both. The objective is to minimize the costs of stockpiling, rehandling, and holding costs, and the constraints include stockpiling, blending, machine working space and allocation, and dynamic destination assignment. The model answers for each of the input blocks the extraction period, the destination, and the allocated excavator, and the time horizon considered has a length of three months, whose periods can be from one day to a week. Making use of the SA's ability to find better solutions even when occluded by a non-improving solution and the capability of TS to avoid returning to preceding recording solutions, the computational experiments in 24 instances showed that hybrid algorithms are capable of achieving the same lower bound that CPLEX did for most cases, even finding better results for some instances. Generally, SA achieves superior results than TS, albeit TS is faster than SA, while TS-SA formulation combines individual features and provides better and more reliable solutions, especially for large instances [183, 184].

Maiti, Pathak, and Samanta (2020) proposed a specific breakpoint algorithm (SBA) to search breakpoints of problems related to parametric maximum flow, being used to solve relaxed Lagrangian PCKP and linear programming relaxed PCKP (LRPCKP and LPRPCKP) for determining the design of pushbacks. The monolithic PCKP model consists of a single pushback, which is a reduced LRPCKP, in which the objective function becomes a linear piecewise convex min-cut function of a graph, which can be resolved through a parametric maximum flow approach, being the SBA employed to search for related breakpoints of the min-cut function efficiently. These pushbacks can be generated by two approaches, the Lagrangian relaxation approach by Byun and Dimitrakopoulos (2013) and the linear relaxation approach by Chicoisne et al. (2012). Since the relaxed solutions must be transformed into feasible integer solutions (0-1), a new heuristic called rounded top sort has been developed for this purpose. Computational experiments in seven MineLib instances for both approaches indicate that LRPCKP-SBA is faster, though it produces lower-quality solutions than the Bienstock–Zuckerberg (BZ) algorithm and CPLEX solutions reaching optimality gaps of 4–71%.

Nevertheless, LPRPCKP-SBA produces solutions comparable to BZ computationally more efficiently, reaching optimality gaps of 2–31%. Furthermore, the rounded top sort heuristic produces better feasible solutions operating in relaxed solutions than other existing techniques, such as the expected-time top sort heuristic from Moreno, Espinoza, and Goycoolea (2010). Most of these papers will be better explained in the DBS topic [185–189].

The last class of optimizers that solve the production scheduling problem to be presented in this topic is simulation optimization, which consists of finding the best values of the input variables among all the possibilities for a predetermined simulation scenario without evaluating each possibility explicitly, thus minimizing the resources spent and maximizing the information obtained in a simulation experiment. Fytas, Hadjigeorgiou, and Collins (1993) also proposed a simulation structure to assess the long-term mine sequencing problem, consisting of a series of subroutines to simulate mining and waste stripping operations, in order to guide the scheduling procedure through periods to achieve a solution that meets the defined restrictions. The model aims to maximize the revenue of each mining period and encompasses hematite and the weight recovery of an iron mine, defining as constraints the tonnage of ore and concentrate, blending targets, uncovered ore tonnage at the beginning of the period, remaining ore tonnage, and waste volume in each bench, the number of benches per period and stripping capacity. A smoothing process is performed in the pit contour generated by simulation at the end of each period to generate practical and viable sequences, followed by a cash flow analysis model. Frimpong, Asa, and Szymanski (1998) and Frimpong, Asa, and Suglo (2001) developed the MULSOPS multivariate pit shell simulator to solve medium-term production schedule problems addressing random processes of the operation. The optimized pits are produced using the MCS/MFNN algorithm proposed by Frimpong and Achireko (1997). Then, the tactical pit shells are modeled as 3D extraction ellipsoids that extend in all directions above the optimized pit contours, limited by the projected pit wall slopes and mimicking their expansions and sequential interactions. Stochastic and numerical modeling techniques address the multivariate random states and provide feasible solutions to the time-dependent geometric models, considering distinct production and economic parameters. Therefore, the models are simulated to reproduce the proper features of the mining cuts, while the variance simulation is employed to

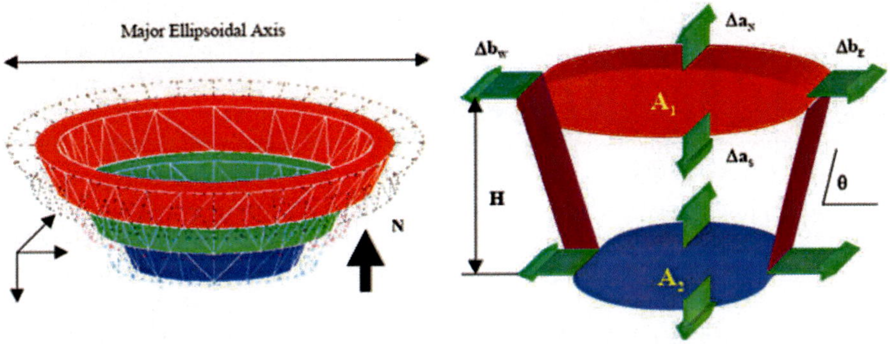

Fig. 9.21 Geometry of the pit shell frustum forming benches and the open-pit expansion dynamics employed by the MULSOPS [191]

supply sensitive stochastic variables to define the input data and tighten the production target bounds. The MULSOPS was used to optimize the tactical schedule of a gold mine for three years, and results are presented. Figure 9.21 presents the geometry of the pit shell frustum forming benches and the open-pit expansion dynamics employed by the MULSOPS [74, 121, 190, 191].

Askari-Nasab, Awuah-Offei, and Frimpong (2004) proposed the open-pit production simulator (OPPS) to solve the tactical production scheduling problem and simultaneously capture the random field and pit expansion dynamics, determining the extraction time and destination for ore, waste, and stockpile materials during each pushback. The algorithm combines the geometrical formulation of open-pit, stochastic simulation, numerical solution of a set of partial differential equations, and a search scheme to return the monetary value in open pit at any specific time instance. Incorporating a geometric pit evolution model, the OPPS employs Taylor series expansion over the dimensions of the geometry of the initial ellipsoidal frustum to provide numerical solutions to the resulting partial differential equations that govern the pushback dynamics so that the sequence that provides the highest NPV is identified. Furthermore, this hybrid stochastic optimization algorithm makes it possible to provide an uncertainty assessment for the sequencing and the results of each scheduling realization, as presented in the computational experiment performed in the Bingham Copper Mine. Askari-Nasab and Szymanski (2005) describe in detail the discrete and continuous simulation processes that occur internally in OPPS, where equally probable realizations are generated through discrete Monte Carlo simulations and only that sequence that maximizes the project's NPV will be subjected to a continuous simulation to describe the process on frequent targets within a shorter time framework. The discrete modeling of open-pit dynamics allows the evolution of pit geometry similar to a sequence of snapshots, managing data representing a finite number of attributes in a finite number of states usually modeled as algebraic equations.

On the other hand, continuous models address a physical problem through partial differential or ordinary equations, which generally cannot be solved by analytical solutions, requiring approximate methods to solve them, such as finite differences, finite elements, finite volumes, and boundary elements. Usually, these methods divide the studied region into a series of small elements, thus generating a discrete approximation of those differential equations for each element and solving them concurrently. Continuous modeling of open-pit dynamics involves characterizing the changes made to the mine geometry and the volume of materials moved by a system of ordinary differential equations, which is solved in OPPS by explicit Runge–Kutta integration scheme, thus providing a practical guide for the short-term production plan. Askari-Nasab and Szymanski (2007a), and later Askari-Nasab, Frimpong, and Szymanski (2008), proposed a case study at an iron ore deposit, in which the LG algorithm and discrete simulation determined the final pit was used at different mining starting points, and then the realization that maximizes NPV served as an input for continuous simulation, reaching a value of 98% of the NPV achieved by LG [32, 192–195].

Askari-Nasab and Szymanski (2007b) introduced the intelligent open-pit optimal production simulator (IOPS) to solve the long-term mine sequencing problem, based on reinforcement learning and the OPPS simulator to address the dynamic processes and random fields. Reinforcement learning can be described by the interaction of the agent and environment entities, in which the first seek to learn and make decisions from the second, while the simulation entity mediates the interactions between both. The agent receives the sensory input from the environment and produces output actions that affect it. In the IOPS framework, the environment is the geological and economic block models, the simulator is the OPPS structure, and the agents are the cash flow of each simulated period outputted by OPPS. The simulated results are transferred to the IOPA so that the long-term scheduling is formulated as a dynamic decision network, where the agent reacts to the open-pit environment by means of simulation and uses a Q-learning algorithm as an engine to maximize the NPV. Therefore, the algorithm starts with the interaction between the environment and the simulation entities. Hence, the OPPS stochastically simulates that the economic pit expansion model returns the next feasible pushbacks with cash flows. The IOPA then employs Q-learning and an epsilon-greedy algorithm to choose the next suitable pushback among all the simulated results, updating the action-value function so that the simulation continues until the simulated pushback design reaches the final pit limits. Askari-Nasab and Awuah-Offei (2009) investigated through a case study using the IOPS the validity of the theorem presented by Caccetta and Hill (2003), who declare that a final pit obtained through a long-term mine sequencing is always a subset of the traditional ultimate pit produced by the LG algorithm.

Moreover, whether the optimal sequence is known, the exploitation time of each block is also known. Therefore, a discount rate can be applied, and the present values of the blocks can be discounted. Using IOPS to define an optimized pushback sequence before defining the ultimate pit contour, this sequence was used to build a discounted economic block model, serving as an input to the LG algorithm. The comparison of discounted and undiscounted models revealed the discrepancies between the two contours, which overlap in most areas, although the traditional ultimate pit model constricts the optimal long-term sequence. However, it is reasonable to define the ultimate pit contour and then schedule the resulting model, although it can overestimate larger deposit reserves with lower annual production goals and conduct the project's overcapacity [32, 90, 196, 197].

Direct Block Scheduling (DBS)

The traditional production scheduling depends on the generation of pushbacks with parameterization of prices and aggregation of the blocks within these pushbacks into benches to produce long-term mine sequences. There should also be an extraction and destination schedule for each block over a set of years. However, this incremental fashion has no success in incorporating operational requirements, such as multiple destinations, multiple capacity, truck hours, blending requirements, and stockpile. Furthermore, once mine sequencing usually relies on the pushbacks' design, bad contours would preclude sequences from reaching a maximum NPV or even obtain a viable solution. An alternative to the generation of pushbacks is the DBS methodology, where the blocks are individually evaluated as to their extraction (or not) and destination. At the same time, the mine production scheduling problem is solved under operational constraints, and the resultant annual sequences are clustered into phases for the pit design. However, DBS methods demand a high computational capacity to be resolved, since it is a formulation of the scheduling problem as an IP, as a MIP, or as MILP, being generally very large considering the number of restrictions and variables in the real application, which has been a disadvantage, despite their benefits. In recent years, the effectiveness of new technologies and algorithms increased the competitiveness of DBS, and new algorithms relying on IP and heuristics emerged with acceptable processing times. In this topic, all the IP, MIP, and MILP formulations that target the long-term production scheduling problem were grouped, even if there are eventually metaheuristics and hybrid algorithms between them. The algorithms that promote the relaxation of the original problems were also addressed here, as long as these formulations have been used in deterministic approaches since the stochastic approaches are mentioned later. Figure 9.22 presents traditional pushback methodology versus direct block scheduling [90, 198, 199].

Johnson (1968) pioneered his work on long-term production scheduling when he subjected each block of a model to linear restrictions of precedence, slope, blending, mining and milling equipment capacity, grade control, and other limitations, seeking to maximize profit in many periods. Although there was no possibility of solving a problem of this magnitude with modern computational tools, it was decomposed into subproblems easier to solve. Even today, solving this monolithic problem is computationally unfeasible when formulated with many restrictions and applied to very large block models. Smith (1978) used a MIP formulation to access the long-term mine sequence of an alluvial diamond deposit. Gershon (1982) formulated a generalized mine scheduling optimization model using MIP, called mine scheduling optimization (MSO), which maximizes NPV and provides optimal long-range plans for each

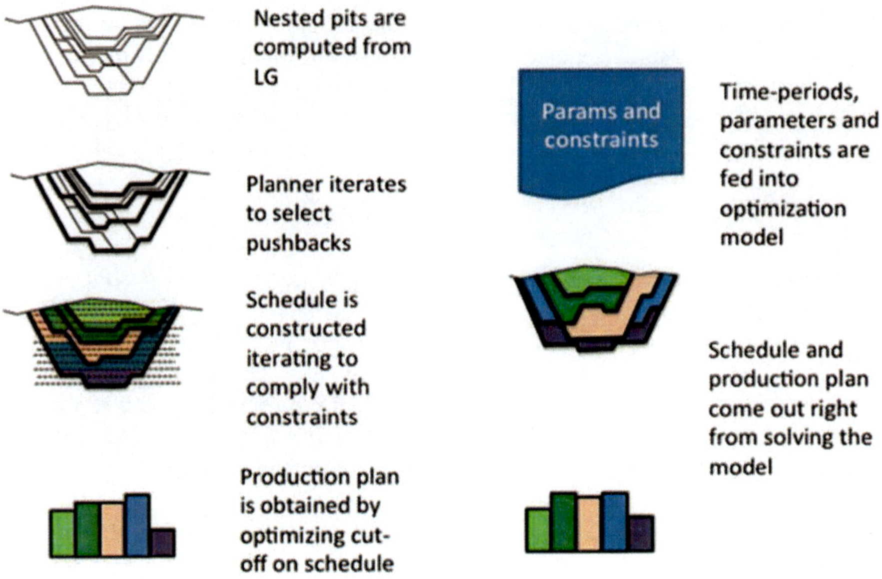

Fig. 9.22 Traditional pushback methodology versus direct block scheduling [199]

operation and has been successfully applied on copper, coal, and cement operations. Gershon (1983b) proposed a MIP formulation to optimize mine sequencing, mill blending simultaneously, and processing requirements, considering the interrelationships between these problems for long-, intermediate-, and short-term scheduling [46, 200–202].

Barbaro, Ramani, and Luckie (1982) proposed a MIP model to optimize the siting of a centralized preparation plant system and the production schedule for each of the multiple coal mines in the system that should supply the plant, aiming to minimize the total discounted cost of the system over the mine life. The requirements regarded encompass production capacities, demand requirements, transportation, blending, and preparation plant performance. At the same time, the model determines the best location of the preparation plant based on minimizing the total discounted cost, production levels for each mine, the clean coal quality, and the total cost of the system. Barbaro and Ramani (1986) presented a generalized multi-period MIP model to optimize the production schedule for multiple coal mines, select and locate new processing facilities, schedule the processed ore blending, and select markets. In addition, the model optimizes them by maximizing the total profit before tax, regarding the revenue from the supply of ore and the associated costs related to the production, haulage, blending, processing, and dumping of waste, being applied to a coal preparation plant system [203–205].

Dagdelen and Johnson (1986) applied the Lagrangian relaxation decomposition approach, where they introduced Lagrangian multipliers to decompose the complex multi-period IP model into reduced single-period models, which can be handled by applying ultimate pit or maximum flow algorithms. By relaxing the mining and processing constraints, adding them into the objective function, and inserting Lagrangian multipliers, the optimum schedule is achieved by adjusting the multiplier using the sub-gradient method. However, there may be no convergence toward an optimal solution when multipliers do not produce viable solutions regarding the constraints imposed, resulting in the gap problem. Klingman and Philips (1988) proposed a 0-1 IP model that incorporates cutoff factors for product grades and processing, composing a decision support system that helps decide which areas of each parcel to mine, which strata to process, and which strata to process should be separated into a phosphate reserve. Caccetta, Kelsey, and Gianinni (1998) applied this proposition in a real orebody for six time periods, obtaining a timeline solution within 5% of the optimal theoretical, but not considering the concept of dynamic cutoff grade in the scheduling solution [4, 57, 89, 206–208].

Akaike and Dagdelen (1999) applied the method named 4D network relaxation to an IP model, implementing a dynamic cutoff grade conception over the sequencing process and the capability to handle the stockpile option. Urbaez and Dagdelen (1999) developed the program "Mine Scheduler" capable of optimizing the NPV of the cash flow in MILP problems. Hoerger, Hoffman, and Seymour (1999) implemented the Newmont in-house production scheduling software "Optimizer," a MILP able to maximize life-of-mine NPV under various operational constraints for open cast and underground mines. Mogi et al. (2001) reconsidered the 4D network relaxation method aiming to diminish the consequences of the gap problem, though not eliminating it. Ramazan (2001) introduced the fundamental tree concept in his thesis, contemplated when there is a conjunction of blocks that (1) can be profitably exploited, (2) meets the slope requirements, and (3) when any subset of the grouped blocks does not meet 1 and 2. Applying this concept to a multi-mineral copper deposit in Peru, first, an LP formulation is used to define a series of fundamental trees in the deposit (decreasing the number of variables demanded sequencing). Then, a MIP formulation is applied to determine the yearly mine schedule from the fundamental trees, considering constraints such as mine production capacity, processing mill capacity, grade of ore, etc. The results point to a high NPV achievement using this methodology than using commercial schedulers. Caccetta and Hill (2003), in turn, presented a MILP model for the scheduling problems along with a BC solution approach, combining breadth-first and depth-first searches to propitiate a variety of feasible pit sequences. The constraints employed were the mining extraction sequence; operational requirements, e.g., maximum vertical depth and minimum bottom width; grade control; logistics; stockpiles; and mining, processing, and refining capacities. Computational experiments were carried out in six instances from 6720 to 209,664 blocks so that the smaller instance reached guaranteed solutions within 0.4% of the optimum in 12 min, while the larger instance obtained a solution with optimality within 8.3% in 20 h [4, 5, 90, 209–214].

Ramazan and Dimitrakopoulos (2003) discussed the size issues in complex multivariate deposits, such as iron and nickel laterite ores, in which both MIP and traditional optimization models may not be able to generate feasible schedules in terms of production requirements. Ramazan and Dimitrakopoulos (2004a) presented several approaches to formulating a MIP model, guiding effective long-term open-pit mine sequencing. They proposed an alternative to decreasing the solution time by defining certain variables as binary or linear and minimizing or even preventing partial block mining. The dependency of the model's tightness in terms of the dataset, objective function, and restrictions for the solution time was discussed, in addition to the size of the model, and it was verified that if the gap of the first feasible solution found were greater than 10%, that would occur a closing problem. Ramazan and Dimitrakopoulos (2004b) studied applying a MIP model to a polymetallic nickel laterite deposit considering the uncertainty of grades and thickness of the layers by means of equally probable simulated orebody models. They verified that the schedules generated, while being the direct MIP outputs or the smoothed models, are quite different and barely resulted in infeasible solutions, far from optimal when feasible. They proposed a modification to re-evaluate the stochastic schedules by maximizing the chance of blocks being excavated in a given period (whose aim was to maximize the NPV), followed by the smoothing for operational reasons. The resultant schedule presented less spreading over the mineralized deposit than a direct MIP, a very important operational issue, and dealt with geological and risk uncertainty while attending to all constraints. Figure 9.23 presents the comparison of four simulated sequences obtained by means of conventional MIP optimization and their smoothed models (left) and one sequence acquired using the suggested MIP approach and its smoothed model (right) [215–217].

Stone et al. (2004) developed a MILP-based software called Blasor to generate and optimize the ultimate long-range pit and schedule for an operation in Western Australia with multiple pits and blended ore. They introduced a process of aggregation of blocks preceding the scheduling by an LP approach due to the magnitude of the block models. To provide selectivity, binning is applied to exploit blocks with similar characteristics together by the optimizer employing a fuzzy clustering algorithm. Fricke (2006) developed strengthened IP and MIP formulations and presented

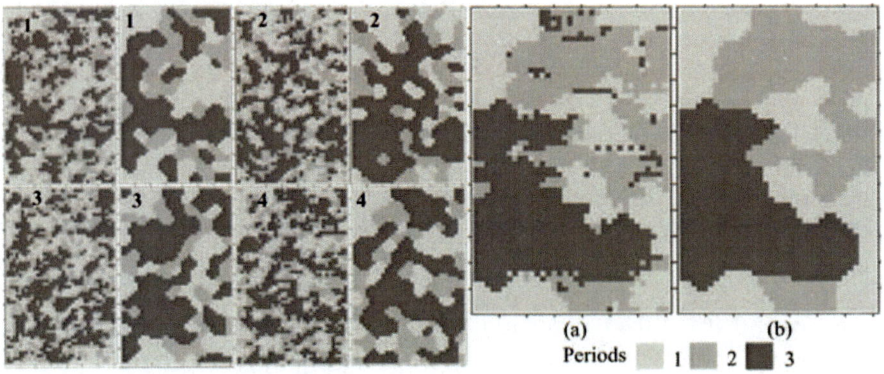

Fig. 9.23 Comparison of four simulated sequences obtained by means of conventional MIP optimization and their smoothed models (left), and one sequence acquired using the suggested MIP approach and its smoothed model (right) [217]

different heuristics for the open-pit mine sequencing problem, even developing a novel polyhedron derived from the continuous PCKP. Two IP-based algorithms were developed, in addition to an adaptation of the floating cone algorithm and a Dantzig–Wolfe reformulation of a binary integer programming (BIP) algorithm, the latter being impractical when compared to other heuristics. Boland, Fricke, and Froyland (2007) presented a strengthened IP formulation in which 0-1 knapsack inequalities (PCKP) are formed by combining precedence and production constraints and then decreasing the computational demands to achieve an optimal solution. Boland et al. (2009) formulated the scheduling problem as a MIP, in which the aggregated blocks are employed for sequencing, while each block is applied in decision-making. Utilizing an iterative disaggregation method, the aggregates obtained in the processing result in an optimal solution equivalent to that obtained using an LP relaxation of the individual blocks in the MIP model. They also suggest tactics for creating fine aggregates for MIP processing, employing duality results, and exploring the problem framework, allowing the resolution of very large instances in an acceptable time with high NPV quality [218–221].

Amaya et al. (2009) proposed a scalable IP-based approach to solve very large problems (with millions of blocks) of the long-term open-pit mine scheduling, in which the usual IP framework combined with a local search algorithm makes it possible to achieve near-optimal solutions in large instances in an acceptable time. Bley et al. (2010) considered the proposition of Caccetta and Hill (2003), although they have combined the corresponding precedence and production constraints in the blocks, considering each period and attribute to create a PCKP. This strategy allows deriving multiple additional constraints, leading to a more rigid linear relaxation and a substantial decrease in the computation time necessary to solve sequencing problems. Bienstock and Zuckerberg (2010) proposed an iterative algorithm (BZ) for solving the LP relaxation problems observing the subjacent combinatorial framework of precedence constrained and capacitated problems of optimization. Their solution employs the primal structure extracted from the solution of subproblems instead of updating the dual information, thus accelerating the convergence and resulting in an algorithm applicable to several types of problems, capable of achieving optimality in instances containing millions of variables and tens of millions of restrictions over twenty iterations. Moreno, Espinoza, and Goycoolea (2010) introduce the critical multiplier algorithm to solve the linear relaxation of multi-period scheduling problems formulated as an IP and an LP-based topological sort heuristic to achieve viable solutions. The critical multiplier algorithm is based on two observations: (1) Solving two single-time period maximum closure problems and obtaining a convex composition of the solutions are enough to solve a single-period PCKP problem; (2) solving a series of single-period problems and grouping them in the correct order are enough to solve a multi-period PCKP. If the precedence relations are represented as a directed graph, it accepts a topological ordering of its nodes, and when weights are given for each item in that ordering, they are considered topologically sorted. This process conducts to different classes of top sort heuristics determining weights in distinct ways. Hence, the authors compared the performance of a greedy formulation, Gershon's heuristic, and a proposition where the LP relaxation inspired the weight function, called the expected-time top sort heuristic. Computational experiments considering 15 periods in the Marvin orebody dataset and three real mines (one including 4,320,480 blocks) show that the critical multiplier algorithm can solve the LP relaxation of large problems in a reasonable time even when the CPLEX algorithm demanded several days or was unable to solve it. A comparison between the topo sort heuristics showed that the values obtained by the greedy and Gershon formulations are very poor for some of the instances, while the expected-time top sort heuristic reached solutions within 6% of optimality in all instances. Finally, it is possible to achieve results with less than 1% of optimality in a few hours by gathering these solutions as a basis for local search algorithm [90, 188, 189, 222, 223].

Askari-Nasab et al. (2011) presented four MILP formulations to solve large multi-period open-pit mine sequencing problems, and the goal was to maximize the NPV while meeting mining and processing capacities, grade blending, and precedence constraints. The models differ from each other by using continuous and/or binary variables to control extraction, process, and precedence and by resolution of blocks in units or aggregates of blocks (mining cuts). The efficiency of each model is analyzed in terms of memory consumption, execution time, an adequate range of scheduling. Cullenbine, Wood, and Newman (2011) introduced a sliding time window heuristic (STWH) to solve an IP model along with a discretized time window and bounded by precedence, ore control, and capacity constraints with approximately 2% of optimality when optimal algorithms could not find solutions. The heuristic recurrently determines, resolves, and partially repairs an approximation model containing repaired variables in initial periods, a rigorous sub-model determined along a horizon in intermediate periods, and a relaxed sub-model in later periods, the Lagrangian relaxation stage being critical for computational efficiency. Including production and processing lower and upper bounds, the formulation solved problems encompassing 15 time periods and approximately 25,000 blocks, the largest instances requiring up to an hour and a quarter to run while exhibiting a 2.4% average optimality gap. Chicoisne et al. (2012) proposed a methodology to solve the IP model of the scheduling problem that consists of (1) solving the LP relaxation with a decomposition method named critical multiplier algorithm; (2) applying a rounding heuristic based on topological sorting to the partial solution achieved so far; and (3) applying a local search heuristic to improve these previous partial solutions. Is required for the first step that the formulation has a unique resource constraint per period, and the proposition solved the LP relaxation of real mine planning instances with approximately 5,000,000 blocks and 20 time periods within 1–6% of optimality [187, 224, 225].

Byun and Dimitrakopoulos (2013) developed a parametric maximum flow algorithm to resolve the LP relaxation of the maximum closure problem with a unique capacity constraint (PCKP) for each period, consisting of a piecewise-linear function of a determined parameter. Computational experiments have shown that LP relaxation can be solved in less than one minute, for instances, containing hundreds of thousands of blocks. Smith and Wicks (2014) developed a MIP application for medium-range mine sequencing considering a 60-month production sequence of Lumwana copper mine, regarding constraints related to shovel location, stockpiling, grade control, mining, and processing capacity. Their MIP model framework optimizes the sequencing problem over two stages: First, a stage accounting for the annual production of the aggregate reserve blocks is applied, followed by a monthly production schedule for sequencing the volumes employed in the current production scheduling. Lambert and Newman (2014) proposed a tailored Lagrangian relaxation algorithm to generate a feasible initial integer solution and then use it to solve the open-pit mine sequencing problem formulated as an IP model. First, the variables that incorporate a value of 0 in the optimized solution are eliminated, and then, a heuristic is employed to generate an initial feasible integer solution to be solved by the BB algorithm. Subsequently, a Lagrangian relaxation utilizes the results obtained during the generation of the initial solution to determine a dualization design that defines the resource constraints that must be dualized and those that must be kept explicit, producing substantial reductions in solution time in determining near-optimal solutions compared to solving the original monolithic problem. The computational experiments were performed on twelve instances obtained from two datasets, reaching an average 59% reduction in solution times for models with 10.000 blocks and 22% on average for models with 18.000 blocks derived from Marvin dataset [186, 226, 227].

Lamghari, Dimitrakopoulos, and Ferland (2015) introduced a two-stage hybrid algorithm to resolve the production scheduling problem. The first stage involves solving a sequence of LP problems to produce an initial solution, while the second phase applies a VND procedure to upgrade the solution. The basic idea of proposing an initial solution is to decompose the IP or MIP model into multiple LP subproblems, each associated with a period and easier to solve, in addition to a fixing heuristic (to turn the subproblem solution into feasible, if necessary) and then combined into a feasible initial solution. Posteriorly, the VND promotes a neighborhood shift to evade local optima based on the movements of exchange, shift-after, and shift-before, while using the best improvement local descent makes it possible to explore different neighborhoods. This hybrid formulation solves the 20 instances (including Marvin and MineLib) for the different proposed constraints with up to 3.2% of average optimality when compared to CPLEX, and in less than 2 min, also providing novel best-known results for these instances and being able to resolve those that other algorithms have found intractable. Morales et al. (2015) confronted the ultimate pits and mine sequences achieved in the block models of MineLib by applying two different methods, the classic LG nested pits, using GEMCOM Whittle algorithm, and DBS methodology, using SimSched and BOS2 routines. They found that although the total NPVs did not diverge so much, the pit geometries (especially in the early years) and total effort to produce the solutions differ considerably, where the traditional method generally takes many hours more to achieve an optimal solution than DBS methodology [144, 199, 228].

Vossen, Wood, and Newman (2016) proposed a variant of MIP with lower bounds on resource consumption in each period beyond the usual upper bounds and allowed fractional block extraction while enforcing the slope restrictions of the pit wall. The algorithm uses the hierarchical Benders decomposition to solve the LP relaxation of the MIP, exploring variables aggregated over time and recursively decomposing a model into the main problem and two subproblems instead of the usual single subproblem. Thus, a BB heuristic generates mixed-integer solutions with high quality, solving medium-size instances to near optimality in minutes. Aras (2018) developed an IP algorithm to solve the block-by-block mine production scheduling problems addressed with multi-capacities, grade uncertainty, blending, variable pit slopes, stockpiles, multi-destinations, and truck hours. Achieving optimality gaps of less than 1.3% in all the large mine sequencing problems tested, he also developed a new cone pattern generation scheme that can handle variable pit slope angles and introduced a new variable substitution methodology to formulate the open-pit mine sequencing problem with the significantly reduced number of sequencing constraints. Mai, Topal, and Erten (2018) proposed a methodology to solve long-term block scheduling problems based on a new approach of block aggregation followed by the formulation of an IP model whose aim is to maximize the NPV, while the constraints considered are related to precedence, grade control, processing, and mining capacity. The Topcon algorithm aggregates the blocks in TopCones that must respect slope and other constraints and must be positive when considering the total economic value of the blocks clustered, which cannot be divided into smaller cones if restrictions are violated and whose units and cone shape can be explicitly controlled. The number of generated TopCones can be controlled by the minimum number of blocks per cone, average grade, total ore or waste tonnage, amount of material, or other constraints, and applying the methodology to a large-scale copper deposit showed that higher NPV solutions were obtained as opposed to commercial software in less than 10 min. Figure 9.24 presents the ultimate pit limit obtained using TopCone aggregation for a 2D copper block model [198, 229–231].

Souza et al. (2018b) proposed combining a MIP model with SA and TS meta-heuristics, thus composing a greedy algorithm that seeks higher-grade blocks in the initial periods, achieving the same NPV of the project in fewer periods by the traditional methodology. Samavati et al. (2018) formulated an IP model, encompassing the precedence and knapsack constraints, using the fractional solutions of the IP

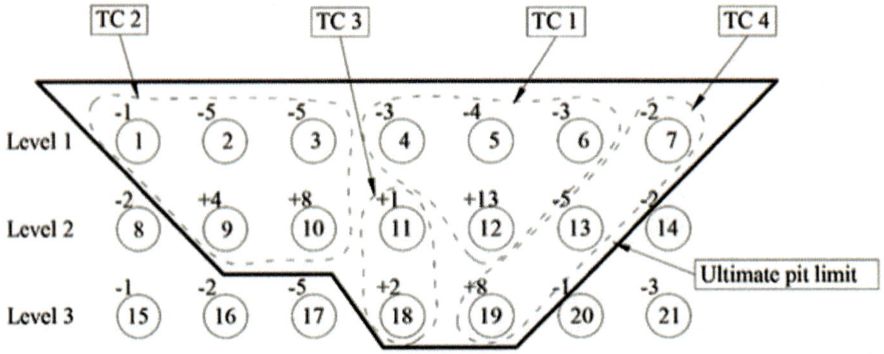

Fig. 9.24 Ultimate pit limit obtained using TopCone aggregation for a network representation of a 2D copper block model [231]

model's LP relaxation to quickly generate nearly optimal feasible (integer) solutions and employing a fast heuristic to compute a viable LP solution. The authors propound the adding cut technique to extract the first property in the LP relaxation solution, which consists of a BB algorithm that iteratively applies the top sort heuristic to each node in the tree until the stopping criteria are reached. An expectation of the exploitation time of the blocks can be obtained by appending some special cuts to the LP relaxation model and solving it. Therefore, the topological sort can obtain a near-optimal schedule. Jélvez, Morales, and Nancel-Penard (2019) and Jélvez et al. (2020) proposed a full-binary formulation and an expected-time incremental heuristic (ETInc), inspired on an STWH and linear relaxation to select in advance a subset of blocks to be sequenced within each time horizon for large precedence-constrained production scheduling problems and formulation even capable of tackling problems with blending constraints. The STWH decomposes the monolithic problem into subproblems on a rolling horizon basis, aiming to diminish periods in each subproblem, similar to the approaches explored by Cullenbine, Wood, and Newman (2011) and Lambert and Newman (2014). Then, a process inspired by the expected exploitation time is used to preselect the subset of blocks to be included in the model by solving the LP relaxation of the monolithic problem, aiming to decrease the number of blocks within each subproblem, as shown by Chicoisne et al. (2012); the authors employed the BZ algorithm at this stage. Hence, a viable solution is built iteratively, and the computational experiments performed on MineLib instances showed that near-optimal feasible solutions were obtained with less than 5% gap in all instances, considering as stopping criterion to return the best viable solution whenever the 1% gap or the time limitation has been reached. The proposed hybrid heuristic could improve the best-known results for several problems in the literature, being the first to generate a viable solution for the only problem with blending constraints in MineLib with a 1.76% optimality gap. Figure 9.25 presents the hybrid algorithm proposed by Jélvez, Morales, and Nancel-Penard (2019) [143, 187, 188, 225, 227, 232–235].

Letelier et al. (2020) introduced a structure inspired by the work of Bienstock and Zuckerberg (2010) to solve the problem with integral variables, with steps including preprocessing, cutting plane, heuristic, and specialized branching techniques, both

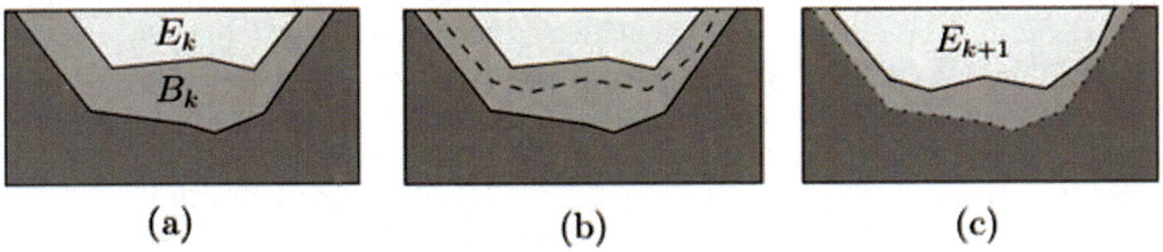

Fig. 9.25 An iteration of the presented algorithm, where E_k is the group of blocks previously extracted and B_k is the group of preselected blocks applied in auxiliary problem. First **a** B_k blocks are calculated considering the expected exploitation time and a tolerance parameter, when **(b)** a subset of B_k is solved for extraction and processing decisions and **(c)** the scheduled blocks are finally updated [234]

considering DBS and bench-phase scheduling problems. The authors tested their solution on real instances whose sizes vary from 20,000 to 5,000,000 blocks and embracing twenty to fifty time periods. Using new heuristics, they could diminish the LP relaxation value by approximately 33% and achieve viable solutions reaching an average gap of 1.52% from the preceding calculated bound. Gholamnejad et al. (2020) introduced a chance-constrained MILP model to tackle the long-range mine sequencing problem by including a single stockpile whose attributes are treated as random variables normally distributed in a probabilistic approach. Firstly, the deterministic alternative of the problem is enunciated, which acts as a basis for the stochastic alternative, where uncertainty is regarded in the average grade of ore sent from stockpile to processing plant. Since the optimization process does not allow the stochastic objective function and constraints to be handled directly, a deterministic variant of the model was performed to convert stochastic linear equations to deterministic linear equivalents. Finally, chance-constrained programming was employed to reach the homologous deterministic solution of the primary model. The methodology implementation to an open-pit gold mine considering ten periods showed that the probabilistic approach to the stockpile provided a better average feed grade for the processing plant, increasing the NPV and the likelihood of meeting grade blending constraints compared to the use of the traditional stockpile deterministic model. Flores-Fonseca, Linfati, and Escobar (2021) also proposed a MILP formulation to solve the scheduling problem in two instances, considering the use or not of a single stockpile. The objective function maximized the NPV, respecting the constraints related to precedence, destination, grade and stockpile control, equipment, and plant capacity. In both instances, one obtained from Newman et al. (2010), the use of stockpiles increased the NPV and flexibility of the operation, and the CPLEX obtained better solutions with less computing time than Gurobi, which even presented a larger gap in one of the instances [5, 155, 188, 236, 237].

Stochastic Mine Planning

Since traditional methodologies for optimizing open-pit mine design and sequencing derive from a unique estimated orebody model, the geological variability is not considered unless a conditional simulation is employed to quantitatively address the resulting uncertainties and generate several equally probable orebody models. Dimitrakopoulos (1998) stated that mining enterprises were mainly subject to uncertainties related to geology (e.g., in situ grade variability and material-type distribution), to technical specifications (e.g., slope constraints, mining, and processing capacity), and economics (e.g., commodity prices, capital, and operating costs), the former being the most influential. This approach provides an appropriate input for stochastic integer programming (SIP), a mathematical programming methodology that achieves the optimal outcome for a given objective (e.g., maximize the NPV or minimize ore grade variation in plants) within the viable solution space delimited by a group of constraints (e.g., grade control, quantity, and quality constraints). A series of orebody realizations as input to a SIP model allows the integration of uncertainty and variability of in situ deposit into the mine sequencing optimization process. SIP can be defined as an extension of the MIP encompassing uncertainty in a single or multiple of the related coefficients, which leads to increase complexity and size of problems compared to MIP-based formulations. This topic will investigate mainly the advances obtained in the scope of open-pit production scheduling under geological uncertainty, considering the plethora of assumptions of uncertainties, operational particularities, complexities, and formulations of the problems, many of which addressed as a two-stage SIP model solved by approximate methods [238, 239].

Ravenscroft (1992) affirmed that the possible error in estimation and spatial variability could be quantified by geostatistical estimation, but not allowing for easy use of this information in sensitivity analysis; moreover, he proposes the use of conditional simulation to permit the incorporation of both estimation error and variability in the distribution in mine scheduling. Dowd (1997) suggests a framework for risk integration in open-pit mining enterprises. Frimpong, Whiting, and Szymanski (1998) developed an IPOP based on the combination of a stochastic optimization annealing process and MFNN, as previously mentioned, and Frimpong, Szymanski, and Narsing (2002) proposed an algorithm capable of incorporating structural, hydrological, and geotechnical elements. Smith and Dimitrakopoulos (1999) suggested a methodology that quantifies uncertainty by defining a set of simulations of the deposit, applying the scheduling algorithm to each realization, determining the chance of a block being exploited in a given period, and using the patterns of probability to develop a risk-based sequence. Finally, Dimitrakopoulos, Farrelly, and Godoy (2002) presented a framework for incorporating quantitative modeling of the uncertainty of grade, tonnage, and geology into open-pit design and planning by generalized sequential Gaussian simulation and direct block simulation. The authors applied the methodologies in a gold deposit. They highlighted the differences between the projections obtained from a single orebody or multiple simulated models, presenting the effects on expectations of NPV, production cost, mill feed grade, and ore tonnage while identifying periods with a high risk of negative discounted cash flow. Figure 9.26 illustrates an uncertainty assessment for a parameter that may, for example, be the ore reserves in a gold mine [75, 78, 239–243].

Ramazan and Dimitrakopoulos (2003) discussed an LP approach that considered grade uncertainty, equipment access, and mobility constraints, relying on the expected grades of block and the chances of different element grades being above the established cutoffs, both information obtained from orebody realizations. As aforementioned, Ramazan and Dimitrakopoulos (2004b) presented effective MIP formulations to produce viable mining patterns of optimized probabilistic mine sequences. Dimitrakopoulos and Ramazan (2004) introduced a risk-based optimization model for long-range open-pit mine sequencing that addresses geological uncertainty regarding grade, ore quantity, and quality, besides risk quantification and usual operational requirements, while evaluating a lateritic nickel deposit with single or multiple realizations. However, the NPV of a traditional LG resolution was slightly higher. Since the objective of this risk-based optimization was not increasing NPV, the probability of not meeting the production requirements in the first period with

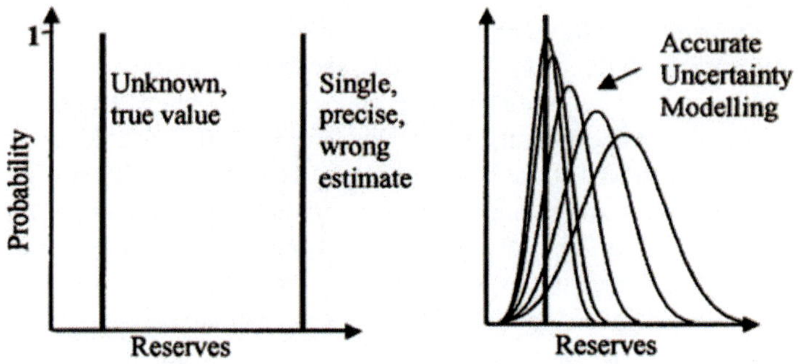

Fig. 9.26 Comparison of a deterministic and a probabilistic approach; search for accurate quantification of uncertainty for the various components that affect mining decisions increases the chances of including real but unknown values in models [243]

the LG approach, was 6% higher. Menabde et al. (2004) developed an algorithm for mine optimization under uncertainty within BHP Billiton, based upon conditionally simulated orebody models and MIP formulation, intending to simultaneously optimize the mine sequence's cutoff grade the pushback design. The objective function aims the maximum NPV subject mining, processing, and market constraints, and in the case studied, there was obtained 4,1% increasing of NPV for ultimate pits with low variance, which indicates that for deposits and ultimate pits with high variance, the results may be substantially higher [217, 244–246].

Godoy and Dimitrakopoulos (2004) presented an optimization framework for mine sequencing relying on efficient waste management and grade uncertainty from multiple equally possible representations of an orebody. Since any long-range mine sequence can be achieved inside a domain of viable solutions between the best and the worst economics, defined by the sum of combinations of ore and waste extracted in each period, multiple domains can be obtained from multiple orebody models. The authors first obtained a stable solution domain (SSD) considering the possibilities for each orebody model and then formulated an LP optimization to maximize the NPV and find mining rations by defining the schedule for ore and waste extraction within the SSD and considering the constraints imposed. Hereafter, an algorithm applies these mining rates to generate a series of production sequences respecting the slope constraints, and, finally, combinatorial optimization is performed to deduce a unique production schedule from all sequences using SA. Applying the methodology in a large gold mine in Western Australia proved to capitalize on postponing waste exploitation and quantifying grade uncertainty to supply a risk-resilient mine sequence while increasing asset value. Figure 9.27 illustrates an optimal solution obtained inside the SSD, derived from a series of simulated resource models [247].

Ramazan and Dimitrakopoulos (2004c) proposed a SIP model to tackle the long-range open-pit mine scheduling, containing substantially fewer binary variables than conventional MIP models and efficient solution time. Being also discussed by Ramazan and Dimitrakopoulos (2003) and by Ramazan et al. (2004) as part of an ARC Linkage Project at the University of Queensland, this approach allows

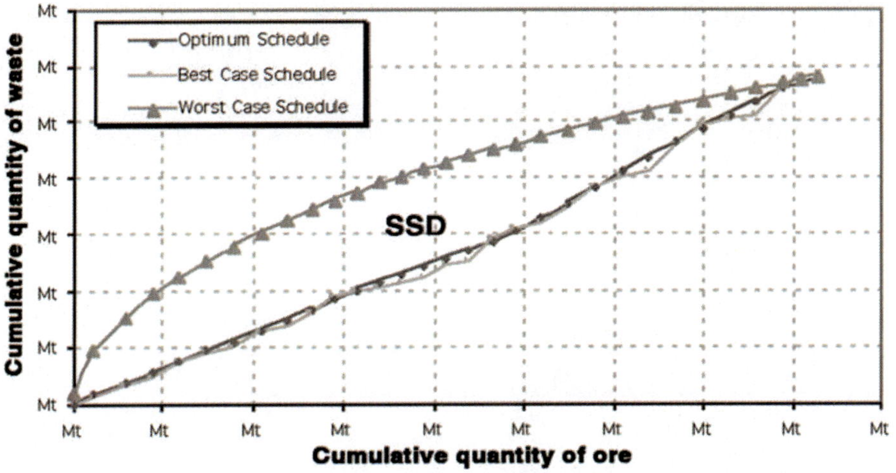

Fig. 9.27 The optimal solution achieved within the SSD was deduced from a set of simulated models [247]

defining a risk profile relying on the uncertainty assessed through orebody simulations, offering the choice of minimizing the risk in each production periods, tolerating the risk in given periods or all periods. Dimitrakopoulos and Ramazan (2008) proposed a SIP model that returns optimal mine sequences from simulated orebody models without averaging their grades (E-type model). Given the objective function of maximizing the NPV, the SIP approach can be applied to minimize the risk of disbanding from production bounds related to ore, metal, and grade blending. A case study on gold and copper deposits highlights that the expected NPV when employing the SIP model is 10 and 25% higher, respectively, than the conventional sequence. Ramazan and Dimitrakopoulos (2013) extended the work in Dimitrakopoulos and Ramazan (2008) by proposing a new two-stage SIP-based optimization model to integrate uncertain supply to production schedules including stockpile and risk profile management. Figure 9.28 illustrates the grade distributions of a 2D deposit in three orebody realizations and the E-type model [248–252].

Jalali, Ataee-pour, and Shahria (2006) introduced an algorithm based on probabilistic logic using the Markov chain process, where the probability of mining a block is proportional to the desirability it may provide, considering the geometrical constraints. Applying this algorithm to a 2D block model, the probability of mining each block was obtained, and then, the optimum pit limits were defined as the pit that provides the highest probability of mining. In comparison with the 2D LG dynamic programming algorithm, results match each other. Golamnejad, Osanloo, and Karimi (2006) proposed a model to solve long-range mine sequencing problems by stochastic chance-constrained binary IP, explicitly accounting for the grade uncertainty. In each block, the probability distribution function of the grade obtained by geostatistical simulation is employed as an input to the optimization structure. A confidence level is defined for each sequencing period for the constraints, and

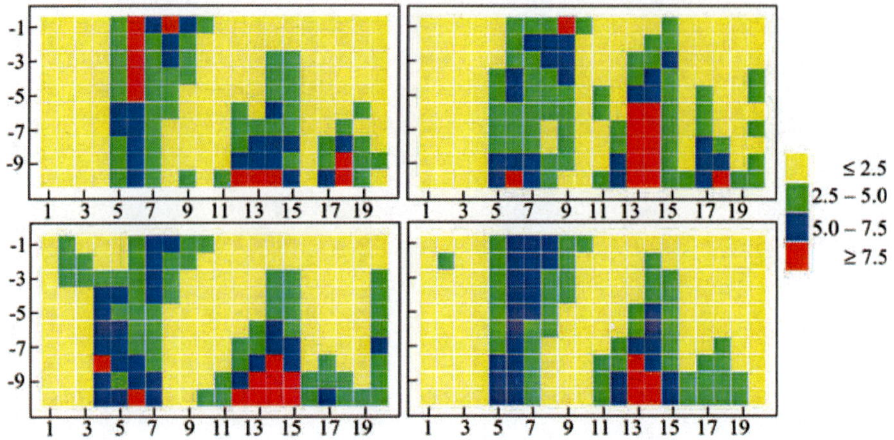

Fig. 9.28 Grade distributions of a hypothetical 2D deposit in three orebody realizations and the E-type model of the 50 simulated realizations (at bottom right) [248]

then, the stochastic objective function and constraints are converted into deterministic nonlinear binary variables to enable the model's resolution. This formulation yields sequences with a high probability of meeting the production targets, while it concomitantly maximizes the expected NPV and minimizes the variance function [253, 254].

Leite and Dimitrakopoulos (2007) proposed an approach to assess the life-of-mine schedule (long-range), generating several stochastically simulated representations of a disseminated copper deposit with relatively low-grade variability and then submitting each realization to a scheduler, submitting all schedules to a process of SA. This approach generated a schedule with an NPV 26% higher than the conventional one while presenting low chances of deviating significantly from targets, conversely to the traditional approach. In this study, for the constraints imposed, the conventional scheduling overestimates the ore tonnages while underestimating the NPV, and even though a generalization can be tricky, regardless of whether it overestimates or underestimates the NPV in different cases, a conventional approach may produce ore tonnages and NPV that are unrealistic and have little chance of being realized. Kent, Peattie, and Chamberlain (2007) compared the grade and risk of conventional scheduling for a gold mine in Namibia that uses uniform conditioning to create a model with the desired selective mining unit to a hundred schedules of stochastic realizations of the orebody reblocked for the same support. The comparison of the approaches showed some risk in the deterministic economic pit, but it also showed some potential for upside, particularly in-depth. A grade control model for reconciliation was already used in the mine based on the blasthole sampling, which usually has levels 10% lower than those registered at the plant. The comparison between the simulated and deterministic orebody and grade control models made it possible to verify a difference in selectivity, especially in areas with high levels. The authors

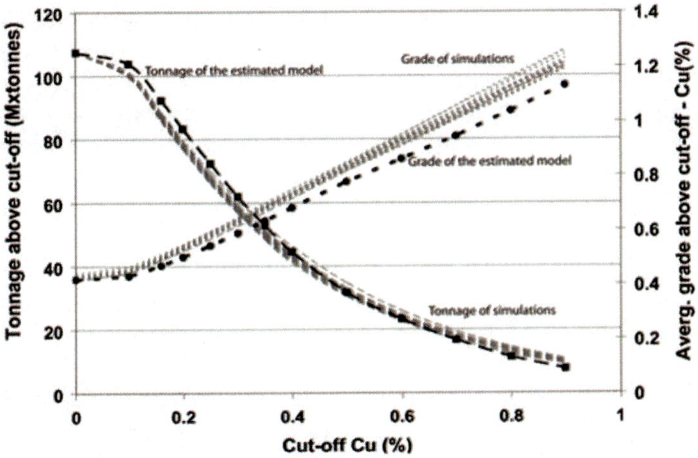

Fig. 9.29 Grade–tonnage curves for estimated and simulated models [255]

did not draw further conclusions. Figure 9.29 presents the grade–tonnage curves for estimated and simulated models in Leite and Dimitrakopoulos (2007) [255, 256].

Dimitrakopoulos, Martinez, and Ramazan (2007a, 2007b) proposed an approach to long-range production scheduling that captures the upside potential of the deposit while minimizing downside risk for key project performance indicators. First, several equally probable stochastic orebody models are simulated to quantify grade uncertainty, followed by the optimization of the schedule of each realization by the LG nested pit methodology, implemented with the heuristic Milawa scheduler of the GEMCOM Whittle algorithm, aiming to maximize the NPV and respect operational constraints, such as slope angle, mill plant, and mining equipment capacity. Hence, with the ultimate pit and the annual extraction sequence, the degree of risk is quantified for each pit design considering the key performance indicators, such as total project NPV, periodic amount of ore to feed the mill, metal production, and cash flows, in a risk analysis similar to that performed by Dimitrakopoulos, Farrelly, and Godoy (2002). Then, a set of discounted cash flows is calculated using each simulated orebody model for a given pushback or year of production. However, this distribution can be generated for any other indicator, such as ore tonnage and metal quantity, considering mill capacity and market demand, thus providing the necessary reference levels. Pit designs that may not meet key performance indicators are discarded. Then, the remaining designs are evaluated considering indicator distributions and benchmarks (e.g., minimum acceptable return, mill demand, market specifications), then calculating the project's upside potential and downside risk. Finally, select the designs that meet the preset decision-making criteria. The authors applied the methodology to a low-grade open-pit gold mine, comparing the risk associated with each pit schedule obtained when applied to each of the 13 orebody realizations, though excluding the application of generated schedule for the realization itself to avoid overestimating the expected profit. Then, the pit among all designs that maximize the expected realized profit or/and other operational criteria can be selected. Figure 9.30 shows the uncertainty in the distribution of a key performance indicator (DCF, discounted cash flow), reward potential, and downside risk concerning a point of reference (MAR, minimum acceptable return) (above), and presents the comparison of risk profiles for ore tonnage within the first pushback for 13 pit designs in each orebody realization, where the arrows at the top indicate selected designs with at least a 70% chance of being above one million tons of ore [32, 243, 257, 258].

Dimitrakopoulos and Abdel Sabour (2007) proposed a framework for assessing the effectiveness of incorporating multiple uncertainties in mine planning using real options valuation (ROV), including market prices, foreign exchange rate, and geological uncertainties, comparing the results predicted by this methodology with those achieved by the NPV approach. For each of the equally probable orebody realizations assessed, a mine schedule with different simulated sequences of exploitation, expected tonnages, operating costs, and metal grades is obtained using a conventional scheduler over all periods, and the only operational flexibility considered is the premise of closing the enterprise any time at no cost. A case study in a small gold mine in Australia with a 3-year lifetime was developed considering 12 orebody

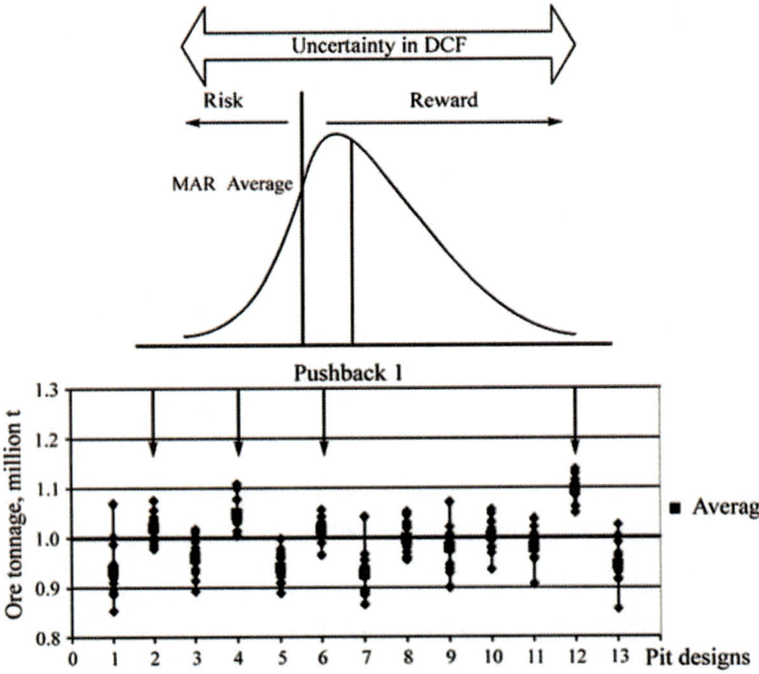

Fig. 9.30 Uncertainty in the distribution of a key performance indicator (DCF), reward or upside potential, and downside risk concerning a point of reference (MAR) (above); comparison of risk profiles for ore tonnage within the first pushback for 13 pit designs

simulations with their respective tonnages, grades and operating costs, in which the schedules were obtained using the LG nested pits algorithm. Metal price values were simulated and foreign exchange rates were determined through a correlation coefficient attached to them, the assessment being developed for both the NPV and the proposed ROV-based methods. Assuming the mine was exploited in the past, the difference between the best and worst designs obtained by the proposed method varied 12% of the range presented by the NPV method, and the scenario pointed out by the ROV, in general, performs better than that indicated by the NPV approach. In addition, designs that maximize NPV do not necessarily maximize value based on the ROV approach, and comparing the net mine value generated for both techniques when selecting a mine design resulted in an 11–18% increase in favor of the ROV approach. Abdel Sabour, Dimitrakopoulos, and Kumral (2008) proposed a multi-criteria ranking system based on ROV to select mine designs and schedules under metal prices, exchange rates, and geological uncertainties. Integrating operational flexibility revises the ultimate pit limits based on new information acquired. With a formulation based in Dimitrakopoulos, Martinez, and Ramazan (2007a, 2007b) to attribute geological uncertainty, and assuming the mines were exploited in the past, the price paths and exchange rates are obtained by stochastic modeling, where for each orebody realization several scenarios of the market variables are assumed in order to generate multiple designs and schedules for each model. Hence, a classification system assigns equal weights to the indicators of upside potential, downside risk, probability of completion, and value statistics to measure the robustness of the proposed designs, indicators obtained from NPV and ROV approach. Therefore, along with the expected values of NPV and ROV, these two sets of uncertainty-based classifications generated for each mine design are compared to rank the design propositions. A case study was developed at a copper mine in Canada and a gold mine in Australia, the first of which contains 10 designs that have been evaluated using 20,000 market simulations over the mine's 8-year life.

In comparison, the second contains 12 designs evaluated using an unrevealed number of market simulations throughout 3 years of mine life. Since it has been assumed that mines have been exploited in the past, the actual market parameters can be used to check among these four classification methods which one indicated more profitable designs, where the ROV multicriteria classification system selected a design with a higher project value, in addition to a significantly lower average mis-ranking classification than the other three methods. The reduction in the average mis-ranking for the copper mine reached 37–58%, while for the gold mine, a reduction of 51–64% was obtained [32, 257–260].

Boland, Dumitrescu, and Froyland (2008) extended a MIP model of the open-pit mine sequencing problem to address geological uncertainty, thus developing a mixed-integer multistage stochastic programming. In their formulation, the resolutions assumed in later periods are dependent on considerations of the geological properties of the material extracted in previous periods, incorporating an endogenous uncertainty by means of multiple orebody realizations, also showing that non-anticipatively can be modeled through linear constraints encompassing variables that are present in the model. Next, Albor Consuegra and Dimitrakopoulos (2009) expanded Leite and Dimitrakopoulos (2007), adding to the life-of-time methodology, which uses a stochastic simulation of the deposit, schedule generation in each orebody simulated and SA of the schedules, sensitivity analysis, and applying it to a copper deposit. When creating 10 deposit realizations, stochastically optimized pit limits were considered to be 17% larger in terms of total tonnage and 10% higher in terms of NPV than conventional optimal pit limits, adding one year to the project. Finally, Meagher, Abdel Sabour, and Dimitrakopoulos (2009) employed ROV and stochastically simulated orebodies to quantify the value of management flexibility to review or change decisions and integrate it with the value assigned to each mining block at the planning time, addressing the concept of value for money over time and integrating the block destination flexibility. In the first stage, the output of this ROV model is an expected value with a time-dependent discount for each block that captures the value of management flexibility instead of a single value. Then, the second stage consists of a parametric minimum-cut algorithm applied to produce single pushback designs, intrinsically leveraging the uncertainty in a set of orebody simulations to produce low-risk open-pit pushback designs, either considering market uncertainty. Albor Consuegra and Dimitrakopoulos (2010) then employed a stochastic simulation approach to integrating uncertainty into a set of pushbacks obtained by the LG nested pit methodology, achieving a 30% increase in NPV when compared to the conventional approach in a porphyry copper deposit, mainly due to differences in scheduling patterns, the waste mining rate, and the extension of the pit limits [261–264].

Lagos et al. (2011) address the production scheduling problem under metal uncertainty, assuming the availability of an independent and identically distributed sample (i.i.d.) of the joint distribution of the ore grades in the deposit. Considering a simplified model where exploitation and processing decisions are subject to capacity constraints for a single-period time horizon, three approaches were evaluated, i.e., minimizing the value-at-risk, minimizing the conditional value-at-risk, and a modulated convex hull model that aims to maximize the worst in-sample profit and maximize the in-sample mean profit. A case study was applied to a vein-type mine in which an i.i.d. sample of 1000 joint ore grades was obtained, where several schedules were obtained for different levels of risk applied to each of the three models proposed. In-sample profits were also obtained for each mine plan from their in-sample ore grades. The value-at-risk approach proved to be unappropriated from a risk-averse perspective, unlike the conditional value-at-risk and modulated convex hull approaches, which attain lower probabilities to the worst profits, although there is less chance of making high profits. Godoy and Dimitrakopoulos (2011) investigated four cases to illustrate the effects of geological uncertainty for pit optimization and design and applied different risk analysis approaches. The first analysis quantifies uncertainty and risk analysis of the ultimate pit limit, including net value, costs, tonnage, grade, and metal. The procedure consists of generating a set of incremental pit shells for the estimated orebody and applying these designs to each equally probable realization of the deposit. The second case identifies areas of upside potential and downside risk in terms of the ultimate pit limits, in other words, where the grade uncertainty had a great impact in defining the ultimate pit limits. Instead of developing a risk analysis on a given set of incremental pit shells, this procedure generates alternative sets of shells, one set for each simulated model. The third case consists of a trade-off analysis to quantify the impact of the grade uncertainty on the tonnage, grades, metal, and net value of two different mining scenarios for a given pushback, considering the depletion of this pushback as single or separate stages. Finally, the fourth analysis quantifies and evaluates the uncertainty and risk analysis related to the ore blocks considering the increment between successive pit shells. Dimitrakopoulos (2011) addressed the advances achieved so far with the use of stochastic simulation and stochastic optimization compared to the deterministic traditional mine planning, highlighting the possibility of increasing the net value of the schedule and the tonnage extracted from the pit through these techniques, which also provide schedules with less risk of not meeting the restrictions imposed [32, 265–267].

Asad and Dimitrakopoulos (2013, advance online publication in 2012) proposed a stochastic network flow approach to developing optimal phase design and ultimate pit limit problem, representing it as a graph and incorporating uncertainty in the metal content and commodity price, thus reflecting potential supply and demand. The authors propose an extension in Meagher, Abdel Sabour, and Dimitrakopoulos (2009) expanding the formulation of the problem as graphs by integrating the changes that Tachefine and Soumis (1997) made in the structure of the maximum closure problem proposed by Picard (1976), which included constraints on resources or production capacity and relaxed the problem using Lagrangian relaxation. Hence, the classical structure of the maximum flow algorithm is respected. Then, a modification in the sub-gradient method is employed to solve the algorithm and create successive phases that contain uniform amounts of material, thus being operationally manageable and addressing the gap problem. An application in a copper deposit in Canada demonstrates that the stochastic approach develops an optimal design with a pit approximately 45% larger than that obtained by the conventional approach, which was employed as a push–relabel maximum flow algorithm to solve the E-type orebody model. Asad and Dimitrakopoulos (2012a) developed a case study on a gold deposit to compare this stochastic network flow approach to the conventional LG algorithm, thus generating a 30% increase in discounted cash flow, a 21% ultimate pit limit larger and about 7% more metal, while also maintaining a better consistency in phase size compared to the traditional approach. Asad and Dimitrakopoulos (2012b) proposed a methodology to determine the optimal production capacities based on a stochastic framework, where multiple grade tonnage curves within the optimal predetermined ultimate pit limit are derived from equally possible simulated orebody realizations. The proposed stochastic model maximizes the NPV over several periods, never less than the initial investment and considering a given concentration capacity. Based on the theory of cutoff grades, this optimization problem seeks those combinations of concentration capacity and the corresponding concentration cutoff grade policy that maximize the return subject to the balancing of mining and refining capacities, whose stockpile implementation can significantly improve the results. The case study on a copper deposit indicates that among the available grade tonnage curves, the one with the minimum operational scale corresponds to the optimum, guaranteeing the neutralization of the initial investment in all scenarios, also corresponding to the point where the difference between the return value and the initial investment is maximum.

Figure 9.31 shows a graph structure in multiple realizations with undiscounted block values (above) and with discounted block values (below) [32, 47, 263, 268–271].

Lamghari and Dimitrakopoulos (2012) presented a two-stage SIP methodology combined with two different hybrid metaheuristic solutions based on TS to solve the open-pit production scheduling problem with metal uncertainty. The first stage determines for each period of the horizon a set of blocks to be mined respecting the precedence and minimum and maximum mining limits and accounting for the grade uncertainty accessed through conditional simulated models. In the second stage, possible violations of processing and metal production constraints in each period can be addressed by adding stockpile management, which entails an additional cost, so the problem consists of identifying a first-stage solution that minimizes the expected cost second-stage solution. The objective function includes two elements to maximize the expected NPV of the mining operation and minimize the expected costs incurred whenever the stochastic constraints are violated due to metal uncertainty. The constraints are related to reserve, slope, mining, processing, and metal production, the last two being stochastic constraints. One of the strategies used to extend the search for a feasible domain combines TS and long-term memory (LTM) of the search history, while the other relies on the VNS method with TS. Numerical results in 10 instances indicate the efficiency of the approaches to produce good solutions in relatively short computational times. Although both methodologies produce robust results in problems with relatively small sizes, the combination of TS-VNS cannot compete with the approach that combines TS-LTM (which produced gaps smaller than 4% in 92.5% of all runs), or even with pure TS. Lamghari and Dimitrakopoulos (2016a) extended their previous work by developing four different heuristics to address the production scheduling problem, accounting for metal's uncertainty and considering multiple destinations and a stockpile. The heuristics employed are a TS, a VND, a very large neighborhood search heuristic based on network flow (NF), and a diversified local search (DLS) that combines VND and NF, the last two approaches unexplored so far. Three different heuristics were employed to obtain initial solutions, a random heuristic (RH), a look-ahead heuristic (LAH), and a BC algorithm, thus resulting in twelve combinations to evaluate each instance in each of the three benchmark datasets used, besides the CPLEX reference. The numerical results show that these solutions are effective and are capable of solving in a few minutes to a few hours instances that CPLEX is unable to solve or requires significantly more time. Both network flow-based heuristics (NF and DLS) surpass TS and VND, achieving more efficient and robust results for tested instances and being less sensitive to initial solutions [272, 273].

Marcotte and Caron (2013) proposed a stochastic optimization formulation based on obtaining the average of the realizations of a simulated deposit whose grades were previously converted into expected profit and compared the results with the expected profit of the E-type of the simulations, exploring 12 deposit scenarios from different spherical covariance model parameters. On average, stochastically optimized pits presented less bias and produced 2.7% more profits than those optimized from the estimated model, although the relative gain can reach 10–25% for poorer deposits. Greater ranges of the variogram model contribute to increasing the

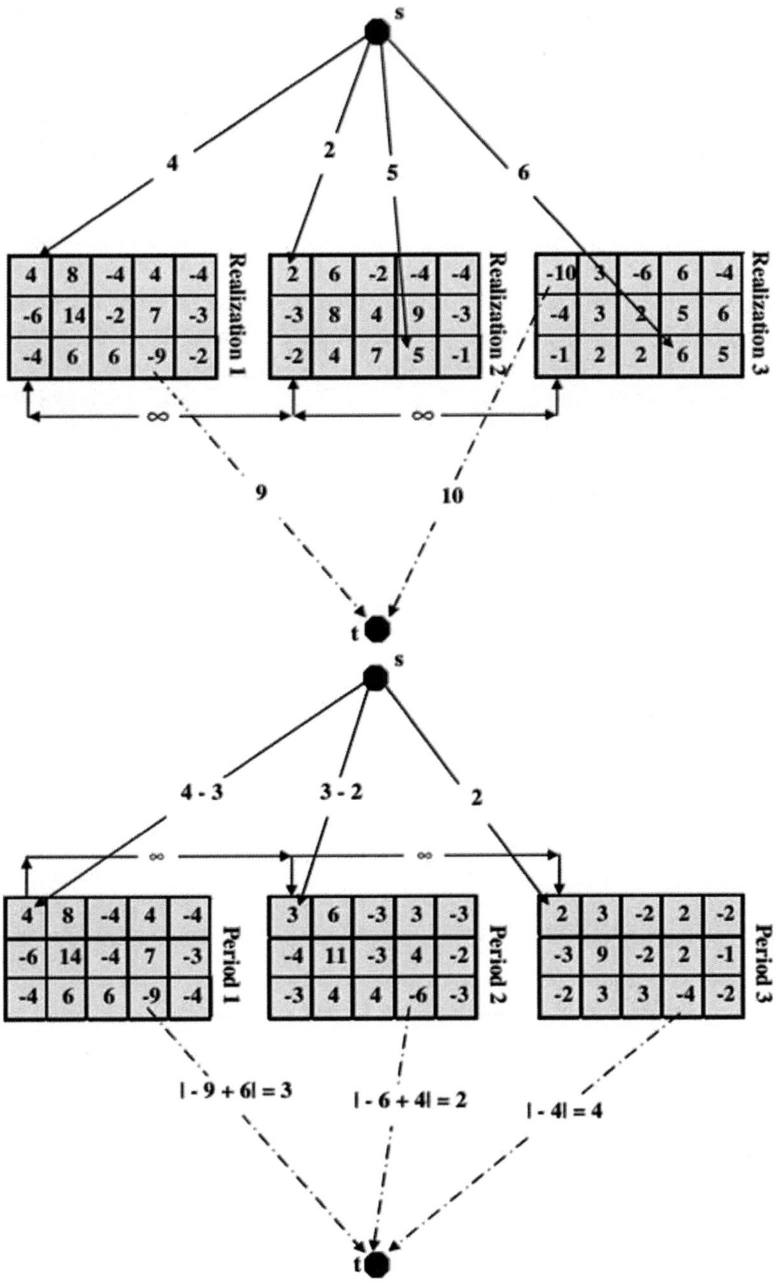

Fig. 9.31 Graph structure in multiple realizations with undiscounted block values (above) and with discounted block values (below) [268]

gains from stochastic optimization over the classical approach, as well as the gains increase with treatment costs, although they tend to decrease with mining costs. Dimitrakopoulos and Jewbali (2013) proposed a multistage stochastic production scheduling approach that integrates short- and long-term mine planning, considering the grade control data. First, the high-density future grade control data is simulated based on the exploration and grade control data available in previously mined parts of the deposit. A conditional simulation by successive residuals is performed to update the preexisting simulated models with this simulated future grade control information. Therefore, a stochastic programming formulation deals with these multiple simulated orebody models to maximize NPV and minimize deviations from expected production targets, meeting constraints related to grade control, ore production, mining capacity, slope, and reserve, followed by quantifying the risk in the generated production schedules. The case study developed in a large gold mine in operation in Australia showed that the approach adds value to the operation and matches with the mine's reconciliation, achieving better results in terms of tons of ore and metal when compared to a SIP that considers only exploration data, also reaching a 41% higher NPV than the traditional production schedule. Benndorf and Dimitrakopoulos (2013) proposed a SIP formulation to address the long-term mine planning problem considering the joint multi-element grade uncertainty, extending the work of Ramazan and Dimitrakopoulos (2004c, 2013) to multi-element deposits. The objective function aims to maximize the NPV of a single pit to define a global schedule and minimize deviations from production targets, assessing the influence that this pit would have on the multiple realizations of the orebody. In addition, the third stage of the objective function controls smooth mining by penalizing the non-mining of adjacent blocks in the same period, the priorities of these three stages being controlled by the magnitude of the corresponding cost parameters assigned to each stage. The constraints assumed are related to reserve, slope, ore tonnage, and grade control considering multiple elements, equipment access, mobility, mining, and processing capacity. The scheduling algorithm Blasor, a MILP-based software proposed by Stone et al. (2004), was used to solve this formulation in deterministic and stochastic scenarios. A case study was developed at an iron ore deposit in Western Australia, whose operational and economic parameters were provided for the next five years, and geological uncertainty was accessed through 20 simulated orebody models comprising the joint simulation of the five elements, while the deterministic model considered in the evaluation was obtained through the E-type of the realizations. The results showed that the stochastic schedule was able to control the risk of deviation from production targets over the periods, in contrast to the deterministic schedule, whose mean values meet the specifications, while the range of values obtained promotes considerable deviations from the targets [218, 248, 252, 274–276].

Goodfellow and Dimitrakopoulos (2013) proposed two formulations to integrate grade and material uncertainty into pushback design in mines with multiple elements, materials, and destinations, aiming to reduce risk and maintain similar pushback sizes. Both propositions require an initial pushback design to define the initial targets, which can be a traditional pushback design, pit shells, or a randomly generated design. The first proposition aims to minimize the average absolute deviation of a target tonnage in all orebody simulations. The target for each processor destination and the pushback is defined as the average tonnage on the available realization and the initial pushback design. As this formulation assesses the linear change in absolute deviation of the tonnage, being very sensitive to drastic differences in the target quantities between the destinations due to the volume variance, the second proposition evaluates the targets' square deviations and standardizes them their target size. As these objective functions are mathematically nonlinear, metaheuristics such as SA and breadth-first search (BFS) algorithm were employed in the resolution. The case study at a copper mine in Chile shows a significant reduction in variability in terms of the amount of material in the pushbacks sent to the processes without displacing the risk to other material destinations or pushback, reaching a reduction in the variance of 35–61%. Montiel and Dimitrakopoulos (2013) proposed an adaptation of the SA-based multistage method. Godoy and Dimitrakopoulos (2004) proposed to solve the long-term risk-based production scheduling problem in mining complexes with multiple rock types and processing streams. The method requires the definition of an initial solution in each stochastic orebody realization obtained by a conventional scheduler, and then, the SA approach generates a final schedule. The objective function minimizes the average deviations from production targets for multiple mining periods, orebody simulations, and different processes. The case study at a complex copper mine in Chile with multiple ore material and processing techniques results in a schedule with less than 5% deviations from the expected mill and waste production targets versus a 20% deviation from a conventional schedule [247, 277, 278].

Lamghari, Dimitrakopoulos, and Ferland (2014) proposed two variants of a VND algorithm based on a two-stage stochastic programming approach to solve the long-term production scheduling problem with metal uncertainty. In order to maximize the expected NPV and minimize the expected recourse costs of violating processing constraints, first, a set of blocks is mined in each period to satisfy the constraints related to reserve, precedence, and mining, and only then the metal content of each block is calculated in the second stage. The stochastic model is then transformed into deterministic by assessing the probability of the metal contained in each realization block when all the realizations are considered. To generate the initial solution, two alternatives were proposed based on a decomposition approach that separates the problem into a series of subproblems representing each period, a rigorous approach, and a heuristic based on a greedy algorithm. Three neighborhood structures (swap, shift-after, and shift-before) try to improve the solutions. Computational tests on three large-scale gold and copper instances result in solutions with an average deviation of less than 3% from optimality obtained in a few minutes to a few hours, while CPLEX was only able to solve the problem by demanding an unviable time. Koushavand, Askari-Nasab, and Deutsch (2014) proposed calculating the cost of uncertainty in a long-term production schedule considering the deviations from the target production. A MILP algorithm based on Askari-Nasab et al. (2011) is formulated to find the sequence of blocks from a predefined pit shell and its respective destinations with two objectives: maximizing the NPV of the estimated model to

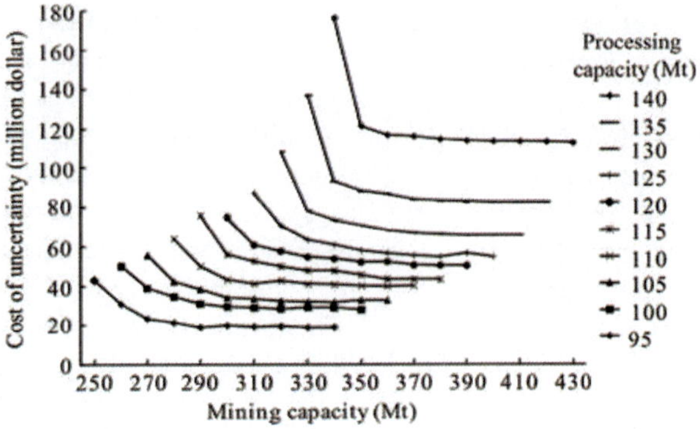

Fig. 9.32 Cost of uncertainty versus different mining and processing capacities in a synthetic case [280]

minimize the cost of the uncertainty of multiple stochastic realizations. The objective function is subject to constraints related to grade blending, mining, processing capacity, reserve, precedence, mining–processing portion, over- and underproduction, and, finally, upper and lower limit for the stockpile, and the cost of uncertainty is modeled as an asymmetric penalty function, where the cost of underproduction is higher than the cost of overproduction. A clustering technique is employed to reduce the number of variables and make the problem tractable. When compared to an LP optimization in an oil sand deposit in Canada, considering 50 realizations of the orebody, the proposed method resulted in a slightly lower NPV due to the need to add new constraints to the problem formulation. However, it reduced the deviation from the target production. The parameters that impact the uncertainty in the optimization problem are studied, and Fig. 9.32 shows the cost of uncertainty versus different mining and processing capacities, where it is possible to infer that mines with high processing and mining capacities have a higher cost of uncertainty and that greater processing capacity requires much greater mining capacity to achieve the minimum cost of uncertainty [224, 279, 280].

Castillo and Dimitrakopoulos (2014) proposed a three-stage model based on ROV to solve the long-term production scheduling problem under the market and geological uncertainty. In the first stage, the ultimate pit and the schedule of a base case considering deterministic economic information and an estimated orebody model are defined. The second stage aims to add flexibility and is divided into two parts. The first is to change the commodity price to generate new nested pits for forecasting an expansion in the base case, while the second part is to generate equally probable orebody simulations and metal price simulations. In the third stage, finally, the model performs a reassessment of the mine's operational status at the end of each period, confronting the expanded base case schedule with multiple orebody models and commodity price paths, which implies changes in the cutoff policy, the possibility of changing destinations of the block and the revision of the option of expanding or contracting the ultimate pit limits. A case study was developed in a gold mine, in which the base case was optimized through the Milawa scheduler considering the E-type of the 20 orebody realizations and a deterministic gold price.

This conventional approach delivered an 11-year schedule, and when the metal price was increased following the rise of two revenue factors in order to obtain nested pits for possible expansions (up to 15 and 16 years), the new assessment showed a decrease in NPV, being neglected by the traditional scheduling framework, that kept the forecast for 11 years. Hence, the stochastic model is optimized using the CPLEX. A case considering only geological uncertainty resulted in an average of 14% less ore, but 21% more NPV, as the estimated model tends to overestimate the medium grade tonnage and underestimate the average ore grade. The first expansion was accepted by this methodology, which would result in 15 years of operation and 22% more NPV when compared to the base case, while the second revenue factor expansion proved to be unprofitable. A case was developed considering the E-type model, the generated schedule, and 20.000 price simulations, demonstrating that the project can increase its NPV by more than 13% and almost 18% if the expansions were considered. Finally, a joint stochastic case was proposed, considering the application of 20.000 price paths for each of the 20 orebody realizations, resulting in 16% and 17% more NPV if both expansions, respectively, were considered, proving that the traditional scheduling framework tends to underestimate the size of the ultimate pit and disregard significant potentially profitable expansions. Figure 9.33 shows 10 price paths generated by a geometric Brownian motion with Poisson jumps model based on parameters obtained from historical data [281].

As aforementioned, Khan (2015) assessed the implementation of the PSO, BA, and DE metaheuristics to solve the open-pit production scheduling problem with grade uncertainty through a set of 15 simulated realizations of a copper deposit. A SIP model based on the formulation of Ramazan and Dimitrakopoulos (2004c) was used as a reference to assess the performance of the metaheuristic techniques. The PSO algorithm variants performed better than bat and DE algorithms in the average percentage gap achieved. On the other hand, the DE algorithm showed a relatively slow convergence rate, although it proved to be more robust. Moosavi and

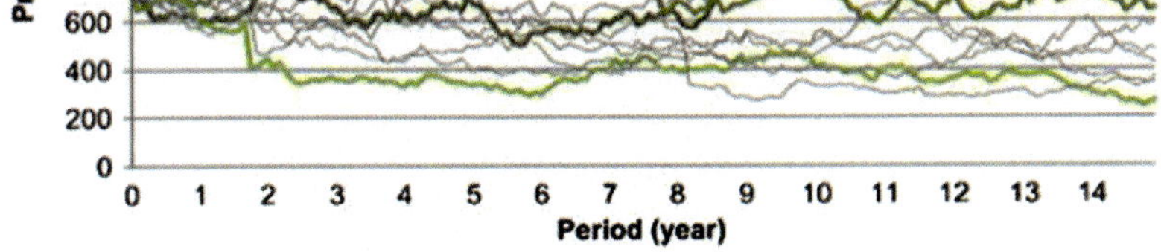

Fig. 9.33 Example of 10 price paths generated by a geometric Brownian motion with Poisson jumps model based on parameters obtained from historical data [281]

Gholamnejad (2015) proposed a MILP formulation for the long-term production scheduling problem under tonnage uncertainty calculated through an ore block probability index, which assigns the probability that each block will be mined in a given period by standard normal cumulative distribution function or indicator kriging. The economic value of a block is updated with this uncertainty-based tonnage. Then, the objective function that aims to maximize the NPV also considers this uncertainty, taking into account restrictions related to the grade control, precedence, and mining and processing equipment capacity. The case study at an iron ore deposit in Iran compared traditional and uncertainty-based models, confirming that the probabilistic approach increases NPV in the first periods by delivering more certain areas of the deposit at the beginning, thereby reducing the probability of deviation from the ore production target. Silva, Dimitrakopoulos, and Lamghari (2015) proposed a SIP model of the long-term production scheduling problem under tonnage uncertainty addressing multiple processing streams and a stockpile, based on the formulation by Lamghari and Dimitrakopoulos (2016a). A hybrid multistage heuristic algorithm is employed to solve the problem, where the initial solutions can be generated by two heuristics based on the principle of divide and conquer, the first one solving exactly each subproblem using BC.

In contrast, the second one is based on a greedy algorithm, and then, this initial feasible solution is improved using a network flow-based algorithm. The case study in a gold mine with two processing streams and a stockpile considered a set of 15 stochastic simulations, and the stochastic implementation was able to control deviations in ore production. Although a greedy heuristic found the initial solution in a few seconds concerning the hours spent by the BC algorithm, the improvement stage was 16% slower, and the NPV was 7.9% lower [170, 248, 273, 282, 283].

Montiel and Dimitrakopoulos (2015) proposed a framework to simultaneously address uncertainty in a mining complex, generating risk-based schedules for mining, processing, and transportation. The model aims to maximize NPV and minimize deviation from mining and metallurgical processing targets, while several constraints have been imposed, such as mining and processing capacity, blending, rehandling, precedence, and penalties. The model can handle various materials with multiple metallurgical properties from different mines, storing them in different stockpiles and sending them to different destinations. Considering the geological uncertainty of different deposits as equally probable stochastic simulations, the method takes an initial solution for each realization using a conventional optimizer. It improves it through three perturbation mechanisms associated with the extraction periods and destinations of the blocks, operational standards, and the transportation system. Then, an SA-based algorithm iteratively integrates the multilevel perturbations. A case study was developed in a mining complex with two copper pits, five types of materials, and five destinations, with the metal recovery of each material being distinct. Twenty stochastic orebody simulations were generated for each deposit, and a base case was developed using Whittle software with the E-types of deposits, presenting a scenario with impractical and large target deviations in terms of capacities and blending.

Nevertheless, the stochastic framework was employed in two cases: The first, considering just the multi-pit multiprocess, obtained 3% more NPV than the base case and low deviation from the targets; the second, considering the three disturbance mechanisms, reached 5% more NPV than the base case and resulted in a more flexible schedule. Montiel, Dimitrakopoulos, and Kawahata (2016) extended this work by proposing a framework to optimize mining complexes with open-pit and underground operations under geological uncertainty while simultaneously dealing with mining, blending, processing, and transportation operations. Based on the SA algorithm that iteratively integrates multilevel perturbations, a case study was developed in a gold complex containing open-pit and underground operations, resulting in 20 orebody realizations each to access uncertainty. The initial solution was obtained through the existing plans provided by the EPS and Milawa scheduler software for underground and open-pit mines, respectively, and the expected NPV obtained by the stochastic approach was 14% higher than that obtained by commercial software, also fulfilling the operational targets for autoclave processing [284, 285].

Del Castillo, Godoy, and Dimitrakopoulos (2015) proposed a MIP formulation to solve long-term mine production scheduling problems under geological uncertainty that considers the SSD proposed by Godoy and Dimitrakopoulos (2004), providing an optimal annual production rate, the optimal utilization of the fleet, and an equipment acquisition program. The objective function minimizes risk, as it deals with the revenue of high-grade ore (sent to the processing plant), the cost of mining the low-grade ore (sent to a stockpile), the cost of mining waste, purchase costs, and the ownership costs. The constraints are related to mining and processing capacity, grade control, interrelation and availability of equipment, and the relationship between waste and metal extracted, in addition to the restrictions inherent in assuming the SSD. A case study on a gold deposit was implemented using the Milawa algorithm, considering 20 equally probable orebody simulations to define the mining rates and an estimated orebody to be evaluated by the proposed optimizer, whose results were compared to those of a traditional case considering the average production rate. The conventional mining rate extracts larger amounts of waste early in the mine's life to obtain a fixed mining rate, requiring unnecessary capital investments in the early years, even when it is not necessary to meet mill demand. The NPV of the proposed approach was 20.7% higher than the traditional case, with 40% of equipment acquisition being delayed by seven years. Del Castillo and Dimitrakopoulos (2016) proposed a multivariate destination policy based on coalition formation clustering to solve the long-term production scheduling problem for polymetallic mines with multiple processing streams under geological uncertainty. Being a line of study of the cooperative game theory, this approach does not require cutoff grades and defines the destination of material considering the value and the relationship of the groups of blocks processed together. Based on the Shapley value mechanism, this methodology allows improving decisions. It simultaneously accounts for the variables that affect the blending and processing requirements, whether metals or deleterious elements, to maximize revenue and minimize deviation from production targets of the process streams. A preprocessing stage using *k*-means++ has been proposed to reduce the processing requirements, in which the scenarios of the deposit are clustered to group blocks into families and, then, processing destinations that are not viable based on the material type of the families are eliminated. A case study was proposed in a copper–gold mine with two deleterious elements and six processing streams. The proposed methodology was applied considering 15 orebody realizations, and a base case was developed considering an estimated model and a traditional scheduling approach. The risk assessment of the base case reached an NPV 4.8% lower than expected, in addition to the occurrence of shortfalls in the tonnage supplied to processing streams over the life of mine, while the proposed methodology presented an expected NPV 5.6% higher than the base case and was able to meet the processing requirements [286, 287].

Gilani and Sattarvand (2016) proposed a stochastic optimization algorithm based on the ACO approach to solving the long-term production scheduling problem under geological uncertainty, considering the ACS formulation to update the pheromone in each iteration and extending Sattarvand's (2009) work. As the ACO does not apply any explicit constraint handling mechanism, a normalization procedure is used to modify the selected depths of each column and convert them into a feasible pit, respecting the slope requirements. The model aims to maximize the NPV and minimize the stripping ratio, and the other constraints are managed by adding a penalty cost to the objective function when a violation occurs. The initial solution was obtained through the deterministic version of ACO in the E-type model of the realizations, so the stochastic ACO updates the risk block model (which indicates the probability of each block being mined in any period); therefore, blocks with a high probability of being exploited in a period receive more pheromone and have better chances of scheduling in that period. Two pheromone update strategies were developed, one considering a single probability value where the blocks with less risk than the threshold are included, and the other considering multiple probability values, where all the blocks contribute to pheromone update. A case study at a copper mine in Iran highlighted that the two stochastic strategies improved NPV compared to the traditional approach, with the multiple probability strategy being the one that produced better schedules. However, the single probability strategy is more practical [160, 288].

Lamghari and Dimitrakopoulos (2016b) developed a two-phase mixed-integer stochastic programming model to solve the production scheduling problem under metal uncertainty based on the progressive hedging (PH) algorithm and STWH. The first phase consists of decomposing the set of equally probable realizations of the orebody in groups selected at random, reducing the time needed to converge and the number of subproblems to be solved compared to the iteration of each realization. Hence, the subproblems related to each group are solved iteratively by the PH to achieve a common solution. As the PH only has guaranteed convergence in convex problems, the algorithm hardly converges (if converges) to a local optimum, which is why a stopping criterion must be applied. Then, the second phase starts in the current subproblem, considerably smaller than the original, where an STWH (based on a fix-and-optimize scheme) is applied. The objective function and constraints are similar to those of Lamghari and Dimitrakopoulos (2012), and the numerical results were obtained from the same datasets then studied, a gold and a copper deposit containing five instances each. This new proposition is compared with the sequential heuristic proposed in Lamghari, Dimitrakopoulos, and Ferland (2013), based on time-decomposition strategy, surpassing it in terms of quality, although it requires more time to reach the solutions. However, it takes considerably less time when compared to using CPLEX to solve the problem directly. It was verified that the STWH is not efficient when employed alone, and if a BC algorithm is used on the second phase instead of the STWH, the quality of the solutions is slightly superior, although the solution time is considerably greater. Matamoros and Dimitrakopoulos (2016) developed a SIP with a recourse model to address multivariate deposits' short-term production scheduling problem considering the metal uncertainty while simultaneously optimizing the fleet allocation and mine scheduling, including stockpiles. The model aims to minimize the total cost of mining and deviations from production targets, so the constraints are divided into two-stage decisions, where the first group does not consider uncertainty and evaluate the estimated orebody model.

In contrast, the second group (recourse actions) considers it and assesses the equally probable realizations of the orebody. The first group includes constraints such as extraction costs, movement of shovels, lack of production, and inadequate mining width; the second group contemplates the possibility of associating resource costs with deviations from quality targets and ore production, hauling costs, and lack of mining with maximum shovel productivity. A case study developed at a multi-element iron mine considers the long-term production schedule of the first year provided previously, where the probabilistic approach considers the realizations of the fleet and the orebody in its solution. At the same time, the deterministic scenario was based on the average values for all related inputs. The resulting stochastic monthly schedules could reduce costs, manage the fleet efficiently, and respect production constraints, while conventional schedules failed to optimize fleet allocation simultaneously and mine scheduling while respecting the constraints [272, 279, 289, 290].

Goodfellow and Dimitrakopoulos (2016) proposed a two-stage SIP model to solve the long-term production scheduling problem of mining complexes under uncertainty, aiming at the first stage to maximize the NPV and at the second stage to minimize the risk of deviation from targets, adapting the initial decisions using recourse variables related to processing stream variables and penalties. The first stage aims to optimize the extraction sequence and destination policies that, instead of relying on cutoff grade policies, decompose each material from the multivariate mines into clusters (*k*-means++ algorithm) by the similarity of attributes and then decide the destination of clusters in each orebody realization, for a processing stream, stockpile, or waste dump. In turn, second-stage resource decisions are employed to optimize the various processing flows in the supply chain. Three combinations of metaheuristics used to solve this problem are assessed, an SA approach, an SA with downstream PSO (SA-D-PSO) approach, and a SA with downstream DE (SA-D-DE) approach; in addition, three perturbation mechanisms are used during annealing, consisting of the extraction sequence, destination policy, and processing stream perturbations. A case study was proposed in a copper–gold mining complex, in which the initialization of the algorithm, which can be based on a predefined sequence or not, was determined to start randomly, and the uncertainty was accessed based on 25 orebody simulations. A

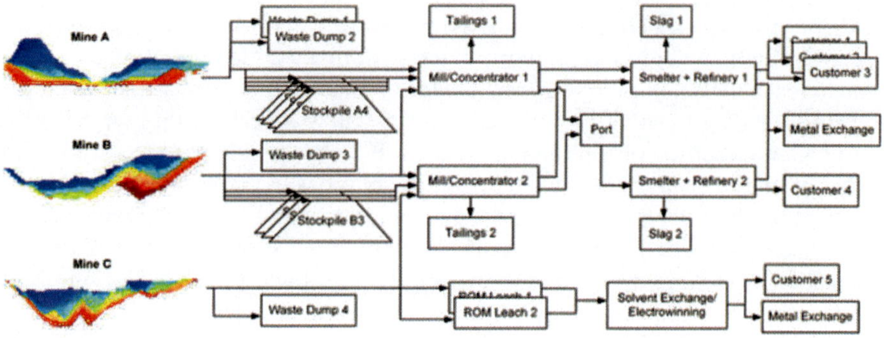

Fig. 9.34 Example of a mining complex with multiple mines, stockpiles, processing streams, and destinations [292]

deterministic assessment was developed using an E-type orebody model, comparing a schedule provided by commercial software and the methodology proposed, with each proposed metaheuristic formulation performing 10 tests. The proposed formulation surpasses the NPV of commercial software by 13.8%, while in the comparison between metaheuristics, the SA-D-PSO and SA-D-DE algorithms increased the NPV by 1.91% and 2.57%, respectively, when comparing to the SA algorithm, although they require 2.4 and 2.9 times more processing time. Hence, 10 trials were performed using the SA-D-DE optimizer for the stochastic approach, whose NPV was 6.6% higher than the deterministic metaheuristic approach and 22.6% higher than that obtained by commercial software, which can also generate designs that minimize the risk of not meeting production targets. Goodfellow and Dimitrakopoulos (2017) extended their previous work and analyzed the blending policy for a nickel laterite complex with multiple stockpiles and blending constraints. Figure 9.34 presents an example of a mining complex with multiple mines, stockpiles, processing streams, and destinations [291, 292].

Montiel and Dimitrakopoulos (2017) proposed a heuristic approach to solve life-of-mine production scheduling problems considering operational alternatives for processing plants and geological uncertainty through iterative random improvements that swap mining block periods and destinations until a final solution is generated. The objective function maximizes the NPV subject to constraints related to mining and processing capacity, grade control, precedence, and blending, and the slope requirements are based on a geotechnical zone file inserted as the initial solution. First, the heuristic assigns periods and destinations to the mining blocks by the initial solution. The set of equally probable orebody simulations are assessed for profits and costs, and the algorithm assigns each block destination based on the overall profitability by available destination accumulating it through simulations. The iterative perturbation stage finally selects blocks at random. It sorts their available destination, bringing the blocks to early periods if the best destination increases the objective function's value or pushes them to later periods otherwise until a stopping criterion is reached. At the same time, a verification mechanism is employed to ensure that constraints are respected. The case study in a copper deposit with 25 orebody realizations used to improve the solution and other 25 simulations to assess the risks shows the robustness of this approach, capable of controlling mining and processing capacities and increasing NPV by 30% compared to the LG algorithm. Montiel and Dimitrakopoulos (2018) proposed a two-stage stochastic MINLP framework based on Montiel and Dimitrakopoulos (2015) to solve long-term production scheduling problems that optimize mining, stockpiling, blending, and processing variables under geological uncertainty. The first stage consists of an SA metaheuristic with multiple neighborhoods and a perturbation mechanism based on the proposition of Montiel and Dimitrakopoulos (2015), while the second stage consists of an adaptation mechanism that ensures the operability of each bench in each pit in the mining complex given the infrastructure, constraints, access, and equipment available. A case study in a gold complex with multiple pits, stockpiles, and processing streams compared the results of the stochastic proposition with a long-term deterministic mine plan adopted by the company, identifying that the first one provides an NPV 7% higher than the deterministic plan, also increasing the probabilities of attending the production and blending targets [32, 284, 293, 294].

Moreno et al. (2017) developed a two-stage stochastic MIP model to solve the long-term production scheduling problem under geological uncertainty that can be reformulated as a PCKP, which can be near-optimally solved using decomposition techniques. In the first stage, decisions are made based on a deterministic scenario, where each region (a block or a phase) of the deposit is assigned to an extraction period. Moreover, in the second stage, recourse decisions are made based on the multiple equally probable orebody realizations, when each block in the region is then reassigned to an extraction period and a destination. The model aims to minimize the cost of extraction (first stage) and maximize the expected value obtained with recourse decisions (second stage), and the constraints can be related to the reserve, precedence, capacity, and slope. A case study was developed in a copper deposit in Chile, where two sets of 50 orebody realizations were composed, the first based on 76 drill holes (high uncertainty), while the second relied on 753 drill holes (low uncertainty). As a result, two problems have been solved: block-level scheduling and bench-phase scheduling, which is based on a predefined set of phases calculated using the Milawa algorithm. Three approaches were used to evaluate each problem and each set of scenarios, i.e., the proposed two-stage stochastic model, the deterministic approach based on the E-type of the simulations and an approach called "Cristal Ball," in which the deterministic model is solved for each of the 50 realizations, which provides upper bounds for the best possible NPV of each one. The NPV value of its linear relaxation was obtained for each approach using a version of the BZ algorithm, reaching a gap of less than 1%, and the results provided by the stochastic approach showed higher NPVs than the deterministic model in all scenarios while capturing the most of the uncertainty of the problem [188, 295].

Richmond (2018) developed a stochastic framework based on a directed graph to solve the long-term production scheduling problem under geological, metal price, and production cost uncertainties using Monte Carlo simulation and a cone-based heuristic. In order to maximize the NPV, a traditional floating cone algorithm was used, which decomposes a full directed graph problem into a series of subproblems, where whenever the sum of the payoffs associated with a new cone is major than the solution previous, this new closure is added to the solution and a new iteration begins until a stopping criterion is reached. A case study considering 25 copper deposit realizations, 25 stochastic copper price simulations, 25 operating cost simulations, and monthly copper recoveries obtained at random from a normal distribution and a discrete five-year range to define the time of initial investment was developed, thus quantifying several aspects of uncertainty in a mine plan. Jamshidi and Osanloo (2018) developed a linear formula to determine block economic value in multivariate deposits, considering the effects of blending and assessing this technique's influence in defining the ultimate pit limit under geological uncertainty. Considering the blocks in an iron deposit whose iron content and beneficial and deleterious elements are within an acceptable range, these blocks are classified as direct-shipping ore, and the block economic value can be calculated directly. Otherwise, they are classified as non-direct-shipping ore and require blending in their block economic value definition. A case study developed at an iron ore mine in Iran compared the deterministic final pit produced using the NPV scheduler using this methodology with the pit produced by traditional block economic value, reaching an increase of 3% in reserves. The uncertainty was assessed using 20 orebody realizations, and the final pits were obtained individually using the NPV scheduler and the proposed methodology for determining the block economic value. It evaluated the probability of each block being extracted considering all the pits [296, 297].

Rimélé, Dimitrakopoulos, and Gamache (2018) proposed a two-stage SIP model to address the long-term production scheduling problem under geological uncertainty, including the disposal of waste and tailings within the exploited parts of a pit, a useful strategy mainly when the deposit has low dip layers and long strike length. At each period, the algorithm defines an upper and a lower strip following the pit dip to delimit the storage zone, the strips between these two being reserved for storage so that the framework aims to define both the extraction sequence and the disposal of materials. The objective function aims to maximize the expected NPV while minimizing deviations from production targets, subject to a reserve, precedence, grade control, and capacity constraints, in addition to several constraints related to the disposal of materials, such as the singularity of the upper and lower strips, permission for storage within a strip, quantity of tailings stored outside the pit, among others. An application was proposed at a multi-element iron ore mine in Canada considering 10 extraction periods, two destinations, 38 strips of storage, and 10 equally probable orebody simulations, the results of which demonstrated the proposal's main aspects and advantages were solved using an STWH. Ajak, Lilford, and Topal (2018) proposed using a predictive data mining algorithm to generate possible real options for the operation to deal with the clay material geological uncertainty, where grade control data is used to train and test the machine learning model. The mine plan data is used to apply the model for ROV analysis. A decision tree based on the ID3 algorithm was selected among other algorithms to deal with the problem of predicting the presence of clay material in the multi-element iron ore before being fed into the processing plant, and a dataset of 832 mining blocks was used to train and test the model, reaching 78.6% of accuracy. Thus, techniques such as the Poisson distribution model and Monte Carlo simulations were applied in order to proceed with the ROV analysis and create a discounted cash flow model, resulting in the prediction that the present value would be reduced considering the effect of feeding the crusher with clay material when compared to the traditional model. Because of this forecast, the options of developing a rehandling material policy, building a bypass circuit, or adding a new mining area were evaluated, reaching increases in the present value in the order of 21%, 16%, and 12%, respectively [298, 299].

Quigley, Dimitrakopoulos, and Grammatikopoulos (2018) proposed a risk-resilient mine production scheduling methodology to deal with the extraction of a deposit with multiple rare earth elements, using multiple-point simulation of geological domains through single normal equation simulation (SNESIM), joint stochastic simulation of the continuous variables using min/max autocorrelation factors (MAF), and a two-stage stochastic recourse model. Issues related to geological modeling and grade estimation are beyond the scope of this chapter, whereas the SIP formulation proposed is based on Ramazan and Dimitrakopoulos (2004c, 2013) and Benndorf and Dimitrakopoulos (2013), where the objective function is to maximize the expected NPV (first stage: anticipative decisions) and minimize the costs of deviating from production targets (second stage: adaptive decisions). To solve this formulation, a multi-neighborhood TS metaheuristic based on Lamghari and Dimitrakopoulos (2012) was used, considering four types of neighborhoods at each iteration, where the one that promotes the biggest change in the objective value is selected, and the solution is then updated. The strategies consist of changing the extraction period of a block, changing its destination and extraction period simultaneously, and exchanging the two blocks' destinations in the same period. A study case in a fictitious rare earth deposit known as the Jupiter mine was developed, considering 15 elements in the orebody and 20 realizations of the domains using SNESIM, each one filled with grades from one joint stochastic simulation using MAF. When comparing the stochastic schedule with the conventional one, generated through Whittle software, it can be seen that the former surpasses the latter by better meeting the targets of mineral grades, achieving greater accuracy and certainty.

On the other hand, the expected NPV is 12% higher in the stochastic approach, which could be even better if the penalty fees for not meeting the customer's agreements in the conventional scenario were considered. Paithankar and Chatterjee (2019) proposed a hybrid of maximum flow and GA to solve SIP models that address long-term production scheduling problems, in which the structure of the graph for maximum flow is created for multiple periods under geological uncertainty, and a GA metaheuristic controls the flow in the arcs to develop the production schedule. The objective function aims to maximize the expected NPV subject to a reserve, slope, mining, processing, and metal production constraints, and the genetic operators employed by the algorithm include random selection, crossover, and mutation, in addition to a parametric diversification strategy used to guide the search that changes parts of the chromosome when operational constraint violations occur. In order to avoid addressing the entire problem at once, a variable chromosome length GA is applied, using the benefits of a multistage strategy, so that the problem is solved for one stage and then the best chromosome for that stage is added as a predecessor to solve the next stage. Numerical results were obtained for three different problems applied to real copper and gold deposits, considering 20 equally probable orebody realizations for each deposit, the average gap obtained in these cases being less than 4%. More numerical results were obtained by applying the proposed methodology to the instances of MineLib, whose results were compared to those obtained by the modified top sort method to solve the Lagrangian relaxation (Espinoza et al. 2013) by the sequential heuristics with the VND approach (Lamghari, Dimitrakopoulos, and Ferland 2015) and by a graph-based algorithm (Liu and Kozan 2016). The results obtained were robust, and the average gap generated is less than 6%, with the exception of one instance [142, 143, 228, 248, 252, 276, 300, 301].

Del Castillo and Dimitrakopoulos (2019) proposed an adapted multistage stochastic programming model to solve a mining complex's long-term production scheduling problem under geological uncertainty, expanding the two-stage structure as multiple recourse stages are solved iteratively, and a scenario tree structure is developed on feasible mine designs. This model optimizes the annual production schedule as a dynamic plan. Investment decisions are made sequentially based on the new information that arises each period, allowing for advance planning of possible configuration changes resulting from new capital investments. By defining a threshold in which a representative portion of the scenarios differs in terms of major investment decisions (such as building a new processing plant), a solution can be divided into feasible branches or unique solutions for each period, which is why minor decisions must be avoided. Otherwise, the model would grow exponentially. The mine plan is optimized as a two-stage SIP based on the formulation by Goodfellow and Dimitrakopoulos (2016), so the iterations of the second stage start at a year t, freezing it and considering all future extraction and processing decisions, and if a representative number of scenarios consider an investment in a given year, two branches are created. Thus, the iteration freezes the previous period and proceeds to the period $t + 1$ until all branches in all periods have been evaluated. An adaptive neighborhood search combined with an SA mechanism was used to solve the problem, in which each decision variable defines a neighborhood, and the solution space is iteratively perturbed. A case study was developed in a copper mine complex composed of two mines, two processing destinations, and a stockpile, considering 100 scenarios for the mining complex, the options of investing in the truck and shovel fleet, and in a secondary crusher. The base case was designed as a standard two-stage simultaneous stochastic optimization of the mining complex, resulting in a unique production sequence and destination policy, and the purchase options for truck, shovel, and secondary crusher were included in the first stage of the model. For a 20-year life of mine, the proposed methodology indicates that there is a 42% chance of investing in the secondary crusher in year 3, which leads to a 3% higher NPV in the P50 value compared to the base case, while an NPV 8% higher can be achieved by considering the P90 value. Figure 9.35 presents an example of including investment alternatives in the mine planning process [291, 302].

Mai et al. (2019) proposed a stochastic framework in which the TopCone algorithm proposed by Mai, Topal, and Erten (2018) is employed in the orebody simulations

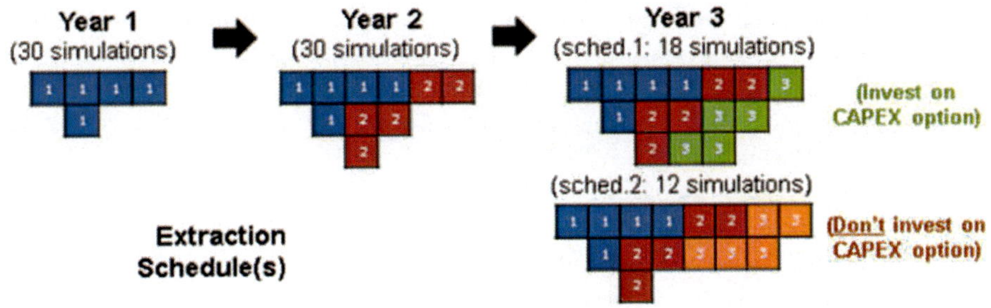

Fig. 9.35 Example of including investment alternatives in the mine planning process [302]

to reduce the scale of the data. Thus, a multistage SIP model works the aggregate blocks to address the long-term production scheduling problem under geological uncertainty over a practical timespan. The Topcon algorithm considers the E-type of the realizations to generate its TopCones, whose stochastic values are evaluated by replacing the realizations in the TopCone set. Hence, the SIP model to maximize the economic values of mining TopCones, maximize the bonuses obtained by meeting the grade requirements, and minimize the cost of not meeting the production targets is employed, subject to constraints related to processing and mining capacity, blending, sequencing, resource, and non-negativity. A case study developed at a large-scale multi-element iron ore deposit in Western Australia considering 20 equally probable orebody realizations and 8 periods compared the scheduling result of the proposed formulation with a deterministic IP model in CPLEX, reaching 6.88% and 2.28% more NPV result with and without considering grade bonus/penalty and uncertainty cost, respectively. Using an ordinary computer, the proposed formulation showed the ability to manage risk and produce solutions in a practical timespan. Saliba and Dimitrakopoulos (2019) proposed adapting and applying the SIP model by Goodfellow and Dimitrakopoulos (2016) to solve the long-term production scheduling problem of a multi-pit and multiprocessor gold mining complex in the USA under simultaneous geological and market uncertainty, considering strict geochemical blending constraints. The model simultaneously optimizes the extraction sequence, destination policy, and processing stream decisions, eliminating the use of a priori cutoff grade optimization, since these decisions are determined by the optimization process considering the operational constraints and the variability of the material, thus reducing the number and complexity of decisions to be made. The multistage model aims to maximize NPV, minimize processing and transportation costs, minimize the cost of deviating from geochemical, processing, and mining capacity targets, and minimize mining cost and the cost of deviating from the schedule smoothness and sink rate constraints. The 5 commodity price simulations were obtained using the geometric Brownian motion with Poisson jump diffusion technique, evaluated along with 15 equally probable orebody realizations, and the metaheuristic approach proposed by Goodfellow and Dimitrakopoulos (2016, 2017) was employed to solve the problem. This simultaneous case was compared to a base case which was also a simultaneous stochastic optimization, although constrained by a conventional cutoff grade procedure and considering a constant gold price instead of simulated ones. The results showed that the joint case extracted 8% more material and reached 3% more NPV in p50 than the base case, allowing more flexibility regarding price changes and a reduction in complexity since the joint case requires 12 stockpiles versus 38 required in the base case [231, 291, 292, 303, 304].

Sepúlveda, Álvarez, and Bedoya (2020) proposed an implicit stochastic optimization model based on GA metaheuristic and Monte Carlo simulation to solve the long-term production scheduling problem of polymetallic deposits under geological, technical, and market uncertainties. The objective function aims to maximize the expected value of the NPV and minimize risks. Constraints were grouped in relation to mining, processing, floatation, capacity, and conditions of decision variables, and a set of pushbacks already determined can be used as an initial solution. Geological uncertainty was considered through a grade distribution function for each block, while economic uncertainty considers forecasts with time series based on historical data to predict prices and costs. The cutoff grade policy is dynamic over time. Multiple processes and destinations are also considered, including stockpiles, in a formulation in which genetic operators employ selection, crossing, and mutation to propagate the characteristics of the best individuals to the next populations. A case study developed in a mine that exploits copper, molybdenum, and gold is performed, and the results obtained are discussed. Quigley and Dimitrakopoulos (2020) expanded the SIP model proposed by Matamoros and Dimitrakopoulos (2016), which addresses the simultaneous optimization of the short-term production scheduling and the allocation of equipment under equipment performance and metal uncertainties, in order to minimize the total mining cost and deviations from production targets. Among the improvements are the adaptation of the model to incorporate mining complexes, the incentive to extract interconnected blocks in the same period to avoid unnecessary shovel movements, the concept of including ramp positions that allow the construction of smoother mining patterns by defining a single horizontal precedence direction for each block, and a new joint simulation of equipment performance parameters using principal component analysis (PCA). To solve the problem, three variants of the earliest start heuristic are employed at the beginning of each STWH iteration, thus eliminating some binary block extraction decisions and reducing the solution time. A case study was developed in a very large copper mining complex containing two pits, the deposit of which was represented as 20 equally probable simulations, the material to be exploited was defined based on a year in the long-term production schedule, and the length of the period was considered to be one month. The numerical results were obtained by applying the proposed stochastic methodology and comparing it with a conventional schedule, where the deposit was estimated by ordinary kriging, and the average performance of equipment is considered. The improvement and consistency of the quantity and quality of the ore delivered to each processing destination in the stochastic approach are evident, reducing the risk of deviation from production targets and even reducing the risk of deviations from the shovel's production, in addition to proposing a schedule with a greater likelihood of being accomplished. Figure 9.36 presents a mine area with its associated ramp design and horizontal precedence sets on the top and bottom benches [290, 305, 306].

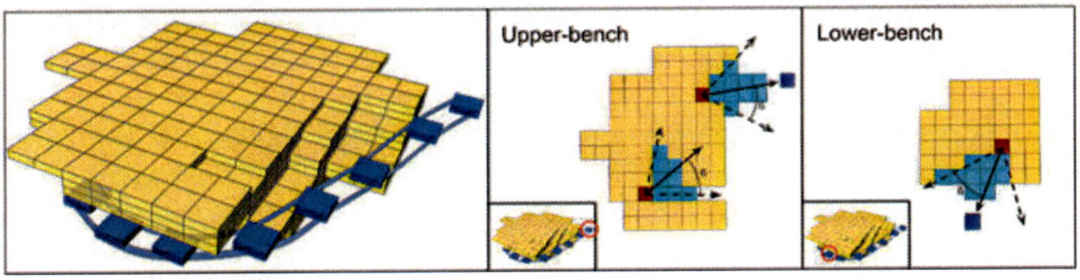

Fig. 9.36 It is a mine area with its associated ramp design and horizontal precedence sets on the top and bottom benches [306]

Lamghari and Dimitrakopoulos (2020) proposed a hyper-heuristic based on reinforcement learning and TS to solve the stochastic open-pit production scheduling problem with one or multiple processing streams and stockpiles, thus consisting of an algorithm that selects or generates heuristics to solve the problem instead of solving it. Three hyper-heuristic approaches are proposed to select among 27 existing simple low-level perturbative heuristics to address the problem and improve a candidate solution at a certain stage of the search process. This decision consists of a score-based learning mechanism that associates each heuristic to a score depending on its previous performance. Two approaches are previously developed formulations, the Tabu search and the choice function hyper-heuristics (HH1 and HH2), while the third is a new proposal for hyper-heuristic based on the ideas of the previous ones (HH3), although it includes new resources to surpass their weaknesses. Numerical results were obtained from applying the three hyper-heuristics proposed on five copper and gold mine test sets, containing 43 instances of different sizes and operational characteristics, most previously studied in Lamghari and Dimitrakopoulos (2012, 2016a). The results were also compared with those obtained by applying the TS heuristic proposed in Lamghari and Dimitrakopoulos (2012) for cases with a single process stream and the DLS heuristic proposed in Lamghari and Dimitrakopoulos (2016a) for cases with multiple process streams and stockpiles. It can be concluded that the HH1 and HH2 are faster but have resulted in solutions of inferior quality; on the other hand, HH3 is faster and more robust when compared to TS and DLS, being substantially better at solving larger and more difficult instances, although the other two may perform slightly better in some instances. Lagos et al. (2020) developed an adaptive framework to solve the multi-period production scheduling under geological uncertainty that can learn from the information that arises as more blocks are exploited, thus producing an operational risk reduction policy. Based on the non-adaptive stochastic programming proposed by Moreno et al. (2017), a rolling horizon policy was added to the structure in order to update the ore grade distribution in each period and generate new scenarios based on the recent observations and then does a two-stage stochastic algorithm solve the problem to set extraction decisions for the current evaluated period.

Moreover, a one-step look-ahead policy based on an approximate dynamic programming approach has also been proposed, an independent formulation of the non-adaptive stochastic programming, which would overcome possible failures to anticipate that extraction decision will be reviewed after each period. Numerical tests were carried out in mines of various sizes with 100 equally probable orebody realizations compared to the proposed approaches. The rolling horizon formulation surpasses non-adaptive stochastic programming in terms of computational time and gap, although both achieved similar convergence results when the number of drill holes was large. On the other hand, the approximate dynamic programming approach does not perform well, consuming more computational time to reach the solutions [272, 273, 295, 307, 308].

Gilani et al. (2020) developed a SIP model to tackle the long-term production scheduling problem under geological uncertainty and then employed a PSO-based algorithm to solve it, comparing four different formulation strategies. The SIP formulation aims to maximize the NPV and minimize risks of not meeting the production targets subject to operational and technical constraints, such as slope, reserve, and penalty cost, the latter related to the violation of mining, processing, and grade constraints. The initial solution can be obtained by applying a traditional scheduling algorithm to the E-type model, in order to create a deterministic case, and to the geological simulations generated, in order to create a stochastic base that will compose the risk block model, which assigns the probability of extracting each block in each period. Similar to the ACO-based algorithm proposed by Gilani and Sattarvand (2016), the 3D surfaces that represent the nested pits are converted into column depths, so these depth sets are fed into the SIP model and solved by the PSO propositions, which iterate, perturb, update the particle positions (depths) and normalize the solution to meet the constraints. The PSO propositions are based on Gbest (GPSO-SRB and GPSO-MRB) and Lbest population topologies (LPSO-SRB and LPSO-MRB) (see the explanation in Khan and Niemann-Delius (2014) for more details). The SRB version includes blocks to contribute in-depth determination process whose probability values obtained from the risk block model are equal to or greater than the risk reference probability, while the MRB version allows all blocks to contribute according to their probability in the risk block model. A case study at a copper mine in Iran with 20 equally probable orebody realizations was developed, in which the GPSO-SRB and GPSO-MRB increased the fitness value of the initial E-type solution by 9.44% and 12.09%, respectively, while LPSO-SRB and LPSO-MRB increased by 9.85% and 13.68%, respectively. In terms of execution time, the MRB version is faster than the SRB and generates better solutions, especially in low-risk scenarios. Finally, the GPSO strategies were more susceptible to trap in local optima than the LPSO. Both and Dimitrakopoulos (2020) proposed a stochastic model to simultaneously optimize short-term production scheduling, shovel relocation, heterogeneous hauling fleet schedule, and destination of materials in mining complexes under geological and equipment performance uncertainties, extending the work of Goodfellow and Dimitrakopoulos (2016) by adding features related to short-term planning. Rather than just minimizing operating costs, the objective function aims to maximize the metal production and profit from the mining complex while also aiming to minimize penalties for deviations from production targets, risk of not reaching shovel targets by mining area, shovel movement cost, risk of running short of truck haulage capacity, truck cost, and a term to impose a smooth mining schedule. A SA metaheuristic is adapted to handle all additional decision variables related to short-term production scheduling, and a case study was developed in a gold mining complex, including two open pits, six mining areas, two processing streams, and a stockpile, considering 15 equally probable orebody realizations to take into account geological uncertainty. A comparison of the proposed methodology with a two-step approach as proposed by Goodfellow and Dimitrakopoulos (2016) resulted in a 56% reduction in costs generated by shovel movements and a 54% reduction in production loss due to shovel relocation, in addition to reducing 3.1% of truck costs and 25% of the required haulage capacity in the most intense periods [169, 288, 291, 309, 310].

Dimitrakopoulos and Senécal (2020) proposed using a parallel multi-neighborhood Tabu search (P-MNTS) approach to solve a multistage SIP model addressing the long-term production scheduling problem with multiple processing streams under mineral supply uncertainty. The objective function consists of four parts, the first aiming to maximize the expected NPV, the second maximizing the value of processing the resources of the stockpile for a fixed grade, the third minimizing the rehandling cost, and the fourth minimizing the costs for not meeting the production targets. The constraints are related to reserve, precedence, mining, processing capacities, stockpiling, and additional constraints. The algorithm defines a feasible initial solution, searches simultaneously for multiple neighborhoods, and then applies the best permissible moves in the solution until a stopping criterion is met, when a diversification strategy as proposed by Lamghari and Dimitrakopoulos (2012) is employed to build a new starting solution, iterating until the maximum time allotted for the algorithm is reached. To prevent searching in entire neighborhoods, a mechanism is used to remove solutions that are far from meeting processing capacity constraints from the search space; on the other hand, in order to search the solution space extensively, another mechanism is used to consider infeasible solutions in terms of mining capacity constraints in the neighborhoods. Two types of move operators have been implemented to generate new neighbors, corresponding to a change in block's extraction period or block's destination, so that four different neighborhoods can be created, using each one of these operators individually, simultaneously, or by swapping the destination of two blocks extracted in the same period. The initial solution's generation was tested using two strategies: an exact method that generates a suboptimal solution and a random heuristic. Two case studies were developed, evaluating three small and medium copper deposits among the ten instances assessed in Lamghari and Dimitrakopoulos (2012). The second applies the proposed methodology to a large gold deposit, comparing the results against the linear relaxation values obtained through CPLEX. The results showed that both initialization methods could generate near-optimal feasible solutions, although the exact method can achieve a higher NPV in a much more expensive computational time. The gap widens as the complexity of the problem increases, and solutions that used parallelism improved their performance than those that did not [272, 311].

Maleki et al. (2020) proposed a mixed SIP model to solve the long-term production scheduling problem under geological uncertainty, simultaneously addressing the rock type and seven quantitative variables in 100 equally probable realizations of an iron ore deposit, in which a plurigaussian approach simulated the domains and the turning bands algorithm simulated the grades. The model aims to maximize the expected NPV and minimize the cost of deviations from production targets, subject to constraints such as precedence, slope, grade control, and mining and processing capacity. First, the ultimate pit limit was obtained for each realization to reduce the size of the problem. Then, a temporal decomposition strategy was employed to divide the monolithic problem into subproblems associated with fewer periods, together with a block preselection procedure based on a linear programming relaxation solution. The proposed methodology was compared to a deterministic one that considered the E-type of the realizations, improving the expected NPV by 6% and the amount of iron ore extracted by 16%. Even though the SIP model allows for deviations, it controls the cost associated with not meeting the targets to maximize the objective function and resulted in a schedule with greater capacity to meet the requirements than the deterministic model. Rimélé, Dimitrakopoulos, and Gamache (2020) proposed a two-stage SIP model combined with a stochastic dynamic programming (SDP) algorithm to address the long-term production scheduling problem under simultaneous geological and commodity price uncertainty. The two-stage SIP model consists of a common framework, which develops a deterministic schedule in the first stage and then applies recourse variables to the realizations of the deposit in order to denote deviations from production targets. Therefore, the objective function maximizes the expected NPV while minimizing the cost of deviating from production targets, subject to reserve, slope, and mine production target constraints.

Nevertheless, in this methodology, the SIP model is also responsible for pre-computing a set of possible subsets of blocks aiming given metal targets, thus building a tree of viable extraction sequences, in which the children nodes correspond to variations in metal targets considering an increase or decrease in an initial root value. Hence, an additional constraint is considered to fix blocks previously extracted at the parent node in the tree. Thus, the SDP algorithm is used to define the best policy (or which subset of blocks should be extracted in the next period) depending on the price evolution using the backpropagation method, beginning at the last period and arriving at the first one. A case study was developed on a small part of a copper orebody, which was scheduled in a maximum of 4 periods and considered the 5% variation in the metal's production target from one period to another. Geological uncertainty was addressed as 20 orebody realizations. In contrast, copper price uncertainty was modeled as a Markovian process, considering five equally probable price variations in each period, the problem being solved by CPLEX and the results highlighting the main aspects and advantages of the proposal. Figure 9.37 presents a tree of metal production targets computing the pool of subsets of blocks [312, 313].

Chatterjee and Dimitrakopoulos (2020) proposed a hybrid method that combines Lagrangian relaxation and a BC algorithm to divide the long-term production scheduling problem under geological uncertainty into subproblems solve it sequentially. The formulation of the problem was based on the two-stage SIP model proposed by Ramazan and Dimitrakopoulos (2013), in which the objective function aims to maximize the expected economic return and minimize the cost of deviating from production targets, subject to constraints related to precedence, reserve,

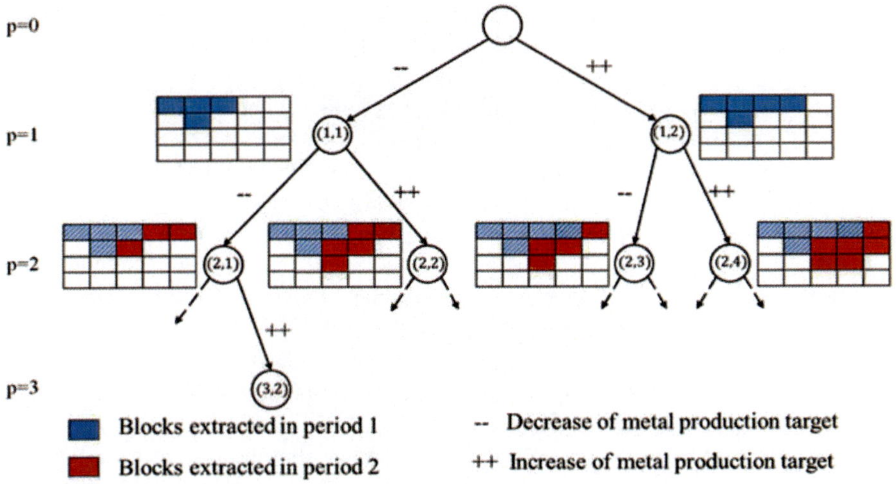

Fig. 9.37 Tree of metal production targets computing the pool of subsets of blocks [313]

grade control, mining, and processing capacity. First, Lagrangian relaxation is used to relax constraints and generate subproblems, which can be seen as a minimum-cut problem of a directed graph, which can be solved by a maximum flow algorithm such as the sub-gradient method in order to represent an upper-bound solution value for the initial problem. To generate the solution of the lower bound, the procedure proposed by Meagher, Abdel Sabour, and Dimitrakopoulos (2009) was used, in which a parametric minimum-cut algorithm is applied to produce a single pushback design that meets all the constraints of the upper bound. Therefore, in order to add new elements to the lower bound, a MIP model is formulated and solved using a branch-and-cut algorithm so that each solution sequentially updates the lower bound of the subproblem until reaching a final solution for a determined period, which is then eliminated from the problem so that the next period updates are obtained from the new smaller subproblem. The proposed methodology was tested using six real datasets of copper and coal mines, and the results are compared with the linear relaxation of the exact formulation of the problem, solved in CPLEX considering the E-type of the 20 equally probable orebody realizations of each deposit. The optimality gap showed reasonable results, within 3.5%, while the computational time was reduced significantly, including the instance in which the exact solution was not reached in four days of processing. A case study was developed in a copper deposit that resulted in a schedule that meets the resource constraints, reaching 11% more NPV than the obtained by the linear relaxation of the exact formulation and in 26% more NPV than that obtained by the Whittle software [252, 263, 314].

Tolouei et al. (2020) proposed three hybrid metaheuristic-based algorithms to solve an IP model of the long-term production scheduling problem under grade uncertainty, combining Lagrangian relaxation with PSO (LR-PSO), BA (LR-BA), and firefly algorithm (LR-FA). With the objective function of maximizing the NPV subject to constraints related to grade blending, reserve, slope, mining, and processing capacity, the explicitness of the constraints on the objective function and the introduction of Lagrange multipliers relax the problem and split it into smaller subproblems. An initial population of Lagrange multipliers is defined, and then, the algorithms update them until a stopping criterion is reached. Considering a GA approach and a traditional sub-gradient as benchmarks, the techniques were employed on an artificial dataset with 150 blocks, in which the LR-FA showed better results from the point of view of the duality gap. Hence, the second pushback data of an iron ore mine from Iran was assessed, in which the average ore grade in the twelve-year schedule using the LR-FA method reached 55.12%, while for the LR-BA, LR-PSO, LR-GA, and LR-SG methods, the grades reached were 54.89%, 54.64%, 54.46%, and 54.31%, respectively. The NPV comparison confirmed the robustness of the LR-FA approach, defeating the LR-BA, LR-PSO, LR-GA, and LR-SG approaches by 1.87%, 2.83%, 4.43%, and 5.43%, respectively. Tolouei et al. (2021a) expanded the work of Tolouei et al. (2020), comparing the use of hybrid approaches combining Lagrangian relaxation and augmented Lagrangian relaxation with BA and FA to solve the long-term production scheduling problem under grade uncertainty, comparing their results to those obtained using GA, a traditional sub-gradient and conventional methods. By applying the techniques to the same pushback data proposed, the results showed that the augmented Lagrangian relaxation versions required less computational time and reached smaller gaps than the Lagrangian relaxation versions, resulting in higher NPVs and a higher average grade of ore. When comparing metaheuristics, the FA algorithms were faster and more robust, followed by BA, the GA, the sub-gradient, and the traditional approaches, similar to that stated by Tolouei et al. (2020), but whose augmented Lagrangian relaxation versions were better for all hybrid methods [315, 316].

Trends for the Future

The interaction among the departments is a sine qua non condition for the success of any mining enterprise, such as the quantity and the quality of the data supplied by each one of them, in this new scenario provided by the emergence of concepts such as the Fourth Industrial Revolution and Mining 4.0, where the integration of operation technology to information technology (OT-IT), Internet of things (IoT), big data, cloud computing, the insertion of artificial intelligence techniques in the mining routine, and the paradigms are completely new. As the discovery of new world-class, massive deposits diminishes, as the average grade of the ore exploited decreases, and as the increase in the scale of the equipment is no longer a solution to enhance productivity, dealing with data and information with a disruptive eye seems to be the solution to increase the results of companies worldwide [317].

Bassan et al. (2008) predicted that the future of mining scenario in 2025 would be composed of data-savvy mineworkers, augmented reality, omnipresent computing, presence, digital labeling, transforming enterprises from a motionless distinct series of functions to a digitally conscious, integrated, and interactive environment with information accessible just in time. In some ways, the authors were accurate. At that time, they predicted that the industry would adopt remote operations centers and IT systems integration using service-oriented architectures (SOA), which, together with automated sensor and control technologies, actually provided a level of integration that improved the capacity and performance of the entire system. Figure 9.38 presents their idea of management evolution, considering the degree of process automation and the level of business integration [318].

The development and implementation of technologies capable of promoting integration in the mining industry are exponential, generating an amount of data that is difficult to analyze and quantify. Figure 9.39 presents the evolution of the implementation of autonomous trucks in mining over the years. As in many other unit operations, it has an exponential trend of modernization and integration, contributing to the generation of more and more data each second.

At the end of the 2010s, Bassan and Knights (2008) highlighted the following topics as relevant for advanced analytics in mine planning:

- concurrent optimization of pit limits and pushbacks;
- incorporation of geometallurgical parameters in mine planning;
- rapid revision of life-of-mine plans;

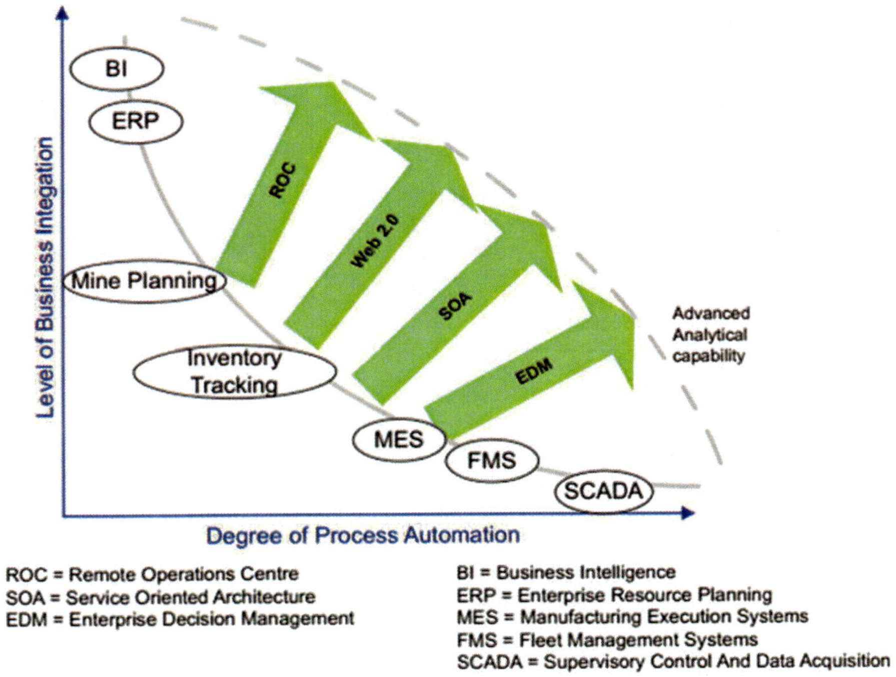

Fig. 9.38 Relationship between business integration and degree of process automation mining systems [318]

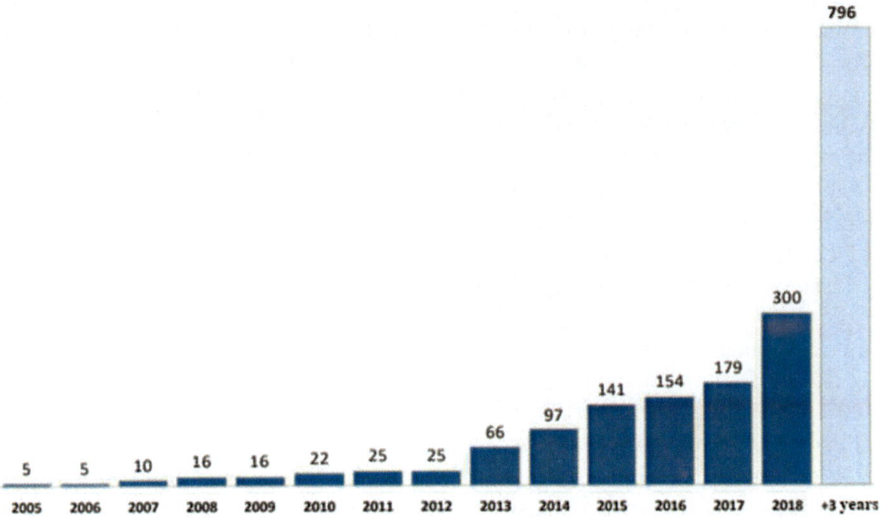

Fig. 9.39 Evolution of implementation of autonomous trucks in mining over the years [319]

- uncertainty quantification;
- rapid reconciliation and adaptive planning;
- collaborative mine planning; and
- modern financial valuation [320].

There was no computational capacity to implement direct block scheduling (DBS) techniques with several constraints or to develop a stochastic mine planning approach in great block models; Fig. 9.40 presents the evolution of microprocessor trend data, which increased considerably from one decade to another. Commercial software also did not allow the implementation of semiautomatic projects for roads and shoulders, and although the evolution in these areas is significant, there is still a long way to make the whole process faster, more accurate, and reduce risks.

However, there is no doubt that as the complexity of the business scenario increases, analytics become more complex, and the use of new technologies and tools such as advanced analytics is becoming a fundamental routine for professionals and companies that seek to connect to what is most modern. These tools are already proving their ability to add value and robustness to the mining industry. Those who do not observe what happens at the forefront of technology tend to lose unique opportunities and market value in companies and employability in mining professionals.

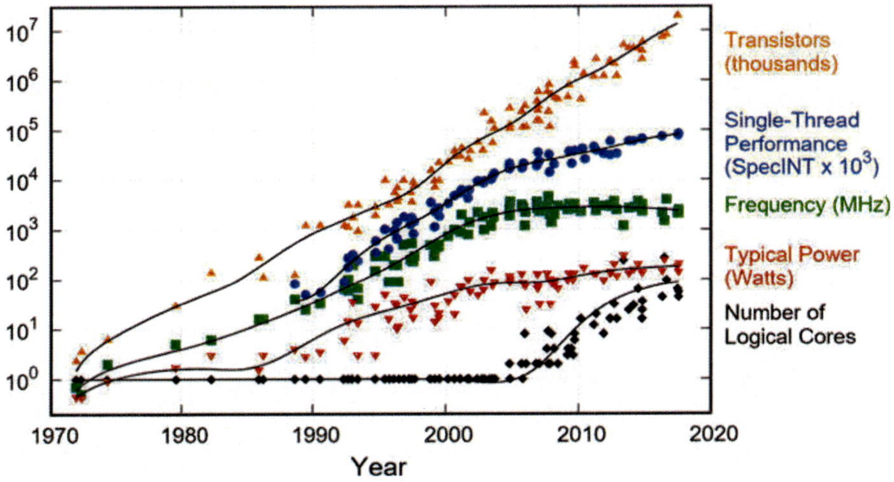

Fig. 9.40 Forty-two years of microprocessor trend data [321]

Conclusion

Mine planning is a fundamental stage in every mining project, whose impacts of unwise and inaccurate decisions can be dreadful to the enterprise since the entire operation phase depends on this and the geological modeling, activities that must be re-evaluated every time new data arises. The popular traditional methods for mine planning cannot support mining companies in the near future when they need to improve the quality of their materials and reduce the final cost of products in the competitive market. New technologies helped the mining companies to collect the required data from different assets. However, there is a long way to convert the raw data to useful information to make better decisions. Advanced analytics and potentially applying artificial intelligence can help researchers and mine managers speed up their progress in this way. However, the lack of knowledge, experience, and reliable references in using advanced analytics for mine planning is a critical challenge in improving the decisions in this field. This chapter tried to cover those above-mentioned scientific and technical gaps to help researchers and mine managers use the advanced analytics approaches to increase the quality of their decisions in mine planning, specifically in surface extraction.

References

1. Cherchenevski, P.K., J.F.C.L. Costa, and R.H. Rubio. 2019. Grade uncertainty embedded in long term scheduling: Stochastic mine planning. *REM International Engineering Journal* 72: 275–284. https://doi.org/10.1590/0370-44672018720119.
2. Abzalov, M. 2016. Conversion resources to reserves. In *Applied Mining Geology*, ed. M.

Abzalov, 365–372. Modern Approaches in Solid Earth Sciences 12. Cham: Springer. https://doi.org/10.1007/978-3-319-39264-6_29.
3. Monkhouse, P.H.L., and G.A. Yeates. 2018. Beyond naïve optimization. In *Advances in Applied Strategic Mine Planning*, ed. R. Dimitrakopoulos, 3–18. Cham: Springer. https://doi.org/10.1007/978-3-319-69320-0.
4. Osanloo, M., J. Gholamnejad, and B. Karimi. 2008. Long-term open pit mine production planning: A review of models and algorithms. *International Journal of Mining, Reclamation and Environment* 22 (1): 3–35. https://doi.org/10.1080/17480930601118947.
5. Newman, A.M., E. Rubio, R. Caro, A. Weintraub, and K. Eurek. 2010. A review of operations research in mine planning. *Interfaces* 40: 222–245. https://doi.org/10.1287/inte.1090.0492.
6. Whittle, J. 1989. *The Facts and Fallacies of Open-Pit Optimization.* Balwyn North, Victoria: Whittle Programming Pty Ltd.
7. Dagdelen, K. 2001. Open pit optimization—Strategies for improving the economics of mining projects through mine planning. In *Proceedings of the 17th International Mining Congress and Exhibition of Turkey*, IMCET 2001, Ankara.
8. Axelson, A.H. 1964. A practical approach to computer utilization in mine planning. In *Proceedings of the 4th International APCOM Symposium*, Golden.
9. Canessa, G., E. Moreno, and B.K. Pagnoncelli. 2020. The risk-averse ultimate pit problem. *Optimization and Engineering*. https://doi.org/10.1007/s11081-020-09545-4.
10. Campos, P.H.A., Arroyo, C.F., and N. Morales. 2018. Application of optimized models through direct block scheduling in traditional mine planning. *Journal-South African Institute of Mining and Metallurgy* 118 (4): 381–386. https://doi.org/10.17159/2411-9717/2018/v118n4a8.
11. Erikson, J.D. 1968. Long-range open-pit planning. Mining Engineering 20 (4): 75–78.
12. Soderberg, A., and D.O. Rausch. 1968. Pit planning and layout. In *Surface Mining*, ed. E.P. Pfleider, 141–165. New York: SME-AIME.
13. Koskiniemi, B.C. 1979. Hand methods. In *Open-Pit Mine Planning and Design*, ed. J.T. Crawford and A. Hustrulid, 187–195. New York: SME/AIME.
14. Hustrulid, W., M. Kuchta, and R. Martin. 2013. *Open Pit Mine Planning and Design*, 3rd ed. Leiden: CRC Press.
15. Luke, K.W. 1972. Functional optimization of open-pit mine design utilizing geologic cross-section data. *Transactions of the Metallurgical Society of AIME* 252: 125–131.
16. Pyrcz, M.J., and C.V. Deutsch. 2014. *Geostatistical Reservoir Modeling*, 2nd ed. New York: Oxford University Press.
17. Gilani, S.O., and J.A. Sattarvand. 2015. A new heuristic non-linear approach for modeling the variable slope angles in open pit mine planning algorithms. *Acta Montanistica Slovaca* 20 (4): 251–259.
18. Kim, Y.C. 1978. Ultimate pit limit design methodologies using computer models, state of the art. *Mining Engineering* 30 (10): 1454–1459.
19. Pana, M.T. 1965. The simulation approach to open-pit design. In *Proceeding of the 5th International APCOM Symposium*, Tucson.
20. Lemieux, M. 1979. Moving cone optimizing algorithm. In *Computer Methods for the 80s in the Mineral Industry*, ed. A. Weiss, 329–345. New York: SME/AIME.
21. Wright, E.A. 1999. Moving cone II—A simple algorithm for optimum pit limits design. In *Proceedings of the 28th International APCOM Symposium*, Colorado.
22. Khalokakaie, R. 2007. Optimum open pit design with modified moving cone II methods. *Journal of Faculty of Engineering* 41 (3): 297–307.
23. Galić, I., B. Janković, and I. Mrakovčić. 2009. An another way for open-pit mine design optimization: Floating slopes method. *Rudarsko-Geolosko-Naftni Zbornik* 21: 103–111.
24. Elahi Zeyni, E., R. Kakaie, and A. Yousefi. 2012. A new algorithm for optimum open pit design: Floating cone method III. *Journal of Mining and Environment* 2 (2): 118–125.
25. Korobov, S. 1974. Method for determining optimal open pit limits. Technical Report (EP74-R-4), Ecole Polytechnique de Montreal.

26. Dowd, P.A., and A.H. Onur. 1992. Optimizing open pit design and sequencing. In *Proceeding of the 23rd International APCOM Symposium*, Tucson.
27. Dowd, P.A., and A.H. Onur. 1993. Open pit optimization—Part 1: Optimal open pit design. Transactions of the Institution of Mining and Metallurgy 102 (A): A95–A104.
28. Khalokakaie, R. 1999. Computer-aided optimal open pit design with variable slope angles. PhD thesis, University of Leeds.
29. Shishvan, M.S., and J.A. Sattarvand. 2012. Modeling of accurate variable slope angles in open pit mine design using spline interpolation. *Archives of Mining Sciences* 57 (4): 921–932. https://doi.org/10.2478/v10267-012-0061-y.
30. Xu, X.C., X.W. Gu, Q. Wang, J.P. Liu, and J. Wang. 2014. Ultimate pit optimization with the ecological cost for open-pit metal mines. *Transactions of the Nonferrous Metals Society of China* 24 (5): 1531–1537. https://doi.org/10.1016/S1003-6326(14)63222-2.
31. Sadeghi, M., H. Dehghani, and B.J. Shokri. 2021. Ultimate pit limit determination using flashlight algorithm. *International Journal of Mining and Geo-Engineering* 55 (1): 41–46. https://doi.org/10.22059/IJMGE.2020.296120.594840.
32. Lerchs, H., and I.F. Grossmann. 1965. Optimum design of open pit mines. *CIM Bulletin* 58: 17–24.
33. Huttagosol, P. 1988. Modified tree graph algorithm for ultimate pit limit analysis. MSc thesis, Colorado School of Mines.
34. Johnson, T.B., and R.W. Sharp. 1971. Three-dimensional dynamic programming method for optimal ultimate pit design. USBM Report of Investigation 7553.
35. Barnes, R.J. 1982. Optimizing the ultimate pit. MSc thesis, Colorado School of Mines.
36. Koenigsberg, E. 1982. The optimum contours of an open-pit mine: An application of dynamic programming. In *Proceedings of the 17th International APCOM Symposium*, Golden.
37. Wilke, F.L., and E.A. Wright. 1984. Determining the ultimate optimal pit for hard rock open pit mines using dynamic programming. *Erzmetall* 37: 138–144.
38. Shenggui, Z., and A.M. Starfield. 1985. Dynamic programming with color graphics smoothing for open-pit design on a personal computer. *International Journal of Mining Engineering* 3: 27–34.
39. Wright, E.A. 1987. The use of dynamic programming for open pit mine design, some practical implications. *Mining Science and Technology* 4: 97–104. https://doi.org/10.1016/S0167-9031(87)90214-3.
40. Yamatomi, J., G. Mogi, A. Akaike, and U. Yamaguchi. 1995. Selective extraction dynamic cone algorithm for three-dimensional open pit designs. In *Proceedings of the 25th International APCOM Symposium*, Brisbane.
41. Bond, G.D. 1995. A mathematical analysis of the Lerch and Grossmann algorithm and the nested Lerch and Grossmann algorithm. PhD thesis, Colorado School of Mines.
42. Vallet, R. 1976. Optimisation mathematique de l'exploitation d'une mine a ciel ouvert ou le problem de l'enveloppe. *Annales des Mines de Belgique* 113–135.
43. Hochbaum, D.S., and A. Chen. 2000. Performance analysis and best implementations of old and new algorithms for the open-pit mining problem. *Operational Research: An International Journal* 48 (6): 894–914. https://doi.org/10.1287/opre.48.6.894.12392.
44. Zhao, Y. 1992. Algorithms for optimum design and planning of open-pit mines. PhD thesis, University of Arizona.
45. Zhao, Y., and Y.C. Kim. 1992. A new ultimate pit limit design algorithm. In *Proceedings of the 23rd International APCOM Symposium*, Tucson.
46. Johnson, T.B. 1968. Optimum open-pit mine production scheduling. PhD thesis, University of California.
47. Picard, J.C. 1976. Maximal closure of a graph and applications to combinatorial problems. *Management Science* 22 (11): 1268–1272. https://doi.org/10.1287/mnsc.22.11.1268.
48. Sahni, J. 1996. Application of push-relabel and heuristics to open-pit mines. MSc thesis, University of British Columbia.
49. Cai, W. 1989. Application of network flow and zero-one programming to open pit mine design problems. PhD thesis, University of Arizona.

50. Faaland, B., K. Kim, and T. Schmitt. 1990. A new algorithm for computing the maximal closure. *Management Science* 36 (3): 315–331. https://doi.org/10.1287/mnsc.36.3.315.
51. Giannini, L.M., L. Caccetta, P. Kelsey, and S. Carras. 1991. PITOPTIM: A new high-speed network flow technique for optimum pit design facilitating rapid sensitivity analysis. *AusIMM Proceedings* 2: 57–62.
52. Yegulalp, T.M., and J.A. Arias. 1992. A fast algorithm to solve the ultimate pit limit problem. In *Proceedings of the 23rd International APCOM Symposium*, Tucson.
53. Caccetta, L., L.M. Giannini, and P. Kelsey. 1994. On the implementation of exact optimization techniques for open pit design. *Asia-Pacific Journal of Operational Research* 11: 155–170.
54. Caccetta, L. 2007. Application of optimization techniques in open-pit mining. In *Handbook of Operations Research in Natural Resources*, vol. 99, ed. A. Weintraub et al., 547–559. International Series in Operations Research & Management. Boston: Springer. https://doi.org/10.1007/978-0-387-71815-6_29.
55. Underwood, R., and B. Tolwinski. 1996. A mathematical programming viewpoint for solving the ultimate pit problem. *European Journal of Operational Research* 107 (1): 96–107. https://doi.org/10.1016/S0377-2217(97)00141-0.
56. Hochbaum, D.S. 2001. A new-old algorithm for minimum-cut and maximum-flow in closure graphs. *Networks* 37 (4): 171–193. https://doi.org/10.1002/net.1012.
57. Hochbaum, D.S. 2008. The pseudo flow algorithm: A new algorithm for the maximum flow problem. *Operations Research* 56 (4): 992–1009. https://doi.org/10.1287/opre.1080.0524.
58. Mwangi, A.D., Z. Jianhua, H. Gang, R.M. Kasomo, and M.M. Innocent. 2020. Ultimate pit limit optimization using Boykov-Kolmogorov maximum flow algorithm. *Journal of Mining and Environment* 12 (1): 1–13. https://doi.org/10.22044/JME.2020.10170.1953.
59. Meyer, M. 1969. Applying linear programming to the design of ultimate pit limits. *Management Science* 16 (2): B121–B135. https://doi.org/10.1287/mnsc.16.2.B121.
60. Gershon, M.E. 1983. Optimal mine production scheduling evaluation of large scale mathematical programming approaches. *International Journal of Mining Engineering* 1 (4): 315–329.
61. Huttagosol, P., and A. Cameron. 1992. A computer design of ultimate pit limit by using transportation algorithm. In *Proceedings of the 23rd International APCOM Symposium*, Tucson.
62. Underwood, R., and B. Tolwinski. 1998. A mathematical programming viewpoint for solving the ultimate pit problem. *European Journal of Operational Research* 107 (1): 96–107.
63. Khodayari, A.A. 2013. A new algorithm for determining ultimate pit limits based on network optimization. *International Journal of Mining and Geo-Engineering* 47 (2): 129–137. https://doi.org/10.22059/ijmge.2013.51334.
64. Matheron, G. 1975. Paramétrage des contours optimaux. Centre de Géostatistique et de Morphologie mathématique. Internal Report N-403; Note géostatistique 128, 54 pp, Fontainebleau.
65. François-Bongarçon, D., and A. Marechal. 1976. A new method for open pit design: Parameterization of the final pit contour. In *Proceedings of the 14th International APCOM Symposium*, New York.
66. François-Bongarçon, D., and D. Guibal. 1982. Algorithms for parameterizing reserves under different geometrical constraints. In *Proceedings of the 17th International APCOM Symposium*, New York.
67. François-Bongarçon, D. 1994. Myth and reality: A status report on computer open-pit optimization algorithms in the 1990s. In *Mining Latin America: Challenges in the Mining Industry*, 77–87. London: Chapman & Hall.
68. Dagdelen, K., and D. François-Bongarçon. 1982. Towards the complete double parameterization of recovered reserves in open pit mining. In *Proceedings of the 17th International APCOM Symposium*, New York.
69. Whittle, J. 1988. Beyond optimization in open pit design. Whittle programming. In *Proceedings of the 1st Canadian Conference on Computer Applications in the Mineral Industry*, 331–337. Rotterdam: Balkema.

70. Coléou, T. 1989. Technical parameterization of reserves for open pit design and mine planning. In *Proceedings of the 21st International APCOM Symposium*, Littleton.
71. Sayadi, A.R., N. Fathianpour, and A.A. Mousavi. 2011. Open pit optimization in 3D using a new artificial neural network. *Archives of Mining Sciences* 56 (3): 389–403.
72. Denby, B., D. Schofield, and S. Bradford. 1991. Neural network applications in mining engineering. *University of Nottingham Department of Engineering Magazine* 43: 13–23.
73. Achireko, P.K., and S. Frimpong. 1996. Open pit optimization using artificial neural networks on conditionally simulated blocks. In *Proceedings of the 26th International APCOM Symposium*, State College.
74. Frimpong, S., and P.K. Achireko. 1997. The MCS/MFNN algorithm for open pit optimization. *International Journal of Surface Mining, Reclamation and Environment* 11: 45–52. https://doi.org/10.1080/09208119708944055.
75. Frimpong, S., J.M. Whiting, and J. Szymanski. 1998. Stochastic-optimization annealing of an intelligent, open pit design. *Mineral Resources Engineering* 7 (1): 15–27.
76. Achireko, P.K. 1998. Application of modified conditional simulation and artificial neural networks to open-pit optimization. PhD thesis, Dalhousie University Caltech.
77. Asa, E. 2002. An intelligent 3-D open pit design and optimization using machine learning: Adaptive logic networks and neuro-genetic algorithms. PhD thesis, University of Alberta.
78. Frimpong, S., J. Szymanski, and A. Narsing. 2002. An intelligent computational algorithm for surface mine layouts optimization. *SIMULATION* 78 (10): 600–611. https://doi.org/10.1177/0037549702078010002.
79. Yang, Z.Y., L.S. Lowndes, and B. Denby. 1999. Genetic algorithm optimization of a large U.K. coal mine ventilation network. In *Proceedings of the 8th US Mine Ventilation Symposium*, Rolla.
80. Sattarvand, J., S.O. Gilani, M.S. Shishvan, and A. Khan. 2015. Application of the metaheuristic approaches in open pit mine planning. In *Proceedings of the 37th International APCOM Symposium*, Fairbanks.
81. Alipour, A., A.A. Khodaiari, A. Jafari, and R. Tavakkoli-Moghaddam. 2020. Production scheduling of open-pit mines using genetic algorithm: A case study. *International Journal of Management Science and Engineering Management* 15 (3): 176–183. https://doi.org/10.1080/17509653.2019.1683090.
82. Rosas, J.F.S. 2009. A genetic optimizer for stochastic problems with applications to orebody uncertainty in mine planning. PhD thesis, Laurentian University.
83. Denby, B., and D. Schofield. 1993. Genetic algorithms: A new approach to pit optimization. In *Proceedings of the 24th International APCOM Symposium*, Wollongong.
84. Denby, B., and D. Schofield. 1994. Open pit design and scheduling using genetic algorithms. *Transactions of the Institution of Mining and Metallurgy* 103 (A): A21–A26.
85. Denby, B., and D. Schofield. 1995. Inclusion of risk assessment in open pit design and scheduling. *Transactions of the Institution of Mining and Metallurgy* 104 (A): A67–A71.
86. Songolo, M. 2010. Pushback design using genetic algorithms. MSc thesis, Curtin University.
87. Denby, B., D. Schofield, G. Hunter. 1996. Genetic algorithms for open pit scheduling—An extension into 3-dimensions. In *Mine Planning and Equipment Selection*, ed. W.T. Hennies, L.A. Ayres Da Silva, and A.P. Chaves, 177–185. Rotterdam: A. A. Balkema.
88. Denby, B., D. Schofield, and T. Surme. 1998. Genetic algorithms for flexible scheduling of open-pit operations. In *Proceedings of the 27th International APCOM Symposium*, London.
89. Saleki, M., R. Kakaie, and M. Ataei. 2019. Mathematical relationship between ultimate pit limits generated by discounted and undiscounted block value maximization in open-pit mining. *Journal of Sustainable Mining* 18: 94–99. https://doi.org/10.1016/j.jsm.2019.03.003.
90. Caccetta, L., and S. Hill. 2003. An application of branch and cut to open-pit mine scheduling. *Journal of Global Optimization* 27: 349–365. https://doi.org/10.1023/A:1024835022186.
91. Meagher, C., R. Dimitrakopoulos, and D. Avis. 2014. Optimized open pit mine design, pushbacks, and the gap problem—A review. *Journal of Mining Science* 50 (3): 508–526. https://doi.org/10.1134/S1062739114030132.

92. Wang, Q., and H. Sevim. 1995. Alternative to parameterization in finding a series of maximum-metal pits for production planning. *Mining Engineering* 178–182.
93. Gershon, M.E. 1987. Heuristic approaches for mine planning and production scheduling. *International Journal of Mining and Geo-Engineering* 5 (1): 1–13. https://doi.org/10.1007/BF01553529.
94. Whittle, J. 1998. Beyond optimization in open pit design. In *Proceedings of the 1st Canadian Conference on Computer Applications in the Mineral Industry*, Quebec.
95. Ramazan, S., and K. Dagdelen. 1998. A new push-back design algorithm in the open it mining. In *Proceedings of the 17th International Symposium on Mine Planning and Equipment Selection*, Calgary.
96. Seymour, F. 1995. Pit limit parametrization from modified 3D Lerchs-Grossmann algorithm. Society of Mining, Metallurgy, and Exploration, Manuscript, 1994.
97. Roman, R.J. 1971. Mine-mill production scheduling by dynamic programming. *The Journal of the Operational Research Society* 22 (4): 319–328. https://doi.org/10.1057/jors.1971.76.
98. Roman, R.J. 1974. The role of the time value of money in determining an open-pit mining sequence and pit limits. In *Proceedings of the 12th International APCOM Symposium*, Golden.
99. Wright, E.A. 1989. Dynamic programming in open pit mining sequence planning a case study. In *Proceedings of the 21st International APCOM Symposium*, New York.
100. Gershon, M.E., and F.H. Murphy. 1989. Optimizing single hole mine cuts by dynamic programming. *European Journal of Operational Research* 38 (1): 56–62. https://doi.org/10.1016/0377-2217(89)90468-2.
101. Onur, A.H., and P.A. Dowd. 1993. Open pit optimization—Part 2: Production scheduling and inclusion of roadways. *Transactions of the Institution of Mining and Metallurgy* 102 (A): A95–A104.
102. Tolwinski, B., and R. Underwood. 1992. An algorithm to estimate the optimal evolution of an open-pit mine. In *Proceedings of the 23rd International APCOM Symposium*, Tucson.
103. Tan, S., and R. Ramani. 1992. Optimization models for scheduling ore and waste production in open-pit mines. In *Proceedings of the 23rd International APCOM Symposium*, Tucson.
104. Elevli, B. 1995. Open pit mine design and extraction sequencing using OR and AI concept. *International Journal of Surface Mining, Reclamation and Environment* 9 (4): 149–153. https://doi.org/10.1080/09208119508964741.
105. Tolwinski, B., and T.S. Golosinski. 1995. Long term open pit scheduler. In *Proceedings of the International Symposium on Mine Planning and Equipment Selection*, Calgary.
106. Tolwinski, B. 1998. Scheduling production for open-pit mines. In *Proceedings of the 27th International APCOM Symposium*, London.
107. Sevin, H., and D.D. Lei. 1998. The problem of production planning in open-pit mines. *INFOR* 36 (1–2): 1–12. https://doi.org/10.1080/03155986.1998.11732339.
108. Erarslan, K., and N. Çelebi. 2001. A simulative model for optimum pit design. *CIM Bulletin* 94 (1055): 59–68.
109. Wang, Q., and H. Sun. 2001. A theorem on open-pit planning optimization and its application. In *Proceedings of the 29th International APCOM Symposium*, Beijing.
110. Gu, X.W., P.F. Wang, Q. Wang, Y.Y. Zhang, J.P. Liu, and B. Chen. 2011. Simultaneous optimization of final pit and production schedule in open-pit coal mines. *Advances in Materials Research* 323: 222–228. https://doi.org/10.4028/www.scientific.net/AMR.323.222.
111. Wang, Q., X.C. Xu, and X.W. Gu. 2013. A dynamic-programming based model for phase-mining optimization in open-pit metal mines. *Applied Mechanics and Materials* 316–317: 896–901. https://doi.org/10.4028/www.scientific.net/AMM.316-317.896.
112. Latorre, E., and T.S. Golosinski. 2011. Definition of economic pit limits considering the time value of money. *CIM Journal* 2 (3): 162–170.
113. Nanjari, E.L., and T.S. Golosinski. 2013. Optimising open pit mine scheduling considering the time value of money and mining restrictions. *International Journal of Mining, Reclamation and Environment* 27 (3): 156–165. https://doi.org/10.1080/09658416.2012.655166.
114. Albach, H. 1967. Long-range planning in open-pit mining. *Management Science* 13 (10): B549–B568. https://doi.org/10.1287/mnsc.13.10.B549.

115. Manula, C.B., and H. Gezik. 1968. Application of linear programming in the crushed stone industry. In *Proceedings of the Annual Meeting of the American Institute of Mining, Metallurgical, and Petroleum Engineers*, New York.
116. Johnson, T.B. 1969. Optimum production scheduling. In *Proceedings of the 8th International APCOM Symposium*, Salt Lake City.
117. Wilke, F., and T. Reimer. 1977. Optimizing the short-term production schedule for an open-pit iron ore mining operation. In *Proceedings of the 15th International APCOM Symposium*, Brisbane.
118. Blom, M., A.R. Pearce, and P.J. Stuckey. 2019. Short-term planning for open-pit mines: A review. *International Journal of Mining, Reclamation and Environment* 33 (5): 318–339. https://doi.org/10.1080/17480930.2018.1448248.
119. Chanda, E., and F. Wilke. 1992. An EPD model of open-pit short term production scheduling optimization for stratiform orebodies. In *Proceedings of the 23rd International APCOM Symposium*, Tucson.
120. Youdi, Z., C. Qingziang, and W. Lixin. 1992. Combined approach for surface mine short term planning optimization. In *Proceedings of the 23rd International APCOM Symposium*, Tucson.
121. Fytas, K., C. Pelley, and P. Calder. 1987. Optimization of open-pit short- and long-range production scheduling. *CIM Bulletin* 80 (904): 55–61.
122. Fytas, K., J. Hadjigeorgiou, and J.L. Collins. 1993. Production scheduling optimization in open-pit mines. *International Journal of Surface Mining, Reclamation and Environment* 7 (1): 1–9. https://doi.org/10.1080/09208119308964677.
123. Sundar, D.K., and D. Acharya. 1995. Blast schedule planning and shift wise production scheduling of an opencast iron ore mine. *Computers and Industrial Engineering* 28 (4): 927–935. https://doi.org/10.1016/0360-8352(94)00221-8.
124. Chanda, E.K.C., and K. Dagdelen. 1995. Optimal blending of mine production using goal programming and interactive graphics systems. *International Journal of Surface Mining Reclamation and Environment* 9 (4): 203–208. https://doi.org/10.1080/09208119508964748.
125. Hu, Q., W. Wei, and S. Fang. 1995. Short-term production scheduling for open-pit mines by PERT networks with resource constraints. In *Proceedings of the International Symposium on Mine Planning and Equipment Selection*, Calgary.
126. Vujić, S., and G. Ćirović. 1996. Production planning in mines using fuzzy linear programming. *Yugoslav Journal of Operations Research* 6 (2): 205–215.
127. Pendharkar, P.C. 1997. A fuzzy linear programming model for production planning in coal mines. *Computers and Operations Research* 24 (12): 1141–1149. https://doi.org/10.1016/s0305-0548(97)00024-5.
128. Djilani, M.C. 1997. Interactive open-pit design using parameterization techniques. PhD thesis, University of Leeds.
129. Halatchev, R. 1993. Stage open-cut exploitation of ore deposit. In *Proceedings of the 24th International APCOM Symposium*, Wollongong.
130. Halatchev, R. 1996. Factor of compromise in the assessment of open pit long-term production plans. In *Proceedings of Surface Mining'96*, Johannesburg.
131. Halatchev, R. 1999. Company strategy—A basis for production scheduling of an open pit complex. In *Proceedings of Strategic Mine Planning Conference.*
132. Halatchev, R. 2002. The time aspect of the optimum long-term open-pit production sequencing. In *Proceedings of the 30th International APCOM Symposium*, Littleton.
133. Halatchev, R., and E. Kozan. 2003. Coal production scheduling optimization of single seam open-cast mining. In *Proceeding of the 5th Large Open Pit Mining Conference*, Kalgoorlie.
134. Kozan, E., and S.Q. Liu. 2011. Operations research for mining: A classification and literature review. *ASOR Bulletin* 30 (1): 2–23.
135. Halatchev, R., and P. Lever. 2004. The optimum production equipment scale for open pit coal mining. In *Proceedings of the SME Annual Meeting*, Denver.
136. Halatchev, R. 2005. A model of discounted profit variation of open-pit production sequencing optimization. In *Application of Computers and Operations Research in the Mineral Industry*,

ed. S.D. Dessureault, R. Ganguli, V. Kecojevic, and J. Girard-Dwyer, 305–323. London: Taylor & Francis Group.

137. Halatchev, R. 2015. The spatial aspect of the long-term open pit mine sequence optimization. In *Proceedings of the SME Annual Meeting*, Denver.
138. Souza, F.R., L. Soares, H. Burgarelli, A. Nader, C. Arroyo, and L. Alberto. 2018. Direct stockpile scheduling mathematical formulation. *DYNA* 85 (204): 296–301. https://doi.org/10.15446/dyna.v85n204.62642.
139. Picard, J.C., and B.T. Smith. 2004. Parametric maximum flows and the calculation of optimal intermediate contours in open pit mine design. *INFOR* 42 (2): 143–153. https://doi.org/10.1080/03155986.2004.11732697.
140. Topal, E., and S. Ramazan. 2012. Strategic mine planning model using network flow model and real case application. *International Journal of Mining, Reclamation and Environment* 26 (1): 29–37. https://doi.org/10.1080/17480930.2011.600827.
141. Amankwah, H., T. Larsson, and B. Textorius. 2014. A maximum flow formulation of a multi-period open-pit mining problem. *Operational Research: An International Journal* 14: 1–10. https://doi.org/10.1007/s12351-013-0140-7.
142. Liu, S.Q., and E. Kozan. 2016. New graph-based algorithms efficiently solve large scale open pit mining optimization problems. *Expert Systems with Applications* 43: 59–65. https://doi.org/10.1016/j.eswa.2015.08.044.
143. Espinoza, D., M. Goycoolea, E. Moreno, and A. Newman. 2013. MineLib: A library of open pit mining problems. *Annals of Operations Research* 206: 93–114. https://doi.org/10.1007/s10479-012-1258-3.
144. Smith, M.L., and T.W. You. 1995. Mine production scheduling for optimization of plant recovery in surface phosphate operations. *International Journal of Surface Mining Reclamation and Environment* 9 (2): 41–46. https://doi.org/10.1080/09208119508964716.
145. Smith, M.L. 1998. Optimizing short-term production schedules in surface mining: Integrating mine modeling software with ANIPL/CPLEX. *International Journal of Surface Mining Reclamation and Environment* 12 (4): 149–155. https://doi.org/10.1080/09208118908944038.
146. Smith, M.L. 1999. Optimizing inventory stockpiles and mine production: An application of separable and goal programming to phosphate mining using AMPL/CPLEX. *CIM Bulletin* 92 (1030): 61–64.
147. Rehman, S.U., and M.W.A. Asad. 2010. A mixed-integer linear programming (MILP) model for short-range production scheduling of cement quarry operations. *Asia-Pacific Journal of Operational Research* 27 (3): 315–333. https://doi.org/10.1142/S0217595910002727.
148. Askari-Nasab, H., H. Eivazy, M. Tabesh, and M.M. Badiozamani. 2011. A mathematical programming model for open pit short-term production scheduling. In *Proceedings of the SME Annual Meeting*, Denver.
149. Ben-Awuah, E., and H. Askari-Nasab. 2011. Oil sands mine planning and waste management using mixed integer goal programming. *International Journal of Mining, Reclamation and Environment* 25 (3): 226–247. https://doi.org/10.1080/17480930.2010.549656.
150. Eivazy, H., and H. Askari-Nasab. 2012. A mixed-integer linear programming model for short-term open pit mine production scheduling. *Mining Technology* 121 (2): 97–108. https://doi.org/10.1179/1743286312Y.0000000006.
151. L'Heureux, G., M. Gamache, and F. Soumis. 2013. Mixed integer programming model for short term planning in open-pit mines. *Mining Technology* 122 (2): 101–109. https://doi.org/10.1179/1743286313Y.0000000037.
152. Yavarzadeh, S., J. Abodallheisharif, and A. Neishabouri. 2014. Modeling of short term production scheduling with the objective grade control. In *Mine Planning and Equipment Selection: Proceedings of the 22nd MPES Conference*, Dresden, Germany, ed. C. Drebenstedt, and R. Singhal, 379–387, October 14th–19th, 2013. Cham: Springer. https://doi.org/10.1007/978-3-319-02678-7_37.
153. Mousavi, A.A., E. Kozan, and S.Q. Liu. 2015. Integrated approach to optimize open-pit mine block sequencing. In *Industrial Engineering Non-traditional Applications in International Settings*, ed. B.Y. Kara, B. Bidanda, and I. Sabuncuoglu, 83–98. Boca Raton: CRC Press.

154. Kozan, E., and S.Q. Liu. 2018. An open-pit multi-stage mine production scheduling model for drilling, blasting, and excavating operations. In *Advances in Applied Strategic Mine Planning*, ed. R. Dimitrakopoulos, 655–668. Cham: Springer. https://doi.org/10.1007/978-3-319-69320-0_38.
155. Flores-Fonseca, C., R. Linfati, and J.W. Escobar. 2021. Exact algorithms for production planning in mining considering the use of stockpiles and sequencing of power shovels in open-pit mines. *Operational Research: An International Journal*. https://doi.org/10.1007/s12351-020-00618-x.
156. Gershon, M.E. 1985. Developments in computerized mine production scheduling. In *Proceedings of SME Fall Meeting*, Albuquerque.
157. Gershon, M.E. 1986. A blending-based approach to mine planning and production scheduling. In *Proceedings of the 19th International APCOM Symposium*, State College.
158. Wang, Q., and H. Sevim. 1992. Enhanced production planning in open pit mining through intelligent dynamic search. In *Proceeding of the 23rd International APCOM Symposium*, Tucson.
159. Wang, Q., and H. Sevim. 1993. Alternative to parameterization in finding a series of maximum-metal pits for production planning. In *Proceeding of the 24th International APCOM Symposium*, Wollongong.
160. Sattarvand, J. 2009. Long-term open-pit planning by ant colony optimization. PhD thesis, Rheinisch-Westfälische Technische Hochschule Aachen University.
161. Sevim, H., and D.D. Lei. 1996. Production planning with working-slope maximum-metal pit sequences. *Transactions of the Institution of Mining and Metallurgy* 105 (A): A93–A98.
162. Kumral, M., and P.A. Dowd. 2002. Short-term mine production scheduling for industrial minerals using multi-objective simulated annealing. In *Proceedings of the 30th International APCOM Symposium*, Littleton.
163. Kumral, M., and P.A. Dowd. 2005. A simulated annealing approach to mine production scheduling. *The Journal of the Operational Research Society* 56: 922–930. https://doi.org/10.1057/palgrave.jobs.2601902.
164. Samanta, B., A. Bhattacherjee, and R. Ganguli. 2005. A genetic algorithms approach for grade control planning in a bauxite deposit. In *Application of Computers and Operations Research in the Mineral Industry*, ed. S.D. Dessureault, R. Ganguli, V. Kecojevic, and J. Girard-Dwyer, 337–342. London: Taylor & Francis Group. https://doi.org/10.1201/9781439833407.ch44.
165. Zhang, M. 2006. Combining genetic algorithms and topological sort to optimize open-pit mine plans. In *Mine Planning and Equipment Selection: Proceedings of the 15th MPES Conference*, Fiordo, Torino, Italy, ed. M. Cardu, R. Cicco, E. Lovera, and E. Michelotti, 1234–1239.
166. Ferland, J.A., J. Amaya, and S. Djuimo. 2007. Application of a particle swarm algorithm to the capacitated open-pit mining problem. In *Autonomous Robots and Agents—Studies in Computational Intelligence*, vol. 76, ed. S.C. Mukhopadhyay, and G.S. Gupta, 127–133. Cham: Springer. https://doi.org/10.1007/978-3-540-73424-6_15.
167. Shishvan, M.S., and J. Sattarvand. 2015. Long term production planning of open-pit mines by ant colony optimization. *European Journal of Operational Research* 240 (3): 825–836. https://doi.org/10.1016/j.ejor.2014.07.040.
168. Myburgh, C., and K. Deb. 2010. Evolutionary algorithms in large-scale open-pit mine scheduling. In *Proceedings of the 12th Annual Conference on Genetic and Evolutionary Computation—GECCO'10*, Portland. https://doi.org/10.1145/1830483.1830696.
169. Khan, A., and C. Niemann-Delius. 2014. Production scheduling of open-pit mines using particle swarm optimization algorithm. *Advances in Operations Research* 2014: 1–9. https://doi.org/10.1155/2014/208502.
170. Khan, A. 2015. Development of new metaheuristic tools for long-term production scheduling of open-pit mines. PhD thesis, Rheinisch-Westfälische Technische Hochschule Aachen University.
171. Asl, M.M., and J. Sattarvand. 2016. An imperialist competitive algorithm for solving the production scheduling problem in open pit mine. *International Journal of Mining and Geo-Engineering* 50 (1): 131–143. https://doi.org/10.22059/IJMGE.2016.57862.

172. Alipour, A., A.A. Khodaiari, A. Jafari, and R. Tavakkoli-Moghaddam. 2017. A genetic algorithm approach for open-pit mine production scheduling. *International Journal of Mining and Geo-Engineering* 51 (1): 47–52. https://doi.org/10.22059/ijmge.2017.62152.
173. Ramezanalizadeh, T., M. Monjezi, A.R. Sayadi, and A. Mousavi. 2020. Development of a MIP model to maximize NPV and minimize adverse environmental impact—A heuristic approach. *Environmental Monitoring and Assessment* 192 (605): 1–15. https://doi.org/10.1007/s10661-020-08550-5.
174. Pendharkar, P.C., and J.A. Rodger. 2000. Nonlinear programming and genetic search application for production scheduling in coal mines. *Annals of Operations Research* 95: 251–267. https://doi.org/10.1023/A:1018958209290.
175. Souza, M.J.F., I.N. Coelho, S. Ribas, H.G. Santos, and L.H.C. Merschmann. 2010. A hybrid heuristic algorithm for the open-pit-mining operational planning problem. *European Journal of Operational Research* 207 (2): 1041–1051. https://doi.org/10.1016/j.ejor.2010.05.031.
176. Tabesh, M., and H. Askari-Nasab. 2011. Two-stage clustering algorithm for block aggregation in open-pit mines. *Mining Technology* 120 (3): 158–169. https://doi.org/10.1179/1743286311y.0000000009.
177. Tabesh, M., and H. Askari-Nasab. 2013. Automatic creation of mining polygons using hierarchical clustering techniques. *Journal of Mining Science* 49 (3): 426–440. https://doi.org/10.1134/S1062739149030106.
178. Tabesh, M., C. Mieth, and H. Askari-Nasab. 2013. Open pit production planning using controlled pushbacks and aggregates. In *Proceedings of the 23rd World Mining Congress*, Montréal.
179. Tabesh, M., C. Mieth, and H. Askari-Nasab. 2014. A multi-step approach to long-term open-pit production planning. *International Journal of Mining and Mineral Engineering* 5 (4): 273–298. https://doi.org/10.1504/IJMME.2014.066577.
180. Tabesh, M., H. Askari-Nasab, and R. Peroni. 2015. A comprehensive approach to strategic open pit mine planning with stockpile consideration. In *Proceedings of the 37th International APCOM Symposium*, Alaska.
181. Blom, M.L., C.N. Burt, A.R. Pearce, and P.J. Stuckey. 2014. A decomposition-based heuristic for collaborative scheduling in a network of open-pit mines. *INFOR* 26 (4): 658–676. https://doi.org/10.1287/ijoc.2013.0590.
182. Blom, M.L., A.R. Pearce, and P.J. Stuckey. 2016. A decomposition-based algorithm for scheduling open-pit networks over multiple periods. *Management Science* 62 (10): 3059–3084. https://doi.org/10.1287/mnsc.2015.2284.
183. Mousavi, A., E. Kozan, and S.Q. Liu. 2016. Open-pit block sequencing optimization: A mathematical model and solution technique. *Engineering Optimization* 48 (11): 1932–1950. https://doi.org/10.1080/0305215X.2016.1142080.
184. Mousavi, A., E. Kozan, and S.Q. Liu. 2016. Comparative analysis of three metaheuristics for short-term open pit block sequencing. *Journal of Heuristics* 22 (3): 301–329. https://doi.org/10.1007/s10732-016-9311-z.
185. Maiti, N., P. Pathak, and B. Samanta. 2021. An efficient algorithm for the precedence constraint knapsack problem concerning large-scale open-pit mining pushback design. *Mining Technology* 1–14. https://doi.org/10.1080/25726668.2020.1866369.
186. Byun, J., and R. Dimitrakopoulos. 2013. An efficient algorithm for the LP relaxation of the maximal closure problem with a capacity constraint. Les Cahiers du GERAD, G-2013-60
187. Chicoisne, R., D. Espinoza, M. Goycoolea, E. Moreno, and E. Rubio. 2012. A new algorithm for the open-pit mine production scheduling problem. *Operations Research* 60 (3): 517–528. https://doi.org/10.1287/opre.1120.1050.
188. Bienstock, D., and M. Zuckerberg. 2010. Solving LP relaxations of large-scale precedence constrained problems. In *Integer Programming and Combinatorial Optimization. IPCO 2010*, vol. 6080, ed. F. Eisenbrand, and F.B. Shepherd, 1–14. Lecture Notes in Computer Science. Berlin, Heidelberg: Springer. https://doi.org/10.1007/978-3-642-13036-6_1.
189. Moreno, E., D. Espinoza, and M. Goycoolea. 2010. Large-scale multi-period precedence constrained knapsack problem: A mining application. *Electronic Notes in Discrete Mathematics* 36: 407–414. https://doi.org/10.1016/j.endm.2010.05.052.

190. Frimpong, S., E. Asa, and J. Szymanski. 1998. MULSOPS: Multivariate optimized pit shells simulator for tactical mine planning. *International Journal of Surface Mining, Reclamation and Environment* 12 (4): 163–171. https://doi.org/10.1080/09208118908944040.
191. Frimpong, S., E. Asa, and R.S. Suglo. 2001. Numerical simulation of surface mine production system using pit shell simulator. *Mineral Resources Engineering* 10 (2): 185–203. https://doi.org/10.1142/S0950609801000609.
192. Askari-Nasab, H., K. Awuah-Offei, and S. Frimpong. 2004. Stochastic simulation of open-pit pushbacks with a production simulator. In *Proceedings of the CIM Mining Industry Conference and Exhibition*, Edmonton.
193. Askari-Nasab, H., and J. Szymanski. 2005. Modelling open pit dynamics using Monte Carlo simulation. In *Proceedings of the 5th Computer Applications in the Minerals Industry*, Banff.
194. Askari-Nasab, H., and J. Szymanski. 2007. Continuous modeling of open-pit dynamics. In *Proceedings of the 16th MPES Conference*, Bangkok.
195. Askari-Nasab, H., S. Frimpong, and J. Szymanski. 2008. Investigating the continuous-time open-pit dynamics. *Journal-South African Institute of Mining and Metallurgy* 108 (2): 61–71.
196. Askari-Nasab, H., and J. Szymanski. 2007. Open pit production scheduling using reinforcement learning. In *Proceedings of the 33rd International APCOM Symposium*, Santiago.
197. Askari-Nasab, H., and K. Awuah-Offei. 2009. Open pit optimization using discounted economic block values. *Mining Technology* 118 (1): 1–12. https://doi.org/10.1179/037178409X12450752943243.
198. Aras, C., K. Dagdelen, and T.B. Johnson. 2019. Generating pushbacks using direct block mine production scheduling algorithm. In *Mining Goes Digital*, ed. C. Mueller et al. London: CRC Press.
199. Morales, N., E. Jélvez, P. Nancel-Penard, A. Marinho, O. Guimarães. 2015. A comparison of conventional and direct block scheduling methods for open-pit mine production scheduling. In *Proceedings of the 37th International APCOM Symposium*, Alaska.
200. Smith, C.E. 1978. The use of mixed-integer programming in planning an alluvial diamond deposit depletion. In *Proceedings of the 10th Annual Conference of the Operations Research Society of South Africa*, Durban.
201. Gershon, M.E. 1982. A linear programming approach to mine scheduling optimization. In *Proceedings of the 17th International APCOM Symposium*, New York.
202. Gershon, M.E. 1983. Mine scheduling optimization with mixed-integer programming. *Mining Engineering* 35: 351–354.
203. Brickey, A.J. 2015. Underground production scheduling optimization with ventilation constraints. PhD thesis, Colorado School of Mines.
204. Barbaro, R.W., R.V. Ramani, and P.T. Luckie. 1982. Optimal location of and production scheduling for a centralized coal preparation plant. Special Research Report Number SR-116, Coal Research Section, The Pennsylvania State University.
205. Barbaro, R.W., and R.V. Ramani. 1986. Generalized multiperiod MIP model for production scheduling and processing facilities selection and location. *Mining Engineering* 38 (2): 107–114.
206. Dagdelen, K., and T.B. Johnson. 1986. Optimum open pit mine production scheduling by Lagrangian parameterization. In *Proceedings of the 19th International APCOM Symposium*, State College.
207. Klingman, D., and N. Phillips. 1988. Integer programming for optimal phosphate-mining strategies. *The Journal of the Operational Research Society* 39 (9): 805–810. https://doi.org/10.1057/jors.1988.140.
208. Caccetta, L., P. Kelsey, and L.M. Giannini. 1998. Open-pit mine production scheduling. In *Proceedings of the 3rd Regional APCOM Symposium*.
209. Akaike, A., and K. Dagdelen. 1999. A strategic production scheduling method for an open-pit mine. In *Proceedings of the 28th International APCOM Symposium*, Littleton.
210. Urbaez, E., and K. Dagdelen. 1999. Implementation of the linear programming model for optimum open pit production scheduling problem. In *Proceedings of the SME Annual Meeting*, Denver.

211. Hoerger, S., L. Hoffman, and F. Seymour. 1999. Mine planning at Newmont's Nevada operations. *Mining Engineering* 51 (10): 3–7.
212. Mogi, G., T. Adachi, A. Akaike, and J. Yamatomi. 2001. Optimum production scale and scheduling of open-pit mines using revised 4D network relaxation method. In *Proceedings of the 17th International Symposium on Mine Planning and Equipment Selection.*
213. Kawahata, K., P. Schumacher, and K. Criss. 2016. Large-scale mine production scheduling optimization with mill blending constraints at Newmont's Twin Creeks Operation. *Mining Technology* 125 (4): 1–5. https://doi.org/10.1080/14749009.2016.1212510.
214. Ramazan, S. 2001. Open pit mine scheduling based on fundamental tree algorithm. PhD thesis, Colorado School of Mines.
215. Ramazan, S., and R. Dimitrakopoulos. 2003. Production scheduling optimization in a nickel laterite deposit: MIP and LP applications and infeasibility in the presence of orebody variability. In *Proceedings of the 12th International Symposium on Mine Planning and Equipment Selection*, Kalgoorlie.
216. Ramazan, S., and R. Dimitrakopoulos. 2004. Recent operations research applications and efficient MIP formulations in open-pit mining. *Transactions of the Society for Mining, Metallurgy and Exploration Inc.* 316: 73–78.
217. Ramazan, S., and R. Dimitrakopoulos. 2004. Traditional and new MIP models for production scheduling with in-situ grade variability. *International Journal of Surface Mining and Reclamation and Environment* 18 (2): 85–98. https://doi.org/10.1080/13895260412331295367.
218. Stone, P., G. Froyland, M. Menabde, B. Law, R. Pasyar, and P. Monkhouse. 2004. Blaser-blended iron-ore mine planning optimization at Yandi. In *Proceedings of the Orebody Modelling and Strategic Mine Planning International Symposium*, Melbourne.
219. Fricle, C. 2006. Applications of integer programming in open-pit mining. PhD thesis, The University of Melbourne.
220. Boland, N., C. Fricke, and G. Froyland. 2007. A strengthened formulation for the open-pit mine production scheduling problem. In *Optimization Online*. Available via DIALOG. http://www.optimization-online.org/DB_FILE/2007/03/1624.pdf. Accessed March 01, 2021.
221. Boland, N., I. Dumitrescu, G. Froyland, and A.M. Gleixner. 2009. LP-based disaggregation approaches to solving the open pit mining production scheduling problem with block processing selectivity. *Computers and Operations Research* 36 (4): 1064–1089. https://doi.org/10.1016/j.cor.2007.12.006.
222. Amaya, J., D. Espinoza, M. Goycoolea, E. Moreno, T. Prevost, and E. Rubio. 2009. A scalable approach to optimal block scheduling. In *Proceedings of the 34th International APCOM Symposium*, Vancouver.
223. Bley, A., N. Boland, C. Fricke, and G. Froyland. 2010. A strengthened formulation and cutting planes for the open-pit mine production scheduling problem. *Computers and Operations Research* 37 (9): 1641–1647. https://doi.org/10.1016/j.cor.2009.12.008.
224. Askari-Nasab, H., Y. Pourrahimian, E. Ben-Awuah, and S. Kalantari. 2011. Mixed integer linear programming formulations for open-pit production scheduling. *Journal of Mining Science* 47 (3): 338–359. https://doi.org/10.1134/S1062739147030117.
225. Cullenbine, C., R.K. Wood, and A.A. Newman. 2011. A sliding time window heuristic for open pit mine block sequencing. *Optimization Letters* 5: 365–377. https://doi.org/10.1007/s11590-011-0306-2.
226. Smith, M.L., and S.J. Wicks. 2014. Medium-term production scheduling of the Lumwana mining complex. *Interfaces* 44 (2): 176–194. https://doi.org/10.1287/inte.2014.0737.
227. Lambert, W.B., and A.M. Newman. 2014. Tailored Lagrangian relaxation for the open pit block sequencing problem. *Annals of Operations Research* 222: 419–438. https://doi.org/10.1007/s10479-012-1287-y.
228. Lamghari, A., R. Dimitrakopoulos, and J.A. Ferland. 2015. A hybrid method based on linear programming and variable neighborhood descent for scheduling production in open-pit mines. *Journal of Global Optimization* 63: 555–582. https://doi.org/10.1007/s10898-10014-10185-z.

229. Vossen, T.W.M., R.K. Wood, and A.M. Newman. 2016. Hierarchical benders decomposition for open-pit mine block sequencing. *Operations Research* 64 (4): 771–793. https://doi.org/10.1287/opre.2016.1516.
230. Aras, C. 2018. A new integer solution algorithm to solve open-pit mine production scheduling problems. PhD thesis, Colorado School of Mines.
231. Mai, N.L., E. Topal, and O. Erten. 2018. A new open-pit mine planning optimization method using block aggregation and integer programming. *Journal-South African Institute of Mining and Metallurgy* 118: 705–714. https://doi.org/10.17159/2411-9717/2018/v118n7a4.
232. Souza, F.R., H.R. Burgarelli, A.S. Nader, C.E.A. Ortiz, L.S. Chaves, L.A. Carvalho, V.F.N. Torres, T.R. Câmara, and R. Galery. 2018. Direct block scheduling technology: Analysis of avidity. *REM: International Engineering Journal* 71 (1): 97–104. https://doi.org/10.1590/0370-44672017710129.
233. Samavati, M., D. Essam, M. Nehring, and R. Sarker. 2018. A new methodology for the open-pit mine production scheduling problem. *Omega* 81: 169–182. https://doi.org/10.1016/j.omega.2017.10.008.
234. Jélvez, E., N. Morales, and P. Nancel-Penard. 2019. Open-pit mine production scheduling: improvements to Minelib library problems. In *Proceedings of the 27th International Symposium on Mine Planning and Equipment Selection, MPES 2018*, ed. E. Widzyk-Capehart, A. Hekmat, and R. Singhal, 223–232. Cham: Springer. https://doi.org/10.1007/978-3-319-99220-4_18.
235. Jélvez, E., N. Morales, P. Nancel-Penard, and F. Cornillier. 2020. A new hybrid algorithm for the precedence constrained production scheduling problem: A mining application. *Omega* 94 (11). https://doi.org/10.1016/j.omega.2019.03.004.
236. Letelier, O.R., D. Espinoza, M. Goycoolea, E. Moreno, and G. Muñoz. 2020. Production scheduling for strategic open pit mine planning: A mixed-integer programming approach. *Operations Research* 68 (5): 1425–1444. https://doi.org/10.1287/opre.2019.1965.
237. Gholamnejad, J., A. Azimi, R. Lotfian, S. Kasmaeeyazdi, and F. Tinti. 2020. The application of a stockpile stochastic model into long-term open pit mine production scheduling to improve the feed grade for the processing plant. *Rudarsko-Geolosko-Naftni Zbornik* 35 (4): 115–129. https://doi.org/10.17794/rgn.2020.4.10.
238. Dimitrakopoulos, R. 1998. Conditional simulation algorithms orebody uncertainty in open pit optimization. *International Journal of Mining, Reclamation and Environment* 12 (4): 173–179. https://doi.org/10.1080/09208118908944041.
239. Ramazan, S., and R. Dimitrakopoulos. 2018. Stochastic optimization of long-term production scheduling for open pit mines with a new integer programming formulation. In *Advances in Applied Strategic Mine Planning*, ed. R. Dimitrakopoulos, 139–153. Cham: Springer. https://doi.org/10.1007/978-3-319-69320-0_11.
240. Ravenscroft, P.J. 1992. Risk analysis for mine scheduling by conditional simulation. *Transactions of the Institution of Mining and Metallurgy* 101 (A): A104–A108.
241. Dowd, P. 1997. Risk in minerals projects: analysis, perception, and management. *Transactions of the Institution of Mining and Metallurgy* 106 (A): A9–A18.
242. Smith, M., and R. Dimitrakopoulos. 1999. The influence of deposit uncertainty on mine production scheduling. *International Journal of Mining, Reclamation and Environment* 13 (4): 173–178. https://doi.org/10.1080/09208119908944244.
243. Dimitrakopoulos, R., C.T. Farrelly, and M. Godoy. 2002. Moving forward from traditional optimization: Grade uncertainty and risk effects in open-pit design. *Transactions of the Institution of Mining and Metallurgy* 111 (A): A82–A88.
244. Ramazan, S., and R. Dimitrakopoulos. 2003. Production scheduling optimization in nickel laterite deposit: MIP and LP application and in the presence of orebody variability mine. In *Proceedings of the 12th MPES Conference*, Kalgoorlie.
245. Dimitrakopoulos, R., and S. Ramazan. 2004. Uncertainty-based production scheduling in open pit mining. *Transactions of the Society for Mining, Metallurgy and Exploration Inc.* 316: 106–112.

246. Menabde, M., G. Froyland, P. Stone, and G. Yeates. 2004. Mining schedule optimization for conditionally simulated orebodies. In *Proceedings of the International Symposium on Orebody Modelling and Strategic Mine Planning: Uncertainty and Risk Management*, Perth.
247. Godoy, M., and R. Dimitrakopoulos. 2004. Managing risk and waste mining in long-term production scheduling of open-pit mines. *Transactions of the Society for Mining, Metallurgy and Exploration Inc.* 316: 43–50.
248. Ramazan, S., and R. Dimitrakopoulos. 2004. Stochastic optimization of long-term production scheduling for open pit mines with a new integer programming formulation. In *Proceedings of the International Symposium on Orebody Modelling and Strategic Mine Planning: Uncertainty and Risk Management*, Perth.
249. Ramazan, S., and R. Dimitrakopoulos. 2003. Stochastic integer programming based modeling for long-term production scheduling of open-pit mines. ARC-Linkage Report N-6002-1. University of Queensland, Brisbane.
250. Ramazan, S., R. Dimitrakopoulos, J. Benndorf, and L. Archambeault. 2004. Extension of SIP modeling for long-term production scheduling with stochastically designed stockpiles and multiple ore processors. ARC-Linkage Report N-6003-1, W H Bryan Mining Geology Research Centre, University of Queensland, Brisbane.
251. Dimitrakopoulos, R., and S. Ramazan. 2008. Stochastic integer programming for optimizing long term production schedules of open-pit mines: Methods, application, and value of stochastic solutions. *Mining Technology* 117 (4): 155–160. https://doi.org/10.1179/174328609X417279.
252. Ramazan, S., and R. Dimitrakopoulos. 2013. Production scheduling with uncertain supply: A new solution to the open-pit mining problem. *Optimization and Engineering* 14 (2): 361–380. https://doi.org/10.1007/s11081-012-9186-2.
253. Jalali, S.E., M. Ataee-pour, and K. Shahria. 2006. Pit limit optimization using stochastic process. *CIM Bulletin* 99 (1096): 1–11.
254. Golamnejad, J., M. Osanloo, and B. Karimi. 2006. A chance-constrained programming approach for open pit long-term production scheduling in stochastic environments. *Journal of the South African Institute of Mining and Metallurgy* 106: 105–114.
255. Leite, A., and R. Dimitrakopoulos. 2007. A stochastic optimization model for open pit mine planning: Application and risk analysis at a copper deposit. *Mining Technology* 116 (3): 109–118. https://doi.org/10.1179/174328607X228848.
256. Kent, M., R. Peattie, and V. Chamberlain. 2007. Incorporating grade uncertainty in the decision to expand the main pit at the Navachab gold mine, Namibia, through stochastic simulation. *AusIMM Spectrum Series* 14: 207–218.
257. Dimitrakopoulos, R., L. Martinez, and S. Ramazan. 2007. A maximum upside/minimum downside approach to the traditional optimization of open-pit mine design. *Journal of Mining Science* 43: 73–82. https://doi.org/10.1007/s10913-007-0009-3.
258. Dimitrakopoulos, R., L. Martinez, and S. Ramazan. 2007. Optimising open pit design with simulated orebodies and Whittle Four-X—A maximum upside/minimum downside approach. In *Australasian Institute of Mining and Metallurgy Publication Series*, 201–206. Perth, Australia: Australasian Institute of Mining and Metallurgy.
259. Dimitrakopoulos, R., and S.A. Abdel Sabour. 2007. Evaluating mine plans under uncertainty: Can the real options make a difference? *Resources Policy* 32 (3): 116–125. https://doi.org/10.1016/j.resourpol.2007.06.003.
260. Abdel Sabour, S.A., R.G. Dimitrakopoulos, and M. Kumral. 2008. Mine design selection under uncertainty. *Mining Technology* 117 (2): 53–64. https://doi.org/10.1179/174328608X343065.
261. Boland, N., I. Dumitrescu, and G. Froyland. 2008. A multistage stochastic programming approach to open-pit mine production scheduling with uncertain geology. http://www.optimization-online.org/DB_HTML/2008/10/2123.html. Accessed January 23, 2021.
262. Albor Consuegra, F.R., and R. Dimitrakopoulos. 2009. Stochastic mine design optimization based on simulated annealing: Pit limits, production schedules, multiple orebody scenarios, and sensitivity analysis. *Mining Technology* 118 (2): 79–90. https://doi.org/10.1179/037178409X12541250836860.

263. Meagher, C., S.A. Abdel Sabour, and R. Dimitrakopoulos. 2009. Pushback design of open-pit mines under geological and market uncertainties. In *Proceedings of the International Symposium on Orebody Modelling and Strategic Mine Planning: Uncertainty and Risk Management*, Perth.
264. Albor Consuegra, F.R., and R. Dimitrakopoulos. 2010. Algorithmic approaches to pushback design based on stochastic programming: Method, application, and comparisons. *Mining Technology* 119 (2): 88–101. https://doi.org/10.1179/037178410X12780655704761.
265. Lagos, G., D. Espinoza, E. Moreno, and J. Amaya. 2011. Robust planning for an open-pit mining problem under ore-grade uncertainty. *Electron Notes in Discrete Mathematics* 37: 15–20. https://doi.org/10.1016/j.endm.2011.05.004.
266. Godoy, M., and R. Dimitrakopoulos. 2011. A risk quantification framework for strategic mine planning: Method and application. *Journal of Mining Science* 47 (2): 235–246. https://doi.org/10.1134/S1062739147020109.
267. Dimitrakopoulos, R. 2011. Stochastic optimization for strategic mine planning: A decade of developments. *Journal of Mining Science* 47 (2): 138–150. https://doi.org/10.1134/S1062739147020018.
268. Asad, M.W.A., and R. Dimitrakopoulos. 2013. Implementing a parametric maximum flow algorithm for optimal open pit mine design under uncertain supply and demand. *The Journal of the Operational Research Society* 64 (2): 185–197. https://doi.org/10.1057/jors.2012.26.
269. Tachefine, B., and F. Soumis. 1997. Maximal closure on a graph with resource constraints. *Computers and Operations Research* 24 (10): 981–990. https://doi.org/10.1016/S0305-0548(97)00008-7.
270. Asad, M.W.A., and R. Dimitrakopoulos. 2012. Performance evaluation of a new stochastic network flow approach to optimal open pit mine design—Application at a gold mine. *Journal of the South African Institute of Mining and Metallurgy* 112 (7): 649–655.
271. Asad, M.W.A., and R. Dimitrakopoulos. 2012. Optimal production scale of open-pit mining operations with uncertain metal supply and long-term stockpiles. *Resources Policy* 37 (1): 81–89. https://doi.org/10.1016/j.resourpol.2011.12.002.
272. Lamghari, A., and R. Dimitrakopoulos. 2012. A diversified Tabu search approach for the open-pit mine production scheduling problem with metal uncertainty. *European Journal of Operational Research* 222 (3): 642–652. https://doi.org/10.1016/j.ejor.2012.05.029.
273. Lamghari, A., and R. Dimitrakopoulos. 2016. Network-flow based algorithms for scheduling production in multi-processor open-pit mines accounting for metal uncertainty. *European Journal of Operational Research* 250: 273–290. https://doi.org/10.1016/j.ejor.2015.08.051.
274. Marcotte, D., and J. Caron. 2013. Ultimate open-pit stochastic optimization. *Computers and Geosciences* 51: 238–246. https://doi.org/10.1016/j.cageo.2012.08.008.
275. Dimitrakopoulos, R., and A. Jewbali. 2013. Joint stochastic optimization of short and long term mine production planning: Method and application in a large operating gold mine. *Mining Technology* 122 (2): 110–123. https://doi.org/10.1179/1743286313y.0000000040.
276. Benndorf, J., and R. Dimitrakopoulos. 2013. Stochastic long-term production scheduling of iron ore deposits: Integrating joint multi-element geological uncertainty. *Journal of Mining Science* 49 (1): 68–81. https://doi.org/10.1134/S1062739149010097.
277. Goodfellow, R., and R. Dimitrakopoulos. 2013. Algorithmic integration of geological uncertainty in pushback designs for complex multiprocess open-pit mines. *Mining Technology* 122 (2): 67–77. https://doi.org/10.1179/147490013x13639459465736.
278. Montiel, L., and R. Dimitrakopoulos. 2013. Stochastic mine production scheduling with multiple processes: Application at Escondida Norte, Chile. *Journal of Mining Science* 49: 583–597. https://doi.org/10.1134/S1062739149040096.
279. Lamghari, A., R. Dimitrakopoulos, and J.A. Ferland. 2014. A variable neighborhood descent algorithm for the open-pit mine production scheduling problem with metal uncertainty. *The Journal of the Operational Research Society* 65 (9): 1305–1314. https://doi.org/10.1057/jors.2013.81.
280. Koushavand, B., H. Askari-Nasab, and C.V. Deutsch. 2014. A linear programming model for long-term mine planning in the presence of grade uncertainty and a stockpile. *International*

Journal of Mining Science and Technology 24 (4): 451–459. https://doi.org/10.1016/j.ijmst.2014.05.006.

281. Del Castillo, F., and R. Dimitrakopoulos. 2014. Joint effect of commodity price and geological uncertainty over mine life and ultimate pit limit. *Mining Technology* 123 (4): 207–219. https://doi.org/10.1179/1743286314y.0000000069.
282. Moosavi, E., and J. Gholamnejad. 2015. Long-term production scheduling modeling for the open-pit mines considering tonnage uncertainty via indicator kriging. *Journal of Mining Science* 51 (6): 1226–1234. https://doi.org/10.1134/S1062739115060526.
283. de Freitas, S.M., R. Dimitrakopoulos, and A. Lamghari. 2015. Solving a large SIP model for production scheduling at a gold mine with multiple processing streams and uncertain geology. *Mining Technology* 124 (1): 24–33. https://doi.org/10.1179/1743286314y.0000000075.
284. Montiel, L., and R. Dimitrakopoulos. 2015. Optimizing mining complexes with multiple processing and transportation alternatives: An uncertainty-based approach. *European Journal of Operational Research* 247 (1): 166–178. https://doi.org/10.1016/j.ejor.2015.05.002.
285. Montiel, L., R. Dimitrakopoulos, and K. Kawahata. 2016. Globally optimizing open-pit and underground mining operations under geological uncertainty. *Mining Technology* 125 (1): 2–14. https://doi.org/10.1179/1743286315Y.0000000027.
286. Del Castillo, M.F., M.C. Godoy, and R. Dimitrakopoulos. 2015. Optimal mining rates revisited: Managing mining equipment and geological risk at a given mine setup. *Journal of Mining Science* 51 (4): 785–798. https://doi.org/10.1134/S1062739115040165.
287. Del Castillo, M.F., and R. Dimitrakopoulos. 2016. A multivariate destination policy for geometallurgical variables in mineral value chains using coalition-formation clustering. *Resources Policy* 50: 322–332. https://doi.org/10.1016/j.resourpol.2016.10.003.
288. Gilani, S.O., and J. Sattarvand. 2016. Integrating geological uncertainty in long-term open pit mine production planning by ant colony optimization. *Computers and Geosciences* 87: 31–40. https://doi.org/10.1016/j.cageo.2015.11.008.
289. Lamghari, A., and R. Dimitrakopoulos. 2016. Progressive hedging applied as a metaheuristic to schedule production in open-pit mines accounting for reserve uncertainty. *European Journal of Operational Research* 253 (3): 843–855. https://doi.org/10.1016/j.ejor.2016.03.007.
290. Matamoros, M.E.V., and R. Dimitrakopoulos. 2016. Stochastic short-term mine production schedule accounting for fleet allocation, operational considerations, and blending restrictions. *European Journal of Operational Research* 255 (3): 911–921. https://doi.org/10.1016/j.ejor.2016.05.050.
291. Goodfellow, R.C., and R. Dimitrakopoulos. 2016. Global optimization of open pit mining complexes with uncertainty. *Applied Soft Computing* 40: 292–304. https://doi.org/10.1016/j.asoc.2015.11.038.
292. Goodfellow, R.C., and R. Dimitrakopoulos. 2017. Simultaneous stochastic optimization of mining complexes and mineral value chains. *Mathematical Geosciences* 49 (3): 341–360. https://doi.org/10.1007/s11004-017-9680-3.
293. Montiel, L., and R. Dimitrakopoulos. 2017. A heuristic approach for the stochastic optimization of mine production schedules. *Journal of Heuristics* 23: 397–415. https://doi.org/10.1007/s10732-017-9349-6.
294. Montiel, L., and R. Dimitrakopoulos. 2018. Simultaneous stochastic optimization of production scheduling at Twin Creeks Mining Complex, Nevada. *Mining Engineering* 70 (12): 48–56. https://doi.org/10.19150/me.8645.
295. Moreno, E., X. Emery, M. Goycoolea, N. Morales, and G. Nelis. 2017. A two-stage stochastic model for open pit mine planning under geological uncertainty. In *Proceedings of the 38h International APCOM Symposium*, Golden.
296. Richmond, A. 2018. Direct net present value open pit optimization with probabilistic models. In *Advances in Applied Strategic Mine Planning*, ed. R. Dimitrakopoulos, 217–228. Cham: Springer. https://doi.org/10.1007/978-3-319-69320-0_15.
297. Jamshidi, M., and M. Osanloo. 2018. UPL determination of multi-element deposits with grade uncertainty using a new block economic value calculation approach. *Journal of Mining and Environment* 9 (1): 61–72. https://doi.org/10.22044/jme.2017.5763.1387.

298. Rimélé, M.A., R. Dimitrakopoulos, and M. Gamache. 2018. A stochastic optimization method with in-pit waste and tailings disposal for open pit life-of-mine production planning. *Resources Policy* 57: 112–121. https://doi.org/10.1016/j.resourpol.2018.02.006.
299. Ajak, A.D., E. Lilford, and E. Topal. 2018. Application of predictive data mining to create mine plan flexibility in the face of geological uncertainty. *Resources Policy* 55: 62–79. https://doi.org/10.1016/j.resourpol.2017.10.016.
300. Quigley, M., R. Dimitrakopoulos, and T. Grammatikopoulos. 2018. Risk-resilient mine production schedules with good quality for rare earth element projects. *Mining Technology* 127 (1): 41–55. https://doi.org/10.1080/14749009.2017.1323172.
301. Paithankar, A., and S. Chatterjee. 2019. Open-pit mine production schedule optimization using a maximum-flow and genetic algorithms hybrid. *Applied Soft Computing* 81. https://doi.org/10.1016/j.asoc.2019.105507.
302. Del Castillo, M.F., and R. Dimitrakopoulos. 2019. Dynamically optimizing the strategic plan of mining complexes under supply uncertainty. *Resources Policy* 60: 83–93. https://doi.org/10.1016/j.resourpol.2018.11.019.
303. Mai, N.L., E. Topal, O. Erten, and B. Sommerville. 2019. A new risk-based optimization method for the iron ore production scheduling using stochastic integer programming. *Resources Policy* 62: 571–579. https://doi.org/10.1016/j.resourpol.2018.11.004.
304. Saliba, Z., and R. Dimitrakopoulos. 2019. Simultaneous stochastic optimization of an open-pit gold mining complex with supply and market uncertainty. *Mining Technology* 128 (4): 216–229. https://doi.org/10.1080/25726668.2019.1626169.
305. Sepulveda, G.F., P.J. Alvarez, and J.B. Bedoya. 2020. Stochastic optimization in mine planning schedule. *Computers and Operations Research* 115. https://doi.org/10.1016/j.cor.2019.104823.
306. Quigley, M., and R. Dimitrakopoulos. 2020. Incorporating geological and equipment performance uncertainty while optimizing short-term mine production schedules. *International Journal of Mining, Reclamation and Environment* 34 (5): 362–383. https://doi.org/10.1080/17480930.2019.1658923.
307. Lamghari, A., and R. Dimitrakopoulos. 2020. Hyper-heuristic approaches for strategic mine planning under uncertainty. *Computers and Operations Research* 115. https://doi.org/10.1016/j.cor.2018.11.010.
308. Lagos, T., M. Armstrong, T. Homem-de-Mello, G. Lagos, and D. Sauré. 2020. A framework for adaptive open-pit mining planning under geological uncertainty. *Optimization and Engineering*. https://doi.org/10.1007/s11081-020-09557-0.
309. Gilani, S.O., J. Sattarvand, M. Hajihassani, and S.S. Abdullah. 2020. A stochastic particle swarm-based model for long term production planning of open-pit mines considering the geological uncertainty. *Resources Policy* 68. https://doi.org/10.1016/j.resourpol.2020.101738.
310. Both, C., and R. Dimitrakopoulos. 2020. Joint stochastic short-term production scheduling and fleet management optimization for mining complexes. *Optimization and Engineering* 21: 1717–1743. https://doi.org/10.1007/s11081-020-09495-x.
311. Dimitrakopoulos, R., and R. Senécal. 2020. Long-term mine production scheduling with multiple processing destinations under mineral supply uncertainty, based on multi-neighborhood Tabu search. *International Journal of Mining, Reclamation and Environment* 34 (7): 459–475. https://doi.org/10.1080/17480930.2019.1595902.
312. Maleki, M., E. Jélvez, X. Emery, and N. Morales. 2020. Stochastic open-pit mine production scheduling: An iron deposit case study. *Minerals* 10 (7): 585. https://doi.org/10.3390/min10070585.
313. Rimélé, A., R. Dimitrakopoulos, and M. Gamache. 2020. A dynamic stochastic programming approach for open-pit mine planning with geological and commodity price uncertainty. *Resources Policy* 65. https://doi.org/10.1016/j.resourpol.2019.101570.
314. Chatterjee, S., and R. Dimitrakopoulos. 2020. Production scheduling under uncertainty of an open-pit mine using Lagrangian relaxation and branch-and-cut algorithm. *International Journal of Mining, Reclamation and Environment* 34 (5): 343–361. https://doi.org/10.1080/17480930.2019.1631427.

315. Tolouei, K., E. Moosavi, A.H. Bangin Tabrizi, P. Afzal, and A.A. Aghajani Bazzazi. 2020. A comprehensive study of several meta-heuristic algorithms for open-pit mine production scheduling problem considering grade uncertainty. *Journal of Mining and Environment* 11 (3): 721–736. https://doi.org/10.22044/jme.2020.9127.1803.
316. Tolouei, K., E. Moosavi, A.H.B. Tabrizi, and P. Afzal. 2021. Application of an improved Lagrangian relaxation approach in the constrained long-term production scheduling problem under grade uncertainty. *Engineering Optimization* 53 (5): 735–753. https://doi.org/10.1080/0305215X.2020.1746295.
317. Humphrays, D. 2019. Mining productivity and the fourth industrial revolution. *Mineral Economics* 33: 115–125. https://doi.org/10.1007/s13563-019-00172-9.
318. Bassan, J., V. Srinivasan, P. Knights, and C. Farrelly. 2008. A day in the life of a mine worker in 2025. In *Proceedings of the First International Future Mining Conference*, Sydney.
319. Guieiro, G.A., E.A. Leitão, E.S. Balbino, and L. van Melis. 2019. Implantação de caminhões fora de estrada autônomos na indústria de mineração. In *Proceedings of 20° Simpósio de Mineração*, São Paulo. https://doi.org/10.5151/2594-357X-33680.
320. Bassan, J., and P. Knights. 2008. Applications of advanced analytics in mining: Safer, smarter, sustainable operations. In *Proceedings of the Australian Mining Technology Conference*, Queensland.
321. Rupp, K. 2018. 42 years of microprocessor trend data. https://www.karlrupp.net/2018/02/42-years-of-microprocessor-trend-data. Accessed September 29, 2018.

Chapter 10
Advanced Analytics for Dynamic Programming

Pritam Biswas, Rabindra Kumar Sinha, and Phalguni Sen

Abstract In open-pit mine planning, the determination of optimum cut-off grade is essential for the mine designing process. Selection of the optimum cut-off grade maximizes the net present value (NPV) of an open-pit mining operation subject to the variable mining capacities, milling, smelting, and refining. The cut-off grade of a metalliferous deposit is dynamic; thus, the dynamic programming approach may be considered one of the suitable methods for solving the cut-off grade determination problem. This chapter analyzes cut-off grade optimization of an open-pit copper mine project considering fixed annual mine, mill, and refinery production. A computer-driven model, namely the cut-off grade optimizer (COGO) derived from a dynamic programming algorithm built-in Visual Basic using C programming language, has been developed. The cut-off grade planner (COGP) software package developed in the present research comprises three modules—input data module, output result module, and graphical module. The algorithm and the associated software have been validated using a portion of the data of the Sarcheshmeh copper mine from a published paper. The results of the analysis are encouraging. The optimum cut-off grade of the copper deposit has been determined to be 0.24% resulting in a maximum NPV of $83,714 million, with an optimized mine life of 23.5 years, and the average mill head grade corresponds to 0.707%.

Keywords Dynamic programming · Cut-off grade · Refining · Optimization · Mining

Introduction and Review of Previous Work Done

The selection of optimum cut-off grade for metalliferous deposits is a common problem faced by the mine planners because of the variation in costs of operation, metallurgical characteristics, and stripping ratios for the different deposits. Cut-off

P. Biswas (✉) · R. K. Sinha · P. Sen
Department of Mining Engineering, IIT(ISM) Dhanbad, Dhanbad 826004, India

R. K. Sinha
e-mail: rksinha@iitism.ac.in

A. Soofastaei (ed.), *Advanced Analytics in Mining Engineering*,
https://doi.org/10.1007/978-3-030-91589-6_10

grade optimization theory supports the fulfillment of the ultimate objective of a mining operation through maximizing the Present Net Value (NPV) [1]. Henning [2] was the first who presented a methodology for the calculation of optimum cut-off grade. According to him, fixed costs are irrelevant in determining the cut-off grade, and the maximization of profits implies a descending order of cut-off grades during the project's life.

However, the optimum cut-off grade at any moment may not necessarily be one of the above three limiting cut-off grades, namely, mining, concentrator, and refinery cut-off grades, as it has been decided based on the interaction of mine, plant, and refinery capacities [3]. Lane [4] optimized the cut-off grade terminology based on dynamic programming, which is dependent on the opportunity cost of deferring the processing of higher-grade ore. Dowd [5] employed a method of general approach using stochastic programming, which predicts the uncertainty of pricing of metals in the future concerning future market conditions in terms of probability. The optimizing criterion is the maximum NPV.

Optimization and determination of optimal cut-off grade using Lane's approach for a single element through a computer-aided application developed by the C ++ /CLI programming language in VC ++ IDE [6]. Yasrebi et al. [7] used a nonlinear programming method to modify the optimum cut-off grade model to maximize the profit value of the mining project. Birch [8] utilized the concept of risk assessment and mixed-integer linear programming using an MS-excel solver for maximization of either profit or NPV. He also integrated the cost of dilution and its impact on ore cost under varying degrees of uncertainty. Using Lane's algorithm Zarshenas and Saeedi [9] demonstrated that with an increase in dilution, there is a reduction in the average grade and thereby increasing the cut-off grade. Therefore, the grade at which the highest NPV is achieved should be accepted as the optimal cut-off grade. However, the optimum cut-off grade depends on various variables like price of the metal, costs of production, cost of processing, grade-tonnage distributions, limiting constraints related to mine, mill, smelter and refinery, total reserve, and mine life [10].

Khan and Asad [11] shared a two-stage stochastic linear programming based on optimization of cut-off grade that accounts for uncertainty in grade and maximizes the NPV subject to the constraint of production capacity over the entire mine life. Franco-Sepulveda et al. [12] used a genetic algorithm to optimize the NPV of a mining project. They had analyzed the metal price from a historic date to the current date and concluded that for 50% of the time, the sale value was below the estimated value. However, this method worked only for a few variables with the average price value or the ore grade. Githiria and Musingwini [13] developed a model using stochastic and dynamic programming extending Lane's algorithm. Biswas and Sen [14]; Ahmadi and Bazzazi [15] stated that the NPV and cut-off grade are the twin critical factors affecting the mine planning, and choosing the optimal cut-off grade maximizes the NPV and the total profit.

This chapter considers that the optimum cut-off grade for a metalliferous deposit is dynamic. A model has been developed to determine cut-off grades using dynamic programming techniques. The model has been developed in Visual C, a sequence

Table 10.1 Possible combinations of eight different models based on constraints

Models	Models	Mill (*C*)	Refinery (*R*)	Restrictions in production capacities
1*	Restricted	Restricted	Restricted	All (mine, mill, and refinery)
2	Restricted			M
3		Restricted		C
4			Restricted	R
5	Restricted	Restricted		M and C
6	Restricted		Restricted	M and R
7		Restricted	Restricted	C and R
8				No restrictions (M, C, and R)

*signifies the model considered for this study among the eight modules. The first model considered for this study

of mathematical and computational techniques based on simple algebraic equations and iterative algorithm steps that fulfill the criteria of dynamic programming for the cut-off grade optimization and NPV maximization subject to mining constraints concentrating and refining capacities. For this research work, we have included a real case study of the Sarcheshmeh copper mine taken from a published paper [16] to demonstrate the functioning of the computer tool.

Operational Model

In metalliferous open-pit mine planning and design, the cut-off grade is mainly affected by three governing capacity constraints of various operations: mine, mill, and refinery. Therefore, eight different possibilities can be considered while calculating the optimum cut-off grade. However, in this part of the study, only the first model was considered here to calculate the optimum cut-off grade, limiting conditions on the three main stages: mine, mill, and refinery capacities, as shown in Table 10.1.

Model Prerequisite

Development of ultimate pit limit and finalizing mineable ore reserves as the grade-tonnage report for fixing the pit limit and mine production, processing and refining capacities, costs of these operations, techno-economic parameters including the price of metal are the prerequisites for deploying the cut-off grade optimization model.

Model Formulation

The objective function of cut-off grade optimization is to optimize the total net present value (PRV) of a given mine design. Therefore, the present value (NPV) of a discrete series of cash flow is given by Eq. (10.1):

$$\text{PVR} = \sum_{n=1}^{N} \frac{P_n}{(1+r)^n} \tag{10.1}$$

For a continuous cash flow series, PRV may be calculated using Eq. (10.2) when the cash flow is the same every year. What follows is the development of a generalized function for (n) for an open-pit mine.

$$\text{PVR} = \frac{cf \times \left\{(1+r)^p - 1\right\}}{r \times (1+r)^p} \tag{10.2}$$

Mining operations generate revenue by excavating and processing material to create a saleable product. Revenue is calculated as the product of recoverable ore grade, tonnes of ore processed, and the selling price of the product. During operation, mine incurs both fixed costs and variable costs. Variable costs can be further broken down into excavating and treating ore and those associated with excavating and handling waste. Typical variable ore costs may include: drilling and blasting, hauling, crushing, floatation, tailings disposal, among other variable waste costs may include drilling and blasting, dumping, remediation, and other fixed costs include general, administrative costs, and any other costs that are independent of production level. Capital costs can be discounted and subtracted from the discounted cash flow to determine NVP using Eq. (10.3).

$$\text{NVP}_{\dot{\upsilon},\delta} = \text{PVR}_{\dot{\upsilon},\delta} - \text{cf}_{\ddot{\upsilon},\partial} \tag{10.3}$$

For obtaining the optimized returns $\left(\text{NPV}_{\dot{\upsilon},\delta}\right) \rightarrow A_{m(\dot{\upsilon},\partial)}, A_{c(\dot{\upsilon},\partial)}, A_{r(\dot{\upsilon},\partial)}$ are the initial decision variables for all the stages. Decision variables are governed by the limiting capacities (constraints). Based on the cut-off grade c_{t^*} and the stage transition $\left(\dot{\upsilon} - 1 \le \gamma \le \partial - 1\right)$, the amount of production in different stages such as mine, mill, and refinery are subsequently determined by any one of the following conditions given in Eqs. 10.4–10.6:

$$A_{m_p} \le \text{MM}_{\dot{\upsilon}} - 1, \gamma)\forall p \tag{10.4}$$

$$A_{c_p} \le \text{MC}_{\dot{\upsilon}} - 1, \gamma)\forall p \tag{10.5}$$

$$A_{r_p} \le \text{MR}_{\dot{\upsilon}} - 1, \gamma)\forall p \tag{10.6}$$

Where:

$cf_{\dot{\upsilon},}$	cash flow;
n	time of cash flow during the year;
N	Life of the project;
r	discounting rate;
pp	period indicator (year);
$MM_{\dot{\upsilon},}$	Maximum capacity of mine production (tonne/year);
$MC_{\dot{\upsilon},}$	Maximum capacity of the mill to process ore (tonne/year);
$MR_{\dot{\upsilon},}$	Maximum capacity of the refinery for metal production (tonne/year);
A_m	Actual production rate of mining (ore and waste) (tonne/year);
A_c	Actual processing rate of mill (tonne/year);
A_r	Actual rate of production in refinery for metal production (tonne/year);
c_{t*}	Cut-off grade;
$a\bar{g}$	Average grade of the material above the cut-off grade;
G_m	Maximum grade in the mineral inventory data;
G_t	Grade interval;
SR	Overall stripping ratio;
y_m	mining recovery;
y_c	mill recovery; and
y_r	refinery recovery.

It is assumed that the distribution of grade tonnage of a mineral inventory comprises of v-classes of grade for example (cg_1, cg_2), (cg_2, cg_3), (cg_3, cg_4) … (cg_{v-1}, cg_v) and for each class of grades, there exist t_t tonnes of material. If 't' signifies grade increment (cg_{t-1}, cg_t) and c_{t*} is taken as the cut-off grade, then the ore quantity is T_o, waste quantity is T_w, and $av\bar{g}$ can be determined from Eqs. 10.7–10.9, respectively:

$$T_o(c_{t^*}) = \sum_{t=t^*}^{T} (t_t) \tag{10.7}$$

$$T_w(c_{t^*}) = \sum_{t=1}^{t^*} (t_t) \tag{10.8}$$

$$a\bar{g}(c_{t^*}) = \frac{\sum_{t=t^*}^{T} \left[t_t\left(\frac{cg_t + cg_{t+1}}{2}\right)\right]}{T_{ore}(cg_t)} \tag{10.9}$$

To determine the total metallurgical recovery (y_{mcr}) comprises of all the three stages,

$$y_{\mathrm{mcr}(\dot{\upsilon},\partial)} = y_m \times Y_c \times y_r \tag{10.10}$$

Life of the project will be the complex life of the Mine-Mill-Refinery, as:

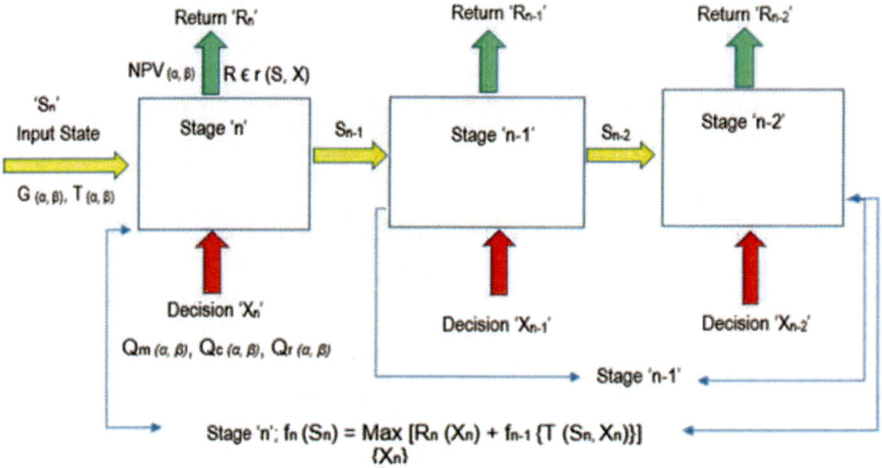

Fig. 10.1 Dynamic programming multi-stage decision process for '*n*' stages

$$\text{Life of the Mine} - \text{Mill} - \text{Refinery complex} = (T_o + T_w)/A_m \tag{10.11}$$

To determine the production rate of all the three stages: mine, mill, and refinery are subsequently estimated using any of the following conditions depending on the cut-off grade c_{t*} values. Determination of the optimum cut-off grade is a Multi-Staged Decision Problem [17], and output from each state is fed as an input to the next. Thus, each stage has several possible states associated with it, as shown in Fig. 10.1.

The results are optimized across all the stages, governed by Bellman's principle of optimality [18]. State variables (Sn(α,β)) and decision variables $A_{m(\dot{\upsilon},\partial)}$, $A_{c(\dot{\upsilon},\partial)}$, $A_{r(\dot{\upsilon},\partial)}$ are having discrete values. Recursive relationship identifies the optimal decision at stage 'n' for the state 'Sn.' It gives the decision for each state at stage (n−1). Here, $f_n(S_n)$ is the optimized objective function value at stage 'n,' and 'X_n' is the decision we are making which maximizes over $\{X_n\}$.

$$f_n(S_n) = \text{Max}\left[R_n(X_n) + f_{n-1}\{T(S_n, X_n)\}\right] \tag{10.12}$$

R_n: Return for decision, 'X_n' in the current state.

$T(S_{n,\,n})$: Transfer function to get the state 'S_{n-1}' corresponding to 'S_n' and 'X_n.'

NPV of the project is calculated by calculating the PV of all the cash flows using Eq. 10.1 and then subtracting the discounted values of all the investments. The flowchart of the software algorithm has been shown in Fig. 10.2.

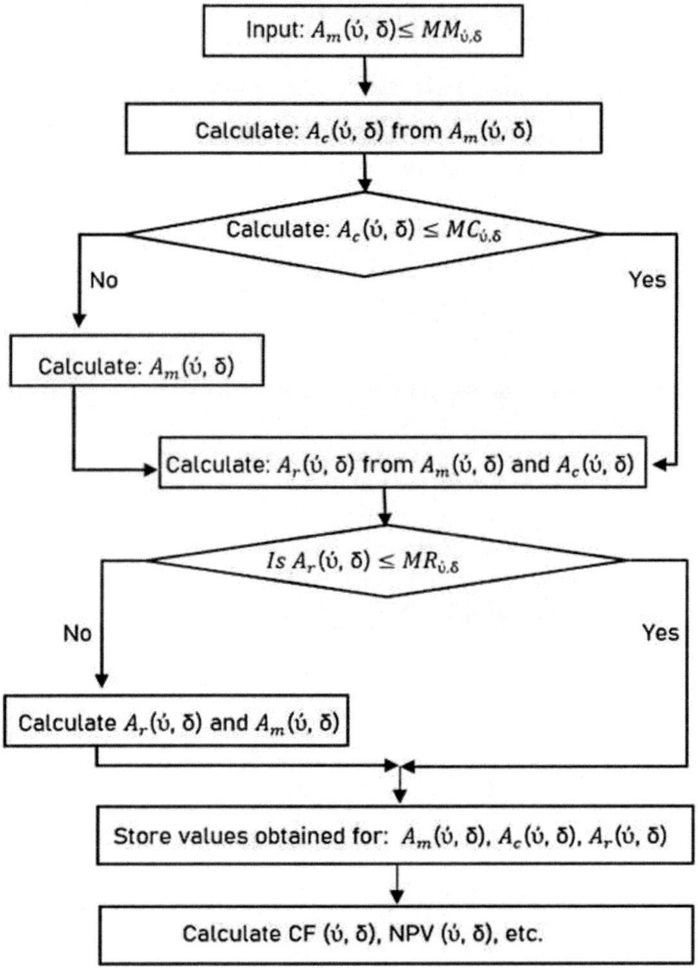

Fig. 10.2 Flowchart of the software algorithm

Development of Computer Model

A computer model has been developed based on the simple programming algorithm to determine the optimum cut-off grade using the formulae as discussed. In addition, the computer model has greater operating flexibility and a graphical user interface (GUI) based on the user interface program. The developed software and the different modules are discussed below.

Brief Details About the Software Tool

The following have been used in the development of this software:

- the language used to develop the computer package is C# 4.0, ideal programming language for the development of scientific and business applications which requires great extent of calculation speed and accuracy of data;
- the framework of development is Microsoft. Net Framework 4.0; it supports the current version and higher (with back version compatibility);
- Intergrated Development Interface (IDE) used in this software is Microsoft Visual Studio 2010;
- the database used in this software is the internal. Net Framework database and dataset. This software package has an entirely GUI-based environment which is very user friendly;
- it supports both 32-bit and 64-bit processor;
- supports all versions of Microsoft Windows (XP/Vista/7/8) and other platforms (LINUX and Mac) which support Microsoft. Net Framework 4.0 compatibility and higher version of the framework;
- size of the software is compact and can be installed in seconds;
- supports the low-speed capacity computer with as minimum RAM as up to 64 MB; and
- results can be printed or exported to the text file, which Microsoft Notepad supports.

Further, detailed specifications and requirements for C# programming language and development framework can be accessed on the website of Microsoft Visual Studio.

Modules of the Software Tool

The software package comprises of three modules:

- input data module;
- output result module; and
- result graphical module.

A brief discussion of the above modules has been given below:

Input Data Module

The input data module consists of all the input functions: the grade range, quantity of material, various fixed costs, and variable costs. The tool is user friendly, and it will ask the user whether the values are correctly entered or not, and then only it processes the data and moves to the next module. The input data module can be categorized into two parts:

1. Mineral inventory data

   The mineral inventory input data of grade class intervals and the tonnage distribution for each given class interval.

   - grade ranges; and
   - quantity of material

After data collection, the data have been validated and rearranged forgiven grade interval for the whole range starting from zero to the maximum value of the grade in the mineral inventory data. Then, this data can be filled in the data-grid-view (GUI form) given on the main window, as shown in Fig. 10.3.

2. Costs and other parameters input

   This part of the input module can be accessed by clicking on the 'Data Input' button on the main form window.The data input wizard consists of five sections of different input data, which are:

   - mine costs input;
   - mill costs input;
   - refinery costs input;
   - additional input; and
   - rates input.

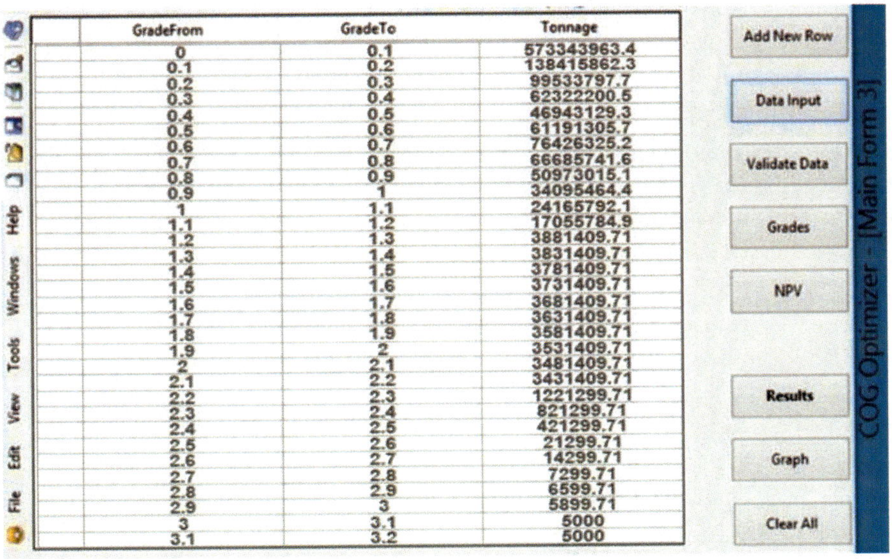

GradeFrom	GradeTo	Tonnage
0	0.1	573343963.4
0.1	0.2	138415862.3
0.2	0.3	99533797.7
0.3	0.4	62322200.5
0.4	0.5	46943129.3
0.5	0.6	61191305.7
0.6	0.7	76426325.2
0.7	0.8	66685741.6
0.8	0.9	50973015.1
0.9	1	34095464.4
1	1.1	24165792.1
1.1	1.2	17055784.9
1.2	1.3	3881409.71
1.3	1.4	3831409.71
1.4	1.5	3781409.71
1.5	1.6	3731409.71
1.6	1.7	3681409.71
1.7	1.8	3631409.71
1.8	1.9	3581409.71
1.9	2	3531409.71
2	2.1	3481409.71
2.1	2.2	3431409.71
2.2	2.3	1221299.71
2.3	2.4	821299.71
2.4	2.5	421299.71
2.5	2.6	21299.71
2.6	2.7	14299.71
2.7	2.8	7299.71
2.8	2.9	6599.71
2.9	3	5899.71
3	3.1	5000
3.1	3.2	5000

Fig. 10.3 Mineral inventory data input window in the software main window

The following input data are the part of above cost parameters and required during calculation:

- fixed annual costs, variable costs;
- investment for mine, plant, smelter, and refinery;
- limiting/maximum capacities of mine, concentrator, smelter, and refinery;
- sale value of the metal;
- discount rate/year;
- yield/recovery in the plant; and
- handling loss.

Output Result Module

This module gave the final optimum cut-off grade of the deposit in addition to the optimized NPV of the entire operation. The output of the result module has been divided into four sections, as shown below:

- mine section;
- mill section;
- refinery section; and
- profits section.

Result Graphical Module

This module is the heart of the computer tool, and it is an application of the GUI. After the computation, it provides the graphical data representation of the result in various graphs between the specified variables. The tool provides the following graphical presentations to the user.

- Grade Distribution versus Tonnage available;
- Cut-off grade versus Cumulative Tonnage; and
- Cut-off grade versus Average Grade;

Results

Case of Study

The grade-tonnage distribution, operational data (mine, mill, and refinery capacities) of the Sarcheshmeh copper mine were used to compute the model. Only financial parameters have been considered as per the Indian scenario (in INR). Mineral inventory is given in Table 10.2.

Table 10.2 Grade tonnage distribution of deposit [16]

Grade from	Grade to	Tonnage (in tonnes)
0.00	0.10	573,343,963.40
0.10	0.20	138,415,862.30
0.20	0.30	99,533,797.70
0.30	0.40	62,322,200.50
0.40	0.50	46,943,129.30
0.50	0.60	61,191,305.70
0.60	0.70	76,426,325.20
0.70	0.80	66,685,741.60
0.80	0.90	50,973,015.10
0.90	1.00	34,095,464.40
1.00	1.10	24,165,792.10
1.10	1.20	17,055,784.90
1.20	1.30	3,881,409.71
1.30	1.40	3,831,409.71
1.40	1.50	3,781,409.71
1.50	1.60	3,731,409.71
1.60	1.70	3,681,409.71
1.70	1.80	3,631,409.71
1.80	1.90	3,581,409.71
1.90	2.00	3,531,409.71
2.00	2.10	3,481,409.71
2.10	2.20	3,431,409.71
2.20	2.30	1,221,299.71
2.30	2.40	821,299.71
2.40	2.50	421,299.71
2.50	2.60	21,299.71
2.60	2.70	14,299.71
2.70	2.80	7299.71
2.80	2.90	6599.71
2.90	3.00	5899.71
3.00	3.10	5000.00
3.10	3.20	5000.00

In addition to the grade-tonnage deposit (mineral inventory) data, the required technical and financial data has been given in Table 10.3, which must be required in the calculation and software input.

Table 10.3 Technical and financial parameters were used for this case study in INR ($)

Mining		
Particulars	Unit	Values
Maximum mining capacity	tonnes/year	55,000,000
Percentage mining recovery	%	100
Specific investment cost for mining	$/tonne/year	200
Unit cost of excavation	$/tonne of ore and waste	100
Fixed mining cost	$/year	1000
Average ore transportation cost	$/tonne/km	5
Average waste transportation cost	$/tonne/km	5
Average ore transportation haul distance	km	0.5
Average waste transportation haul distance	km	2.0
Mill		
Maximum mill capacity	tonnes/year	23,000,000
Mill concentrate grade	%	23
Percentage mill recovery	%	86
Specific investment cost for mill	$/tonne/year	400
Unit cost of ore processing	$/tonne	300
Fixed mill cost	$/year	2400
Average concentrate transportation cost	$/tonne/km	5
Average mill tailing transportation cost	$/tonne/km	5
Average concentrate transportation haul distance	km	2
Average mill tailing transportation haul distance	km	1
Refinery		
Maximum refinery capacity	tonnes/year	150,000
Percentage refinery recovery	%	84
Specific investment cost for refinery	$/tonne/year	1500
The unit cost of refining	$/tonne of metal produced	8000
Fixed refinery cost	$/year	4800
Average metal transportation cost	$/tonne/km	5
Average slag transportation cost	$/tonne/km	5
Average metal transportation haul distance	km	15
Average slag transportation haul distance	km	1
Additional costs		
Other costs including general administration cost (percentage of the total operating cost)	%	25
Rates		
Selling price of metal	$/tonne	500,000

(continued)

Table 10.3 (continued)

Mining		
Particulars	Unit	Values
Tax rate	%	40
The interest rate of investment	%	20
Discount rate	%	15
Grade interval precision		
Grade Interval increment	%	0.01

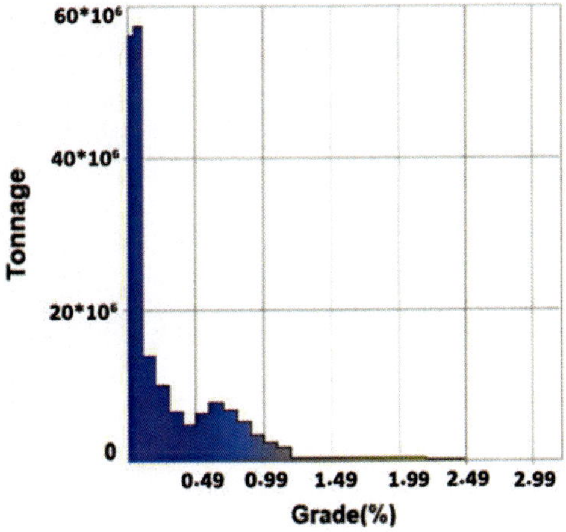

Fig. 10.4 Tonnage versus grade distribution (%)

Graphs of the Results Output

As mentioned in the graphical result module, the output data module for the above results output has been shown through graphs depicted in Figs. 10.4, 10.5, 10.6 and 10.7.

Discussions of Results

In this first model of cut-off grade optimization, after running the software tool, the cut-off grade optimization sheet is generated, and the summarized statistics of the result data of possible NPV variations of the projects, as an example, has been shown in Fig. 10.8. Therefore, it is very clear from the results we got after the simulation is that at Mine-Mill-Refinery complex Life (years) of 23.5 and average grade of 0.7068, we are getting the optimum cut-off grade of 0.24 with a maximum NPV of $83,713,640,025.55 and after achieving its peak value NPV starts declining and reaches its lowest value by the end of mine life with depletion of the reserves by

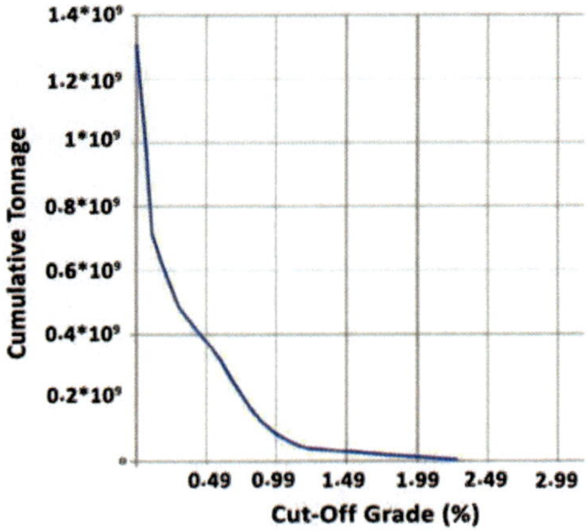

Fig. 10.5 Cum. tonnage versus COG (%)

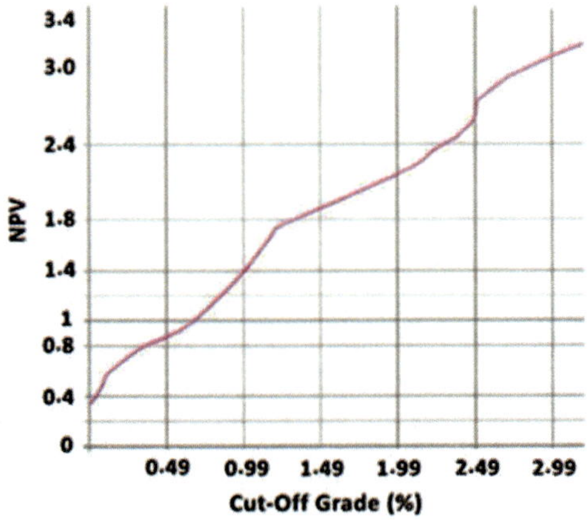

Fig. 10.6 Avg. grade (%) versus COG (%)

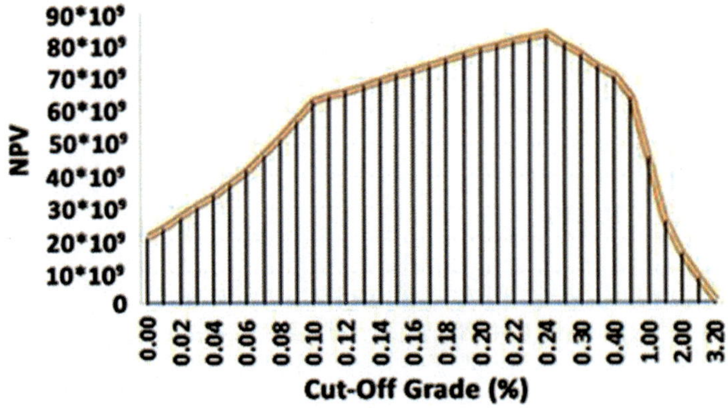

Fig. 10.7 NPV ($) versus COG (%)

Results

MineLife	CutGrade	AveGrade	NPV	Ore	Waste	Conc	Tail	Metal	Revenue
38.6480	0.07	0.477995	46066018359.999...	888905002.630	401340774.380	411075.583502	22588924.416498	79419.802733	39709901366.336...
36.1552	0.08	0.505780	51022435356.736...	831570606.260	458675170.720	434970.970998	22565029.029002	84036.391597	42018195798.384...
33.6624	0.09	0.536940	56496365019.868...	774236209.920	516009567.060	461768.546995	22538231.453005	89213.683279	44606841639.714...
31.1696	0.10	0.572284	62658629432.288...	716901813.580	573343963.400	492164.621181	22507835.378819	95086.204812	47543102406.099...
30.5678	0.11	0.581484	64214930768.046...	703060227.350	587185549.630	500076.381213	22499923.618787	96614.756850	48307378425.194...
29.9660	0.12	0.590853	65774160988.773...	689218641.120	601027135.860	508133.211285	22491866.788715	98171.336420	49085668210.155...
29.3642	0.13	0.600400	67421119712.565...	675377054.890	614868722.090	516344.030857	22483655.969143	99757.666762	49878833380.787...
28.7624	0.14	0.610138	68999022471.354...	661535468.660	628710308.320	524718.505891	22475281.494109	101375.615338	50687807669.035...
28.1606	0.15	0.620078	70684833861.877...	647693882.430	642551894.550	533267.128617	22466732.871383	103027.209249	51513604624.374...
27.5588	0.16	0.630234	72282151662.714...	633852296.200	656393480.780	542001.307751	22457998.692249	104714.652658	52357326328.780...
26.9570	0.17	0.640620	73876342064.476...	620010709.970	670235067.010	550933.470798	22449066.529202	106440.346558	53220173279.064...
26.3552	0.18	0.651253	75629847264.143...	606169123.740	684076653.240	560077.180364	22439922.819636	108206.911246	54103455623.150...
25.7534	0.19	0.662148	77241129344.743...	592327537.510	697918239.470	569447.266789	22430552.733211	110017.211944	55008605971.788...
25.1516	0.20	0.673326	79049444399.644...	578485951.280	711759825.700	579059.979809	22420940.020191	111874.388099	55937194049.548...
24.7188	0.21	0.681525	80152165206.352...	568532571.510	721713205.470	586111.155478	22413888.844522	113236.675238	56618337619.161...
24.2861	0.22	0.689838	81469132502.810...	558579191.740	731666585.240	593260.378108	22406739.621892	114617.905051	57308952525.253...
23.8533	0.23	0.698271	82530414755.431...	548625811.970	741619965.010	600512.984117	22399487.015883	116019.108531	58009554265.674...
23.4590	0.24	0.706831	83713640025.545...	538672432.200	751573344.780	606877.768747	22355401.433860	117240.784922	58624392460.988...

Fig. 10.8 Statistics of the results output with all probable NPV variations

time. Thus, the cut-off grade optimization maximizes undiscounted profit for a given mining operation, thus increasing investment return. In general, the discounted cash flow initially starts with a low value, or even it may be negative. After that, however, it starts increasing and reaches a maximum value when mine produces its targeted production and will decline gradually.

Conclusion

It has been observed that the cut-off grade affects the economy of the mining operation. This model offers a definite conceptual advantage and will overcome the shortcomings of other methodologies. The developed computer tool has been designed to be user friendly and can work in any *Windows®* based platform. The computer tool computes the optimum cut-off grade based on maximum NPV and gives the various costs associated with mine, mill, and refinery. The computer can produce the result in seconds, and thus it will be an influential industry-oriented tool.

The proposed model conforms well to the large open-pit mines such as the Sarcheshmeh copper mine with diversified production constraints. The model does not consider uncertainty in economic parameters, especially the uncertainty and fluctuations on metal prices, stockpiling or blending, mining dilution, multi-mineral deposits, and mine rehabilitation. This model intends to introduce a general method for the solution. This method is capable of handling technically more acceptable relationships and definitions. So further modifications to cut-off grade optimization are essential to incorporate the above parameters and then further optimize NPV values as per the scenario. Therefore, incorporating the above uncertainty parameters, there is a scope of modification in the cut-off grade optimization according to the requirement. Thus, any material, which is waste today can be considered ore in the future or vice versa depending upon the market dynamics.

References

1. Asad, M. 2007. Optimum cut-off grade policy for open pit mining operations through net present value algorithm considering the metal price and cost escalation. *Engineering Computations*.
2. Henning, U. 1963. Calculation of cut-off grade. *Canadian Mining Journal* 84 (3): 54–57.
3. Lane, K.F. 1964. Choosing the optimum cut-off grade.
4. Lane, K.F. 1988. The economic definition of ore: Cut-off grades in theory and practice.
5. Dowd, P. 1976. Application of dynamic and stochastic programming to optimize cut-off grades and production rates.
6. Githiria, J., C. Musingwini, and J. Muriuki. 2016. Development of a computer-aided application using lane's algorithm to optimize cut-off grade. *Journal of the Southern African Institute of Mining and Metallurgy* 116 (11): 1027–1035.
7. Yasrebi, A.B., et al. 2015. Determination of optimized cut-off grade utilizing non-linear programming. *Arabian Journal of Geosciences* 8 (10): 8963–8967.
8. Birch, C. 2017. Optimization of cut-off grades considering grade uncertainty in narrow, tabular gold deposits. *Journal of the Southern African Institute of Mining and Metallurgy* 117 (2): 149–156.
9. Zarshenas, Y., and G. Saeedi. 2017. Determination of optimum cut-off grade with considering dilution. *Arabian Journal of Geosciences* 10 (7): 165.
10. Biswas, P., et al. 2020. Determination of optimum cut-off grade of an open-pit metalliferous deposit under various limiting conditions using a linearly advancing algorithm derived from dynamic programming. *Resources Policy* 66: 101594.
11. Khan, A., and M.W.A. Asad. 2019. A method for optimal cut-off grade policy in open pit mining operations under uncertain supply. *Resources Policy* 60: 178–184.
12. Franco-Sepulveda, G., G. Jaramillo, and J. Del Rio. 2009. Use of genetic algorithms for optimization of open-pit mining operations with geological and market uncertainty. In Proceedings of the 27th international symposium on mine planning and equipment selection-MPES 2018. Springer.
13. Githiria, J., and C. Musingwini. 2019. A stochastic cut-off grade optimization model to incorporate uncertainty improved project value. *Journal of the Southern African Institute of Mining and Metallurgy* 119 (3): 217–228.
14. Biswas, P., and P. Sen. 2019. Development of a computer aided tool for determination of optimum cut-off grade using dynamic programming based on limited mine capacity of open pit metalliferous deposits. *Journal of Mines, Metals and Fuels* 67 (10): 461–468.
15. Ahmadi, M.R., and A.A. Bazzazi. 2020. Application of meta-heuristic optimization algorithm to determine the optimal cut-off grade of open-pit mines. *Arabian Journal of Geosciences* 13 (5): 1–12.
16. Maleki, B., E. Mozaffari, and M. Mahdavipour. 2016. Optimizing the cut-off grade in sarcheshmeh copper mine using lane quartet model. *Journal of Mining and Metallurgy A: Mining* 52 (1): 27–35.
17. Li, S., and C. Yang. 2012. An optimum algorithm for cut-off grade calculation using multi-stage stochastic programming. *Journal of Theoretical and Applied Information Technology* 45 (1): 117–122.
18. Bellman, R. 1972. A rand corporation research study. In *Dynamic programming*, 1–365. Princeton University Press.

Chapter 11
Advanced Analytics for Drilling and Blasting

Ali Siamaki

Abstract The mining operation is associated with uncertainties that are not limited to ore characteristics and market demand; operational costs, operational performance, regulations are all a part of these uncertainties. Managing these uncertainties effectively and consistently through the mine life helps miners to maintain sustainability in their operations. Drilling and blasting as a key activity in the mining cycle have a key role in operational risk management. Many empirical methods have been developed to determine blast design parameters or the prediction of blast outcomes. However, these methods cannot cover all associated parameters and factors in design or prediction. Applications of advanced analytical methods, including AI-capable engineers, develop powerful tools that consider multiple parameters to predict blasting results or determine design parameters to achieve desired results. This chapter shows how analytical models can optimize blast patterns according to the real bench geometry and discusses advanced analytical solutions for predicting blast fragmentation and vibration by considering affecting parameters on the blast outcomes. In addition, it presents developed models based on recorded data in a mine site that can be used to determine blast design parameters to achieve desired blast outcomes. Discussed models in this chapter are based on AI and machine learning.

Keywords Blasting · Fragmentation · Ground vibration · Artificial intelligence · Optimization · Advanced analytics · Prediction

Blast Mechanism

In rock blasting in mines, achieving desired fragmentation is the main objective of blasting. Therefore, a small portion of the explosive energy is used to fragment rock mass, and a considerable part of the energy is consumed for flyrock, airblast, ground vibration, fragmented rock movement, and heat. Table 11.1 shows some of the components that absorb explosive energy during rock blasting [1].

A. Siamaki (✉)
Business Development Manager at GKM Consultants, Toronto, Canada
e-mail: asiamaki@gkmconsultants.com

A. Soofastaei (ed.), *Advanced Analytics in Mining Engineering*,
https://doi.org/10.1007/978-3-030-91589-6_11

Table 11.1 Components of explosive energy absorption in rock blasting [1]

Immediately in and around borehole	In the surrounding rock mass	Lost into the atmosphere (outside immediate blasting zone)
• The chambering of the borehole • Crushing (fines) • Friction • Heat	• Elastic and plastic deformation from tensile and compression forces • Micro-cracks • Macro-cracks • Formation of large fragments (greater than 1 mm) • Formation of fines (smaller than 1 mm) • Backbreaker • Friction and impact between fragments • Movement of fragments	• Heat • Fumes • Airblast • Ground vibration

Under well-controlled conditions, Ouchterlony et al. [2] assessed energy partitioning of limestone production blasts and computed in-hole losses, seismic energy, the kinetic energy of throw, and fragmentation energy. Their search revealed the energy partitioning of a blast with bulk emulsion explosive would be [2]

- Crushing and other losses in the rock mass; 20–40%
- The kinetic energy of throw; 10–20%
- Seismic energy; 5–10%
- Fragmentation energy; 0.1–2%

Therefore, it can be concluded that only 60–70% of the explosive energy is transmitted to the rock mass, which less than 1–2% of that is responsible for fragmentation [1]. Therefore, explosive type and characteristics, rock mass characteristics, and blast geometry define how the energy is partitioned into the mentioned categories.

Figure 11.1 shows affecting parameters on blast geometry.

Detonation of explosives confined in a blast hole results in applying shock waves to the borehole wall, and producing a large amount of very high temperature and pressure gases is a concise portion of time. Therefore, a better understanding of this mechanism and rock fracturing, fragmentation, and seismic propagation help mining engineers optimize their blasting operations and prevent unwanted blasting results, including ground vibration.

A deeper look at the interaction of the shock wave and borehole/rock mass shows in a plan view of a single borehole in Fig. 11.2 [4]. The shock wave pressure is more than the dynamic compressive strength of the rock, which results in crushed and fractured zones immediately around the borehole. The attenuation rate of the shock wave is high; therefore, the crushed zone extends around the borehole cannot exceed more than four times of borehole radius. At this time, the tangential component of the seismic wave is more than the dynamic tensile strength of the rock, which results

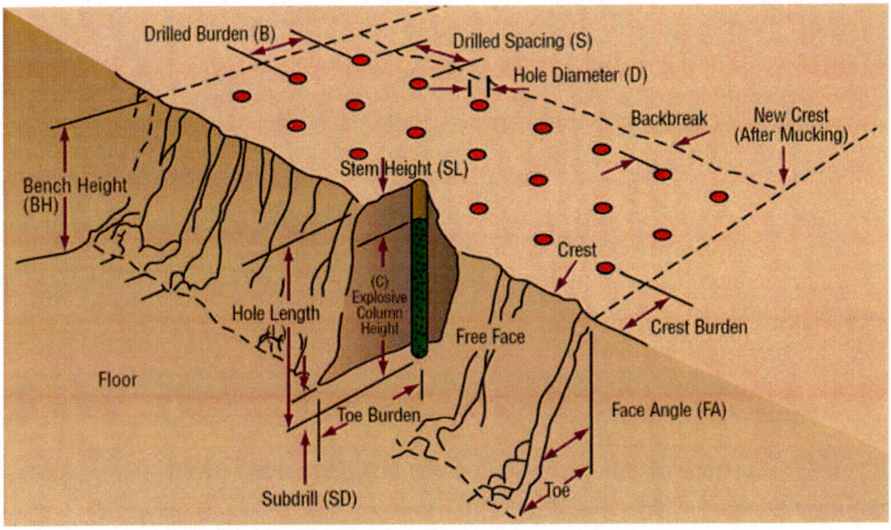

Fig. 11.1 Design parameters in drilling and blasting [3]

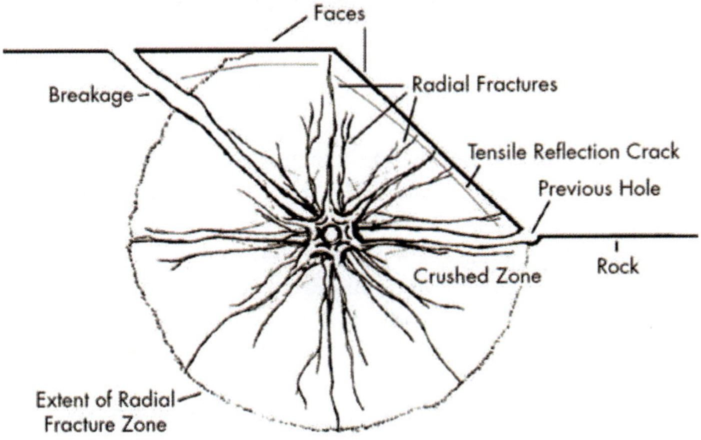

Fig. 11.2 Generation of crushed and fractured zones during blasting [8]

in the generation of radial fractures around the borehole. This fracture zone extends ten times of borehole diameter to 25 times [5–7]. By arriving the incident wave to the bench-free face, the wave is ultimately reflected. If the pressure of the reflected wave from the free surface is still more than the dynamic tensile strength of the rock, spalling has happened, and tensile cracks will be generated parallel to the free face. Seismic wave stress falls off rapidly with distance from the borehole. Consequently, rock behavior changes from plastic deformation to elastic deformation till the seismic wave attenuates completely. This latter situation is called ground vibration.

Blast Design

Rock blasting objectives in mining are achieving desired fragmentation size for ore and required size for waste at minimum cost, controlling and minimizing ore dilution to increase the value of the extracted ore, and minimizing unwanted blasting results, including ground vibration and airblast. Therefore, proper blast design is essential for a mining operation to achieve and maintain these objectives.

Blast design parameters and, as a result, overall blast performance are controlled by a combination of controllable and uncontrollable characteristics. Identifying these parameters or factors, considering them in blast design, and adjusting controllable ones lead to reaching blast objectives. Figure 11.3 shows controllable and uncontrollable parameters involved in blast design.

Uncontrollable parameters are related to rock mass characteristics, structural discontinuities and fracture networks, local geology, water, weather, and regulations. In comparison, controllable parameters are all the parameters related to drilling and blast design, explosive, timing, etc. [10].

Following the importance of uncontrollable parameters on blasting outcomes, Cunningham (1987) used Lilly's (1986) "blast ability index" to quantify the effect of rock parameters on fragmentation and incorporated it in his Kuz-Ram model. He defined rock factor A and highlighted how density, mechanical strength, elastic characteristics, and fractures should be considered while blasting.

$$A = 0.06(\mathrm{RMD} + \mathrm{JF} + \mathrm{RDI} + \mathrm{HF}) \tag{11.1}$$

where RMD is the mass description, JF is the joint factor, RDI is the rock density influence, and HF is the hardness factor [11, 12].

The primary objective of a production blasting for ore is optimizing fragmentation to improve downstream efficiency. However, fragmentation of waste rock should be just enough for efficient loading and hauling. Osanloo and Hekmat used Rosin–Rammler–Bennett expression to investigate the relationship between fragmentation size and shovel productivity. They discovered that a 50% increase in particle size results in a 20% reduction in shovel output [13].

Bench height is fixed by many factors, including block model size, slope stability, and loading equipment. Other blast design parameters should be determined by considering rock mass properties and explosive type. While typical blast patterns are square or staggered equilateral, staggered patterns provide a better energy distribution of explosives between boreholes. According to the bench width and minimizing shovel moves, most blasts are drilled four to five rows. Table 11.2 presents typical blasting parameters in open-pit mines [14].

Figure 11.4 shows a list of input parameters in the most important developed and suggested equations for burden calculation. These parameters can be categorized into blast design parameters or geometry, explosive parameters, and rock mass characteristics. Drill hole diameter has been considered in all suggested equations. Considering multiple parameters makes these equations more complicated, and in

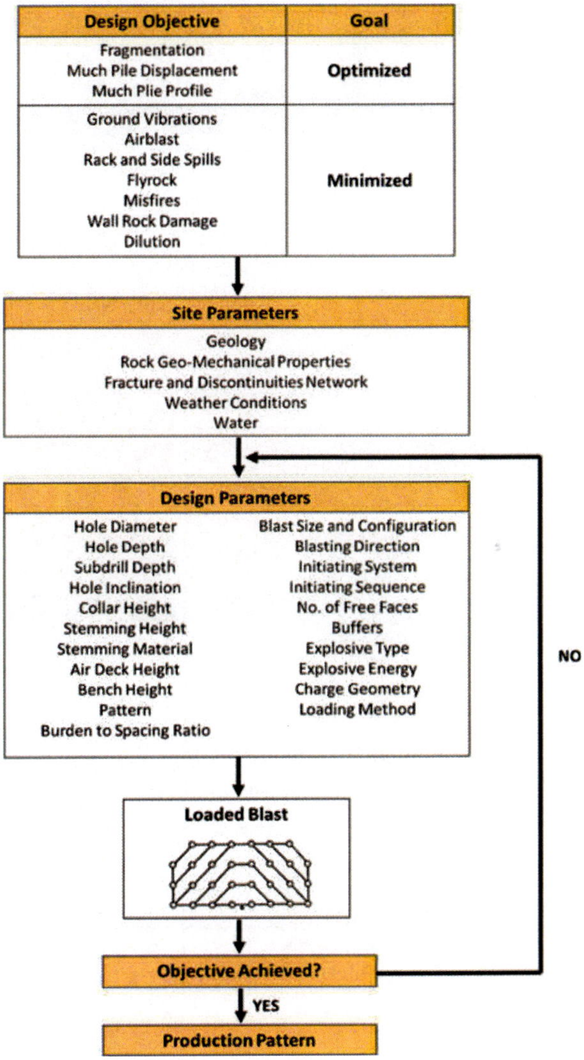

Fig. 11.3 Blast design flowsheet logic (modified after Hustrulide et al.) [9]

most cases, offered equations are suitable for a specific rock type or mine site. It reveals the limitation of empirical models in prediction and highlights the need for using artificial intelligence (AI) models in blast design.

Following is some of the common relations for the first estimation of blasting parameters in open-pit mines [3]:

- Blast hole diameter (D) in mm $\leq 15 \times$ Bench height in m;
- Bench height $\geq$ (Blast hole diameter in mm)/15;
- Burden (B) $= (25–40) \times D$;
- Spacing (S) $= 1.15 \times B$;

Table 11.2 Open-pit production blast parameters [14]

Parameters	Range
Burden	3.66–9.14 m
Spacing	5.49–10.06 m
Face height	10.69–15.24 m
Stemming	6.10–7.62 m
Hole diameter	251–432 mm
Subdrilling	0.91–2.13 m
Powder factor	0.24–1.19 kg/m^3
Weight of explosive/8 ms delays	454–4536 kg
Back break	7.62–13.72 m

		1952 Anderson	1952 Fraenkel	1955 Pearse	1959 Hino	1960 Allsman	1963 Ash	1963 Langefors & Kihlstrom	1967 Hansen	1972 Ucar	1972 Konya	1980 Foldesi	1980 Praillet	1980 Lopez Jimeno	1983 Konya	1985 Berta	1985 Carr	1990 Olofsson	1990 Rustan
Geometry	Drill hole/charge diameter	×	×	×	×	×	×	×	×	×	×	×	×	×	×	×	×	×	×
Geometry	Bench height								×	×			×					×	
Geometry	Drill hole length	×	×																
Geometry	Stemming												×						
Geometry	Subdrilling												×						
Geometry	Charge length		×						×	×									
Geometry	Drill hole inclination							×							×			×	
Geometry	Burden/spacing ratio							×											
Explosive	Explosive density						×	×	×	×	×	×	×	×	×	×		×	
Explosive	Detonation velocity						×					×	×	×	×	×			
Explosive	Detonation pressure			×	×	×													
Explosive	Binomial rock-explosive constant				×														
Explosive	Explosive strength							×										×	
Explosive	Loading equipment												×						
Rock	Rock density				×	×					×			×		×	×		
Rock	Rock strength		×	×	×							×	×		×				
Rock	Rock constants/factors	×		×				×										×	
Rock	Rock mass seismic velocity													×	×	×	×		

Fig. 11.4 Comparison of considered parameters in burden calculation in different suggested formulas [15]

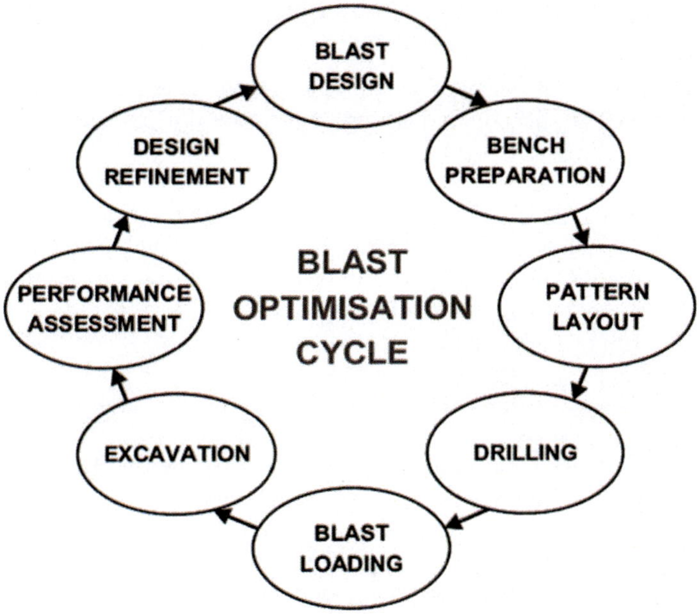

Fig. 11.5 Blast optimization cycle [10]

- Subdrill = (3–15) × *D*;
- Charge length (*C*) ≥ 20*D*;
- Stemming ≥ 20 × *D* or (0.7–1.2) × *B*; and
- Stemming material size = *D*/10–*D*/20 (angular material with minimum fines).

By reviewing different references, it is concluded that blast design equations provide a general guideline for blast design. These design parameters should be modified and optimized during blasting by monitoring drilling and charging performance and recording and analyzing blasting outcomes. Recording pre and post-blasts data provide a data bank that AI models can use for blast design and prediction. Figure 11.5 demonstrates the blast optimization cycle [10].

Blast Prediction

Blasting plays a key role in mining operations, and its results can directly affect downstream operations, including crushing and milling. Therefore, predicting a blast performance helps engineers control blast fragmentation and control and minimize negative effects of blasting like ground vibration and flyrock. This section presents suggested prediction models for blast fragmentation and ground vibration.

Fragmentation

The primary objective of rock blasting in mines is achieving desired fragmentation, which is to prevent any secondary blasting, improve loading and hauling efficiency, increase mineral liberation, and most importantly achieving desired fragmentation size, which can reduce downstream energy consumption and costs. Fragmentation of blasted material increases by an increase in the burden to hole diameter ratio decreases if the spacing to burden ratio increases, increases if stemming length to burden ratio increases, decreases with the increase in explosive charge and decreases by an increase in bench height to burden ratio [16].

There is no complete theoretical solution for the degree of fragmentation prediction. However, many empirical and size distribution models have been developed by many researchers [17–20]. Most of these models offer equations to calculate the average fragment size (x_{50}) as well as the entire fragment size distribution curve.

The Kuz-Ram model is an empirical fragmentation model based on the Kuznetsov and Rosin–Rammler equations and a Cunningham method that generates the Rosin–Rammler equation's coefficient of uniformity from blasting parameters [21, 22]. This model obtains the mean fragment size (X_{50}) based on the rock factor, the rock volume is broken per blasthole and the explosive weight.

$$X_{50} = A\left(\frac{V}{Q}\right)^{0.8} Q^{0.167} \tag{11.2}$$

Where X_{50} is the mean fragment size (cm); A is the rock factor (7 for medium hard rocks, 10 for hard but highly fissured rocks, 13 for very hard, weakly fissured rocks); V is the rock volume (m^3), and Q is the weight of TNT (kg) equivalent in energy to the explosive charge in one borehole.

The role of the Rosin–Rammler equation is to estimate the complete fragmentation distribution. For example, for a given mesh size or screen opening, X, this equation can estimate the percentage of fragments retained [17].

$$R_x = \exp\left(\frac{x}{x_c}\right)^n \tag{11.3}$$

where R_x is the proportion of fragments larger than the mesh size X (cm), and X_c is the characteristic fragment size (cm).

Ouchterlony [23] developed the Swebrec function and replaced the original Rosin–Rammler equation to predict rock fragment size distribution. This model has been called the Kuznetsov-Cunningham-Ouchterlony (KCO) model. The ability of this model in fines prediction is better than previous ones. This model considers the fraction smaller than size $x(P(x))$ obtained based on the sieve size (x), the maximum in situ block size (x_{max}) and the 50% passing size (x_{50}) [17]:

$$P(x) = \frac{1}{[1 + f(x)]} \tag{11.4}$$

$$f(x) = \left[\frac{\ln\left(\frac{X_{max}}{X}\right)}{\ln\left(\frac{X_{max}}{X_{50}}\right)}\right]^b \tag{11.5}$$

Where P(x) is the percentage of fragments passing a given mesh size, X; X_{max} is the upper limit of fragment size; X_{50} is the mean fragment size; b is the curve undulation parameter.

Ground Vibration

Ground vibrations from blasting is one of the significant drawbacks of blasting, which cause damage to close by commercial and residential buildings. In addition, high amplitude vibrations may stimulate unstable wedges, planes, or faults on the pit slope. All researchers agreed that peak particle velocity (PPV) is the best representative of blast-induced vibration. Most of the presented equations for predicting PPV are consideringldistance from blast and maximum weight per delay as the influencing parameters. United States Bureau of Mining (USBM) has done extensive research on the affecting parameters on PPV in surface mines. As a part of this research, Duvall and Fogelson [24] showed that the PPV is scaled with the square of the charge weight. Later on, Ambrasey and Hendron [25] considered the cube root of the maximum charge. Currently, the cube root of the maximum charge is considered in a sensitive building near the blasting or underground blasting. Langefors and Khistrom [7] presented a relation for predicting the PPV based on Q/D3/2, which is acceptable for different charge weights. Ghosh and Deamen [26] stated that numerous non-elastic factors cause a reduction in the energy of the wave during its propagation; this non-elastic effect causes a decrease in the wave range due to its geometric propagation. They modified the relation suggested by the USBM based on the non-elastic damping factor [6, 27–29].

$$\mathrm{PPV} = K\left(D/Q^{1/2}\right)^{-\beta} \quad \text{Duvall and Fogelson [46]} \tag{11.6}$$

$$PPV = K\left(D/Q^{1/3}\right)^{-\beta} \quad \text{Ambrasey and Hendron [47]} \tag{11.7}$$

$$\mathrm{PPV} = K\left(\sqrt{D/Q^{2/3}}\right)^{\beta} \quad \text{Langefors and Khilstrom [7]} \tag{11.8}$$

$$\begin{aligned} \mathrm{PPV} &= K\left(D/Q^{1/2}\right)^{-\beta} e^{P\cdot D} \\ \mathrm{PPV} &= K\left(D/Q^{1/3}\right)^{-\beta} e^{P\cdot D} \end{aligned} \quad \text{Ghosh and Deamen [50]} \tag{11.9}$$

In these relations, PPV is the peak particle velocity, D is the distance from the explosion point, Q is the charge per delay, and K, β are the area parameters. In the relations proposed by Ghosh and Daemen, P is the non-elastic damping factor.

Advanced Analytics in Blasting

As discussed earlier, much different rock, explosive, and design parameters or factors control blasting results. Therefore, considering controllable and uncontrollable parameters, including rock mass parameters in blast design and result prediction, provides more realistic solutions for this part of the mining cycle, benefiting Mine to Mill Optimization goals. Furthermore, modeling and design software and digital technologies become part of day-to-day mining operations, provide a better understanding of the relations between these parameters and make blast optimization more achievable.

The empirical models for blast design and predictions are widely favored in blasting operations since they have simple functions and are easily parameterized. However, oversimplification and considering limited factors make these models less accurate.

As part of digital technologies solutions, data analysis and analytics tools play a key role in data collection, storage, data cleaning, and analysis in mining operations. These tools, including AI, help miners to take into account different parameters in their predictions, identify patterns in the mining dataset and, between those parameters, make predictions for the new conditions with the aim of those patterns. AI solutions in blasting are mainly based on machine learning and neural networks algorithms.

While pre-blast datasets, including blast design and rock mass characteristics, are used in an AI model to make predictions about post-blast results, blasting outcomes data including fragmentation, vibration, and muck pile shape can be used to train an AI model to recommend blast design parameters based on desired outcomes. By adding pre-blast and post-blast data to the AI model, each mine site has its model perfected overtime to give the best possible predictions and solutions tailored to that site (Fig. 11.6) [30].

In the last two decades, AI techniques have been used in blasting operations to predict blasting outcomes, including ground vibrations. Many researchers investigated the effects of different parameters on ground vibration from blasting by using neural networks models [31–34]. Some researchers used other machine learning techniques, including fuzzy models, to predict ground vibration and blast air overpressure [35–38]. Despite conventional models in predicting ground vibration, these new models were capable of many different parameters from rock characteristics to blast design and geometry to improve prediction efficiency. All these neural networks and machine learning models are using supervised learning algorithms. The supervised learning method uses a dataset of parameters as a training set to develop a model that can predict outputs for future data.

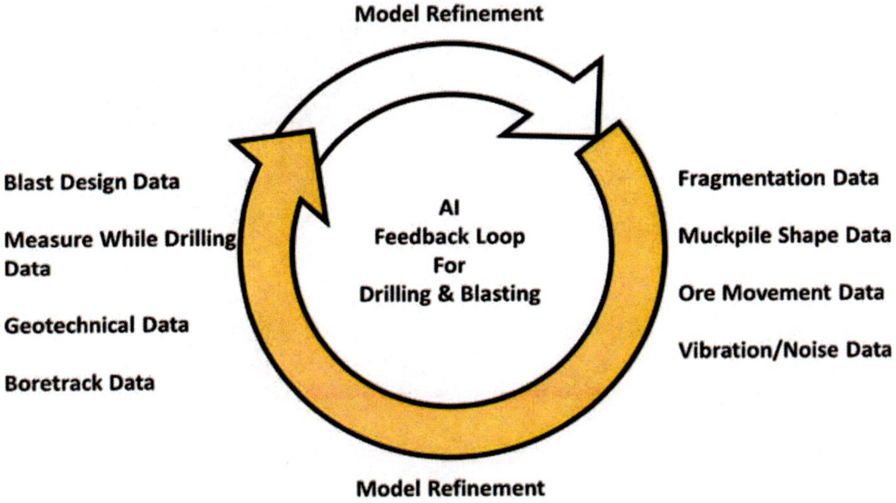

Fig. 11.6 Feedback loop for optimizing AI model based on pre- and post-blast data [20]

These AI models do not limit the number of input parameters is a significant advantage over empirical models. For instance, some of them affecting parameters in blast vibration are powder factor, rock properties, size of the blast, deck blasting, geometric parameters of blast including blast hole diameter and bench height, distance from the blast, burden and spacing, delay intervals, etc. [15]. AI developed models for ground vibration prediction consider these parameters to improve predictions for a specific site and rock type. In contrast, the USBM formula, a widely accepted method in predicting ground vibration, only considers distance from blasting, the maximum charge per delay, and equation constants that reflect environmental parameters will be obtained by regression [39]. This section will review the application of data analytics solutions and AI in blasting for blast design, blast operations, and blast results prediction.

Burden Optimization

Blast design optimization requires a 3D profile of the blasting bench to define actual burden, rock mass characteristics, fracture network, actual drill hole measurements, the actual weight of charge, and length of stemming per borehole, and recorded post-blast data. Strayos provides a comprehensive toolkit for blast optimization based on AI. This toolkit starts by capturing a bench profile by a UAV, defining bench toe and crest, and determining the discontinuities network (Fig. 11.7). It follows by gathering drill hole and blast design data and predict muckpile shape and blast fragmentation based on pre and post-recorded data from the previous blast and new blast data [40, 41].

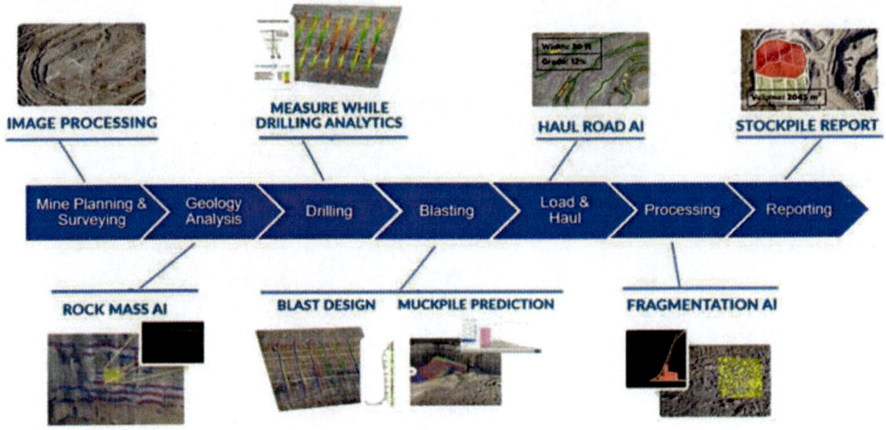

Fig. 11.7 Using AI in mine to mill optimization [41]

As discussed in the blast design section, burden plays a key role in determining other blast parameters. Rai stated that the burden is one of the most crucial blast design parameters for better fragmentation [42]. Improving blast outcomes for a constant burden needs attention to the front row burden. Bench face and locations of bench toe and crest effects on the front row burden. Choudhary and Arora showed blasting results, including fragmentation, muckpile shape, and final wall profile, which can be improved by considering a front row burden range of 0.5–0.7 of the designed burden [43].

Burden measurement is essential to determine locations of blast holes, reduce flyrock, improve fragmentation, and reduce ground vibration. Therefore, bench profiling is getting more attention in blasting operations. New technologies, including drones and photogrammetry, provide accurate profiling, which gives the engineers better data for the most efficient blast results. Figure 11.8 shows the effect of inaccurate burden measurement and locating boreholes in the wrong location on the blast performance [44].

Research at the University of Missouri of Science and Technology in a small pit showed that the 3D profile of a blast bench is the most accurate measurement to find and load burden. A drone considers all the points. Drones take a full visual view of the complete bench, and Strayos software, using AI, stitches the pictures into a map used for blast design. The 3D minimal burden is the optimal measurement. It is repeatable and gives the blaster the least line of the explosive line [44].

Blast Fragmentation Prediction

Improving blast fragmentation prediction methods needs considering a wide-ranging of affecting parameters on the fragmentation performance. While empirical methods

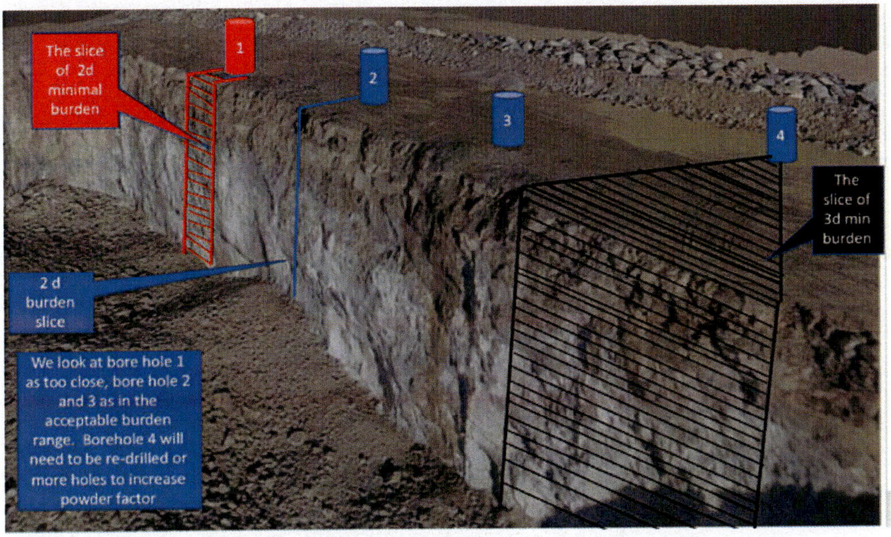

Fig. 11.8 Hypothetical location of blast holes with different burdens [44]

have limitations in the number of parameters, AI can include different controllable and uncontrollable parameters to improve prediction accuracy. This section will discuss two developed methods based on machine learning and neural network in detail.

Fragmentation results from explosive and rock mass interaction and blast design parameters govern fragmentation distribution over a production blast. For example, Badroddin et al. [45] considered blast design and rock mass parameters from 37 recorded blasts in a copper mine to predict 20% and 80% passing size in the fragmentation distribution curve. In this research, input parameters were axial compressive strength, cohesion strength, and elastic modulus considered for rock mass. Furthermore, design parameters were considered delay interval, specific charge, stemming, the burden to spacing ratio, blast ability index, and hole diameter. Table 11.3 presents input and output ranges for this AI model [44].

The training procedure consists of a sequential data feed into the network, followed by the comparative evaluation of the corresponding output provided by the NN and the actual result. This network is called feed-forward network. The network adjusts the weighting of the connection links in a continuous effort and bases on the error between target and network response to produce the results that best correspond to the training dataset. Usually, a significant number of iterations is required to converge below a pre-specified threshold for the residual error. Figure 11.9 illustrates a feed-forward neural network model [46].

Many training algorithms are available for training the NN models to achieve desired error and update neurons weights through each iteration. However, the Levenberg–Marquardt algorithm was designed to approach second-order training speed without computing the Hessian matrix. Instead, the Hessian matrix is approximated

Table 11.3 Input and output parameters and their range

Variable		Range
Input	Axial compressive strength (MPa)	16.6–39.9
	Cohesion strength (MPa)	0.9–10.30
	Elastic modulus (GPa)	8.3–34.4
	Delay (ms)	20–70
	Specific charge (kg/m^3)	0.2–0.59
	Stemming (m)	2.36–4.72
	B/S	0.63–0.89
	Blastibility index	64.25–87.5
	Hole diameter (mm)	127–165
Output	X_{20} (mm)	4.8–9.8
	X_{80} (mm)	19.1–51.4

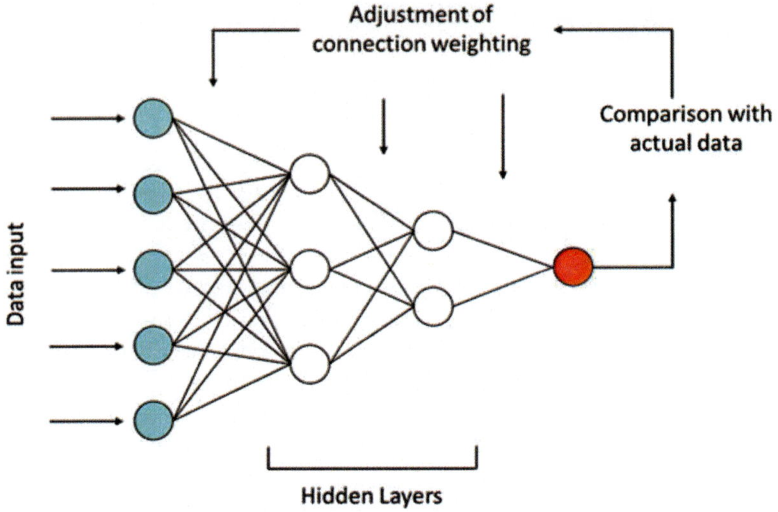

Fig. 11.9 Feed-forward neural network structure [46]

as the multiplying of the Jacobian matrix (which contains the first derivatives of the network error) and its transposed matrix. Consequently, this algorithm appears to be the fastest method for training moderate-sized feed-forward neural networks used in this research.

The developed NN model structure in the MATLAB software includes four hidden layers with the structure of {16, 14, 12, 10}. As discussed earlier, Levenberg–Marquardt has been used as the learning algorithm in these hidden layers, and tan-sigmoid was considered as the transfer function. For quality control purposes, 8 datasets were used to test the trained network. Figure 11.10 compares obtained X_{20}

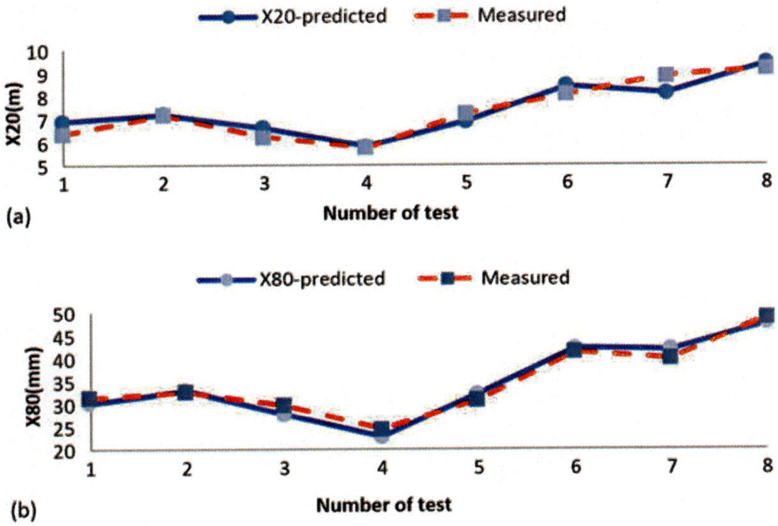

Fig. 11.10 Comparison of measured fragmentation and predicted fragmentation by NN: **a** X_{20} and **b** X_{80} [44]

and X_{80} from the NN model with real data in the dataset. X_{20} gained a correlation coefficient of 0.88, and X_{80} gained 0.97.

Amoako and Jha also developed an artificial neural network (ANN) model using Keras Python libraries. They considered recorded data for 102 blasts to develop a 50% passing size prediction model in the fragmentation distribution curve. Table 11.4 shows the range of considered inputs and corresponding outputs in the AI models [47].

Different layer structures were considered to obtain the most effective prediction model by using Keras regarding the ANN model. In all networks, hyperparameter tuning is performed using the Bayesian Optimization object in Keras. Bayesian Optimization saves time by automating the search process to find the best combination of neurons for many hidden layers. According to the test datasets correlation with

Table 11.4 Input and output parameters and their ranges for the X_{50} fragmentation prediction models [47]

Variable		Range
Input	S/B	1–1.75
	H/B	1.33–6.82
	B/D	17.98–39.47
	T/B	0.5–4.67
	Pf (kg/m^3)	0.22–1.26
	Xb (m)	0.29–2.35
	E (Gpa)	9.57–60
Output	X_{50} (m)	0.12–0.96

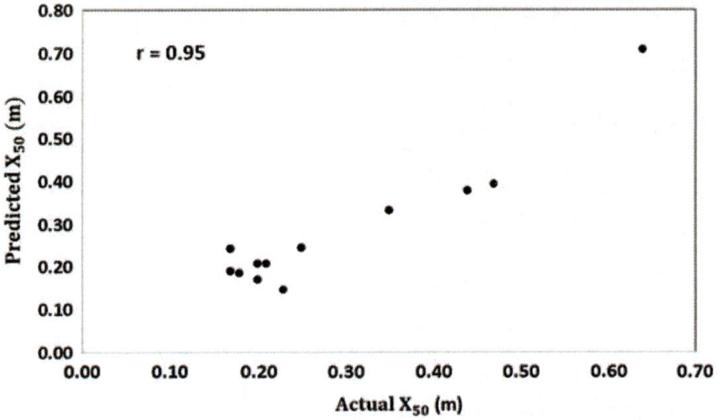

Fig. 11.11 Actual versus predicted values for the ANN developed model

desired output data, the NN with two hidden layers had the best performance and was chosen for the modeling. The achieved correlation coefficient for this model is 0.95 (Fig. 11.11).

Blast Design

Many empirical relations have been developed to determine blast design parameters. For instance, the explosive specifications and strength of the rock mass were considered in the presented relations by Allsman [48], and Speath [49]. Since the proper choice of the type and the amount of the explosive and rock was not easy. In 1968 Ash presented a simple relation for determining burden based on blast hole diameter [6]. According to Fig. 11.6, most equations were developed to determine the burden based on rock characteristics, blast hole diameter, and explosive properties. Blasting desired outcomes have not been considered in the blast design equations. While blasting, design parameters should be chosen according to blast objectives.

In another attempt, Bakhshandeh et al. [50] trained an ANN model to determine burden, spacing, and total charge weight in the blast pattern based on the desired vibration from blasting according to the adjacent buildings. For this reason, 41 sets of data from a copper mine blasting have been considered for the ANN model training and 10 datasets for testing the model. Desired PPV, the distance between the measuring point and blasting location, explosive density, and volume of extraction block were selected as the network input parameters. Rock mass density has not been considered an input parameter since it was the same for different blasts. Other designed parameters were limited by bench height and drill hole diameters. Therefore, blast hole depth is 15 m, out of which 3.5 m is sub-drilling, and stemming

Table 11.5 Input and output parameters and their ranges for the blast design parameters determination

Parameter		Range
Output	Peak particle velocity (mm/s)	0.83–10.99
	Distance from the blasting (m)	740–3525
	Explosive density	0.8–1.5
	Blasting block volume (m^3)	29,760–102,000
Input	Burden (m)	7.0–7.5
	Spacing (m)	7.0–9.5
	Charge weight (tom)	7.00–27.14

varies between 6 and 7 m [51]. Table 11.5 presents the input and output parameters and their ranges in the training process.

The network was trained using the backpropagation algorithm in MATLAB software. However, algorithms do not always converge to the absolute minimum in machine learning and ANN models, and they might stop at a local minimum. Therefore, performing multiple models is required to ensure achieving the best model performance. In this model, the mean square error function has been chosen as the performance function. This function controls network errors by controlling errors between the output and the actual values through the training process.

The ANN model consists of four hidden layers by the structure of {16, 14, 12, 10, 3}. The higher number of layers and neurons in each layer improves the network-ability in training, and in contrast, it increases training time. The sigmoid tangent function was used as the transfer function in the hidden layers, which can scale the response in a span of [−1, 1]. Also, the linear transfer function was considered for the output layer. The algorithm used in this network is that of Levenberg–Marquardt. Correlation coefficients of 0.76, 0.96, and 0.65 have been achieved for burden, spacing, and charge weight, respectively, for the test datasets.

Badroddin et al. [45] used blast ability index (BI), uniaxial compression strength, blasthole diameter and specific weight of the rock, rock quality designation (RQD), cohesion strength to train an ANN model to obtain the desired burden for the blast pattern [50]. In this study, ANN results compared with actual data and also results from a burden determination equation that was developed by Khoshrou et al. [52] by nonlinear multiple regression analysis for the same mine:

$$B = \frac{e^{-5.577} \cdot \sigma_c^{0.735} D^{0.13} \cdot C^{0.052} \cdot \mathrm{BI}^{1.75}}{\gamma_r^{0.354} \cdot \mathrm{RQD}^{0.177} \cdot q^{0.433}} \tag{11.10}$$

- B is the burden (m);
- σ_c is the uniaxial compressive strength (MPa);
- D is the hole diameter (mm);
- C is the cohesion strength (MPa);
- BI is the blast ability index;
- γ_r is the specific weight of rock (gr/cm^3);

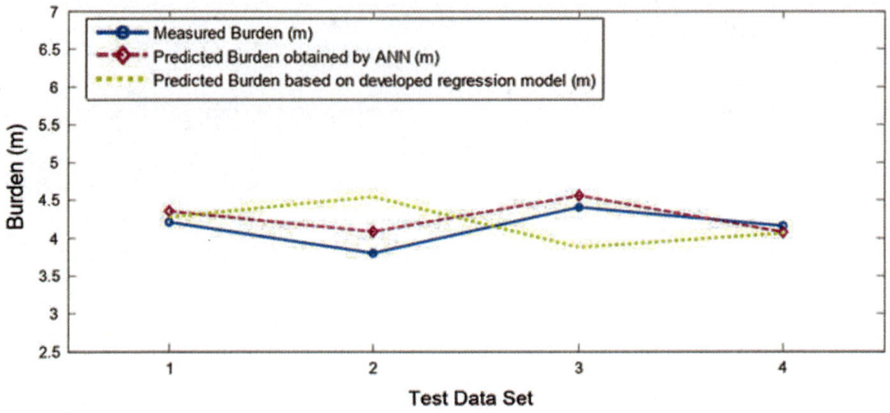

Fig. 11.12 Comparison of measured and predicted burden for four different artificial neural network models

- RQD is the rock quality designation (percent); and
- q is the powder factor (kg/m^3).

Four different feed-forward network architecture has been used to find the best model for burden determination. First, the network performance was evaluated using mean square error (MSE), which compared the output and the actual data with the structure of $\{19, 16, 10, 1\}$.

A coefficient of determination (CoD) and mean absolute error (MAE) between the predicted and observed values is taken as the measure of performance of the ANN model and suggested equation. The prediction capability of the models was found to be strong and better than the most applied fragmentation prediction model. Figure 11.12 compares the results data for the test dataset for the ANN model, regression model, and real data.

Conclusion

Digital technologies have been a part of the mining industry. While these technologies reduce operational uncertainties, they help mining engineers to make more accurate decisions based on the provided insights from digital technologies. Drilling and blasting operations benefit from digital, and AI tools can help engineers predict blast results by considering many controllable and uncontrollable parameters. In traditional methods, only limited controllable parameters have been considered, and most uncontrollable ones have been neglected. By developing advanced analytical methods, mining and blasting engineers can locate boreholes on the blasting block according to actual burden and define charge weight and stemming length to optimize fragmentation or reduce blasting environmental issues, including vibration. Some of the data analytics applications in blast operations, according to this chapter, are:

- Predictive models for blast fragmentation that can predict d50 and d80 sizes based on geotechnical properties of rock mass, blast design, and geometry parameters;
- Predictive models for blast-induced vibration based on geotechnical data of rock mass and charge type and weight; and
- Determine blast design and geometry parameters based on desired blast fragmentation and vibration according to the safety standards and downstream requirements.

References

1. Calnan, J.T. 2015. Determination of explosive energy partition values in rock blasting through small-scale testing. University of Kentucky.
2. Ouchterlony, F., et al. 2004. Where does the explosive energy in rock blasting rounds go? *Science and Technology of Energetic Materials* 65 (2): 54–63.
3. Nobel, D. 2010. Blasting and explosives quick reference guide. Kalgoorlie: Dyno Nobel Asia Pacific Pty Limited.
4. Zhang, Z.-X. 2016. Rock fracture and blasting: Theory and applications. Butterworth-Heinemann.
5. Shadabfar, M., et al. 2021. Estimation of damage induced by single-hole rock blasting: A review on analytical, numerical, and experimental solutions. *Energies* 14 (1): 29.
6. Bhandari, S. 1997. Engineering rock blasting operations.
7. Langefors, U., and B. Kihlström. 1963. *The modern technique of rock blasting*, vol. 405. New York: Wiley.
8. Darling, P. 2011. SME mining engineering handbook. Vol. 1. SME.
9. Hustrulid, W.A., M. Kuchta, and R.K. Martin. 2013. Open pit mine planning and design, two-volume set and CD-ROM pack. CRC Press.
10. Hustrulid, W.A. 1999. Blasting principles for open pit mining: General design concepts. Balkema.
11. Lilly, P.A. 1986. An empirical method of assessing rock mass blast ability.
12. Cunningham, C. 1987. Fragmentation estimations and the Kuz-Ram model-four years on. In *Proceedings of 2nd international symposium on rock fragmentation by blasting.*
13. Osanloo, M., and A. Hekmat. 2005. Prediction of shovel productivity in the Gol-e-Gohar iron mine. *Journal of Mining Science* 41 (2): 177–184.
14. International Society of Explosives, E. 2011. *ISEE blasters' handbook*. Cleveland, Ohio: International Society of Explosives Engineers.
15. Jimeno, C.L., et al. 1995. *Drilling ǎnd blasting of rocks.*
16. Singh, P., et al. 2016. Rock fragmentation control in opencast blasting. *Journal of Rock Mechanics and Geotechnical Engineering* 8 (2): 225–237.
17. Rosin, P. 1933. Laws governing the fineness of powdered coal. *Journal of Institute of Fuel* 7: 29–36.
18. Koshelev, E., et al. 1971. Statistics of the fragments forming with the destruction of solids by the explosion. *Journal of Applied Mechanics and Technical Physics* 12 (2): 244–256.
19. Kuznetsov, V. 1973. The mean diameter of the fragments formed by blasting rock. *Soviet Mining* 9 (2): 144–148.
20. Ouchterlony, F. 2016. The case for the median fragment size as a better fragment size descriptor than the mean. *Rock Mechanics and Rock Engineering* 49 (1): 143–164.

21. Cunningham, C. 1983. The Kuz-Ram model for production of fragmentation from blasting. In *Proceedings of 1st symposium on rock fragmentation by blasting*, Lulea.
22. Lizotte, Y.C. 1990. Empirical procedures for prediction of rock fragmentation by blasting. Canada Centre for Mineral and Energy Technology, Mining Research Laboratories.
23. Ouchterlony, F. 2005. The Swebrec© function: Linking fragmentation by blasting and crushing. *Mining Technology* 114: 29–44.
24. Duvall, W.I., and D.E. Fogelson. 1962. *Review of criteria for estimating damage to residences from blasting vibration,* vol. 5968. US Bur Mines.
25. Ambraseys, N.R., A.J. Hendron, K.G. Stagg, and O.C. Zienkiewicz, eds. 1968. Dynamic behaviour of rockmasses. In *Rock mechanics in rock mechanics in engineering practice*, pp. 203–207. London: Wiley.
26. Ghosh, A., and J.K. Daemen. 1983. A simple new blast predictor of ground vibrations ınduced predictor. In *Proceedings of the 24th US Symposium on Rock Mechanics*, 20–23 June. Texas, USA
27. Balbas Anton, M. and J. Diaz Garcia. 1995. Spatial relation between laws of vibration from blasting. *International Journal of Surface Mining and Reclamation* 9(4): 161–164.
28. Singh, S.R., and R.D. Lamond. 1993. Prediction and measurements of blast vibrations. *International Journal of Surface Mining and Reclamation* 7(4): 149–154.
29. Roy, P.P. 1991. Prediction and control of ground vibration due to blasting. *Colliery Guardian* (United Kingdom) 239(7).
30. Sahu, R. 2020. AI in drill and blast.
31. Mohamed, M.T. 2009. Artificial neural network for predicting and controlling blasting vibrations in Assiut (Egypt) limestone quarry. *International Journal of Rock Mechanics and Mining Sciences* 46 (2): 426–431.
32. Amnieh, H.B., M. Mozdianfard, and A. Siamaki. 2010. Predicting blasting vibrations in Sarcheshmeh copper mine by a neural network. *Safety Science* 48 (3): 319–325.
33. Petr, V., M. Simoes, and T. Rozgonoyi. 2003. Future development of neural network prediction for blasting design parameter of production blasting. *Explosive and Blasting Technique* 625–630. Holmberg.
34. Soofastaei, A. 2020. *Data analytics applied to the mining industry*. CRC Press.
35. Xue, X. 2019. Neuro-fuzzy based approach for prediction of blast-induced ground vibration. *Applied Acoustics* 152: 73–78.
36. Shahnazar, A., et al. 2017. A newly developed approach for predicting ground vibration using a hybrid PSO-optimized ANFIS-based model. *Environmental Earth Sciences* 76 (15): 1–17.
37. Zhou, J., et al. 2020. Novel approach for forecasting the blast-induced AOP using a hybrid fuzzy system and firefly algorithm. *Engineering with Computers* 36 (2): 703–712.
38. Martins, P., and A. Soofastaei. 2020. Process analytics. In *Data analytics applied to the mining industry*, 131–148. CRC Press.
39. Duvall, W.I. and B. Petkoff. 1959. Spherical propagation of explosion-generated strain pulses in rock. US Department of the Interior, Bureau of Mines.
40. 2020. Visual AI platform for quarry and mining.
41. Jha, A., et al. 2020. Detection of geological features using aerial image analysis and machine learning. In *Proceedings of the 46th annual conference on explosives and blasting technique*, USA: Denver, CO.
42. Rai, P. 2002. Evaluation of the effects of some blast design parameters on fragmentation in opencast mines. Ph. D. thesis, Banaras Hindu University, Varanasi.
43. Choudhary, B., and R. Arora. 2018. Influence of front row burden on fragmentation, muckpile shape, excavator cycle time, and back break-in surface limestone mines. *Iranian Journal of Earth Sciences* 10 (1): 1–10.
44. 2020. Burden profiling. Do you measure what you think you are?
45. Badroddin, M., H. Khoshrou, and A. Siamaki. 2012. Prediction of burden at the sungun copper mine by artificial neural network. In *Applications of Computers and Operations Research in the Mineral Industry Symposium Issue (35th APCOM)*.

46. Badroddin, M., H. Khoshrou, and A. Siamaki. 2013. Prediction of fragment size distribution from blasting: Artificial neural networks approach.
47. Benardos, A., and D. Kaliampakos. 2004. Modelling TBM performance with artificial neural networks. *Tunneling and Underground Space Technology* 19 (6): 597–605.
48. Allsman, P.L. 1960. Analysis of explosive action in breaking rock. *Transactions AIME* 217: 475–476.
49. Speath, G.L. 1960. Formula for proper blasthole spacing. *Engineering News Record*218(3): 53.
50. Amnieh, H.B., A. Siamaki, and S. Soltani. 2012. Design of blasting pattern in proportion to the peak particle velocity (PPV): Artificial neural networks approach. *Safety Science* 50 (9): 1913–1916.
51. Amoako, R.A.J., A. 2020. Rock fragmentation prediction using machine learning.
52. Khoshrou, H., M. Badroddin, and E. Bakhtavar (2010) Determination of the practicable burden in Sungun open-pit mine, Iran. In *Rock Fragmentation by Blasting*, ed. J.A. Sanchidrián, pp. 271–276. London: Taylor & Francis Group.

Chapter 12
Advanced Analytics for Rock Fragmentation

Paulo Martins and Ali Soofastaei

Abstract Applied advanced analytics is not a new term in the main industries, but the mining industry has recently started using innovative analytical projects. Nevertheless, there is a big gap between the current analytical situation and desired level of maturity that is needed to use advanced analytics in mining practically. The mentioned gap is not just in the technological development in the mine sites; the significant research gap also has been identified by reviewing the published papers and books in this field. Rock fragmentation is one of the most important mining operations, and other related operations can be improved by optimizing the fragmentation process. This chapter shows some examples of new analytical approaches to improve operators' performance in drilling and blasting. The other safety-related topics, such as ground vibration and flyrock problems, is also discussed in this chapter.

Keywords Drilling · Blasting · Rock fragmentation · Advanced analytics · Prediction

Introduction

Rock blasting is an essential process to determine mine productivity, energy efficiency, and process optimization. Often, poor management practices at different stages in the entire blasting process will be observed. Although there is a large amount of data available from various sources in the drilling and blasting operations, it has always been challenging and time-consuming to pull it all together and interpret it. Rock fragmentation, an essential outcome of the blasting process, directly impacts mucking productivity, fleet efficiency, plant throughput, and energy consumption. Moreover, some studies estimate that less than 30% of the explosive energy is effectively used to fragment the rock, and the remaining 70% is adversely

P. Martins (✉)
Institute for Mineral and Energy Resource, University of Adelaide, Adelaide, Australia

A. Soofastaei
AI Program Leader, Vale, Brisbane, Australia
URL: http://www.soofastaei.net

A. Soofastaei (ed.), *Advanced Analytics in Mining Engineering*,
https://doi.org/10.1007/978-3-030-91589-6_12

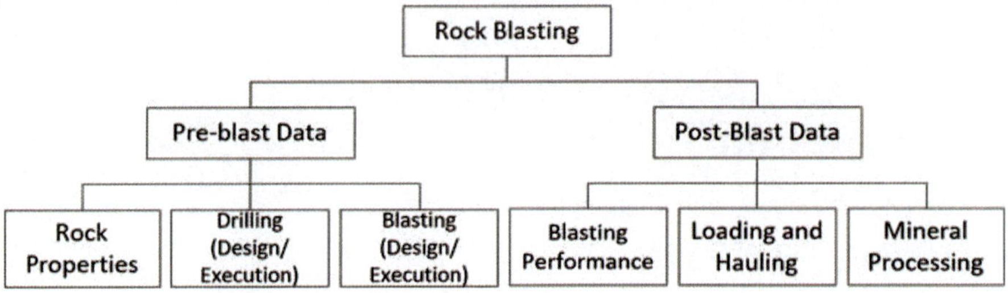

Fig. 12.1 Sources of data associated with rock blasting

released in the surrounding environment, resulting in dust, ground vibration, fumes, flyrock, air overpressure [1]. Due to its interconnected characteristic, an integrated strategy to collect data considering the different stages of the mining value chain is essential to analyze further and enable advanced analytics methods for performance assessment, monitoring, prediction, and optimization.

For a comprehensive analytical study of rock breakage performance and impact, the data collection approach can be divided into two major groups: pre-blast and post-blast. Figure 12.1 shows the major groups and their related domains:

The integrated analysis of this data provides the ability to streamline the process, predict performance, optimize the operations, and enable people to make informed decisions. The following sessions will describe the most relevant data types found in each pre-blast domain and the blast performance domain related to the post-blast group. This chapter will not explore the data collected in the loading and hauling and mineral processing domains and their relationship with rock breakage and the upstream process.

Rock Properties

Despite the significant technological advances, blasting remains as much an art as a science. Much progress has been made to modeling the fracturing process and the propagation of stress waves in an isotropic or anisotropic medium. However, the natural variability of the rock mass usually distorts the complex blasting design strategies, and, consequently, much of the engineering work ends up relying on relatively simple 'rules of thumb' from global experience as part of the documentation and dissemination of knowledge typical of the mining industry [2].

Various geological factors play an essential role in rock fragmentation and its environmental impacts. Rock properties can be classified into two groups: physical and structural. Physical rock properties are rock strength, modulus of elasticity, density, longitudinal wave velocity. Structural rock properties are joint spacing and orientation, bedding planes orientation and thickness, fracture frequency, mud or clay seams, presence of voids or cavities.

These properties often dictate the performance of explosives and, consequently, the size of the fragmented blocks and the direction and distance of the blast-induced movement.

Drilling Data

The natural geological and geotechnical characteristics of the rock are unchangeable. However, the drill and blast design specifications, such as burden and spacing, depth, diameter, and amount/type of explosive used, can be customized. The ability to modify these parameters dynamically in response to drilled information is critical to achieving good fragmentation. Considering the scale and frequency of blasting in a mining operation, characterizing rock mass at a more granular level is labor-intensive and time-consuming. However, drilling performance data can provide viable means of characterizing the rock mass systematically, routinely, and in real time [3]. A common practice in mining operations is monitoring while drilling (MWD) to access drilling parameters such as the penetration rate, rotational speed, flushing pressure, and GPS-based drill hole positioning data [4]. Significant efforts have been made to use drilling information to characterize rock properties and rock mass conditions, including the definition of hardness domains for blast design optimization.

Nowadays, soft computing methods are successfully applied as a powerful tool to solve complex engineering problems. Over the last years, many published works have focused on estimating rock properties through artificial intelligence techniques, and drilling measured data.

In [5], the authors summarize a range of recent research devoted to applying MWD data for different applications, including characterization of rock properties, identification of geological delineation, hardness domains, and rock mass strength. Table 12.1 describes some recent studies.

Table 12.1 Summary of measure while drilling (MWD) data applied for rock properties estimation

Application	Related publications
Rock hardness definition	[4]
Rock mass strength (contour map)	[6]
Rock weathering grades	[7]
Classification of rock types	[8–12]
Detection of geological boundaries	[13, 14]
Prediction of geo-mechanical properties	[15]
Rock quality designation (RQD)	[16]
Fracture frequency and RQD	[17]
Fracture frequency	[18]
Rock mass rating (RMR)	[12]

It is essential to consider that the raw data obtained from field MWD systems is commonly inappropriate for direct analysis. Analyzing and interpreting the data is as important as recording the data. Some sources of inconsistency, such as different operators' skills, unexpected situations in the operational environment, or equipment conditions, could generate errors in the data that do not represent the entire rock property or structural discontinuity under investigation. Site-specific rules and statistical analysis are necessary to manage these variations.

Indeed, a high return can be achieved by an extensive assessment of MWD data quality and the application of algorithms to clean and prepare the data. In addition, this consistent and reliable database can interpret physical and structural rock properties and develop prediction and optimization models that embed explosives characteristics, fragmentation, and environmental impacts from blasting.

Blasting Design Data

The blasting design activity is an essential component of the continuous improvement loop that promotes efficient rock fragmentation. Therefore, the ability to interpret the relevance of the multiple parameters involved and dynamically respond to new information from the drilling and downstream process is critical for mining control and optimization.

Top performance operations maintain a robust database of historical blast designs and their monitored performance. Improvement opportunities are investigated based on the database, and modern advanced analytics techniques can be applied to uncover considerable value. Commercial blast management platforms can be implemented to ensure efficiency, traceability, integration, and quick access to stored blasts. A centralized blast management tool can also provide scalability and enable multiple data types, including video records, maps, seismography results, simulation, and graphics.

The most necessary drilling and blasting design parameters are:

- blast-hole diameter;
- bench height considered in the mining plan;
- the burden for blast-holes;
- spacing between blast-holes;
- stemming length;
- deviation and alignment of blast-holes;
- drilling pattern (staggered or rectangular);
- properties of the explosive(s) used;
- powder factor;
- maximum charge per delay; and
- delay patterns.

Whenever possible, the blast performance assessment should consider the actual design applied for each blast. During a blast operation, deviation from the original

Fig. 12.2 Desirable and undesirable outcomes of rock blasting. Adapted from [19]

plan is typical, demanding clear procedures to monitor, control, and report unexpected changes.

Blast Performance Data

According to [19], blast performance can be divided into desirable and undesirable outcomes (Fig. 12.2).

Although only three desirable parameters are listed, the positive effect of an adequate fragmentation pervades the entire value chain, potentially increasing throughput and metallurgical performance and reducing costs. Several studies have investigated the application of advanced analytics methods to predict and optimize both desirable and undesirable outcomes. Figure 12.3 adapted from [19] compares the most common input parameters for prediction models developed to assess the blast impact using empirical models, artificial neural networks (ANN), and support vector machine (SVM).

The following sessions will review the most recent prediction and optimizations based on advanced analytics for fragmentation, ground vibration, and flyrock.

Rock Fragmentation Models

Rock fragmentations results are directly linked to all the downstream processes' performance in terms of costs and efficiency. Over the last decades, multiple models have been developed to quantify the rock breakage obtained after blasting execution. Popular empirical models applied for rock fragmentation are listed in [19]. The practical application of these models resulted in a holistic approach known as Mine to Mill (M2M) optimization, first introduced by the JKMRC—Julius Kruttschinitt Mineral Research Centre. Several M2M studies and cases have been developed worldwide to optimize energy consumption and throughput in the crushing and grinding process. An equally important field of study is the control of blasting, which includes ground vibration, flyrock, back break, and overpressure.

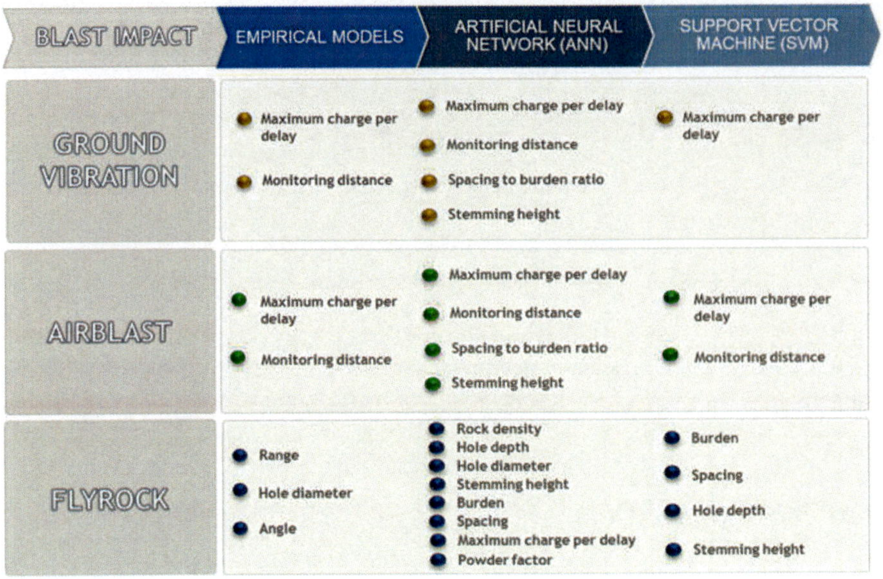

Fig. 12.3 Predominate parameters for predicting blast impacts

The modern application of modern meta-heuristics for rock fragmentation such as support vector machines (SVM), artificial neural networks (ANN), Fuzzy Systems (FS), and adaptive neuro-fuzzy inference systems (ANFIS) has consistently presented more excellent accurate results than traditional empirical models. As synthesized in [20], ANN models present significant advantages in flexibility, nonlinearity, higher faulty tolerance, and adaptive learning compared to regression models. In addition, ANN models can embed nonlinear effects and interactions between models' variables and find meaning derived from complex and uncertain data. The work [21] used the most common type of ANN trained with Multilayer Perceptron algorithms (MLP) and considered input parameters some commonly monitored variables involved in the drilling and blasting design and geo-mechanical properties of the rock mass. The output selected to create the network were the P_{80}, P_{50}, and P_{20} fragmentation results, defined with 80%, 50%, and 20% of the throughput size in the crusher within open-pit mine, were used for the output data formation of a network. Most of the prediction models reported in the literature use a set of selected fragmentation ranges as the target variable. Other output metrics include the mean fragmentation size and the uniformity index.

A review of the literature also demonstrates that hybrid approaches have been widely explored in the research field. For example, optimization algorithms have been successfully applied to adjust and improve the performance of the prediction algorithm. Table 12.2 shows the most recent publications and their respective methods. Satisfactory performance—R^2 results above 0.9, was reached in most of the referred studies.

Table 12.2 Summary of AA-based techniques and related publications for rock fragmentation

Technique	Publications
Cat swarm optimization	[21]
Multivariate regression	[22, 23]
Fuzzy inference system	[24]
Adaptive neuro-fuzzy inference system (ANFIS)	[25–27]
Artificial neural network (ANN)	[28–30]
Multivariate regression analysis	[22, 29]
Support vector regression (SVR)	[25]
Decision-making trial and evaluation laboratory (DEMATEL)	[22]
Neuro-fuzzy method	[31]
Imperialist competitive algorithm (ICA)	[32]
Artificial neural network (ANN) with artificial bee colony	[33]
Artificial neural network (ANN) with firefly algorithm (FA)	[34, 35]
Particle swarm optimization (PSO)	[26, 32]
ANFIS combined with particle swarm optimization (PSO)	[26, 36]
ANFIS combined with firefly algorithm (FA)	[27, 36, 37]
ANFIS combined with genetic algorithm (GA)	[37]
Gaussian process regression (GPR)	[35, 38]
Firefly (FA) with boosted generalized additive model (BGAM)	[39]
Firefly (FA) with gradient boosting machine (GBM)	[35]
Firefly (FA) with support vector machine (SVM)	[35]
Firefly (FA) with Gaussian process (GP)	[35]
Support vector machine (SVM)	[36, 40]
Ant colony optimization (ACO) with boosted regression tree (BRT)	[36]
Hyperbolic tangent function	[41]

The most common performance indicators for model assessment are the root-mean-squared error (RMSE), mean square error (MSE), mean absolute error (MAE), and determination coefficient (R^2). In [39], a Taylor diagram was also plotted to evaluate the performance of six different models. In [35, 36], a color-scale ranking was also considered to evaluate the models visually.

Although significant performance results are demonstrated in recent publications, the number of blasts and the operational parameters (inputs) considered in each study is worth discussing. Moreover, a sensitive part of the methodology is related to the method and procedure used to collect and evaluate rock fragmentation output parameters. The most common practice is the 2D image analysis of different photographs taken from the blasted muck pile. The most common software packages for this method are Gold-Size, Split-Desktop, and WipFrag, and the rock delineation procedure can be manual or automatic, depending on the software used. In order to standardize the analytical process, an overall strategy to represent the blasted muck pile should be considered. Some important aspects are the number and positions of the images and the muck pile sections to be covered during the collection stage.

Considering that the performance of the models significantly depends on the input variables, some of the described studies also developed sensitivity analysis using different approaches. In [35], the Sobol sensitivity method was applied and showed that width (*W*), spacing (*S*), and burden (*B*) are the parameters that significantly affect the accuracy of the simulation process of rock fragmentation in bench blasting. The bench height was analyzed in [36]. However, no strong effects were reported. The analysis performed in [22] based on the DEMATEL method suggested that the blast-hole pattern (BHP) is the most influential parameter for mean fragmentation size.

Although advanced analytics applications have been introduced for predicting the size of rock or particle size distribution, they are only applicable to the original test quarry, mine, or area and cannot represent different real-world conditions. Therefore, a comprehensive assessment of the influence of input variables, especially rock properties, is also a significant concern for future studies. Moreover, with the development of modern infrastructure and connectivity technologies such as the Internet of things (IoT) and cloud computing, new and advanced sensors such as those applied to monitor the drilling operation and high-resolution rock images can be widely adopted as new features to improve fragmentation models.

Ground Vibration Models

According to some estimations, only 20–30% of the explosive energy is effectively applied to fragment the rock. The other amount is released in different forms, generating adverse effects such as ground vibration, flyrock, fumes, tremendous noise, and a large amount of released heat [42]. Ground vibration is one of the most detrimental impacts induced by rock blasting. Its destructive effects can affect the stability of adjacent benches, slopes, underground structures, roads, water pipelines and cause severe damages to surrounding facilities. In addition, ground vibration is often a significant source of complaints and litigations from nearby communities settled around a mine operation. For small mining operations, economic factors may restrict the continuous monitoring of ground vibration during rock blasting. In this regard, the estimation of ground vibration through predictive models can support pre-blast planning activities and management decisions. Therefore, a considerable value is provided by predicting the vibration level before operation [43].

According to [44], the intensity of ground vibration is associated with multiple parameters, classified into two groups: controllable and uncontrollable. The controllable parameters comprise the explosives specifications and drill and blast design parameters. The uncontrollable parameters refer to the inherent condition of the rock mass, such as geology, lithology, hardness, density, geo-mechanical properties.

The most common parameters measured when monitoring ground vibrations are frequency and peak particle velocity (PPV). The maximum value defines PPV in wave propagation and vertical, transverse, and longitude waveforms in a rocky environment. Similar to fragmentation models, several empirical models have been developed to estimate PPV value. The two independent parameters usually considered in these models are the maximum charge per delay and the distance between the blast bench and the monitored point. However, their accuracy is inadequate for estimating the safety zone required to protect people and assets from the blasting effects. Therefore, statistical methods for predicting the PPV are also referred to in the literature. Contrarily to empirical models that only use the distance and maximum charge per delay as predictors, statistical models have the advantage of incorporating more features, such as explosives properties, rock mass characteristics, drilling, and blasting design. Nevertheless, the performance of statistical models can decline significantly when new data available are different from the data used to create the model [45].

Over the last decade, advanced analytics approaches have been widely adopted to develop practical solutions for the challenge of predicting ground vibration originated from rock blasting. The recent survey published in [46] indicates that more than 50% of the machine learning applications in blast-induced impact predictions explore algorithms for ground vibration. According to the studies investigated in [47], most publications are dedicated to predicting and optimizing PPV applied artificial neural networks (ANN). Figures 12.4 and 12.5 present the corresponding figures for each of the eight groups described by the authors.

Flyrock induced by rock blasting is one of the most critical elements that must be accurately predicted in open-pit mines. The undesired projection of rock fragments by explosive energy beyond the safety zone of the explosion blocks has been a significant challenge faced by blast professionals and mining managers. Flyrock events lead to fatalities, injuries, and physical damage. Therefore, predicting the projection range and size of the rocks thrown is a remarkable step in reducing and controlling blasting accidents in the operations [48].

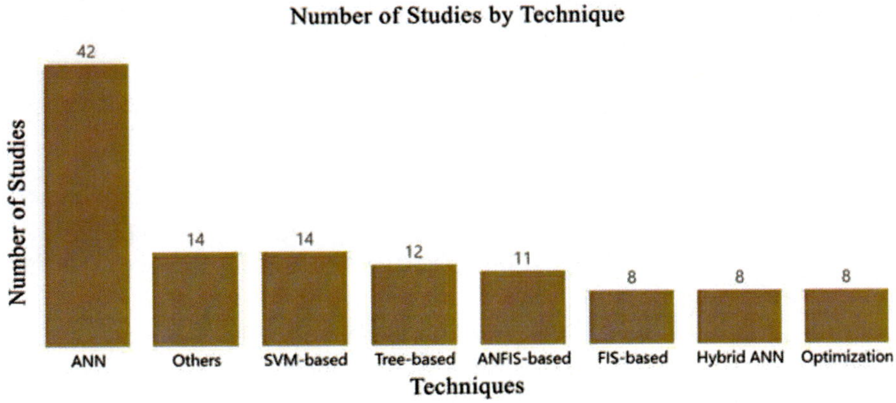

Fig. 12.4 Models applied for PPV prediction and optimization—quantity per group [45]

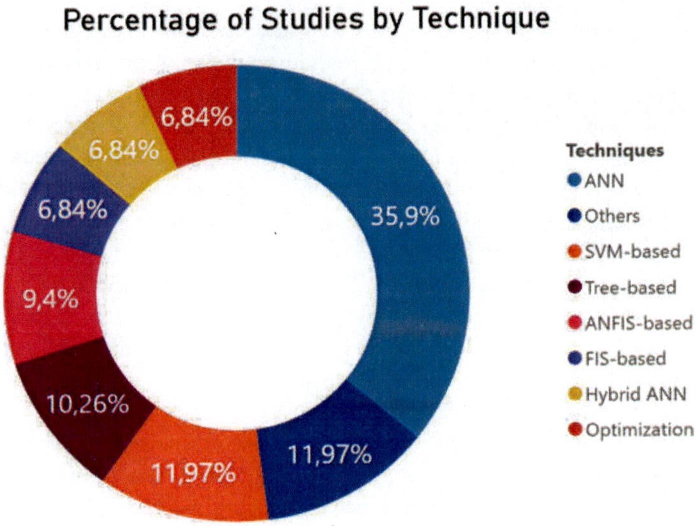

Fig. 12.5 Models applied for PPV prediction and optimization—percentage per group [45]

Following the same trend demonstrated in the previous sessions, a significant number of scientific papers have successfully explored the potential of advanced artificial and intelligence and machine learning algorithms to predict the flyrock distance. In order to reach such satisfactory results, most authors considered geological structures and rock mass properties in their models. According to the review presented in [19], rock density is an essential factor in predicting flyrock distance—the lighter the rock, the greater its projection. The drill and blast design parameters are the most frequently used, burden, stemming length, and hole depth. Flyrock distance is inversely proportional to burden and stemming length, while it increases with the increase in hole depth. The survey shows that powder factor, the maximum charge per delay, and specific charge are widely used to develop the flyrock prediction models when it comes to explosives characteristics.

The publications listed in Table 12.3 are just a recent search and by no means represent an exhaustive list of related publications in this field. Nevertheless, the exponential growth of advanced analytics methods is noticeable to predict rock fragmentation and environmental impacts related to blasting. The most common accuracy criteria described in the above publications are the determination coefficient (R^2), variance account for (VAF), root-mean-squared error (RMSE), and mean absolute error (MAE). In [47], a Taylor diagram was also used to demonstrate the performance of five different models.

The study developed in [68] considered 12 new hybrid models to optimize the support vector regression (SVR) model. The authors selected different evolutionary algorithms, including particle swarm optimization (PSO), genetic algorithm (GA), imperialist competitive algorithm (ICA), and artificial bee colony (ABC). For each

Table 12.3 Summary of AA-based techniques and related publications for ground vibration

Technique	Publications
Random forest (RF)	[43, 45, 49, 50]
Bayesian network (BN)	[45]
Gradient boost regression (GBR)	[43]
Extreme gradient boosting (XGBoost)	[43]
Cat boost (CatB)	[43]
Fuzzy inference system (FIS)	[51, 52]
Support vector machine (SVM)	[49, 50, 53–55]
Classification and regression trees (CART)	[49]
Chi-squared automatic interaction detection (CHAID)	[49]
General regression neural network (GRNN)	[54]
Recurrent adaptive neuro-fuzzy inference system (RANFIS)	[47]
Artificial neural network (ANN)	[47, 49, 52, 56–59]
Artificial neural network with K-nearest neighbors (ANN-KNN)	[57]
Artificial neural network with particle swarm optimization (ANN-PSO)	[60]
Artificial neural network with imperialist competitive algorithm (ANN-ICA)	[61]
Artificial neural network and genetic algorithm (ANN-GA)	[62]
Adaptive neuro-fuzzy inference system (ANFIS)	[47, 63–65]
ANFIS with particle swarm optimization (ANFIS-PSO)	[63]
Gene expression programming (GEP)	[66]
Support vector regression (SVR)	[67, 68]
Fuzzy C-means clustering (FCM)	[67]
Particle swarm optimization with extreme learning machine (PSO-ELM)	[69]
Autonomous groups PSO with extreme learning machine (AGPSO-ELM)	[69]
Mini-max probability machine regression (MPMR)	[69]
Gaussian process regression (GPR)	[69]
Least square support vector machine (LS-SVM)	[69]
Gray wolf optimization with extreme gradient boosting (GWO-XGBoost)	[43]
Whale optimization algorithm with extreme gradient boosting (WOA-XGBoost)	[43]
Bayesian optimization algorithm with extreme gradient boosting (BO-XGBoost)	[43]
Particle swarm optimization with extreme gradient boosting machine	[70]
Artificial bee colony with artificial neural network (ABC-ANN)	[30]

(continued)

Table 12.3 (continued)

Technique	Publications
Extreme learning machines with grasshopper optima. algorithm (GOA-ELM)	[71]
Extreme learning machines with Harris hawks optimization (HHO-ELM)	[71]
Chaos recurrent adaptive neuro-fuzzy inference system and PSO (CRANFIS-PSO)	[47]

evolutionary algorithm, three forms of kernel function, linear (*L*), radial basis function (RBF), and polynomial (*P*), were determined and developed. The comparison showed the highest accuracy for the GA-SVR-RBF model ($R^2 = 0.991$). The relevance of the input variables was assessed through the Hilbert–Schmidt Independence Criterion (HSIC) method, which indicated that *W* (explosive charge per delay) and *R* (monitoring distance) are the most relevant parameters for predicting PPV.

Flyrock Prediction Models

Flyrock induced by rock blasting is one of the most critical elements that must be accurately predicted in open-pit mines. The undesired projection of rock fragments by explosive energy beyond the safety zone of the explosion blocks has been a significant challenge faced by blast professionals and mining managers. Flyrock events lead to fatalities, injuries, and physical damage. Therefore, predicting the projection range and size of the rocks thrown is a remarkable step in reducing and controlling blasting accidents in the operations [48].

Following the same trend demonstrated in the previous sessions, a significant number of scientific papers have successfully explored the potential of advanced artificial and intelligence and machine learning algorithms to predict the flyrock distance. In order to reach such satisfactory results, most authors considered geological structures and rock mass properties in their models. According to the review presented in [19], rock density is an essential factor in predicting flyrock distance—the lighter the rock, the greater its projection. The drill and blast design parameters are the most frequently used, burden, stemming length, and hole depth. Flyrock distance is inversely proportional to burden and stemming length, while it increases with the increase in hole depth. The survey shows that powder factor, the maximum charge per delay, and specific charge are widely used to develop the flyrock prediction models when it comes to explosives characteristics.

Not surprisingly, most of the AA approaches presented in Table 12.4 found outstanding performance on their models, reaching high accuracy results, $R^2 > 0.9$. In [88], a sensitivity analysis was performed to evaluate the impact of each variable on the output (flyrock distance). The method concluded that hole diameter and powder factor have higher effects than the others.

Table 12.4 Summary of AA-based techniques and related publications for flyrock distance

Technique	Publications
Artificial neural network (ANN)	[24, 46, 72–81]
Support vector machine (SVM)	[72, 82]
Fuzzy inference system (FIS)	[83, 84]
Particle swarm optimization (PSO)	[85]
Support vector regression (SVR)	[86]
Least squares support vector machines (LS-SVM)	[86]
Extreme learning machine (ELM)	[81]
Outlier robust ELM (ORELM)	[81]
Adaptive neuro-fuzzy inference system (ANFIS)	[76, 79]
Artificial neural network and genetic algorithm (ANN-GA)	[1, 77, 87, 88]
Artificial neural network and artificial bee colony (ANN-ABC)	[87]
Artificial neural network and particle swarm optimization (ANN-PSO)	[60, 87–89]
ANN and imperialist competitive algorithm (ANN-ICA)	[87, 88, 90]
ANN network and ant colony optimization (ANN-ACO)	[74]
Artificial neural network and firefly algorithm (ANN-FA)	[34, 87]
Deep neural network and whale optimization algorithm (WOA)	[75]
SVR and lasso and elastic-net regularized genre. linear model	[91]
Recurrent fuzzy neural network and genetic algorithm (RFNN-GA)	[77]
Recurrent fuzzy neural network and PSO (RFNN-GA)	[92]
Gene expression programming and cuckoo optimization algorithm	[93]

The study presented in [75] showed that a deep learning model compared with the best model of ANN provided much better results for flyrock prediction. The authors applied a developed Whale Optimization Algorithm (WOA) to minimize the flyrock distance.

The research field of advanced analytics applied for rock blasting still demands integrated models for rock fragmentation and other desirable and undesirable impacts. In [19], a neural network with 7-13-3 architecture predicted rock fragmentation, ground vibrations, and air blasts. The developed model proved to be more accurate and efficient compared to empirical equations and multivariate regression.

By developing more holistic and robust models, perhaps using deep learning, significant resources can be saved, and people involved in the blasting process will investigate the relation of the input parameters on the blast performance in one attempt [19]. Furthermore, as mentioned in previous sections, geological properties and new data sources (e.g., advanced sensors such as MWD) can be incorporated to leverage the prediction performance of future models. Moreover, technical teams can utilize modern connectivity infrastructure and enable real-time data exchange from upstream and downstream processes to boost data-driven decisions in rock blasting and the entire mining value chain.

Conclusion

This chapter tried to show a new way to use advanced analytics in mining, especially in rock fragmentation. There are different effective parameters on the quality of the rock fragmentation process—the parameters such as rock properties, drilling data, blasting design, and performance data. Therefore, the selected analytical model can also play an important role in rock fragmentation performance. Furthermore, applying intelligent approaches such as prediction and optimization can help the operators improve their performance quality. This chapter clarified how the mine managers could use the new technological analytics approach to better rock fragmentation. This chapter also discussed the application of advanced analytics to improve mining safety by solving the ground vibration and flyrocks concerns during the drilling and blasting.

References

1. Monjezi, M., H.A. Khoshalan, and A.Y. Varjani. 2012. Prediction of flyrock and back break in open pit blasting operation: A neuro-genetic approach. *Arabian Journal of Geosciences* 5 (3): 441–448.
2. Silva, J., J. Amaya, and F. Basso. 2017. Development of a predictive model of fragmentation using drilling and blasting data in open-pit mining. *Journal of the Southern African Institute of Mining and Metallurgy* 117 (11): 1089–1094.
3. Rai, P., et al. 2015. An overview on measurement-while-drilling technique and its scope in excavation industry. *Journal of the Institution of Engineers (India): Series D* 96(1): 57–66.
4. Segui, J., and M. Higgins. 2002. Blast design using measurement while drilling parameters. *Fragblast* 6 (3–4): 287–299.
5. Park, J., and K. Kim. 2020. Using drilling performance to improve rock-breakage efficiencies: A part of mine-to-mill optimization studies in a hard-rock mine. *International Journal of Mining Science and Technology* 30(2): 179–188.
6. Ghosh, R., H. Schunnesson, and U. Kumar. 2014. Evaluation of rock mass characteristics using measurement while drilling in Boliden minerals Aitik copper mine, Sweden. In *Mine planning and equipment selection*, 81–91. Springer.
7. Chen, J., and Z. Yue. 2015. Ground characterization using breaking-action-based zoning analysis of rotary-percussive instrumented drilling. *International Journal of Rock Mechanics and Mining Sciences* 75: 33–43.
8. Hatherly, P., et al. 2015. Drill monitoring results reveal geological conditions in blast-hole drilling. *International Journal of Rock Mechanics and Mining Sciences* 78: 144–154.
9. Kadkhodaie-Ilkhchi, A., et al. 2010. Rock recognition from MWD data: A comparative study of boosting, neural networks, and fuzzy logic. *IEEE Geoscience and Remote Sensing Letters* 7 (4): 680–684.
10. Zhou, H., et al. 2012. Automatic rock recognition from drilling performance data. IEEE.
11. Basarir, H., and C. Karpuz. 2016. Preliminary estimation of rock mass strength using diamond bit drilling operational parameters. *International Journal of Mining, Reclamation, and Environment* 30 (2): 145–164.
12. Klyuchnikov, N., et al. 2019. Data-driven model for identifying the rock type at a drilling bit. *Journal of Petroleum Science and Engineering* 178: 506–516.
13. Ahsan, N., et al. 2015. Adaptive sampling applied to blast-hole drilling in surface mining. *International Journal of Rock Mechanics and Mining Sciences* 75: 244–255.

14. Leung, R., and S. Scheduling. 2015. Automated coal seam detection using a modulated specific energy measure in a monitor-while-drilling context. *International Journal of Rock Mechanics and Mining Sciences* 75: 196–209.
15. Kumar, C.V., H. Vardhan, and C.S. Murthy. 2021. Artificial neural network for predicting rock properties using acoustic frequencies recorded during rock drilling operations. *Modeling Earth Systems and Environment* 1–21.
16. Basarir, H., et al. 2017. The use of soft computing methods to predict rock properties based on measurement while drilling data. In *Proceedings of the Eighth International Conference on Deep and High-Stress Mining*, Australian Centre for Geo-mechanics.
17. He, M., Z. Zhang, and N. Li. 2021. Prediction of fracture frequency and RQD for the fractured rock mass using drilling logging data. *Bulletin of Engineering Geology and the Environment* 80 (6): 4547–4557.
18. Babaei Khorzoughi, M. 2013. Measurement while drilling techniques for improved rock mass characterization in open-pit mines. University of British Columbia.
19. Dumakor-Dupey, N.K., S. Arya, and A. Jha. 2021. Advances in blast-induced impact prediction—a review of machine learning applications. *Minerals* 11 (6): 601.
20. Dhekne, P., M. Pradhan, and R. Jade. 2014. Artificial intelligence and prediction of rock fragmentation. In *Mine planning and equipment selection*, 891–898. Springer.
21. Huang, J., et al. 2020. A new auto-tuning model for predicting the rock fragmentation: A cat swarm optimization algorithm. *Engineering with Computers* 1–12.
22. Babaeian, M., et al. 2019. A new framework for evaluating rock fragmentation in open-pit mines. *Journal of Rock Mechanics and Geotechnical Engineering* 11 (2): 325–336.
23. Hudaverdi, T., P. Kulatilake, and C. Kuzu. 2011. Prediction of blast fragmentation using multivariate analysis procedures. *International Journal for Numerical and Analytical Methods in Geo-mechanics* 35 (12): 1318–1333.
24. Monjezi, M., M. Rezaei, and A.Y. Varjani. 2009. Prediction of rock fragmentation due to blasting in Gol-E-Gohar iron mine using fuzzy logic. *International Journal of Rock Mechanics and Mining Sciences* 46 (8): 1273–1280.
25. Esmaeili, M., et al. 2015. Application of PCA, SVR, and ANFIS for modeling of rock fragmentation. *Arabian Journal of Geosciences* 8 (9): 6881–6893.
26. Hasanipanah, M., et al. 2018. Feasibility of PSO–ANFIS model to estimate rock fragmentation produced by mine blasting. *Neural Computing and Applications* 30 (4): 1015–1024.
27. Mojtahedi, S.F.F., et al. 2019. Proposing a novel hybrid intelligent model to simulate particle size distribution resulting from blasting. *Engineering with Computers* 35 (1): 47–56.
28. Sayadi, A., et al. 2013. A comparative study on applying various artificial neural networks to simultaneous prediction of rock fragmentation and back break. *Journal of Rock Mechanics and Geotechnical Engineering* 5 (4): 318–324.
29. Enayatollahi, I., A.A. Bazzazi, and A. Asadi. 2014. Comparison between neural networks and multiple regression analysis to predict rock fragmentation in open-pit mines. *Rock mechanics and rock Engineering* 47 (2): 799–807.
30. Taheri, K., et al. 2017. A hybrid artificial bee colony algorithm-artificial neural network for forecasting the blast-produced ground vibration. *Engineering with Computers* 33 (3): 689–700.
31. Salimi, A., et al. 2012. A neuro-fuzzy approach for predicting rock fragmentation in open-pit mines. In *Proceedings 21st International Symposium on Mine Planning and Equipment Selection (MPES)*, India: New Delhi.
32. Sayevand, K., H. Arab, and S.B. Golzar. 2018. Development of the imperialist competitive algorithm in predicting the particle size distribution after mine blasting. *Engineering with Computers* 34 (2): 329–338.
33. Ebrahimi, E., et al. 2016. Prediction and optimization of back-break and rock fragmentation using an artificial neural network and a bee colony algorithm. *Bulletin of Engineering Geology and the Environment* 75 (1): 27–36.
34. Asl, P.F., et al. 2018. Optimization of flyrock and rock fragmentation in the Tajareh limestone mine using meta-heuristics method of firefly algorithm. *Engineering with Computers* 34 (2): 241–251.

35. Xie, C., et al. 2021. Predicting rock size distribution in mine blasting using various novel soft computing models based on meta-heuristics and machine learning algorithms. *Geoscience Frontiers* 12(3): 101108.
36. Zhang, S., et al. 2020. Prediction of rock size distribution in mine bench blasting using a novel ant colony optimization-based boosted regression tree technique. *Natural Resources Research* 29 (2): 867–886.
37. Zhou, J., et al. 2019. Performance evaluation of hybrid FFA-ANFIS and GA-ANFIS models to predict a muck-pile particle size distribution after blasting. *Engineering with Computers* 1–10.
38. Gao, W., et al. 2018. Developing GPR model for forecasting the rock fragmentation in surface mines. *Engineering with Computers* 34 (2): 339–345.
39. Fang, Q., et al. 2021. modeling rock fragmentation by firefly optimization algorithm and boosted generalized additive model. *Neural Computing and Applications* 33 (8): 3503–3519.
40. Shi, X.-Z., et al. 2012. Support vector machines approach to mean particle size of rock fragmentation due to bench blasting prediction. *Transactions of Nonferrous Metals Society of China* 22 (2): 432–441.
41. Bahadori, M. and H. Bakhshandeh Amnieh. 2018. Implementation of hyperbolic tangent function to estimate size distribution of rock fragmentation by blasting in open-pit mines. *International Journal of Mining and Geo-Engineering* 52(2): 187–197.
42. Jiang, N., et al. 2015. Damage characteristics of surrounding rock subjected to VCR mining blasting shock. *Shock and Vibration* 2015.
43. Qiu, Y., et al. 2021. Performance evaluation of hybrid WOA-XGBoost, GWO-XGBoost and BO-XGBoost models to predict blast-induced ground vibration. *Engineering with Computers* 1–18.
44. Parida, A., and M. Mishra. 2015. Blast vibration analysis by different predictor approaches—a comparison. *Procedia Earth and Planetary Science* 11: 337–345.
45. Zhou, J., et al. 2020. Prediction of ground vibration induced by blasting operations through the Bayesian network and random forest models. *Soil Dynamics and Earthquake Engineering* 139: 106390.
46. Monjezi, M., A. Bahrami, and A. Yazdian Varjani. 2010. Simultaneous prediction of fragmentation and flyrock in blasting operation using artificial neural networks. *International Journal of Rock Mechanics and Mining Sciences* (1997) 47(3): 476–480.
47. Zhu, W., H.N. Rad, and M. Hasanipanah. 2021. A recurrent chaos ANFIS optimized by PSO to predict ground vibration generated in rock blasting. *Applied Soft Computing* 108: 107434.
48. Rosales-Huamani, J.A., et al. 2020. Design of a predictive model of rock breakage by blasting using artificial neural networks. *Symmetry* 12 (9): 1405.
49. Zhang, H., et al. 2020. A combination of feature selection and random forest techniques to solve a problem related to blast-induced ground vibration. *Applied Sciences* 10 (3): 869.
50. Longjun, D., et al. 2011. Comparisons of random forest and support vector machine for predicting blasting vibration characteristic parameters. *Procedia Engineering* 26: 1772–1781.
51. Fişne, A., C. Kuzu, and T. Hüdaverdi. 2011. Prediction of environmental impacts of quarry blasting operation using fuzzy logic. *Environmental Monitoring and Assessment* 174 (1): 461–470.
52. Mohamed, M.T. 2011. Performance of fuzzy logic and artificial neural network in predicting ground and air vibrations. *JES Journal of Engineering Sciences* 39 (2): 425–440.
53. Hasanipanah, M., et al. 2015. Feasibility of indirect determination of blast-induced ground vibration based on support vector machine. *Measurement* 75: 289–297.
54. Mohamadnejad, M., R. Gholami, and M. Ataei. 2012. Comparison of intelligence science techniques and empirical methods for predicting blasting vibrations. *Tunneling and Underground Space Technology* 28: 238–244.
55. Khandelwal, M., P. Kankar, and S. Harsha. 2010. Evaluation and prediction of blast-induced ground vibration using support vector machine. *Mining Science and Technology (China)* 20 (1): 64–70.
56. Monjezi, M., M. Ghafurikalajahi, and A. Bahrami. 2011. Prediction of blast-induced ground vibration using artificial neural networks. *Tunneling and Underground Space Technology* 26 (1): 46–50.

57. Amiri, M., et al. 2016. A new combination of artificial neural network and K-nearest neighbors models to predict blast-induced ground vibration and air-overpressure. *Engineering with Computers* 32 (4): 631–644.
58. Monjezi, M., et al. 2006. Prediction and analysis of blast parameters using artificial neural network. *Noise and Vibration Worldwide* 37 (5): 8–16.
59. Yu, Z., et al. 2020. A new multi-kernel relevance vector machine based on the HPSOGWO algorithm for predicting and controlling blast-induced ground vibration. *Engineering with Computers* 1–16.
60. Armaghani, D.J., et al. 2014. Blasting-induced flyrock and ground vibration prediction through an expert artificial neural network based on particle swarm optimization. *Arabian Journal of Geosciences* 7 (12): 5383–5396.
61. Hajihassani, M., et al. 2015. Ground vibration prediction in quarry blasting through an artificial neural network optimized by the imperialist competitive algorithm. *Bulletin of Engineering Geology and the Environment* 74 (3): 873–886.
62. Xu, S., et al. 2021. Blasting vibration control using an improved artificial neural network in the Ashlee copper mine. *Shock and Vibration* 2021.
63. Shahnazar, A., et al. 2017. A newly developed approach for predicting ground vibration using a hybrid PSO-optimized ANFIS-based model. *Environmental Earth Sciences* 76 (15): 1–17.
64. Iphar, M., M. Yavuz, and H. Ak. 2008. Prediction of ground vibrations resulting from the blasting operations in an open-pit mine by adaptive neuro-fuzzy inference system. *Environmental Geology* 56 (1): 97–107.
65. Armaghani, D.J., et al. 2015. The feasibility of the ANFIS model is to predict ground vibrations resulting from quarry blasting. *Environmental earth sciences* 74 (4): 2845–2860.
66. Faradonbeh, R.S., et al. 2016. Prediction of ground vibration due to quarry blasting based on gene expression programming: A new model for peak particle velocity prediction. *International Journal of Environmental Science and Technology* 13 (6): 1453–1464.
67. Sheykhi, H., et al. 2018. Forecasting ground vibration due to rock blasting: A hybrid intelligent approach using support vector regression and fuzzy C-means clustering. *Engineering with Computers* 34 (2): 357–365.
68. Nguyen, H., et al. 2020. Predicting blast-induced ground vibration in open-pit mines using vibration sensors and support vector regression-based optimization algorithms. *Sensors* 20 (1): 132.
69. Armaghani, D.J., et al. 2020. A novel approach for forecasting ground vibrations resulting from blasting: Modified particle swarm optimization coupled extreme learning machine. *Engineering with Computers* 1–15.
70. Zhang, X., et al. 2020. Novel soft computing model for predicting blast-induced ground vibration in open-pit mines based on particle swarm optimization and XGBoost. *Natural Resources Research* 29 (2): 711–721.
71. Yu, C., et al. 2021. Optimal ELM–Harris hawks optimization and ELM–grasshopper optimization models to forecast peak particle velocity resulting from mine blasting. *Natural Resources Research* 30 (3): 2647–2662.
72. Amini, H., et al. 2012. Evaluation of flyrock phenomenon due to blasting operation by support vector machine. *Neural Computing and Applications* 21 (8): 2077–2085.
73. Monjezi, M., et al. 2013. Evaluation of the effect of blast design parameters on flyrock using artificial neural networks. *Neural Computing and Applications* 23 (2): 349–356.
74. Saghatforoush, A., et al. 2016. Combination of neural network and ant colony optimization algorithms for prediction and optimization of flyrock and back-break induced by blasting. *Engineering with Computers* 32 (2): 255–266.
75. Guo, H., et al. 2021. Deep neural network and whale optimization algorithm to assess flyrock induced by blasting. *Engineering with Computers* 37 (1): 173–186.
76. Trivedi, R., T. Singh, and N. Gupta. 2015. Prediction of blast-induced flyrock in opencast mines using ANN and ANFIS. *Geotechnical and Geological Engineering* 33 (4): 875–891.
77. Rad, H.N., et al. 2020. Prediction of flyrock in mine blasting: A new computational intelligence approach. *Natural Resources Research* 29 (2): 609–623.

78. Zhou, J., et al. 2020. A Monte Carlo simulation approach for effective assessment of flyrock based on the intelligent system of neural network. *Engineering with Computers* 36 (2): 713–723.
79. Armaghani, D.J., et al. 2016. Evaluation and prediction of flyrock resulting from blasting operations using empirical and computational methods. *Engineering with Computers* 32 (1): 109–121.
80. Ghasemi, E., et al. 2014. Application of artificial intelligence techniques for predicting the flyrock distance caused by the blasting operation. *Arabian Journal of Geosciences* 7 (1): 193–202.
81. Lu, X., et al. 2020. ORELM: A novel machine learning approach for predicting flyrock in mine blasting. *Natural Resources Research* 29 (2): 641–654.
82. Manoj, K., and M. Monjezi. 2013. Prediction of flyrock in open pit blasting operation using the machine learning method. *International Journal of Mining Science and Technology* 23 (3): 313–316.
83. Rezaei, M., M. Monjezi, and A.Y. Varjani. 2011. Development of a fuzzy model to predict flyrock in surface mining. *Safety Science* 49 (2): 298–305.
84. Shams, S., et al. 2015. Application of fuzzy inference system for predicting rock fragmentation induced by blasting. *Arabian Journal of Geosciences* 8 (12): 10819–10832.
85. Hasanipanah, M., et al. 2017. Application of PSO to develop a powerful equation for prediction of flyrock due to blasting. *Neural Computing and Applications* 28 (1): 1043–1050.
86. Rad, H.N., et al. 2018. Developing the least squares support vector machine for estimating the blast-induced flyrock. *Engineering with Computers* 34 (4): 709–717.
87. Li, D., M. Koopialipoor, and D.J. Armaghani. 2021. A combination of fuzzy Delphi method and ANN-based models to investigate factors of flyrock induced by mine blasting. *Natural Resources Research* 30 (2): 1905–1924.
88. Koopialipoor, M., et al. 2019. Three hybrid intelligent models in estimating flyrock distance resulting from blasting. *Engineering with Computers* 35 (1): 243–256.
89. Zhou, J., et al. 2019. Use intelligent methods to design effective pattern parameters of mine blasting to minimize flyrock distance. *Natural Resources Research* 1–15.
90. Marto, A., et al. 2014. A novel approach for blast-induced flyrock prediction based on the imperialist competitive algorithm and artificial neural network. *The Scientific World Journal* 2014.
91. Guo, H., et al. 2021. A new technique to predict fly-rock in bench blasting based on an ensemble of support vector regression and GLMNET. *Engineering with Computers* 37 (1): 421–435.
92. Kalaivaani, P., et al. 2020. A novel intelligent approach to simulate the blast-induced flyrock based on RFNN combined with PSO. *Engineering with Computers* 36 (2): 435–442.
93. Dehghani, H., and M. Pourzafar. 2021. Prediction and minimization of blast-induced flyrock using gene expression programming and cuckoo optimization algorithm. *Environmental Earth Sciences* 80 (1): 1–17.

Chapter 13
Advanced Analytics for Rock Blasting and Explosives Engineering in Mining

Jorge Luiz Valença Mariz and Ali Soofastaei

Abstract Blasting is a fundamental operation in the mining of metal and non-metal mine sites. There is a historical story of using explosive material in both underground and surface mines sites. The development of new technology to collect the data from the blasting operation provides an opportunity for mining companies to think about applying machine learning (ML) and artificial intelligence (AI) methodologies for prediction and optimization. There are many effective parameters on blasting performance, and the main part of them is controllable. Using ML and AI helps the operators to find the best quantity and quality of explosive materials. Moreover, intelligence methods can be useful to solve the related explosion environment problems such as vibration, air overpressure, and fly rocks. It has been evidenced that using the new technologies and AI approaches are valuable approaches to improving blasting efficiency and safety. However, applying these intelligence tools in mining is very new, and there are few experiences in this field. The lack of this information can see in academic research as well. This chapter tries to cover the earlier mentioned gap by providing some material regarding the explosive and initiating systems, characterization of explosives, rock fragmentation modeling, and blasting environmental concerns. In addition, this chapter reviews many related case studies to help the researchers who are aiming to study and work in the field of using advanced analytics in blasting mining operations.

Keywords Advanced analytics · Blasting · Mining · Prediction · Optimization

Introduction

The use of explosives in the mining industry dates back to the seventeenth century when Caspar Weindel employed black powder (known by the Chinese since the

J. L. V. Mariz (✉)
Universidade Federal Do Rio Grande Do Sul (UFRGS), Porto Alegre, Brazil

A. Soofastaei
Vale, Brisbane, Australia
URL: http://www.soofastaei.net

A. Soofastaei (ed.), *Advanced Analytics in Mining Engineering*,
https://doi.org/10.1007/978-3-030-91589-6_13

second century B.C. and used for war purposes since the twelfth century) to break a rock in an underground mine in Schemnitz (Banská Štiavnica), in current Slovakia. He used a fuse to trigger the powder and, before the spread of this technique in mines in Europe, introduced for civilian purposes by Bertold Schwartz since the fourteenth century, miners used only hand tools in mineral exploitation [1].

Between the seventeenth and nineteenth centuries, black powder was applied in rock breaking until dynamite was discovered, dominated, and produced on a large scale. Italian chemist Ascanio Sobrero accidentally discovered nitroglycerin (NG) in 1846 when he tried to heat a drop of liquid in a test tube, which exploded and threw glass fragments everywhere, severely injuring his face and hands. Fearing the destructive power of his discovery, he chose not to proceed with the research. Between 1864 and 1867, the ingenious Alfred Nobel studied NG and was the first to produce it on a scale. During the testing phase, several explosions, including one that killed his brother and four others, convinced the authorities that new experiments should be banned "on the outskirts of Stockholm," so Nobel moved his laboratory to a boat anchored in a lake. Finally, he discovered how to stabilize the mixture with diatoms and turn it into a paste, molding it into rods suitable for the holes and naming it dynamite, starting its commercial production. He also created an igniter: a wooden plug filled with black powder, which could be detonated by lighting a fuse, causing the surrounding NG to explode [2].

Nitroglycerin was hegemonic until the middle of the twentieth century when an unfortunate accident brought the perception that a known fertilizer could be used as an explosive. In 1947, there was a fire on the ship SS Grandcamp, containing 2341 tons of fertilizer (ammonium nitrate) anchored in the port of Texas City. To not damage the cargo, the firefighters decided not to use water to fight the fire but to muffle it, without success, since the NH_4NO_3 does not need oxygen to start combustion. The fire spread in the cellars, causing an explosion that was heard from 240 km away, producing a 600-m incendiary mushroom that destroyed even two small planes passing by. There was a chain reaction when flaming debris hit a ship containing a cargo of sulfur and crude fuel deposits located in the port, causing further explosions. After that, 576 people died in the explosion, and approximately 4000 were injured. The explosions destroyed a residential area of 500 houses. The accident drew attention to the explosive power of ammonium nitrate, which today is the basis of most explosives used in the industry [3–5].

Regardless of the type of explosive used, rock breaking in underground and open-pit mines employs mostly blasting, so that an efficient operation needs to master the technique and be aware of the risks, especially when all subsequent operations depend on the result of the blasting. In addition to meeting the fragmentation optimization criteria and geomechanical stability, as new mines are operated in already established urban centers and as populations start to occupy the surroundings of the mines, the control of ground motion (seismic waves), air blast, and the frocks play a fundamental role in many operations, even preventing the recovery of specific deposit sectors. Although the breaking of rocks with explosives is still a brute force operation, the manufacture of more stable explosives, the development of new ignition technologies, blasting simulation, and the use of advanced analytics to minimize the impact of environmental constraints seem to be the trend of this operation in the context of Industry 4.0.

Explosives and Initiation Systems

Characterization of Explosives

Explosives are a combination of elements that, when properly stimulated, produce a fast reduction–oxidation reaction involving fuels and oxidizing agents. In the explosion process, the fuels (such as carbon and hydrogen) are oxidized, and the oxidizers (such as ammonium nitrate) are reduced. In the proper condition and proportion, even if the elements are not explosives in themselves, the mixture detonates, or the explosion front spreads along a column of explosive; under other conditions, the mixture will deflagrate (burn) at a very rapid rate. An example of such a reaction involving the mixture of ammonium nitrate (AN) and fuel oil (FO) is shown in Eq. (13.1) [6, 7].

$$3NH_4NO_3 + CH_2 \rightarrow 7H_2O + CO_2 3N_2 + Heat \text{ of Explosion} \tag{13.1}$$

This is the ideal oxygen balanced reaction, where there is just enough oxygen present to oxidize the fuels (hydrogen and carbon) in the mixture to H_2O and CO_2, and all the nitrogen forms N_2. The amount of heat liberated by this exothermic reaction is the cited heat of the explosion. The breaking of the molecules on the left side of the equation demands an amount of energy, called heat of breakdown, while the union of atoms on the right side releases another amount, called heat of formation. The resultant heat of explosion is the difference between the heat of formation of the products and the heat of breakdown of the explosive ingredients [7].

This understanding of enthalpy and balancing equations is essential for defining the proportion of components of the explosive because if the oxygen is not balanced, there will be the generation of toxic gases and loss of energy (therefore, loss of efficiency in rock breaking). For example, in the case of ammonium nitrate and fuel oil (ANFO), if there is a deficiency of oxygen, deadly carbon monoxide gas will be formed; on the other hand, if there is an excess of oxygen, toxic nitrogen dioxide arises [7].

Several parameters are essential in the selection of explosives. First, the composition is related to the density, stabilization of the components, sensitivity, and the balance of the reduction–oxidation reaction; the addition of aluminum powder incorrect particle size and proportion to the ANFO (ALANFO), for instance, increases the total energy of the explosive without changing its sensitivity and water resistance, but this process must be accurate to achieve the goal. Sensitivity, in contrast, is a measure of the ease with which an explosive can be detonated by heat, friction, impact, or shock and, although high sensitivities increase the potential of the wave to travel along the column, it reduces the safety of handling those explosives. Finally, Water resistance is essential for the selection and, since there are explosives with much resistance (water gel) or no resistance (ANFO) to humid environments, regardless of the type of explosive used, the period of exposure must be minimal. Otherwise, it can reach a point where the detonation wave cannot propagate through the explosive column [6].

Still, regarding the main parameters of explosives, the velocity of detonation (VOD) is the speed with which a detonation wave propagates through a column of explosive and, generally, the higher the VOD, the greater the brisance (shattering effect). Furthermore, the weight strength is the explosive energy per unit weight, while the bulk strength is explosive energy per unit volume, the second being more related to drilling diameter; a P–V curve diagram is another way to measure the energy of an explosive. Finally, the detonation pressure is the pressure of the gas directly at the detonation front, while the explosion pressure denotes the gas pressure applied to the borehole wall just after the detonation, being the second approximately half of the amount of the first [6].

Despite the number of parameters to be evaluated, the main objective is to employ an explosive suitable for the rock type to be blasted. Hard and massive rock types require creating new fractures for adequate fragmentation, and the shock energies required for the new generation of fractures are associated with high explosion pressures, high VOD, and high density. Other rock types, which are already cracked, depend more upon the heaving/displacing action provided by gas pressure for breaking, which is generally related to explosives with lower VOD and density. Although two explosives can have the same amount of total energy (same areas under P–V curves), their behavior can be completely different, as one may have a higher peak of pressure, but the other maintains a higher gas pressure while expanding. Therefore, explosives with high brisance/low gas pressure are recommended for hard, brittle rocks, though low-brisance/high-gas explosives are indicated for softer/jointed rocks. Figure 13.1 presents a P–V diagram for two different explosives, explosive A related to higher peak pressure, while explosive B is related to higher gas pressure during expansion [6].

Types of Explosives

Before introducing the main types of explosives used in industry, it is important to explore the concepts of blasting agents and blasting explosives. A blasting agent is a chemical mixture that contains no ingredient that is in itself an explosive and cannot be initiated by a No. 8 detonator in an unconfined state; on the other hand, a blasting explosive contains an ingredient that is in itself an explosive and/or can be initiated by a No. 8 detonators. These conditions have a major impact on shipment, storage, and handling procedures [6].

The combination of ammonium nitrate (AN) and fuel oil (FO) in the 94/6 weight ratio makes up one of the most common blasting agents applied in industry, the

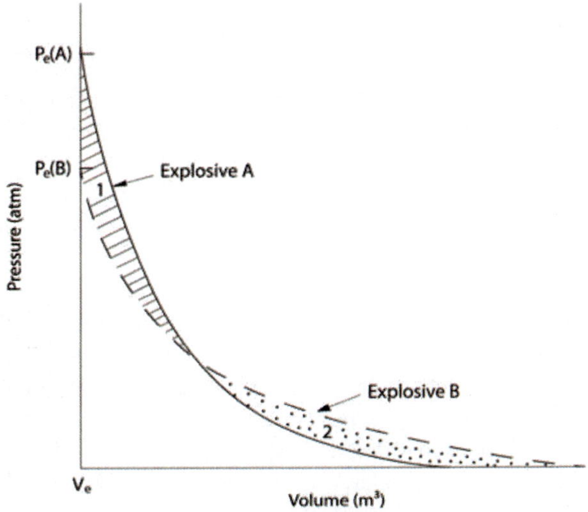

Fig. 13.1 Representation of the P–V curves for two explosives with the same energy [6]

Fig. 13.2 ANFO ingredients by weight [8]

ANFO. Maximum strength is achieved when about 5.5% of fuel oil by weight is added; as stated before, when little fuel oil is added, energy falls quickly, and several nitrous oxides (NO_x) are formed, while excess fuel oil also causes energy to decreases, forming carbon monoxide (CO). ANFO is marketed in porous prills, with granulometry in the range of 0.84–3.36 mm, and although the mixture of AN and FO can be prepared in the mines, usually in trucks for bulk loading, it can be supplied as an ANFO package ready for use. A major advantage of ANFO is its low cost per unit of energy delivered, while a major disadvantage is the lack of water resistance due to ammonium nitrate dissolving easily in water, even with fuel oil; low density and low energy range are additional disadvantages. Figure 13.2 presents ANFO ingredients by weight [6].

When paint grade aluminum with specified size and purity density is added as a fuel to the standard ANFO, ALANFO is created, whose total explosive energy is substantially greater than the first, although there is no change in its sensitivity and water resistance. The most common compositions contain between 5 and 15 percent aluminum by weight, and as more aluminum is added to the mixture, the proportion of fuel oil should decrease. This explosive can be used as the toe load in bench blasting [6].

Notwithstanding, there are situations, such as when dealing with the back break, where bulk agents with less strength are desired to reduce the explosion pressure in the blast holes. The addition of low-density materials, such as polystyrene, bagasse, sawdust, or peanut skin, can reduce the explosive's density and energy concentration. In order to minimize the segregation between these materials and the prills, they must have the same particle size. Although up to 90% of polystyrene has been successfully used in mixtures in larger diameters, the 50/50 ratio achieved the desired reduction of VOD and density for all cited materials without compromising the results in different studies for different studies diameters [6].

Modern dynamite products are detonator-sensitive mixtures that contain nitroglycerin (NG) as a sensitizer or as a primary means for developing energy that decomposes at a stable VOD when properly initiated, although they differ from the original Nobel formula. It consists of a mixture of liquid nitrate esters (NG and nitroglycol), a gelling agent, oxidizing salts, fuel, and materials called dopes. Nitrocellulose acts as a gelling agent and provides resistance to water while prevents fluid leakage from the cartridges. Fuels provide energy, and the oxidizing salts (AN and sodium nitrate) provide the oxygen for the detonation, while dopes provide another useful property [8].

Energy can vary among dynamite grades, but dynamite tends to be high brisance energy explosive. Besides energy, the sensitivity, the VOD, and the fume class of the product must be consulted with manufacturers for each application. The four basic types of dynamites are granular, semigelatin, gelatin, and permissible, the latter being those approved by Mine Safety and Health Administration (MSHA) for application in underground coal mines. Dynamites also differ in the material used to provide their main energy source, being "straight" those based on NG, or "extra" dynamites, those whose large part of NG has been replaced by AN. Figure 13.3 presents standard cartridge dynamite packages and sizes [8].

Another type of explosives is slurries, bulk explosives compatible with water that can be either blasting agents or explosives, depending on the ingredients used and on the degree of cap-sensitivity, and can have any desired consistency from a free-flowing liquid to a cohesive gel. Although the use of water as an explosive constituent seems unlikely, a certain amount is necessary to provide the required consistency and texture of a water gel. If more water is used than necessary, the weight strength decreases; on the other hand, if little water is used, the liquid phase is insufficient, and the resulting high viscosity makes pumping difficult. The water

Fig. 13.3 Standard cartridge dynamite packages and sizes [8]

content, which ranges from 5 to 40% by weight with an average of 15%, also reduces hazardous sensitivity associated with fire, friction, and impact [6].

Since common oxidizers (such as ammonium nitrate and sodium nitrate) are soluble, the water resistance of a water gel depends on its physical state. Guar gam (polysaccharide) is generally used to thicken the mixture, and then cross-linking agents are employed to establish an impenetrable barrier that prevents the leaching of soluble salts by water permeating through the borehole. The presence of air bubbles or void space in the mixture is necessary for efficient detonation and to define the appropriate sensitivity level, detonation velocity, and critical drilling diameter. Aerating agents or gas formers can be used, as well as physical agitation, and this methodology allowed the substitution of molecular explosive additives (such as TNT), smokeless powder, organic salts (like amine nitrates and perchlorates), or aluminum powder as sensitizers [6].

Slurries can be divided into three categories: those containing high-explosive fuel sensitizers; those containing metallic fuel sensitizers; and those that do not contain any high-energy ingredients. Originally, water gels were composed of AN-TNT-H_2O. However, later different ingredients were tested, divided into fuel sensitizers (explosives, such as TNT, or non-explosives, such as aluminum), oxidizers (ammonium nitrate or sodium nitrate), cross-linking agents (such as potassium dichromate), gelling agents (guar gum, starch, or acrylamide polymers), gas formers (such as peroxides or sodium bicarbonate), and aerating agents (such as perlite). The ingredients of the slurries can be selected as desired to vary the energy, sensitivity, oxygen balance, rheology, and stability of the final product. Figure 13.4 presents the rheology of a pourable water gel (left) and examples of typical water gel in cartridges (right) [6].

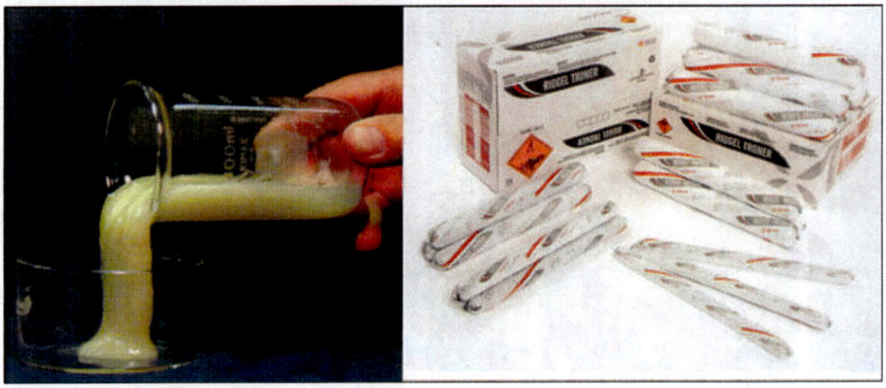

Fig. 13.4 Rheology of a pourable water gel (left) and typical water gel in cartridges (right) [8]

The type of explosive that has gained more popularity in recent decades is the emulsion, which consists of a two-phase system in which an inner or dispersed phase is distributed as very small droplets in an outer or continuous phase. The importance of particle size of its components lies in increasing the rate and efficiency of reactions as the proximity of an oxidizing unit to a fuel unit becomes more intimate. Despite the other explosives listed in this chapter, both the oxidizer and the fuel are liquid; instead of being dry prills, the emulsion oxidizer is a highly concentrated solution of ammonium nitrate and/or other salts. Using emulsifiers, the droplet size of this solution was reduced to microscopic proportions (2 to 10 microns), where there is a film of oil around each droplet. The result is a mixture of fuel and oxidizers similar to other explosives, but the particle size is intended to mimic the combination found in molecular explosives such as NG or TNT [6].

Depending on the internal phased ratio, without adding thickeners, a relatively high viscosity can be obtained just by the nature of the system. Due to oxygen balance constraints, a large amount of oxidizer salt solution is required to react with a relatively small amount of fuel oil, such as a ratio of approximately 90/10 from the aqueous phase to fuel phase on a volumetric basis. The main sensitization of emulsions is provided by the intimacy between the fuel and the oxidizer achieved when the oxidizing solution is dispersed in very small droplets; even so, small air bubbles must be added as additional sensitizers. Due to the absence of a cross-linked gel network to assist in maintaining the air bubbles, however, the addition of very small glass microballoons is necessary to control sensitization and density. Various types of oils and/or waxes can be used in the fuel phase to adjust the product's fluidity from a pourable consistency to a firm, moldable solid. The aluminum content can also be varied to produce a series of energies, just as it was done with the slurries. Figure 13.5 presents examples of typical cartridges emulsions with different consistency [6].

Another ANFO species is that air space among the prills is filled by emulsion, which propitiates an explosive denser and more water-resistant than regular ANFO

Fig. 13.5 Consistency of typical cartridges emulsion [8]

with relatively little added cost, called heavy ANFO (HANFO). The cited parameters can be tailored by changing the emulsion level. Depending upon the AN prill type and the matrix density (emulsion), all spaces among prills will be filled in the 35–40% matrix range. The proper selection of the blend ratio depends on factors such as borehole environment (wet or dry) and diameters, such as the tradeoff between gas energy and shock energy (brisance). Blends using more than 50% emulsion are not recommended due to the lack of sensitivity of the product [6].

Boosters, finally, are a key component of the primer, which together with an initiating device provide the required energy to initiate the explosive charge. Since the development of ANFO products, water gels, and non-sensitized emulsions, the need for high-velocity and high-energy boosters has arisen. In small diameter applications, an initiating device can serve as the priming mechanism, although in most applications, the booster is primordial to providing enough energy to the explosive charge to create full detonation as quickly as possible. The booster must exceed the detonation pressure of the explosive being primed to reach its full brisance. Since most primer systems are placed at the bottom of boreholes, the water resistance of the boosters should be enough to withstand the exposure for the time required in the blasting situation. Figure 13.6 presents examples of various cast booster sizes and shapes [8].

Initiation Systems

Initiation systems were created to control explosive energy release in quantities and in a sequence that ensures productivity and safety, timing the multiple explosive charges to control breakage, rock movement, and adverse environmental effects such as vibration, noise (air blast), and flyrock. Several detonator categories are available, with different physical properties, performance characteristics, time precision (from

Fig. 13.6 Various cast booster sizes and shapes [8]

milliseconds to seconds), and application uses, making the selection and correct employ of these components a key factor for the success of these components the detonation. The most common detonators are electronic, shock tubes, electric and detonating cords [8].

The electronic system provides precise and accurate timing, allowing significant improvement in blasting results in fragmentation, vibration, and air blast compared with former methods. In addition, the possibility of simultaneous initiation of two or more primers in a single column has improved fragmentation due to collision shock waves. Despite the many benefits, some limitations are the purchase cost; the communication within the wired system, which can cause failure in command due to static electricity, stray current or current leakage; and the time required for testing, programming, arming and firing procedures, which may exceed the expected, demanding high battery levels. In addition, troubleshooting can be complicated and require extensive knowledge of electric circuits and electronics to fix it. Shock and pressure sensitivity may be an issue too. Figure 13.7 presents typical electronic system components [8].

The shock tube system is one of the most widely used systems globally due to its ease of use, durable construction, intuitive layout, and electrical insensitivity, preventing premature activation by extraneous electrical energy even in high static or high stray current environments. At the same time, the tubes are also resistant to load and fire. The principal deficiency of non-electric systems is the lack of circuit test capability, which makes the blaster dependent on the manufacturer's quality control to ensure proper and complete filling of the shock tubes to avoid dispersion of the firing times of the system. When it occurs, a borehole charged behind another

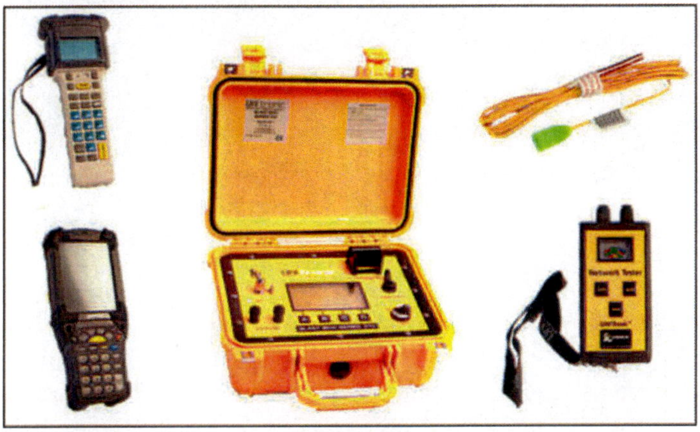

Fig. 13.7 Typical electronic system components [8]

borehole can be activated first, which would cause damage to other tubes and the rock mass. Figure 13.8 presents a shock tube detonator with a J-connector [8].

Electrical initiation systems use an electrical power source to initiate detonators connected to circuits, where the voltage level is sufficient to generate ample current to ignite the detonators. Provided with delayed powders inside detonators, this type of system became dominant from the 1940s to the 1970s, when the non-electric shock tubes appeared. The main benefit of electric detonators is the possibility of thoroughly testing the circuit continuity and resistance before initiation. This step was not possible until its popularization, although this benefit is no longer unique since the appearance of electronic systems. Its susceptibility to initiation by stray current, static electricity, or direct electrical current is the primary limitation of these systems. Caution must also be taken to preclude any lead-in-lines or connecting wire from coming into contact with the suspended electrical lines due to the movement

Fig. 13.8 Shock tube detonator with J-connector [8]

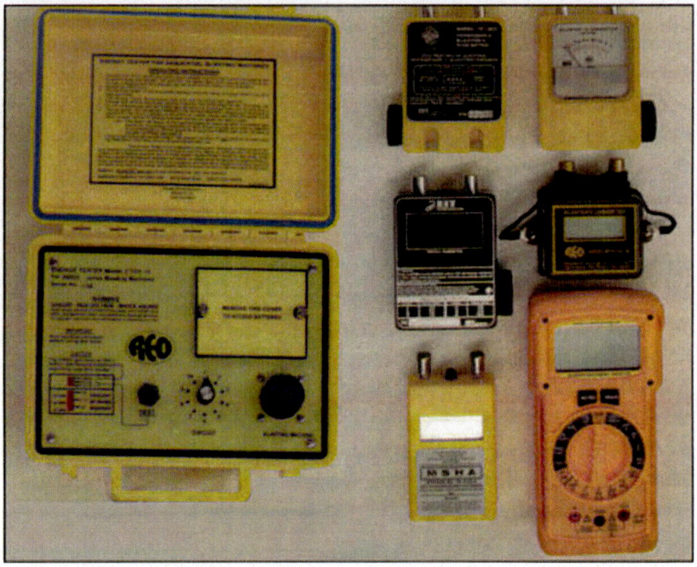

Fig. 13.9 Typical electric initiation system instruments [8]

of the ground during the blast, which could cause injury or death to the blaster due to current return. Figure 13.9 presents typical electric initiation system instruments [8].

The detonating cord is a flexible cord containing a center core of a high-velocity, detonator sensitivity explosive (generally PETN, pentaerythritol tetranitrate) used to detonate other high explosives by contact or transmit the detonation signal to another cord or a shock tube. It is relatively resistant to accidental detonation from impact, shock, friction, or extraneous electricity, but since detonating cords are high explosives, it has been detonated accidentally on rare occasions. When used in conjunction with millisecond connectors, delays between rows and boreholes are achieved, but with less precision and variety of times than other systems. The ability to fire row by row with boreholes firing very close together provides excellent movement and muck pile loosening, presenting excellent pre-splitting results. A disadvantage of detonating cord downline is that its explosion adversely affects the main charge of explosives in the borehole. Noise generation is a limiting factor in its widespread use, mainly due to exposed detonating cords, which create an air blast. The ability to donate initiation to another cord through a knot may be reduced due to inappropriate or loose knots. Detonating cords used to make primers can cause low-order detonations of ANFO products in small diameter holes, and the cord cores loaded with PETN are susceptible to reduced sensitivity when wet. Figure 13.10 presents various detonating cords and connectors [8].

Finally, one recent improvement in initiation systems is the wireless system developed by Orica, the WebGen, which includes wireless in-hole primers initiated by a firing command that communicates through rock, water, and air via ultra-low

Fig. 13.10 Various detonating cords and connectors [8]

frequency signals called electromagnetic induction. The blasting sequence is stored in the primer during encoding, which is performed when charging the blast. All primers require the correct unique signal to "wake up" and then the correct signal to fire. Figure 13.11 presents typical wireless initiation system instruments [9].

Fig. 13.11 Typical wireless initiation system instruments [9]

Rock Fragmentation Modeling: Empirical and Advanced Analytical Approaches

A fragmentation model may be defined as a set of equations that define (i) the position and the shape of the fragment size distribution and (ii) how these parameters are influenced by factors like the amount of explosive used, explosive strength, blasting geometry, and rock properties. This dependence is usually of the multiplicative form, e.g., for the median fragment size, x_{50}, such as the Eq. (13.2)

$$x_{50} = f_1(\text{rock}) \times f_2(\text{geometry}) \times f_3(\text{explosives}) \tag{13.2}$$

where f_1, f_2, and f_3 denote data functions that describe the argument [10].

Before the advent of image analysis software, the fragment size distribution was formerly determined by the mass weights of the remaining material sieve decks, followed by creating a histogram and a cumulative distribution function (CDF), which could be smoothed by polynomial interpolation. The CDF or mass passing function P has the form of a continuous, piecewise linear function which is defined over the interval $0 \leq x \leq x_{\max}$, whose most widely used CDF is the Rosin–Rammler (RR) distribution, which may be written as the Eq. (13.3)

$$P_{\text{RR}}(x) = 1 - e^{-(x/xc)n} = 1 - 2^{-(x/x50)n} = 1 - 5^{-(x/x80)n} \tag{13.3}$$

where x_c is the characteristic size, while x_{50} and x_{80} are the 50% and 80% passing size, respectively.

Figure 13.12 presents large-size manufactured sieves (up) and an example of the methodology employed to take pictures of the muck pile (down) for further image analysis, where basketballs serve as a size reference, both approaches aiming to describe the fragment size distribution [10–14].

Koshelev et al. (1971) proposed Eq. (13.4).

$$x_c \approx \{x\}10Q^{1/6}(V_0/Q)^{0.8} \tag{13.4}$$

where $\{x\}$ is the mean fragment size, Q is the amount of TNT explosive in the blast hole (kg), and V_0 is the blasted volume (m^3), which implies $q = Q/V_0$ is the specific charge (kg/m^3). After applying this proposition to different materials, Kuznetsov (1973) adapted it to insert a rock factor based on Protodyakonov's (1962) stamp hardness f of the blasted rock, resulting in Eq. (13.5).

$$\{x\} \approx AQ^{1/6}(V_0/Q)^{0.8}\theta^{-2/3} \tag{13.5}$$

where θ means the TNT equivalent of charge, Q is the amount of actual explosive used (kg), and A is the rock factor, which can assume the values 7 for medium hard rocks, 10 for hard and high fissured rocks, or 13 for very hard and weakly fissured rock [10, 15–17].

Fig. 13.12 Large-size manufactured sieves (up) and an example of the methodology used to take pictures of the muck pile (down) for further image analysis [13, 14]

There were some other propositions of size distribution models before it, but the most widespread (until today!) was the Kuz-Ram model created by Cunningham (1983), who improved the former works to obtain the average size of the fragment x and the uniformity index n for the muck pile whose sieving curve follows the RR function. The proposed equations were

$$\begin{aligned} x &= \left[AQ^{1/6}(115/E)^{19/30}\right]/q^{0.8}(\text{in cm}) \text{ and} \\ n &= (2.2 - 14B/d)(1 - W/B)[1 + (S/B - 1)/2]L/H \end{aligned} \tag{13.6}$$

where $E = \text{RWS} = S_{\text{ANFO}}$ is the relative weight strength of the explosive concerning ANFO (%), A is given by the criteria of Protodyakonov (1962), as proposed by Kuznetsov (1973), d is the diameter of the blast hole (mm), W is the standard deviation of drilling accuracy (m), L is the charge length above grade level (m), B and S are the burden and the spacing of the blast holes, respectively (m), and H is the bench height (m) [10, 16–18].

Spathis (2004) identified an inconsistency in Cunningham's (1983) proposition since he considered x^{-} equal to x_{50}. This flaw, however, ceases to be valid when the revised model proposed in Cunningham (1987) modified the rock factor A to adopt Lilly's (1986) blast ability index, where

$$A = 0.06(\text{RMD} + \text{JF} + \text{RDI} + \text{HF}) \tag{13.7}$$

making it possible to predict the entire sieving curve for a wider range of rock conditions than before. RMD describes the rock mass, assuming 10 if the rock is powdery or friable, JF value if there are vertical joints, or 50 if it is a massive rock. JF is the joint factor, calculated by

$$\mathrm{JF} = \mathrm{JPS} + \mathrm{JPA} \tag{13.8}$$

where JPS and JPA are the joint plane spacing and the angle parts, respectively. JPS is 10 if S_j (average joint spacing (m)) is less than 0.1 m, 20 if S_j has a value between 0.1 and x_O (the defined oversize (m)), and 50 if S_j is greater than x_O. JPA, in turn, assumes a value of 20 if the dip is out of the face, 30 if the strike is perpendicular to the face, and 40 if it dips into the face. RDI is the rock density influence, calculated by the equation $\mathrm{RDI} = 25\rho - 50$, where ρ is the rock density (t/m^3). Finally, HF stands for hardness factor and can assume the value of $E/3$ if $E < 50$ GPa or $\sigma_c/5$ if $E > 50$ GPa, with σ_c being the uniaxial compressive strength (MPa). Cunningham (1983) also expanded the uniformity index n, which became

$$n = (2.2 - 14B/d)(1 - W/B)\sqrt{(1 + S/B)/2}$$
$$(|L_b - L_c|/L_{\text{tot}} + 0.1)^{0.1} L_{\text{tot}}/H \tag{13.9}$$

where L_b and L_c denote the bottom and column charge lengths (m), respectively, and $L_{\text{tot}} = L_b + L_c$ [10, 19–21].

Despite their flaws and limitations, these propositions were the basis for fragmentation modeling, motivating decades of research after them. Stagg et al. (1990) and Otterness et al. (1991), from the United States Bureau of Mines (USBM), concluded that there appear to be two zones of breakage, with the finer material generated mainly around the borehole, and proposed a RR function for finer material and a standard normal distribution for coarser material. Djordjevic (1999), Kanchibotla et al. (1999), and Thornton et al. (2001), from the Julius Kruttschnitt Mineral Research Centre (JKMRC), created the crush zone model (CZM) and the two-component model (TCM), which were developed to overcome the frequent underestimation of the Kuz-Ram model concerning the number of fines in muck piles, both of which are special sub-models for generating fines from a circular compressive failure or crushed zone around the blast hole, while Kuz-Ram model should still model the coarse material. Figure 13.13 presents a CZM crushed zone concept (left) and a star-shaped breakage zone (right) [10, 22–26].

When analyzing the Less Fines Project data, researchers realized that the so-called Swebrec fragment size distribution fitted sieved fragmentation data better than the RR function. The transformation $P(x) = 1/[1 + f(x)]$ with $f(x_{\max}) = 0, f(x_{50}) = 1$ ensures that $x_{\max}$ and x_{50} are fixed points on the curve and a suitable choice for $F(x)$ is

$$F(x) = [\ln(x_{\max}/x)/\ln(x_{\max}/x_{50})]^b \tag{13.10}$$

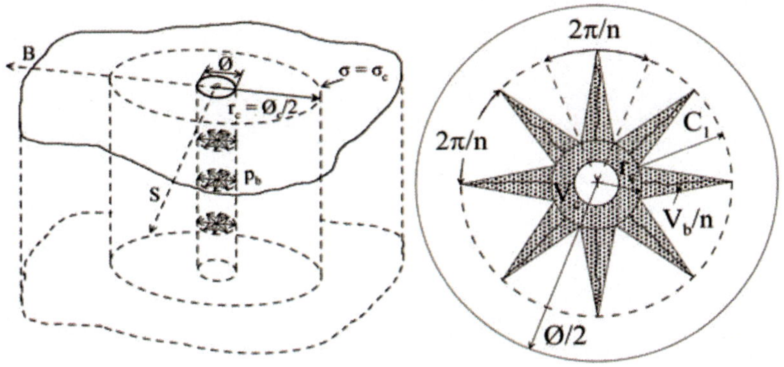

Fig. 13.13 CZM crushed zone concept (left) and star-shaped breakage zone (right) [10]

where b is the undulation exponent or natural breakage characteristics exponent. These results motivated the proposition of replacing the RR distribution in the Kuz-Ram model by Ouchterlony (2005), who called the resulting equation as the Kuznetsov–Cunningham–Ouchterlony (KCO) model, with median fragment size equals to

$$X_{50} = g(n)AQ^{1/6}(115/\text{RWS})^{19/30}/q^{0.8} \tag{13.11}$$

with x_{50} in mm and $g(n)$ being the possible mean to median conversion factor, later established as 1. This model was improved and calibrated in several latter works, where b was defined as

$$b = 0.4(B_{\text{ref}}/B)^{0.25}x_{50}^{0.25}\ln(x_{\max}/x_{50}) \tag{13.12}$$

where B_{ref} is a reference burden of about 4 m, although it can be tuned [10, 27–36].

Cunningham (2005) presented the third generation of his Kuz-Ram models. He removes some deficiencies of the previous models and considers new blast parameters, such as initiation and delay, making it possible to predict the effect of using electronic delay detonators instead of pyrotechnic ones. The RR function is still used, and there are explicit calibration factors for application in different blasting locations. The equation for the mean particle size x_m, specified this time as the median fragment size x_{50}, is calculated as

$$x_{50} = A_t A\left[Q^{1/6}(115/\text{RWS})^{19/30}/q^{0.8}\right]C(A) \tag{13.13}$$

where $C(A)$ is a multiplicative factor (generally between 0.5 and 2). The (delay) timing factor A_t is given by

$$0.66\tau^3 - 0.13\tau^2 - 1.58\tau + 2.1 \quad \text{if} \quad \tau \le 1 \tag{13.14}$$

or

$$0.9 + 0.1(\tau - 1) \tag{13.15}$$

otherwise, being $\tau = \Delta T/T_{\max}$, where ΔT is the (nominal) in-row delay (ms) and $T_{\max} = 15.6\ B/c_p$, where c_p is the P wave velocity (m/ms). The uniformity index n, in turn, is given by

$$n = n_s\sqrt{}(2 - 30B/d)\sqrt{}[(1 + S/B)/2]$$
$$(1 - W/B)(L/H)^{0.3}(A/6)^{0.3}C(n) \tag{13.16}$$

where $C(A)$ is a multiplicative factor. n_s is a factor that depends on the dispersion or standard deviation s_t of the in-row delay time ΔT, expressed using the scatter ratio R_s, where

$$n_s = 0.206 + (1 - Rs/4)^{0.8} \tag{13.17}$$

with $R_s = 6\ s_t/\Delta T$. Rock factor A is given by

$$A = 0.06(\mathrm{RMD} + \mathrm{RDI} + \mathrm{HF}) \tag{13.18}$$

without the term JF, although RMD, RDI, and HF remain the same. The joint factor JF used to calculate RMD is redefined using a joint condition factor JCF, being calculated as

$$\mathrm{JF} = (\mathrm{JCF})(\mathrm{JPS}) + \mathrm{JPA} \tag{13.19}$$

where JCF assumes the value 1 for tight, 1.5 for relaxed, and 2 for gouge-filled joints. JPS is called the vertical joint plane spacing factor and is related to the joint spacing S_j (m) and to the "reduced pattern" $\sqrt{}$BS (m), assuming value 10 when $S_j < 0.1$ m, 20 when 0.1 m $< S_j <$ 0.3 m, 80 when 0.3 m $< S_j < 0.95\sqrt{}$BS m, and 50 when $S_j > \sqrt{}$BS. In turn, JPA assumes 40 if the dip is out of the face, 30 if the strike is perpendicular to the face, and 20 if it dips into the face, which is the reverse of the previous proposition. After testing the effects of using 25 ms pyrotechnic delays or 9 ms electronic delays, Cunningham (2005) found that shorter intervals decreased the average size x_{50} while eliminating dispersion improved uniformity n (¼ steeper sieving curve), resulting in the elimination of material with an excess dimension of 1 m and a significant decrease in fine material with a dimension of 1 mm [10, 37].

Sanchidrian and Ouchterlony (2017a, b), based on the discovery of the fragmentation-energy fan and the previous dimensional analysis (DA), developed a basic dimensionless prediction equation for the fragment size x_P for an arbitrary percentage value P, called the x_P-frag model. Every fragment size is given by

$$x_p = L_c k\left[\min\left(s_j/L_j, a_s\right) + a_o j_o\right] k_2^h \left(\sigma^-/qeL_c^\lambda\right)^k f_t(\Pi_t) \tag{13.20}$$

with $\sigma^- = {\sigma_c}^2/(2E)$, where $L_c = \sqrt{HS}$ is a characteristic size, $k_2 = B/(\sqrt{HS}\cos\theta_i)$ is a bench shape factor, in which θ_i is the inclination angle to the vertical of the blast holes. σ^- is the stored elastic energy at compressive failure, e is the energy of the explosive per unit mass (J/kg), and the parameters k, h, k, and λ can be determined by experimental data. For the influence of jointing, the classical approach was used, being $J_F = J_S + J_o$, where J_S is a non-dimensional ratio s_j/L_j, in which s_j is the mean discontinuity spacing and L_j is a characteristic length, capped by a limiting value as for large joint spacings. Jo is defined as $a_o j_o$ in which j_o is Lilly's (1986) joints orientation index normalized to one, being j_o 0.25 if horizontal, 0.5 when dipping out of the face, 0.75 if sub-vertical striking normal to the face, and 1 when dipping into the face or no visible jointing. The dimensionless delay factor Π_t is equal to $c_p \Delta t/L_t$, where Δt is the interhole delay and c_p is the P-wave speed. The delay time correction factor function $f_t(\Pi_t)$, in turn, is given by

$$\delta 1 + (1 - \delta_1 - \delta_2 \Pi_t)e^{-\delta 3 \Pi t} \tag{13.21}$$

where δ_1, δ_2, and δ_3 are P-dependent constants to be determined from the fitting. The authors confirm that the average median expected error of the percentile size prediction for x_P-frag is about 20%, whereas the corresponding number for the Kuz-Ram model and the CZM is about 60%, as Fig. 13.14 presents the boxplot of distributions

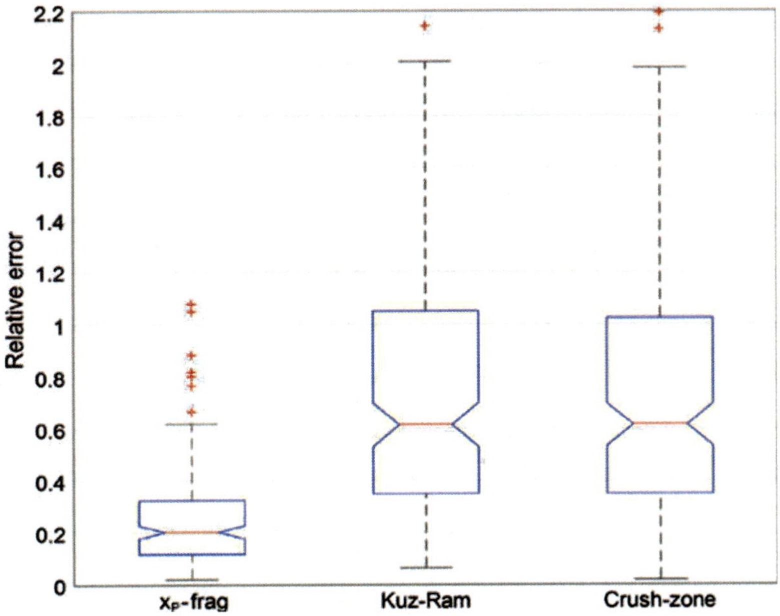

Fig. 13.14 Boxplot of distributions of the medians of the absolute errors for all percentiles of each dataset for xP-frag, Kuz-Ram, and CZM models [10]

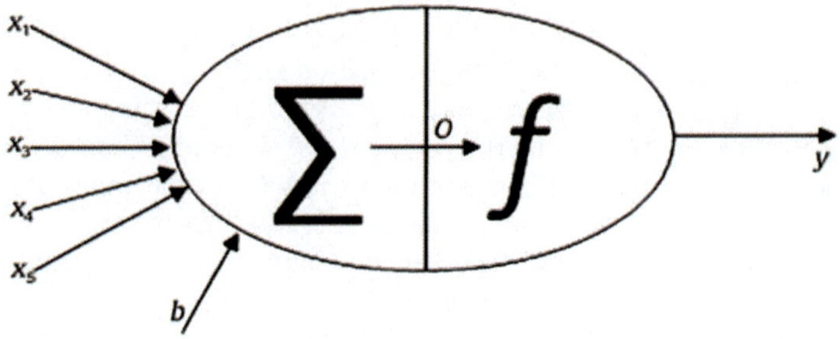

Fig. 13.15 A neuron structure with inputs, bias, transfer function, threshold, and output [40]

of the medians of the absolute errors for all percentiles of each dataset [10, 21, 38, 39].

Besides these mathematical-empirical models presented, advanced analytics techniques, such as artificial neural networks (ANN) and machine learning, combined with metaheuristics or optimization methods, increase the possibilities of analyzing rock breakage using explosives. An ANN is a computer system inspired by the functioning of the neural network, encompassing a series of interconnected layers of nodes (neurons) that have the ability to map input vectors and generate an output target from this information. The internal connections of neurons are called synaptic weights, which are adjusted as the learning process proceeds through sample analysis until the entire network reaches generalizability to predict the results for other samples. A single artificial neuron interferes with the signal travel through the hidden layers according to the received signal weights (a nonlinear function of the sum of the inputs) and as long as a threshold is surpassed for the selected transfer function. This technique's capabilities are generalizing and transforming independent variables into dependent ones, calculating arithmetic and logical functions, nonlinear processing, parallel computing, imprecise or fuzzy information management, pattern recognition, and function approximation. Figure 13.15 presents an example of neuron structure with inputs, bias, transfer function, threshold, and output [40].

Monjezi, Rezaei, and Yazdian Varjani (2009) employed a fuzzy inference system (FIS) to predict rock fragmentation from blasts in an iron mine in Iran, comparing its result to that obtained by multivariate regression analysis (MVRA). Fuzzy modeling is a flexible and robust artificial intelligence subsystem that uses fuzzy logic to process inaccurate information, choosing a membership function that determines whether an element belongs or not to a given set, thus receiving a membership value between 0 and 1, a procedure called fuzzification. Linear and nonlinear shapes can be used as membership functions, and rule-based models (fuzzy "if–then" rules) capable of combining numerical data and specialized knowledge must be created to be applied in a FIS. Despite the database determining the membership functions and the rule base, FIS also requires a mechanism to aggregate these individual rules, followed by a defuzzification process to achieve representative values from the fuzzy set. A total of 300 "if–then" rules, an aggregation mechanism based on the Mamdani algorithm, and the centroid of area defuzzification method were employed in this case study. On the other hand, the MVRA is utilized to define the relationship between distinct variables (dependent or independent ones) in order to achieve predictive variables based on inputs. A dataset with 415 blasts containing information about rock mass properties, explosive properties (only ANFO was utilized), and blast geometry was used, with 115 instances reserved for model testing. Fragmentation quality was defined in terms of 80% passing size (D_{80}), assessed by the image analysis method. The coefficient of determination (R^2) and root mean square error (RMSE) indices for the MVRA were 0.80 and 6.83, respectively, while these indices for the fuzzy model were 0.96 and 3.26, respectively, corroborating the robustness of the latter. Kulatilake et al. (2010) employed an MVRA and a backpropagation neural network (BPNN) with a single hidden layer to a blasting database gathered by Hudaverdi, Kulatilake, and Kuzu (2011) to foresee the mean particle size of rock blasts. This database consists of information on 91 blasts worldwide using ANFO, covering blast design and explosive parameters, in situ block size, and modulus of elasticity. A hierarchical cluster analysis divided the dataset into two concerning intact rock stiffness, the last parameter. Considering approximately 10% of the data for testing in each cluster, the mean square error (MSE) and the training cycles were evaluated for four training methods: BFGS Quasi-Newton (BFGSQN), Levenberg–Marquardt (LM), gradient descent (GD), and gradient descent momentum (GDM) algorithms, with the LM algorithm providing more stability and learning speed. On the other hand, the multiple regression analysis provided a prediction equation for each cluster, reaching an R^2 of 0.9407 for the ANN approach, 0.8197 for the multivariate regression, and 0.5698 for the Kuznetsov's equation for both clusters, corroborating that the proposed methods are more efficient than one of the methodologies most used by the industry. Figure 13.16 presents the FIS procedure [14, 16, 41, 42].

Monjezi, Bahrami, and Yazdian Varjani (2010) proposed a BPNN to predict rock fragmentation and blast-induced flyrocks in an iron ore mine in Iran simultaneously, employing a dataset with 250 instances from which 10% were set aside for testing. For this purpose, the quality of the fragmentation was defined in terms of 80% passing size (D_{80}), assessed by the image analysis method, and the flyrock was verified using a measuring tape. Several ANN architecture propositions were evaluated, and it was

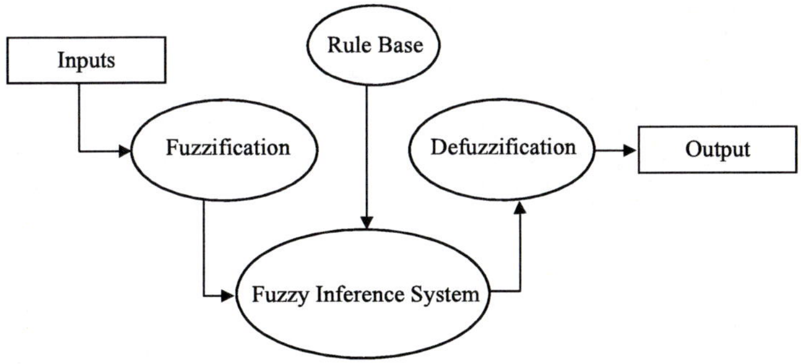

Fig. 13.16 Fuzzy inference system (FIS) procedure [14]

verified that an 8-3-3-2 architecture achieved better results in RMSE, mean absolute error (MAE), and mean relative error (Er). R^2 equal to 0.9785 and 0.9883 were obtained for fragmentation and flyrock, respectively, reiterating the robustness of the proposed solution. A sensitivity analysis using relative strength of effects (RSE) was performed, identifying that the charge per delay, burden to spacing ratio, rock density, and the number of blasting rows are, respectively, the most influential parameters on fragmentation, while hole diameter, stemming, charge per delay and powder factor is the most influential on flyrocks. Taking these results into account in further blasts reduced the flyrock distance from 110 to 30 m while also allowing for improvements in terms of fragmentation. Bahrami et al. (2011) employed a BPNN to address the fragmentation prediction at an iron ore mine in Iran, comparing the results to those obtained by MVRA. Several architectures were compared, and a 10-9-7-1 structure was selected from analyzing their root mean square of error values. The dataset consists of 220 instances of blasting operations at that mine, and 10% of the data has been set aside for testing. The root means square of error and the R^2 were obtained for the MVRA and BPNN techniques, reaching respectively 0.701 and 1.958 for the MVRA, and 0.97 and 0.56 for the ANN, proving that the latter is better predicted the fragmentation. A sensitivity analysis was also develop using the cosine amplitude method (CAM), unraveling that the main parameters in descending order among the ten selected as input are blast ability index, the charge per delay, burden, SMR, and powder factor. Figure 13.17 presents the structure of the BPNN proposed by Monjezi, Bahrami, and Yazdian Varjani (2010) [43, 44].

Shi et al. (2012) employed a support vector machine (SVM) to foresee the mean particle size of rocks from blasts, comparing the results to those obtained by MVRA, ANN, and Kuznetsov's proposition. Support vector regression (SVR) traces the data nonlinearly in a space with multiple dimensions in convex quadratic programming (QP) problems. It thus solves a linear regression problem in this space, is obtained

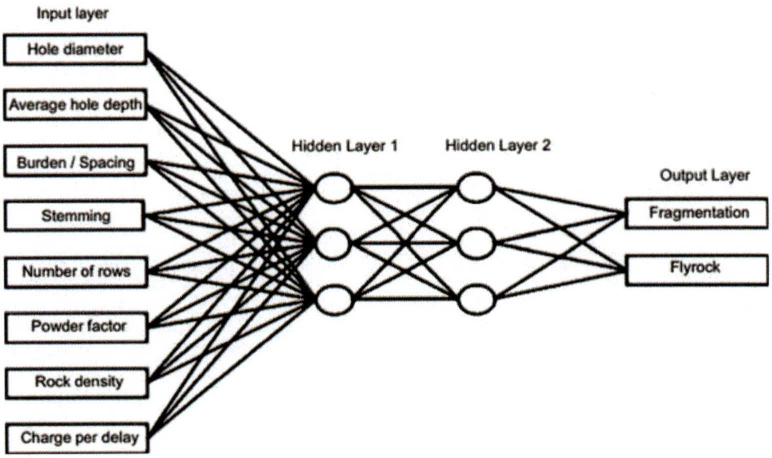

Fig. 13.17 Structure of the backpropagation neural network (BPNN) [43]

in this study the optimal parameters of the model through K-fold cross-validation and a grid search method (GSM) approach. Considering seven input parameters and a database with 102 rock blasting instances from several mines worldwide, most obtained from Hudaverdi, Kulatilake, and Kuzu (2011), 12 of these instances were employed for validation. The Gaussian kernel function was selected to train the samples in the SVR, and after 90 training sets, a mean square error of 0.0144 and a square correlation coefficient of 0.9263 were obtained. When comparing the parameters R^2, MAE, and RMSE of the results, Kuznetsov's proposition reached 0.614, 0.344, and 0.025, respectively; the MVRA reached 0.815, 0.170, and 0.017; the ANN reached 0.941, 0.107, and 0.009; and the SVM, finally, reached the values 0.962, 0.066 and 0.006, respectively, for the test data. The indexes showed that the SVR approach resulted in more robust results than the other methodologies, predicting values less dispersed and close to reality. Gao and Fu (2013) applied a BPNN to predict the rock size distribution from blasts in a mining face in a tantalum-niobium mine in China, using a dataset with 14 events, from which 11 were employed for training and 3 to test the model. The ANN has 10 nodes in the input layer (one for each influence factor regarded), 16 nodes in the hidden layer, and 4 in the output layer, returning passing percentages it certain thresholds, with a 10-16-4 architecture. The residual fitting error was 0.001 for the epoch 54,393, resulting in a robust prediction, even to the few data evaluated, reaching 1.32% of average relative error. Figure 13.18 presents an example of hyper-plane classification using SVM [16, 41, 45–47].

Sayadi et al. (2013) employed an MVRA, BPNN, and radial basis function neural network (RBFNN) to predict rock fragmentation and back break in a set of limestone mines in Iran, using a dataset with 103 instances of blasts from these mines, containing 6 parameters as input and setting aside 10% of instances for testing. A specific MVRA model was developed for each problem, and all parameters were considered in the formulations. Unlike BPNN, the RBFNN structure has a unique hidden layer, reducing the computation time and requiring a Gaussian transfer function to achieve its results. After some trial-and-error procedures, it was found that the 6-10-2 and 6-36-2 architectures were better in terms of minimizing error and maximizing the accuracy of BPNN and RBFNN, respectively, with a spreading factor of 0.79 being used in the latter. The quality of the fragmentation also was defined in terms of 80% passing size (D_{80}). The approaches were compared in terms of R^2, RMSE, value account for (VAF), and maximum relative error, and the results

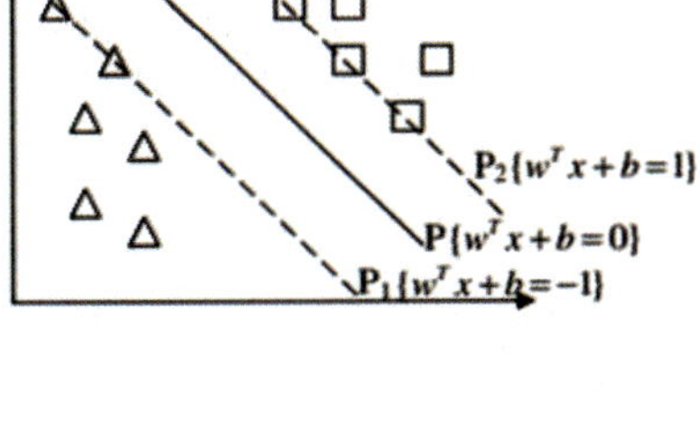

Fig. 13.18 Example of hyper-plane classification using support vector machine (SVM) [46]

showed BPNN achieved better results in both problems than the other methodologies. A sensitivity analysis was performed using the CAM, reaching that both stemming and burden are the most influential parameters on fragmentation and back break. Shi et al. (2013) compared the results of an MVRA, a regular BPNN, and an ANN combined with a genetic algorithm (GA) in predicting the mean particle size of rocks from blasts. The GA approach is an evolutionary metaheuristic inspired by processes such as inheritance, crossover, and mutation, which can be observed in natural selection, aiming to achieve a robust output for predicting or adapting patterns in complex problems, although not guaranteeing optimality. Therefore, an initial population of chromosomes (strings of fixed length) is randomly defined. Those with best-fitting values are more likely to remain in the next generation than others, even though crossover and mutation procedures are generally implemented to maintain generality and avoid local optima. The ANN architecture was 7-4-1, addressing the seven defined parameters, while the GD algorithm is used to achieve the minimum error. The initial weights of the ANN-GA are chosen randomly for a given range and written as a code string. The fitness function is updated using roulette wheel operations, in which the genes with high fitness are more likely to be chosen. Genetic crossovers and variations are employed to preserve population diversity. The dataset consists of 20 blasting information instances, divided into two groups of 10, the first using 8 instances for training, while the second using 7 instances. The mean relative errors of MVRA, BPNN, and ANN-GA were, respectively, 11.15%, 7.38%, and 3.09%, proving the robustness of the proposed solution. Figure 13.19 presents an example of a GA flowchart [40, 48, 49].

Karami and Afiuni-Zadeh (2013) compared the performances of an RBFNN and an adaptive neuro-fuzzy inference system (ANFIS) to predict the rock fragmentation from blasts in an iron mine in Iran. ANFIS consists of a FIS implemented on an ANN architecture, taking advantage of both FIS's flexibility and ability to work with ill-defined systems and ANN's learning procedures. An ANFIS system with 5 layers and 2 inputs was created, as shown in Fig. 13.20, in which the circles depict nodes with fixed parameters, while the squares depict nodes with adaptive parameters, which must be determined by training. The first layer's nodes are fuzzy sets in fuzzy

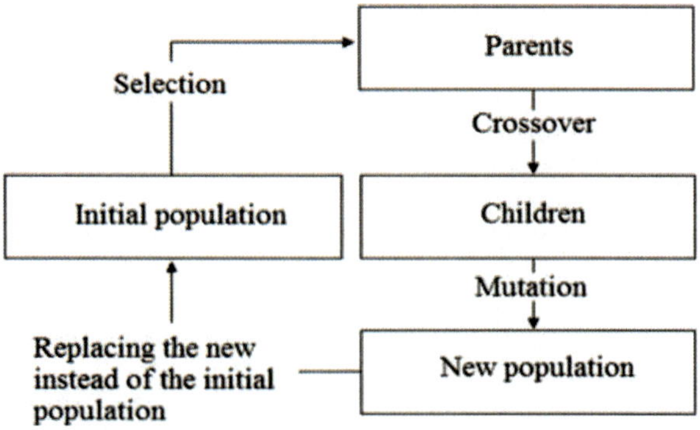

Fig. 13.19 Example of a genetic algorithm (GA) flowchart [48]

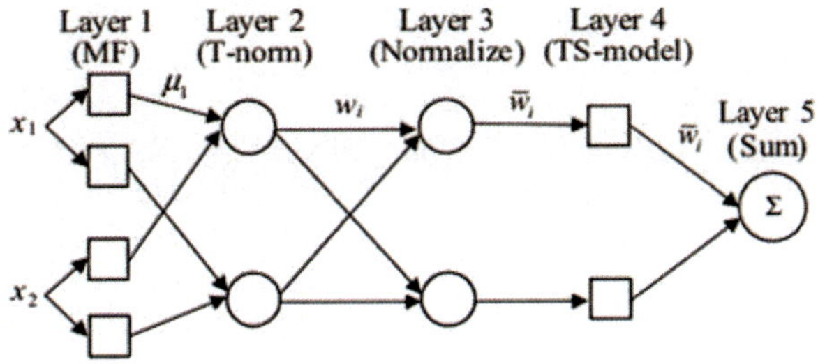

Fig. 13.20 Adaptive neuro-fuzzy inference system (ANFIS) architecture [50]

rules and contain the premise parameters, while the second and third layers calculate the product of their inputs and normalize these products. The fourth layer's adaptive nodes are first-order Takagi–Sugeno (TS) FIS models containing consequent parameters, while the last layer sums the inputs from the fourth layer to calculate the overall output. The ANFIS model can be trained through the least squares estimation, determining the consequent parameters for the fixed parameters assumed in the first layer, thus fixing the parameters in the fourth layer, and using backpropagation to adjust the first layer's premise parameters. A dataset with 30 blast instances containing 7 parameters was employed, and the fragmentation quality was defined in terms of 80% passing size (D_{80}), obtained through an image processing system. Distinct ANFIS models were built by applying distinct combinations of the seven input parameters, and both approaches were subjected to different train–test strategies, where the best ANFIS reached 0.882 and 2.851 for R^2 and RMSE, respectively, and the best RBFNN reached 0.746 and 3.391 for the same indexes, proving the robustness of the former. Enayatollahi, Bazzazi, and Asadi (2014) developed a case study on large iron ore in Iran in which an ANN, linear and nonlinear MVRA, and the Kuz-Ram approaches were assessed in terms of their ability to predict the rock fragmentation from blasts. The dataset comprises 70 instances of mine blasts, containing 12 input information and the result of fragmentation in each instance, of which 10 instances were separated for testing. A trial-and-error methodology was applied to define the ANN architecture by evaluating the RMSE, reaching the 12-15-11-1 architecture as the best. Multiple linear regression included five parameters in the equation (i.e., specific drilling, hole depth, water depth, powder factor, and tensile strength), considering the coefficient of correlation achieved, while multiple nonlinear regression included six parameters (i.e., specific drilling, hole depth, powder factor, water depth, stemming, and burden) for the same reason. The results achieved by these three methodologies were compared with those provided by a commercial software that uses the Kuz-Ram model and, in addition to the R^2, RMSE, bias, standard error of prediction (SEP), accuracy factor, and the Nash–Sutcliffe coefficient of efficiency were employed for comparison. The verified values corroborate the superiority of the ANN compared to other approaches to predict blast fragmentation. A CAM was also employed to identify the most sensitive factors, with the stemming being the most influential; thus, for a new blast, the stemming was adjusted, and the resulting fragmentation was evaluated, showing that the results improved. Figure 13.20 presents the ANFIS architecture proposed by Karami and Afiuni-Zadeh (2013) [18, 20, 50, 51].

Bakhtavar, Khoshrou, and Badroddin (2015) employed multidimensional regression analysis (MDRA) and nonlinear MVRA to foresee the mean particle size of blasting operations in a copper mine in Iran, comparing the results with a traditional sieve analysis, image processing, Kuz-Ram, and modified Kuz-Ram model proposed by Gheibie et al. (2009). The DA technique restructures a series of dimensionless products out of the dimensional variables of the problem, and the mass and force systems are generally used, in which the first considers mass (M), length (L), and time (T) as basic units, and the latter considers length (L) and time (T). A dataset containing detonation information considering 10 independent parameters and the mean particle size as a dependent parameter was employed. Once the force system was chosen, the units of these parameters should be expressed as an arrangement of the base units aforementioned. Thus, a dimensional matrix is arranged, the determinant is achieved, and Buckingham's π-theorem is considered to determine the number of dimensionless products of the variables and the number of homogeneous linear equations derived. Finally, since Buckingham's π-theorem suggests that a single equation relating the dependent and independent parameters can be defined, this equation can be rewritten in terms of the independent dimensionless products. Three samples were analyzed, and the MDRA achieved the best results compared to the real sieving, with predictions differing by 20.4, 6.8, and 9.1%.

On the other hand, the image processing differed 65.3, 22.4, and 27.4%; the Kuz-Ram model, 210.6, 102.3, and 174%; and the modified Kuz-Ram model, 289.3, 153.1, and 243.3%. Ebrahimi et al. (2016) applied an ANN to predict rock fragmentation and back break in a lead and zinc mine in Iran, in addition to an artificial bee colony (ABC) algorithm to optimize blasting pattern parameters. The ABC is a metaheuristic based on the behavior of the swarms of honeybees that does not guarantee optimality, being a colony separated into three classes: employees, onlookers, and explorers. The solution begins with the explorer bees looking for a food source on which the employees will work. While the onlooker bees continually seek new food sources in each iteration, the explorer bees communicate relying on the information about the quality of the food sources (amount of nectar) provided by the employees in the "dancing area," and the process continues until the optimization finds the best food source until a stopping criterion is reached. The dataset contains information from 34 blasts, and 30% of the instances were set aside for testing. The ANN was defined as a 5-5-4-2 architecture, and the quality of the fragmentation was defined in terms of 80% passing size (D_{80}), being evaluated by the R^2, RMSE, and VAF indices, reaching, respectively, 0.78, 2.76, and 77.73 for fragmentation and 0.77, 0.53, and 77.25 for back break analysis. In turn, the outputs of the ABC algorithm were evaluated by the Rastrigin function, providing a suggestion for the 5 blasting parameters considered to reach the optimal result of 29 cm and 3.25 m for fragmentation and back break, respectively. This proposition was also compared to commercial software that employs the Kuz-Ram model, which achieved a fragmentation result of 33.5 cm for the input defined by the ABC algorithm, a result similar to the previous 29 cm obtained. Figure 13.21 shows an example of the "dancing area" mechanism employed in the ABC algorithm [13, 18, 20, 52, 53].

Fig. 13.21 Example of the "dancing area" mechanism employed in the artificial bee colony algorithm (ABC) [52]

Als et al. (2018) proposed the use of an ANN and firefly algorithm (FFA) to simultaneously predict flyrock and rock fragmentation from blasts in a limestone mine in Iran, using a dataset with 200 instances of blasts that contain information from 8 independent parameters, of which 20% of the instances were used for testing purposes. The FFA is based on the social behavior of fireflies, whose flashes depict a way of attracting other fireflies. Hence, the attractiveness ($\beta(r)$) of a firefly is directly linked to its light intensity ($I(r)$), which is determined in the algorithm by a fitness function, and which decreases by increasing the square of the distance (r^2). Fragmentation quality was defined in terms of 80% passing size (D_{80}), assessed by the image analysis method. After distinct network architectures were tested, the 8-8-6-2 was chosen as the one with the best MSE, and the R^2, RMSE, and Er presented for fragmentation were 0.94, 0.1, and 26.12%, respectively, while for flyrock, they were 0.93, 0.09 and 7.2%, respectively. Hence, the FFA was employed to identify the best blasting pattern, indicating independent variable values that would reach 25.7 cm and 77.3 m for fragmentation and flyrock, respectively. Experiments were carried out, and the values measured in the new blasts were 29.8 and 91.2 m, while the average values in the mine were 38.3 and 135.3 m, corroborating the ability of the ANN-FFA methodology to result in gains of quality and safety.

Furthermore, ANN-FFA even predicted fragmentation better than the Kuz-Ram model and predicted flyrock better than models proposed in the literature, such as Lundborg (1975) and Gupta (1980) models. Finally, a sensitivity analysis using the CAM indicated that the GSI factor and burden are the most influential parameters in fragmentation and flyrock. Figure 13.22 presents parameter relationships in the firefly algorithm [18, 20, 54–56].

Hasanipanah et al. (2018a) proposed the combination of ANFIS and particle swarm optimization (PSO) to foresee rock fragmentation, comparing the results of this ANFIS-PSO approach with those obtained using SVM, ANFIS, and nonlinear MVRA models. PSO is an evolutionary metaheuristic in which each swarm (or particle) that represents a feasible solution in space is subject to a local search and perturbation mechanisms so that at each iteration, the swarms tend to converge to a

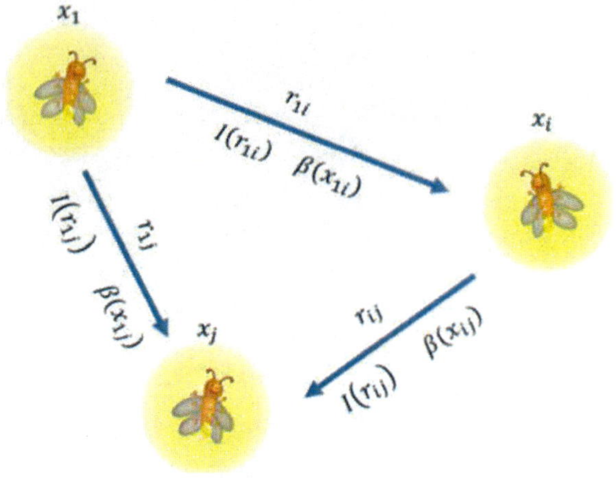

Fig. 13.22 Parameter relationships in the firefly algorithm (FFA) [57]

solution that results in the optimum fitness function. Several topologies can be implemented in a PSO algorithm, aiming for the best global (gbest) or best local (lbest) positions. The proposed ANFIS-PSO approach is similar to the ANFIS described in Karami and Afiuni-Zadeh (2013), but instead of a backpropagation mechanism to train the network, a PSO approach is employed. The dataset consists of 72 blasting instances, from which 14 instances were used for testing, and the five parameters considered as input are specific charge, spacing, burden, stemming, and maximum charge used per delay. The ANFIS-PSO approach reached 0.89 and 1.31 for R^2 and RMSE, respectively, while SVM reached 0.83 and 1.66, ANFIS reached 0.81 and 1.78, and nonlinear MVRA reached 0.57 and 3.93, proving the best quality solution from the first. Finally, the sensitivity analysis by the CAM indicated the maximum charge used per delay as the most relevant parameter and the burden as having the least effect on rock fragmentation. Figure 13.23 shows examples of particle topologies in PSO: (a) gbest star, (b) lbest ring, (c) von Neumann, and (d) four clusters topologies [50, 58].

Mojtahedi et al. (2019) compared ANFIS-FFA and MVRA solutions to predict the rock fragmentation from blasts in two quarry mines in Iran, considering a dataset with

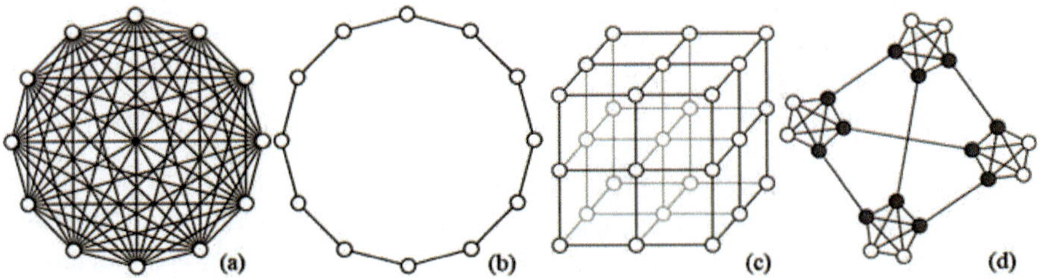

Fig. 13.23 Examples of particle topologies in particle swarm optimization (PSO): **a** gbest star, **b** lbest ring, **c** Von Neumann and **d** four clusters topologies [59]

72 instances, encompassing 5 independent variables and the dependent fragmentation variable, separating 14 instances for testing. Fragmentation quality was defined as 80% passing size (D_{80}) and was assessed by the image analysis method. An initial ANFIS network is first generated from the training data. Thus, the FFA stage updates its parameters by trial-and-error considering the RMSE index when it finally updates the premise and consequent parameters of the fuzzy rules in the ANFIS stage until a stopping criterion is reached, aiming to optimize the ANFIS-FFA model. In order to increase the performance of the ANFIS stage, six combinations of independent variables were assessed as the ANFIS input, in which the scenario that applied the 5 variables was the best. To evaluate the performance of the propositions, the R^2, VAF, RMSE, and Nash and Sutcliffe (NS) indexes were used, and the ANFIS-FFA approach reaches values of 0.980, 97.64, 0.52, and 0.976, respectively, while the MVRA reached values of 0.669, 60.96, 1.48, and 0.581, respectively, assuring the best accuracy of the former. Valencia et al. (2019) proposed a framework based on image processing and machine learning to identify deviations in the allocation of drill holes and propose, before the blast's execution, the charge's adjustment to guarantee an effective explosive energy distribution (EED). First, a drone flies over the bench at a small height and takes pictures of the drill hole pattern, which must be processed through photogrammetry software to build a digital elevation model. Then, several machine learning algorithms and parameters were tested to define those employed to automatically recognize the drill hole locations; the aggregate channel features (ACF) algorithm was selected for training with the 136 drill hole samples provided. Thus, the detected drill holes can be compared to their intended locations, while the case study presented up to a 1-m deviation from the plan. Hence, a block model of the bench is built, and a 4D dynamic calculation is applied to derive the energy concentration of each block. Finally, a binary optimization model based on the powder factor is developed to optimize the amount of charge in each drill hole subject to a cost function to penalize deviations from the average energy concentration. This problem can be solved by a metaheuristic algorithm, although none have been proposed. Figure 13.24 presents a scheme with the planned drill hole pattern, resulting in an ideal EED (a); the actual EED resulting from an error in the positioning of drill hole D6 (b); and the charge adjustment to balance the drilling error (c) [60, 61].

Yaghoobi et al. (2019) proposed an algorithm combining pattern recognition and machine learning techniques, exploring both supervised ANN and the features extraction method to describe the size distribution of fragments from detonations in an iron mine in Iran, using a dataset with 226 images from which 26 were used for testing. The algorithm's first steps are similar to those employed in commercial image processing software, including preprocessing, image enhancement, scales extraction, perspective elimination, and normalization. Hence, the images must be transformed into a reduced series of features (a feature vector) to diminish the processing time, using discrete Fourier transform, discrete wavelet transform, Gabor filters, and combinations. Thus, considering these feature vectors supervised ANN inputs, the values from F10 to F100 were determined as ANN targets, is obtained by applying the manual mode of the Split-Desktop software to the training images and considering

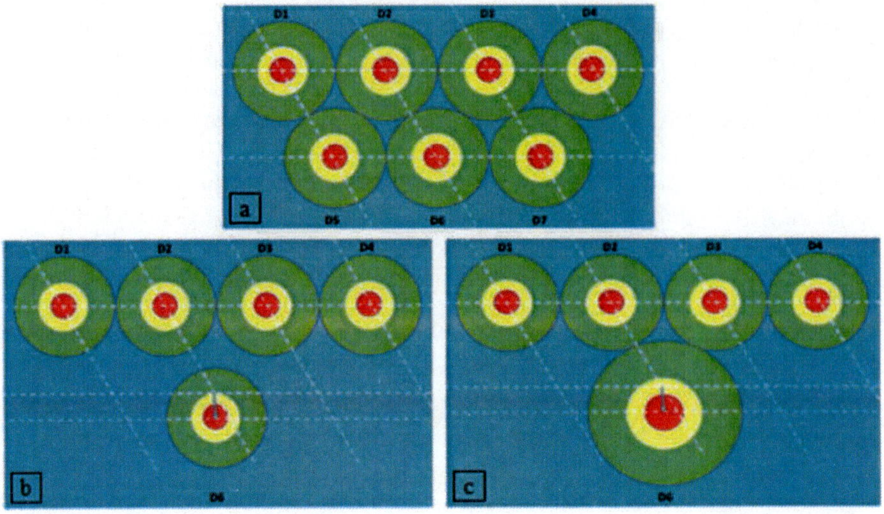

Fig. 13.24 Planned drill hole pattern, resulting in an ideal EED (**a**); the actual EED resulting from an error in the positioning of drill hole D6 (**b**); and the charge adjustment to balance the drilling error (**c**) [61]

the mean relative error (MRE) to compare feature extraction techniques. The ideal network architecture was tested in terms of RMSE, and a 40-15-10-10 architecture was chosen. To benchmark the predictions, the automatic mode of the Split-Desktop software was also used on the test images, and the results were compared in terms of MRE improvement between each technique, applying this index to each feature extraction technique. The techniques that resulted in more improvements from F10 to F100 were Fourier transform, Gabor filters, and discrete wavelet methods, reaching 67%, 57%, and 48%, respectively. In contrast, for fine to medium particles (F10 to F50), the Fourier transforms, Gabor filters, and Fourier–Gabor methods reached improvements of 52%, 40%, and 32%, respectively. For more information on size distribution in muck piles, see studies such as Luerkens (1986), Bootlinger and Kholus (1992), Lin, Ken, and Miller (1993), Barron, Smith, and Prisbrey (1994), Kemeny (1994), and Yen, Lin, and Miller (1998), in addition to other applications (e.g., classification) and locations (e.g., conveyor belt) analyzed in other studies using image processing techniques [62–68].

Before presenting the next studies, it is useful to explain some concepts and techniques. First, the k-nearest neighbors (k-NN) algorithm is a nonparametric classification or regression model that remembers the values and location of neighbors in the training dataset rather than learning from them when making predictions. In contrast, the optimal number of neighbors is its main parameter, which a grid search technique can determine. Second, principal component analysis (PCA) is a procedure that converts a series of possibly correlated variables into values of linearly uncorrelated variables using vector orthogonalization (provided the data is normally distributed) and can be determined by decomposing a covariance matrix in eigenvectors, whose larger eigenvalues better explain the variance of the data. Therefore, principal component regression (PCR) is a PCA-based technique in which the principal components with larger variances are usually selected as regressors, and the number of components employed in the regression is the main parameter for adjusting the algorithm. Finally, in a Gaussian process (GP), every finite set of random variables (since it is a stochastic process) or linear combination results in a multivariate normal distribution, deriving a joint distribution of all random variables over functions with a continuous domain. The definition of the covariance function implies the behavior of the process, as long as the stationarity, smoothness, isotropy, and periodicity of the GP are respected. Figure 13.25 presents (a) an example of a k-NN classifier with different values of k and (b) the principal components of a dataset [69].

Zhang et al. (2020a) proposed a combination of ant colony optimization (ACO). They boosted the regression tree (BRT) to foresee the size distribution of blasts in a quarry mine in Vietnam from a dataset with 136 instances, considering 6 independent variables as input and the dependent size distribution as output, while reserving 25 instances for testing. The ACO algorism is a metaheuristic inspired by the behavior of foraging ants, in which the ants in search of food sources deposit pheromones in their paths, which attracts other ants to those paths with more pheromone deposited, so the entire colony tends to converge on the shortest path (best feasible solution) as it receives more pheromone over time. The path's pheromone level is periodically reduced to reflect evaporation, and usually, a local search phase is employed to improve the final solution after the construction phase. On the other hand, the BRT framework employs two algorithms to achieve its solution, a regression tree technique based on decision trees, which do not demand transformation or removal of outliers, and a boosting technique that combines multiple decision trees to upgrade the accuracy of the model, thus generating an optimal regression tree. First, the feeding of the ACO-BRT model with the training dataset starts with an initial BRT.

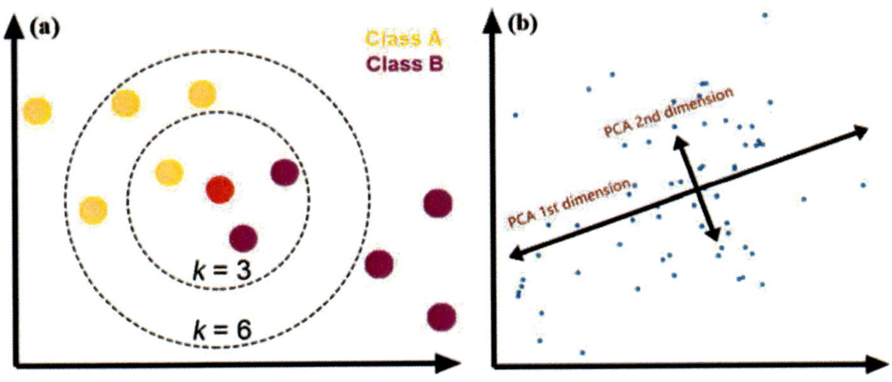

Fig. 13.25 **a** An example of a k-nearest neighbors classifier (k-NN) with different values of *k* and **b** the principal components of a dataset in a principal component analysis (PCA) [70, 71]

Thus the ACO hyper-parameters are adjusted by a trial-and-error procedure, and, after this definition, the hyper-parameters of the BRT phase are adjusted through the ACO phase until an RMSE target is reached, the final ACO-BRT is determined, and the testing dataset is used. The proposed framework was benchmarked with ANFIS-PSO, ANFIS-FFA, ANN-FFA, SVM, k-NN, PCR, and GP models, in addition to the traditional Kuz-Ram model assessed employing commercial software. The methodologies were evaluated in terms of R^2, MAE, and RMSE, and the ACO-BRT model outperforms the other methodologies, reaching 0.963, 1.388, and 1.964 in the training dataset, respectively, and 0.962, 1.075, and 1.643 in the testing dataset, respectively, results with high accuracy. ANFIS-PSO, ANFIS-FFA, and ANN-FFA reached R^2 values of 0.929–0.952, MAE values of 1.306–1.607, and RMSE values of 1.801–2.279, while AI techniques such as SVM, k-NN, PCR, and GP reached values of 0.892–0.940 for R^2, 1.492–1.893 for MAE and 2.137–2.741 for RMSE. Finally, sensitivity analysis indicates that the predictions were impacted by the explosive charge per delay, spacing, and burden. Figure 13.26 presents ants traversing two paths from their nest to a food source and deciding on the shortest after a while in an ACO approach, while Fig. 13.27 shows an example of BRT, in which an ensemble model with three trees results in a single optimal tree [69].

Huang et al. (2020) compared the performances of a cat swarm optimization (CSO) and PSO algorithms to predict the rock fragmentation from blasting operations at two quarry mines in Iran from a dataset with 75 instances, considering 6 independent

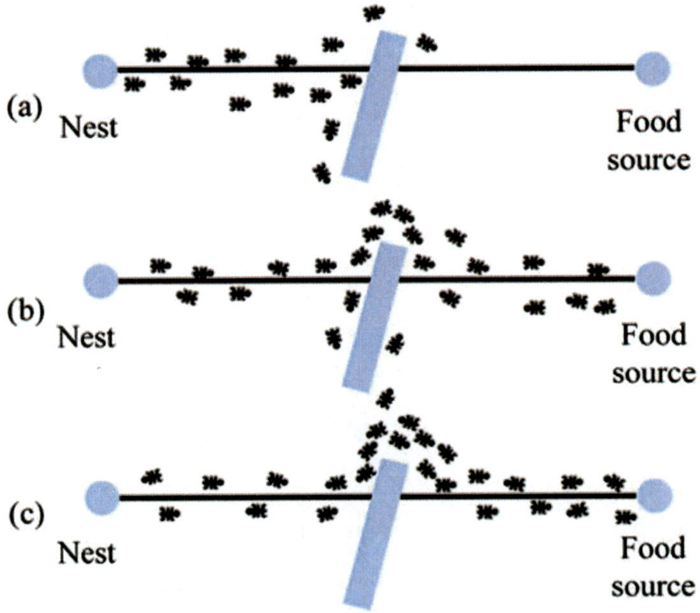

Fig. 13.26 Ants traversing two paths from their nest to a food source and deciding on the shortest after a while in an ant colony optimization (ACO) approach [72]

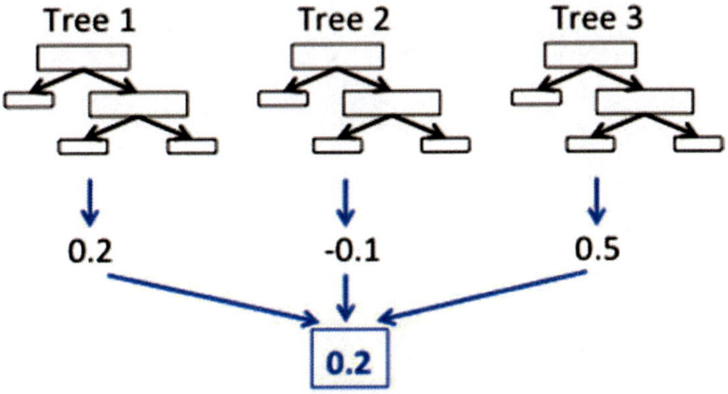

Fig. 13.27 An example of a boosted regression tree (BRT) is an ensemble model with three trees, resulting in a unique optimal tree [73]

variables as input and the dependent fragmentation as output, reserving 15 instances for testing. Fragmentation quality was determined in terms of 80% passing size (D_{80}), being assessed by the image analysis method. The CSO algorithm is an auto-tuning metaheuristic inspired by felines behavior, combining PSO and ACO principles in a stochastic framework and encompassing exploration (seeking mode) and exploitation (tracing mode) phases, relying on a mixture ratio parameter to adjust both modes. Since each cat has its position, velocity, and fitness function value, a brief explanation is that the population is randomly divided into seeking and tracing cats. In the first place, when seeking mode creates copies of the seeking cat's positions, randomly perturbs these positions (except for the best cat), calculates their fitness function, and chooses the best candidates to move.

Moreover, the tracing mode depicts the quick movements that cats make when hunting a target, so these cats' velocities are updated and adjusted to be between the upper and lower bounds when their new locations and fitness functions are computed when new iterations start until a stopping criterion is met. Both CSO and PSO were applied in their linear and power form, and the values of R^2, RMSE, VAF, and mean absolute bias error (MABE) indexes were computed and compared. The R^2 values of CSO in its power and linear form, respectively, were 0.985 and 0.964, while those of PSO in its power and linear form, respectively, were 0.962 and 0.955 so that the CSO power model was elected the best. Finally, sensitivity analyses indicated that stemming was the most influential parameter in this dataset. Figure 13.28 presents the structure of the CSO algorithm [74].

Fang et al. (2021) proposed a hybrid model of the boosted generalized additive model (BGAM) with FFA to predict the rock size distribution from blasting operations in a limestone mine in Vietnam, considering a dataset with 136 instances, 20% of which were used for testing, encompassing 6 independent variables as input and the dependent size distribution as output. A commercial software using the image analysis method assesses the size distribution, and the ANN-FFA, ANFIS-FFA, SVM,

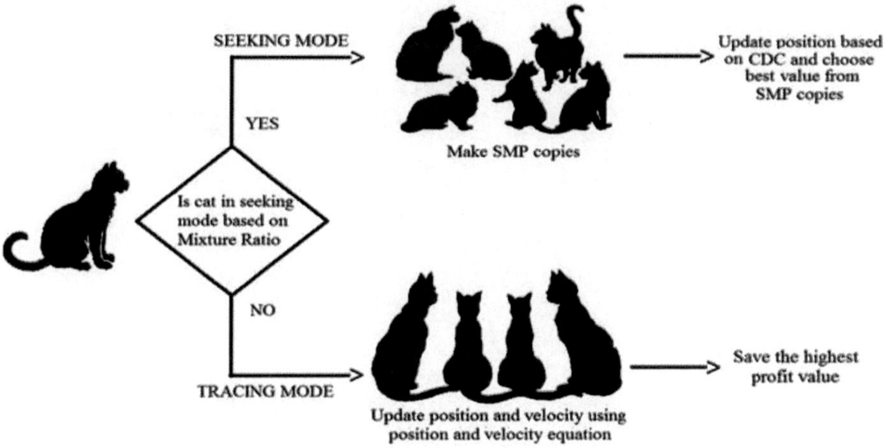

Fig. 13.28 Structure of the cat swarm optimization (CSO) algorithm [75]

GP, and k-NN models were also used to assess the prediction. BGAM is an algorithm based on generalized additive models capable of clarifying data properties and the relationships between input and output variables in nonlinear problems, in addition to a boost stage, which is considered the primary model in this proposition. At the same time, the FFA aims to optimize its hyper-parameters. The BGAM-FFA approach reached an R^2 value of 0.980, while the ANN-FFA, ANFIS-FFA, SVM, GP, and k-NN models reached 0.967, 0.968, 0.972, 0.940, and 0.963, respectively, highlighting the accuracy of the proposed methodology. Xie et al. (2021) proposed four combinations of the FFA algorithm and machine learning techniques (i.e., FFA with gradient boosting machine (FFA-GBM), FFA-SVM, FFA-GP, and ANN-FFA) to predict the size distribution of rocks from blasting operations in a limestone mine in Vietnam, considering a dataset with 136 instances with 6 independent variables as input and the dependent size distribution as output. A commercial software using the image analysis method assesses the size distribution, while 25 instances of the dataset have been reserved for testing. GBM is a stepwise tree-based model that improves as new tree-based models aiming to compensate for the disadvantages of former tree-based models (weak learners) are created so that a combination of weak models generates an improved final model with high performance. Among machine learning techniques, the ANN architecture must be defined before optimization, so a trial-and-error procedure resulted in a 6-14-7-1 architecture. These hybrid algorithms are initialized with the training data, so FFA (with predefined parameters) is used first to improve its performance. Thus, it optimizes the entire framework iteratively until a termination criterion is reached. The models were evaluated for R^2, RMSE, and MAE indexes, reaching values of 0.996, 0.591, and 0.392 for FFA-GBM, respectively; values of 0.979, 1.290, and 0.950 for FFA-SVM, respectively; values of 0.979, 1.213, and 0.991 for ANN-FFA, respectively; and 0.940, 2.137, and 1.573 values for

FFA-GB, respectively, attesting to the superiority of the first when compared to the others [76, 77].

Vu et al. (2021) employed a mask region-based convolutional neural network (mask R-CNN) to automatically measure blast fragmentation from a dataset with more than 200 images of blasting muck piles in a coal mine in Vietnam, 10% of which were used for testing purposes. CNN is regularized network generally used for visual imaging purposes due to its ability to exploit the hierarchical pattern in data by convolutional operations and thus reunite patterns of increasing complexity from smaller and simpler patterns recorded in their kernels (or filters, the vector of weights and bias). Furthermore, the receptive field of different neurons performs a partial overlap to extend over the entire visual field, whereas the network optimizes the kernels through automated learning. In this mask R-CNN proposition, given an image of a blasted muck pile, the model aims to locate the position of the rock fragments and classify them, first splitting the image into overlapping patches, then training the model for rock fragment segmentation, reuniting the patches into a unique image and, finally, calculating a CDF curve to measure the fragmentation. The proposed mask R-CNN architecture includes a convolutional backbone for feature extraction, a region proposal network to generate regions of interest that must be classified for class prediction, and a bounding-box regressor to refine the regions of interest, culminating in a fully convolutional network with processes to predict the pixel-accurate mask from each region of interest. The authors drew rock fragments manually on the images to achieve results similar to muck pile sieving and, compared to Split-Desktop predictions, the proposed model achieved fragmentation results much closer to reality than the commercial software, with an average precision of 92% and 83% for bounding-box and segmentation masks, respectively. Figure 13.29 presents an image of the muck pile (a); colored rock texture image by mask R-CNN (b); fragment delineation by Split-Desktop (c); and comparison of rock fragmentation predictions (d) [78].

Esmaeili et al. (2015) used PCA to verify the influencing parameters. Thus, SVR and ANFIS models were applied to predict the fragmentation of blasting operations in an iron mine in Iran, reaching R^2 values of 0.89 0.83, respectively. Shams et al. (2015) applied FIS and MVRA to predict the rock fragmentation in a copper mine in Iran, reaching R^2 values of 0.922 and 0.738, respectively. Trivedi, Singh, and Raina (2016) used a BPNN with 8-10-10-2 architecture and an MVRA to simultaneously predict fragmentation and flyrock in four limestone mines in India, reaching R^2 values of 0.962, 0.77 for rock fragmentation, respectively. Gao et al. (2018) compared GP performances with five different kernels to predict rock fragmentation from the dataset used in Hasanipanah et al. (2018a), reaching R^2 values from 0.936 to 0.948. Finally, Zhou et al. (2021a) compared the performances of ANFIS-FFA, ANFIS-GA, ANFIS, SVR, and ANN techniques to predict the fragmentation of blasting operations at two quarry mines in Iran, reaching R^2 values of 0.981, 0.989, 0.956, 0.924, and 0.922, respectively. For more information, see Trivedi et al. (2014) and Dumakor-Dupey, Arya, and Jha (2021) reviews [18, 20, 58, 79–85].

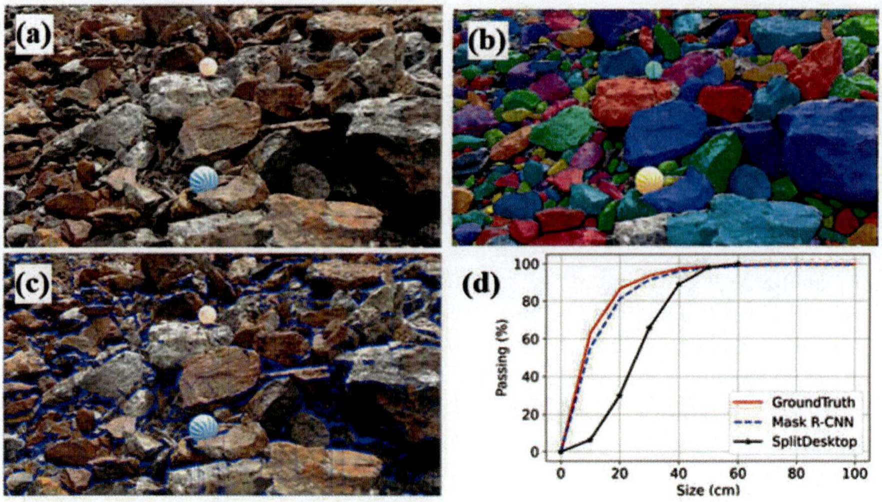

Fig. 13.29 An image of the muck pile (**a**); colored rock texture image by mask R-CNN (**b**); fragment delineation by Split-Desktop (**c**); and comparison of rock fragmentation predictions (**d**) [78]

Environmental Concerns

There are two forms of energy released when an explosive charge is detonated: shock waves or high gas pressure. The initial burst of energy crushes the rock around the borehole to approximately twice its radius. In contrast, beyond the crush zone, radial crashes are formed by the interaction of the shock wave to existing discontinuities, extending to 20 or more times the borehole diameter. The gases then crack and heave the fractured rock mass, and the amount of energy that was not consumed in rock blasting, rock breakage, and movement promotes the ground vibration, which travels elastically as rock particles are temporarily displaced and return to their original position. Air overpressures (AOP, air blast) are generated if any energy is liberated into the atmosphere. Flyrocks are fragments that throw beyond the planned clearance zone and have an enormous risk associated with their occurrence due to the deadly consequences it may bring about. Figure 13.30 presents blast vibrations, such as surface, body, and airwaves [8].

This section analyzes these negative environmental effects, presenting classic empirical methodologies and modern advanced analytics techniques for predicting these adverse effects. For the sake of brevity, studies using techniques already presented in this chapter are quickly cited, while those using as-yet-undisclosed techniques are more detailed. Additional information should be obtained from the publications themselves.

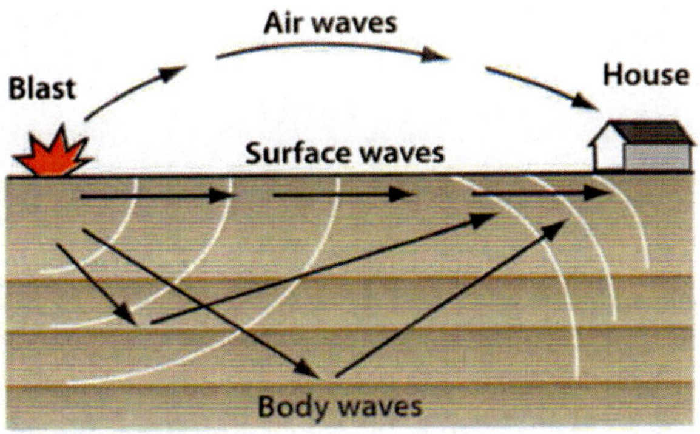

Fig. 13.30 Blast vibration types [8]

Vibration (Seismic Waves)

When a blast is detonated, ground waves radiate in all directions. Whose intensity of the waves (amplitude, frequency, and duration) in a given location depends on the size of the blast, the displacement routes, and site characteristics. Vibrations travel with the earth and its surface and are generally composed of many different families of waves. Some of the most important types are P waves (primary or compression), S waves (secondary or shear waves), and R waves (Rayleigh or transverse waves). P waves transmit energy or sound through rock at a seismic velocity from 1,800 to 6,100 m/second, traveling in the direction of propagation. S waves follow the P waves, traveling at about 0.6 times the velocity of the other and vibrating perpendicularly to the direction of the wave's course. R waves travel about 0.9 times the velocity of S waves at speeds ranging from 920 to 1500 m/s, conversely to others. They travel on the surface. As ground waves travel through the earth, they cause each particle of the earth to vibrate in three dimensions and are generally described as harmonic waves. The motion of vibrating particles can be described by their displacement, velocity, acceleration, and frequency. Phase, in its turn, is the relationship among displacement, velocity, and acceleration at the same instant of time, which consists of an important concept to evaluate the interaction among constructive and destructive waves [8].

Blasting vibrations are generally recorded and reported as velocity–time histories in three-component directions (x, y, and z), designated as radial, transverse and vertical. These time histories (or blasting waveforms) make it possible to deduct the wave propagation velocity over time. The highest zero to the peak value in the waveform is the peak particle velocity (PPV), and waves tend to have higher frequency energy in the initial part of the waveform and lower frequency energy in the last part due to attenuating factors. Wave amplitudes (PPV) and frequency at a location are strongly related to the size of a blast (single charge) and distance from the location, where the amplitude of vibration increases as the charge weight increases and decreases when the distance increases. Figure 13.31 presents three components of ground vibration in a typical measurement and PPV detection [8].

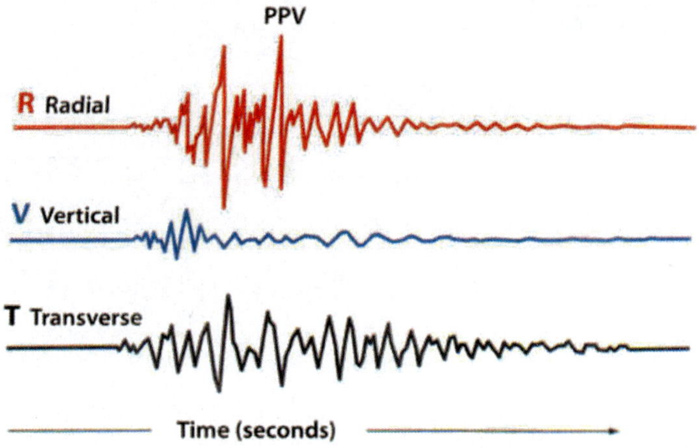

Fig. 13.31 Three components of ground vibration and detection of PPV [8]

The natural frequency of buildings usually ranges from 5 to 10 Hz, and if the frequency in the ground induced by the blast is equal to or close to the frequency of buildings, residents feel uncomfortable as the buildings resonate. Therefore, up to a certain limit directly related to frequency, the PPV value does not cause damage to structures; when these limits are exceeded, damage occurs, with the frequency range between 5 and 10 Hz being the most critical. The first blast-based damage criteria to consider the effect of blast-induced PPV and frequency on buildings was proposed by Siskind et al. (1980a), who evaluated the frequency range from 1 to 100 Hz and the correlation of measured PPV with building damage, thus proposing a safe zone based on these two factors. Figure 13.32 presents the USBM RI 8507 blast-based damage criteria proposed by Siskind et al. (1980a) [86, 87].

Several experiments were carried out to obtain general equations for predicting amplitudes (PPV) and frequency, encompassing different parameters in their construction, such as different rocks types, explosives, monitoring distances, etc. Considering the PPV in mm/s, W as the maximum charge per delay, in kg, D as the distance between the blast face and vibration monitoring point, in meters, and site constants such as $K, \beta, A, n,$ and α, Table 13.1 presents empirical blast-induced ground vibration predictors proposed over decades. Although, in many situations, they make predictions far from reality, especially when compared to modern advanced analytics techniques [85, 88].

The ANN approach has been widely used to predict ground vibration for the past few decades. Similar to the methodology applied to predict rock fragmentation, in which several parameters related to blast design, rock mechanics, explosive parameters, among others, are used as input in order to reach the mean particle size or the 80% passing size, here the inputs are explored to reach seismic wave parameters as output, generally expressed as PPV. Singh et al. (2004) used an ANN with 6-5-2

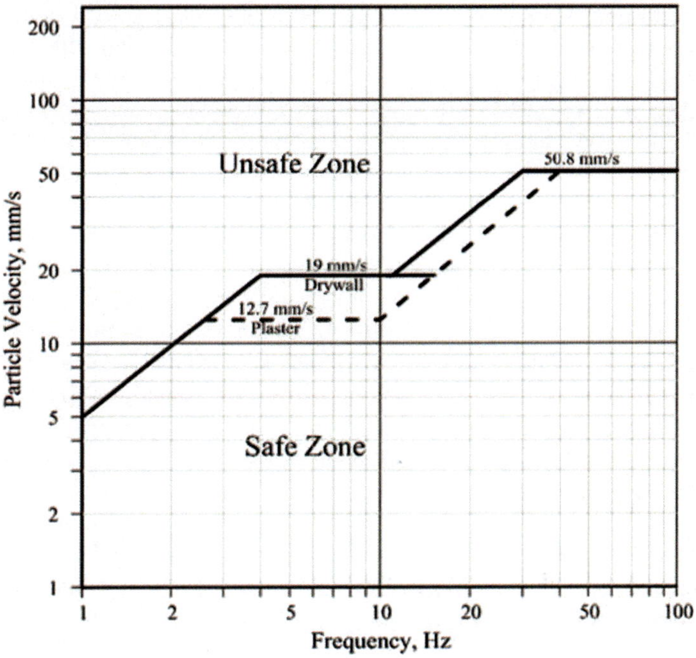

Fig. 13.32 USBM RI 8507 blast-based damage criteria, proposed by Siskind et al. (1980a) [86, 87]

Table 13.1 Most used empirical blast-induced ground vibration predictors [89–98]

Empirical PPV predictors	Equations
USBM: Duvall–Petkof (1959)	$\text{ppv} = K[D/\sqrt{W}]^{-\beta}$
Langefors–Kihlstrom (1963)	$\text{ppv} = K[\sqrt{(W/D^{3/2})}]^{\beta}$
General: Davies–Farmer–Attewell (1964)	$\text{ppv} = K\,D^{-\beta}\,(W)^{A}$
Ambraseys–Hendron (1968)	$\text{ppv} = K[D/(W)^{1/3}]^{-\beta}$
Bureau of Indian Standard (1973)	$\text{ppv} = K[(W^{2/3}/D)]^{\beta}$
Ghosh–Daemen I (1983)	$\text{ppv} = K[D/\sqrt{W}]^{-\beta}\,e^{-\alpha D}$
Ghosh–Daemen II (1983)	$\text{ppv} = K[D/(W)^{1/3}]^{-\beta}\,e^{-\alpha D}$
Gupta et al. I (1987)	$\text{ppv} = K[W/(D)^{3/2}]^{\beta}\,e^{\alpha D}$
Gupta et al. II (1987)	$\text{ppv} = K[W/(D)^{2/3}]^{\beta}\,e^{\alpha D}$
Gupta et al. III (1988)	$\text{ppv} = K[D/\sqrt{W}]^{-\beta}\,e^{\alpha(D/W)}$
CMRS: Roy (1993)	$\text{ppv} = n + K[\sqrt{W}/D]$
Rai–Singh (2004)	$\text{ppv} = K\,D^{-\beta}\,WA\,e^{-\alpha D}$

architecture to predict rock anisotropy and blast-induced p-wave velocity in sandstone mines, reaching an R^2 of 0.9481 and 0.9987, respectively, while used another ANN with 10-4-2 architecture to predict the same output for the marble mines, reaching the values of 0.9149 and 0.687, respectively. Singh and Singh (2005) used an ANN with a 9-6-1 architecture to predict the blast-induced frequency in coal measure sandstones from a dataset with 15 blast instances, reaching an R^2 value of 0.905, while an MVRA approach developed for comparison achieved an R^2 value of 0.716. Khandelwal and Singh (2006) used an ANN with 13-8-2 architecture to predict the blast-induced PPV and frequency in a coal operation in India, using 150 blast instances for training and 20 for testing the models, reaching R^2 values of 0.9994 and 0.9868, respectively, while an MVRA approach developed for comparison only achieved R^2 values of 0.4971 and 0.0356, respectively. Khandelwal and Singh (2007) employed an ANN with 2-5-1 architecture to predict blast-induced PPV in a magnesite mine in India, comparing the results with four empirical predictors, i.e., USBM, Ambraseys–Hendron, Langefors–Kihlstrom, and Indian Standard predictors, where only the ANN approach was able to predict the PPV values with acceptable accuracy. Khandelwal and Singh (2009) used an ANN with 10-15-2 architecture to predict the PPV and frequency in a coal operation in India, reaching R^2 values of 0.9864 and 0.9086, respectively, while an MVRA approach developed for comparison achieved R^2 values of 0.3508 and 0.098, respectively. Seven conventional empirical predictors were also tested, while the PPV achieved through these methods reached R^2 values between 0.1306 and 0.5409, results considerably inferior to those obtained by the ANN approach. Figure 13.33 presents a graph comparing measured and predicted PPV by ANN and different empirical approaches in Khandelwal and Singh (2007) [89, 90, 92, 93, 99–103].

In addition to ANN approaches, many advanced analytics and machine learning techniques have been applied to predict ground vibration, such as ANFIS, SVM, random forest (RF), several metaheuristics, and evolutionary algorithms, among others chronologically described in this section. Iphar, Yavuz, and Ak (2008) applied an ANFIS model to predict blast-induced PPV in a magnesite mine in Turkey and compared the results with those obtained using an empirical regression model developed for this purpose, achieving R^2 values of 0.99 and 0.90 for ANFIS and the regression model, respectively. Singh, Dontha, and Bhardwaj (2008) compared the performances of ANFIS and MVRA models to simultaneously predict blast-induced PPV and frequency in a coal mine in India, considering 192 blast instances, 10% of which were used to test the models, comprising 6 independent variables as input. The ANFIS model reached R^2 values of 0.9986 and 0.9988 for PPV and frequency, respectively, while the MVRA indexes were not presented, although the errors were considered high. Mohamed (2009) assessed the influence of individually inputting the scaled distance, simultaneously inputting the measured distance and maximum explosives per delay, and concomitantly inputting 15 parameters in ANNs with different architectures to predict the blast-induced PPV in a limestone mine in Egypt, comparing the results to the general predictor. Nevertheless, architectures 1-7-7-1, 2-6-6-1 and 15-5-7-1 were employed, reaching MSE values of 0.0227, 0.0114 and 0.0092, respectively, and R^2 values of 0.92, 0.96, and 0.97, respectively. The average

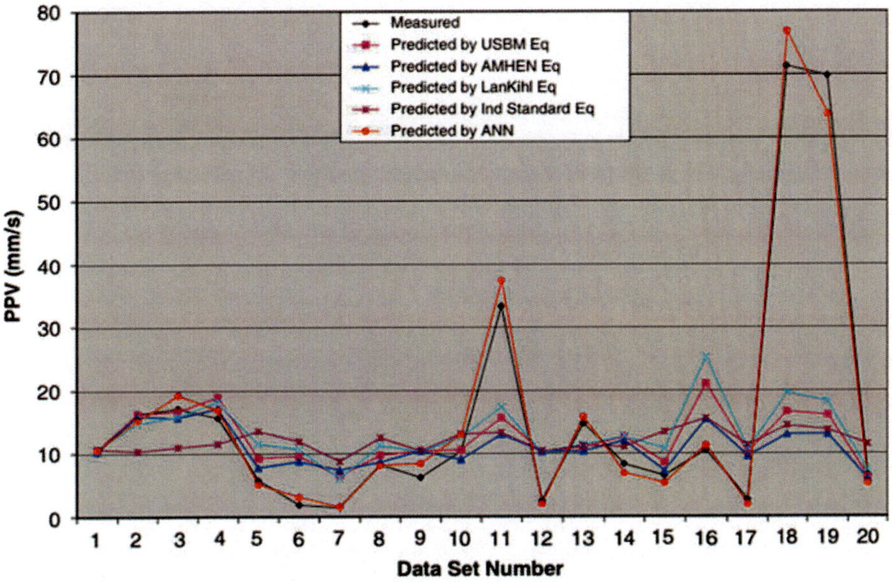

Fig. 13.33 Comparison of measured and predicted PPV by ANN and different empirical approaches in Khandelwal and Singh (2007) [101]

standard deviations of the results were 0.2993, 0.1956, and 0.1473, respectively, while the general empirical predictor reached the value of 0.5499, an inferior result, mainly when compared to simultaneously inputting 15 variables in the prediction. Bakhshandeh Amnieh, Mozdianfard, and Siamaki (2010) used an ANN with 20-17-15-10-1 architecture to predict the blast-induced PPV at a copper mine in Iran, achieving an R^2 value of 0.9935 [104–107].

Monjezi et al. (2010) compared the performances of different types of ANN to predict the blast-induced PPV in a copper mine in Iran. Therefore, they evaluated a multilayer perceptron neural network (MLPNN), which is a BPNN with one or multiple hidden layers that have been addressed in many of the studies aforementioned, an RBFNN and a general regression neural network (GRNN), which relies on four layers, i.e., input, pattern, summation, and output layers, and do not require training parameter arrangement. The MLPNN with 6-20-1 architecture achieved the best results, with an R^2 value of 0.954, while a sensibility analysis by CAM showed that the distance from the blast face was the most influential parameter. Khandelwal, Kankar, and Harsha (2010) used an SVM model to predict blast-induced PPV in a magnesite mine in India, using a dataset with 170 blast instances, while 20 of these instances were used to test the model, considering the distance from the blast face and explosive charge per delay as input. The SVM methodology reached an R^2 value of 0.955, while 4 conventional predictors achieved poor values, from 0.163 to 0.337. Mohammadi, Bakhshandeh Amnieh, and Bahadori (2011) used ANFIS models to predict blast-induced PPV in a copper mine in Iran from a dataset with 46 blast instances, 22 of which using ANFO and 24 using emolan explosive (a mixture of ANFO and emulite that has VOD up to 4000–5000 m/s), taking 28 instances for model training, while the remaining instances were used to validate and test them. Considering the distance from the blast face and explosive charge per delay as input, both models reached R^2 values of 0.97, and the sensitivity analysis indicated that the PPVs from blasts with ANFO are 90% dependent on the distance from the blast face, while the PPVs from blasts with emolan are almost equally influenced by the two parameters. Dehghani and Ataee-pour (2011) used an ANN with 9-25-1 architecture to predict the blast-induced PPV at a copper mine in Iran, achieving an RMSE value of 0.0245. Thus, from the sensitivity analysis performed, a prediction equation was developed through the DA approach using the force system, obtaining R^2 and RMSE values of 0.775 and 3.49, respectively, a better result than those obtained by conventional predictors [108–111].

Monjezi, Ghafurikalajahi, and Bahrami (2011) used an ANN with 4-10-5-1 architecture to predict tblast-induced PPV in Iranian tunnel blasting operations, achieving an R^2 value of 0.9493, while an MVRA approach achieved a value of 0.6932. Conventional predictors were also used, with USBM being the best, with an R^2 value of 0.7987, while Langefors-Kihlstrom was the worst, with an R^2 value of 0.3760. A sensitivity analysis performed using the CAM revealed that the distance from the blasting face was the most influential parameter in the PPV, while the stemming was the least influential. Khandelwal, Kumar, and Yellishetty (2011) used an ANN with 2-5-1 architecture to predict blast-induced PPV in surface coal mines in Iran, considering the distance from the blast face and explosive charge per delay as input, achieving an R^2 value of 0.919. Five empirical predictors were also assessed, reaching only R^2 values from 0.225 to 0.591. Fisne, Kuzu, and Hudaverdi (2011) employed a FIS approach to predict the blast-induced PPV in a quarry mine in Turkey, considering maximum charge per delay and distance from the blast face, comparing the results to a regression model developed for this purpose. The FIS model reached R^2, VAF, and RMSE values of 0.96, 0.91, and 5.31, respectively, while the regression model achieved 0.82, 0.59, and 11.32, respectively. Longjun et al. (2011) employed the SVM and RF approach to predict blast-induced PPV, frequency simultaneously, and duration of blasts in a copper mine in China from a dataset with 108 instances, considering 9 independent variables as input and those three dependent variables as output, reserving 15 instances for testing. The RF regressor is composed of decision trees that depend on a random vector to assign numerical values so that these independent trees are randomly perturbed and can explore the space of possibilities.

In contrast, the numerical output values of the random trees are averaged to produce a single output value for the forest. The results show that both methodologies achieved predictions with a low error ratio. Besides, the average error rate of the SVM is lower than that of the RF. Figure 13.34 presents an example of an RF flowchart [89, 90, 112–115].

Verma and Singh (2011) used a GA approach to predict the blast-induced PPV in a coal mine in India, achieving an R^2 and mean average percentage error (MAPE) values of 0.997 and −0.088, respectively, while a comparison with 8 empirical predictors resulted in MAPE values from -3,736.42 to 44.48. Furthermore, the MAPE values

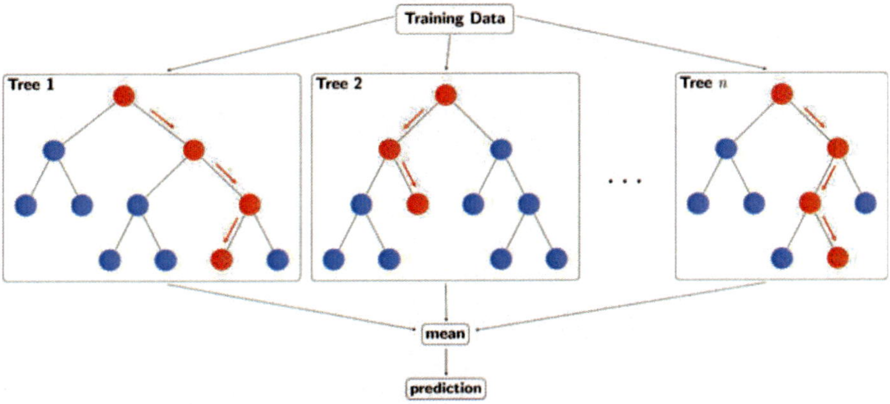

Fig. 13.34 Example of a random forest (RF) flowchart

of an MVRA model and an ANN with 7-4-1 architecture were -0.184 and -0.160, respectively, reinforcing the superiority of the GA approach. Khandelwal (2011) used two SVM models to predict blast-induced PPV at three coal mines in India, considering a scaled distance (single input) or both distance from the blast face and explosive charge per delay as two inputs, reaching R^2 values of 0.867 and 0.964, respectively, while five conventional predictors reached values from 0.356 to 0.655. Mohamed (2011) proposed an ANFIS model to predict the blast-induced PPV and the air overpressure of blasts in a limestone mine in Egypt, considering maximum charge per delay and distance from the blast as input, achieving R^2 values of 0.95 and 0.93, respectively, while a regression model build for this purpose reached R^2 values of 0.66 and 0.55, respectively. Shuran and Shujin (2011) developed a BPNN with 3-6-1 architecture to predict blast-induced PPV in an iron mine, considering 134 blast instances, 10 of which were used for training, encompassing 3 independent variables as input, reaching an R-value of 0.999, while an empirical predictor reached a value of 0.854 [116–119].

Mohamadnejad, Gholami, and Ataei (2012) compared the performance of GRNN and SVM in predicting PPV from blasting operations nearby a dam in Iran, considering the maximum charge per delay and distance from the blast face as input, achieving R^2 values of 0.658 and 0.946, respectively, while the best R^2 of five conventional predictors (Bureau of Indian Standard) was 0.658. Álvarez-Vigil et al. (2012) used an ANN with 11-5-2 architecture to predict blast-induced PPV and frequency in a limestone mine in Spain, reaching R^2 values of 0.98 and 0.95, respectively, while an MVRA approach developed for comparison achieved R^2 values of 0.5 and 0.15, respectively. Hudaverdi (2012) employed a hierarchical cluster approach to predict the PPV from blast operations in a sandstone quarry in Turkey, considering a dataset with 88 instances with 8 independent variables. The hierarchical cluster analysis consists of a technique that creates relatively homogeneous groups of objects (blasts in this case) based on their similar characteristics, represented as the distance from other objects, where each object represents a separate cluster that will be combined

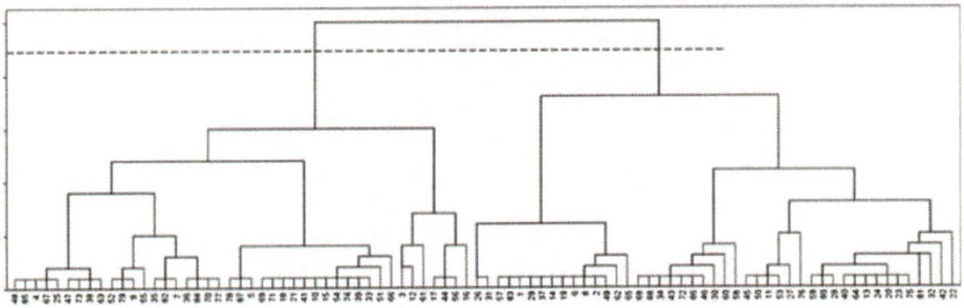

Fig. 13.35 Hierarchical classification dendrogram in Hudaverdi (2012) [122]

with the others until only one remains, using a distance criterion. This study considered the Pearson correlation distance to define the distance between clusters and the dendrogram. There are two main groups of blasts, combined based on their calculated relative distance, in which blasts with high similarity are placed together. Therefore, of the 88 instances, 43 blasts incorporated group 1 and the remain 45 incorporated groups 2, whose main variables responsible for this separation are the ratio of burden to hole diameter and powder factor, in other words, blasts that have high values for one of these variables have low values for the other. A discriminant analysis was also used to ensure that the groups are correctly defined. Thus, a regression analysis was performed for each group, where the R^2 values of groups 1 and 2 reached 0.908 and 0.891, respectively, while a regression analysis considering the entire dataset reached the R^2 value of 0.774. Figure 13.35 presents the hierarchical classification dendrogram developed in Hudaverdi (2012) [93, 120–122].

Li, Yan, and Zhang (2012) used SVM to predict blast-induced PPV in tunnel construction in China, considering the distance from the blast face and explosive charge per delay as input, reaching an R^2 value of 0.954, while a regression analysis reached a value of 0.655. Bakhshandeh Amnieh, Siamaki, and Soltani (2012) used an ANN with 4-8-3 architecture to predict the design of the blasting pattern, i.e., explosive weight, burden, and spacing, at a copper mine in Iran, considering blast-induced PPV, blast face distance, block volume and explosive density as input, reaching R^2 values of 0.651, 0.77 and 0.963, respectively. Mohamad et al. (2012) used an ANN with 9-4-4-2 architecture to predict blast-induced PPV and frequency in a quarry mine in Malaysia, achieving R^2 values of 0.997 and 0.989, respectively. Three empirical predictors were also assessed, which presented errors from 32 to 63%, with the USBM predictor being the best, while ANN presented an error of 0.01% for velocity and 4% for frequency. Mohammadnejad et al. (2012) employed SVM models to predict the blast-induced PPV from two limestone mines, considering 26 blast instances, 40% of which were used to validate and test the model, comprising distance from the blast face and maximum charge per delay as input, thus reaching an R^2 value of 0.944 [89, 123–126].

Ataei and Kamali (2013) employed an ANFIS to predict the blast-induced PPV from the excavations to build a power plant and a dam in Iran, considering 29 blast instances, 20% of which were used for testing. Comprising the distance from the blast face and maximum charge per delay as input, the ANFIS model achieved an R^2 value of 0.9897, while the USBM predictor achieved a value of 0.9079. Gorgulu et al. (2013) employed an ANN with 11-10-1 architecture to predict the blast-induced PPV in a boron mine in Turkey, considering 36 blast instances with 11 independent variables as input, achieving an R^2 value of 0.9553, while a regression analysis reached a value of 0.55. The authors also tried to incorporate the directional influence in the analysis, collecting vibration data in straight lines, whose regressions reached superior values compared to the entire data, with R^2 values from 0.77 to 0.88. Monjezi, Hasanipanah, and Khandelwal (2013) used an ANN with 3-4-4-1 architecture to predict blast-induced PPV in a dam adjacent to two mines in Iran, achieving an R^2 value of 0.927. Five empirical predictors were also evaluated, three of them achieving poor results, while the R^2 values achieved by the USBM and Roy predictors were 0.945 and 0.949, respectively, an unusual occasion where conventional methods were more accurate than the ANN approach. However, other indices show the robustness of this approach at the expense of empirical ones. Finally, a sensitivity analysis showed that the distance from the blast face is the most influential parameter. Ghasemi, Ataei, and Hashemolhosseini (2013) used a FIS model to predict PPV from blast operations at a copper mine in Iran, taking 10 independent variables as input, achieving an R^2 value of 94.59, while an MVRA approach only achieved a value of 34.55. A comparison was made with 10 ground vibration predictors, whose R^2 values ranged from 45.40 to 65.25 [89, 97, 127–130].

Verma and Singh (2013) used an ANFIS model to predict the blast-induced P wave velocity in detonations on three different rock types, i.e., marble, travertine, and granite. Using a dataset with 136 blast instances, of which 26% were used to test and validate the model, and considering 6 independent parameters as input, the ANFIS model achieved R^2 and MAPE values of 0.9253 and 0.51%, respectively. Saadat, Khandelwal, and Monjezi (2014) used an ANN with 4-11-5-1 architecture to predict blast-induced PPV at an iron mine in Iran, achieving an R^2 value of 0.957 MVRA approach developed for comparison achieved an R^2 value of 0.276. Four empirical predictors were also assessed, reaching only R^2 values from 0.139 to 0.446. Verma, Singh, and Maheshwar (2014) compared the performance of an ANN with 6-4-1 architecture, an ANFIS model, an MVRA approach, and an SVM model in predicting blast-induced P waves velocity in detonations in three different types of rocks, i.e., marble, travertine, and granite. Using a dataset with 150 blast instances, of which 25% were used to test and validate the model, and considering 6 independent parameters as input, the methodologies reached MAPE values of 0.258, 0.309, 0.583, and 0.769 for SVM, ANFIS, ANN, and MVRA, respectively. Vasović et al. (2014) used an ANN with 3–6-1 architecture to predict blast-induced PPV in a limestone mine in Serbia, using a dataset with 32 blast instances, of which 50% were reserved for model validation and testing, considering the distance from the blast face, the explosive charge per delay and total charge as input, reaching an R-value of 0.89978. For comparative purposes, 5 empirical predictors were also evaluated, reaching R values from 0.69 to 0.88, while empirical AOP predictors were also evaluated, although no advanced analytical technique was employed to benchmark these results [131–134].

Jahed Armaghani et al. (2014) proposed to use an ANN-PSO approach to simultaneously predict blast-induced flyrock and PPV at three quarry mines in Malaysia, using a dataset with 44 instances of blast operations that contain information from 10 independent parameters, of which 20% of the instances were used for testing purposes. Sensitivity analysis determined PSO parameters, such as defining 300 particles as the swarm size, while the 10-15-2 network architecture was defined by trial and error. In this ANN-PSO approach, each particle depicts a series of weights and biases, while the objective function of the PSO is to minimize MSE in the network, returning the gbest value after a stopping criterion is reached. The Lundborg predictor was used to compare the results for flyrock, while 6 empirical ground vibration predictors were used for the same purpose, the ANN-PSO approach being superior to all methodologies used as a benchmark, reaching an R^2 of 0.93025 for the testing dataset. A sensitivity analysis using CAM identified sub-drilling and charge per delay as the most effective parameters in ground vibration, while powder factor and charge per delay were the most influential on flyrock distance. Lapcevic et al. (2014) employed an ANN with 5–20-1 architecture to predict blast-induced PPV in a copper mine in Serbia, considering 42 blast instances, 50% of which were used to validate and test the model, encompassing 5 independent variables as input. The ANN approach reached an R^2 value of 0.916, while 5 empirical regressors reached 0.13 to 0.31. Finally, sensitivity analysis by hierarchical analysis indicated that the blast face's distance and delay time were the most influential parameters. Xue and Yang (2014) proposed a GRNN model to predict blast-induced PPV and frequency in a coal mine in India, using 20 blast instances presented by Khandelwal and Singh (2006), of which 15 instances were used for training and 5 were used for testing, encompassing 13 independent variables as input. The GRNN trained with 15 instances reached R^2 values of 1.0000 and 0.9992 for PPV and frequency, respectively, while the ANN with 13-8-2 architecture trained by Khandelwal and Singh (2006) with 150 instances reached R^2 values of 0.9994 and 0.9868, respectively [55, 101, 135–137].

Gorgulu et al. (2015) extended the work in Gorgulu et al. (2013) using an ANN with 10-10-2 architecture to predict blast-induced PPV and frequency in one coal mine and two boron mines in Turkey, considering 431 blast instances with 10 independent variables as input. Regression analyzes were developed independently for the three datasets, reaching R^2 values for PPV of 0.55, 0.66, and 0.70. The proposed ANN reached R^2 values for PPV and frequency 0.976 and 0.836, respectively, while MVRA functions reached 0.45 and 0.17. Hajihassani el at. (2015a) employed a similar ANN-PSO approach to simultaneously predict blast-induced PPV and air blast at a quarry mine in Malaysia, using a 9-12-2 architecture network and achieving average R^2 and MSE values of 0.89 and 0.038, respectively. A comparison was made with 6 ground vibration predictors and 5 AOP predictors, in which the ANN-PSO algorithm reached values with lower average errors compared to the empirical equations. The sensitivity analysis indicated that sub-drilling and maximum charge per delay were the most influential parameters on PPV while stemming and maximum charge per delay was the most influential parameters on AOP [128, 138, 139].

Hajihassani et al. (2015b) proposed a combination of ANN and imperialist competitive algorithm (ICA) to predict blast-induced PPV in a quarry mine in

Malaysia, using a dataset with 95 blast instances that contain information from 6 independent parameters, of which 20% of these instances were used for testing the model. ICA is an evolutionary algorithm inspired by the socio-political experience, in which each element of the population portrays a country whose best ones (considering the minimum error) are elected as imperialists. At the same time, the other depicts the colonies of these empires. As imperialist countries spread their characteristics to their colonies over time, some revolutions prevent the ICA from being trapped in local optima, where the weakest colony in each iteration is replaced by a new random one, which can be reassigned to a stronger empire. This procedure remains until the most powerful empire prevails over the other countries. Similar to other combinations of ANN and metaheuristics, the ANN architecture is previously defined, the initial metaheuristic parameters are then chosen, and the ICA is used to optimize the ANN weights and biases to achieve a minimum error. Using 135 countries and 11 imperialists, this ANN-ICA model achieved a R^2 value of 0.976, while the ANN alone reached 0.911. On the other hand, a regression analysis developed for this purpose reached an R^2 value of 0.856, while an MVRA approach reached a value of 0.878, assuring the robustness of the ANN-ICA approach. Figure 13.36 presents an example of an imperialist competitive algorithm (ICA) flowchart [140].

Jahed Armaghani et al. (2015a) employed an ANFIS model to predict blast-induced PPV at a quarry mine in Malaysia, using a dataset with 109 blast instances and considering the distance from the blast face and explosive charge per delay as input, of which 20% of these instances were used to test the model. This ANFIS model achieved an R^2 value of 0.969, while an ANN alone achieved 0.902. On the other hand, a regression analysis developed for this purpose reached an R^2 value

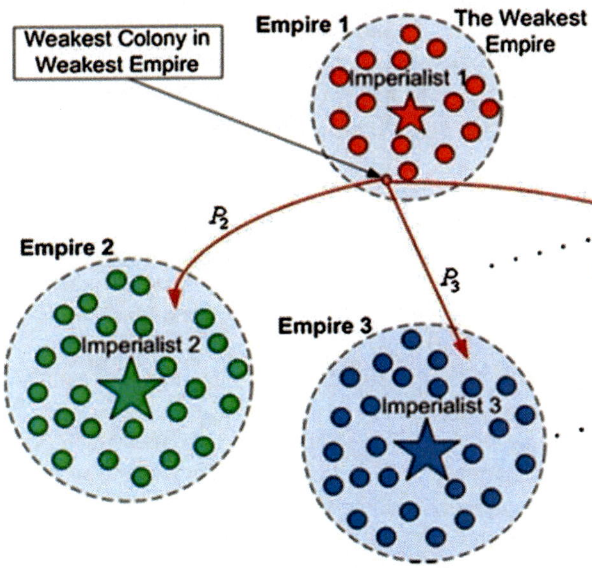

Fig. 13.36 Example of imperialist competitive algorithm (ICA) flowchart [140]

of 0.836, while 5 empirical predictors reached values from 0.315 to 0.834. Jahed Armaghani et al. (2015b) compared the performance of the ANN and ANFIS models in predicting blast-induced PPV, AOP, and flyrock at four quarry mines in Malaysia, using a dataset with 166 blast instances, of which 20% of these instances were used to test the model. To predict the PPV and AOP, the distance from the blast face and explosive charge per delay were used as input to the model, while stemming, the maximum charge per delay, burden to spacing ratio, and distance from the blast face were used to predict the flyrock. Various modifications in training parameters and datasets were performed in both methodologies, and the best results achieved in predicting PPV, AOP, and flyrock using ANN were R^2 values of 0.771, 0.864, and 0.834, respectively.

In contrast, the R^2 values obtained using ANFIS were 0.939, 0.947, and 0.959, respectively, highlighting the robustness of the second methodology. Mohebi, Shirazi, and Tabatabaee (2015) used an ANFIS model to predict blast-induced PPV at a quarry mine in Turkey, using a dataset with 102 blast instances, of which 14 instances were used to test the model, also selecting 4 among the 8 independent measured parameters as input by trial-and-error approach. For comparative purposes, a hierarchical cluster analysis was used, and two main clusters were created, whose regressions reached R^2 values of 0.8331 and 0.8473, while the value obtained by ANFIS was 0.9275. Hasanipanah et al. (2015) employed an SVM model to predict blast-induced PPV in tunneling operations near a dam in Iran, using a dataset with 80 blast instances and considering the distance from the blast face and explosive charge per delay as input, of which 20 of these instances were used to test the model. The SVM approach reached an R^2 value of 0.95, while 4 empirical predictors used as a benchmark achieved values from 0.001 to 0.684 [141–144].

Dindarloo (2015a) used an SVM model to predict blast-induced PPV at an iron mine in Iran, using a dataset with 100 blast instances, 20% of which was used to test the model, and taking 12 independent variables as input. This approach achieved an R^2 value of 0.99, while a partial least squares regression (PLSR) employed for comparison achieved a value of 0.94. Dindarloo (2015b) used a gene expression programming (GEP) approach to predict blast-induced frequency in coal measure sandstones from the same dataset employed in Singh and Singh (2005), considering 15 blast instances and 9 independent parameters as input. Genetic programming (GP) is an extension of GA. While the latter applies the Darwinian principle of survival to binary strings, which aims to find optimal values for a given series of variables, the first acts on tree structures, aiming to derive these optimal values and the structure for that purpose simultaneously, thus varying size and shape. In GP, all feasible solutions are evaluated by a fitness function, while the best nodes in the tree survive and generate an offspring population subjected to genetic operations (i.e., mutation and crossover), where other nodes or subtrees replace random nodes in the tree. Therefore, each iteration performs a new fitness function evaluation and offspring generation until a stopping criterion is reached. On the other hand, gene expression programming (GEP) is also based on tree structures and aims to simultaneously derive the optimal values for a given series of variables and the structure for that purpose. However, these entities are encoded as chromosomes, likewise in the GA

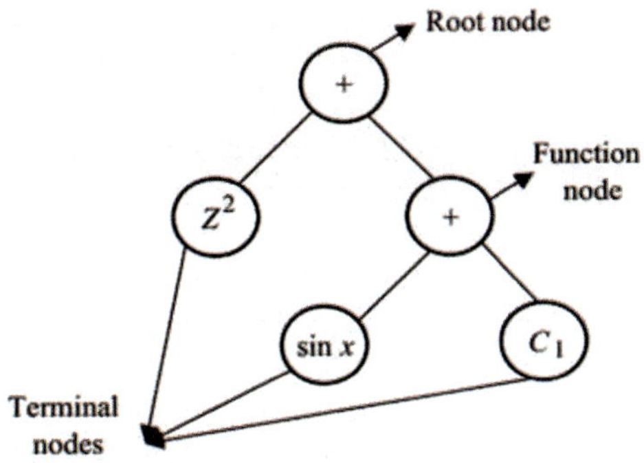

Fig. 13.37 Genetic programming (GP) tree representation of the function $y = z^2(\sin x + c1)$ [146]

approach, simplifying the creation of genetic diversity and evolution of complex programs. The comparison of the proposed GEP with an ANN approach reached R^2 values of 0.91 and 0.81, respectively, while the MAPE values were 4.7 and 9.3, respectively. Figure 13.37 presents the GP tree representation of the function [100, 145, 146].

$$y = z^2(\sin x + c_1) \tag{13.22}$$

Shirani Faradonbeh et al. (2016a) proposed a GEP model to predict blast-induced PPV in a quarry mine in Malaysia, considering 102 blast instances and 6 independent parameters as input, of which 20% of these instances were used to test and validate the model. Using genetic operators such as mutation, inversion, transposition, and recombination, 5 GEP models were developed, from which the best, with 32 chromosomes, reached an R^2 value of 0.874, while a nonlinear MVRA performed for comparison purposes reached the value of 0.790. A sensitivity analysis using CAM identified the maximum charge per delay and distance from the blast face as the most effective parameters in ground vibration. Monjezi et al. (2016) modified the USBM ground vibration predictor by adding the effect of water and comparing the results with techniques such as GEP, linear and nonlinear MVRA. A dataset with 35 blasting operations in an iron mine in Iran was employed, of which 20% were used to test the models, and the distance from the blast face and explosive charge per delay, the water condition of the blast holes was also considered. The conventional USBM predictor reached an R^2 value of 0.614, while the other 3 predictors reached 0.465–0.642. Moreover, the modified USBM predictor reached an R^2 value of 0.752, corroborating the influence of the water condition on the PPV estimate. Furthermore, the GEP, linear, and nonlinear MVRA techniques reached R^2 values of 0.918, 0.797, and 0.814, respectively, reaffirming advanced analytical and machine learning techniques [89, 147, 148].

Singh et al. (2016) extended the work of Verma, Singh, and Maheshwar (2014) by including GA in the analysis, besides the ANN with 6-4-1 architecture, ANFIS, MVRA, GA, and SVM models to predict blast-induced P waves velocity in detonations in three types of rocks, i.e., marble, travertine, and granite. Using a dataset with 150 blast instances, of which 25% were used to test and validate the model, and considering 6 independent parameters as input, the methodologies reached MAPE values of 0.258, 0.309, 0.583, 0.769, and 0.198, respectively, for SVM, ANFIS, ANN, MVRA, and GA, the latter being the one with the best result. Ghoraba et al. (2016) compared the performance of an ANN with 2-4-1-architecture and an ANFIS model in predicting blast-induced PPV in an iron mine in Iran, using a dataset with 115 blast instances, of which 20% were used to test the model, considering the distance from the blast face and explosive charge per delay as input. The USBM predictor was assessed for comparative purposes, reaching an R^2 value of 0.749, while the ANN and ANFIS models reached 0.888 and 0.952, respectively. Amiri et al. (2016) proposed an ANN-kNN model to simultaneously predict blast-induced PPV and AOP at two quarry mines in the vicinity of a dam in Iran, using a dataset with 75 blast instances 20% were used to test the model. Considering the distance from the blast face and explosive charge per delay as input, the data is normalized, and k-means clustering is an unsupervised algorithm that divides a dataset into a predefined number of clusters considering their centroids creates two clusters. Hence, a k-NN is employed in the prediction, whereas an ANN with 2-6-2 architecture was defined by trial and error to also predict the values in each cluster, from which a weighted combination of the results of both methods is achieved, in addition to a padding stage, when the final result is achieved. An ANN with a 2-4-2 architecture was defined by trial and error for comparison purposes, in addition to the USBM empirical predictors, which were also applied. The R^2 values achieved for PPV were 0.81, 0.82, and 0.88 for the USBM, ANN, and ANN-kNN methods, respectively, while the values achieved for AOP were 0.89, 0.93, and 0.95, respectively. Figure 13.38 presents an example of k-means with two clusters [89, 133, 149–151].

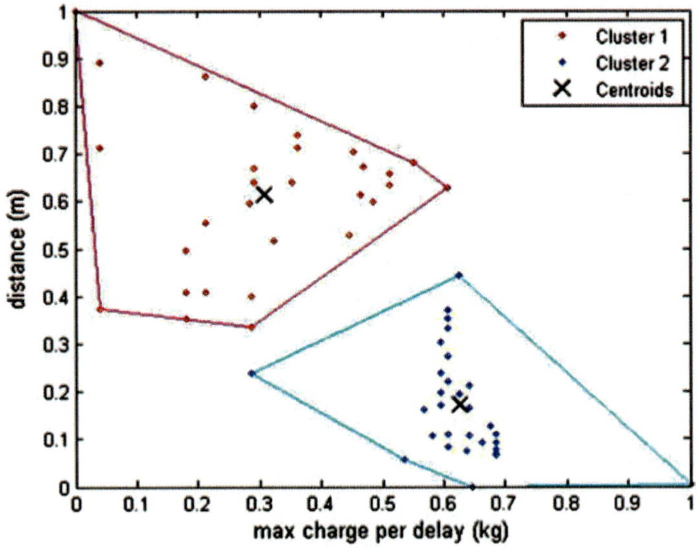

Fig. 13.38 Example of k-means with two clusters [151]

Ghasemi, Kalhori, and Bagherpour (2016) employed an ANFIS-PSO model to predict blast-induced PPV at a copper mine in Iran, considering 120 blast instances, 20% of which were used for testing, encompassing 6 independent variables as input. This ANFIS-PSO approach reached an R^2 value of 0.957, while an SVR and a USBM model reached 0.924 and 0.729, respectively. Koçaslan et al. (2017) employed an ANFIS model to predict blast-induced PPV at two coal mines and two boron mines in Turkey, considering 521 blast instances, 50% of which were used to validate and test the model, encompassing 5 independent variables as input. ANFIS achieved an impressive R^2 value of 1.00, while predictors based on scaled distance developed for each of the four mines reached values from 0.57 to 0.81. Finally, ram Chandar, Sastry, and Hegde (2017) compared the performance of an ANN with 7-4-1 architecture, linear and nonlinear MVRA models in predicting blast-induced PPV in three limestones dolomite and coal mines in India, considering 168 blast instances and comprising 7 independent variables as input. Models were developed for each of the three mines and the entire dataset, and the R^2 values achieved by ANN, linear and nonlinear MVRA models for the combined data were 0.878, 0.842, and 0.738, respectively, while a regression based on the scaled distance for the combined data achieved an R^2 value of 0.746. Shahnazar et al. (2017) proposed an ANFIS-PSO model to predict blast-induced PPV in a quarry mine in Malaysia, considering 81 blast instances, and the data were randomly divided into 5 groups for training and testing, covering the distance from the blast face and explosive charge per delay as input. The ANFIS-PSO approach reached an R^2 value of 0.9840, while an ANFIS and an empirical model based on the USBM predictor achieved 0.9654 and 0.9232, respectively [89, 152–155].

Samareh et al. (2017) developed ANN, nonlinear MVRA, and PSO-GA models to predict blast-induced PPV in a copper mine in Iran, considering 95 blast instances, 30% of which were used to test the model, encompassing 11 independent variables as input. First, a sensitivity analysis performed by the cosine frequency method derived that 4 out of 11 variables would better predict the PPV, from which the ANN architecture was defined as 4-5-1, reaching an R^2 value of 0.7378. Hence, power, exponential, and second-order nonlinear versions of the nonlinear MVRA were built, verifying that the power version was superior, from which an R^2 value of 0.6707 was achieved. Furthermore, the 8 coefficients in the nonlinear MVRA power version that considered the main 4 independent variables were optimized by a PSO-GA algorithm, thus reaching an R^2 value of 0.7507. Next, Xue, Yang, and Li (2017) used an ANFIS model to predict blast-induced PPV on three different rock types, i.e., marble, travertine, and granite, using the same database applied in Verma and Singh (2013), considering 25 blast instances, 20% of which were used to validate and test the model, comprising 7 independent variables as input. The ANFIS model was trained by a gradient descent method and least squares algorithm, reaching an MSE value practically equal to zero since its significant figure is in the sixth-place digit. Taheri et al. (2017) proposed an ANN-ABC approach to predict blast-induced PPV at a copper mine in Iran, using a dataset with 89 blast instances, of which 20% were used to test the model, considering the distance from the blast face and explosive charge per delay as input. The ANN architecture was previously defined as 2-5-1.

Hence, the initial ABC parameters were chosen, and this stage was used to optimize the ANN weights and biases, reaching an R^2 value of 0.92, while the ANN alone reached a value of 0.86. For comparative purposes, 4 empirical predictors were also evaluated, reaching R^2 values from 0.58 to 0.70 [131, 156–158].

Shirani Faradonbeh and Monjezi (2017) proposed a GEP model to predict blast-induced PPV in an iron mine in Iran, considering 115 blast instances and 8 independent parameters as input, of which 20% of these instances were used to test and validate the model. Moreover, a cuckoo optimization algorithm (COA) approach is employed to optimize the blast pattern. COA is a metaheuristic inspired by the behavior of this family of birds, who never build their nests, thus ensuring the continuity of the species through the nests of other birds, where they enter, eliminate an egg from the host and replace it with their egg, which the unsuspecting host must hatch. First, a cuckoo population is generated in a habitat, of which each bird lays several eggs in the host nests within a radial distance of the habitat, besides eggs less similar (less fitness value) to the host's eggs have more chance to be eliminated, while the remaining eggs would generate the cuckoo chicks. Then, K-means cluster these chicks, so the cluster with the highest average fitness is chosen by cuckoos from other groups as the next destination, where new habitat is built and another iteration proceeds, aiming to find the settlement where eggs are more likely to survive, and food resources are available. The GEP parameters were defined, and the result was compared to that of a nonlinear MVRA, reaching R^2 values of 0.874 and 0.579, respectively, while 5 conventional predictors reached values from 0.620 to 0.742. Sensitivity analysis identified the distance from the blast face, the maximum charge per delay, and powder factor as the most influential parameters in ground vibration. Finally, after choosing the GEP equation as the objective function and the COA parameters, 8 strategies were proposed, and the corresponding blast patterns were obtained, reaching a maximum decrease in the ground vibration of up to 55.33%. Figure 13.39 presents an example of COA, representing two nests close to the cuckoo habitat, where cuckoo's eggs (starry) are more similar to the host's eggs in the Nest b [159].

Khandelwal et al. (2017) used a classification and regression tree (CART) model to predict blast-induced PPV in a coal mine in India, considering 51 blast instances, 30% of which were used to test the model, comprising distance from the blast face and maximum charge per delay as input. CART is a nonparametric tree algorithm that can identify the significant variables and eliminate non-significant ones, not requiring transformation or removal of outliers for its application. This algorithm selects a variable as the root, applying rules to divide the data across nodes by evaluating the value of a variable when a "yes or no" answer is returned to a question, thus generating branches across sub-nodes. The process continues splitting the data until a stopping criterion is reached when the end nodes have their predicted values calculated. As a result, the proposed methodology reached an R^2 value of 0.9, while the MVRA reached a value of 0.58, and 3 empirical predictors reached values from 0.67 to 0.84. Hasanipanah et al. (2017a) also employed a CART model to predict blast-induced PPV in a copper mine in India, considering 86 blast instances, 20% of which were used to test the model, encompassing distance from the blast face and

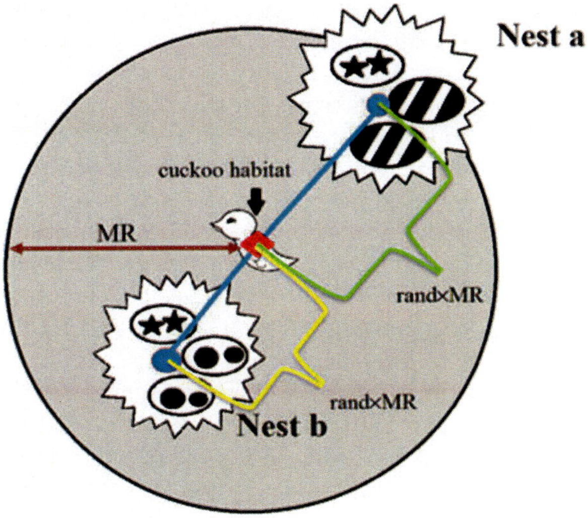

Fig. 13.39 Example of cuckoo optimization algorithm (COA), representing two nests close to the cuckoo habitat, where cuckoo's eggs (starry) are more similar to the host's eggs in the Nest b [160]

maximum charge per delay as input. The CART methodology reached an R^2 value of 0.950, while the MVRA reached 0.883, and 4 empirical predictors reached values from 0.796 to 0.851. Figure 13.40 presents the tree structure of the CART model used in Khandelwal et al. (2017), where the nodes are divided by ranges of values presented when the data is asked about the distance from the blast face and maximum charge per delay [161, 162].

Hasanipanah et al. (2017b) employed two GA models to predict blast-induced PPV from tunneling operations in the vicinity of a dam in Iran, considering 85 blast instances, comprising distance from the blast face and maximum charge per

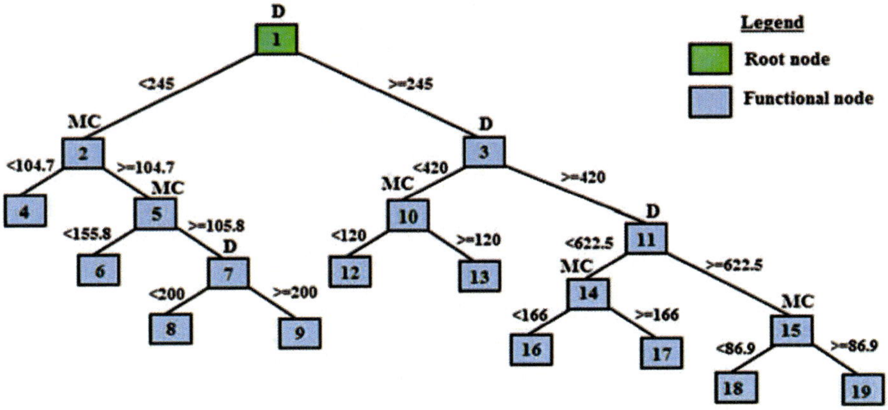

Fig. 13.40 Tree structure of the classification and regression tree (CART) model was used in Khandelwal et al. (2017) [162]

delay as input. GA's linear and power forms reached R^2 values of 0.873 and 0.920, respectively, while 7 empirical predictors reached values from 0.4 to 0.826, with Roy's predictor being the most accurate among them, although both GA models achieved better predictions. On the other hand, Hasanipanah et al. (2017c) used two PSO models to predict the blast-induced PPV from two mines in the vicinity of a dam in Iran, considering 80 blast instances, encompassing distance from the blast face and maximum charge per delay as input. The linear and power forms of PSO reached R^2 values of 0.901 and 0.938, respectively, while an MVRA and the USBM predictor reached 0.869 and 0.867, respectively. Fouladgar, Hasanipanah, and Bakhshandeh Amnieh (2017) employed the COA to predict blast-induced PPV from blasts in a copper mine in Iran, considering 85 blast instances, comprising distance from the blast face and maximum charge per delay as input. The COA reached an R^2 value of 0.957, while 7 empirical predictors reached 0.894 to 0.925. Jahed Armaghani et al. (2018a) used two ICA models to predict the blast-induced PPV from three quarry mines in Malaysia, considering 73 blast instances, encompassing distance from the blast face and maximum charge per delay as input. The power and quadratic forms of ICA reached R^2 values of 0.93 and 0.94, respectively, while 4 empirical predictors reached values from 0.875 to 0.897. In turn, Behzadafshar et al. (2018) employed three ICA models to predict blast-induced PPV from operations in two mines near a dam in Iran, considering 76 blast instances, 20% of which were used to test the models. Comprising 5 independent parameters as input, the linear, power, and quadratic forms of the ICA achieved R^2 values of 0.939, 0.935, and 0.920, respectively, while an ANN with 5-6-1 architecture reached a value of 0.911 and 4 empirical predictors reached values from 0.865 to 0.904 [89, 97, 163–167].

Abbaszadeh Shahri and Asheghi (2018) used a regular multilayer perceptron BPNN and a generalized feed-forward neural network (GFNN) to predict the blast-induced PPV in tunneling operations near a dam in Iran, considering 37 blast instances, 45% of which were used to test and validate the models. A GFNN is a generalization of the regular multilayer perceptron, although its connections can jump over one or more layers, making it capable of solving problems in considerably fewer epochs than a BPNN with the same number of neurons. Considering the total charge, distance from the blast face, and maximum charge per delay as input, the BPNN with 3-3-2-3-1 architecture and the GFNN with 3-4-3-1 architecture reached R^2 values of 0.932 and 0.954, respectively, while 6 empirical predictors reached values from 0.440 to 0.860. Mokfi et al. (2018) used a group method of data handling (GMDH) model to predict blast-induced PPV at a quarry mine in Malaysia, the same dataset used in Shirani Faradonbeh et al. (2016a), considering 102 blast instances, 20% of which were used to test and validate the model. GMDH is an ANN with a self-organizing heuristic that finds its optimal structure, using a percentage of the training dataset by a least squares technique to find polynomials (since the GMDH derives a relationship between input and output in polynomial form) and the remaining training data to assess the number of neurons that would result in the optimal solution. Encompassing 6 independent variables as input, the optimal GMDH parameters were defined by trial and error, reaching a 6-16-20-20-1 architecture and an R^2 value of 0.911, superior to the previously evaluated GEP and

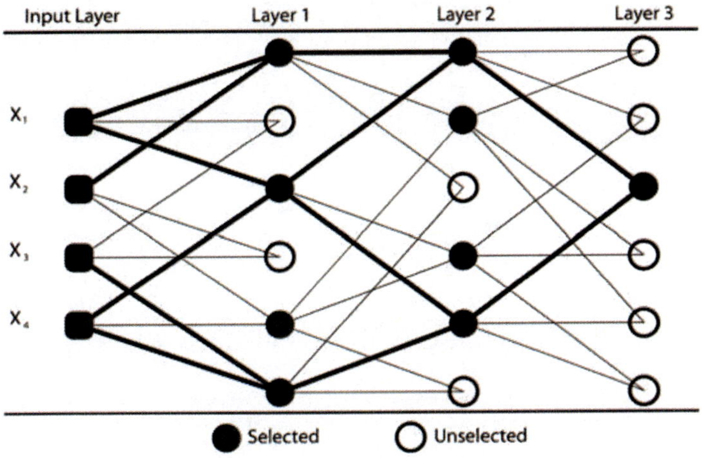

Fig. 13.41 Example of group method of data handling (GMDH) model deriving its architecture [169]

nonlinear MVRA models, which reached R^2 values of 0.874 and 0.790, respectively. Figure 13.41 presents a GMDH model deriving its architecture [147, 168, 169].

Garai et al. (2018) developed distinct models based on ANN, RF, and empirical scaled distance regression to predict the blast-induced PPV in a coal mine in India, aiming to develop predictions for three initiation systems, i.e., detonating cord with cord relay, nonelectric detonators, and electronic detonators. For this purpose, 128 blast instance data were collected, and the distance from the blast face and maximum charge per delay were used as input into the models. For the detonating cord, nonelectric and electronic detonators, the ANN models reached R^2 values of 0.966, 0.966, and 0.980, respectively; the RF models reached 0.953, 0.957, and 0.979, respectively; and the regression models reached values of 0.942, 0.943, and 0.942, respectively. Sheykhi et al. (2018) employed combinations of SVM, USBM empirical predictor, and fuzzy C-means clustering (FCM) to predict blast-induced PPV in a copper mine in Iran, considering 120 blast instances, 15% of which were used to test the models, also encompassing 6 independent variables as input. FCM (also called soft clustering) is a clustering technique that splits the data by applying fuzzy logic, in which each data point (observation) is assigned to one or more clusters, thus achieving similarity measures (e.g., distance and connectivity) aiming group members with high similarity to each other and high dissimilarity to members of other clusters. In the hybrid formulations, the FCM was defined to generate 3 clusters of training data considering 6 input variables. Hence, the SVM or USBM regressor was employed to predict the PPV values for each cluster, only considering the distance from the blast face and maximum charge per delay to make them comparable. Therefore, the testing dataset is also grouped by FCM and has PPV values predicted by SVM or USBM regressor. The FCM-SVM and FCM-USBM achieved R^2 values of 0.853 and 0.494, respectively, while the SVM and USBM approach reached 0.708 and 0.354,

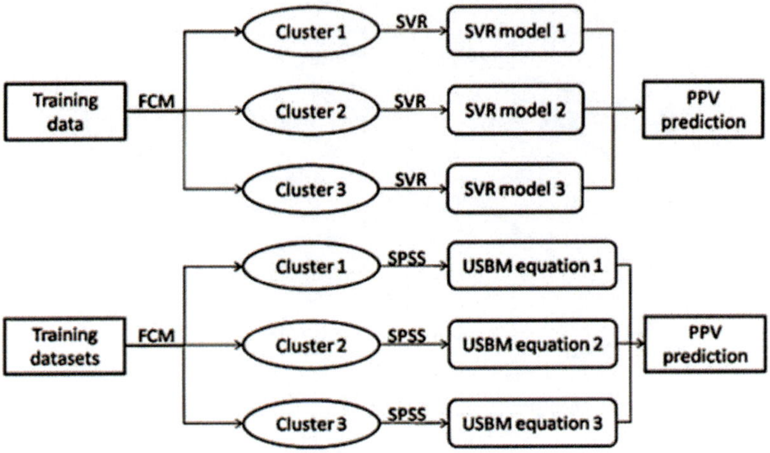

Fig. 13.42 Scheme of the fuzzy C-means-support vector machine (FCM-SVM) and fuzzy C-means-USBM (FCM-USBM) approaches [171]

respectively, indicating that the use of FCM increased the accuracy of predictions. Figure 13.42 presents the scheme of the FCM-SVM and FCM-USBM approaches [89, 170, 171].

Zhongya and Xiaoguang (2018) compared the performances of a BPNN, MVRA, and an extreme learning machine (ELM) when optimized by dimensionality reduction using factor analysis and mean impact value (FA-MIV) in predicting blast-induced PPV of a dataset with 108 blast instances, 10% of which were used to test the models, comprising 9 independent variables as input. ELM is a least squares-based ANN with a single hidden layer applied to solve regression and classification problems, which randomly defines the input weights and hidden biases so that the parameters of the hidden nodes do not require adjustment, providing high speed of learning high generalization ability. FA is an extension of PCA, in which common factors are expressed as a linear combination of variables and their factor scores are computed, even though they can be rotated to alter the distribution of the different factors when the common factors have little significance. Hence, the factors that represent most of the information from the original variables must be extracted (those who represent 85% of the total variance explained), and the principal components are represented as a linear combination of common factors, thus decreasing repetitive information in high-dimensional space to transform it into a low-dimensional space. In turn, the MIV is an index that determines the effect exerted by the input neurons on the output neurons, and in this FA-MIV approach aims to rank the weights of each principal component in the output parameter. Therefore, the two principal components with minimum weights can be eliminated, allowing the PPV to be explained in a low-dimensional space with four input characteristic parameters. BPNN reached R^2 values of 0.7614 and 0.5513 with or without the use of FA-MIV; similarly, the MVRA reached values of 0.5811 and 0.4047, while the ELM reached R^2 values of

0.9604 and 0.7856 with the use or not of the FA-MIV, corroborating the robustness and accuracy of the ELM and FA-MIV techniques [172].

Hasanipanah et al. (2018b) proposed a FIS-ICA approach to predict blast-induced PPV at a copper mine in Iran, considering 50 blast instances, 20% of which were used for testing, encompassing distance from the blast face and maximum charge per delay as input. The FIS-ICA approach reached an R^2 value of 0.942, while 4 empirical predictors reached values from 0.519 to 0.638. Prashanth and Nimaje (2018) compared the performances of a BPNN with architecture 8-70-1, an RBFNN with architecture 8-80-1, a cascade forward neural network (CFNN) with architecture 8-70-1, and an SVM model to predict PPV in a coal mine in India, considering 121 blast instances, 30% of which were used for testing, encompassing 8 independent variables as input. A CFNN is similar to a feed-forward BPNN, although its inputs computed after each hidden layer are backpropagated to the input layer, and the weights are updated since there are connections from this type of ANN input all previous layers to the following layers. The BPNN, RBFNN, CFNN, and SVM models achieved R^2 values of 0.9581, 0.9918, 0.9300, and 0.9043, respectively. Ragam and Nimaje (2018) employed a GRNN with 2-14-1 architecture to predict blast-induced PPV at an explosives manufacturer's site in India, considering 14 blast instances, 30% of which were used to test the model, comprising distance from the blast face and maximum charge per delay as input. The GRNN reached an R^2 value of 0.9988, while 5 empirical predictors reached values from 0.31097 to 0.69816. Iramina et al. (2018) compared the performances of an ANN with 1-3-3-1 architecture and the Geological Strength Index (GSI) and Rock Quality Designation (RQD) geomechanical parameter relationships to predict blast-induced PPV in a quarry mine in Brazil, using the equations proposed by Kumar, Choudhury, and Bhargava (2016). The RMSE values for an empirical USBM-based equation, an ANN, RQD, and GSI approaches were 3.63, 3.25, 7.81, and 4.36, respectively, indicating that the geomechanical parameter relationships had a lower predictive capacity [89, 173–177].

Nguyen et al. (2019a) developed 5 different ANN architectures to predict blast-induced PPV in a coal mine in Vietnam, considering 68 blast instances, 20% of which were used to test the models, comprising distance from the blast face maximum charge per delay as input. The models achieved R^2 values from 0.936 to 0.973, with the 2-10-8-5-1 architecture was superior, while the Ambraseys–Hendron equation reached 0.768. Nguyen et al. (2019b) extended their previous work, evaluating 9 different ANN architectures to predict blast-induced PPV in a coal mine in Vietnam, considering 136 blast instances, 15% of which were used to test the models, encompassing 7 independent variables as input. The models reached R^2 values from 0.839 to 0.980, with the 7-10-8-5-1 architecture being the superior, while the USBM and MVRA models reached R^2 values of 0.816 and 0.947, respectively. Sensitivity analysis by standardized rank regression coefficients (SRRC) indicated that the maximum charge per delay was the most influential parameter. Nguyen et al. (2019c) proposed an eXtreme gradient boosting (XGBoost) algorithm to predict blast-induced PPV in a coal mine in Vietnam, comparing their results to those of the SVM, RF, and k-NN methodologies, considering 146 blast instances, 20% of which were used to test the models, comprising 9 independent variables as input. XGBoost builds boosted trees and operates in parallel to solve classification and regression problems, implementing machine learning algorithms in a gradient boosting structure and aiming to optimize the value of an objective function generally consisting of a training loss function, employed to evaluate the model's performance, and a regularization term, which controls the complexity of the model. The SVM, RF, and k-NN approaches reached R^2 values of 0.934, 0.947, and 0.859, respectively, while the XGBoost model reached 0.952. A sensitivity analysis showed that the distance from the blast face and the maximum charge per delay was the most influential parameters [89, 92, 178–180].

Nguyen et al. (2019d) developed a hybrid model based on hierarchical K-means clustering (HKM) and cubist algorithm (CA) to predict blast-induced PPV in a coal mine in Vietnam, considering 136 blast instances, 20% of which were used to test the models, comprising 7 independent variables as input. The HKM clustering technique aims to restart the k-means clustering several times for better results since k-means models set their initial centroids randomly, so all instances are defined as clusters, and their centroids are recorded, these clusters being grouped recursively hierarchically until a predefined number of iterations and final clusters are reached. On the other hand, CA is a robust tree-based algorithm that can solve regression problems for thousands of input attributes by first branching all the data into a tree, then building a regression model on each node, pruning the tree to avoid overfitting and smoothing it to amend for discontinuities due to data splitting. Therefore, for each of the two clusters created by the HKM algorithm from the data, a CA model is created to predict the PPV in the proposed HKM-CA approach. An empirical model based on the USBM predictor reached an R^2 value of 0.816, while the RF, SVM, CART, and CA models developed for comparison purposes reached values of 0.984, 0.960, 0.968, and 0.989, respectively, the HKM-CA model being the superior, as it reached an R^2 value of 0.995. A sensitivity analysis showed that the elevation from the blast face to the monitoring point was the most influential parameter. Nguyen, Bui, and Moayedi (2019) developed an ANN with architecture 2-6-8-6-1, k-NN, SVM, and CART models to predict blast-induced PPV in a coal mine in Vietnam, considering 68 blast instances, 20% of which were used to test the models, encompassing maximum charge per delay and distance from the blast face as input. Three empirical regressors were applied, reaching R^2 values from 0.529 to 0.933. The Ghosh–Daemen model was superior, while the ANN, k-NN, SVM, and CART models reached 0.981 0.737, 0.886, and 0.618, respectively. Nguyen (2019) compared three types of kernel function (linear, polynomial, and RBF) in SVR models to predict blast-induced PPV in a coal mine in Vietnam, considering 181 blast instances, 15% of which were used to test the models, comprising the distance from the blast face and the maximum charge per delay as input. The models reached R^2 values from 0.915 to 0.924, with RBF being the best kernel function, and an equation based on the USBM predictor reached the value of 0.643 [89, 94, 181–183].

Nguyen et al. (2019e) compared the performances of an SVM, an ANN with 6-8-6-1 architecture, and BGAMs models to predict blast-induced PPV in a coal mine in Vietnam, considering 79 blast instances, 20% of which were used for testing the models, encompassing 6 independent variables as input. The SVM, BGAMs, and ANN models reached R^2 values of 0.975, 0.990, and 0.970, respectively, while an equation based on the Ambraseys and Hendron predictor reached 0.787. Chen et al. (2019) combined three metaheuristics, i.e., FFA, GA, and PSO, with an ANN with 6-5-1 architecture and an SVR model to predict blast-induced PPV in a quarry mine in Malaysia, considering 95 blast instances, 20% of which were used to test the models, comprising 6 independent variables as input. The three hybrid ANN models reached R^2 values from 0.924 to 0.937, while the SVR models reached values from 0.957 to 0.977, with the SVR-PSO model being superior, highlighting the superiority of SVR models over ANN models. Furthermore, a modified FFA algorithm was proposed to improve the SVR-FFA model, in which instead of a full pairwise search, the parameters are adjusted to avoid this search in fireflies that are very distant from each other, thus reaching an R^2 value of 0.984, a higher value when compared to the other propositions. Das, Sinha, and Ganguly (2019) developed an ANN with 15-4-1 architecture to predict blast-induced PPV in three coal mines in India, considering 248 blast instances, 30% of which were used to test the models, comprising 15 independent variables as input. The ANN and MVRA models reached R^2 values of 0.97 and 0.80, respectively, while 4 empirical predictors reached values from 0.63 to 0.74. A sensitivity analysis indicated that the distance from the blast face and the maximum charge per delay was the most influential parameters. Shang et al. (2019) compared the performance of ANN-FFA, k-NN, CART, and SVM models in predicting blast-induced PPV at a quarry mine in Vietnam, considering 83 blast instances, 20% of which were used to test the models, encompassing 5 independent variables as input. The k-NN, CART, and SVM models reached R^2 values of 0.851, 0.899, and 0.927, respectively, while the ANN-FFA model, which has a 5-16-20-1 architecture, reached the value of 0.966. A sensitivity analysis indicated that the maximum charge per delay and the distance from the blast face were the most influential parameters [92, 184–187].

Torres et al. (2019) proposed an ANN with 13-15-1 architecture to predict blast-induced peak vector sum (PVS) at an iron mine in Brazil, considering 133 blast instances, 30% of which were used to validate and test the model, encompassing 13 independent variables as input, reaching an R^2 value of 0.9242. Xue (2019) proposed combinations of subtractive clustering algorithm (SCA) and FCM with an ANFIS model to predict blast-induced PPV in a limestone mine in Iran, from a dataset used in Hosseini and Baghikhani (2013), considering 79 blast instances, 29 of which were used to test the model. SCA is an unsupervised clustering algorithm that considers all data points as potential cluster centers and measures the density at each data point so that data points with many neighbors have higher potential values. The data point with higher potential is elected as the first cluster center, then the density and potential for each data point are recalculated, and a new cluster center is defined, a procedure that continues until all cluster centers are found. When combined with an ANFIS, the SCA and FCM act by generating tuned membership functions according to the dataset's domain, thus improving the ANFIS performance. Encompassing the distance from the blast face, the maximum charge per delay, and scaled distance as input, the SCA-ANFIS, and FCM-ANFIS models achieved R^2 values of 0.9231 and 0.9299, respectively, while 3 empirical predictors achieved values from 0.6183 to 0.7052. Zhang et al. (2019) developed an XGBoost-PSO model to predict blast-induced PPV in a quarry mine in Vietnam, considering 175 blast instances, 20% of which were used to test the model, encompassing 5 independent variables as input. The XGBoost-PSO model reached an R^2 value of 0.968, while 3 empirical predictors reached values from 0.545 to 0.686. Jiang et al. (2019) employed an ANFIS model to predict blast-induced PPV in the vicinity of a dam in Iran, considering 90 blast instances, 20% of which were used to test the model, comprising distance from the blast face and maximum charge per delay as input. The ANFIS reached an R^2 value of 0.983, while an MVRA reached 0.876 [188–192].

Arthur, Temeng, and Ziggah (2019) compared the performance of a wavelet neural network (WNN) with 5-3-1 architecture to a BPNN with 5-1-1 architecture, RBFNN with 5-9-1 architecture, GRNN, and GMDH in predicting blast-induced PPV in a manganese mine in Ghana, considering 210 blast instances, 80 of which were used to test the model, encompassing 5 independent variables as input. WNN is similar to a BPAA, although it uses a wavelet function as the activation function instead of sigmoid or hyperbolic functions, usually employing the Gaussian, Mexican hat, and Morlet wavelet functions. The R^2 values obtained from the WNN, BPNN, RBFNN, GRNN, and GMDH methodologies were 0.7120, 0.7288, 0.7185, 0.6869, and 0.6870, respectively, while 4 empirical predictors reached values from 0.5574 to 0.6136. Yang et al. (2019) used ANFIS, ANFIS-PSO, and ANFIS-GA models to predict blast-induced PPV at two quarry mines in Iran, considering 86 blast instances, 20% of which were used to test the models, comprising 6 independent variables as input. The R^2 achieved by ANFIS, ANFIS-PSO, and ANFIS-GA were 0.884, 0.966, and 0.979, respectively, while 2 empirical predictors reached values of 0.845 to 0.874, in addition to a sensitivity analysis that indicated the maximum charge per delay as the most influential parameter. Bui et al. (2019a) employed a k-NN-PSO approach with three different kernel functions (i.e., quartic, tri weight, and cosine) to predict blast-induced PPV in a coal mine in Vietnam, considering 152 blast instances, 20% of which were used to test the models, using distance from the blast face and explosive charge per delay as input. In this proposition, the PSO is used to optimize the k-NN hyper-parameters, and the use of quartic, tri weight, and cosine kernels reached R^2 values of 0.964, 0.977, and 0.960, respectively, while an empirical regressor based on the USBM, RF, and SVR techniques reached values of 0.579, 0.952, and 0.944, respectively. Azimi, Khoshrou, and Osanloo (2019) used an ANN-GA model with 2-8-16-1 architecture to predict blast-induced PPV in a copper mine in Iran, considering 70 blast instances, 30% of which were used to validate and test the model—comprising maximum charge per delay and 3 different types of measuring distance as input verified that using a modified radial distance results in better predictions. The ANN-GA and ANFIS models reached R^2 values of 0.9883 and 0.9203, respectively, while 5 empirical models using this modified radial distance reached 0.804–0.853 [193–196].

Nguyen et al. (2020a) employed four different combinations of evolutionary algorithms and SVR to predict blast-induced PPV in a limestone mine in Vietnam, i.e., SVR-PSO, SVR-GA, SVR-ICA, and SVR-ABC, using three types of the kernel function, i.e., linear, polynomial and RBF, thus developing 12 different hybrid models.

One hundred twenty-five blast instances were assessed, 20% of which were used to test the models, comprising 4 independent variables as input. The models reached R^2 values from 0.901 to 0.991, with RBF the best kernel function and SVR-GA-RBF the best-evaluated model. Nguyen et al. (2020b) proposed an HKM-ANN approach to predict blast-induced PPV in a limestone mine in Vietnam, considering 185 blast instances, 30% of which were used to validate and test the model encompassing 5 independent variables as input. In this proposition, the HKM was used to define 2 clusters, then two ANN models with 5-10-6-1 and 5-8-7-1 architectures were used to predict the PPV in each cluster, reaching an R^2 value of 0.988, while an HKM-SVR reached a value of 0.940. The FCM strategy was also used for comparison purposes so that the FCM-ANN and FCM-SVR models reached R^2 values of 0.976 and 0.942, respectively. An SVR and empirical approaches reached R^2 values of 0.956 and 0.781, respectively, while 4 ANNs with different architectures reached values from 0.955 to 0.965. Shakeri, Shokri, Dehghani (2020) used an ANN with 5-20-1 architecture, GEP, and MVRA models to predict blast-induced PPV in a copper mine in Iran, considering 113 blast instances 30% of which were used to test the model, comprising 5 independent parameters as input. The MVRA, ANN, and GEP models reached R^2 values of 0.7020, 0.8862, and 0.9114, respectively, and a sensitivity analysis showed that the distance from the blast face and the explosive charge per delay were the most influential variables [197–199].

Arthur, Temeng, and Ziggah (2020a) assessed the performance of 13 BPNN training algorithms in predicting blast-induced PPV in a manganese mine in Ghana in terms of the number of epochs, MSE and R indexes, considering 210 blast instances, 80 of which were used to test the models, encompassing 5 independent variables as input. ANN's optimal architecture was defined as 5-1-1, and the Broyden–Fletcher–Goldfarb–Shanno Quasi-Newton algorithm achieved the highest R values and lowest MSE values Levenberg–Marquardt algorithm proved to be the fastest, these being the two algorithms with the best overall performance. Arthur, Temeng, and Ziggah (2020b) employed multivariate adaptive regression splines (MARS) to predict blast-induced PPV in a manganese mine in Ghana, considering 210 blast instances, 80 of which were used to test the models, encompassing 5 independent variables as input. MARS consists of a nonparametric multivariate technique based on recursive partition and adaptive regression splines method that can automatically map the nonlinear relationship between input and output variables through a series of coefficients and piecewise polynomials derived from the regression data. First, a forward selection iteratively adds basis function pairs to reach the lowest training error in the construction phase. Then, a backward deletion phase iteratively removes the least contributing basis functions to avoid overfitting, and a generalized cross-validation criterion achieves the final MARS model. The MARS model reached an R^2 value of 0.7074, while BPNN with 5-1-1 architecture, GRNN, and RBFNN with 5-13-1 architecture models reached values of 0.6879, 0.6869, and 0.6762, respectively, while 4 empirical predictors reached values from 0.5574 to 0.6136. Figure 13.43 presents two datasets with the true functions (green) and the MARS functions (red), (a) having small error variance and (b) having high error variance [200, 201].

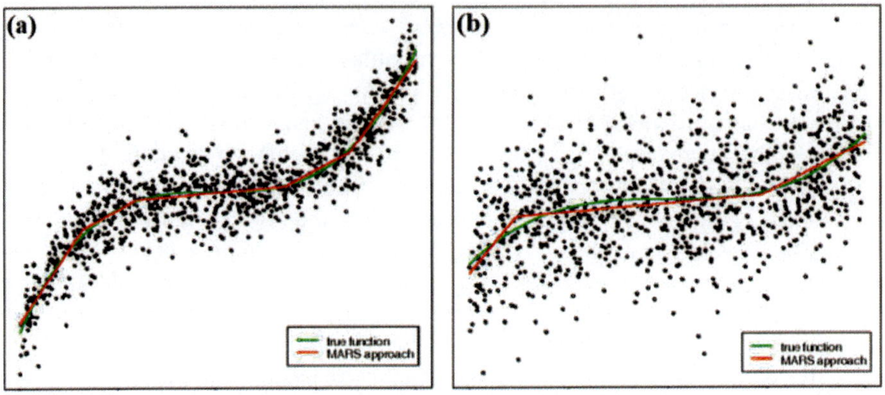

Fig. 13.43 Two datasets with the true functions (green) and the multivariate adaptive regression splines (MARS) approaches (red), **a** having small error variance and **b** having high error variance [202]

Arthur and Kaunda (2020) proposed a combination of the Paretosearch algorithm and goal attainment method to determine blast design parameters that would minimize the blast-induced ground vibration and maximize the production at a quarry mine in the USA in a multiobjective optimization model, encompassing 51 blast instances, 18% of which were used for testing, considering 8 independent variables as input. Paretosearch technique is a pattern search algorithm that iteratively finds optimal non-dominant Pareto solutions in multiobjective problems, projecting an initial series of points in a linear subspace of linearly feasible points by solving a linear programming problem, thus using a polling algorithm (e.g., mesh adaptive direct search) with a related mesh size to discover the non-dominant points in each iteration. If a non-dominated point is in the polled points, the poll is a success, and the algorithm doubles the mesh size repeatedly in the successful directions, extending the poll until a dominated point is produced. Otherwise, the algorithm continues to poll until a non-dominated point is found or until it runs out of points, when the poll is considered unsuccessful and Paretosearch halves its mesh size. After polling, the algorithm measures the volume of the non-dominated points in the hyper-volume of the objective function and the spread of the Pareto set, being these two possible stopping criteria, and this procedure theoretically converges to points close to the true Pareto front.

On the other hand, goal attainment is a multiobjective nonlinear optimization method that minimizes a series of objectives related to a series of goals expressed in vectors, using a slack variable to express the trade-offs between objectives and sequential quadratic programming (SQP) in its solution. Therefore, the proposed method consists of executing the goal attainment method at the solutions points of Paretosearch, trying to find blast parameters that satisfy the objective functions and constraints, with production maximization modeled as a nonlinear regression

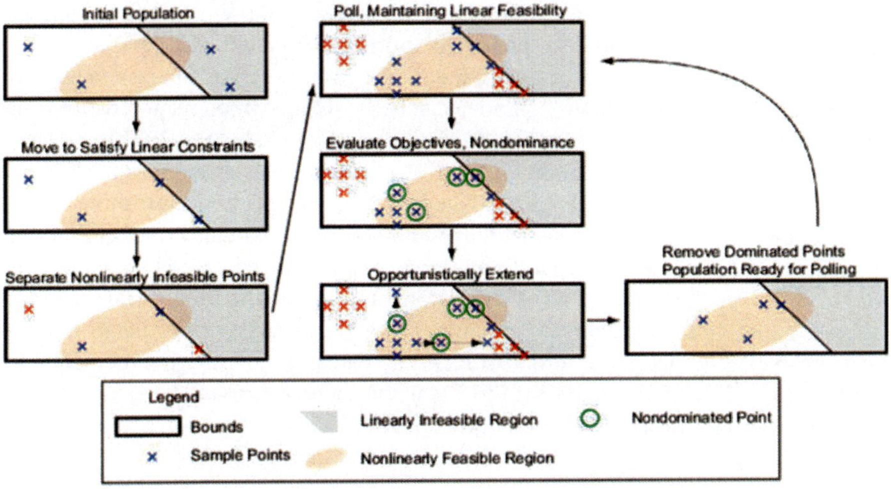

Fig. 13.44 Paretosearch algorithm scheme [204]

model and blast-induced ground vibration minimization modeled as a linear regression model. Compared to a Paretosearch model, the hybrid Paretosearch-goal attainment method performed slightly better, optimizing blast parameters according to the constraints. Figure 13.44 presents a Paretosearch algorithm scheme [203].

Bayat et al. (2020) used an ANN with 4-23-1 architecture to predict blast-induced PPV in a limestone mine in Iran and an FFA with 151 fireflies to optimize the blast design in order to minimize the vibration, considering 151 blast instances, 20% of which were used to test the model, comprising 4 independent parameters as input. The ANN reached an R^2 value of 0.977, in addition to the FFA approach, which reduced the PPV by 60%, increasing the burden by 1%, reducing spacing by 10%, and maximum charge per delay by 88%. Sensitivity analysis by the CAM method indicated that the maximum charge per delay was the most influential parameter. Fang et al. (2020a) proposed an M5Rules-ICA model to predict blast-induced PPV in a quarry mine in Vietnam, considering 125 blast instances, 20% of which were used for testing, encompassing 5 independent variables as input. M5Rules is a rule-based technique that first builds an M5 tree model for training data, prunes the tree, and generates a series of if–then rules through a partial regression tree algorithm. Hence, the best leaf is turned into a rule. The tree is discarded, dataset instances are removed and replaced by the generated rule. Thus, the remaining instances are recursively submitted to this process until all have been included in the rules when a complete tree is built rather than a partially explored tree. The M5Rules-ICA model reached an R^2 value of 0.995, while an M5Rules, RF, and SVM models reached 0.971, 0.988, and 0.983, respectively, while 2 empirical models reached values of 0.696 and 0.774. A sensitivity analysis showed that the maximum charge per delay was the most influential parameter. Lawal and Idris (2020) used an ANN with 3-3-1 architecture to predict blast-induced PPV in a sandstone quarry in Turkey, considering the 88 blast instances Hudaverdi (2012) used for training, besides another 14 instanced for validating and testing, comprising 3 independent parameters as input. The ANN reached an R^2 value of 0.88, while the hierarchical classification by Hudaverdi (2012) and an MVRA model reached 0.84 and 0.68, respectively, while 4 empirical predictors reached values from 0.10 to 0.45 [122, 205–207].

Amiri, Hasanipanah, and Amnieh (2020) proposed a combination of ANN and itemset mining (IM) to predict blast-induced PPV in andesite and a tuff mine in Iran, considering 92 blast instances, using leave-one-out cross-validation and comprising 6 independent parameters as input. Frequent itemset mining is based on data mining, which is a process that extracts knowledge from raw data, so IM aims to extract groups of objects that often co-occur in a database, thus developing rules related to those frequent itemsets extracted in order to facilitate decision making. IM methodology employed in this study uses the Apriori algorithm, which extracts frequent itemsets by an iterative level-wise search technique. The FindSupport function calculates each candidate's support and evaluates those that are frequent or not the function CandidateGeneration, which generates the frequent candidates. The dataset was normalized and discretized, so 20% were used to extract frequent and confident itemsets by IM. Hence, according to the extracted patterns, an ANN with 6-4-1 architecture was trained considering the training instances chosen based on the confident frequent itemsets. The ANN-IM model reached an R^2 value of 0.944, while the ANN reached a value of 0.898, and 3 empirical predictors reached values from 0.810 to 0.883, highlighting the accuracy of the proposed methodology. Yang et al. (2020) proposed SVR models combined with metaheuristics to predict blast-induced PPV in two quarry mines in Iran, considering 90 blast instances, 30% of which were used for testing purposes, encompassing 5 independent parameters as input. Therefore, the SVR, ANN-FFA, SVR-FFA, SVR-PSO, and SVR-AG models reached R^2 values of 0.969, 0.946, 0.992, 0.979, and 0.971, respectively, while 5 empirical predictors reached values from 0.853 to 0.909 [208, 209].

Wei et al. (2020) evaluated the performance of a nested-ELM model in predicting blast-induced PPV from four different datasets, from a set with few instances and few input parameters to another with many instances and many input parameters comparing their results to ELM, ELM-PSO, and ELM-SA models. Since the ELM's hidden node parameters are random, little or no adjustment is made to the hidden nodes, requiring more hidden nodes compared to a regular BPNN, thus increasing the complexity and reducing the algorithm's efficiency. Therefore, the nested-ELM aims to optimize the number of hidden nodes and parameters, taking the random input weights and biases generated by ELM to the hidden nodes, defining the fitness function, using roulette wheel selection to define the parameters that meet the fitness function requirements and then optimizing the network structure, adjusting the number of hidden nodes and their parameters. Both datasets with few instances (35) and with many instances (105) had 5 separate instances for testing and the rest for training, and it was verified that the nested-ELM achieved the lowest mean training time, MAPE and RMSE values, using 10 or 61 hidden nodes, depending on the dataset evaluated. Although ELM had the lowest mean training than the other hybrid algorithms, its MAPE and RMSE tend to be higher. Jahed Armaghani et al. (2020a) proposed a GMDH and a generalized structure GMDH (GS-GMDH) models to predict PPV in the vicinity of a dam in Iran, considering 96 blast instances, 30% of which were used to test the model, encompassing three independent parameters as input. While the GMDH algorithm allows two variables as inputs for each neuron, which are selected from adjacent neurons, and presents a second-order polynomial structure, the proposed GS-GMDH algorithm allows two or three variables as inputs for each neuron, and the inputs can be selected from adjacent or non-adjacent layers, presenting second and third-orders polynomial structures. Four variants of each algorithm were developed, considering the three variables as input or combinations of two variables, models with three inputs being the superior, in which the best GS-GMDH and GMDH reached R^2 values of 0.942 and 0.908, respectively [210, 211].

Jahed Armaghani et al. (2020b) proposed combining ELM algorithm with PSO and autonomous groups particles swarm optimization (AGPSO) to predict blast-induced PPV in a quarry mine in Malaysia, considering the 102 blast instances used in Shirani Faradonbeh et al. (2016a), encompassing 6 independent parameters as input. Minimax probability machine regression (MPMR), least squares support vector machine (LS-SVM), and GP models were also evaluated for comparison purposes. AGPSO is a generalization of PSO based on individual diversity in termite colonies that allow particles to have different cognitive and social parameter strategies when searching within a search space, whereas in PSO, all particles have the same parameters. Therefore, the AGPSO aims to overcome the fact of being trapped in local minima and the slow convergence rates in solving high-dimensional problems presented by the PSO, and three distinct types of AGPSO were employed in this study in hybridization with ELM, looking for optimizing ELM input weights and hidden biases, likewise for ELM-PSO. On the other hand, MPMR assumes knowledge of statistics of predictors and the predicted value and then maximizes the minimum probability of the regression model considering all feasible distributions with known mean and covariance matrix, directly computing the probability within the $\pm\varepsilon$ bounds on the created regression surface. LS-SVM is an extension of SVM that converts inequality constraints into linear constraints and directly transforms two programming problems into linear equations, increasing the computational speed and the classification ability. The ELM-APSO model reached an R^2 value of 0.84, and the three versions of the ELM-AGPSO reached 0.87 to 0.9, while the MPMR, LS-SVM, and GP models reached values of 0.8, 0.87, and 0.84, respectively. Figure 13.45 presents a comparison between a sync function with Gaussian noise and an MPMR model (a) and a two-spiral classification problem with two classes solved by LS-SVM (b) [147, 212].

Mahdiyar et al. (2020) used a GEP model to predict blast-induced PPV in four quarry mines in Malaysia and Monte Carlo simulations (MCS) to assess the possible range of PPV values, considering 149 blast instances, 20% of which were used to test the model, encompassing 5 independent parameters as input and achieving an R^2 value of 0.6823 in the prediction. Computing the minimum and maximum values of the variables, defining their related distribution functions, and carrying out 10,000 simulations, the results indicated that the PPV could assume values from 1.13 to

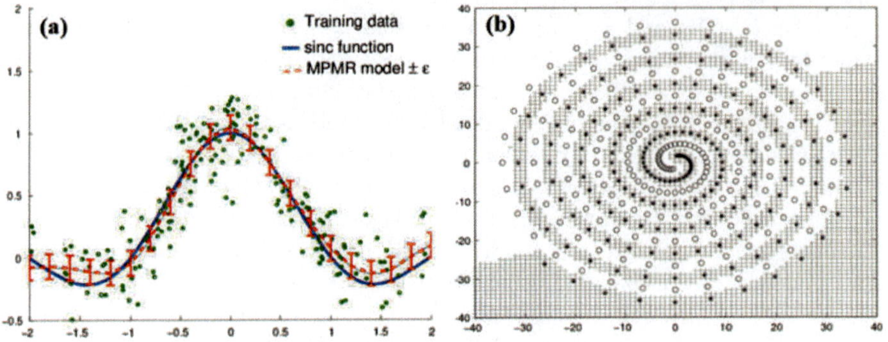

Fig. 13.45 A comparison between a sync function with Gaussian noise and a minimax probability machine regression (MPMR) model (**a**) and a two-spiral classification problem with two classes solved by least squares support vector machine (LS-SVM) (**b**) [213, 214]

34.58 mm/s. Li et al. (2020) proposed a combination of biogeography-based optimization (BBO) with ANN (ANN-BBO), an ANN-DIRECT, and three other models (i.e., ANN-PSO, MPMR, and ELM) to predict blast-induced PPV at two mines nearby a dam in Iran, considering the 80 blast instances, 30% of which were used in testing, encompassing 7 independent parameters as input. BBO is an evolutionary algorithm inspired by concepts of biogeography such as migration of species from a habitat (island) to another, emergence and extinction of inhabitants in these habitats. The migration process is the chief operator, which involves rates of immigration and emigration to modify the characteristics of habitats, maintaining the population of one candidate while combining the population of the others, iterating up to a given measure of quality or a certain number of iterations were achieved. DIRECT is a deterministic optimization algorithm that uses a robust direct search approach to find the global optimum of the objective function without requiring any further information. In the proposed formulations, both BBO and DIRECT were used to optimize ANN's weights and biases. The ANN-BBO and ANN-DIRECT models reached R^2 values of 0.988 and 0.981, respectively, while the ANN-PSO, MPMR, and ELM models reached 0.972, 0.971, and 0.965, respectively, and 3 empirical predictors reached values from 0.724 to 0.799. A sensitivity analysis showed that the maximum charge per delay was the most influential parameter. Figure 13.46 presents a BBO algorithm training an ANN [215, 216].

Ding et al. (2020a) proposed an XGBoost, XGBoost-ICA, XGBoost-PSO, GBM, an ANN with 7-12-15-1 architecture and SVM models to predict blast-induced PPV at a coal mine in Vietnam, considering 136 blast instances, 20% of which were used to test the model, comprising 7 independent parameters as input. The R^2 values achieved by XGBoost-ICA, XGBoost-PSO, XGBoost, SVM, ANN, and GBM models were 0.988, 0.977, 0.965, 0.965, 0.945, and 0.935, respectively, and a sensibility analysis by Sobol method showed that elevation from the blasting point was the most influential parameter in predicting PPV with XGBoost-ICA. Ding et al. (2020b) compared the performances of a bagged support vector regression with FFA (BSVR-FFA),

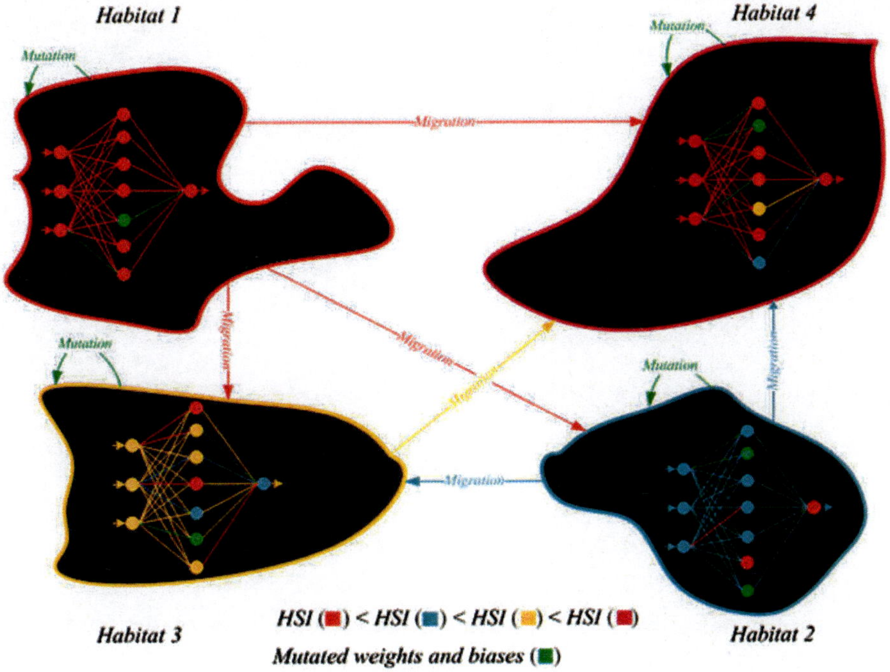

Fig. 13.46 A biogeography-based optimization (BBO) algorithm training an ANN [217]

a BPNN with 2-4-1 architecture, and RBFN models to predict blast-induced PPV nearby a dam in Iran, considering 87 blast instances, 20% of which were used to test the model, encompassing distance from the blast face and explosive charge per delay as input. Bagging is an ensemble algorithm that relies on bootstrapping and aggregating, in which randomly sampled training sets are generated from the original training set and, in the proposed model, the hybrid BSVR had its weights modified by FFA to increase its performance. As a result, the BSVR-FA, BPNN, and RBFN models achieved R^2 values of 0.996, 0.896, and 0.828, respectively, while a sensibility analysis showed that explosive charge per delay was the most influential parameter [218, 219].

Yu et al. (2020a) developed a hybrid RF with Harris hawks optimization (RF-HHO) model to predict blast-induced PPV in a copper mine in China, considering 137 blast instances, 20% of which were used to test the model, comprising 7 independent parameters as input. HHO algorithm is inspired by the social behavior of Harris hawks, which are cooperative predators, as they commonly locate, assault, and share prey with other individuals, and consists of three phases, i.e., exploration, the transition from exploration to exploitation, and exploitation phases. In the first phase, Harris hawks explore and find prey in the desert, and the individuals' position and an average position are computed and, according to the escaping energy displayed by the prey, they decide to remain in the transition phase or attack. Then,

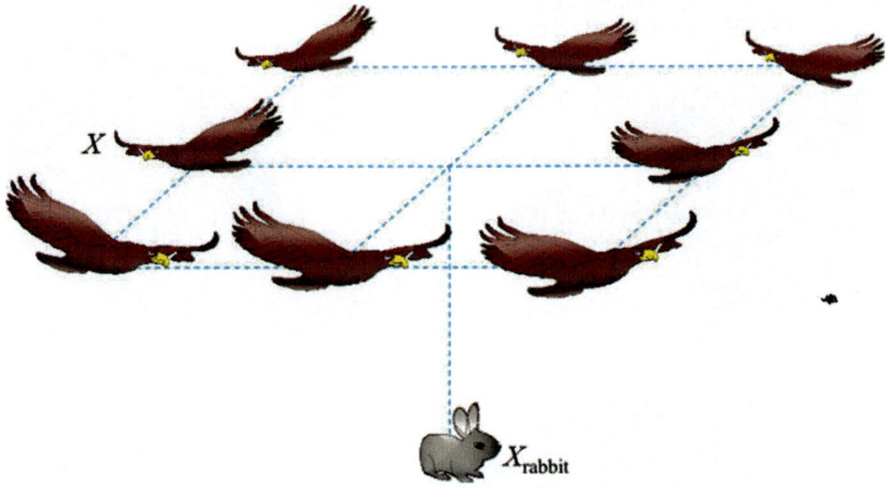

Fig. 13.47 Relationship between Harris Hawks and Prey in the Harris Hawks optimization algorithm (HHO) [220]

in the exploitation phase, distinct situations are modeled based on the assessment of the possibility of escape by the prey using a range of random values from 0 to 1, reflecting escaping behaviors distinct from the prey, from high energy, when the probability of a successful attack by the Harris hawks is low, to exhaustion, when this probability is high. In the proposed RF-HHO model, the HHO algorithm searches the optimal combination of variables to grow each try and the number of trees in RF, which is used to predict the PPV, reaching an R^2 value of 0.94, while an RF model reached a value of 0.90. In addition, an MCS approach was performed, resulting in an average PPV value of 0.98 cm/s and indicating that the PPV value does not exceed 1.95 cm/s with 90% probability, whereas a sensitivity analysis showed that the distance from the blast face is the most influential parameters. Figure 13.47 presents the relationship between Harris hawks and prey in the HHO algorithm [220].

Yu et al. (2020b) proposed a hybrid relevance vector machine (RVM) with PSO and gray wolf optimizer (GWO) model, comparing the performances of three single kernel functions and three multikernel functions in order to predict blast-induced PPV in a copper mine and a coal mine in China. The copper dataset considered 137 blast instances, from which 20 were removed as outliers and 20% of the remainder were used to test the models, encompassing 7 independent parameters as input. In contrast, the coal dataset considered 50 blast instances, 20% of which were used to test the models, encompassing 11 independent parameters as input. RVM is a machine learning technique with a function similar to SVM, but as it is based on Bayesian structure, it provides parsimonious solutions for regression and probabilistic classification, using a learning method based on expectation maximization. GWO algorithm is a metaheuristic inspired by the behavior of gray wolves, whose pack is divided into four groups: the alpha (α), which is the leader, responsible for making decisions; the beta (β) individuals, which reinforce the alpha's orders throughout the pack; the delta (δ) group, composed of scouts, sentinels, hunters, etc.; and the omega (ω) group, whose function is act as the scapegoat. Thus, when prey is located, the pack uses the processes of encircling, hunting, and attacking, so the GWO algorithm employs coefficient vectors, the current position vectors of the prey, and wolves in each iteration until the prey is attacked, or the optimal solution is reached. In the proposed hybrid formulation, the PSO improves the optimization capability of the GWO, which is used to optimize the RVM parameters, and a tenfold cross-validation stage is also employed. In the copper dataset, the RVM-PSO-GWO model using 6 different kernel functions reached R^2 values from 0.4878 to 0.9179, with the RBF function being the best, while an ANN with 7-9-1 architecture reached a value of 0.7973 and 7 empirical predictors reached values from 0.1587 to 0.4338. In turn, in the coal dataset, the 6 RVM-PSO-GWO versions reached R^2 values from 0.9079 to 0.9708, with the hybrid RBF with the sigmoid function being the best, while an ANN with 7-4-1 architecture reached a value of 0.9858 and 7 empirical predictors reached values from 0.3950 to 0.9896. Figure 13.48 presents the four groups of a pack of wolves in the GWO algorithm [221].

Zhang et al. (2020b) compared the performances of 5 machine learning classifiers, i.e., CART, chi-squared automatic interaction detection (CHAID), RF, ANN, and SVM, to predict blast-induced PPV in a quarry mine, considering 102 blast instances, 30% of which were used to test the models, comprising 6 independent parameters as input. Like CART, CHAID is a nonparametric decision tree algorithm that identifies significant variables and eliminates non-significant ones while automatically pruning the tree during its generation. However, its combinations and splits are based on a chi-square test, and more settings were required to produce a CHAID. A feature selection (FS) phase was used to reduce the dimensionality of the data, as this technique identifies the correlation between input and output variables to select those input with the highest correlations, so 5 independent variables were chosen as input to the models. In addition, the powder factor was the most important

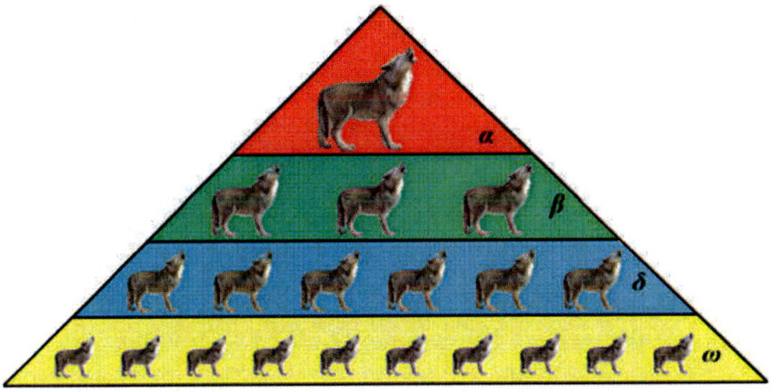

Fig. 13.48 Four groups of a pack of wolves in the gray wolf optimizer (GWO) algorithm [221]

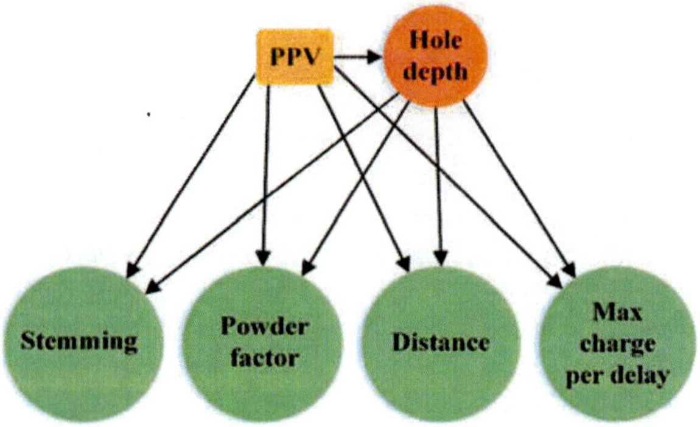

Fig. 13.49 Bayesian network (BN) graph to evaluate the probability of PPV values, where the hole depth was chosen to form the binary compositions [223]

variable in the CHAID, RF, and SVM models, whereas the distance from the blast face was the most important variable in the CART and ANN models. The R^2 values achieved by the CART, CHAID, RF, ANN, and SVM models were 0.56, 0.68, 0.83, 0.84, and 0.85, respectively. Zhou et al. (2020a) proposed hybrid FS-RF and FS with Bayesian Network (FS-BN) models to predict blast-induced PPV in a quarry mine, considering 102 blast instances, 30% of which were used to test the models, encompassing 6 independent parameters as input. BN is a graphical model that represents multivariate probability distributions, in which the network nodes depict a series of random variables, while the directed arcs represent the causal relationships between these variables. The tree augmented naive Bayes (TAN) method can address the interdependencies among the variables. Of the 6 measured input variables, FS chose 5 to feed the models so that the level of accuracy achieved by FS-RF and FS-BN models were 90.32% and 87.09%, respectively, while 5 empirical predictors reached R^2 values from 0.275 to 0.654 and an equation developed for this case study reached an R^2 value of 0.672. Figure 13.49 presents the BN graph to assess the probability of PPV values, where the hole depth was chosen to form the binary compositions [222, 223].

Bui et al. (2020a) proposed a combination of FCM and quantile regression neural network (FCM-QRNN) to predict blast-induced PPV in a quarry mine in Vietnam, considering 83 blast instances, 20% of which were used to test the models, comprising 8 independent parameters as input. QRNN combines ANN with quantile regression principles, which calculates the conditional median of the target on distinct resource values, as opposed to linear regressions, which calculates the conditional mean of the target, resulting in a robust ANN with a single hidden layer able to calculate conditional quantiles of the predictive distribution by employing asymmetric weights to positive/negative errors. The author used an unmanned aerial vehicle (UAV) to assist in data collection. It was verified that the generation of two clusters by FCM, with 34 and 37 blast instances, and the 8-2-1 architecture for QRNN resulted in a better prediction in the training phase, where the FCM-QRNN model reached an R^2 value of 0.952 in the testing phase considering clusters with 4 and 8 instances. On the other hand, an ANN with 8-8-12-6-1 architecture, a QRNN, and an RF model reached R^2 values of 0.946, 0.920, and 0.894, respectively, while an empirical model based on the USBM predictor reached the value of 0.658. Lawal (2020) used an ANN with 2-5-1 architecture and an MVRA model to predict blast-induced PPV at five quarry mines in Nigeria, considering a dataset with 100 blast instances, 20% of which were used for testing, encompassing the distance from the blast face and the maximum charge per delay as input, achieving R^2 values of 0.988 and 0.738, respectively [89, 224, 225].

Zhou et al. (2021b) proposed a GEP model to predict blast-induced PPV in a quarry mine, considering 102 blast instances, 20% of which were used to test the model, encompassing 6 independent parameters as input, reaching R^2 values from 0.864 to 0.880 among the 5 models generated. Moreover, the extraction and multiplication of the expression trees of the best GEP model result in a PPV prediction equation, from which 10,000 MCS scenarios were performed to evaluate the risk in operation, where PPV range between 0 mm/s and 10.8 mm/s was verified, with 99% probability of being less than 8 mm/s. A sensitivity analysis showed that the distance from the blast face was the most influential parameter. Lawal, Kwon, and Kim (2021) proposed an ANN with 6-5-1 architecture, a hybrid ANN with moth-flame optimization algorithm (ANN-MFO), GEP and MVRA models to predict blast-induced PPV from tunnel construction in South Korea, considering 56 blast instances, 30% of which were used to validate and test the models, comprising 6 independent parameters as input. MFO is a metaheuristic inspired by the transverse orientation mechanism that moths use when flying toward a distant light source and the spiral movement they perform when flying around a nearby artificial light, thus changing their movement from 1 to 3D by adjusting their position vectors. MFO is a three-tuple-based algorithm composed of a function that generates a random population of moths and the related fitness values, another function that moves the moths in the search space, and a stopping function to respond when the termination criteria is reached, or when the global optimization is achieved. Once the ANN architecture is defined and the algorithm started in the proposed formulation, the MFO optimizes its weights and biases. The ANN, ANN-MFO, GEP, and MVRA models achieved adjusted R^2 values of 0.6914, 0.9577, 0.4991, and 0.6266, respectively, while the Langefors–Kihlstrom and Bureau of Indian Standard predictors reached values of 0.4469 and −0.2573, respectively. The sensitivity analysis by the weight partitioning method showed that the maximum charge per delay was the most influential variable. Figure 13.50 presents the transverse orientation of moths (left) and their spiral movement (right) [89, 93, 226, 227].

Zhu, Nikafshan Rad, and Hasanipanah (2021) proposed a combination of chaos recurrent adaptive neuro-fuzzy inference system (CRANFIS) and PSO (CRANFIS-PSO) to predict blast-induced PPV in operations nearby a dam in Iran, comparing the results to those obtained by an ANN with 6-7-1 architecture, ANFIS, recurrent ANFIS (RANFIS), and CRANFIS models. Considering the Embedding Theorem, even if the source and dynamics of the chaotic behaviors in a system are inaccessible, a chaotic system achieves and stores information about the hidden states of a dynamical system, thus ensuring equivalent dynamics of the system and reducing their complexity and dimension. A dataset can be considered chaotic by measuring the fractal dimension

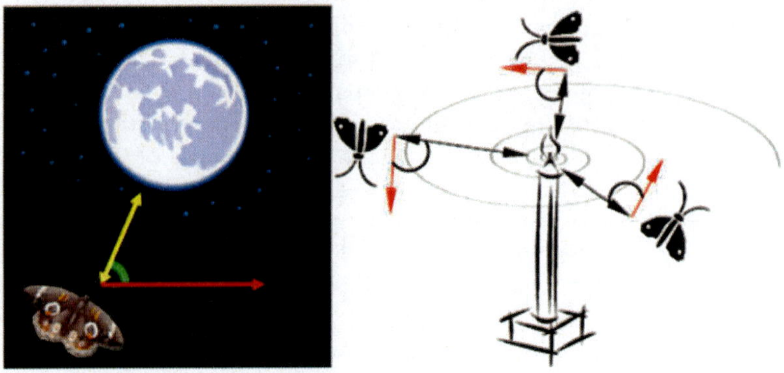

Fig. 13.50 Transverse orientation of moths (left) and their spiral movement (right) [228]

using a mutual information function and the correlation dimension algorithm. In addition, the Lyapunov exponents' algorithm can evaluate the system response to small perturbations, which measures its degree of convergence or divergence. RANFIS is a machine learning method capable of generating estimates by receiving feedback from its prediction model, so once a dataset is proved to be chaotic, its fractal dimensions (lags) are measured and adjusted to feed a RANFIS, thus composing a CRANFIS. Both the RANFIS and CRANFIS models were implemented using the FCM strategy, and the CRANFIS-PSO version uses the PSO algorithm to improve the value of the parameters in membership functions. Considering 84 blast instances, 20% of which were used to test the models, and comprising 6 independent parameters as input, CRANFIS-PSO, ANN, ANFIS, RANFIS, and CRANFIS models achieved R^2 values of 0.997, 0.775, 0.882, 0.958, and 0.967, respectively, while 4 empirical predictors reached values from 0.587 to 0.806. A sensitivity analysis showed that the burden was the most influential parameter [229].

Qiu et al. (2021) employed combinations of the GWO, whale optimization algorithm (WOA), and Bayesian optimization algorithm (BO) with XGBoost models in order to predict blast-induced PPV in a coal mine in India, considering 150 blast instances, 20% of which were used to test the model, encompassing 13 independent parameters as input. WOA is a population-based metaheuristic inspired by the predatory behavior of humpback whales, whose hunting method (bubble-net feeding) consists of first encircling the prey (a fish school), evaluating its location by echo, thus updating their position by contractive enveloping and spiral mechanisms while trapping the prey in the bubble net, culminating in capture as their movement converges. The BO algorithm is a robust technique used for hyper-parameter tuning when the unknown goal function is modeled as a multidimensional Gaussian distribution by GP, thus assuming a high-dimensional normal distribution form, so the algorithm learns the results of the objective function from the previous sampling points to achieve a combination of hyper-parameter that will result in the global optimum. In addition, the authors compared these results to those generated by CatBoost (CatB, Categorical Boosting), gradient boosting regression (GBR), XGBoost, and RF models, with CatB being a gradient boosting framework that uses ordered boosting methodology in binary decision trees to make its predictions. The GWO, WOA, and BO approaches were employed to optimize the parameters in XGBoost, and the R^2 values achieved by the XGBoost-GWO, XGBoost-WOA, XGBoost-BO, CatB, GBR, XGBoost, and RF models were 0.9751, 0.9757, 0.9727, 0.9208, 0.9656, 0.9592, and 0.9386, respectively. Figure 13.51 presents the spiral motion strategy and the bubble-net mechanism used in the WOA (a), the surrounding shrink mechanism (b), and the spiral updating position mechanism (c) [230].

Lawal et al. (2021) proposed a combination of sine cosine algorithm (SCA) and an ANN with 4-5-1-1 architecture (ANN-SCA), a GEP, and an ANFIS model to predict blast-induced PPV at five quarry mines in Nigeria, considering 100 blast instances, encompassing 4 independent parameters as input. SCA is a population-based metaheuristic inspired by the mathematical function that localizes the optimal solution in the loop of these two curves. The algorithm generates multiple initial random candidate solutions that float through the search space toward the global optimum using a mathematical model of sine and cosine functions. SCA comprises

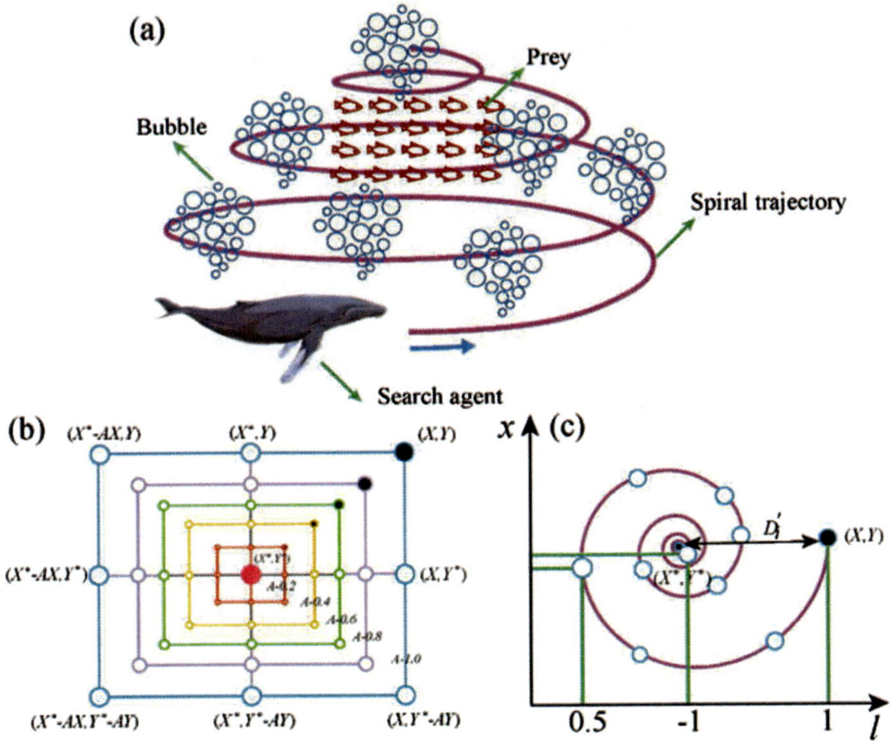

Fig. 13.51 Spiral motion strategy and the bubble-net mechanism are used in the whale optimization algorithm (WOA) **a**, the surrounding shrink mechanism **b**, and the spiral updating position mechanism **c** [230]

exploration and exploitation phases, distinguished by the degree of randomness in each phase, and in the proposed ANN-SCA model, the SCA phase is used to optimize ANN weights and biases. The SCA-ANN, GEP, and ANFIS models reached R^2 values of 0.999, 0.989, and 0.997, respectively, while the USBM predictor reached 0.679. Yu et al. (2021) assessed the performances of an ELM model and its combination with HHO and grasshopper optimization algorithm (GOA) metaheuristics to predict blast-induced PPV in a quarry mine in Malaysia, considering 166 blast instances, 20% of which were used for testing, comprising 6 independent parameters as input. GOA is a metaheuristic inspired by the herding pattern of grasshoppers, whose individuals in a swarm are subject to three types of influences that determine their behavior, i.e., social relationship, gravity force, and wind advection. Naturally, there is a strong repulsion between two grasshoppers if the distance between them is in the range from 0 to 2.079, the latter being the comfort zone; otherwise, if they are far apart, in the range from 2.079 to 4, there is a strong attraction, decaying beyond this value asymptotically. In addition, a shrinkage coefficient is multiplied to reduce the comfort, repulsion, and attraction zones to balance the exploration and exploitation phases, aiming to converge the swarm toward a solution. The gravitational force vector points toward the center of the earth, while the wind advection points toward a target (food source), and the sum of these three factors (including the social relationship) multiplied by random numbers ranging from 0 to 1 result in the location of the swarm toward the space of feasible solutions. The ELM, ELM-HHO, and ELM-GOA models achieved R^2 values of 0.8501, 0.9001, and 0.9105, respectively. Figure 13.52 presents a GOA scheme and its attraction and repulsion forces [231, 232].

Bui et al. (2021) compared the performances of an ANN with 6-18-16-6-1 architecture, an ANN-CSO, SVM, and tree-based ensemble models to predict blast-induced PPV in a quarry mine in Vietnam, considering 118 blast instances for training and 3 new blasting events for validation and testing, encompassing 6 independent parameters as input. The ANN-CSO, ANN, SVM, and tree-based ensemble models achieved R^2 values of 0.987, 0.990, 0.960, and 0.932, respectively, while two empirical predictors achieved 0.229 and 0.227, and a sensitivity analysis showed that the

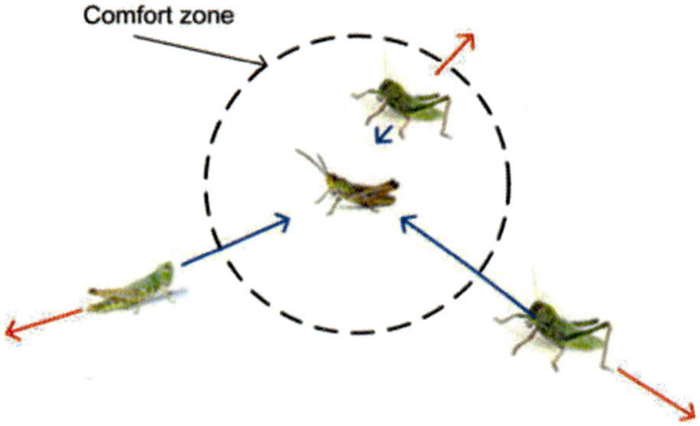

Fig. 13.52 Grasshopper optimization algorithm (GOA) scheme and its attraction and repulsion forces [233]

distance from the blast face was the most influential parameter. Nguyen and Buy (2021) proposed combining an ANN with 9-10-5-1 architecture with hunger games search (HGS), GOA, FFA, and PSO metaheuristics to predict blast-induced PPV in a coal mine in Vietnam, considering 252 blast instances, 30% of which were used for training, comprising 9 independent parameters as input. HGS is a two-stage meta-heuristic inspired by the cooperative behavior of animals to find food sources as they struggle to survive against starvation. The first stage consists of the behavior of the individuals (who have different levels of team spirit) in search of food sources in the feasible space and communicating between them when food is found, with random numbers controlling the hunting pattern of the animals to avoid trapping in local optima. Hence, the second stage controls each individual's hunger level since satisfied animals have no desire to hunt, unlike starving animals, and random numbers are also employed to perturb the system. The ANN-HGS, ANN-GOA, ANN-FFA, and ANN-PSO models achieved R^2 values of 0.922, 0.921, 0.899, and 0.893, respectively, and a sensitivity analysis showed that burden, distance from the blast face, and the number of blasting groups were the most influential parameter [234, 235].

Over the past decade, more than 100 papers have been published aiming to employ advanced analytics and machine learning techniques to predict blast-induced ground vibration, many of which have been covered in this topic, and as new methods emerge, they are rapidly incorporated into the list of feasible techniques by researchers. For more information, see Trivedi et al. (2014) and Dumakor-Dupey, Arya, and Jha (2021) reviews [84, 85].

Air Overpressure (Air Blast)

When a charge is detonated, air vibrations radiate in all directions, which frequency, amplitude, and duration of the vibrations at a given place depending on the detonation's size, blast confinement, and atmospheric conditions. Since fluids (like air) have no shear strength, there are no S waves in the atmosphere, so only P waves are considered in this kind of analysis. Air overpressure (AOP) is a transient pressure change in a fluid medium generated by explosions, in which the pressure suddenly increases and decreases beyond average pressure. The AOP measurement computes the absolute value of the largest pressure change, regardless of whether it was positive or negative, while its frequency is determined by the number of cycles occurring in one second. Figure 13.53 presents the most relevant factors affecting air overpressure [8, 236].

Siskind et al. (1980b) proposed a blast-based air overpressure criteria considering measuring equipment with different features and the maximum acceptable noise level achieved, being one of the most used references worldwide in terms of AOP. Table 13.2 presents the USBM RI 8485 blast-based air overpressure criteria proposed by Siskind et al. (1980b) [238].

Furthermore, several experiments were carried out to obtain general equations for AOP prediction, comprising different parameters in their construction, such as

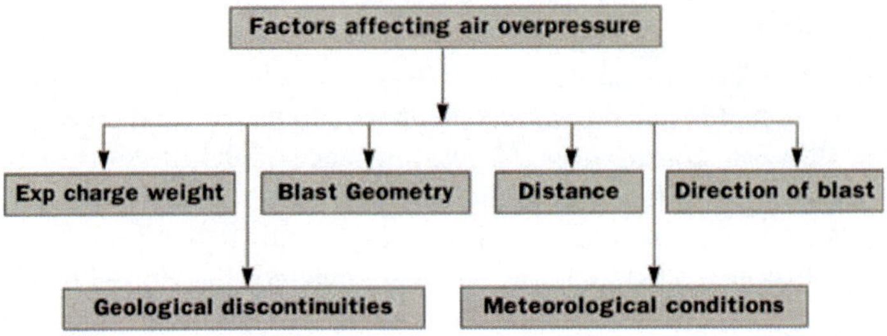

Fig. 13.53 Most relevant factors affecting air overpressure [237]

Table 13.2 USBM RI 8485 blast-based air overpressure criteria, proposed by Siskind et al. (1980b) [238]

Lower frequency limit of measuring system	Maximum level
0.1 Hz high pass system (flat response)	134 dB
2 Hz high pass system (flat response)	133 dB
6 Hz high pass system (flat response)	129 dB
C-weighted system in events with less than 2 s (slow response)	105 dB

distinct rock types, weather conditions, explosives, monitoring distances, etc. Considering the AOP in dB, W as the maximum charge per delay, in kg, D as the distance between the blast face and vibration monitoring point, in meters, and site constants such as k, β_1, and β_2, Table 13.3 presents empirical blast-induced AOP predictors. Similar to the blast-induced PPV empirical predictors, in many situations, these equations can make forecasts far from reality, chiefly when compared to machine learning and advanced analytics techniques.

Khandelwal and Singh (2005) employed an ANN with 2–5-1 architecture to predict blast-induced AOP in two limestone mines in India, using a dataset with 41 blast instances, while 15 instances from a magnesite mine were used to validate the model, considering the distance from the blast face and explosive charge per delay as input. The ANN model reached an R^2 value of 0.9574, while the values

Table 13.3 Most used empirical blast-induced air overpressure predictors [238–243]

Empirical AOp predictors	Equations
Holmberg–Persson (1979)	$\text{AOp} = 0.7\,\text{k}\,\text{W}^{1/3}/D$ (mbar)
USBM: Siskind et al. (1980b)	$\text{AOp} = \beta_1[D/W^{1/3}]^{\beta_2}$
NAASRA (1982)	$\text{AOp} = 140[(\text{W}/200)^{1/3}]/\text{D}$ (kPa)
Mckenzie (1990)	$\text{AOp} = 165 - 24\log D/W^{1/3}$ (dB)
Ollofson (1990); Persson–Holmberg–Jaimin (1994)	$\text{AOp} = 0.7\,W^{1/3}/D$ (mbar)

obtained by an equation based on the USBM predictor and MVRA were 0.3811 and 0.5258, respectively. Sawmliana et al. (2007) used an ANN with 4-7-1 architecture to predict blast-induced AOP at four mines in India, using a dataset with 70 blast instances, while 25 instances from two other mines were used to validate the model, considering 4 independent variables as input. The ANN model reached an R^2 value of 0.931, while the value obtained by the USBM predictor was 0.867. A sensitivity analysis indicated that the distance from the blast face was the most influential parameter. Khandelwal and Kankar (2011) used an SVM to predict blast-induced AOP at two limestones and dolomite mine and magnesite mines in India, using a dataset with 75 blast instances, of which the 55 instances of the limestone mines were used for training, while the remaining 20 instances of the magnesite mine were used to test the model. Considering the distance from the blast face and explosive charge per delay as input, the SVM model reached an R^2 value of 0.855, while the USBM predictor reached 0.587. As aforementioned, Mohamed (2011) developed an ANFIS to predict blast-induced PPV and AOP in a limestone mine in Egypt, reaching R^2 values of 0.95 and 0.93, respectively [118, 237, 238, 244, 245].

Hajihassani el at. (2014) used an ANN-PSO model with 9-12-1 architecture and a swarm size of 275 particles to predict blast-induced AOP at four quarry mines in Malaysia, using a dataset with 62 blast instances, of which 20% were used for testing. Considering 9 independent variables as input, the ANN-PSO model reached an R-value of 0.91, while four empirical equations were also assessed, although their indexes were not presented. As mentioned before, Jahed Armaghani et al. (2015b) employed ANN and ANFIS models to predict blast-induced PPV, AOP, and flyrock at four quarry mines in Malaysia, achieving R^2 values of 0.864 and 0.947, respectively, for the predicted AOP. Jahed Armaghani et al. (2015c) used an ANN-ICA model with 2-4-1 architecture, 45 countries, and 5 imperialists to predict blast-induced AOP in a quarry mine in Malaysia, using a dataset with 95 blast instances. Encompassing distance from the blast face and explosive charge per delay as input, the ANN-ICA model reached an R^2 value of 0.982, while an empirical equation reached 0.724. Finally, Jahed Armaghani et al. (2015d) proposed an ANFIS model to predict blast-induced AOP at three quarry mines in Malaysia, using a dataset with 128 blast instances, 20% of which were used for testing, comprising 5 independent variables as input. The ANFIS model reached an R^2 value of 0.971, while an ANN with 5-4-1 architecture and an MVRA model reached 0.914 and 0.766, respectively, while 3 empirical predictors reached values from 0.634 to 0.689. Sensitivity analysis by CAM indicated that the maximum charge per delay and distance from the blast face was the most influential parameter. As aforementioned, Hajihassani el at. (2015a) used an ANN-PSO model to predict blast-induced PPV and AOP at a quarry mine in Malaysia, achieving average R^2 and MSE values of 0.89 and 0.038, respectively [139, 142, 246–248].

Hasanipanah et al. (2016) compared the performances of several empirical, ANN, FIS, and ANFIS models to predict blast-induced AOP in a copper mine in Iran, using a dataset with 77 blast instances, 20% of which were used for testing, encompassing distance from the blast face and explosive charge per delay as input. The best ANFIS model reached the R^2 value of 0.954, while the best FIS, ANN, and empirical models reached 0.878, 0.797, and 0.650, respectively. Tonnizam Mohamad et al. (2016) compared the performances of several empirical, ANN-GA and ANN models to predict blast-induced AOP in a quarry mine in Malaysia, using a dataset with 76 blast instances, 20% of which were used to test the models, comprising the distance from the blast face and the explosive charge per delay as input. The best ANN-GA model reached the R^2 value of 0.974, while the best ANN and empirical models reached 0.902 and 0.782, respectively. Jahed Armaghani, Hasanipanah, and Tonnizam Mohamad (2016) compared the performances of several empirical, ANN-ICA and ANN models to predict blast-induced AOP from two mines in the vicinity of a dam in Iran, using a dataset with 70 blast instances, 20% of which were used for testing, encompassing distance from the blast face and explosive charge per delay as input. The best ANN-ICA model reached the R^2 value of 0.961, while the best ANN and empirical models reached 0.882 and 0.886, respectively. As aforementioned, Amiri et al. (2016) employed an ANN-k-NN model to simultaneously predict blast-induced PPV and AOP at two quarry mines near a dam in Iran, reaching an R^2 value of 0.95 for AOP [149, 249–251].

Hasanipanah et al. (2017d) employed three SVM-PSO models to predict blast-induced AOP from operations at two mines near a dam in Iran, considering 67 blast instances, 20% of which were used to test the models, comprising distance from the blast face and explosive charge per delay as input. The SVM-PSO approach's linear, quadratic, and RBF variants reached R^2 values of 0.9695, 0.9865, and 0.9968, respectively, while an MVRA reached 0.8675. Shirani Faradonbeh et al. (2018a) employed GP and GEP models to predict blast-induced AOP in a copper mine in Iran, considering 92 blast instances, 20% of which were used for testing, encompassing distance from the blast face and maximum charge per delay as input. The GP and GEP models reached R^2 values of 0.928 and 0.941, respectively, while an MVRA reached 0.903 and 3 empirical predictors reached values from 0.901 to 0.908. Jahed Armaghani et al. (2018b) proposed an ANN with 2-5-1 architecture and an ANN-GA model to predict blast-induced AOP in a quarry mine in Malaysia, considering 97 blast instances, 20% of which were used to test the models, comprising distance from the blast face and maximum charge per delay as input. The ANN and ANN-GA models reached R^2 values of 0.857 and 0.965, respectively, while an MVRA and the USBM predictor reached 0.82 and 0.774, respectively. Alel et al. (2018) compared the performances of 8 ANN-PSO models, using PSO with genetic operators (i.e., crossover and mutation) and multiswarm optimization (MSO) approaches to predict blast-induced AOP in a mine in Malaysia, considering 76 blast instances, 30% of which were used to validate and test the models. MSO is a variant of PSO that uses multiple sub-swarms in its solution, preventing the swarm from converging to local optima, and the main difference in the formulation is the inclusion of a term related to the swarm optima. Encompassing distance from the blast face and maximum charge per delay as input, the ANN-PSO models achieved R^2 values from 0.9619 to 0.9729, with ANN-MSO using linear decreasing inertia weight and genetic operators being the best, while the USBM predictor achieved a value of 0.8450. Figure 13.54 presents the MSO mechanism [238, 252–255].

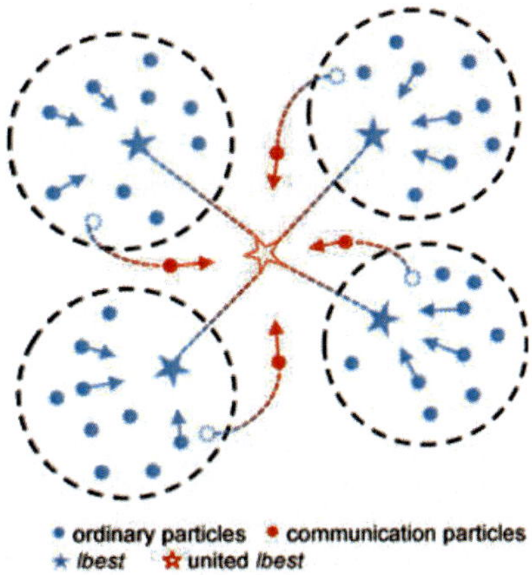

Fig. 13.54 Multiswarm optimization (MSO) mechanism [256]

AminShokravi et al. (2018) three PSO models to predict the blast-induced AOP of operations near a dam in Iran, considering 80 blast instances, 20% of which were used to test the models, comprising 3 independent parameters as input. The linear, power, and quadratic forms of PSO reached R^2 values of 0.960, 0.923, and 0.926, respectively, while an ANN with 3-4-1 architecture reached 0.897, and the USBM predictor reached the value of 0.872. Mahdiyar, Marto, and Mirhosseinei (2018) proposed an MVRA to predict the blast-induced AOP in a quarry mine in Malaysia, considering 76 blast instances and encompassing 8 independent parameters as input, thus employing 10,000 MCS scenarios to evaluate the risk in operation. The MVRA achieved the R^2 value of 0.901, while the AOP range was verified between 67.17 and 134.24 dB, with a 99% probability of being less than 134 dB. A sensitivity analysis showed that the distance from the blast face was the most influential parameter. Nguyen and Bui (2019) compared the performance of 5 ANN architectures, an RF model, and an ANN-RF in predicting blast-induced AOP in a coal mine in Vietnam, considering 114 blast instances, 20% of which were used for testing, comprising 7 independent variables as input. In this ANN-RF proposition, a stacking technique was applied to ensemble the five ANN models using RF algorithm, reaching an R^2 value of 0.985, while the 5 ANN models reached values from 0.942 to 0.966, the RF model reached a value of 0.939, and an empirical predictor based on USBM equation reached an R^2 value of 0.429 [238, 257–259].

Bui et al. (2019b) proposed a lasso and elastic-net regularized generalized linear model (GLMNET) to foresee blast-induced AOP in a coal mine in Vietnam, considering 108 blast instances, 20% of which were used for testing, encompassing 5 independent variables as input. GLMNET is an algorithm that continuously optimizes the objective function in each parameter, fixing the other input parameters in each optimization and applying cyclic coordinate descent, running until convergence while aiming to minimize the loss function of elastic-net regularized regression technique. A lasso regression, in turn, is a technique used to increase the robustness against collinearity of the ordinary least squares regression through the minimization of a cost function, composed of the residual sum of squares and a regularization penalty by the $L1$ norm ($\lambda|\beta|$). The best GLMNET model among 1000 developed in this study reached the R^2 value of 0.975, while an empirical equation based on the USBM regressor reached 0.838. Keshtegar et al. (2019) proposed a general regression analysis employing the modified Fletcher and Reeves (FR) conjugate gradient method with a limited scalar factor and dynamic step size to calibrate 8 nonlinear functions that aim to predict blast-induced AOP in two mines nearby a dam in Iran, considering 81 blast instances, 20% of which were used for testing. Comprising the maximum charge per delay and distance from the blast face as input, the models reached MAE values from 3.007 to 2.936 in the training dataset, while the USBM predictor reached 3.088.

Furthermore, 3 empirical models were derived from these functions, reaching MAE values from 3.79 to 3.52 in the testing dataset, while the USBM predictor reached 4.05. Finally, Gao et al. (2020) introduced a GMDH-GA model to predict blast-induced AOP in the vicinity of a dam in Iran, considering 84 blast instances, 20% of which were used for testing, comprising 4 independent variables as input. In this proposition, GA was used to optimize the GMDH parameters, predicting the AOP, reaching the R^2 value of 0.988 [238, 260–262].

Bui et al. (2020b) compared the performances of 7 artificial intelligence models to predict blast-induced AOP in a coal mine in Vietnam, considering 113 blast instances, 20% of which were used for testing, encompassing 7 independent variables as input. The RF, SVR, GP, BART, BRT, k-NN models, and an ANN with 7-9-7-5-1 architecture achieved R^2 values of 0.943, 0.948, 0.949, 0.945, 0.898, 0.890, and 0.961, respectively, while an empirical equation based on the USBM predictor reached the value of 0.930. Nguyen and Bui (2020) proposed a combination of boosted smoothing spline (BSTSM) and GA (BSTSM-GA) to predict blast-induced AOP in a coal mine in Vietnam, considering 121 blast instances, 20% of which were used for testing, comparing their results with those of the CART, k-NN, an ANN with 7-13-1 architecture, Bayesian ridge regression (BRR), SVR, and GP models. BSTSM is a phased additive modeling strategy based on smoothing splines, which are function estimates derived from a series of noisy observations of a target, to balance the target fitness value using a measure derived from the smoothness of the splines, thus resulting in noisy data smoothing means. The boosting phase extends the additive strategy in stages, similar to minimizing the descent into the function space, and if the independent variables evaluated are continuous, the BSTSM generates an additive model with univariate smoothing splines. In turn, the BRR technique is based on Bayesian inference and ridge regression, which, similar to lasso regression, is employed to increase the robustness of the ordinary least squares regression against collinearity. However, instead of using a regularization penalty by the $L1$ norm ($\lambda|\beta|$), it uses the $L2$ norm ($\lambda\beta^T\beta$). Comprising 7 independent variables as input, the R^2 values achieved by the BSTSM-GA, CART, k-NN, ANN, BRR, SVR, and GP models were 0.971, 0.903, 0.941, 0.957, 0.927, 0.949, and 0.956, respectively, while an empirical equation based on the USBM predictor reached the value of 0.466. A sensitivity analysis showed that, in addition to distance from the blast point and maximum charge per delay, relative humidity and wind speed have a significant influence on AOP prediction [238, 263, 264].

Nguyen et al. (2020c) proposed an ANN with 9-12-8-6-1 architecture, Bayesian regularized neural networks (BRNN) with 9-2-1 architecture, and ANFIS models to predict blast-induced AOP in a coal mine in Vietnam, considering 146 blast instances, 20% of which were used for testing, encompassing 9 independent variables as input. BRNN incorporates the Bayesian structure for the ANN learning process, accounting for the uncertainty in the weight vector by addressing a probability distribution for the weights that represent relative degrees of belief in the different values. Bayesian regularization employs ridge regression instead of the usual nonlinear regression, avoiding the need for backpropagation to adjust their weights and resulting in an ANN with few neurons in hidden layers, difficult to overtrain and overfit. The R^2 values achieved by the ANN, BRNN, and ANFIS models were 0.982, 0.946, and 0.801, respectively, and as the air humidity presented a positive correlation with AOP of 0.81 when analyzed by the Pearson coefficient, it must be carefully measured and considered in the predictions. Nguyen et al. (2020d) employed CA, RF, and GBM models to predict blast-induced AOP in a coal mine in Vietnam, considering 146 blast instances, 20% of which were used for testing, comprising 9 independent variables as input. The CA, RF, and GBM models reached R^2 values of 0.956, 0.953, and 0.950, respectively, while an empirical equation based on the USBM predictor reached 0.872. A sensitivity analysis showed that the stemming length, spacing, and air humidity significantly influence the AOP prediction and distance from the blast point and maximum charge per delay. Fang et al. (2020b) proposed a CA-GA model to predict blast-induced AOP in a quarry mine in Vietnam, comparing its results to those produced by GP, conditional inference tree (CIT), PCA, ANFIS, and k-NN models, considering 164 blast instances, 20% of which were used for testing, encompassing 6 independent variables as input. CIT is a nonparametric tree-based algorithm that employs recursive splitting of dependent variables in a conditional inference structure based on their correlation values using a significance test performed at each algorithm iteration. In the proposed framework, the CA initial solution is obtained. Furthermore, the GA approach improves the CA model hyper-parameters to reach an adequate fitness value for the AOP prediction. The R^2 values achieved by the CA-GA, GP, CIT, HYFIS, PCA, and k-NN models were 0.968, 0.958, 0.954, 0.806, 0.957, and 0.864, respectively, and a sensitivity analysis showed that in addition to the maximum charge per delay and distance from the blast face, the powder factor and stemming length were influential variables in the AOP prediction [238, 265–267].

Zhou et al. (2020b) proposed a FIS-FFA model to predict blast-induced AOP in operations near a dam in Iran, considering 86 blast instances, 20% of which were used to test the models and comprising 3 independent variables as input, reaching an R^2 value of 0.977. Temeng, Ziggah, and Arthur (2020) developed a brain-inspired emotional neural network (BENN) to predict blast-induced AOP in a mine in Ghana, considering 171 blast instances, 20% of which were used to test the models, comprising 4 independent parameters as input. BENN is an advanced class of ANNs inspired by the limbic system theory of emotion, including the thalamus, sensory cortex, orbitofrontal cortex, and amygdala, elements responsible for producing a response to external information through the interaction between the orbitofrontal cortex and the amygdala. The thalamus receives data from the external environment and calculates an inaccurate output, sent directly to the amygdala via the shortest route of sensory information transfer and to the sensory cortex for further processing. Moreover, the sensory cortex also sends its output to the amygdala inputs. However, the orbitofrontal cortex inputs also receive this information to adjust the final response output due to its ability to estimate uncertainty, restrict incorrect amygdala responses, send information to it, and receive information as bias. The proposed BENN approach reached an R-value of 0.8249, while an ANN with 4-16-1 architecture, an SVM, and a GMDH models achieved values of 0.8172, 0.8134, and 0.5878, respectively, all using GA to adjust their weights, while the USBM and McKenzie equations achieved values of 0.7196 and 0.7208. Temeng, Ziggah, and Arthur (2021) extended the previous work, using a BENN to predict blast-induced AOP in a mine in Ghana, considering 324 blast instances, 30% of which were used to test the models, comprising 6 independent parameters as input. The proposed BENN reached the R-value of 0.911, while an ANN with 7-4-1 architecture, an RBFNN with 7-22-1 architecture, GRNN, GMDH, LS-SVM, SVM, and MVRA models reached values of 0.914, 0.919, 0.874, 0.905, 0.910, 0.909, and 0.821, respectively, in which all machine learning techniques used GA to adjust their weights. Figure 13.55 presents a BENN architecture [238, 241, 268–270].

Zhou et al. (2021c) combined a parametric linear model to 4 nonparametric models to predict blast-induced AOP in a quarry mine in Malaysia, such as CHAID, ANN with 6-4-1 architecture, k-NN, and SVM models, considering 62 blast instances and encompassing 9 independent variables as input. When evaluating the possible subsets, the linear model selected 6 inputs as the most important, which are chosen to feed the CHAID, ANN, k-NN, and SVM models, which achieved R^2 values of 0.83, 0.92, 0.30, and 0.04, respectively. Zeng et al. (2021) proposed a CFNN with 7-12-11-9-1 architecture trained by Levenberg–Marquardt algorithm, an ELM with

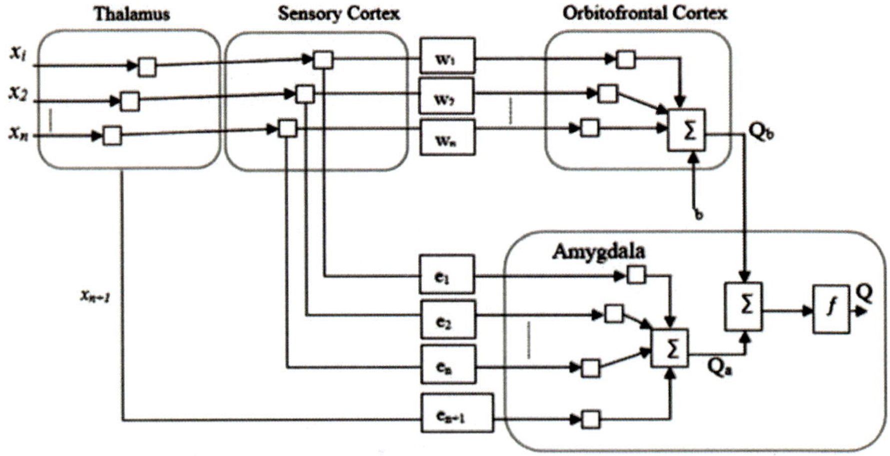

Fig. 13.55 Brain-inspired emotional neural network (BENN) architecture [270]

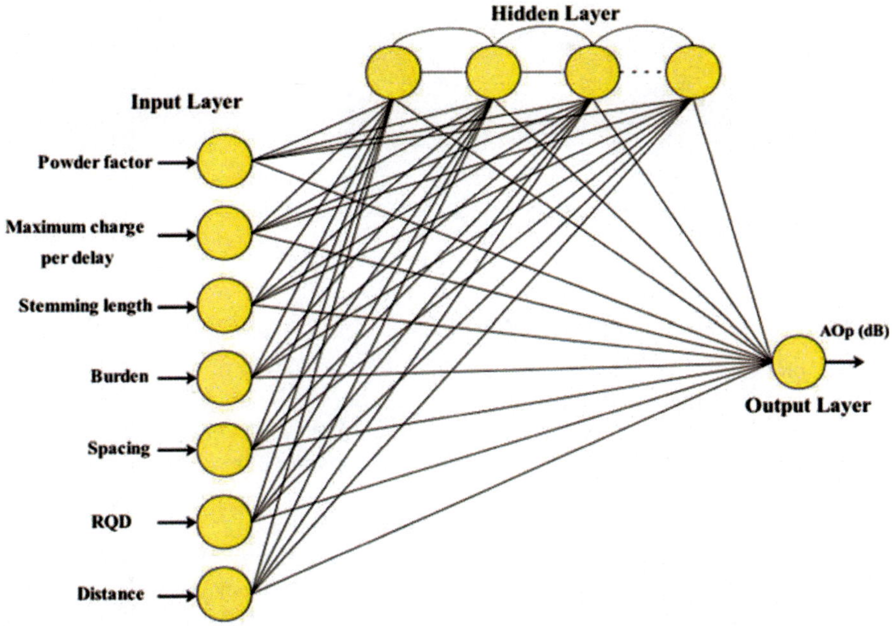

Fig. 13.56 Cascade forward neural network (CFNN) architecture [272]

7-6-1 architecture, and a GRNN model to predict blast-induced AOP in four quarry mines in Malaysia, considering 62 blast instances, 20% of which were used to train the models, comprising 7 independent variables as input. The R^2 values achieved by the CFNN, ELM, and GRNN models were 0.9272, 0.7134, and 0.8041, respectively, while a sensitivity analysis showed that the maximum charge per delay and the powder factor were the most influential variables. Figure 13.56 presents a CFNN architecture [271, 272].

For more information, see Trivedi et al. (2014) and Dumakor-Dupey, Arya, and Jha (2021) reviews [84, 85].

Flyrocks

Flyrock is a loose rock fragment propelled from the blast site beyond the blast area that can reach hundreds of meters away and has the potential to damage property or cause injury and death to people and animals in the vicinity of the blast site. The Mine Safety and Health Administration (MSHA) cited factors for considering when determining the blast area the geology, blast pattern, burden depth, diameter and angle of the boreholes, delay system, type and amount of explosive employed, type and amount of stemming, among other factors. The main mechanisms that generate flyrock in open-pit mines are facing bursting (ejection on the blast face

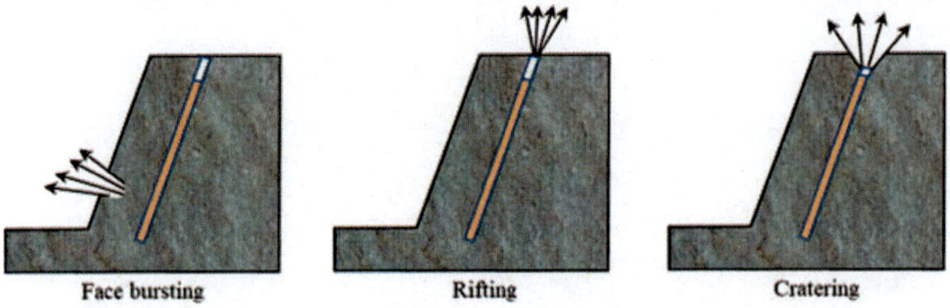

Fig. 13.57 Main mechanisms that generate flyrock in open-pit mines [85]

due to geology-related weakness zones or inadequate burden), rifting (ejection of collar rock when the stemming material is insufficient), and cratering (ejection of the weakened stemming region due to previous explosions on the bench above). Figure 13.57 presents the main mechanisms that generate flyrock in open-pit mines [8, 85].

Similar to the investigations of vibration and air overpressure, several experiments were carried out to obtain general equations for the prediction of flyrock. The main aspects generally covered in the equations were L as the flyrock distance, in meters, and d as the hole diameter, in inches. Furthermore, q was also considered as the specific charge, in kg/m^3, T_b as flyrock fragment size, St as stemming length and B as a burden, all in meters; R_1 as horizontal distance and R_2 as total distance traveled, in meters; V_0 as the initial velocity, in m/s, θ as the departure angle and g as the gravitational constant. Table 13.4 presents the most used empirical blast-induced flyrock predictors, although likewise blast-induced PPV and AOP empirical predictors. In many situations, these equations can make forecasts far from reality, particularly when compared to machine learning and advanced analytics techniques.

Monjezi et al. (2006) employed an ANN with 6-3-4-3 architecture to simultaneously predict the ratio of muck pile, flyrock, and total explosive in a bauxite mine in Iran, considering a dataset with 92 blast instances, of which 18% were used for testing, comprising 6 independent variables as input. Three algorithms were employed in the training phase, i.e., resilient backpropagation, one-step secant, and Powell–Beale conjugate gradient algorithms. The first function achieved R^2 values of 0.9906, 0.9849, and 0.9968 for muck pile, flyrock, and total explosive, respectively, while the other two reached 0.9571, 0.8177, and 0.998 and 0.9805, 0.9122, and

Table 13.4 Most used empirical blast-induced flyrock predictors [55, 56, 273, 274]

Empirical AOp predictors	Equations
Lundborg (1973)	$L = 143d\ (q - 0.2)$
Lundborg et al. (1975)	$L = 260d^{2/3} \| T_b = 0.1d^{2/3}$
Gupta (1980)	$L = [(\mathrm{St}/B)/155.2]^{-0.73}$
Chiapetta et al. (1983)	$R_1 = V_0(2\sin 2\theta/g)$ $R_2 = V_0 \cos(V_0 \sin\theta + 2V_0 \sin\theta + 2gH)/g$

0.9975, respectively. Aghajani-Bazzazi, Osanloo, and Azimi (2009) used MVRA to predict blast-induced flyrock in a phosphate mine in Iran, considering a dataset with 15 blast instances, comprising 3 independent variables as input and reaching an R-value of 0.915, while the power, logarithmic, and polynomial versions reached values of 0.918, 0.914, and 0.928, respectively. As aforementioned, Monjezi, Bahrami, and Yazdian Varjani (2010) employed an ANN with 8-3-3-2 architecture to simultaneously predict rock fragmentation and blast-induced flyrock in an iron mine in Iran, reaching an R^2 value of 0.9883 for flyrock, from which the blast parameters were adjusted and the flyrock distance diminished from 110 to 30 m. Monjezi et al. (2011) used an ANN with 9-13-1 architecture to predict blast-induced flyrock in an iron mine in Iran, considering a dataset with 192 blast instances, 10% of which were used to test the model, comprising 9 independent variables as input and reaching an R^2 value of 0.9705. Sensitivity analysis by the CAM method showed that blast ability index, the maximum charge per delay, hole diameter, stemming length, and powder factor were the most influential parameters, and further adjustments in blast parameters reduced the flyrock distance from 165 to 25 m [43, 275–277].

Rezaei, Monjezi, and Yazdian Varjani (2011) employed a FIS model to predict blast-induced flyrock in an iron mine in Iran, considering a dataset with 490 blast instances, 20% of which were used to test the model, comprising 8 independent variables as input. The FIS model reached an R^2 value of 0.9845, while an MVRA reached a value of 0.7011. Sensitivity analysis by the CAM method showed that powder factor and stemming length were the most influential parameters. Monjezi, Amini Khoshalan, and Yazdian Varjani (2012) proposed an ANN-GA model with 9-16-2 architecture to simultaneously predict blast-induced flyrock and back break in a copper mine in Iran, considering a dataset with 195 blast instances, 25% of which were used to test the model, encompassing 9 independent variables as input. The ANN-GA model achieved R^2 values of 0.976 and 0.956 for flyrock and back break, respectively, while the two MVRA models achieved 0.582 and 0.427, respectively. Sensitivity analysis by the CAM method showed that stemming length and powder factor were the most influential variables in flyrock, whereas stemming length and maximum charge per delay were the most influential variables in the back break. Amini et al. (2012) compared the performances of an ANN with 7-12-1 architecture and an SVM model to predict blast-induced flyrock in a copper mine in Iran, considering a dataset with 245 blast instances, 30% of which were used to test the model, comprising 7 independent variables as input. The ANN reached an R^2 value of 0.92, while the SVM reached a value of 0.97 [278–280].

Tonnizam Mohamad et al. (2012) used an ANN with 8-15-15-1 architecture to predict the flyrock induced by the boulder blasting in a quarry mine in Malaysia, considering a dataset with 16 blast instances, encompassing 8 independent variables as input, reaching an R^2 value of 0.919. In turn, Lundborg et al. (1975) empirical equation and Lundborg (1981) graph reached R^2 values of 0.211 and 0.217, respectively, and sensitivity analysis by the max–min method showed that powder factor, stemming length, and charge length were the most influential parameters. Tonnizam Mohamad et al. (2013) used an ANN with 9-20-2 architecture to simultaneously predict blast-induced flyrock and the size of thrown rock at a quarry mine in Malaysia, considering a dataset with 39 blast instances, 20% of which were used for testing, comprising 9 independent variables as input, reaching an *R*-value of 0.9525. Khandelwal and Monjezi (2013) used SVM and MVRA models to predict blast-induced flyrock in a copper mine in Iran, considering a dataset with 234 blast instances, 20% of which were used for testing, encompassing 6 independent variables as input, achieving R^2 values of 0.948 and 0.440, respectively.

Raina, Murthy, and Soni (2013) proposed an ANN with 7-20-14-8-1 architecture to address the influence of shape (sphericity), size of rocks, and other parameters on blast-induced flyrock, by blasting 75 concrete models of $18'' \times 18'' \times 6''$ with single holes and distinct parameters, resulting in 136 blast instances, 10 of which were used for testing. Using a high-speed camera to identify the fragments' initial velocity and launch angle and physical measurements to determine their ranges, weights, and dimensions, the fragments' sphericity was calculated, and these data fed the ANN, considering 7 independent parameters as input. No index was shown for the ANN test phase, and a sensitivity analysis showed that the initial fragment velocity, launch angle, and weight were the most influential parameters [55, 281–284].

Monjezi et al. (2013) used an ANN with 9-5-2-1 architecture and an MVRA model to predict blast-induced flyrock in a copper mine in Iran, considering a dataset with 310 blast instances, 10% of which were used for testing, encompassing 9 independent variables as input. The ANN and MVRA models achieved R^2 values of 0.980 and 0.883, respectively, while the Lundborg (1973) and Lundborg et al. (1975) empirical predictors reached R^2 values of 0.038 and 0.377, respectively. A sensitivity analysis showed that the most influential parameters were the powder factor, hole diameter, stemming length, and maximum charge per delay. As afore said, Jahed Armaghani et al. (2014) proposed an ANN-PSO model to simultaneously predict blast-induced flyrock and PPV at three quarry mines in Malaysia, reaching an R^2 value of 0.93025, while the Lundborg predictor showed low accuracy, with powder factor and maximum charge per delay being the most influential variables in predicting flyrock. Trivedi, Singh, and Raina (2014) used an ANN with 6-10-10-1 architecture and an MVRA model to predict blast-induced flyrock at four limestone mines in Iran, considering a dataset with 125 blast instances, 25% of which were used for testing, comprising 6 independent variables as input, reaching R^2 values of 0.983 and 0.815, respectively. A sensibility analysis showed that burden and stemming length were the most influential variables in $d = 165$ mm blast holes, whereas unconfined compressive strength and RQD were the most influential variables in $d = 115$ mm blast holes [55, 135, 273, 285, 286].

Ghasemi et al. (2014) employed an ANN with 6-9-1 architecture and a FIS model to predict blast-induced flyrock in a copper mine in Iran, considering a dataset with 230 blast instances, 20% of which were used for testing, encompassing 6 independent variables as input, reaching R^2 values of 0.939 and 0.957, respectively. Marto et al. (2014) compared the performances of an ANN with 7-8-1 architecture, an ANN-ICA, and an MVRA models to predict blast-induced flyrock in a quarry mine in Malaysia, considering a dataset with 113 blast instances, 20% of which were used for testing, comprising 7 independent variables as input and reaching R^2 values of 0.981, 0.919, and 0.743, respectively. An empirical equation reached an R^2 value of 0.118. In contrast, sensitivity analysis by CAM showed that the powder factor and the maximum charge per delay were the most influential parameters, which were used individually to develop power form predictors, which reached R^2 values of 0.565 and 0.544, respectively. Trivedi, Singh, and Gupta (2015) compared the performances of an ANN with 6-10-10-1 architecture, an ANFIS, and an MVRA model to predict blast-induced flyrock in four limestone mines in India, considering a dataset with 125 blast instances, 20% of which were used for testing, encompassing 6 independent variables as input and reaching R^2 values of 0.95, 0.98, and 0.72, respectively [287–289].

As aforementioned, Jahed Armaghani et al. (2015b) used ANN and ANFIS models to predict blast-induced PPV, AOP, and flyrock at four quarry mines in Malaysia, achieving R^2 values of 0.834 and 0.959, respectively, for the predicted flyrock distance. Jahed Armaghani et al. (2016) used an ANN with 2-5-1 architecture and an ANFIS model to predict blast-induced flyrock at five quarry mines in Malaysia, considering a dataset with 232 blast instances, 20% of which were used for testing, comprising maximum charge per delay and powder factor as input, achieving R^2 values of 0.925 and 0.964, respectively. Saghatforoush et al. (2016) employed an ANN-ACO model with 5-5-5-2 architecture and 30 ants to simultaneously predict blast-induced flyrock and back break in an iron mine in Iran, considering a dataset with 97 blast instances, 30% of which were used for testing, encompassing 5 independent variables as input, reaching R^2 values of 0.994 and 0.832, respectively. Yari et al. (2016) used an ANN with 12-8-8-1 architecture to predict blast-induced flyrock in a copper mine in Iran, considering a dataset with 334 blast instances, 20% of which were used for validation and testing, comprising 12 independent variables as input, reaching an R-value of 0.989, while 4 empirical predictors reached values from 0.206 to 0.693. As aforementioned, Trivedi, Singh, and Raina (2016) used a BPNN with 8-10-10-2 architecture and an MVRA model to simultaneously predict fragmentation and flyrock in four limestone mines in India, reaching R^2 values of 0.985 and 0.81 for flyrock, respectively [81, 142, 290–292].

Shirani Faradonbeh, Jahed Armaghani, and Monjezi (2016) proposed a GP model to predict blast-induced flyrock in six quarry mines in Malaysia, considering a dataset with 262 blast instances, 20% of which were used for testing, encompassing 6 independent variables as input, achieving an R^2 value of 0.908, while an MVRA model reached the value of 0.816. Shirani Faradonbeh et al. (2016b) used GP and GEP models to predict blast-induced flyrock in an iron mine in Iran, considering a dataset with 97 blast instances, 30% of which were used for testing, comprising 5 independent variables as input, achieving R^2 values of 0.893 and 0.987, respectively. Stojadinović et al. (2016) employed an ANN with 19-8-6-1 architecture to predict launch velocity of blast-induced flyrocks in limestone and two copper mines in Serbia, considering a dataset with 905 blast instances, 15% of which were used for validating, encompassing 19 independent variables as input. The acquisition of the flyrock launch velocity was performed using a high-speed camera, ranging from 8 to 237 m/s and considering ±10% of error, with the proposed model extrapolating the error margin only in 6% of the validation dataset. A new dataset with 237 blast instances was obtained for testing purposes, 36 of which generated flyrock, and among these, 77% of the predictions complied with ±10% of error [293–295].

Hasanipanah et al. (2017e) employed a PSO model to develop a power equation capable of predicting blast-induced flyrock in three quarry mines in Malaysia, considering a dataset with 76 blast instances, comprising 5 independent variables as input and achieving an R^2 value of 0.96, while an MVRA model reached the value of 0.88. A sensitivity analysis showed that rock density was the most influential parameter. Hasanipanah et al. (2017f) employed a CART model to predict blast-induced flyrock in a quarry mine in Malaysia, considering a dataset with 65 blast instances, 20% of which were used for testing, encompassing 6 independent variables as input and achieving an R^2 value of 0.872, while an MVRA model reached the value of 0.860. A sensitivity analysis showed that the powder factor was the most influential parameter. Dehghani and Shafaghi (2017) proposed a combination of differential evolution (DE) and DA algorithm to predict blast-induced flyrock in a copper mine in Iran, considering a dataset with 300 blast instances, comprising 9 independent variables as input. DE is a random evolutionary optimization algorithm that relies on genetic operators such as mutation, crossover, and selection, whose concept is similar to GA. However, the latter represents the data as string chromosomes, while DE actually deals with real-valued vectors (real numbers).

Furthermore, DE differs in the generation of new population and genetic operator procedures in the search for the global optimum, although not guaranteeing optimality. The DE-DA model reached the R^2 value of 90.8%, while a DA model by the mass system and a nonlinear MVRA model reached values of 55.7% and 75.5%, respectively, with Lundborg (1973), Lundborg et al. (1975), and Gupta (1980) empirical predictors reaching values of 33.4%, 25.7%, and 0%, respectively. A sensibility analysis showed that stemming length and powder factor were the most influential variables [55, 56, 273, 296–298].

Bakhtavar, Nourizadeh, and Sahebi (2017) proposed a DA-FIS model to predict blast-induced flyrock in a copper mine in Iran, considering a dataset with 320 blast instances, 10% of which were used for testing, encompassing 6 independent fuzzified variables as input and employing the force system in the DA, reaching an R^2 value of 0.9761. As mentioned before, Als et al. (2018) employed an ANN-FFA model to simultaneously predict blast-induced flyrock and rock fragmentation in a limestone mine in Iran, reaching an R^2 value of 0.93 for flyrock, while a sensitivity analysis showed that the GSI factor and burden were the most influential parameters in both fragmentation and flyrock predictions. Shirani Faradonbeh et al. (2018b) used a GEP model to develop an equation capable of predicting blast-induced flyrock at three quarry mines in Malaysia, considering a dataset with 76 blast instances, 30% of which were used to test the model, comprising 5 independent variables as input, reaching an R^2 value of 0.924. Hence, an FFA model was employed to optimize the blasting parameters to minimize the flyrock distance, reducing the order of 34%, from 60.0 to 39.8 m. Nikafshan Rad et al. (2018) compared the performances of the LS-SVM and SVR models to predict the blast-induced flyrock in an iron mine in Iran, considering a dataset with 90 blast instances, 20% of which were used to test the models, encompassing 7 independent variables as input, reaching R^2 values of

0.969 and 0.945, respectively. A sensitivity analysis showed that the powder factor and rock density were the most influential parameters [54, 299–301].

Hasanipanah et al. (2018c) developed a rock engineering systems (RES) methodology to predict blast-induced flyrock while assessing and modeling the level of risk associated with it at a quarry mine in Malaysia, considering a dataset with 62 blast instances, comprising 11 independent variables as input. RES is a technique that characterizes the influential parameters and interaction mechanisms in rock engineering problems, based on an interaction matrix, in which the most influential parameters in the system are placed on the main diagonal, while the values of the influence that a parameter exerts on the other are placed in off-diagonal positions. The sum of each row and column are termed cause and effect, respectively, and this RES formulation used the expert semi-quantitative method for numerically encode the interaction matrix, where the influence values range from 0 to 4. The 11 inputs were plotted on an effect–cause diagram. Therefore, the dominant and subordinate parameters were identified, so the evaluation of the 6 parameters chosen as dominant reveals that the level of risk in this dataset ranges from 24.82 to 42.02, with an overall value of 32.95, considered low to medium. This methodology also enables the development of a regression analysis, whose R^2 value reached was 0.894, while an MVRA model reached the value of 0.641, while a sensitivity analysis showed that powder factor and RMR were the most influential parameters. Figure 13.58 presents an interaction matrix with two parameters (left) and an example of coding an interaction matrix (right) [302].

Nguyen et al. (2019f) employed an ensemble of five ANN models (EANN) to predict blast-induced flyrock at a quarry mine in Vietnam, considering a dataset with 210 blast instances, 20% of which were used to test the models, encompassing 5 independent variables as input. First, individually trained and tested ANNs with 5-7-5-1, 5-10-8-1, 5-14-9-1, 5-18-13-1, and 5-21-16-1 architectures. Therefore, an EANN with 5-25-21-15-1 architecture was trained and tested, relying upon a new

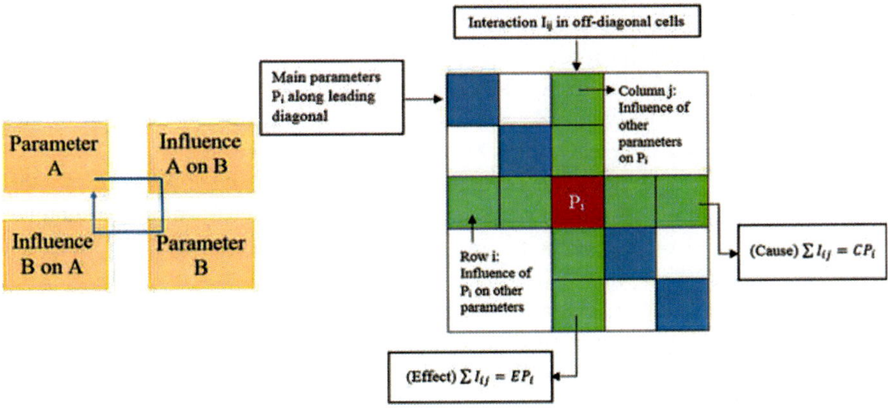

Fig. 13.58 An interaction matrix with two parameters (left) and an example of coding an interaction matrix (right) [302]

dataset composed of the flyrock predictions generated by the individual ANN models, culminating in new training and testing phases using the original data. The five ANN models reached R^2 values from 0.973 to 0.975, while the EANN reached 0.986. An ANN with the same 5-25-21-15-1 architecture as the EANN was also developed for comparison purposes, achieving an R^2 value of 0.975, a tower presented by EANN. A sensitivity analysis showed that maximum charge per delay and stemming length were the most influential parameters. Koopialipoor et al. (2019) compared the performances of ICA, PSO, and GA metaheuristics when combined with an ANN with 6-9-1 architecture to predict blast-induced flyrock at six quarry mines in Malaysia, considering a dataset with 262 blast instances, 20% of which were used to test the models, encompassing 6 independent variables as input. The ANN-ICA, ANN-PSO, and ANN-GA hybrid models achieved R^2 values of 0.943, 0.958, and 0.930, respectively, and a sensitivity analysis showed that hole diameter and powder factor were the most influential parameters [303, 304].

Lu et al. (2020) proposed an ANN with 5-7-1 architecture, ELM, and outlier robust ELM (ORELM) models to predict blast-induced flyrock in three quarry mines in Malaysia, considering a dataset with 82 blast instances, 30% of which were used for testing, comprising 5 independent variables as input. ORELM is an EML variant developed to be robust to outlier datasets, including the $L1$ Norm as loss function and the $L2$ Norm for evaluating the output weight in an ELM framework, simultaneously minimizing output weight limit and training errors. Thus, the ORELM not only addresses dispersion change but also minimizes the overall convergence range. The R^2 values achieved by ANN, ELM, ORELM, and MVRA models were 0.912, 0.955, 0.958, and 0.883, respectively, and an evaluation in which one of the five inputs was excluded from the training phase for each machine learning model demonstrated that the exclusion of rock density resulted in a significant decrease inaccuracy. However, this is the most influential parameter. Jahed Armaghani et al. (2020) compared the performances of the SVR, MARS, and PCR models to predict the blast-induced flyrock at six quarry mines in Malaysia, considering a dataset with 262 blast instances, 20% of which were used to test the models, encompassing 6 independent variables as input, reaching R^2 values of 0.9392, 0.8845, and 0.6427, respectively. Hence, a GWO model was used to optimize the blasting parameters to minimize the flyrock distance, achieving a reduction of the order of 4%, from 75.0 to 72.4 m [305, 306].

Zhou et al. (2020c) developed an ANN with 6-14-1 architecture to predict blast-induced flyrock in a quarry mine in Malaysia, considering a dataset with 65 blast instances, comprising 6 independent variables as input, reaching an R^2 value of 0.906. Hence, a 50-particle PSO model was employed to optimize the blasting parameters to minimize the flyrock distance, achieving a theoretical flyrock distance of 34 m and a practical distance of 109 m considering the maintenance of some engineering specifications. Zhou et al. (2020d) proposed an ANN with 6-9-1 architecture and an MVRA model to predict blast-induced flyrock in a quarry mine in Malaysia, considering 260 blast instances, encompassing 6 independent parameters as input, achieving R^2 values of 0.944 and 0.819, respectively. Moreover, 10,000 MCS scenarios were developed to assess the risk in operation, based on the correlation between input variables obtained from the ANN weights, resulting in a flyrock distance range between 55 and 367 m, with a 90% probability of being less than 331 m. A sensitivity analysis showed that maximum charge per delay and hole diameter was the most influential parameter. Han et al. (2020) employed an RF model to identify and exclude the most influential among 6 input parameters and then developed a BN model on the remaining parameters in order to perform a probabilistic prediction of blast-induced flyrock in six quarry mines in Malaysia, considering 262 blast instances, 30% of which were used for testing. As the BN models work with categorical data, the flyrock distance output was categorized in four ranges, and the data fed the RF model, which identified that the maximum charge per delay was the most influential parameter. The remaining variables fed the BN model, which identified that among these input variables, the hole diameter was the most influential since it was present in the four probability conditions, with values between 97.5 and 127.5 mm, while the combination of hole diameter and hole depth appeared in three probability predictions, suggesting this as an influential combination of factors [307–309].

Hasanipanah et al. (2020) proposed an ANN with 5-7-1 architecture combined with an adaptive dynamical harmony search (ANN-ADHS) approach to predict blast-induced flyrock at three quarry mines in Malaysia, considering a dataset with 82 blast instances, 20% of which were used for testing purposes, encompassing 5 independent variables as input. Harmony search (HS) is a metaheuristic inspired by the improvisation process that jazz musicians employ in the search for harmony, in which each musician aims to tune their instruments and improve their performance to achieve a pleasing result, whose quality is measured by aesthetic estimation and achievement requires several attempts. In HS, analogy, musicians are agents, and the harmony is the objective function to be improved by a series of practices (iterations), and the mechanism considers that each musician can sound notes (decision variables) at random, from their memory notes close to those in their memory. When a new harmony is improvised in HS metaheuristic, it may or may not be accepted into the harmony memory depending on its fitness function, iterating until a stopping criterion is met and the best-stored harmony is returned. On the other hand, two random adjusting procedures were employed in ADHS methodology in each iteration, comprising an adjustment of the harmony elements based on the maximum and minimum values for each random variable in the harmony memory and a dynamic pitch procedure. The R^2 values achieved by the ANN-ADHS, ANN-HS, ANN-PSO, and ANN models were 0.930, 0.871, 0.832, and 0.831, respectively, while a sensitivity analysis showed that spacing and powder factor was the most influential parameters. Figure 13.59 presents an analogy between improvisation and optimization in HS metaheuristic [310].

Hasanipanah and Bakhshandeh Amnieh (2020) proposed a fuzzy rock engineering system (FRES) methodology to predict blast-induced flyrock while assessing and modeling the level of risk associated with it at a quarry mine in Malaysia, considering a dataset with 62 blast instances, comprising 11 independent variables as input. In order to numerically encode the interaction matrix, the FRES proposition used a fuzzy expert semi-quantitative method, and for the 5 parameters elected as dominant using an effect–cause diagram, the overall risk value achieved was 33.31, considered medium to high. Furthermore, the regression analysis coefficients obtained through

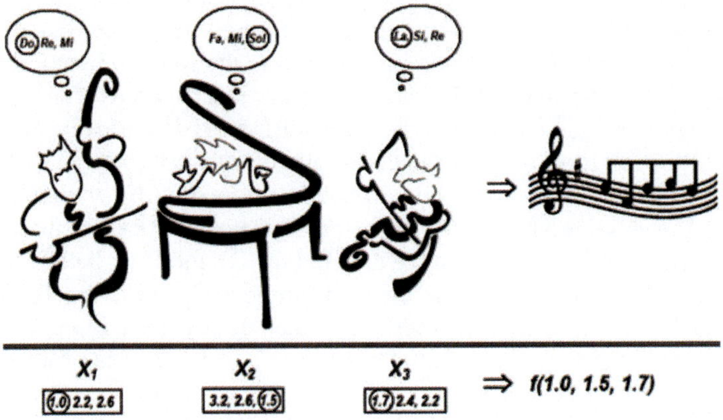

Fig. 13.59 Analogy between improvisation and optimization in harmony search (HS) metaheuristic [311]

this FRES methodology were optimized by PSO, GA, and ICA models, reaching R^2 values of 0.981, 0.984, and 0.983, respectively. Nguyen et al. (2020e) developed four variants of SVM-WOA models to predict blast-induced flyrock in a quarry mine in Vietnam, considering a dataset with 210 blast instances, 20% of which were used to test the models, comprising 5 independent variables as input, comparing their results to those presented by an ANN with 5-16-11-1 architecture, ANFIS, GBM, RF, and CART models. The SVM-WOA variants used linear, polynomial, RBF, and hyperbolic tangent kernel functions, reaching R^2 values of 0.937, 0.976, 0.977, and 0.972, respectively, while the ANN, ANFIS, GBM, RF, and CART models reached values of 0.971, 0.965, 0.971, 0.972, and 0.973 [312, 313].

Nikafshan Rad et al. (2020) combined a recurrent fuzzy neural network (RFNN) with GA (RFNN-GA) to predict blast-induced flyrock in two quarry mines nearby a dam in Iran, considering a dataset with 70 blast instances, 20% of which were used to test the models, encompassing 4 independent variables as input. An RFNN is a learning algorithm with dynamic architecture capable of adding delays in a recurrent fashion (feedback loops) to a normal input–output mapping process, thus mapping not only the current inputs but also the previous ones and possibly previous outputs. Developed in an ANFIS structure, layer 0 of RFNN corresponds to input variables, layer 1 has a recurrent connection and calculates the values of Gaussian membership of the input variables, layers 2 and 3 are related to fuzzy rules and normalization, while layers 4 and 5 are related to the consequent parameters and the overall output. The R^2 values achieved by an ANN with 4-3-1 architecture, the ANN-GA, and RFNN-GA models were 0.866, 0.944, and 0.966, respectively, while a sensitivity analysis showed that maximum charge per delay was the most influential parameter. Baghat et al. (2021) proposed a CART model to predict the flyrock distance of boulder blasting operations in railway construction in India, considering a dataset with 61 blast instances, 20% of which were used for testing comprising 11 independent variables as input. Different combinations of inputs generated different CART and

MVRA models, and those with superior results achieved R^2 values of 0.9555 and 0.7938, respectively, while a sensitivity analysis showed that specific charge and specific drill density were the most influential parameters [314, 315].

Guo et al. (2021a) proposed an ensemble model combining 6 SVR variants with a GLMNET approach (SVR-GLMNET) to predict blast-induced flyrock in a quarry mine in Vietnam, considering a dataset with 210 blast instances, 30% of which were used for validation and test, encompassing 5 independent variables as input. First, 100 SVR models were developed by the trial-and-error methodology, from which the 6 superior models (with RBF kernel function) reached R^2 values from 0.968 to 0.972, while 100 GLMNET models were also developed, with the best reaching an R^2 value of 0.920. Therefore, the predictions of these six SVR models in the 40 instances of the validating dataset were used to generate a new training dataset with 6 variables as input, thus, used to train the GLMNET so that the SVR-GLMNET model achieved an R^2 value of 0.993. Sensitivity analysis by the Sobol method showed that stemming length, the maximum charge per delay, and powder factor were the most influential parameters. Guo et al. (2021b) compared the performances of a deep neural network (DNN) with 6-6-18-12-1 architecture and an ANN model to predict blast-induced flyrock in a quarry mine in Malaysia, considering a dataset with 240 blast instances, 20% of which were used for testing, comprising 6 independent variables as input. To overcome the tendency of converging to local optima that regular ANNs with multiple hidden layers present, the DNN methodology was implemented, which includes a supervised fine-tuning and an unsupervised layer-wise pre-training to reduce nonlinear dimensions, treating each consecutive pair of layers like a restricted Boltzmann machine (RBM) with joint probability. In this pre-education phase, the initial system weights are created and transferred to the next layer, followed by a backpropagation error algorithm phase to refine these weights, consisting of a procedure to be repeated in all layers up to the output layer, when supervised fine-tuning stage and mapping function will be applied to minimize the MSE in the training phase. The R^2 values achieved by the DNN and ANN models were 0.9093 and 0.8539, respectively, and a WOA approach was employed to optimize the blasting parameters to minimize the flyrock distance, achieving a reduction of the order of 10%, from 71.68 to 64.80 m. Figure 13.60 shows the unsupervised layerwise pre-training mechanism by RBM in DNNs [316, 317].

Li, Koopialipoor, and Jahed Armaghani (2021) developed 5 models combining an ANN with 4-8-1 architecture with different metaheuristic (i.e., GA, PSO, ICA, ABC, and FFA) to predict blast-induced flyrock in a quarry mine in Malaysia, considering the dataset used by Marto et al. (2014), with 113 blast instances, 20% of which were used for testing. This dataset was submitted to the fuzzy Delphi method (FDM) to filter the most influential variables to serve as input, which consists of a questionnaire to be answered by experts, whose opinions are computed as fuzzy triangular numbers and the geometric mean of the responses are considered as the degree of membership of fuzzy numbers. Moreover, the values must be defuzzified, and a threshold must be defined so that in this research, the original dataset encompassing 7 independent variables as input was reduced to 4 input variables. The R^2 values achieved by the ANN-GA, ANN-PSO, ANN-ICA, ANN-ABC, and ANN-FFA models were 0.9466,

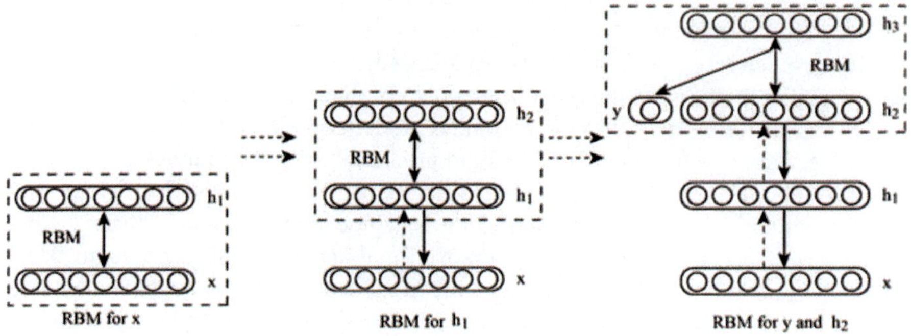

Fig. 13.60 Unsupervised layerwise pre-training mechanism by restricted Boltzmann machine (RBM) in deep neural networks (DNNs) [317]

0.9608, 0.9598, 0.9666, and 0.9719, respectively, and even though Marto et al. (2014) achieved an R^2 value of 0.98 using their ANN-ICA model, the reduction from 7 to 4 inputs is significant in terms of effort in collecting new data. Fattahi and Hasanipanah (2021) proposed combining an ANFIS model with GOA and cultural algorithm (CTA) metaheuristics to predict blast-induced flyrock in three quarry mines in Malaysia dataset with 80 blast instances, 20% of which were used for testing, comprising 5 independent variables as input. CTA is an evolutionary knowledge-based algorithm inspired by the updating of knowledge in society over time as the population evolves and, in turn, uses that knowledge to guide society in solving new problems. Based on two levels of evolution, i.e., population and cultural beliefs, once a solution to a problem is accepted by society, the belief space is updated, so the next iteration (population) will be subject to the influence of the belief space updated with new knowledge. Using the CTA and GOA algorithms to improve the parameter values in the membership functions, the R^2 values achieved by ANFIS-CTA and ANFIS-GOA models were 0.9529 and 0.9741, respectively [288, 318, 319].

Masir, Ataei, and Motahedi (2021) combined MCDM with fuzzy fault tree analysis (FFTA-MCDM) approaches to predict blast-induced flyrock in surface mines. First, the authors considered the literature review and expert opinion using a questionnaire to assess the risk of flyrock occurrence, computing the probabilities using FFTA. Then, in an FTA, a set of basic events (BEs) is combined with logical gates (AND and OR gates) from top to bottom to assess the probability of an undesirable event (TE) occurring, also considering intermediate events (IEs) in its construction, reaching that three main factors are responsible for the occurrence of flyrock: design error, human error, and natural error. Hence, the fuzzy set theory was employed to evaluate the BEs' probabilities and consequently the TE's probability, the fuzzy numbers were defuzzified, and each class is attached to a probability when the decision-making trial and evaluation laboratory (DEMATEL) method was used to assess the interdependence of the events. Furthermore, a fuzzy analytic network process (FANP) was employed to evaluate the consequence severities of flyrock events, determining the weighting of clusters by pairwise comparison, culminating in the number of risk

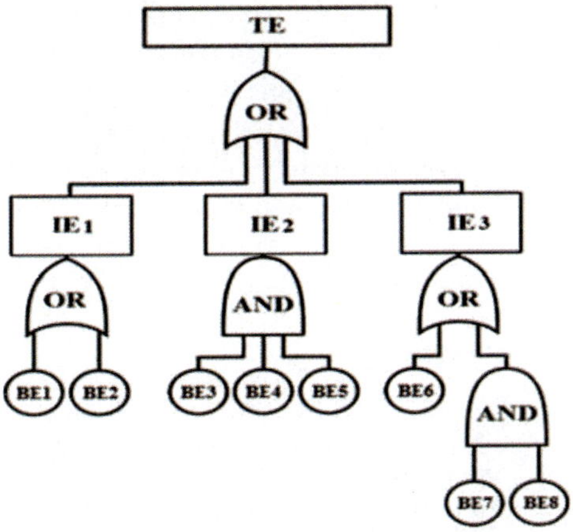

Fig. 13.61 An example of fault tree analysis (FTA) [320]

events by multiplying the probability and severity of the consequences, thus building a risk matrix. Finally, from the evaluation of this risk matrix, it was verified that the risk numbers for the occurrence of flyrock related to design errors, human errors, and natural influence were 12, 6, and 2, respectively, guiding engineers in setting proper values of factors such as burden, spacing, delays, and hole diameter as vital to preventing flyrock occurrence. Figure 13.61 presents an example of FTA [320].

Ye et al. (2021) used GP and RF models to predict blast-induced flyrock in six quarry mines in Malaysia, considering 262 blast instances, 20% of which were used for testing purposes, encompassing 6 independent parameters as input, achieving R^2 values of 0.9083 and 0.9046, respectively. Moreover, more than 1000 MCS scenarios were developed to assess the risk in operation, resulting in a flyrock distance range between 78 and 420 m, with an average distance of 160 m, reaching a 99% probability of being less than 340 m and 90% probability of being less than 290 m. For more information, see Trivedi et al. (2014) and Dumakor-Dupey, Arya, and Jha (2021) reviews [84, 85, 321].

Conclusion

This chapter reviewed the explosive and initiation systems, including the characterization and types of explosives. Furthermore, this chapter explained advanced data analytics methods to extract intelligent insights from the huge amount of blasting collected data. These intelligent insights help the operators and managers make better decisions to control the effective parameters to optimize the process. Blasting operation as a main process in the mine value chain plays a critical role in having successful mining. These days, new technology has equipped mining companies to collect critical data from the blasting operation. The collected data can potentially convert to useful information by using ML and AI in the mining. This chapter reviewed many successful case studies and business cases to highlight the power of these new approached for optimized basting. This chapter also presented the intelligent solutions for the blasting environmental and safety concerns. More studies and practical experiences need to reach a validated methodology to use ML and AI in blasting, but this chapter helped cover the research gap in advanced analytics for blasting operation in surface and underground mine sites.

References

1. Hammelmann, F. and P. Reindes. 2003. Eletronic blasting and blast management. In *Explosives and Blasting Technique: Proceedings of the EFEE 2nd World Conference, Prague, Czech Republic*, ed. R. Holmberg, 10–12 Sept 2003. Rotterdam: A.A. Balkema
2. Wisniak, J. 2008. The development of dynamite. From Braconnot to Nobel. *Educación Química* 19 (1): 71–81.
3. The Editors of Encyclopaedia Britannica. 2008. *Texas City explosion of 1947*. https://www.britannica.com/event/Texas-City-explosion-of-1947/additional-info#history. Accessed 11 Nov 2020
4. History.com Editors. 2009. Fertilizer explosion kills 581 in Texas. https://www.history.com/this-day-in-history/fertilizer-explosion-kills-581-in-texas. Accessed 11 Nov 2020
5. Leighton, H. and C. Hlavaty. 2018. *See historic, rare footage of the aftermath of the deadly 1947 explosion in Texas City*. https://www.chron.com/neighborhood/bayarea/article/Historic-footage-deadly-1947-Texas-City-explosion-12482007.php. Accessed 11 Nov 2020
6. Hustrulid, W. 1999. *Blasting principles for open pit mining. General design concepts*, vol. 1. Rotterdam: A.A. Balkema
7. Hustrulid, W. 1999. *Blasting principles for open pit mining. Theoretical foundations*, vol. 2. Rotterdam: A.A. Balkema
8. International Society of Explosives Engineers. 2011. *ISEE Blasters' handbook*, 18th ed. Ohio: International Society of Explosives Engineers.
9. Orica Limited (2020) *WebGen™ is the world's first wireless initiating system for mining—A significant step in the evolution of blast initiation*. https://www.orica.com/Products-Services/Blasting/Wireless/How-it-works/how-it-works#.X_S8d9hKhPY. Accessed 05 Jan 2021
10. Ouchterlony, F., and J.A. Sanchidrián. 2019. A review of development of better prediction equations for blast fragmentation. *J Rock Mech Geotech Eng* 11 (5): 1094–1109. https://doi.org/10.1016/j.jrmge.2019.03.001.
11. Rosin, P., and E. Rammler. 1933. The laws governing fineness of powdered coal. *J Inst Fuel* 7: 29–36.
12. Bennett, J.G. 1936. Broken coal. *J Inst Fuel* 10: 22–39.
13. Bakhtavar, E., H. Khoshrou, and M. Badroddin. 2015. Using dimensional-regression analysis to predict the mean particle size of fragmentation by blasting at the Sungun copper mine. *Arabian Journal of Geosciences* 8: 2111–2120. https://doi.org/10.1007/s12517-013-1261-2.
14. Monjezi, M., M. Rezaei, and A. Yazdian Varjani. 2009. Prediction of rock fragmentation due to blasting in Gol-E-Gohar iron mine using fuzzy logic. *International Journal of Rock Mechanics and Mining Sciences* 46 (8): 1273–1280. https://doi.org/10.1016/j.ijrmms.2009.05.005.
15. Koshelev, E.A., V.M. Kuznetsov, S.T. Sofronov, and A.G. Chernikov. 1971. Statistics of the fragments forming with the destruction of solids by explosion. *Journal of Applied Mechanics and Technical Physics* 12 (2): 244–256.
16. Kuznetsov, V.M. 1973. The mean diameter of the fragments formed by blasting rock. *Soviet Mining Science* 9 (2): 144–148.

17. Protodyakonov MM (1962) Mechanical properties and drillability of rocks. In *Proceedings of the fifth US symposium on rock mechanics*, Minneapolis.
18. Cunningham, C.V.B. 1983. The Kuz-Ram model for prediction of fragmentation from blasting. In *Proceedings of the 1st International Symposium on Rock Fragmentation by Blasting*, Luleå
19. Spathis, A.T. 2004. A correction relating to the analysis of the original Kuz-Ram model. *Fragblast* 8 (4): 201–205. https://doi.org/10.1080/13855140500041697.
20. Cunningham, C.V.B. 1987. Fragmentation estimations and the Kuz-Ram model—four years on. In *Proceedings of the 2nd International Symposium on Rock Fragmentation by Blasting*, Keystone.
21. Lilly, P.A. 1986. An empirical method of assessing rock mass blastability. In *Proceedings of the Large Open Pit Mine Conference*, Carlton.
22. Stagg, M.S., Rholl, S.A., Otterness, R.E., Smith, N.S. 1990. Influence of shot design parameters on fragmentation. In *Proceedings of the 3rd International Symposium on Rock Fragmentation by Blasting*, Carlton.
23. Otterness, R.E., Stagg, M.S., Rholl, S.A., Smith, N.S. 1991. Correlation of shot design parameters to fragmentation. In *Proceedings of the 7th annual conference on explosives and blasting technique*, Las Vegas.
24. Djordjevic, N. 1999. Two-component model of blast fragmentation. In *Proceedings of the 6th International Symposium on Rock Fragmentation by Blasting*, Johannesburg.
25. Kanchibotla, S.S., W. Valery, S. Morell. 1999. Modelling fines in blast fragmentation and its impact on crushing and grinding. In *Proceedings of the Explo 1999*, Carlton.
26. Thornton, D.M., S.S. Kanchibotla, J.S. Esterle. 2001. A fragmentation model to estimate ROM size distribution of soft rock types. In *Proceedings of the 28th Annual Conference on Explosives and Blasting Techniques*, Nashville.
27. Moser, P. 2003. Less fines production in aggregate and industrial minerals industry. In *Proceedings EFEE 2nd Conf. on Explosives & Blasting Techniques*, ed. R. Holmberg, 335–343. Rotterdam: Balkema
28. F. Ouchterlony. 2003. 'Bend it like Beckham' or a wide-range yet simple fragment size distribution for blasted and crushed rock. Less Fines. *Technical Report 78*, EU project GRD-2000-25224.
29. Ouchterlony, F. 2005. The Swebrec function, linking fragmentation by blasting and crushing. *Mining Technology* 114: 29–44. https://doi.org/10.1179/037178405x44539.
30. Ouchterlony, F. 2005. What does the fragment size distribution from blasting look like? In *Proceedings of the 3rd EFEE World Conference on Explosives and Blasting*, Brighton.
31. Ouchterlony, F., Olsson, M., Nyberg, U., Andersson, P., Gustavsson, L. 2006. Constructing the fragment size distribution of a bench blasting round, using the new Swebrec function. In *Proceedings of the 8th International Symposium on Rock Fragmentation by Blasting*, Santiago.
32. Ouchterlony, F. 2009. Fragmentation characterization; the Swebrec function and its use in blast engineering. In *Proceedings of the 9th International Symposium on Rock Fragmentation by Blasting*, London.
33. Ouchterlony, F., N. Paley. 2013. A reanalysis of fragmentation data from the Red Dog mine—Part 2. *Blasting and Fragmentation Journal* 7 (3): 139–172
34. Ouchterlony, F., P. Bergman, U. Nyberg. 2013. Fragmentation in production rounds and mill throughput in the Aitik copper mine, a summary of development projects 2002–2009. In *Proceedings of the 10th International Symposium on Rock Fragmentation by Blasting*, London.
35. Ouchterlony, F., U. Nyberg, M. Olsson, K. Vikström, P. Svedensten. 2015. Effects of specific charge and electronic delay detonators on fragmentation in an aggregate quarry, building KCO design curves. In *Proceedings of the 11th International Symposium on Rock Fragmentation by Blasting*, Carlton.
36. Ouchterlony, F., J.A. Sanchidrián, and P. Moser. 2017. Percentile fragment size predictions for blasted rock and the fragmentation-energy fan. *Rock Mechanics and Rock Engineering* 50 (4): 751–779. https://doi.org/10.1007/s00603-016-1094-x.

37. Cunningham, C.V.B. 2005. The Kuz-Ram fragmentation model—20 years on. In *Proceedings of the 3rd European Federation of Explosives Engineers (EFEE) World Conference on Explosives and Blasting*, Brighton.
38. Sanchidrián, J.A., and F. Ouchterlony. 2017. A distribution-free description of fragmentation by blasting based on dimensional analysis. *Rock Mechanics and Rock Engineering* 50 (4): 781–806. https://doi.org/10.1007/s00603-016-1131-9.
39. J.A. Sanchidrián, F. Ouchterlony. 2017. xP-frag, a distribution-free model to predict blast fragmentation. In *Proceedings of the 43rd Annual Conference on Explosives and Blasting Techniques*, Orlando.
40. Sayadi, A., M. Monjezi, N. Talebi, and M. Khandelwal. 2013. A comparative study on the application of various artificial neural networks to simultaneous prediction of rock fragmentation and backbreak. *Journal of Rock Mechanics and Geotechnical Engineering* 5 (4): 318–324. https://doi.org/10.1016/j.jrmge.2013.05.007.
41. Kulatilake, P.H.S.W., W. Quiong, T. Hudaverdi, and C. Kuzu. 2010. Mean particle size prediction in rock blast fragmentation using Neural Networks. *Engineering Geology* 114: 298–311. https://doi.org/10.1016/j.enggeo.2010.05.008.
42. Hudaverdi, T., P.H.S.W. Kulatilake, and C. Kuzu. 2011. Prediction of blast fragmentation using multivariate analysis procedures. *International Journal for Numerical and Analytical Methods in Geomechanics* 35 (12): 1318–1333. https://doi.org/10.1002/nag.957.
43. Monjezi, M., A. Bahrami, and A. Yazdian Varjani. 2010. Simultaneous prediction of fragmentation and flyrock in blasting operation using Artificial Neural Networks. *International Journal of Rock Mechanics and Mining Sciences* 47 (3): 476–480. https://doi.org/10.1016/j.ijrmms.2009.09.008.
44. Bahrami, A., M. Monjezi, K. Goshtasbi, and A. Ghazvinian. 2011. Prediction of rock fragmentation due to blasting using Artificial Neural Network. *Engineering Computations* 27: 177–181. https://doi.org/10.1007/s00366-010-0187-5.
45. Shi, X., J. Zhou, B.-B. Wu, D. Huang, and W. Wei. 2012. Support Vector Machines approach to mean particle size of rock fragmentation due to bench blasting prediction. *Transactions of the Nonferrous Metals Society of China* 22 (2): 432–441. https://doi.org/10.1016/S1003-6326(11)61195-3.
46. Khandelwal, M. 2011. Blast-induced ground vibration prediction using support vector machine. *Engineering Computations* 27 (3): 193–200. https://doi.org/10.1007/s00366-010-0190-x.
47. Gao, H., and Z.L. Fu. 2013. Forecast of blasting fragmentation distribution based on BP Neural Network. *Advanced Material Research* 619: 3–8. https://doi.org/10.4028/www.scientific.net/AMR.619.3.
48. Shi, X., D. Huang, J. Zhou, and S. Zhang. 2013. Combined ANN prediction model for rock fragmentation distribution due to blasting. *Journal of Information and Computing Science* 10 (11): 3511–3518. https://doi.org/10.12733/jics20101979.
49. Azarafza, M., M.-R. Feizi-Derakhshi, and A. Jeddi. 2018. Blasting pattern optimization in open-pit mines by using the Genetic Algorithm. *Journal of Geotechnical Geology* 13 (2): 75–81.
50. Karami, A., and S. Afiuni-Zadeh. 2013. Sizing of rock fragmentation modeling due to bench blasting using adaptive neuro fuzzy inference system (ANFIS). *International Journal of Mining Science and Technology* 23 (6): 809–813. https://doi.org/10.1016/j.ijmst.2013.10.005.
51. Enayatollahi, I., A.A. Bazzazi, and A. Asadi. 2014. Comparison between neural networks and multiple regression analysis to predict rock fragmentation in open-pit mines. *Rock Mechanics and Rock Engineering* 47 (2): 799–807. https://doi.org/10.1007/s00603-013-0415-6.
52. Gheibie, S., H. Aghababaei, S.H. Hoseinie, and Y. Pourrahimian. 2009. Modified Kuz-Ram fragmentation model and its use at the Sungun copper mine. *International Journal of Rock Mechanics and Mining Sciences* 46 (6): 967–973. https://doi.org/10.1016/j.ijrmms.2009.05.003.
53. Ebrahimi, E., M. Monjezi, M.R. Khalesi, and D.J. Armaghani. 2016. Prediction and optimization of back-break and rock fragmentation using an artificial neural network and a bee colony

algorithm. *Bulletin of Engineering Geology and the Environment* 75 (1): 27–36. https://doi.org/10.1007/s10064-015-0720-2.

54. Asl, P.F., M. Monjezi, J.K. Hamidi, and D.J. Armaghani. 2018. Optimization of flyrock and rock fragmentation in the Tajareh limestone mine using metaheuristics method of firefly algorithm. *Engineering Computations* 34 (2): 241–251. https://doi.org/10.1007/s00366-017-0535-9.
55. Lundborg, N., N. Persson, A. Ladegaard-Pedersen, and R. Holmberg. 1975. Keeping the lid on flyrock in open pit blasting. *Engineering and Mining Journal* 176: 95–100.
56. Gupta, R.N. 1980. Surface blasting and its impact on environment. In *Impact of mining on environment*, ed. N.J. Trivedy and B.P. Singh, 23–24. New Delhi: Ashish Publishing House.
57. Louzazni, M., A. Khouya, K. Amechnoue, A. Gandelli, M. Mussetta, and A. Crăciunescu. 2018. Metaheuristic algorithm for photovoltaic parameters: Comparative study and prediction with a firefly algorithm. *Applied Sciences* 8 (3): 339. https://doi.org/10.3390/app8030339.
58. Hasanipanah, M., H.B. Amnieh, H. Arab, and M.S. Zamzam. 2018. Feasibility of PSO-ANFIS model to estimate rock fragmentation produced by mine blasting. *Neural Computing and Applications* 30 (4): 1015–1024. https://doi.org/10.1007/s00521-016-2746-1.
59. Bastos-Filho, C.J.A., D.F. Carvalho, M.P. Caraciolo, P.B.C. Miranda, E.M.N. Figueiredo. 2009. Multi-ring particle swarm optimization. In *Evolutionary computation*, ed. W.P. Santos, 523–540. Viena: IntechOpen. https://doi.org/10.5772/9597
60. Mojtahedi, S.F.F., I. Ebtehaj, M. Hasanipanah, H. Bonakdari, and H.B. Amnieh. 2019. Proposing a novel hybrid intelligent model for the simulation of particle size distribution resulting from blasting. *Engineering Computations* 35 (1): 47–56. https://doi.org/10.1007/s00366-018-0582-x.
61. Valencia, J., R. Battulwar, M. Zare Naghadehi, J. Sattarvand. 2019. Enhancement of explosive energy distribution using UAVs and machine learning. In *Mining goes digital*, C. ed. Mueller et al. London: Taylor & Francis Group
62. Yaghoobi, H., H. Mansouri, M.A.E. Farsangi, and H. Nezamabadi-Pour. 2019. Determining the fragmented rock size distribution using textural feature ex-traction of images. *Powder Technology* 342: 630–641. https://doi.org/10.1016/j.powtec.2018.10.006.
63. Luerkens, D.W. 1986. Surface representation derived from a variational principle 1: The gray level function. *Particulate Science and Technology* 4 (4): 361–369. https://doi.org/10.1080/02726358608906467.
64. Bottlinger, M., R. Kholus. 1992. *Characterizing particle shapes and knowledge based image analysis of particle samples*. Quavkenbruck, F.D.R: Deutsches Institut fur lebensmitteltechik e.V. 4750.
65. C.L. Lin, Y.K. Ken, J.D. Miller. 1993. Evaluation of a PC, image-based, on-line coarse particle size analyser. In *Proceedings of the emerging computer techniques for the mineral industry symposium*, Utah.
66. Barron, L., M.L. Smith, and K. Prisbrey. 1994. Neural network pattern recognition of blast fragment size distributions. *Particulate Science and Technology* 12 (3): 235–242. https://doi.org/10.1080/02726359408906653.
67. Kemeny, J. 1994. Practical technique for determining the size distribution of blasted benches, waste dumps and heap leach sites. *Mining Engineering* 46 (11): 1281–1284.
68. Yen, Y.K., C.K. Lin, and J.D. Miller. 1998. Particle overlap and segregation problems in on-line coarse particle size measurement. *Powder Technology* 98 (1): 1–12. https://doi.org/10.1016/S0032-5910(97)03405-0.
69. Zhang, S., X.N. Bui, T. Nguyen-Trung, H. Nguyen, and H.B. Bui. 2020. Prediction of rock size distribution in mine bench blasting using a novel ant colony optimization-based boosted regression tree technique. *Natural Resources Research* 29: 867–886. https://doi.org/10.1007/s11053-019-09603-4.
70. José, I. 2018. *KNN (K-Nearest Neighbors) #1*. https://medium.com/brasil-ai/knn-k-nearest-neighbors-1-e140c82e9c4e. Accessed 10 June 2021.
71. Ngo, L. 2018. *Principal component analysis explained simply*. https://blog.bioturing.com/2018/06/14/principal-component-analysis-explained-simply/. Accessed 10 June 2021

72. Tang, Z., M. Sonntag, and H. Gross. 2019. Ant colony optimization in lens design. *Applied Optics* 58: 6357–6364. https://doi.org/10.1364/AO.58.006357.
73. Bradley, J., M. Amde. 2015. *Random Forests and Boosting in MLlib.* https://databricks.com/blog/2015/01/21/random-forests-and-boosting-in-mllib.html. Accessed 10 June 2021
74. Huang, J., P.G. Asteris, S.M.K. Pasha, A.S. Mohammed, and M. Hasanipanah. 2020. A new auto-tuning model for predicting the rock fragmentation: A cat swarm optimization algorithm. *Engineering Computations.* https://doi.org/10.1007/s00366-020-01207-4.
75. Selvakumar, K., K. Vijayakumar, and C.S. Boopathi. 2017. CSO based solution for load kickback effect in deregulated power systems. *Applied Sciences* 7 (11): 1127. https://doi.org/10.3390/app7111127.
76. Fang, Q., H. Nguyen, X.N. Bui, T. Nguyen-Thoi, and J. Zhou. 2021. Modeling of rock fragmentation by firefly optimization algorithm and boosted generalized additive model. *Neural Computing and Applications* 33: 3503–3519. https://doi.org/10.1007/s00521-020-05197-8.
77. Xie, C., H. Nguyen, X.N. Bui, Y. Choi, J. Zhou, and T. Nguyen-Trang. 2021. Predicting rock size distribution in mine blasting using various novel soft computing models based on meta-heuristics and machine learning algorithms. *Geoscience Frontiers* 12 (3): 101108. https://doi.org/10.1016/j.gsf.2020.11.005.
78. Vu, T., T. Bao, Q.V. Hoang, C. Drebenstetd, P.V. Hoa, H.H. Thang. 2021. *Measuring blast fragmentation at Nui Phao open-pit mine, Vietnam using the Mask R-CNN deep learning model.* Mining Technology. https://doi.org/10.1080/25726668.2021.1944458
79. Esmaeili, M., A. Salimi, C. Drebenstedt, M. Abbaszadeh, and A. Aghajani Bazzazi. 2015. Application of PCA, SVR, and ANFIS for modeling of rock fragmentation. *Arabian Journal of Geosciences* 8: 6881–6893. https://doi.org/10.1007/s12517-014-1677-3.
80. Shams, S., M. Monjezi, V. Johari Majd, and D. Jahed Armaghani. 2015. Application of fuzzy inference system for prediction of rock fragmentation induced by blasting. *Arabian Journal of Geosciences* 8: 10819–10832. https://doi.org/10.1007/s12517-015-1952-y.
81. Trivedi, R., T.N. Singh, and A.K. Raina. 2016. Simultaneous prediction of blast-induced flyrock and fragmentation in opencast limestone mines using back propagation neural network. *International Journal of Mining and Mineral Engineering* 7 (3): 237–252. https://doi.org/10.1504/IJMME.2016.078350.
82. Gao, W., M. Karbasi, M. Hasanipanah, X. Zhang, and J. Guo. 2018. Developing GPR model for forecasting the rock fragmentation in surface mines. *Engineering Computations* 34 (2): 339–345. https://doi.org/10.1007/s00366-017-0544-8.
83. Zhou, J., C. Li, C.A. Arslan, M. Hasanipanah, and H.B. Amnieh. 2021. Performance evaluation of hybrid FFA-ANFIS and GA-ANFIS models to predict particle size distribution of a muck-pile after blasting. *Engineering Computations* 37: 265–274. https://doi.org/10.1007/s00366-019-00822-0.
84. Trivedi, R., T.N. Singh, K. Mudgal, and N. Gupta. 2014. Application of artificial neural network for blast performance evaluation. *International Journal of Research in Engineering and Technology* 3 (5): 564–574. https://doi.org/10.15623/ijret.2014.0305104.
85. Dumakor-Dupey, N.K., S. Arya, and A. Jha. 2021. Advances in blast-induced impact prediction—A review of machine learning applications. *Minerals* 11 (6): 601. https://doi.org/10.3390/min11060601.
86. Siskind, D.E., M.S. Stagg, J.W. Kopp, C.H. Dowding. 1980. *Structure response and damage produced by ground vibration from surface mine blasting.* USBM, Report of Investigations 8507.
87. Uysal, O., and M. Cavus. 2013. Effect of a pre-split plane on the frequencies of blast induced ground vibrations. *Acta Montan Slovaca* 18 (2): 101–109.
88. Hudaverdi, T., and O. Akyildiz. 2019. Evaluation of capability of blast-induced ground vibration predictors considering measurement distance and different error measures. *Environment and Earth Science* 78: 421. https://doi.org/10.1007/s12665-019-8427-5.
89. Duvall, W.I., B. Petkof. 1959. *Spherical propagation of explosion generated strain pulses in rock.* USBM, Report of Investigation 5483.

90. Langefors, U., and B. Kihlstrom. 1963. *The modern technique of rock blasting*. New York: Wiley.
91. Davies, B., I.W. Farmer, and P.B. Attewell. 1964. *Ground vibrations from shallow sub-surface blasts*, 553–559. London: The Engineer.
92. Ambraseys, N.R., and A.J. Hendron. 1968. Dynamic behaviour of rock masses. In *Rock mechanics in engineering practice*, ed. K.G. Stagg and O.C. Zienkiewicz, 203–207. London: Wiley.
93. Indian Standard Institute. 1973. Criteria for safety and design of structures subjected to underground blast. ISI Bullet IS-6922.
94. Ghosh, A., J.K. Daemen. 1983. A simple new blast vibration predictor. In *Proceedings of the 24th US symposium on rock mechanics*, College Station.
95. Gupta, R.N., P.P. Roy, A. Bagachi, and B. Singh. 1987. Dynamic effects in various rock mass and their predictions. *Journal of Mines, Metals & Fuels* 35: 55–462.
96. Gupta, R.N., P.P. Roy, B. Sing. 1988. On a blast induced blast vibration predictor for efficient blasting. In *Proceedings of the 22nd International Conference of Safety in Mines*. Beijing.
97. Roy, P.P. 1993. Putting ground vibration predictors into practice. *Colliery Guardian* 241 (2): 63–67.
98. Rai, R., and T.N. Singh. 2004. A new predictor for ground vibration prediction and its comparison with other predictors. *Indian Journal of Engineering and Materials Science* 11: 178–184.
99. Singh, T.N., R. Kanchan, K. Saigal, and A.K. Verma. 2004. Prediction of P-wave velocity and anisotropic properties of rock using artificial neural networks technique. *Journal of Scientific & Industrial Research* 63 (1): 32–38.
100. Singh, T.N., and V. Singh. 2005. An intelligent approach to prediction and control ground vibration in mines. *Geotechnical and Geological Engineering* 23: 249–262. https://doi.org/10.1007/s10706-004-7068-x.
101. Khandelwal, M., and T.N. Singh. 2006. Prediction of blast induced ground vibrations and frequency in opencast mine—A neural network approach. *Journal of Sound and Vibration* 289 (4–5): 711–725. https://doi.org/10.1016/j.jsv.2005.02.044.
102. Khandelwal, M., and T.N. Singh. 2007. Evaluation of blast-induced ground vibration predictors. *Soil Dynamics and Earthquake Engineering* 27 (2): 116–125. https://doi.org/10.1016/j.soildyn.2006.06.004.
103. Khandelwal, M., and T.N. Singh. 2009. Prediction of blast-induced ground vibration using artificial neural network. *International Journal of Rock Mechanics and Mining Sciences* 46 (7): 1214–1222. https://doi.org/10.1016/j.ijrmms.2009.03.004.
104. Iphar, M., M. Yavuz, and H. Ak. 2008. Prediction of ground vibrations resulting from the blasting operations in an open-pit mine by adaptive neuro-fuzzy inference system. *Environmental Geology* 56: 97–107. https://doi.org/10.1007/s00254-007-1143-6.
105. Singh, T.N., L.K. Dontha, and V. Bhardwaj. 2008. Study into blast vibration and frequency using ANFIS and MVRA. *Mining Technology* 117 (3): 116–121. https://doi.org/10.1179/037178409X405741.
106. Mohamed, M.T. 2009. Artificial neural network for prediction and control of blasting vibrations in Assiut (Egypt) limestone quarry. *International Journal of Rock Mechanics and Mining Sciences* 46 (2): 426–431. https://doi.org/10.1016/j.ijrmms.2008.06.004.
107. Bakhshandeh Amnieh, H., M.R. Mozdianfard, and A. Siamaki. 2010. Predicting of blasting vibration in Sarcheshmeh copper mine by neural network. *Safety Science* 48 (3): 319–325. https://doi.org/10.1016/j.ssci.2009.10.009.
108. Monjezi, M., M. Ahmadi, M. Sheikhan, A. Bahrami, and A. Salimi. 2010. Predicting blast-induced ground vibration using various types of neural networks. *Soil Dynamics and Earthquake Engineering* 30 (11): 1233–1236. https://doi.org/10.1016/j.soildyn.2010.05.005.
109. Khandelwal, M., P.K. Kankar, and S.P. Harsha. 2010. Evaluation and prediction of blast-induced ground vibration using support vector machine. *Mining Science and Technology* 20 (1): 64–70. https://doi.org/10.1016/S1674-5264(09)60162-9.

110. Mohammadi, S.S., H. Bakhshandeh Amnieh, and M. Bahadori. 2011. Prediction ground vibration caused by blasting operations in Sarcheshmeh copper mine considering the charge type by adaptive neuro-fuzzy inference system (ANFIS). *Archives of Mining Sciences* 56 (4): 701–710.
111. Dehghani, H., and M. Ataee-pour. 2011. Development of a model to predict peak particle velocity in a blasting operation. *International Journal of Rock Mechanics and Mining Sciences* 48 (1): 51–58. https://doi.org/10.1016/j.ijrmms.2010.08.005.
112. Monjezi, M., M. Ghafurikalajahi, and A. Bahrami. 2011. Prediction of blast-induced ground vibration using artificial neural networks. *Tunnelling and Underground Space Technology* 26 (1): 46–50. https://doi.org/10.1016/j.tust.2010.05.002.
113. Khandelwal, M., D.L. Kumar, and M. Yellishetty. 2011. Application of soft computing to predict blast-induced ground vibration. *Engineering Computations* 27 (2): 117–125. https://doi.org/10.1007/s00366-009-0157-y.
114. Fişne, A., C. Kuzu, and T. Hudaverdi. 2011. Prediction of environmental impacts of quarry blasting operation using fuzzy logic. *Environmental Monitoring and Assessment* 174: 461–470. https://doi.org/10.1007/s10661-010-1470-z.
115. Longjun, D., L. Xibing, X. Ming, and L. Qiyue. 2011. Comparisons of random forest and support vector machine for predicting blasting vibration characteristic parameters. *Procedia Engineering* 26: 1772–1781. https://doi.org/10.1016/j.proeng.2011.11.2366.
116. Verma, A.K., and T.N. Singh. 2011. Intelligent systems for ground vibration measurement: A comparative study. *Engineering Computations* 27: 225–233. https://doi.org/10.1007/s00366-010-0193-7.
117. Khandelwal, M. 2011. Blast-induced ground vibration prediction using support vector machine. *Engineering Computations* 27: 193–200. https://doi.org/10.1007/s00366-010-0190-x.
118. Mohamed, M.T. 2011. Performance of fuzzy logic and artificial neural network in prediction of ground and air vibrations. *International Journal of Rock Mechanics and Mining Sciences* 48 (5): 845–851. https://doi.org/10.1016/J.IJRMMS.2011.04.016.
119. Shuran, L., and L. Shujin. 2011. Applying BP neural network model to forecast peak velocity of blasting ground vibration. *Procedia Engineering* 26: 257–263. https://doi.org/10.1016/j.proeng.2011.11.2166.
120. Mohamadnejad, M., R. Gholami, and M. Ataei. 2012. Comparison of intelligence science techniques and empirical methods for prediction of blasting vibrations. *Tunnelling and Underground Space Technology* 28: 238–244. https://doi.org/10.1016/j.tust.2011.12.001.
121. Álvarez-Vigil, A.E., C. González-Nicieza, F. López Gayarre, and M.I. Álvarez-Fernández. 2012. Predicting blasting propagation velocity and vibration frequency using artificial neural networks. *International Journal of Rock Mechanics and Mining Sciences* 55: 108–116. https://doi.org/10.1016/j.ijrmms.2012.05.002.
122. Hudaverdi, T. 2012. Application of multivariate analysis for prediction of blast-induced ground vibrations. *Soil Dynamics and Earthquake Engineering* 43: 300–308. https://doi.org/10.1016/j.soildyn.2012.08.002.
123. Li, D.T., J.L. Yan, and L. Zhang. 2012. Prediction of blast-induced ground vibration using support vector machine by tunnel excavation. *Applied Mechanics and Materials* 170: 1414–1418. https://doi.org/10.4028/www.scientific.net/AMM.170-173.1414.
124. Bakhshandeh Amnieh, H., A. Siamaki, and S. Soltani. 2012. Design of blasting pat-tern in proportion to the peak particle velocity (PPV): Artificial neural networks approach. *Safety Science* 50 (9): 1913–1916. https://doi.org/10.1016/j.ssci.2012.05.008.
125. Mohamad, E.T., S.A. Noorani, D.J. Armaghani, and R. Saad. 2012. Simulation of blasting induced ground vibration by using artificial neural network. *Electronic Journal of Geotechnical Engineering* 17: 2571–2584.
126. Mohammadnejad, M., R. Gholami, A. Ramazanzadeh, and M.E. Jalali. 2012. Prediction of blast-induced vibrations in limestone quarries using Support Vector Machine. *Journal of Vibration and Control* 18 (9): 1322–1329. https://doi.org/10.1177/1077546311421052.

127. Ataei, M., and M. Kamali. 2013. Prediction of blast-induced vibration by adaptive neuro-fuzzy inference system in Karoun 3 power plant and dam. *Journal of Vibration and Control* 19 (12): 1906–1914. https://doi.org/10.1177/1077546312444769.
128. Gorgulu, K., E. Arpaz, A. Demirci, A. Kocaslan, M.K. Dilmac, and A.G. Yuksek. 2013. Investigation of blast-induced ground vibrations in the Tulu boron open pit mine. *Bulletin of Engineering Geology and the Environment* 72: 555–564. https://doi.org/10.1007/s10064-013-0521-4.
129. Monjezi, M., M. Hasanipanah, and M. Khandelwal. 2013. Evaluation and prediction of blast-induced ground vibration at Shur River Dam, Iran, by artificial neural network. *Neural Computing and Applications* 22 (7–8): 1637–1643. https://doi.org/10.1007/s00521-012-0856-y.
130. Ghasemi, E., M. Ataei, and H. Hashemolhosseini. 2013. Development of a fuzzy model for predicting ground vibration caused by rock blasting in surface mining. *Journal of Vibration and Control* 19 (5): 755–770. https://doi.org/10.1177/1077546312437002.
131. Verma, A.K., and T.N. Singh. 2013. A neuro-fuzzy approach for prediction of longitudinal wave velocity. *Neural Computing & Applications* 22: 1685–1693. https://doi.org/10.1007/s00521-012-0817-5.
132. Saadat, M., M. Khandelwal, and M. Monjezi. 2014. An ANN-based approach to predict blast-induced ground vibration of Gol-E-Gohar iron ore mine, Iran. *Journal of Rock Mechanics and Geotechnical Engineering* 6 (1): 67–76. https://doi.org/10.1016/j.jrmge.2013.11.001.
133. Verma, A.K., T.N. Singh, and S. Maheshwar. 2014. Comparative study of intelligent prediction models for pressure wave velocity. *International Journal of Geomatics and Geosciences* 2 (3): 130–138. https://doi.org/10.12691/jgg-2-3-9.
134. Vasović, D., S. Kostić, M. Ravilić, and S. Trajković. 2014. Environmental impact of blasting atDrenovac limestone quarry (Serbia). *Environment and Earth Science* 72 (10): 3915–3928. https://doi.org/10.1007/s12665-014-3280-z.
135. Jahed Armaghani, D., M. Hajihassani, E. Tonnizam Mohamad, A. Marto, and S. Noorani. 2014. Blasting-induced flyrock and ground vibration prediction through an expert artificial neural network based on particle swarm optimization. *Arabian Journal of Geosciences* 7 (12): 5383–5396. https://doi.org/10.1007/s12517-013-1174-0.
136. Lapcevic, R., S. Kostic, R. Pantovic, and N. Vasovic. 2014. Prediction of blast-induced ground motion in a copper mine. *International Journal of Rock Mechanics and Mining Sciences* 69: 19–25. https://doi.org/10.1016/j.ijrmms.2014.03.002.
137. Xue, X., and X. Yang. 2014. Predicting blast-induced ground vibration using general regression neural network. *Journal of Vibration and Control* 20 (10): 1512–1519. https://doi.org/10.1177/1077546312474680.
138. Gorgulu, K., E. Arpaz, O. Uysal, Y.S. Duruturk, A.G. Yuksek, A. Kocaslan, and M.K. Dilmac. 2015. Investigation of the effects of blasting design parameters and rock properties on blast-induced ground vibrations. *Arabian Journal of Geosciences* 8: 4269–4278. https://doi.org/10.1007/s12517-014-1477-9.
139. Hajihassani, M., D.J. Armaghani, M. Monjezi, E.T. Mohamad, and A. Marto. 2015. Blast-induced air and ground vibration prediction: A particle swarm optimization-based artificial neural network approach. *Environment and Earth Science* 74 (4): 2799–2817. https://doi.org/10.1007/s12665-015-4274-1.
140. Hajihassani, M., D. Jahed Armaghani, A. Marto, and E. Tonnizam Mohamad. 2015. Ground vibration prediction in quarry blasting through an artificial neural network optimized by imperialist competitive algorithm. *Bulletin of Engineering Geology and the Environment*. https://doi.org/10.1007/s10064-014-0657-x.
141. Jahed Armaghani, D., E. Momeni, S.V.A.N.K. Abad, and M. Khandelwal. 2015. Feasibility of ANFIS model for prediction of ground vibrations resulting from quarry blasting. *Environment and Earth Science* 74 (4): 2845–2860. https://doi.org/10.1007/s12665-015-4305-y.
142. Jahed Armaghani, D., M. Hajihassani, M. Monjezi, E. Tonnizam Mohamad, A. Marto, and M.R. Moghaddam. 2015. Application of two intelligent systems in predicting environmental impacts of quarry blasting. *Arabian Journal of Geosciences* 8: 9647–9665. https://doi.org/10.1007/s12517-015-1908-2.

143. Mohebi, J., A.Z. Shirazi, and H. Tabatabaee. 2015. Adaptive-neuro fuzzy inference system (ANFIS) model for prediction of blast-induced ground vibration. *Science International (Lahore)* 27 (3): 2079–2091.
144. Hasanipanah, M., M. Monjezi, A. Shahnazar, D.J. Armaghani, and A. Farazmand. 2015. Feasibility of indirect determination of blast induced ground vibration based on support vector machine. *Measurement* 75: 289–297. https://doi.org/10.1016/j.measurement.2015.07.019.
145. Dindarloo, S.R. 2015. Peak particle velocity prediction using support vector machines: A surface blasting case study. *Journal of the South African Institute of Mining and Metallurgy* 115 (7): 637–643. https://doi.org/10.17159/2411-9717/2015/V115N7A10.
146. Dindarloo, S.R. 2015. Prediction of blast-induced ground vibrations via genetic programming. *International Journal of Mining Science and Technology* 25 (6): 1011–1015. https://doi.org/10.1016/j.ijmst.2015.09.020.
147. Shirani Faradonbeh, R., D. Jahed Armaghani, M.Z. Abd Majid, M.M.D. Tahir, B. Ramesh Murlidhar, M. Monjezi, and H.M. Wong. 2016. Prediction of ground vibration due to quarry blasting based on gene expression programming: A new model for peak particle velocity prediction. *International Journal of Environmental Science and Technology* 13 (6): 1453–1464. https://doi.org/10.1007/s13762-016-0979-2.
148. Monjezi, M., M. Baghestani, R. Shirani Faradonbeh, M. Pourghasemi Saghand, and D. Jahed Armaghani. 2016. Modification and prediction of blast-induced ground vibrations based on both empirical and computational techniques. *Engineering Computations* 32 (4): 717–728. https://doi.org/10.1007/s00366-016-0448-z.
149. Singh, J., A.K. Verma, H. Banka, T.N. Singh, and S. Maheshwar. 2016. A study of soft computing models for prediction of longitudinal wave velocity. *Arabian Journal of Geosciences* 9: 224. https://doi.org/10.1007/s12517-015-2115-x.
150. Ghoraba, S., M. Monjezi, N. Talebi, D.J. Armaghani, and M.R. Moghaddam. 2016. Estimation of ground vibration produced by blasting operations through intelligent and empirical models. *Environment and Earth Science* 75: 1137. https://doi.org/10.1007/s12665-016-5961-2.
151. Amiri, M., H.B. Amnieh, M. Hasanipanah, and L.M. Khanli. 2016. A new combination of artificial neural network and K-nearest neighbors models to predict blast-induced ground vibration and air-overpressure. *Engineering Computations* 32 (4): 631–644. https://doi.org/10.1007/s00366-016-0442-5.
152. Ghasemi, E., H. Kalhori, and R. Bagherpour. 2016. A new hybrid ANFIS-PSO model for prediction of peak particle velocity due to bench blasting. *Engineering Computations* 32 (4): 607–614. https://doi.org/10.1007/s00366-016-0438-1.
153. Koçaslan, A., A.G. Yuksek, K. Gorgulu, and E. Arpaz. 2017. Evaluation of blast-induced ground vibrations in open-pit mines by using adaptive neuro-fuzzy inference systems. *Environment and Earth Science* 76: 57. https://doi.org/10.1007/s12665-016-6306-x.
154. Ram Chandar, K., V.R. Sastry, and C. Hegde. 2017. A critical comparison of regression models and artificial neural networks to predict ground vibrations. *Geotechnical and Geological Engineering* 35 (2): 573–583. https://doi.org/10.1007/s10706-016-0126-3.
155. Shahnazar, A., H. Nikafshan Rad, M. Hasanipanah, M.M. Tahir, D. Jahed Armaghani, and M. Ghoroqi. 2017. A new developed approach for the prediction of ground vibration using a hybrid PSO-optimized ANFIS-based model. *Environment and Earth Science* 76 (15): 527. https://doi.org/10.1007/s12665-017-6864-6.
156. Samareh, H., S.H. Khoshrou, K. Shahriar, M.M. Ebadzadeh, and M. Eslami. 2017. Optimization of a nonlinear model for predicting the ground vibration using the combinational particle swarm optimization-genetic algorithm. *Journal of the African Earth Sciences* 133: 36–45. https://doi.org/10.1016/j.jafrearsci.2017.04.029.
157. Xue, X., X. Yang, and P. Li. 2017. Evaluation of ground vibration due to blasting using fuzzy logic. *Geotechnical and Geological Engineering* 35 (3): 1231–1237. https://doi.org/10.1007/s10706-017-0162-7.
158. Taheri, K., M. Hasanipanah, S.B. Golzar, and M.Z.A. Majid. 2017. A hybrid artificial bee colony algorithm-artificial neural network for forecasting the blast-produced ground vibration. *Engineering Computations* 33: 689–700. https://doi.org/10.1007/s00366-016-0497-3.

159. Shirani Faradonbeh, R., and M. Monjezi. 2017. Prediction and minimization of blast-induced ground vibration using two robust metaheuristic algorithms. *Engineering Computations* 33 (4): 835–851. https://doi.org/10.1007/s00366-017-0501-6.
160. Ameryan, M., M.R. Akbarzadeh Totonchi, S.J. Seyyed Mahdavi. 2014. Clustering based on cuckoo optimization algorithm. In *Proceedings of the 2014 Iranian Conference on Intelligent Systems (ICIS)*, Bam. https://doi.org/10.1109/IranianCIS.2014.6802605.
161. Khandelwal, M., D.J. Armaghani, R.S. Faradonbeh, M. Yellishetty, M.Z.A. Majid, and M. Monjezi. 2017. Classification and regression tree technique in estimating peak particle velocity caused by blasting. *Engineering Computations* 33: 45–53. https://doi.org/10.1007/s00366-016-0455-0.
162. Hasanipanah, M., R.S. Faradonbeh, H.B. Amnieh, D.J. Armaghani, and M. Monjezi. 2017. Forecasting blast-induced ground vibration developing a CART model. *Engineering Computations* 33 (2): 307–316. https://doi.org/10.1007/s00366-016-0475-9.
163. Hasanipanah, M., S.B. Golzar, I.A. Larki, M.Y. Maryaki, and T. Ghahremanians. 2017. Estimation of blast-induced ground vibration through a soft computing framework. *Engineering Computations* 33 (4): 951–959. https://doi.org/10.1007/s00366-017-0508-z.
164. Hasanipanah, M., R. Naderi, J. Kashir, S.A. Noorani, and A.Z.A. Qaleh. 2017. Prediction of blast-produced ground vibration using particle swarm optimization. *Engineering Computations* 33 (2): 173–179. https://doi.org/10.1007/s00366-016-0462-1.
165. Fouladgar, N., M. Hasanipanah, and H. Bakhshandeh Amnieh. 2017. Application of cuckoo search algorithm to estimate peak particle velocity in mine blasting. *Engineering Computations* 33: 181–189. https://doi.org/10.1007/s00366-016-0463-0.
166. Jahed Armaghani, D., M. Hasanipanah, H.B. Amnieh, and E.T. Mohamad. 2018. Feasibility of ICA in approximating ground vibration resulting from mine blasting. *Neural Computing and Applications* 29 (9): 457–465. https://doi.org/10.1007/s00521-016-2577-0.
167. Behzadafshar, K., F. Mohebbi, M. Soltani Tehrani, M. Hasanipanah, and O. Tabrizi. 2018. Predicting the ground vibration induced by mine blasting using imperialist competitive algorithm. *Engineering Computations* 35 (4): 1774–1787. https://doi.org/10.1108/EC-08-2017-0290.
168. Abbaszadeh Shahri, A., and R. Asheghi. 2018. Optimized developed artificial neural network-based models to predict the blast-induced ground vibration. *Innovative Infrastructure Solutions* 3: 1–10. https://doi.org/10.1007/s41062-018-0137-4.
169. Mokfi, T., A. Shahnazar, I. Bakhshayeshi, A.M. Derakhsh, and O. Tabrizi. 2018. Proposing of a new soft computing-based model to predict peak particle velocity induced by blasting. *Engineering Computations* 34 (4): 881–888. https://doi.org/10.1007/s00366-018-0578-6.
170. Garai, D., H. Agrawal, A.K. Mishra, and S. Kumar. 2018. Influence of initiation system on blast-induced ground vibration using random forest algorithm, artificial neural network, and scaled distance analysis. *Mathematical Modelling of Engineering Problems* 5 (4): 418–426. https://doi.org/10.18280/mmep.050419.
171. Sheykhi, H., R. Bagherpour, E. Ghasemi, and H. Kalhori. 2018. Forecasting ground vibration due to rock blasting: A hybrid intelligent approach using support vector regression and fuzzy C-means clustering. *Engineering Computations* 34 (2): 357–365. https://doi.org/10.1007/s00366-017-0546-6.
172. Zhongya, Z. 2018. Xiaoguang J (2018) Prediction of peak velocity of blasting vibration based on artificial neural network optimized by dimensionality reduction of FA-MIV. *Mathematical Problems in Engineering* 8473547: 1–12. https://doi.org/10.1155/2018/8473547.
173. Hasanipanah, M., H. Bakhshandeh Amnieh, H. Khamesi, D. Jahed Armaghani, S. Bagheri Golzar, and A. Shahnazar. 2018. Prediction of an environmental issue of mine blasting: An imperialistic competitive algorithm-based fuzzy system. *International Journal of Environmental Science and Technology* 15 (3): 551–560. https://doi.org/10.1007/s13762-017-1395-y.
174. Prashanth, R., and D. Nimaje. 2018. Estimation of ambiguous blast-induced ground vibration using intelligent models: A case study. *Noise & Vibration Worldwide* 49 (4): 147–157. https://doi.org/10.1177/0957456518781858.

175. Ragam, P., and D. Nimaje. 2018. Assessment of blast-induced ground vibration using different predictor approaches-a comparison. *Chemical Engineering Transactions* 66: 487–492. https://doi.org/10.3303/CET1866082.
176. Iramina, W.S., E.C. Sansone, M. Wichers, S. Wahyudi, S.M.D. Eston, H. Shimada, and T. Sasaoka. 2018. Comparing blast-induced ground vibration models using ann and empirical geomechanical relationships. *International Engineering Journal* 71 (1): 89–95. https://doi.org/10.1590/0370-44672017710097.
177. Kumar, R., D. Choudhury, and K. Bhargava. 2016. Determination of blast-induced ground vibration equations for rocks using mechanical and geological properties. *Journal of Rock Mechanics and Geotechnical Engineering* 8 (3): 341–349. https://doi.org/10.1016/j.jrmge.2015.10.009.
178. Nguyen, H., X.N. Bui, Q.H. Tran, T.Q. Le, N.H. Do, and L.T.T. Hoa. 2019. Evaluating and predicting blast-induced ground vibration in open-cast mine using ANN: A case study in Vietnam. *SN Applied Sciences* 1: 125. https://doi.org/10.1007/s42452-018-0136-2.
179. Nguyen, H., X.N. Bui, Q.H. Tran, Q.L. Nguyen, D.H. Vu, V.H. Pham, Q.T. Le, and P.V. Nguyen. 2019. Developing an advanced soft computational model for estimating blast-induced ground vibration in Nui Beo open-pit coal mine (Vietnam) using artificial neural network. *Inzynieria Mineraina* 21: 58–73. https://doi.org/10.29227/IM-2019-02-58.
180. Nguyen, H., X.-N. Bui, H.-B. Bui, and D.T. Cuong. 2019. Developing an XGBoost model to predict blast-induced peak particle velocity in an open-pit mine: A case study. *Acta Geophysica* 67 (2): 477–490. https://doi.org/10.1007/s11600-019-00268-4.
181. Nguyen, H., X.-N. Bui, Q.-H. Tran, and N.-L. Mai. 2019. A new soft computing model for estimating and controlling blast-produced ground vibration based on hierarchical K-means clustering and cubist algorithms. *Applied Soft Computing* 77: 376–386. https://doi.org/10.1016/j.asoc.2019.01.042.
182. Nguyen, H., X.-N. Bui, and H. Moayedi. 2019. A comparison of advanced computational models and experimental techniques in predicting blast-induced ground vibration in openpit coal mine. *Acta Geophysica* 67: 1025–1037. https://doi.org/10.1007/s11600-019-00304-3.
183. Nguyen, H. 2019. Support vector regression approach with different kernel functions for predicting blast-induced ground vibration: A case study in an open-pit coal mine of Vietnam. *SN Applied Sciences* 1: 283. https://doi.org/10.1007/s42452-019-0295-9.
184. Nguyen, H., X.N. Bui, Q.H. Tran, and H. Moayedi. 2019. Predicting blast-induced peak particle velocity using BGAMs, ANN and SVM: A case study at the Nui Beo open-pit coal mine in Vietnam. *Environment and Earth Science* 78: 479. https://doi.org/10.1007/s12665-019-8491-x.
185. Chen, W., M. Hasanipanah, H. Nikafshan Rad, D. Jahed Armaghani, and M.M. Tahir. 2019. A new design of evolutionary hybrid optimization of SVR model in predicting the blast-induced ground vibration. *Engineering Computations* 37: 1455–1471. https://doi.org/10.1007/s00366-019-00895-x.
186. Das, A., S. Sinha, and S. Ganguly. 2019. Development of a blast-induced vibration prediction model using an artificial neural network. *Journal of the South African Institute of Mining and Metallurgy* 119 (2): 187–200. https://doi.org/10.17159/2411-9717/2019/v119n2a11.
187. Shang, Y., H. Nguyen, X.N. Bui, Q.H. Tran, and H. Moayedi. 2019. A novel artificial intelligence approach to predict blast-induced ground vibration in open-pit mines based on the firefly algorithm and artificial neural network. *Natural Resources Research* 29: 723–737. https://doi.org/10.1007/s11053-019-09503-7.
188. Torres, N., J.A. Reis, P.L. Luiz, J.H.R. Costa, L.S. Chaves. 2019. Neural network applied to blasting vibration control near communities in a large-scale iron ore mine. In *Proceedings of the 27th International Symposium on Mine Planning and Equipment Selection—MPES 2018*, ed. E. Widzyk-Capehart, A. Hekmat, R. Singhal, 81–91. Cham: Springer. https://doi.org/10.1007/978-3-319-99220-4_7.
189. Xue, X. 2019. Neuro-fuzzy based approach for prediction of blast-induced ground vibration. *Applied Acoustics* 152: 73–78. https://doi.org/10.1016/j.apacoust.2019.03.023.

190. Hosseini, M., and M.S. Baghikhani. 2013. Analysing the ground vibration due to blasting at AlvandQoly limestone mine. *International Journal of Mining Engineering and Mineral Processing* 2 (2): 17–23. https://doi.org/10.5923/j.mining.20130202.01.
191. Zhang, X., H. Nguyen, X.-N. Bui, Q.-H. Tran, D.-A. Nguyen, D.T. Bui, and H. Moayedi. 2019. Novel soft computing model for predicting blast-induced ground vibration in open-pit mines based on particle swarm optimization and XGBoost. *Natural Resources Research* 29: 711–721. https://doi.org/10.1007/s11053-019-09492-7.
192. Jiang, W., C.A. Arslan, M.S. Tehrani, M. Khorami, and M. Hasanipanah. 2019. Simulating the peak particle velocity in rock blasting projects using a neuro-fuzzy inference system. *Engineering Computations* 35: 1203–1211. https://doi.org/10.1007/s00366-018-0659-6.
193. Arthur, C.K., V.A. Temeng, and Y.Y. Ziggah. 2019. Soft computing-based technique as a predictive tool to estimate blast-induced ground vibration. *Journal of Sustainable Mining* 18 (4): 287–296. https://doi.org/10.1016/j.jsm.2019.10.001.
194. Yang, H., M. Hasanipanah, M.M. Tahir, and D.T. Nui. 2019. Intelligent prediction of blasting-induced ground vibration using ANFIS optimized by GA and PSO. *Natural Resources Research* 29: 739–750. https://doi.org/10.1007/s11053-019-09515-3.
195. Bui, X.N., P. Jaroonpattanapong, H. Nguyen, Q.H. Tran, and N.Q. Long. 2019. A novel hybrid model for predicting blast-induced ground vibration based on k-nearest neighbors and particle swarm optimization. *Science and Reports* 9: 13971. https://doi.org/10.1038/s41598-019-50262-5.
196. Azimi, Y., S.H. Khoshrou, and M. Osanloo. 2019. Prediction of blast induced ground vibration (BIGV) of quarry mining using hybrid genetic algorithm optimized artificial neural network. *Measurement* 147: 106874. https://doi.org/10.1016/j.measurement.2019.106874.
197. Nguyen, H., Y. Choi, X.N. Bui, and T. Nguyen-Thoi. 2020. Predicting blast-induced ground vibration in open-pit mines using vibration sensors and support vector regression-based optimization algorithms. *Sensors* 20 (1): 132. https://doi.org/10.3390/s20010132.
198. Nguyen, H., C. Drebenstedt, X.-N. Bui, and D.T. Bui. 2020. Prediction of blast-induced ground vibration in an open-pit mine by a novel hybrid model based on clustering and artificial neural network. *Natural Resources Research* 29: 691–709. https://doi.org/10.1007/s11053-019-09470-z.
199. Shakeri, J., B.J. Shokri, and H. Dehghani. 2020. prediction of blast-induced ground vibration using gene expression programming (GEP), artificial neural networks (ANNs), and linear multivariate regression (LMR). *Archives of Mining Sciences* 65 (2): 317–335. https://doi.org/10.24425/ams.2020.133195.
200. Arthur, C.K., V.A. Temeng, and Y.Y. Ziggah. 2020. Performance evaluation of training algorithms in backpropagation neural network approach to blast-induced ground vibration prediction. *Ghana Mining Journal* 20 (1): 20–33. https://doi.org/10.4314/gm.v20i1.3.
201. Arthur, C.K., V.A. Temeng, and Y.Y. Ziggah. 2020. Multivariate adaptive regression splines (MARS) approach to blast-induced ground vibration prediction. *International Journal of Mining, Reclamation and Environment* 34 (3): 198–222. https://doi.org/10.1080/17480930.2019.1577940.
202. Kriner, M. 2007. *Survival analysis with multivariate adaptive regression splines*. Ph.D. Thesis, Ludwig-Maximilians-Universität Munchen.
203. Arthur, C.K., and R.B. Kaunda. 2020. A hybrid paretosearch algorithm and goal attainment method for maximizing production and reducing blast-induced ground vibration: A blast design parameter selection approach. *Mining Technology* 129 (3): 151–158. https://doi.org/10.1080/25726668.2020.1790262.
204. MathWorks Help Center. 2021. *Paretosearch Algorithm*. https://la.mathworks.com/help/gads/paretosearch-algorithm.html. Accessed 27 June 2021
205. Bayat, P., M. Monjezi, M. Rezakhah, and D. Jahed Armaghani. 2020. Artificial neural network and firefly algorithm for estimation and minimization of ground vibration induced by blasting in a mine. *Natural Resources Research* 29: 4121–4132. https://doi.org/10.1007/s11053-020-09697-1.

206. Fang, Q., H. Nguyen, X.N. Bui, and T. Nguyen-Thoi. 2020. Prediction of blast-induced ground vibration in open-pit mines using a new technique based on imperialist competitive algorithm and M5Rules. *Natural Resources Research* 29: 791–806. https://doi.org/10.1007/s11053-019-09577-3.
207. Lawal, A.I., and M.A. Idris. 2020. An artificial neural network-based mathematical model for the prediction of blast-induced ground vibrations. *International Journal of Environmental Studies* 77 (2): 318–334. https://doi.org/10.1080/00207233.2019.1662186.
208. Amiri, M., M. Hasanipanah, and H. Bakhshandeh Amnieh. 2020. Predicting ground vibration induced by rock blasting using a novel hybrid of neural network and itemset mining. *Neural Computing & Applications* 32: 14681–14699. https://doi.org/10.1007/s00521-020-04822-w.
209. Yang, H., H. Nikafshan Rad, M. Hasanipanah, H. Bakhshandeh Amnieh, and A. Nekouie. 2020. Prediction of vibration velocity generated in mine blasting using support vector regression improved by optimization algorithms. *Natural Resources Research* 29: 807–830. https://doi.org/10.1007/s11053-019-09597-z.
210. Wei, H., J. Chen, J. Zhu, X. Yang, and H. Chu. 2020. A novel algorithm of nested-ELM for predicting blasting vibration. *Engineering Computations*. https://doi.org/10.1007/s00366-020-01082-z.
211. Jahed Armaghani, D., M. Hasanipanah, H. Bakhshandeh Amnieh, D. Tien Bui, P. Mehrabi, and M. Khorami. 2020. Development of a novel hybrid intelligent model for solving engineering problems using GS-GMDH algorithm. *Engineering Computations* 36: 1379–1391. https://doi.org/10.1007/s00366-019-00769-2.
212. Jahed Armaghani, D., D. Kumar, P. Samui, M. Hasanipanah, and B. Roy. 2020. A novel approach for forecasting of ground vibrations resulting from blasting: Modified particle swarm optimization coupled extreme learning machine. *Engineering Computations*. https://doi.org/10.1007/s00366-020-00997-x.
213. Strohmann, T.R., G.Z. Grudic. 2003. *Robust minimax probability machine regression*. https://citeseerx.ist.psu.edu/viewdoc/download?doi=10.1.1.2.3972&rep=rep1&type=pdf. Accessed 28 June 2021.
214. Suykens, J.A.K., and J. Vandewalle. 1999. Least squares support vector machine classifiers. *Neural Processing Letters* 9 (3): 293–300. https://doi.org/10.1023/a:1018628609742.
215. Mahdiyar, A., D. Jahed Armaghani, M. Koopialipoor, A. Hedayat, A. Abdullah, and K. Yahya. 2020. Practical risk assessment of ground vibrations resulting from blasting, using gene expression programming and Monte Carlo simulation techniques. *Applied Sciences* 10 (2): 472. https://doi.org/10.3390/app10020472.
216. Li, G., D. Kumar, P. Samui, H. Nikafshan Rad, B. Roy, and M. Hasanipanah. 2020. Developing a new computational intelligence approach for approximating the blast-induced ground vibration. *Applied Sciences* 10: 434. https://doi.org/10.3390/app10020434.
217. Mirjalili, S. 2021. *Biogeography-based optimizer (BBO) for training multi-layer perceptron (MLP)*. https://www.mathworks.com/matlabcentral/fileexchange/45804-biogeography-based-optimizer-bbo-for-training-multi-layer-perceptron-mlp. Accessed 01 July 2021
218. Ding, Z., H. Nguyen, X.N. Bui, J. Zhou, and H. Moayedi. 2020. Computational intelligence model for estimating intensity of blast-induced ground vibration in a mine based on imperialist competitive and extreme gradient boosting algorithms. *Natural Resources Research* 29: 751–769. https://doi.org/10.1007/s11053-019-09548-8.
219. Ding, Z., M. Hasanipanah, H. Nikafshan Rad, and W. Zhou. 2020. Predicting the blast-induced vibration velocity using a bagged support vector regression optimized with firefly algorithm. *Engineering Computations*. https://doi.org/10.1007/s00366-020-00937-9.
220. Yu, Z., X. Shi, J. Zhou, X. Chen, and X. Qiu. 2020. Effective assessment of blast-induced ground vibration using an optimized random forest model based on a Harris hawks optimization algorithm. *Applied Sciences* 10 (4): 1403. https://doi.org/10.3390/app10041403.
221. Yu, Z., X. Shi, J. Zhou, Y. Gou, X. Huo, J. Zhang, and D. Jahed Armaghani. 2020. A new multikernel relevance vector machine based on the HPSOGWO algorithm for predicting and controlling blast-induced ground vibration. *Engineering Computations*. https://doi.org/10.1007/s00366-020-01136-2.

222. Zhang, H., J. Zhou, D. Jahed Armaghani, M.M. Tahir, B.T. Pham, and V.V. Huynh. 2020. A combination of feature selection and random forest techniques to solve a problem related to blast-induced ground vibration. *Applied Sciences* 10 (3): 869. https://doi.org/10.3390/app10030869.
223. Zhou, J., P.G. Asteris, D. Jahed Armaghani, and B.T. Pham. 2020. Prediction of ground vibration induced by blasting operations through the use of the Bayesian Network and random forest models. *Soil Dynamics and Earthquake Engineering* 139: 106390. https://doi.org/10.1016/j.soildyn.2020.106390.
224. Bui, X.N., Y. Choi, V. Atrushkevich, H. Nguyen, Q.H. Tran, N.Q. Long, and H.T. Hoang. 2020. Prediction of blast-induced ground vibration intensity in open-pit mines using unmanned aerial vehicle and a novel intelligence system. *Natural Resources Research* 29: 771–790. https://doi.org/10.1007/s11053-019-09573-7.
225. Lawal, A.I. 2020. An artificial neural network-based mathematical model for the prediction of blast-induced ground vibration in granite quarries in Ibadan, Oyo State, Nigeria. *Scientific African* 8: e00413. https://doi.org/10.1016/j.sciaf.2020.e00413.
226. Zhou, J., C. Li, M. Koopialipoor, D. Jahed Armaghani, and B. Thai Pham. 2021. Development of a new methodology for estimating the amount of PPV in surface mines based on prediction and probabilistic models (GEP-MC). *International Journal of Mining, Reclamation and Environment* 35 (1): 48–68. https://doi.org/10.1080/17480930.2020.1734151.
227. Lawal, A.I., S. Kwon, and G.Y. Kim. 2021. Prediction of the blast-induced ground vibration in tunnel blasting using ANN, moth-flame optimized ANN, and gene expression programming. *Acta Geophysica* 69: 161–174. https://doi.org/10.1007/s11600-020-00532-y.
228. Mirjalili, S. 2015. Moth-flame optimization algorithm: A novel nature-inspired heuristic paradigm. *Knowledge-Based Systems* 89: 228–249. https://doi.org/10.1016/j.knosys.2015.07.006.
229. Zhu, W., H. Nikafshan Rad, and M. Hasanipanah. 2021. A chaos recurrent ANFIS optimized by PSO to predict ground vibration generated in rock blasting. *Applied Soft Computing* 108: 107434. https://doi.org/10.1016/j.asoc.2021.107434.
230. Qiu, Y., J. Zhou, M. Khandelwal, H. Yang, P. Yang, C. Li. 2021. Performance evaluation of hybrid WOA-XGBoost, GWO-XGBoost and BO-XGBoost models to predict blast-induced ground vibration. *Engineering with Computers*. https://doi.org/10.1007/s00366-021-01393-9.
231. Lawal, A.I., S. Kwon, O.S. Hammed, and M.A. Idris. 2021. Blast-induced ground vibration prediction in granite quarries: An application of gene expression programming, ANFIS, and sine cosine algorithm optimized ANN. *International Journal of Mining Science and Technology* 31 (2): 265–277. https://doi.org/10.1016/j.ijmst.2021.01.007.
232. Yu, C., M. Koopialipoor, B. Ramesh Murlidhar, A.S. Mohammed, D. Jahed Armaghani, E. Tonnizam Mohamad, and Z. Wang. 2021. Optimal ELM-Harris hawks optimization and ELM-grasshopper optimization models to forecast peak particle velocity resulting from mine blasting. *Natural Resources Research* 30 (3): 2647–2662. https://doi.org/10.1007/s11053-021-09826-4.
233. Potnuru, D., A.S. Tummala. 2019. Implementation of grasshopper optimization algorithm for controlling a BLDC motor drive. In *Soft Computing in Data Analytics. Proceedings of International Conference on SCDA 2018*, ed. by J. Nayak, A. Abraham, B. Krishna, G. Chandra Sekhar, A. Das. Singapore: Springer. https://doi.org/10.1007/978-981-13-0514-6_37
234. Bui, X.N., H. Nguyen, Q.H. Tran, D.A. Nguyen, and H.B. Bui. 2021. Predicting ground vibrations due to mine blasting using a novel artificial neural network-based cuckoo search optimization. *Natural Resources Research* 30 (3): 2663–2685. https://doi.org/10.1007/s11053-021-09823-7.
235. Nguyen, H., and X.N. Bui. 2021. A novel hunger games search optimization-based artificial neural network for predicting ground vibration intensity induced by mine blasting. *Natural Resources Research*. https://doi.org/10.1007/s11053-021-09903-8.
236. Revey Associates, Inc. 2013. *Vibration and air-overpressure*. https://higherlogicdownload.s3.amazonaws.com/SMENET/d1f74698-76c6-4c73-8ced-5de57b15be03/UploadedImages/UCA-YM/TAC%20-%202013%20VIBRATION%20AND%20AIR-OVERPRESSURE%20HANDOUT%20-%20OCTOBER%202013.pdf. Accessed 11 July 2021

237. Khandelwal, M., and T.N. Singh. 2005. Prediction of blast induced air overpressure in opencast mine. *Noise & Vibration Worldwide* 36: 7–16. https://doi.org/10.1260/0957456053499095.
238. Siskind, D.E., M.S. Stagg, J.W. Kopp, C.H. Dowding. 1980. Structure response and damage produced by airblast from surface mining. USBM, Report of Investigations 8485.
239. Holmberg, R., P.A. Persson. 1979. Design of tunnel perimeter blasthole patterns to prevent rock damage. In *Proceedings of the IMM tunnelling'79 conference*, London.
240. National Association of Australian State Road Authorities. 1982. *Explosives in roadworks: User's guide*, 2nd ed. Sydney: NAASRA.
241. Mckenzie, C. 1990. Quarry blast monitoring: Technical and environmental perspectives. *Quarry Management* 17: 23–34.
242. Ollofson, S.O. 1990. *Applied explosives technology for construction and mining*. Arla: Applex Publisher.
243. Persson, P.A., R. Holmberg, and L. Jaimin. 1994. *Rock and explosives engineering*. Boca Raton: CRC Press.
244. Sawmliana, C., P.P. Roy, R.K. Singh, and T.N. Singh. 2007. Blast induced air overpressure and its prediction using artificial neural network. *Mining Technology* 116 (2): 41–48. https://doi.org/10.1179/174328607X191065.
245. Khandelwal, M., and P.K. Kankar. 2011. Prediction of blast-induced air overpressure using support vector machine. *Arabian Journal of Geosciences* 4: 427–433. https://doi.org/10.1007/s12517-009-0092-7.
246. Hajihassani, M., D. Jahed Armaghani, H. Sohaei, E. Tonnizam Mohamad, and A. Marto. 2014. Prediction of airblast-overpressure induced by blasting using a hybrid artificial neural network and particle swarm optimization. *Applied Acoustics* 80: 57–67. https://doi.org/10.1016/j.apacoust.2014.01.005.
247. Jahed Armaghani, D., M. Hajihassani, A. Marto, R.S. Faradonbeh, and E. Tonnizam Mohamad. 2015. Prediction of blast-induced air overpressure: A hybrid AI-based predictive model. *Environmental Monitoring and Assessment* 187: 666. https://doi.org/10.1007/s10661-015-4895-6.
248. Jahed Armaghani, D., M. Hajihassani, H. Sohaei, E. Tonnizam Mohamad, A. Marto, H. Motaghedi, and M.R. Moghaddam. 2015. Neuro-fuzzy technique to predict air-overpressure induced by blasting. *Arabian Journal of Geosciences* 8: 10937–10950. https://doi.org/10.1007/s12517-015-1984-3.
249. Hasanipanah, M., D. Jahed Armaghani, H. Khamesi, H. Bakhshandeh Amnieh, and S. Ghoraba. 2016. Several non-linear models in estimating air-overpressure resulting from mine blasting. *Engineering Computations* 32 (3): 441–455. https://doi.org/10.1007/s00366-015-0425-y.
250. Tonnizam Mohamad, E., D. Jahed Armaghani, M. Hasanipanah, B.R. Murlidhar, and M.N.A. Alel. 2016. Estimation of air-overpressure produced by blasting operation through a neuro-genetic technique. *Environment and Earth Science* 75 (2): 174. https://doi.org/10.1007/s12665-015-4983-5.
251. Jahed Armaghani, D., M. Hasanipanah, and E. Tonnizam Mohamad. 2016. A combination of the ICA-ANN model to predict air overpressure resulting from blasting. *Engineering Computations* 32: 155–171. https://doi.org/10.1007/s00366-015-0408-z.
252. Hasanipanah, M., A. Shahnazar, H.B. Amnieh, and D.J. Armaghani. 2017. Prediction of air-overpressure caused by mine blasting using a new hybrid PSO–SVR model. *Engineering Computations* 33 (1): 23–31. https://doi.org/10.1007/s00366-016-0453-2.
253. Shirani Faradonbeh, R., M. Hasanipanah, H. Bakhshandeh Amnieh, D. Jahed Armaghani, and M. Monjezi. 2018. Development of GP and GEP models to estimate an environmental issue induced by blasting operation. *Environmental Monitoring and Assessment* 190 (6): 351. https://doi.org/10.1007/s10661-018-6719-y.
254. Jahed Armaghani, D., M. Hasanipanah, A. Mahdiyar, M.Z.A. Majid, H. Bakhshandeh Amnieh, and M.M. Tahir. 2018. Airblast prediction through a hybrid genetic algorithm-ANN model. *Neural Computing and Applications* 29 (9): 619–629. https://doi.org/10.1007/s00521-016-2598-8.

255. Alel, M.N.A., M.R.A. Upom, R.A. Abdullah, M.H.Z. Abidin. 2018. Optimizing blasting's air overpressure prediction model using swarm intelligence. Journal of Physics: Conference Series 995:012046.
256. Ye, W., W. Feng, and S. Fan. 2017. A novel multi-swarm particle swarm optimization with dynamic learning strategy. *Applied Soft Computing* 61: 832–843. https://doi.org/10.1016/j.asoc.2017.08.051.
257. AminShokravi, A., H. Eskandar, A.M. Derakhsh, H. Nikafshan Rad, and A. Ghanadi. 2018. The potential application of particle swarm optimization algorithm for forecasting the air-overpressure induced by mine blasting. *Engineering Computations* 34 (2): 277–285. https://doi.org/10.1007/s00366-017-0539-5.
258. Mahdiyar, A., A. Marto, and A.S. Mirhosseinei. 2018. Probabilistic air-overpressure simulation resulting from blasting operations. *Environment and Earth Science* 77: 123. https://doi.org/10.1007/s12665-018-7293-x.
259. Nguyen, H., and X.N. Bui. 2019. Predicting blast-induced air overpressure: A robust artificial intelligence system based on artificial neural networks and random forest. *Natural Resources Research* 28: 893–907. https://doi.org/10.1007/s11053-018-9424-1.
260. Bui, X.N., H. Nguyen, Q.H. Tran, H.B. Bui, Q.L. Nguyen, D.A. Nguyen, T.T.H. Le, and V.V. Pham. 2019. A lasso and elastic-net regularized generalized linear model for predicting blast-induced air over-pressure in open-pit mines. *Inzynieria Mineralna* 21 (2/2): 8–20. https://doi.org/10.29227/IM-2019-02-52.
261. Keshtegar, B., M. Hasanipanah, I. Bakhshayeshi, and M.E. Sarafraz. 2019. A novel nonlinear modeling for the prediction of blast-induced airblast using a modified conjugate FR method. *Measurement* 131: 35–41. https://doi.org/10.1016/j.measurement.2018.08.052.
262. Gao, W., A.S. Alqahtani, A. Mubarakali, D. Mavaluru, and S. Khalafi. 2020. Developing an innovative soft computing scheme for prediction of air overpressure resulting from mine blasting using GMDH optimized by GA. *Engineering Computations* 36: 647–654. https://doi.org/10.1007/s00366-019-00720-5.
263. Bui, X.N., H. Nguyen, H.A. Le, H.B. Bui, and N.H. Do. 2020. Prediction of blast-induced air over-pressure in open-pit mine: Assessment of different artificial intelligence techniques. *Natural Resources Research* 29: 571–591. https://doi.org/10.1007/s11053-019-09461-0.
264. Nguyen, H., and X.N. Bui. 2020. Soft computing models for predicting blast-induced air over-pressure: A novel artificial intelligence approach. *Applied Soft Computing* 92: 106292. https://doi.org/10.1016/j.asoc.2020.106292.
265. Nguyen, H., X.-N. Bui, H.-B. Bui, and N.-L. Mai. 2020. A comparative study of artificial neural networks in predicting blast-induced air-blast overpressure at Deo Nai open-pit coal mine, Vietnam. *Neural Computing and Applications* 32: 3939–3955. https://doi.org/10.1007/s00521-018-3717-5.
266. Nguyen, H., X.N. Bui, Q.H. Tran, P.V. Hoa, D.A. Nguyen, L.T.T. Hoa, Q.T. Le, N.H. Do, T.D. Bao, H.B. Bui, and H. Moayedi. 2020. A comparative study of empirical and ensemble machine learning algorithms in predicting air over-pressure in open-pit coal mine. *Acta Geophysica* 68: 325–336. https://doi.org/10.1007/s11600-019-00396-x.
267. Fang, Q., H. Nguyen, X.N. Bui, and Q.H. Tran. 2020. Estimation of blast-induced air over-pressure in quarry mines using cubist-based genetic algorithm. *Natural Resources Research* 29: 593–607. https://doi.org/10.1007/s11053-019-09575-5.
268. Zhou, J., A. Nekouie, C.A. Arslan, B.T. Pham, and M. Hasanipanah. 2020. Novel approach for forecasting the blastinduced AOp using a hybrid fuzzy system and firefly algorithm. *Engineering Computations* 36: 703–712. https://doi.org/10.1007/s00366-019-00725-0.
269. Temeng, V.A., Y.Y. Ziggah, and C.K. Arthur. 2020. A novel artificial intelligent model for predicting air overpressure using brain inspired emotional neural network. *International Journal of Mining Science and Technology* 30 (5): 683–689. https://doi.org/10.1016/j.ijmst.2020.05.020.
270. Temeng, V.A., Y.Y. Ziggah, and C.K. Arthur. 2021. Blast-induced noise level prediction model based on brain inspired emotional neural network. *Journal of Sustainable Mining* 20 (1): 3. https://doi.org/10.46873/2300-3960.1043.

271. Zhou, X., D. Jahed Armaghani, J. Ye, M. Khari, and M.R. Motahari. 2021. Hybridization of parametric and non-parametric techniques to predict air over-pressure induced by quarry blasting. *Natural Resources Research* 30: 209–224. https://doi.org/10.1007/s11053-020-09714-3.
272. Zeng, J., M. Jamei, M. Nait Amar, M. Hasanipanah, and P. Bayat. 2021. A novel solution for simulating air overpressure resulting from blasting using an efficient cascaded forward neural network. *Engineering Computations*. https://doi.org/10.1007/s00366-021-01381-z.
273. Lundborg, N. 1973. *The calculation of maximum throw during blasting*. SveDeFo Rep DS, 4.
274. Chiapetta, R.F., A. Bauer, P.J. Dailey, S.L. Burchell. 1983. The use of high-speed motion picture photography in blast evaluation and design. In *Proceedings of the 9th Conference on Explosives and Blasting Techniques*, Dallas.
275. Monjezi, M., T.N. Singh, M. Khandelwal, S. Sinha, V. Singh, and I. Hosseini. 2006. Prediction and analysis of blast parameters using artificial neural network. *Noise & Vibration Worldwide* 37 (5): 8–16. https://doi.org/10.1260/095745606777630323.
276. Aghajani-Bazzazi, A., M. Osanloo, Y. Azimi. 2009. Flyrock prediction by multiple regression analysis in Esfordi phosphate mine of Iran. In *Proceedings of the 9th International Symposium on Rock Fragmentation by Blasting*, Granada.
277. Monjezi, M., A. Bahrami, A. Yazdian Varjani, and A.R. Sayadi. 2011. Prediction and controlling of flyrock in blasting operation using artificial neural network. *Arabian Journal of Geosciences* 4: 421–425. https://doi.org/10.1007/s12517-009-0091-8.
278. Rezaei, M., M. Monjezi, and A. Yazdian Varjani. 2011. Development of a fuzzy model to predict flyrock in surface mining. *Safety Science* 49 (2): 298–305. https://doi.org/10.1016/j.ssci.2010.09.004.
279. Monjezi, M., H. Amini Khoshalan, and A. Yazdian Varjani. 2012. Prediction of flyrock and backbreak in open pit blasting operation: A neurogenetic approach. *Arabian Journal of Geosciences* 5 (3): 441–448. https://doi.org/10.1007/s12517-010-0185-3.
280. Amini, H., R. Gholami, M. Monjezi, S.R. Torabi, and J. Zadhesh. 2012. Evaluation of flyrock phenomenon due to blasting operation by support vector machine. *Neural Computing & Applications* 21 (8): 2077–2085. https://doi.org/10.1007/s00521-011-0631-5.
281. Tonnizam Mohamad, E., D. Jahed Armaghani, S.A. Noorani, R. Saad, S.V. Alavi Nezhad Khaili Abad. 2012. Prediction of flyrock in boulder blasting by using artificial neural network. Electronic Journal of Geotechnical Engineering 17: 2585–2595
282. Tonnizam Mohamad, E., D. Jahed Armaghani, M. Hajihassani, K. Faizi, and A. Marto. 2013. A simulation approach to predict blasting-induced flyrock and size of thrown rocks. *Electronic Journal of Geotechnical Engineering* 18: 365–374.
283. Khandelwal, M., and M. Monjezi. 2013. Prediction of flyrock in open pit blasting operation using machine learning method. *International Journal of Mining Science and Technology* 23 (3): 313–316. https://doi.org/10.1016/j.ijmst.2013.05.005.
284. Raina, A.K., V.M.S.R. Murthy, and A.K. Soni. 2013. Relevance of shape of fragments on flyrock travel distance: An insight fro concrete model experiments using ANN. *Electronic Journal of Geotechnical Engineering* 18: 899–907.
285. Monjezi, M., A. Mehrdanesh, A. Malek, and M. Khandelwal. 2013. Evaluation of effect of blast design parameters on flyrock using artificial neural networks. *Neural Computing & Applications* 23: 349–356. https://doi.org/10.1007/s00521-012-0917-2.
286. Trivedi, R., T.N. Singh, and A. Raina. 2014. Prediction of blast-induced flyrock in Indian limestone mines using neural networks. *Journal of Rock Mechanics and Geotechnical Engineering* 6 (5): 447–454. https://doi.org/10.1016/j.jrmge.2014.07.003.
287. Ghasemi, E., H. Amini, M. Ataei, and R. Khalokakaei. 2014. Application of artificial intelligence techniques for predicting the flyrock distance caused by blasting operation. *Arabian Journal of Geosciences* 7: 193–202. https://doi.org/10.1007/s12517-012-0703-6.
288. Marto, A., M. Hajihassani, D. Jahed Armaghani, E. Tonnizam Mohamad, and A.M. Makhtar. 2014. A novel approach for blast-induced flyrock prediction based on imperialist competitive algorithm and artificial neural network. *The Scientific World Journal* 2014: 643715. https://doi.org/10.1155/2014/643715.

289. Trivedi, R., T.N. Singh, and N.I. Gupta. 2015. Prediction of blast-induced flyrock in opencast mines using ANN and ANFIS. *Geotechnical and Geological Engineering* 33: 875–891. https://doi.org/10.1007/s10706-015-9869-5.
290. Jahed Armaghani, D., E. Tonnizam Mohamad, M. Hajihassani, S.V. Alavi Nezhad Khalil Abad, A. Marto, M.R. Moghaddam. 2016. Evaluation and prediction of flyrock resulting from blasting operations using empirical and computational methods. Engineering with Computers 32: 109–121. https://doi.org/10.1007/s00366-015-0402-5
291. Saghatforoush, A., M. Monjezi, R. Shirani Faradonbeh. 2016. Combination of neural network and ant colony optimization algorithms for prediction and optimization of flyrock and back-break induced by blasting. *Engineering with Computers* 32: 255–266 (2016). https://doi.org/10.1007/s00366-015-0415-0
292. Yari, M., R. Bagherpour, S. Jamali, and R. Shamsi. 2016. Development of a novel flyrock distance prediction model using BPNN for providing blasting operation safety. *Neural Computing and Applications* 27 (3): 699–706. https://doi.org/10.1007/s00521-015-1889-9.
293. Shirani Faradonbeh, R., D. Jahed Armaghani, and M. Monjezi. 2016. Development of a new model for predicting flyrock distance in quarry blasting: A genetic programming technique. *Bulletin of Engineering Geology and the Environment* 75: 993–1006. https://doi.org/10.1007/s10064-016-0872-8.
294. Shirani Faradonbeh, R., D. Jahed Armaghani, M. Monjezi, and E. Tonnizam Mohamad. 2016. Genetic programming and gene expression programming for flyrock assessment due to mine blasting. *International Journal of Rock Mechanics and Mining Sciences* 88: 254–264. https://doi.org/10.1016/j.ijrmms.2016.07.028.
295. Stojadinović, S., N. Lilić, I. Obradović, R. Pantović, and M. Denić. 2016. Prediction of flyrock launch velocity using artificial neural networks. *Neural Computing and Applications* 27: 515–524. https://doi.org/10.1007/s00521-015-1872-5.
296. Hasanipanah, M., Jahed Armaghani, D., H. Bakhshandeh Amnieh, M.Z. Abd Majid, M.D.M. Tahir. 2017. Application of PSO to develop a powerful equation for prediction of flyrock due to blasting. *Neural Computing and Applications* 28: 1043–1050. https://doi.org/10.1007/s00521-016-2434-1
297. Hasanipanah, M., R. Shirani Faradonbeh, D. Jahed Armaghani, H. Bakhshandeh Amnieh, and M. Khandelwal. 2017. Development of a precise model for prediction of blast-induced flyrock using regression tree technique. *Environment and Earth Science* 76: 27. https://doi.org/10.1007/s12665-016-6335-5.
298. Dehghani, H., and M. Shafaghi. 2017. Prediction of blast-induced flyrock using differential evolution algorithm. *Engineering Computations* 33: 149–158. https://doi.org/10.1007/s00366-016-0461-2.
299. Bakhtavar, E., H. Nourizadeh, and A. Sahebi. 2017. Toward predicting blast-induced flyrock: A hybrid dimensional analysis fuzzy inference system. *International Journal of Environmental Science and Technology* 14 (4): 717–728. https://doi.org/10.1007/s13762-016-1192-z.
300. Shirani Faradonbeh, R., D. Jahed Armaghani, H. Bakhshandeh Amnieh, and E. Tonnizam Mohamad. 2018. Prediction and minimization of blast-induced flyrock using gene expression programming and firefly algorithm. *Neural Computing and Applications* 29 (6): 269–281. https://doi.org/10.1007/s00521-016-2537-8.
301. Nikafshan Rad, H., M. Hasanipanah, M. Rezaei, and A.L. Eghlim. 2018. Developing a least squares support vector machine for estimating the blast-induced flyrock. *Engineering Computations* 34 (4): 709–717. https://doi.org/10.1007/s00366-017-0568-0.
302. Hasanipanah, M., D. Jahed Armaghani, H. Bakhshandeh Amnieh, M. Koopialipoor, and H. Arab. 2018. A risk-based technique to analyze flyrock results through rock engineering system. *Geotechnical and Geological Engineering* 36 (4): 2247–2260. https://doi.org/10.1007/s10706-018-0459-1.
303. Nguyen, H., X.N. Bui, T. Nguyen-Thoi, P. Ragam, and H. Moayedi. 2019. Toward a state-of-the-art of fly-rock prediction technology in open-pit mines using EANNs model. *Applied Sciences* 9 (21): 4554. https://doi.org/10.3390/app9214554.

304. Koopialipoor, M., A. Fallah, D. Jahed Armaghani, A. Azizi, and E. Tonnizam Mohamad. 2019. Three hybrid intelligent models in estimating flyrock distance resulting from blasting. *Engineering Computations* 35 (1): 243–256. https://doi.org/10.1007/s00366-018-0596-4.
305. Lu, X., M. Hasanipanah, K. Brindhadevi, H. Bakhshandeh Amnieh, and S. Khalafi. 2020. ORELM: A novel machine learning approach for prediction of flyrock in mine blasting. *Natural Resources Research* 29: 641–654. https://doi.org/10.1007/s11053-019-09532-2.
306. Jahed Armaghani, D., M. Koopialipoor, M. Bahri, M. Hasanipanah, and M.M. Tahir. 2020. A SVR-GWO technique to minimize flyrock distance resulting from blasting. *Bulletin of Engineering Geology and the Environment* 79: 1–17.
307. Zhou, J., M. Koopialipoor, B.R. Murlidhar, S.A. Fatemi, M.M. Tahir, D. Jahed Armaghani, and C. Li. 2020. Use of intelligent methods to design effective pattern parameters of mine blasting to minimize flyrock distance. *Natural Resources Research* 29: 625–639. https://doi.org/10.1007/s11053-019-09519-z.
308. Zhou, J., N. Aghili, E.N. Ghaleini, D.T. Bui, M.M. Tahir, and M. Koopialipoor. 2020. A Monte Carlo simulation approach for effective assessment of flyrock based on intelligent system of neural network. *Engineering Computations* 36: 713–723. https://doi.org/10.1007/s00366-019-00726-z.
309. Han, H., D. Jahed Armaghani, R. Tarinejad, J. Zhou, and M.M. Tahir. 2020. Random forest and Bayesian network techniques for probabilistic prediction of flyrock induced by blasting in quarry sites. *Natural Resources Research* 29: 655–667. https://doi.org/10.1007/s11053-019-09611-4.
310. Hasanipanah, M., B. Keshtegar, D.K. Thai, and N.T. Troung. 2020. An ANN-adaptive dynamical harmony search algorithm to approximate the flyrock resulting from blasting. *Engineering Computations*. https://doi.org/10.1007/s00366-020-01105-9.
311. Geem, Z.W. 2009. *Music-inspired harmony search algorithm—Theory and application*. Springer, Berlin. https://doi.org/10.1007/978-3-642-00185-7
312. Hasanipanah, M., and H. Bakhshandeh Amnieh. 2020. A fuzzy rule based approach to address uncertainty in risk assessment and prediction of blast-induced flyrock in a quarry. *Natural Resources Research* 29: 669–689. https://doi.org/10.1007/s11053-020-09616-4.
313. Nguyen, H., X.N. Bui, Y. Choi, C.W. Lee, and D. Jahed Armaghani. 2020. A novel combination of whale optimization algorithm and support vector machine with different kernel functions for prediction of blasting-induced fly-rock in quarry mines. *Natural Resources Research* 30: 191–207. https://doi.org/10.1007/s11053-020-09710-7.
314. Nikafshan Rad, H., I. Bakhshayeshi, W.A. Wan Jusoh, M.M. Tahir, and L. Kok Foong. 2020. Prediction of flyrock in mine blasting: A new computational intelligence approach. *Natural Resources Research* 29: 609–623. https://doi.org/10.1007/s11053-019-09464-x.
315. Bhagat, N.K., A. Rana, A.K. Mishra, M.M. Singh, A. Singh, and P.K. Singh. 2021. Prediction of fly-rock during boulder blasting on infrastructure slopes using CART technique. *Geomatics, Natural Hazards and Risk* 12 (1): 1715–1740. https://doi.org/10.1080/19475705.2021.1944917.
316. Guo, H., H. Nguyen, X.N. Bui, and D.J. Armaghani. 2021. A new technique to predict fly-rock in bench blasting based on an ensemble of support vector regression and GLMNET. *Engineering with Computers* 37: 421–435. https://doi.org/10.1007/s00366-019-00833-x.
317. Guo, H., J. Zhou, M. Koopialipoor, D. Jahed Armaghani, and M.M. Tahir. 2021. Deep neural network and whale optimization algorithm to assess flyrock induced by blasting. *Engineering Computations* 37: 173–186. https://doi.org/10.1007/s00366-019-00816-y.
318. Li, D., M. Koopialipoor, and D. Jahed Armaghani. 2021. A combination of fuzzy Delphi method and ANN-based models to investigate factors of flyrock induced by mine blasting. *Natural Resources Research* 30 (2): 1905–1924. https://doi.org/10.1007/s11053-020-09794-1.
319. Fattahi, H., and M. Hasanipanah. 2021. An integrated approach of ANFIS-grasshopper optimization algorithm to approximate flyrock distance in mine blasting. *Engineering Computations*. https://doi.org/10.1007/s00366-020-01231-4.

320. Masir, R.N., M. Ataei, and A. Motahedi. 2021. Risk assessment of flyrock in surface mines using a FFTA-MCDM combination. *Journal of Mining and Environment* 12 (1): 191–203. https://doi.org/10.22044/JME.2020.9107.1799.

321. Ye, J., M. Koopialipoor, J. Zhou, D. Jahed Armaghani, and X. He. 2021. A novel combination of tree-based modeling and monte carlo simulation for assessing risk levels of flyrock induced by mine blasting. *Natural Resources Research* 30: 225–243. https://doi.org/10.1007/s11053-020-09730-3.

Chapter 14
Advanced Analytics for Rock Breaking

Mustafa Erkayaoğlu

Abstract Rock-breaking processes, including drilling and blasting, are the most economical techniques to achieve rock fragmentation in mines and quarries. The main objective of these processes is to achieve fragmentation in the desired size, conforming to the limitations defined for ground vibration, air overpressure, flying rocks, and stability. Fragmentation is a key aspect of downstream processes, such as haulage and mineral processing. The success of rock-breaking-related activities impacts equipment selection for mining and mineral processing, operational efficiency, energy consumption, and recovery. Advanced analytics can aid these processes in various ways by providing a better understanding of the operation. An advanced analytics solution can support decision-making by utilizing data management throughout the rock-breaking process in mining. Additionally, data generated during the drilling and blasting operations can be stored in an integrated database, which can be used by other operational data generating processes in mining. The power of advanced analytics can be used to evaluate the data, identify key performance variables, implement management changes, and evaluate the outcome. Advanced analytics creates an opportunity to leverage the data generated during the rock-breaking process such that improvements in cost and fragmentation can be achieved. This chapter discusses the role of advanced analytics in mechanical rock breaking, drilling, and blasting.

Keywords Rock breaking · Mining · Advanced analytics · Artificial intelligence

Introduction

Mining production is fundamentally related to breaking rock material by either utilizing various equipment or blasting operations. The methods used for rock breaking in underground and surface mining can be classified into mechanical rock cutting equipment and blasting explosives. The basic principle behind mechanical

M. Erkayaoğlu (✉)
Mining Engineering Department, Middle East Technical University (METU), Ankara, Turkey
e-mail: emustafa@metu.edu.tr

A. Soofastaei (ed.), *Advanced Analytics in Mining Engineering*,
https://doi.org/10.1007/978-3-030-91589-6_14

rock cutting for soft and hard rock conditions is the implementation of wedges. Strong rock conditions govern the selection of equipment for rock breaking and limit the application of mechanical rock cutting equipment in a hard rock environment that is commonly present for metalliferous deposits. Underground mining operations that produce seam-type deposits such as coal commonly break rock material by mechanical cutting that operates on cyclic movements. Another common usage of mechanical rock breaking is the tunnel boring machine (TBM) utilized for mining or tunneling activities. The data analytics related to mechanical rock breakage is dominated by the sensors that the cutting machines are equipped with. Usage of sensors leads to a data analytics capability that can be considered similar to process type of data and is commonly based on time series, regression analysis, and other methods.

Surface mining operations commonly initialize production activities by drilling and blasting as rock breakage methods. These rock breakage stages can be considered the initial stages of mineral processing operations as the broken rock material is fed to crushers, mills, and other equipment. Available technology in mobile equipment used for drilling and blasting enables data analytics based on Global Positioning System (GPS) and other computer-aided systems. Drill monitoring systems provide operational data together with GPS data aiming to increase accuracy. Fleet management systems (FMS) track mobile haulage and loading equipment to collect data that can be utilized to develop business intelligence (BI) tools, reports, and advanced analytics tools, such as data mining. This chapter covers the advanced data analysis related to rock breakage by providing insight into data collected by various equipment.

Rock Breaking with Mechanical Equipment

Rock breaking is a fundamental stage of production in the mining industry for both surface and underground operations. Mechanical rock breaking is performed through different machines that operate basically on cutting processes. The principle of using wedges to penetrate the rock is considered as the primary means of mechanical rock breaking. The commonly used equipment, especially in underground mining, is designed to benefit from components that act as wedges in the form of indenters or drag bits. The conventional equipment that can be regarded as mechanical rock breakage tools is long-wall shearers, continuous miners, road headers, raise boring machines, and TBMs. Like other production stages of mining, the performance of equipment utilized in mechanical rock breaking is closely related to various characteristics of the rock material such as strength, texture, temperature, and others. The performance of mining equipment is therefore evaluated based on different operational conditions by analyzing operational data.

Data generated by equipment that cuts rock material is classified as process type of data. This data type can be recorded in high frequency and small size due to its properties similar to sensor readings. The following sections will provide information about the data analytics related to equipment cutting rock with picks or disks and impact breakers.

Data Analytics of Cutting Systems with Picks

The majority of mining machines utilized in mechanical rock breaking are designed to use various types of wedges. Design variables of pick cutting systems can be summarized as pick rake angle, back clearance angle, pick edge angle, pick width, depth of cut, cutting speed, pick spacing, and pick shape. Researchers investigated the performance of shearers, road headers, continuous miners, and other equipment by analyzing data similar to sensor readings. Although different infrastructures can collect and provide access to equipment data, programmable logic controllers (PLCs) and OLE for process control (OPC) servers are common hardware used, especially in underground coal mining. The performance measures of pick cutting systems are the norm and cutting forces, breakout angle, yield, and specific energy.

Long-wall mining is an underground mining method that depends intensively on equipment performance as multiple machines are operated synchronously. Long-wall shearers are the main production units that can be considered as mechanical cutting systems with picks. The analysis of operational data is a vital target for mine management [1, 2]. The first data analysis attempts were based on studying sensor data of shearers aiming to understand optimum production conditions [3]. The increasing amount and variety of data transformed the decision-making process in mining, where advanced analytics has become an essential part of production. The management perspective of relying on human experience evolved into data-driven decision-making [4, 5]. Availability and utilization are used to evaluate equipment and be analyzed in detail by classifying downtime events. Trend analysis of sensor data is a common way to build prediction models for this purpose, and there is a limited amount of research focused on long-wall shearers [6]. New technologies enabled the collection of more data utilizing better network connectivity, which is also essential for automating long-wall equipment [7–11] or increasing data utilization.

BI tools, such as dashboards or balanced scorecards built on data, act as enablers of cultural change management. Key performance indicators (KPI) are part of performance assessments that summarize a time range that could also provide insight into predetermined targets. Measures defined as leading and lagging performance indicators are commonly provided on balanced scorecards, whereas performance can be tracked on a short-term basis from dashboards. For example, a dashboard design for a long-wall manager can be seen in Fig. 14.1.

Data analysis has to be summarized in BI tools to support the decision-making process of engineers. For example, a dashboard might reveal the timeline of total shearer production and durations related to unplanned maintenance events as KPIs. However, in contrast to this data analysis, it is also possible to perform predictive analysis using leading KPIs, such as cutting time, number of cutting passes per shift, and other parameters to help engineers and managers estimate production amount [12]. Therefore, it is important to invest in the required information technology (IT) infrastructure to integrate different data sources for analysis and cut systems with commonly equipped picks with various sensors.

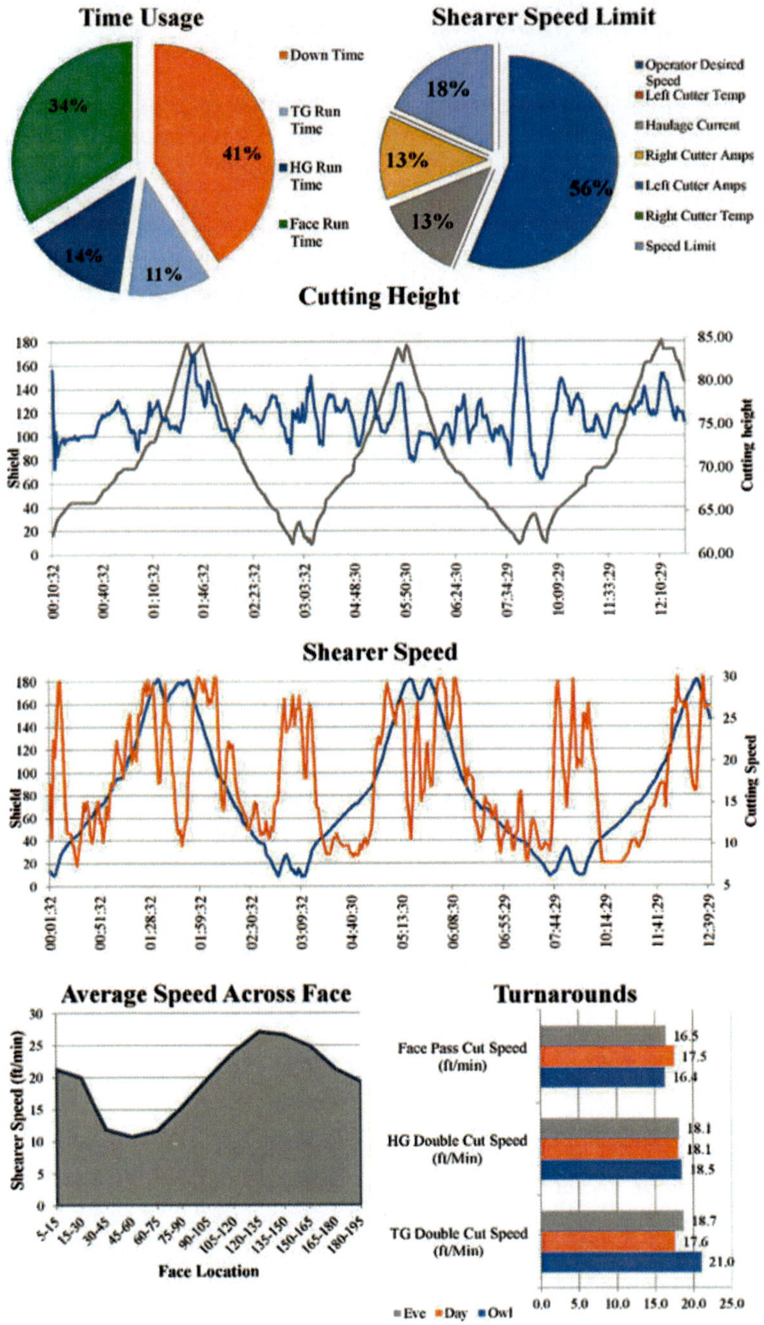

Fig. 14.1 Dashboard for long-wall manager

A common way to measure shearer production is based on tracking the movement of the equipment. The production cycle can be categorized into events that require data analysis to identify the duration coal was cut efficiently with picks. Raw data collected by sensors equipped on the shearer and shields can be analyzed by algorithms that mimic the operator's understanding. The uncertainty related to geological conditions and incidents that are not captured by any sensor increases the complexity of required data analysis for some cases.

Advanced analytics of sensor data for cutting systems with picks requires logic rules that could be enhanced by human input. Systems that provide sensor data can be equipped with onboard PLCs that will enable access by OPC servers. Regardless of the IT infrastructure, collected data has to be preprocessed to detect outliers and data validation. However, the integration of other related data sources requires data warehousing.

Cutting events of systems with picks can be interpreted by using limit values defining boundaries. An example of this practice is the shield number utilized as the location of the shearer during production. An example of the raw data collected from a long-wall shearer is illustrated in Fig. 14.2. The production cycle covers different events, such as entering/leaving headgate or tailgate zone, turnaround, cutting, and pullback movements if required.

Sensors placed on shields can provide data about the location of the shearer triggered when the shearer passes. These events can be evaluated for production-related measures as an example of the advanced data analysis for cutting systems with picks. In addition, raw data can be provided by the shearer itself that is equipped with a PLC. Performance measurement of shearers is based on the definition of events and monitoring of time durations. Operator performance is another measure that can be evaluated by analyzing the time spent between specific shields, such as turnaround zone, which might represent a non-productive event during the change of cutting direction. This analysis can be enhanced if cutting energy data is available and categorized for different rock types cut by the system.

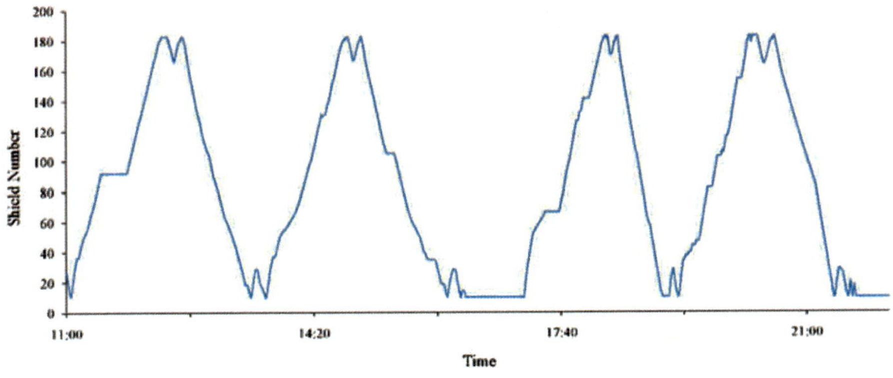

Fig. 14.2 Long-wall movement detection from raw data

A significant contribution of the data collected from cutting systems with picks is the development of BI tools. Dashboards that can utilize real-time data can provide valuable information to decision-makers. The comparison of equipment and operators per shift can be performed if shields predefine each production-related event. Various technologies can be used for data collection in underground mining, and new technologies will continue to enable the utilization of real-time data more commonly. The amount of data generated by underground mining equipment is significant, and another piece of equipment that should be considered for advanced data analysis is cutting systems with disks.

Data Analytics of Cutting Systems with Disks

Mechanical rock-breaking equipment designed to utilize disks, such as raise boring machines (RBM) and TBMs, operates on the action of wedges. The design parameters of disk cutting systems are disk diameter, disk penetration, disk edge angle, cutting speed, and disk spacing. In addition, researchers investigated the performance measures of thrust force, rolling force, yield, and specific energy by laboratory-scale setups and equipment used on site.

Advanced analytics in cutting systems with disks is led by the development of predictive models regarding operational performance. In this field, a fundamental research study developed a predictive equation for TBM performance by utilizing data collected from a TBM tunnel and concluded that rock mass properties had a considerable impact on the analysis [13]. This outcome points out the importance of rock mechanics-related information used for the data analysis of rock breakage. The primary data source is the cutting system with disks that provide sensor data; however, this information has to be enhanced with reliable laboratory or in situ testing of different types of rocks. A model was developed for TBM performance using a database including 158 tunnel sections of selected projects [14]. It is stated that the available data for different geological conditions enabled the model to be considered applicable for different rock types. The construction of databases is essential to create access to operational data for research purposes. The scale and availability of such data sources enable advanced analysis in this field. The predictive performance of various models developed for TBM performance was studied by utilizing a database covering more than 300 TBM projects for statistical analysis. It was concluded that the amount of data should be increased to achieve more reliable models [15]. As is the case for all data science projects, the amount of data is a crucial parameter that especially affects the predictive analysis of cutting performance. Another perspective on evaluating the TBM cutting performance is using numerical analysis and photogrammetric measurement compared to prediction techniques where the data analysis is based on rolling force measurements of a linear cutting machine [16]. It is a common practice to develop laboratory-scale experimental setups to represent cutting systems aiming to overcome the challenge of collecting data from underground operations.

Another common objective of advanced analytics related to cutting systems with disks is the prediction of geological structure.

Researchers investigated the applicability of a wireless system to predict geological conditions for TBM excavation [17]. The amplitude readings from sensors are interpreted to predict rock mass class and strength based on processing the data collected by various sensors. The amount and variety of data provided by sensors enable advanced data analysis tools and be investigated from a geo-statistical perspective. A sample case is the geo-statistical evaluation of drill logging data and TBM operation data to predict geological conditions [18]. The TBM data used in the analysis was thrust load, cutter torque, penetration, cutter rpm, and other commonly assessed time series measures. Drill-hole loggings and geophysical measurements are other data sources utilized for analysis. The evaluation of data from ground-penetrating radar (GPR) images and sensors was studied for event detection and tracking method [19]. It was stated that an extensive database of actual data from tunneling projects, including TBM parameters, would be required to use machine learning for advanced analytics. Training of machine learning algorithms requires a large amount of reliable data, which might be achieved by data preparation based on different methodologies such as transformation. Another study aims to predict TBM performance employing penetration rate which developed a Bayesian prediction model of TBM penetration rate using rock mass parameters [20]. The data sets used in the analysis collected the TBM operation in New York for three years. A similar data set was used to predict TBM performance by particle swarm optimization [21]. It was concluded that the amount of data needs to be increased for more reliable models. Other researchers investigated the potential of nonlinear regression and AI-based algorithms to predict TBM performance [22]. However, the amount of data was limited, and an expanded database was considered for future studies.

Similarly, a prediction model was developed for TBM penetration rate using support vector regression and a database consisting of 150 data points [23]. Data analytics of TBM operational data, mainly cutter head speed, cutter head torque, thrust, and the advance rate, was also utilized to propose a novel methodology [24]. It is stated that additional data is planned to be utilized for prediction purposes. Figure 14.3 represents the sensor data that might be used for advanced data analysis related to TBMs.

Sensor data can be processed with various methodologies before training prediction algorithms for geological conditions. The prediction of TBM performance was also studied using rock mass rating (RMR) by implementing different regression models [26]. The database used in the prediction consists of 46 records and is planned to be extended to develop a more generic prediction model. Rock breakage with mechanical rock cutting systems is commonly performed for underground operations, and data analysis of related equipment is based on sensor data. Another rock breakage method is explosives in blast holes drilled as part of surface mining and underground mining activities.

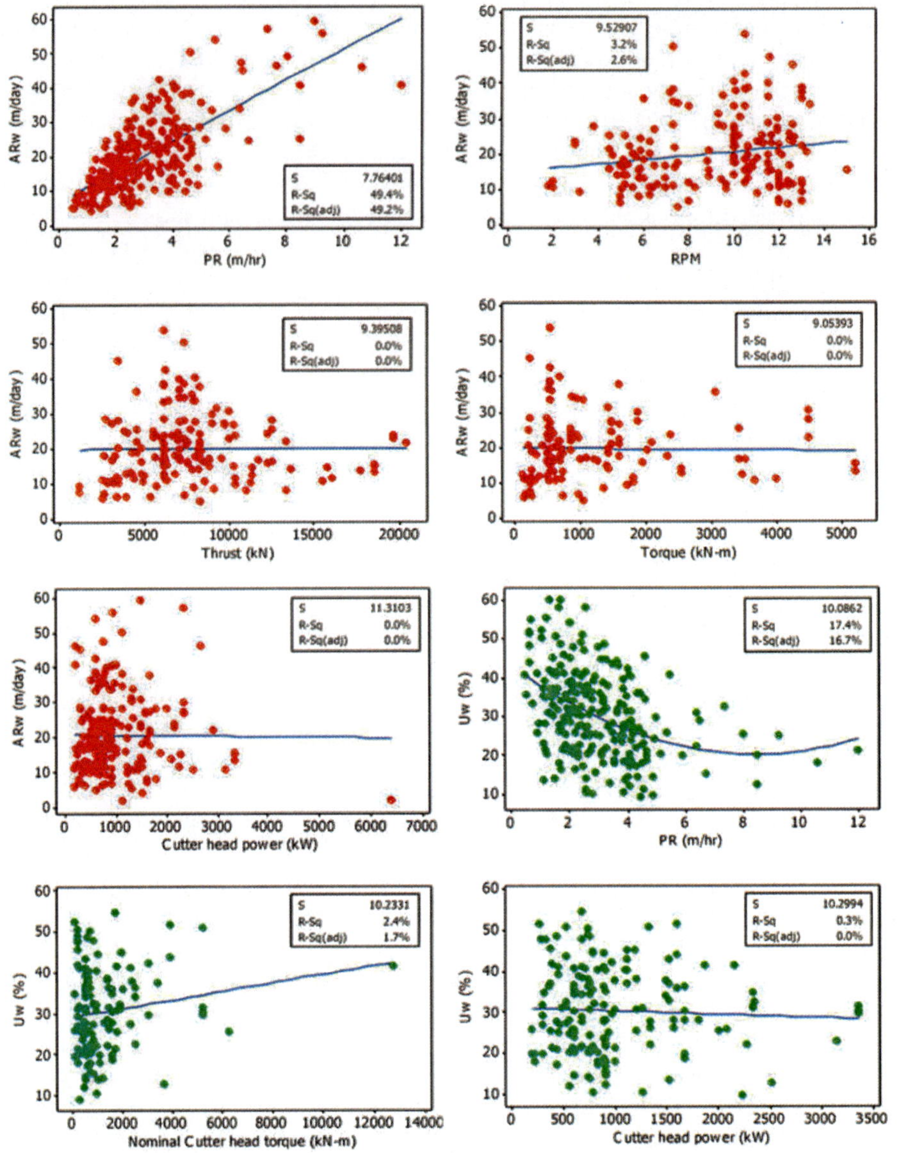

Fig. 14.3 Analysis of TBM performance parameters versus advance rate and utilization [25]

Rock Breaking with Drilling and Blasting

Rock breakage by using explosives is a primary operational stage of mining for surface and underground operations. Available technology enables monitoring drilling and explosives loading equipment by GPS data and sensors, whereas fragmentation analysis is based on digital images taken at different loading and haulage operations stages. The mining and blasting operations in the mining industry significantly impact downstream processes, such as mineral processing. This concept can be defined from a perspective where the output of mineral processing is started to be produced on the mine site. The primary data source for drilling and blasting is the equipment used in drilling, explosive loading, haulage, and loading that FMS commonly monitors. The basic characteristic of drilling and blasting-related data analysis is that information is relational that can be analyzed in multi-dimensional arrays and integrated with other data sources representing downstream processes. The following sections will provide information about the data analytics related to equipment used in drilling and blasting operations.

Blast-hole Drilling

Rock breakage by using explosives is a primary stage of mining activities that are initiated by drilling operations. The mining value chain is majorly affected by the performance of upstream processes such as drilling and blasting. Accuracy of drilling operations can be monitored by GPS locations available in drill monitoring systems or FMS and compared to the short-term production plans. The difference between the planned locations and the actual blast holes is commonly used as a performance measure to evaluate drill operator. Scorecards and other BI tools are developed to integrate drill navigation systems, drill monitoring systems, and FMS to summarize daily production. Drill navigation systems provide visual information to the operator based on the data collected by transducers and sensors available on drilling equipment. This data mainly provides information about the rotation, torque, cooling medium consumption, and depth collected during drilling that can also be subject to ETL processes as part of data warehousing practices.

Various researchers have studied the concept of measurement while drilling (MWD), aiming to predict rock conditions and optimize drilling efficiency [27]. It was pointed out that the penetration rate parameter is not sufficient to provide information about varying conditions, and it should be integrated with other drilling variables. The implementation of different machine learning methods for rock recognition was by using MWD data [28]. It was concluded that additional drilling equipment operated in various geological conditions should be investigated to extend the data set of 28 blast holes. A similar approach was followed by other researchers who proposed a rock recognition methodology based on MWD data and adjusted penetration rate [29, 30]. Different rock types drilled through are categorized by clustering and provide a better understanding of the lithology. The accuracy of the prediction-based research depends on the availability of reliable and multiple sources of data. An advanced MWD system that utilizes high-precision GPS, PLC, and various sensors that collect operational data was developed, and rock properties were predicted by

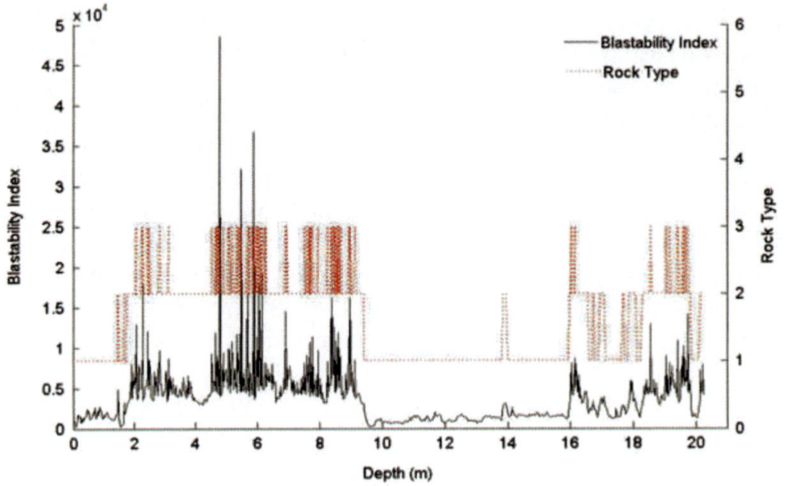

Fig. 14.4 Blast ability index and rock type versus depth for a blast hole in a coal mine [33]

using torque, axial compression, and drilling rate-related data [31]. The geological information that is aimed to be predicted can be interpreted in different ways, especially for coal mining. A specific energy measure to detect coal seams using MWD data of rotary drill rigs without relying on geophysical measurements is an example application [32]. Similarly, MWD data can be evaluated for the blast ability conditions at a coal mine, as shown in Fig. 14.4.

The integration of geophysical measurements and drilling data was investigated for coal mining, and the advanced analysis of drilling-related data could provide accurate information in a shorter time for decision-making purposes [34]. The investigation of rock conditions is crucial information for underground mining operations and tunneling activities. Accurate estimation of RMR with MWD data was also achieved at tunnel excavations [35]. The second stage of rock breakage by drilling and blasting starts charging the drilled blast holes with explosives.

Rock Blasting

Advanced analysis of blasting-related data is commonly focused on fragmentation. The size distribution achieved as a result of blasting activities directly impacts the efficiency of mineral processing. Therefore, the performance of drilling and blasting is generally reported together. Some of the critical measures used to evaluate drilling and blasting performance are drilling accuracy, material passing size of 50%, 80%, etc. Data integration between drilling, blasting, loading, and haulage enables understanding the mining value chain until mineral processing. This highlights the importance of drilling and blasting performance for better recovery at downstream processes.

Fragmentation analysis is based on analyzing digital images of blasted material to understand blasting performance. This is an essential piece of information required for mine-to-mill or mine-to-market concepts. Image processing algorithms are implemented to assess blasting performance where earlier versions of fragmentation analysis were challenged by various environmental conditions such as light amount or dust generation [36]. The recent development in available technology enabled researchers to utilize different equipment for fragmentation analysis. Researchers studied the implementation of high-speed video analysis to detect errors in blasting practice [37]. It is stated that measures such as fragmentation can be analyzed by capturing images representing blast performance. A similar approach was followed by using a portable device to take 3D pictures of broken rock material to perform fragmentation analysis [38]. UAVs with a high-resolution image and video capturing capability can be considered applications that will become more common in the mining industry for blasting performance assessment.

The blasting properties of various types of ore were investigated by a review of blast fragmentation models [39]. It was stated that required data sources are geotechnical parameters obtained from laboratory testing of drill cores, geological interpretation of the field, geophysics, and, most importantly, drill monitoring systems. Data integration is an essential part of advanced data analysis for drilling and blasting, where geological information about the blasted rock material can be utilized. An enhanced geological model that supported better blasting practices resulting in improvement of fragmentation was introduced by researchers [40]. Similarly, indices from underground drilling data were defined to identify stopes with poor fragmentation performance [41]. It was concluded that a more extensive set of geological data would be required to achieve blasting with desired fragmentation.

The prediction of environmental impacts of blasting was also considered as a method to evaluate blasting performance where vibration and overpressure data were utilized together with heave modeling [42]. Therefore, it is of key importance to have geological data available for advanced analytics purposes. The outcomes of blasting operations were investigated as ore loss and dilution by evaluating rock movement [43]. Multiple linear regression was implemented to predict rock movement by utilizing geological information and blast design parameters. Methodologies such as principle component analysis (PCA), support vector regression (SVR), and adaptive neural-fuzzy inference systems (ANFIS) were compared based on rock fragmentation modeling by using a blasting-related database that contains 80 variables at an iron mine [44]. An example of the prediction of blast fragmentation by the artificial neural network (ANN) method can be seen in Fig. 14.5.

The advanced data analysis related to drilling and blasting should be discussed together within the scope of their impact on loading, haulage, and mineral processing. Data collected from drill monitoring, explosives loading equipment, and FMS can be integrated to evaluate the impact of drilling and blasting performance [46]. Figure 14.6 represents a mine-to-mill toolkit developed by using an integrated data warehouse.

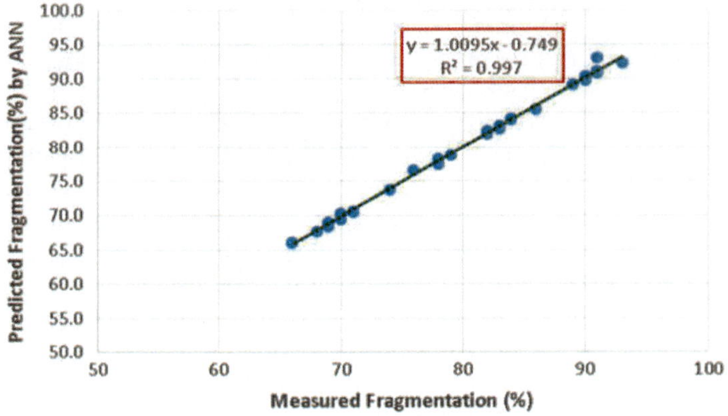

Fig. 14.5 Evaluation of predicted and measured fragmentation by ANN [45]

Measures defined for drilling and blasting can be tracked down in other production stages where locations with comparably lower penetration rates indicate challenging rock material through drilling. Although there might be different reasons for such inefficient drilling practices, these locations are also characterized by coarse fragmentation regardless of the charged explosive amount. The coarse fragmentation of the same location will become the reason for observing lower dig rates indicating operational difficulties at the loading stage. This analysis can be implemented on real-time data to support decision-making at different stages of production.

Conclusion

Rock breakage is a fundamental production stage of mining that has a significant impact on downstream processes. Any potential improvement in this stage can be considered cost efficient than the possible modifications in mineral processing-related activities. The amount of data generated by equipment used for rock cutting, drilling, and blasting is continuously increasing with the available technologies. Prediction of geological conditions is crucial for underground mining operations or tunneling activities that depend on efficient rock cutting systems with picks or disks. Surface mining operations challenged by uncertainty originated from drilling and blasting activities need to provide accurate information such as fragmentation of the feed material for mineral processing. Data is an irreplaceable asset for mining activities and will become more crucial in the future.

Automation and advanced production methods that will become more common in all stages of the minerals industry increase the necessity of using advanced analytics. Data-driven solutions that enable engineers and managers to support decision-making are fundamental pieces of modern mines. Utilization of integrated data on a semantic

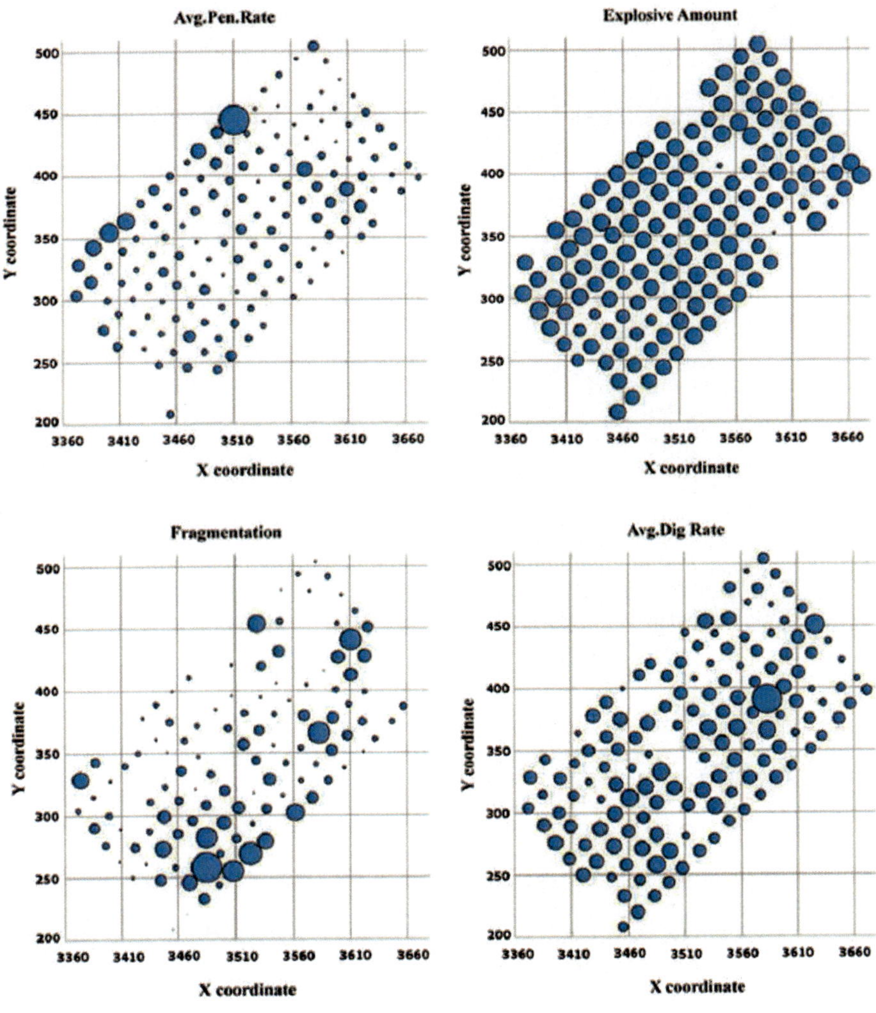

Fig. 14.6 Mine-to-mill toolkit [47]

layer is the key to continuous improvement in all fields of the minerals industry, regardless of the type of equipment or the IT infrastructure used.

References

1. Velzeboer, M.D. 1988. Implications of high productivity long-wall systems.
2. Reid, P.B., et al. 2010. Real-world automation: New capabilities for underground long-wall mining.
3. Price, D.L. and I.C. Jeffrey. 1992. Prediction of shearer cutting performance.

4. Miano, M.P., R.L. Grayson, and S.Q. Yuan. 1993. Society for mining, metallurgy, and exploration, Inc.
5. Martins, P., and A. Soofastaei. 2020. Making decisions based on analytics. In *Data analytics applied to the mining industry*, 193–221. CRC Press.
6. Ramani, R.V., A. Bhattaacherjee, and R.J. Pawlikowski. 1988. Reliability, maintainability and availability analysis of long-wall mining systems.
7. Ralston, J.C., and A.D. Strange. 2013. Developing selective mining capability for long-wall shearers using thermal infrared-based seam tracking. *International Journal of Mining Science and Technology* 23 (1): 47–53.
8. Nienhaus, K., F. Mavroudis, and M. Warcholik. 2010. Coal bed boundary detection using infrared technology for long-wall shearer automation.
9. Skirde, J. and M. Schmid. 2008. Underground information technology infrastructure—a sound basis for efficient future mining operations.
10. Ralston, J.C., et al. 2015. Long-wall automation: Delivering enabling technology to achieve safer and more productive underground mining. *International Journal of Mining Science and Technology* 25 (6): 865–876.
11. Soofastaei, A. 2020. *Data analytics applied to the mining industry.* CRC Press.
12. Erkayaoglu, M., and S. Dessureault. 2017. Using integrated process data of long-wall shearers in data warehouses for performance measurement. *International Journal of Oil, Gas and Coal Technology* 16 (3): 298–310.
13. Yagiz, S. 2008. Utilizing rock mass properties for predicting TBM performance in hard rock condition. *Tunnelling and Underground Space Technology* 23 (3): 326–339.
14. Hassanpour, J., J. Rostami, and J. Zhao. 2011. A new hard rock TBM performance prediction model for project planning. *Tunnelling and Underground Space Technology* 26 (5): 595–603.
15. Farrokh, E., J. Rostami, and C. Laughton. 2012. Study of various models for estimation of penetration rate of hard rock TBMs. *Tunnelling and Underground Space Technology* 30: 110–123.
16. Cho, J.W., et al. 2013. Evaluation of cutting efficiency during TBM disc cutter excavation within a Korean granitic rock using linear-cutting-machine testing and photogrammetric measurement. *Tunnelling and Underground Space Technology* 35: 37–54.
17. Yokota, Y., et al. 2016. Evaluation of geological conditions ahead of TBM tunnel using wireless seismic reflector tracing system. *Tunnelling and Underground Space Technology* 57: 85–90.
18. Yamamoto, T., et al. 2017. Evaluation of the geological condition ahead of the tunnel face by geo-statistical techniques using TBM driving data. *Modern Tunnelling Science and Technology* 1: 213–218.
19. Wei, L., D.R. Magee, and A.G. Cohn. 2017. An anomalous event detection and tracking method for a tunnel look-ahead ground prediction system. *Automation in Construction* 2018 (91): 216–225.
20. Adoko, A.C., C. Gokceoglu, and S. Yagiz. 2017. Bayesian prediction of TBM penetration rate in rock mass. *Engineering Geology* 226 (January): 245–256.
21. Yagiz, S., and H. Karahan. 2011. Prediction of hard rock TBM penetration rate using particle swarm optimization. *International Journal of Rock Mechanics and Mining Sciences* 48 (3): 427–433.
22. Salimi, A., et al. 2016. Application of non-linear regression analysis and artificial intelligence algorithms for performance prediction of hard rock TBMs. *Tunnelling and Underground Space Technology* 58: 236–246.
23. Mahdevari, S., et al. 2014. A support vector regression model for predicting tunnel boring machine penetration rates. *International Journal of Rock Mechanics and Mining Sciences* 72: 214–229.
24. Zhang, Q., Z. Liu, and J. Tan. 2018. Prediction of geological conditions for a tunnel boring machine using big operational data. *Automation in Construction* 2019 (100): 73–83.
25. Farrokh, E. 2013. Study of utilization factor and advance rate of hard rock TBMs.
26. Khademi Hamidi, J., et al. 2010. Performance prediction of hard rock TBM using rock mass rating (RMR) system. *Tunnelling and Underground Space Technology* 25 (4): 333–345.

27. Rai, P., et al. 2015. An overview on measurement-while-drilling technique and its scope in excavation industry. *Journal of the Institution of Engineers (India): Series D* 96(1): 57–66.
28. Kadkhodaie-Ilkhchi, A., et al. 2010. Rock recognition from MWD Data: A comparative study of boosting, neural networks, and fuzzy logic. *IEEE Geoscience and Remote Sensing Letters* 7 (4): 680–684.
29. Zhou, H., et al. 2012. Automatic rock recognition from drilling performance data. IEEE.
30. Zhou, H., et al. 2009. Spectral feature selection for automated rock recognition using Gaussian process classification.
31. Duan, Y., et al. 2015. Advanced technology for setting out of blast holes and measurement while drilling.
32. Leung, R., and S. Scheding. 2015. Automated coal seam detection using a modulated specific energy measure in a monitor-while-drilling context. *International Journal of Rock Mechanics and Mining Sciences* 75: 196–209.
33. Beattie, N. 2019. Monitoring-while-drilling for open-pit mining in a hard rock environment: An investigation of pattern recognition techniques applied to rock identification.
34. Hatherly, P., et al. 2015. Drill monitoring results reveal geological conditions in blast hole drilling. *International Journal of Rock Mechanics and Mining Sciences* 78: 144–154.
35. Galende-Hernández, M., et al. 2018. Monitor-while-drilling-based estimation of rock mass rating with computational intelligence: The case of tunnel excavation front. *Automation in Construction* 93 (May): 325–338.
36. Kemeny, J.M., et al. 1993. Analysis of rock fragmentation using digital image processing. *Journal of Geotechnical Engineering* 119 (7): 1144–1160.
37. Adermann, D., et al. 2015. High-speed video—an essential blasting tool. In *AusIMM Bulletin*, 24–26.
38. Sameti, B., et al. 2015. A portable device for mine face: Rock fragmentation analysis. In *Mining engineering*, 16–23.
39. Scott, A. and I. Onederra. 2015. Characterising the blasting properties of iron ore.
40. Vieira, L. and J.C. Koppe. 2015. Geophysical techniques applied to blasting design.
41. Jackson, J. and E. Sellers. 2017. Blasting related indices as the key to underground blasting improvements in a challenging rock mass.
42. Goswami, T., et al. 2015. A holistic approach to managing blast outcomes.
43. Domingo, J., et al. 2014. Dilution, ore grade and blast movement calculation model.
44. Esmaeili, M., et al. 2015. Application of PCA, SVR, and ANFIS for modeling of rock fragmentation. *Arabian Journal of Geosciences* 8 (9): 6881–6893.
45. Tiile, R.N. 2016. Artificial neural network approach to predict blast-induced ground vibration, airblast and rock fragmentation.
46. Erkayaoğlu, M., and S. Dessureault. 2019. Improving mine-to-mill by data warehousing and data mining. *International Journal of Mining, Reclamation and Environment* 33 (6): 409–424.
47. Erkayaoğlu, M. 2015. A data driven mine-to-mill framework for modern mines.

Chapter 15
Advanced Analytics for Mineral Processing

Danish Ali

Abstract Mineral processing involves methods and technologies with which valuable minerals can be separated from gangue or waste rock in an attempt to produce a more concentrated material. Crushing, grinding, and milling circuits are used to reduce the ore size to a specific range at which the mineral concentration, with procedures like gravity separation and flotation, can be maximized. Advanced data analytics (ADA) techniques including but not limited to machine learning (ML), artificial intelligence (AI), and computer vision-based pattern recognition algorithms, can be used to enhance, optimize, and automate all the activities and procedures involved in these operations. It can be applied toward the design, construction, maintenance, control, performance monitoring, and operation optimization of processes like crushing, grinding, milling, classification (by screens and cyclones), gravity concentration, medium-heavy separation, froth flotation, magnetic and electrostatic separation, and dewatering. This chapter includes brief details on each of these aforementioned processes, followed by practical instances of advanced intelligence-based data-driven frameworks and technologies being applied in these areas. Details on how these ore beneficiation processes can be improved in terms of efficiency, effectiveness, and safety with the application of these innovative data modeling and analytic techniques are also included.

Keywords Advanced analytics · Mining · Rock breaking · Optimization · Productivity · New technology

Introduction to Mineral Processing

Mineral processing is a field that contends with procedures and technologies used for separating valuable minerals from gangue or waste rock. It is a process that converts the extracted ore through mining activity into a more concentrated material, which serves as an input for the extractive metallurgy. Figure 15.1 shows the general

D. Ali (✉)
Mining Engineering and Management, South Dakota School of Mines and Technology, Rapid City, SD, USA
e-mail: Danish.Ali@sdsmt.edu

A. Soofastaei (ed.), *Advanced Analytics in Mining Engineering*,
https://doi.org/10.1007/978-3-030-91589-6_15

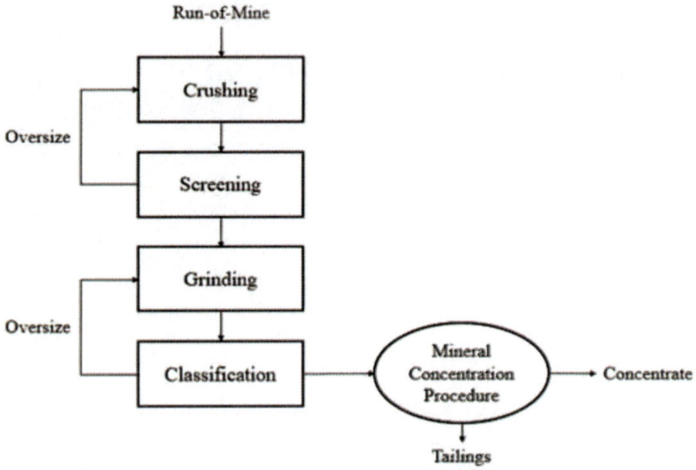

Fig. 15.1 General procedures and unit operations for mineral processing

procedures involved for any mineral processing setup. The process involves mineral liberation, separation, and final concentration before moving to smelting and refining, which falls under metallurgical processes [1].

Surface mining operations commonly initialize production activities by drilling and blasting as rock breakage methods. These rock breakage stages can be considered the initial stages of mineral processing operations as the broken rock material is fed to crushers, mills, and other equipment. Available technology in mobile equipment used for drilling and blasting enables data analytics based on Global Positioning System (GPS) and other computer-aided systems. Drill monitoring systems provide operational data together with GPS data aiming to increase accuracy. Fleet management systems (FMS) track mobile haulage and loading equipment to collect data that can be utilized to develop business intelligence (BI) tools, reports, and ADA tools, such as data mining. This chapter covers the ADA related to rock breakage by providing insight into data collected by various equipment.

Mineral processing is a field that contends with procedures and technologies used for separating valuable minerals from gangue or waste rock. It is a process that converts the extracted ore through mining activity into a more concentrated material, which serves as an input for the extractive metallurgy. Figure 15.1 shows the general procedures involved for any mineral processing setup. The process involves mineral liberation, separation, and final concentration before moving to smelting and refining, which falls under metallurgical processes [1].

Crushers are used in an initial size reduction process to ensure a narrow size range product. Next, screens are used to allow the material to bypass as soon as it falls within the product range, based on the grinding equipment, with the oversize material going back into the crusher. Finally, grinding is done to reduce the size of the material further, generally using either a rod mill or a ball mill. Since grinding reduces the size to fines, at which stage only classifiers can work efficiently enough to ensure high productivity. This series of mineral liberation and separation techniques allows the size of the material to be reduced, allowing an effective concentration, which produces the final concentrate product with all the gangue/waste removed as tailings [2].

There are a series of complex processes and operations required at any mineral processing plant. It is the job of engineers, managers, and executives who design, run, and maintain these operations to ensure the utmost efficiency, effectiveness, and safety. ADA can be utilized for monitoring, analyzing, controlling, or simply automating each of these activities. A few of those tasks have already been automated using ADA like performance monitoring of crushing, grinding, and milling circuits, grain size monitoring and analysis of ore stream, foreign and oversize material detection on belt conveyors, image analysis and performance monitoring of classification process, modeling the performance of gravity concentration process, initial settling rate modeling for a dewatering process, and concentrate grade and recovery monitoring for a flotation process. These applications have covered some areas; however, the considerable room is still available for further improvement and innovation. These existing applications can be implemented and further could be designed, depending upon the need, to improve the operational efficiency and effectiveness in mineral processing.

This chapter aims to provide comprehensive details on applying ADA techniques for various operations in mineral processing. It will demonstrate the value of ADA in terms of descriptive and predictive analysis using advanced intelligence-based techniques like ML, AI, and computer vision-based image analysis tools to improve the operational efficiency and effectiveness of mineral processing. It will provide guidelines for engineers, managers, and executives regarding the inclusion of ADA toward design, construction, maintenance, control, performance monitoring, and operation optimization of crushing, grinding, milling, classification (by screens and cyclones), gravity concentration, medium-heavy separation, froth flotation, magnetic and electrostatic separation, and the process of dewatering. Every section of this chapter provides a high-level summary of the particular mineral processing operation/component, identifying the challenges and the problems faced, and explaining the application of ADA for solving those issues with practical case studies of it.

Crushing, Milling, and Grinding

A combination of the jaw, cone, and impact crushers is used to reduce the size of run-of-mine (ROM) material. Crushers can reduce the size of ROM up to 1 cm. A combination of ball and rod mills is used to reduce the size during grinding further. Steel balls with diameters ranging from 1 to 10 cm are used as the grinding media for the ball mills, whereas for the rod mill, rods with similar diameters and lengths slightly lesser than the rod mill length are used. As a result, mills can produce products with fines being less than 20 μm in size.

A crusher is the first component in the mineral processing work chain; thus, it is crucial to ensure the efficiency of the whole process. Operational data for a mining and processing plant can be used to develop intelligence-based frameworks, which can optimize the crushing process. Baek and Choi [3] demonstrated applying the deep neural network (DNN) model for modeling and improving the crusher utilization in a limestone mine. Deep learning is a specialized application of AI which allows pattern recognition and knowledge acquisition in complex tasks without human assistance. It is based on artificial neural networks, which make it a supervised learning methodology. Therefore, data, usually big data, is required for model development using the deep learning algorithm. The data by Baek and Choi [3] were collected in a limestone mine that had an advanced mine safety management system equipped with information and communication technology (ICT). The data were collected over one month, which was later used to train the intelligence-based model. The deep learning model consisted of a combined two deep neural network models. Figure 15.2 shows the architecture of the deep learning model developed in this study. The first layer was the input layer with 26 nodes, with one node for each explanatory variable. The operating conditions of the truck haulage system were represented by 8 of those nodes, and the remaining 18 nodes represented the truck cycle times. Table 15.1 provides the details for each of those input variables used in this study. The truck's working time was further divided into three sub-categories providing the travel time to the loading area, actual loading time, and time for the truck to get back to the monitoring point, which was a wireless access point. The model consisted of two output nodes, one for each deep neural network, in the final output layer. Multiple

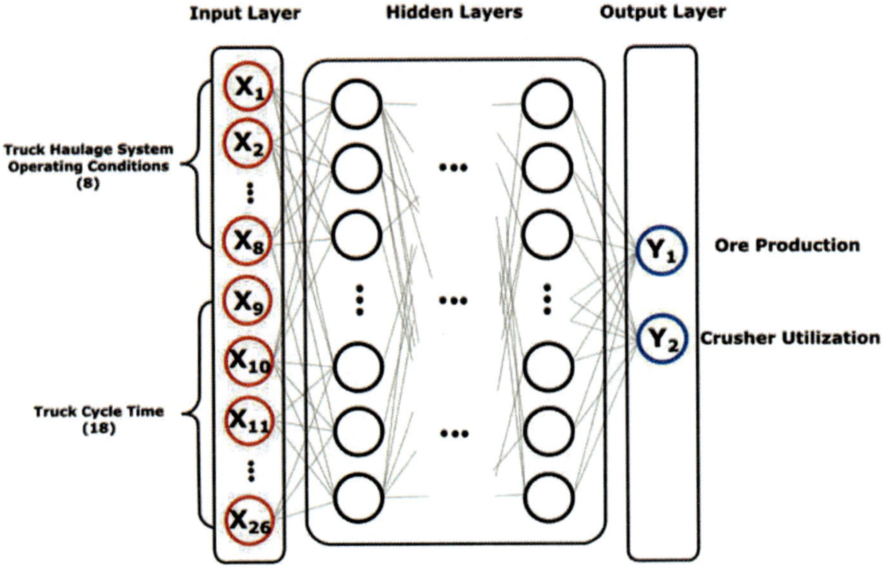

Fig. 15.2 Architecture of the deep learning model developed by Baek and Choi [3]

Table 15.1 Explanatory variables for the deep learning model [3]

Category	Variable description
Operating condition of truck haulage system	Total operating time for truck haulage
	Daily work time
	Initial time for truck dispatch
	Truck dispatch interval
	Total number of dispatched trucks
	Loading capacity of the truck
	Probability of loading 1 occurring
	Probability of loading 2 occurring
Truck cycle time	Travel time for an empty truck
	Working time of the trucks at the loading area
	Loaded truck travel time
	Working time of trucks at crushing area

hidden layers containing multiple information processing units called neurons were placed between the input and output layers. Rectified linear units (ReLU) were used as the activation functions for each of those layers. A training dataset consisting of 6804 data points and 26 input variables was used for model development. A total of 1000 epochs were used during model training with a patience level of 200. The model's accuracy was evaluated using the coefficient of determination (R^2) and mean absolute percentage error (MAPE). Adam was used as the algorithm to optimize the model connection weights.

The optimal deep learning model consisted of five hidden layers for the DNN predicting the ore production and four hidden layers for the DNN predicting the crusher utilization, each with 40 hidden neurons [3]. The developed final deep learning model was evaluated for its ore production and crusher utilization prediction performance over ten days at the mine. Table 15.2 shows the performance results for the optimal deep learning model during training and testing. Figures 15.3 and 15.4 show the real versus predicted plots of the training and testing phase—the

Table 15.2 Deep learning model performance results [3]

Category	Training		Testing	
	R^2	MAPE (%)	R^2	MAPE (%)
Ore production	0.99	2.45	0.99	2.80
Crusher utilization	0.99	2.30	0.99	2.49

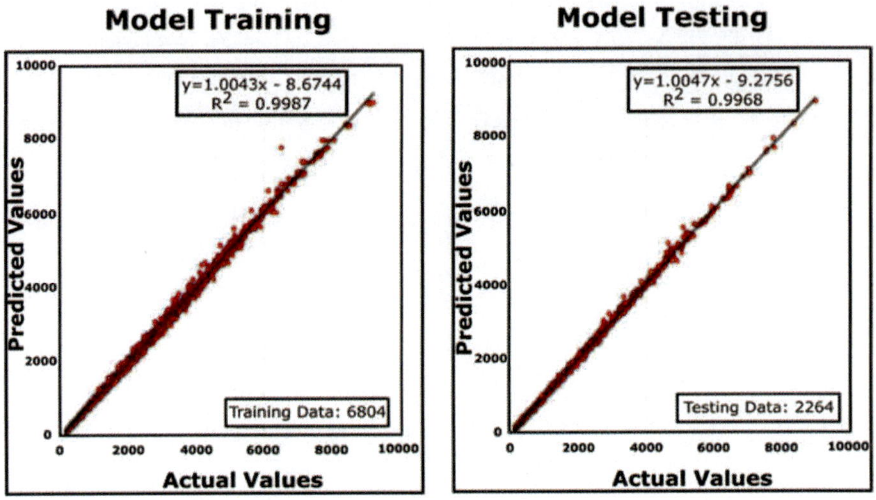

Fig. 15.3 Real versus predicted plots for ore production using the deep learning model [3]

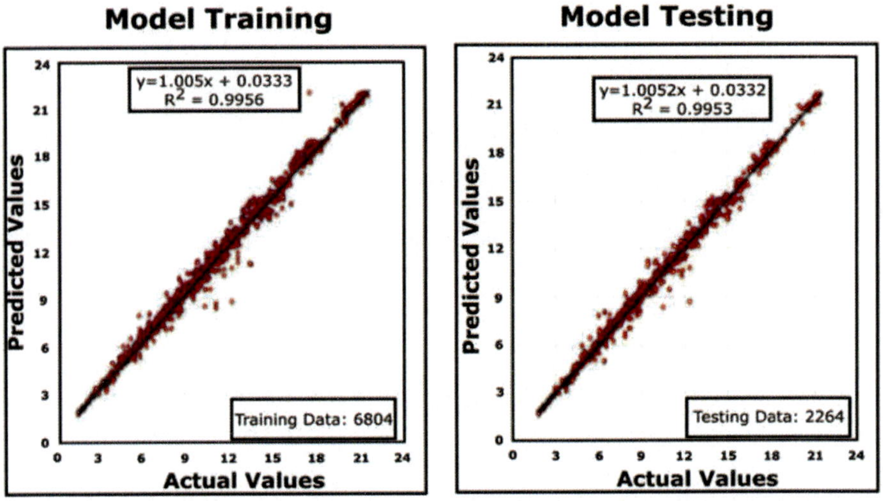

Fig. 15.4 Real versus predicted plots for crusher utilization using the deep learning model [3]

closer the points to the correlation line, the better the prediction. The model successfully analyzed the changing patterns during the process and predicted the future ore production and crusher utilization variation, thus accounting for any operational change and ensuring an optimized process.

Milling is one of the most energy-intensive processes; thus, forecasting the energy needs is critical to ensure smooth processing flow and efficient procedures. Real-time variables, like feed tonnage, rotational speed, and mineralogical features, can be used to design efficient energy forecasting with ML or deep learning. For example, Avalos et al. [4] showed the superiority of recurrent neural networks (RNNs) to achieve the desired forecasting accuracy in an energy consumption prediction task using real-world mining and mineral processing datasets.

ADA can be used to model the mill performance and control the essential variables that affect the mill's efficiency and effectiveness. For example, Freeport-McMoRan implemented a custom-designed AI-based model using three years of operational data from one of their copper mines [5]. The model was designed to understand the mill behavior and tweak the operational parameters, increasing copper production. In addition, the data analytics-based model optimized the feed rate with an adjustment in the ore delivery from the mine and the crushing station, which significantly improved the overall mill production and maintained exceptional mill efficiency.

Grinding is a large part of the total operating cost for any mineral processing plant. The grinding circuit for a typical copper concentrator makes up 47% of the total cost per ton [6]. Controlling the mill circuits completely by using intelligence-based data-driven frameworks can significantly optimize the process because a slight improvement can lead to huge savings and increased productivity. Hellen [7] demonstrated the use of reinforcement learning (RL) to design a system for controlling a grinding mill circuit. The intelligence-based controller showed higher profitability compared to existing control strategies. Moreover, RL-based controllers have the ability to self-learn and adapt to any operational changes. Thus, any modification in the process can be taken into account almost instantly and without manual interference requirements.

The whole objective of these processes—crushing, milling, and grinding—is to optimize the ore particle size for maximizing the mineral/metal recovery in the later processing. Thus, controlling the grain size composition is of paramount importance during all of these ore preparation stages. Therefore, computer vision-based technologies connected with the advanced ML and AI algorithms need to be designed and implemented, which would allow careful monitoring of grain size distribution. For example, Figs. 15.5 and 15.6 show the belt metrics technology [8] designed to monitor the ore stream, which can be implemented after every stage of ore processing.

Figure 15.7 shows the AI and ML algorithm detecting the grain size distribution of an ore stream. This would allow real-time analysis of the material data resulting in onsite decision-making with complete and accurate information for every process in the circuit. Moreover, any oversized material needs to be removed from the ore stream to protect the integrity of the machines and ensure efficient and unhindered processing operation. Figure 15.8 shows the advanced intelligence-based technology, designed by Motion Metrics [8], to detect any oversized material passing through the circuit. Any foreign material entering the circuit could damage the machine and hamper the whole operation performance. Therefore, it needs to be detected and removed. Figure 15.9 shows the computer vision and ML-based technology for detecting a foreign object. These technologies can detect the oversized or foreign material and send the signal to an appropriate mechanism with precise location information of the object to be removed, thus allowing an efficient automated object detection and removal process.

Fig. 15.5 Belt metrics: computer vision-based ore stream monitoring technology [8]

Fig. 15.6 Belt metrics technology with data analysis tool [8]

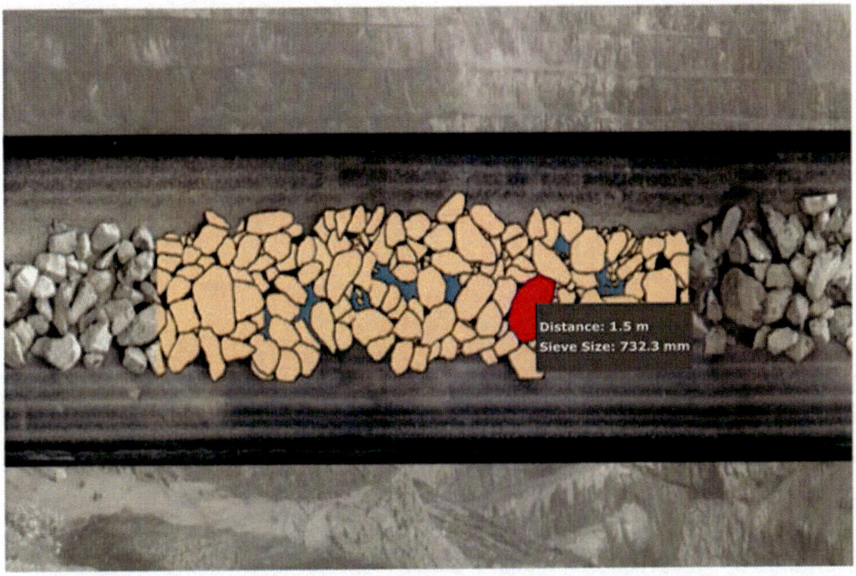

Fig. 15.7 Computer vision-based grain size distribution monitoring technology

ADA can thus be used to develop strategies and technologies to model and monitor the performance of crushing, milling, and grinding circuits. Additionally, it can find ways to optimize the performance, thus improving the system design, equipment utilization, and overall process efficiency in real time.

Classification by Screens and Cyclones

Classification and screening are done to ensure a close-sized feed for the process, reducing the effect of particle mass and allowing the process to rely on a particular physical characteristic solely. Classification is a process of particle sizing that depends on two physical parameters, particle sizes and particle specific gravity (SG). It is generally a wet process and involves a moving fluid that transports the material as the product is separated. Hydrocyclones and hydrosizers are the two main classifiers used in mineral processing. The mechanism works as the feed slurry is separated into two components, with fine and low SG particles forming the overflow component, whereas coarse and high SG particles making the underflow particles. When classification is done through screens, the process differs slightly because the separation depends solely on particle size. Grizzly screens are one of the coarsest screens. Mechanically vibrating screens with single or double deck configurations are the most common for wet processes where oversize material is retained and removed.

Fig. 15.8 Computer vision and AI-based technology for large object detection [8]

The profitability of these operations relies heavily on effective equipment control. However, these machines, especially cyclones with complex configurations and difficulty in measuring the internal flow parameters, are challenging to monitor and control. That is where ADA techniques using the underflow image data of a cyclone can be used to monitor and control the performance to ensure maximum process efficiency. Uahengo [9] demonstrated the intelligent real-time monitoring of particle size and solids concentration within the underflow of the hydrocyclone. Figure 15.10 shows the schematic for the hydrocyclone setup that was used to capture the dataset by Uahengo [9]. The hydrocyclone was suspended on a mixing tank, and both the overflow and underflow streams were exiting in the same mixing tank. A centrifugal pump was used to allow the mixed stream to flow back into the hydrocyclone. A digital camera was used to capture the images of the underflow discharge. The experiments were conducted with the variable feed size range of 41–1189 μm particle size (50%

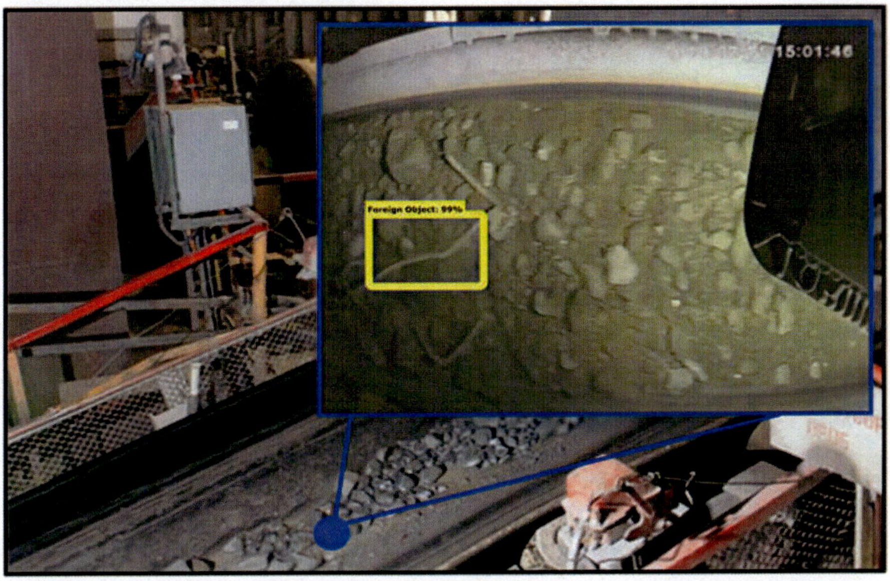

Fig. 15.9 Computer vision and machine learning-based technology for detection of foreign object

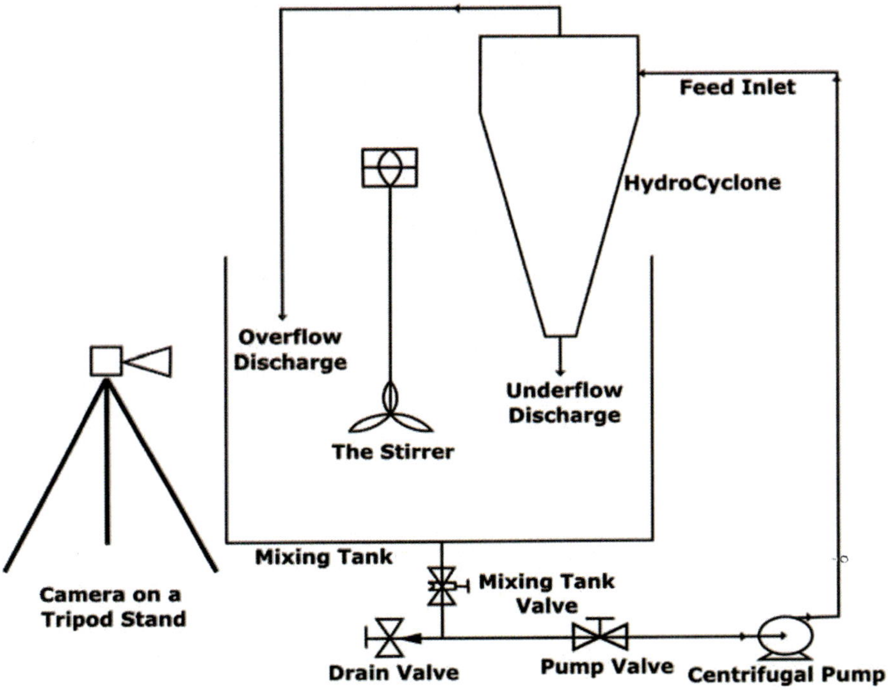

Fig. 15.10 Schematic for the hydrocyclone setup by Uahengo [9]

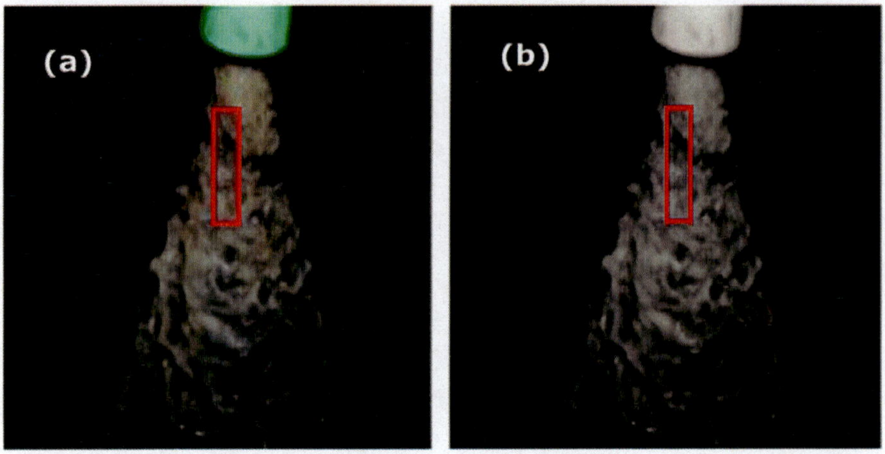

Fig. 15.11 Image feature extraction through GLCM by Uahengo [9] **a** original RBD image, **b** transformed gray-scale image

passing) and the slurry solid weight range of 10.6–16.3%. For each experiment, the samples (both underflow and overflow) were collected, dried, and the sampling time and mass were recorded and particle size distribution (PSD) analysis was performed. The captured images, along with the results of PSD analysis, served as the input data for ML modeling. Background from the images and the noise were removed to improve the image analysis and modeling process. Two main features were extracted from the captured underflow images, including the image texture and the underflow width. Gray-level co-occurrence matrix (GLCM) was used to extract the textual features, whereas the underflow width was captured through manual visualization, as shown in Figs. 15.11 and 15.12, respectively.

The complete dataset consisted of 7 explanatory variables and two response variables with 340 observations. Later, the data were divided into three sub-sets, i.e., training, validation, and testing for neural network model development and optimization. Table 15.3 provides the details on the explanatory and response variables. Ore type was the categorical response variable with mean particle size (MPS) and percentage solid amount in underflow stream being the two continuous response variables in this study. A classification-based neural network model was developed for the ore type, as it was a categorical-type response variable, whereas a regression-based neural network model was developed for MPS and percentage solids since they were of continuous type. The neural network algorithm was applied to train both models using the training dataset comprising of 7 explanatory variables. The sum of squares was used as the cost function with BFGS as the weight optimization algorithm in the classification-based neural network model training. Tangent hyperbolic function (tanh) and SoftMax were used as the activation functions for hidden and output layers, respectively. The optimum neural network consisted of a single hidden layer with 12 neurons (Fig. 15.13) evaluated for its performance to classify the ore type using the testing and validation dataset.

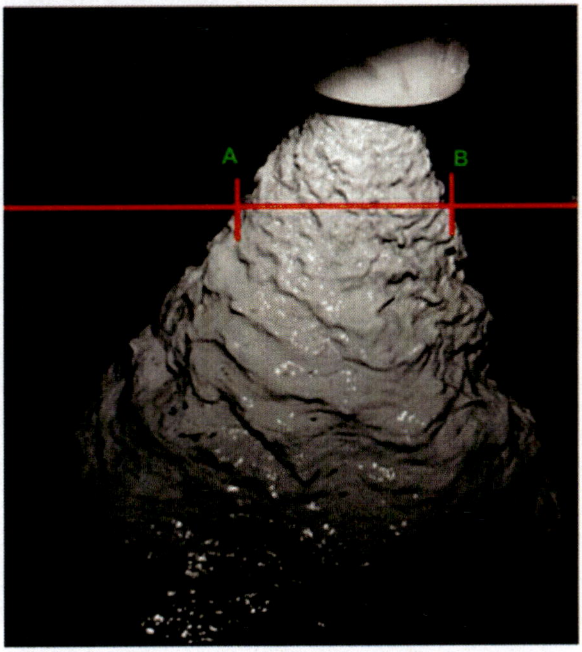

Fig. 15.12 Underflow width extraction [9]

Table 15.3 Details of explanatory and response variables [9]

Category	Type	Variable description
Response	Categorical	Ore type
	Continuous	MPS (μm)
	Continuous	Solids (%)
Explanatory	Continuous	Underflow width
	Continuous	Homogeneity
	Continuous	Entropy
	Continuous	Correlation
	Continuous	Standard deviation of pixel intensity
	Continuous	Energy
	Continuous	Contrast

A linear function was used for a regression-based neural network as the activation function, and 240 images were used during the model training, with 60 images being used for model testing. The optimized neural network consisted of a single hidden layer with ten neurons (Fig. 15.14) evaluated on validation and test dataset. Coefficient of determination (R^2) and mean square error (MSE) were used as the two primary metrics for evaluating the performance of the regression-based neural network model.

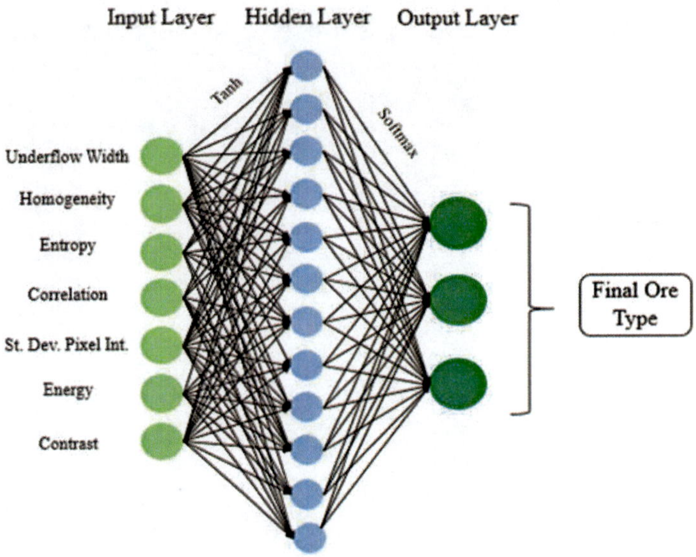

Fig. 15.13 Classification-based neural network model developed by Uahengo [9]

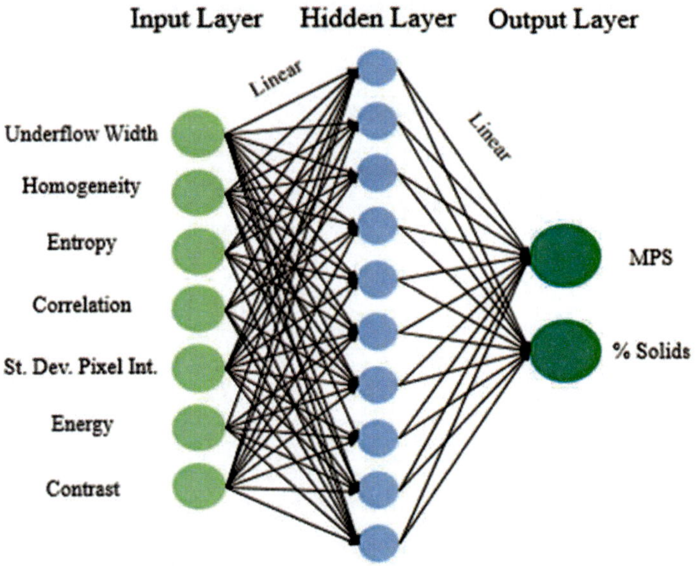

Fig. 15.14 Regression-based neural network model developed by Uahengo [9]

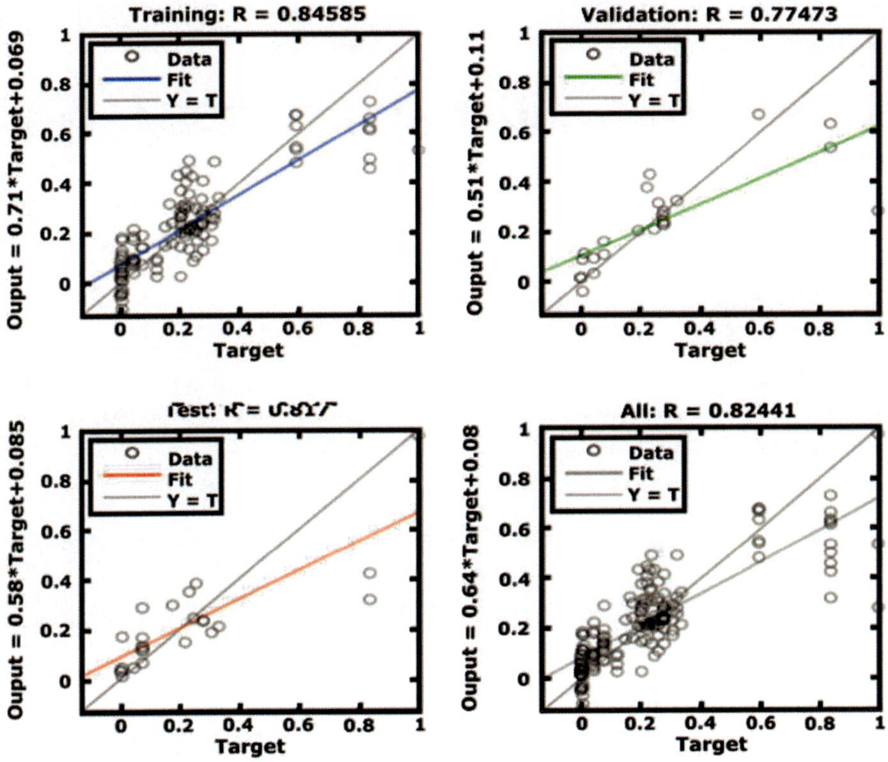

Fig. 15.15 Real versus predicted plots for MPS as the response variable [9]

Figures 15.15 and 15.16 show the real vs. predicted plots for both MPS and solids percentage, respectively, during the training, validation, and testing phase. Table 15.4 provides the performance results for both classification-based and regression-based neural network models during the training, testing, and validation phases. The effectiveness of the neural network model for particle size classes was tested by applying it over a uranium ore processing plant. The ore-type classification model was not tested on the industrial plant due to a lack of ore type (same uranium ore). Such advanced data-driven models can be used to monitor and control the performance of classification processes involving hydrocyclones. As the performance parameters deviate from the expected range, based on the data-driven model, the process variables can be modified and re-optimized in real time to ensure that the equipment performs with desired effectiveness and efficiency at all times.

In combination with AI and ML algorithms, such advanced image analysis and computer vision techniques can be used to model the performance of any classification process and ensure operational efficiency with optimized processing parameters and features. Moreover, data-driven ML and AI can be used to monitor the equipment's health for detecting and predicting the machine breakages and deviations

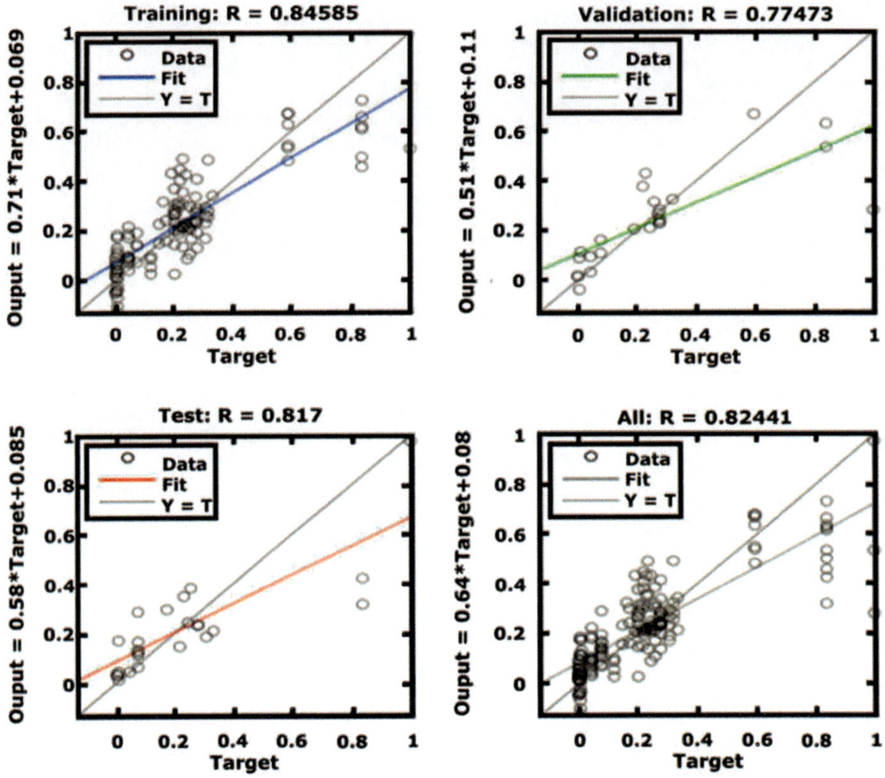

Fig. 15.16 Real versus predicted plots for percentage solid as the response variable [9]

Table 15.4 Performance results of neural network models developed by Uahengo [9]

Model type	Training	Validation	Testing
	%	%	%
Classification-based NN	90.0	60.0	77.8
Regression-based NN (MPS)	84.6	77.5	81.7
Regression-based NN (% solids)	61.3	60.2	67.3

from optimal operations. Predictive maintenance is necessary and heavily relies on sensor data, so a sensor network can be developed to capture all the operational and control data, such as flow parameters, various levels, operating pressures, system temperatures, etc., for the classification process using screens or cyclones. Then, the collected dataset can be used to train the analytics-based ML and AI algorithms to monitor the health and performance of the machine. Upon implementation of such data-trained models, machine health and operating performance can be gauged in real time. Using intelligence monitoring, on-time preventive maintenance or operational change can be done, which will prevent machine downtime and make sure that the equipment performs up to its full potential all the time.

Gravity Concentration and Heavy Media Separation

Gravity concentration and medium-heavy separation deal with separating the minerals based on specific gravities [10]. Compared to other separation processes, it has a relatively low unit cost with lesser power needs, no need for chemical reagents, and the ability to handle large tonnages. Therefore, heavy media separation is one of the most widely used gravity concentration methods applicable to cleaning coal and mineral recovery. Shaking table is another widely employed gravity concentration method/equipment primarily used for recovering coal and certain metal oxides, like tin, tungsten, tantalum, and chrome [11]. Other commonly used gravity separation methods/equipment include jigs, spirals, and cones.

A definitive difference in the specific gravity between the various components of the ore mixture is essential for a successful implementation of any of the aforementioned gravity concentration methods. Equation 15.1 gives the decision criteria for the utility and value of gravity separation, and Table 15.5 provides the particle size guidelines based on the defined ratio [11]. If the ratio is higher than 2.5, gravity separation is favored, and separation of particles up to 200 mesh is possible. With a ratio of 1.75, separation of 65 mesh particles is possible. Materials with a mesh size of 10 can be separated with a ratio of 1.5. Once the ratio drops below 1.25, separation of material with only particles higher than 12.5 mm size is possible, and fluid other than water has to be used.

$$\text{Ratio} = \frac{\text{Heavier Component SG} - \text{SG of Fluid}}{\text{Lighter Component SG} - \text{SG of Fluid}} \tag{15.1}$$

Like any other mineral beneficiation process, performance modeling of gravity concentration is essential to understand the process dynamics, identify the weak links, and thus modify the concerned parameters to ensure process efficiency is maintained. Chaurasia and Nikkam [12] demonstrated the neural network application for modeling the performance of the gravity concentration process for iron ore fines. The trained model could capture, understand, and predict the grade, recovery, and

Table 15.5 Particle size limitation for an efficient gravity separation process [11]

Ratio value	Particle size limit for separation
2.5 or greater	200 mesh (0.075 mm)
1.75	65 mesh (0.212 mm)
1.5	10 mesh (1.7 mm)
1.25 or below	12.5 mm or above

separation efficiency of the process. Thus, such data analytic techniques can be used to monitor the process. Using the performance indicators, the operational parameters can be controlled in real time, with the primary objective of ensuring an efficient gravity concentration process.

Dewatering

Dewatering involves procedures for removing water, which may be required for proper product handling or dry processing. Large amounts of contaminated water require removal during any wet mineral beneficiation process, thus making the dewatering an essential unit operation. Apart from hydrocyclones, standard dewatering methods include dewatering stockpiles, dewatering screens, thickeners, and filters (vacuum belt and pressure type). Dewatering stockpiles are used for sand-sized solids, with the stockpile base requiring sufficient drainage to ensure efficiency. Dewatering screens are vibrating screens with holes allowing through only water or very fine solids. Thickeners consist of large tanks used to dewater fine solids. A vacuum belt filter works with the slurry fed onto the belt, which has a vacuum at the bottom to produce a low-moisture product. Pressure filters are used for very fine solids, which operate with the slurry being fed into multiple filters subjected to pressure for removing the water.

ADA-based techniques can be used for an in-depth real-time performance modeling of the dewatering process. For example, Qi et al. [13] demonstrated the application of particle swarm optimization (PSO) and adaptive neuro-fuzzy inference system (ANFIS) for predicting the initial settling rate, which is the key performance metric for any dewatering process. Flocculation-settling experiments were conducted using the anionic polyacrylamide polymers to gather the dataset. Characteristics and solid content of the mineral processing tailings, along with the type and dosages of the polymer, were selected as the primary explanatory variable, with the initial settling rate (ISR) as the only response variable by Qi et al. [13]. Figure 15.17 shows the computation of the ISR, which is calculated as the slope of the midline-settling time curve and is widely used to characterize the performance of the dewatering process.

Dataset consisted of 102 instances with 17 explanatory variables, so principal component analysis (PCA) was used to reduce the number of explanatory variables to 9 (with PCA99) and 7 (using PCA95) [13]. This reduces the computational requirement during ML modeling. The hybrid algorithm was then trained using the 500 Monte Carlo simulations, and 30 Gaussian membership functions were used for the input variables. Model performance was evaluated using critical metrics like the coefficient of correlation (*R*), mean absolute error (MAE), and root mean square error (RMSE) along with the standard error and slope. Figures 15.18 and 15.19 show the convergence results for the model during the development and testing phase, respectively, with the Monte Carlo simulations. During the training, with the raw dataset, it took about 50 simulations for the model to reach convergence, whereas

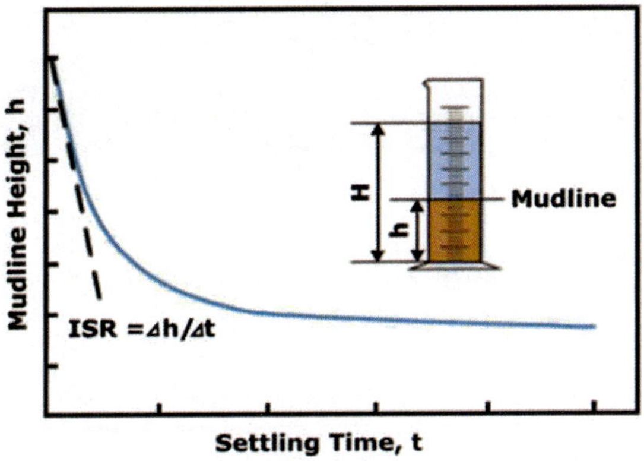

Fig. 15.17 Computation of initial settling rate (ISR) during dewatering process

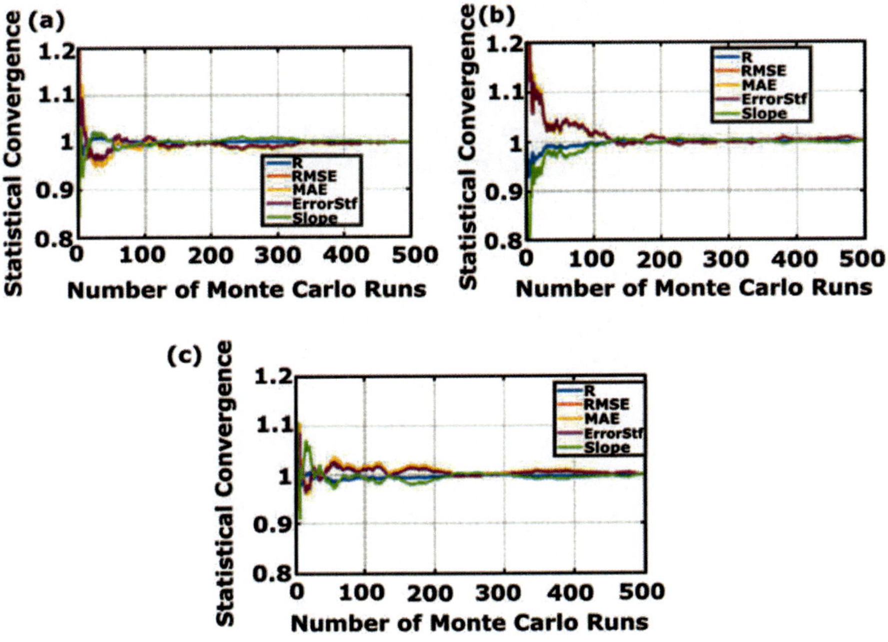

Fig. 15.18 Convergence of model performance during training phase **a** raw dataset, **b** PCA99 dataset, **c** PCA95 dataset [13]

it took about 120 and 220 simulations to reach convergence when using PCA99 and PCA95 datasets, respectively. During the testing, with the raw dataset, it took about 150 simulations for the model to reach convergence, whereas it took about 100 and 220 simulations to reach convergence when using PCA99 and PCA95 datasets, respectively. Figure 15.20 shows the R, MAE, and RMSE results of the model testing phase using raw, PCA99, and PCA95 datasets. The ML hybrid model's performance

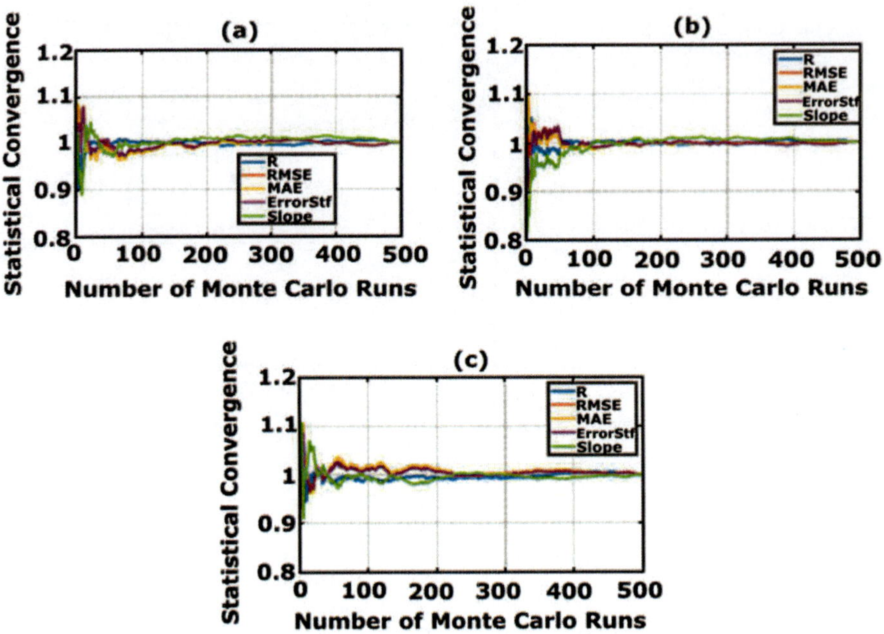

Fig. 15.19 Convergence of model performance during testing phase **a** raw dataset, **b** PCA99 dataset, **c** PCA95 dataset [13]

improved using PCA, characterized by higher *R*-value and lower MAE and RMSE values than the model trained with the raw dataset.

Such automated data-driven models can be used to improve system efficiency and allow cost optimization of the dewatering process. Moreover, the self-learning capability of such frameworks puts them at the cutting edge of innovation, as the operational fluctuations can be tracked, and the model can immediately adapt to any changes without requiring any manual interference, thus making it self-sustainable.

Froth Flotation

Froth flotation is one of the most widely used and most efficient ore beneficiation methods to effectively process complex metallic ores [14] and produce clean coal [15–19]. For coal processing, the goal is to reduce the ash and sulfur content [20]. In this process, various chemical reagents are used to modify the surface wettability of coal and the associated mineral matter. Chemical agents, called collectors, are used to render the coal particles hydrophobic selectively. This surface modification results in coal particles attaching themselves to the air bubbles and floating to the surface, as the concentrate product and the rest of the gangue minerals are left behind as tailings [20]. In addition, chemicals such as pH modifiers, dispersants, and depressants are

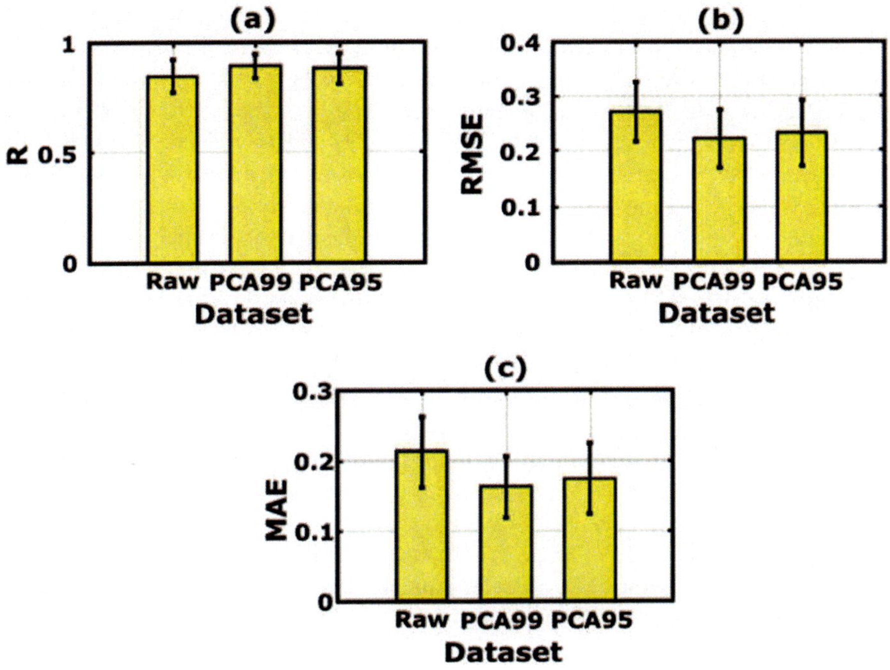

Fig. 15.20 Model performance results during testing phase [13]

usually added to the flotation pulp to enhance the floatability of coal particles further and inhibit the flotation of waste mineral matter [20].

For metallic ore processing, the goal involves separating metals, namely copper, lead, and zinc, from minor minerals, primarily pyrite (FeS_2) [14]. Pyrite present in the final concentrate can result in overall grade degradation and cause problems in the downstream processes [21]. Non-toxic polymers, such as PAM along with its derivatives [22–25], and biodegradable polymers, like chitosan [21, 26], have been introduced recently and are being commonly used as an efficient pyrite depressant in the froth flotation of sulfide ores. Chemical reagents—such as sodium isopropyl xanthate, methyl isobutyl carbinol (commonly known as MIBC), sodium cyanide, and zinc sulfate—are often used as collector, frother, pyrite depressant, and sphalerite depressant, respectively.

Some of the essential variables for a flotation process involve impeller speed, airflow rate, flotation time, and the dosages of chemical reagents, such as the collector, frother, and depressant. Therefore, predicting the concentrate grade and recovery with precision in real time for a given setup and obtaining the optimum values for all the important involved variables is essential for designing and running an efficient flotation setup either for coal or complex metallic ores mixtures.

Based on AI and ML, ADA techniques can be used for smart monitoring of the metallurgical performance, which is given by the concentrate grade and recovery, for any flotation process. Jorjani [27] and Al-Thyabat [28] demonstrated the use of the multi-layered feed-forward artificial neural network (ANN) for predicting the sulfur reductions in coal using mixed culture microorganisms and evaluating the effect of feed size, collector dosage, and impeller speed on the flotation performance for siliceous phosphate, respectively. Mohanty [29] and Jorjani [30] presented the multi-layered ANN model for handling the interface level in a flotation column using multiple values of tailing valve opening and interface level data and predicting the combustible recovery of coal after flotation using the proximate analysis (moisture, volatile matter, and ash) and group macerals (liptinite, fusinite, vitrinite, and ash) data, respectively. Cheng [31] presented the idea of using a single layer artificial neural network to predict the solid concentration of coal-water slurry (CWS) by utilizing the datasets with hard grove grindability index (HGI), moisture content, and degree of parent coal coalification data. Ze-lin [32] demonstrated using a digital image processing dataset with 13 image feature parameters for developing an RBF neural network to predict the washability curve of the coal washing process. Bekat [33] used the datasets with information on moisture, ash content, and coal heating values for constructing a multi-layered ANN model for predicting the bottom ash quantity in a coal-fired power plant. Feng [34] demonstrated the use of proximate analysis data for developing various ML models, support vector machine (SVM), alternating conditional expectation (ACE), and ANN for monitoring the gross calorific value of coal. Pusat [35] developed an adaptive neuro-fuzzy inference system (ANFIS) using the dataset containing the parameters like dying air temperature, drying air velocity, bed height, and sample size for estimating the moisture content for a coal drying process. Khodakarami [36] employed a multi-layered ANN for predicting the flotation performance of clayey coal processing in the presence of hybrid polymer aids. Ali et al. [20] recently demonstrated the use of five different AI algorithms, developed through flotation process datasets for predicting the flotation behavior of fine high-ash coal in the presence of a novel hybrid ash depressant: $Al(OH)_3$-PAM (Al-PAM).

Froth images of the flotation process, along with basic input parameters, were used by Marais [37] for developing the multi-layered ANN and random forest (RF) model to monitor the platinum concentrate grade. Nakhaei et al. [38, 39] demonstrated the use of multi-layered ANN and multivariate nonlinear regression (MNLR) for predicting the grade and recovery of copper and molybdenum using the flotation column plant data. The development of a controller for maintaining the performance by using the fuzzy logic model for a copper floatation plant was presented by Saravani et al. [40]. Using the dataset with parameters like particle size, iron, phosphor, sulfur, and iron oxide percentage contents of run-on-mine, Hosseini and Samanipour [41] developed a multi-layered ANN for predicting the iron, phosphorus, sulfur, and iron oxide recoveries for an iron ore floatation plant. A Mamdani fuzzy logic (MFL)-based model using the processed dataset with parameters like operational method, bacteria type, and process time for monitoring the iron and copper recoveries for a copper flotation plant was developed by Ahmadi and Hosseini [42]. Allahkarami et al. [43] demonstrated the use of a multi-layered ANN algorithm for predicting the grade and recovery of copper and molybdenum using the dataset with collector dosage, frother dosage, F-oil dosage, pH of pulp, particle size, moisture content, solid percentage, and copper, molybdenum, and iron grade in feed as the main parameters. Jahedsaravani et al. [44] presented the use of ANN and ANFIS models for predicting the metallurgical performance of a copper flotation plant, using the dataset with a flow rate of the gas–solid percentage, pH of the slurry, frother, and collector dosages as the input parameters.

Studies have thus demonstrated that the data can be processed in real time using ADA techniques. ML and AI can then further be utilized for designing an algorithm/model which, based on the objective of the plant, includes optimal flotation conditions involving all the critical variables and could maximize the processing plant objective. Furthermore, upon implementation at any flotation setup by using micro-controllers, these intelligent models will monitor and control the input conditions in real time, thus making sure that the metallurgical performance of the flotation plant is never hampered.

Magnetic and Electrostatic Separation

Magnetic separation is an ore beneficiation technique that undertakes mineral separation based on the differences in magnetic properties. It takes advantage of the natural magnetic properties among the mineral constituents within the ore feed [45]. The minerals separated through the magnetic separation technique must exhibit one of the following magnetic properties: ferromagnetic (magnetite, pyrrhotite, etc.); paramagnetic (monazite, rutile, chromite, hematite, etc.); or diamagnetic (plagioclase, calcite, zircon, etc.) [45]. Magnetic separation units on a commercial level involve establishing a magnetic field (low or high) with a continuously moving stream of dry or wet particles through it. Most common types of magnetic separators include drum, cross-belt, roll, high-gradient magnetic separation (HGMS), high-intensity magnetic separation (HIMS), and low-intensity magnetic separation (LIMS) [45].

Electrostatic separation is a processing technique that considers the differences in conductivity between different minerals to achieve the separation [46]. Electrostatic separation is a three-step process: charging of particles, separation at the grounded surface, and separation caused by the trajectory of the particles [2]. The common commercial units are high-tension plates and electrostatic screen separators [47]. The working of electrostatic plate separators involves passing a particle stream over a charged anode. As the electrostatic minerals lose electrons to the plate, they are pulled away from other particles with the induced attraction to the anode [47]. It is primarily used to separate minerals such as monazite, spinel, sillimanite, tourmaline, garnet, zircon, rutile, and ilmenite from placer sand [47]. The typical use of electrostatic separation for rare earth mineral processing involves separating monazite and xenotime from gangue minerals with specific gravity and magnetic properties [48]. The electrostatic technique is widely used in Australia, Indonesia, Malaysia, and India for mineral sand separation [47].

ADA can be used to monitor the metallurgical performance of the separation plant and for studying and optimizing the processing parameters to ensure an efficient and effective magnetic and electrostatic separation process. Lishchuk et al. [49] demonstrated the use of ADA techniques and ML to improve spatial modeling during the magnetic separation process. Two-step data modeling was done to achieve the study objective. The first step involved deploying the process properties into a geological database with the development of multiple non-spatial process models. This was followed by extracting the estimated process properties in the geological database and their coordinates and iron grades. Finally, the transformed process dataset was utilized with the decision tree algorithm to develop an intelligence-based model for integrating and optimizing the process properties (iron recovery and Davis tube mass pull, iron recovery and mass pull of the wet low-intensity magnetic separation, iron oxides liberation, and P80 size) for an iron ore separation plant.

Lai et al. [50] presented AI-based algorithms for modeling the characteristics of an electrostatic separator. First, datasets consisting of all the relevant system parameters—such as DC voltage level, the rotational speed of roller, ambient temperature, and configuration of the electrode—was recorded. Then, using the complete plant dataset, an artificial neural network (ANN) model was trained for accurate and efficient modeling of the electrostatic separation process. The advanced techniques for data modeling, processing, and analyzing can thus be used, along with ML algorithms, for monitoring and optimizing the magnetic and electrostatic mineral separation processes for ensuring operational accuracy, efficiency, and effectiveness.

Conclusion

Mineral processing involves procedures and technologies with which the valuable minerals are separated from gangue or waste rock. It is a three-step procedure involving mineral liberation, separation, and final concentration, after which the product is shifted to metallurgical processing (smelting and refining).

A crusher is the first component in the mineral processing procedure, which allows for the size reduction of the run-of-mine material. Next, a further size reduction is performed through a combination of ball and rod mills in a series of milling and grinding operations. Computer vision-based technologies can be developed through ADA techniques for modeling the performance of crushing, milling, and grinding circuits, allowing for the optimization of the performance, resulting in improved system design, equipment utilization, and overall process efficiency in real time.

Classification and screening involve particle sizing based on two physical parameters, particle size, and particle specific gravity. The two main types of classifiers used for mineral processing include hydrocyclones and hydrosizers. Classification through screens is carried out using grizzly screens or mechanically vibrating screens (single or double deck configuration). ADA techniques, using the underflow image data for these machines, would allow for intelligent monitoring and performance control resulting in maximum process efficiency. In addition, data-driven ML and AI can be utilized to monitor the equipment's health and allow for machine breakages to be detected and predicted in advance, thus allowing for an on-time preventive maintenance or operational change that would ensure that equipment performs up to its full potential all the time.

Gravity concentration and heavy medium separation involve mineral separation based on the difference in specific gravities. The most commonly used gravity separation methods/equipment include shaking tables, jigs, spirals, and cones. Dewatering is a unit operation that allows for water removal for proper product handling and dry processing. The most commonly used dewatering methods include dewatering stockpiles, dewatering screens, thickeners, and filters (vacuum belt and pressure type). One of the most widely used ore beneficiation methods is froth flotation which allows for effective processing of complex metallic ores and coal. The primary goal during coal processing involves reducing the ash and sulfur content, whereas during metallic ore processing, it includes the separation of metals like copper, lead, and zinc from the minor minerals, primarily pyrite (FeS_2). The other two commonly used ore beneficiation techniques include magnetic and electrostatic separation. Magnetic separation involves the separation of minerals based on the differences in magnetic properties. At the same time, electrostatic separation deals with mineral separation by using the differences in conductivity between different minerals.

Each of these mineral beneficiation processes requires proper monitoring of the metallurgical performance and examining and optimizing the processing parameters for ensuring an efficient and effective separation process. ADA techniques can process and analyze all the plant data and utilize ML and AI for process modeling. An intelligence-based algorithm/model can be designed for optimizing the separation process conditions based on the objective of the plant. Upon implementation for any separation process setup through advanced micro controllers, these data-driven intelligent models would allow for intelligent monitoring and parametric control in real time, thus ensuring that the mineral processing plant always operates with utmost efficiency and effectiveness.

References

1. Haldar, S.K. 2017. *Mineral Exploration—Principles and Applications*. Elsevier.
2. Kelly, E.G. 2003. Mineral processing. In *Encyclopedia of Physical Science and Technology*, 29–57.
3. Baek, J., and Y. Choi. 2019. Deep neural network for ore production and crusher utilization prediction of truck haulage system in underground mine. *Applied Sciences* 9: 4180.
4. Avalos, S., W. Kracht, and J.M. Ortiz. 2020. Machine learning and deep learning methods in mining operations: A data-driven SAG mill energy consumption prediction application. *Mining, Metallurgy & Exploration* 37: 1197–1212.
5. Conger, R., H. Robinson, and R. Sellschop. 2018. *Inside a Mining Company's AI Transformation*. McKinsey & Company. https://www.mckinsey.com/industries/metals-and-mining/how-we-help-clients/inside-a-mining-companys-ai-transformation. Accessed 9 Apr 2020.
6. Wills, B.A. 2011. *Wills' Mineral Processing Technology: An Introduction to the Practical Aspects of Ore Treatment and Mineral Recovery*.

7. Hallen, M. 2018. *Comminution Control Using Reinforcement Leaning*. Sweden: Umeå University.
8. MotionMetrics. 2021. *Artificial Intelligence and Computer Vision Based Technologies*. https://www.motionmetrics.com/technologies/. Accessed 15 Jan 2021.
9. Uahengo, F.D.L. 2014. *Estimating Particle Size of Hydrocyclone Underflow Discharge Using Image Analysis*. Stellenbosch University.
10. Burt, R.O. Gravity concentration methods. In *Mineral Processing Design*. Dordrecht: Springer.
11. Gill, C. B. 1991. Gravity concentration. In *Materials Beneficiation. Materials Research and Engineering*. New York, NY: Springer.
12. Chaurasia, R.C., and S. Nikkam. 2017. Application of artificial neural network to study the performance of multi-gravity separator (MGS) treating iron ore fines. *Particulate Science and Technology* 35: 93–102.
13. Qi, C., H.B. Ly, Q. Chen et al. 2020. Flocculation-dewatering prediction of fine mineral tailings using a hybrid machine learning approach. *Chemosphere* 244: 125450.
14. Hayat, M.B. 2018. *Performance, Mitigation of Environmental Hazards of Sulfide Mineral Flotation with an Insight into Froth Stability and Flotation.*
15. Han, C. 1983. *Coal Cleaning by Froth Flotation*. Iowa State University.
16. Erol, M., C. Colduroglu, and Z. Aktas. 2003. The effect of reagents and reagent mixtures on froth flotation of coal fines. *International Journal of Mineral Processing* 71: 131–145. https://doi.org/10.1016/S0301-7516(03)00034-6.
17. Keys, R. 1986. Promoters for froth flotation of coal. US Pat 4,589,980.
18. Demirbaş, A. 2002. Demineralization and desulfurization of coals via column froth flotation and different methods. *Energy Conversion and Management* 43: 885–895. https://doi.org/10.1016/S0196-8904(01)00088-7.
19. Honaker, R.Q., and M.K. Mohanty. 1996. Enhanced column flotation performance for fine coal cleaning. *Minerals Engineering* 9: 931–945. https://doi.org/10.1016/0892-6875(96)00085-4.
20. Ali, D., M.B. Hayat, L. Alagha, and O. Molatlhegi. 2018. An evaluation of machine learning and artificial intelligence models for predicting the flotation behavior of fine high-ash coal. *Advanced Powder Technology* 29: 3493–3506.
21. Hayat, M.B., L. Alagha, and S.M. Sannan. 2017. Flotation behavior of complex sulfide ores in the presence of biodegradable polymeric depressants. *International Journal of Polymer Science* 2017: 1–9. https://doi.org/10.1155/2017/4835842.
22. Boulton, A., D. Fornasiero, and J. Ralston. 2001. Selective depression of pyrite with polyacrylamide polymers. *International Journal of Mineral Processing* 61: 13–22. https://doi.org/10.1016/S0301-7516(00)00024-7.
23. Guévellou, Y., C. Noïk, J. Lecourtier, and D. Defives. 1995. Polyacrylamide adsorption onto dissolving minerals at basic pH. *Colloids Surfaces A Physicochemical Engineering and Aspects* 100: 173–185. https://doi.org/10.1016/0927-7757(95)03156-8.
24. Zhang, J., Y. Hu, D. Wang, and J. Xu. 2004. Depressing effect of hydroxamic polyacrylamide on pyrite. *Journal of Central South University of Technology* 11: 380–384. https://doi.org/10.1007/s11771-004-0079-1.
25. Huang, P., L. Wang, and Q. Liu. 2014. Depressant function of high molecular weight polyacrylamide in the xanthate flotation of chalcopyrite and galena. *International Journal of Mineral Processing* 128: 6–15. https://doi.org/10.1016/j.minpro.2014.02.004.
26. Huang, P., M. Cao, and Q. Liu. 2013. Selective depression of pyrite with chitosan in Pb-Fe sulfide flotation. *Minerals Engineering* 46–47: 45–51. https://doi.org/10.1016/j.mineng.2013.03.027.
27. Jorjani, E., S. Chehreh Chelgani, and S. Mesroghli. 2007. Prediction of microbial desulfurization of coal using artificial neural networks. *Minerals Engineering* 20: 1285–1292. https://doi.org/10.1016/j.mineng.2007.07.003.
28. Al-Thyabat, S. 2008. On the optimization of froth flotation by the use of an artificial neural network. *Journal of China University of Mining and Technology* 18: 418–426. https://doi.org/10.1016/S1006-1266(08)60087-5.

29. Mohanty, S. 2009. Artificial neural network based system identification and model predictive control of a flotation column. *Journal of Process Control* 19: 991–999. https://doi.org/10.1016/j.jprocont.2009.01.001.
30. Jorjani, E., H. Asadollahi Poorali, A. Sam, et al. 2009. Prediction of coal response to froth flotation based on coal analysis using regression and artificial neural network. *Minerals Engineering* 22: 970–976. https://doi.org/10.1016/j.mineng.2009.03.003.
31. Cheng, J., Y. Li, J. Zhou, et al. 2010. Maximum solid concentrations of coal water slurries predicted by neural network models. *Fuel Processing Technology* 91: 1832–1838. https://doi.org/10.1016/j.fuproc.2010.08.007.
32. Ze-lin, Z., Y. Jian-guo, W. Yu-ling, et al. 2011. A study on fast predicting the washability curve of coal. *Procedia Environmental Sciences* 11: 1580–1584. https://doi.org/10.1016/j.proenv.2011.12.238.
33. Bekat, T., M. Erdogan, F. Inal, and A. Genc. 2012. Prediction of the bottom ash formed in a coal-fired power plant using artificial neural networks. *Energy* 45: 882–887. https://doi.org/10.1016/j.energy.2012.06.075.
34. Feng, Q., J. Zhang, X. Zhang, and S. Wen. 2015. Proximate analysis based prediction of gross calorific value of coals: A comparison of support vector machine, alternating conditional expectation and artificial neural network. *Fuel Processing Technology* 129: 120–129. https://doi.org/10.1016/j.fuproc.2014.09.001.
35. Pusat, S., M.T. Akkoyunlu, E. Pekel, et al. 2016. Estimation of coal moisture content in convective drying process using ANFIS. *Fuel Processing Technology* 147: 12–17. https://doi.org/10.1016/j.fuproc.2015.12.010.
36. Khodakarami, M., and O. Molatlhegi, and L. Alagha. 2017. Evaluation of ash and coal response to hybrid polymeric nanoparticles in flotation process: Data analysis using self-learning neural network. *International Journal of Coal Preparation and Utilization*: 1–20.https://doi.org/10.1080/19392699.2017.1308927.
37. Marais C (2010) *Estimation of Concentrate Grade in Platinum Flotation Based on Froth Image Analysis*. University of Stellenbosch.
38. Nakhaei, F., M.R. Mosavi, A. Sam, and Y. Vaghei. 2012. Recovery and grade accurate prediction of pilot plant flotation column concentrate: Neural network and statistical techniques. *International Journal of Mineral Processing* 110–111: 140–154. https://doi.org/10.1016/j.minpro.2012.03.003.
39. Nakhaeie, F., A. Sam, and M.R. Mosavi. 2013. Concentrate grade prediction in an industrial flotation column using artificial neural network. *Arabian Journal for Science and Engineering* 38: 1011–1023. https://doi.org/10.1007/s13369-012-0350-y.
40. Saravani, A.J., N. Mehrshad, and M. Massinaei. 2014. Fuzzy-based modelling and control of an industrial flotation column. *Chemical Engineering Communications* 201: 896–908. https://doi.org/10.1080/00986445.2013.790815.
41. Hosseini, S.H., and M. Samanipour. 2015. Prediction of final concentrate grade using artificial neural networks from Gol-E-Gohar iron ore plant. *American Journal of Mining Metallurgy* 3: 58–62. https://doi.org/10.12691/AJMM-3-3-1.
42. Ahmadi, A., and M.R. Hosseini. 2015. A fuzzy logic model to predict the bioleaching efficiency of copper concentrates in stirred tank reactors. *International Journal of Nonferrous Metallurgy* 4: 1–8. https://doi.org/10.4236/ijnm.2015.41001.
43. Allahkarami, E., O.S. Nuri, A. Abdollahzadeh, et al. 2016. Estimation of copper and molybdenum grades and recoveries in the industrial flotation plant using the artificial neural network prediction of grade and recovery, artificial neural network, copper flotation, copper concentrator plant. *International Journal of Nonferrous Metallurgy* 5: 23–32. https://doi.org/10.4236/ijnm.2016.53004.
44. Jahedsaravani, A., M.H. Marhaban, and M. Massinaei. 2016. Application of statistical and intelligent techniques for modeling of metallurgical performance of a batch flotation process. *Chemical Engineering Communications* 203: 151–160. https://doi.org/10.1080/00986445.2014.973944.

45. Haldar, S. 2013. Chapter 12—Mineral processing. In *Mineral Exploration—Principles and Applications*, 223–250. Elsevier.
46. Higashiyama, Y.A. 1998. Recent progress in electrostatic separation technology. *Particulate Science and Technology* 16: 77–90.
47. Haldar, S. 2018. Chapter 13—Mineral processing. In *Mineral Exploration—Principles and Applications*, 259–290. Elsevier.
48. Zhang, J.E. 2012. A review of rare earth mineral processing technology. In *44th Annual Meeting of The Canadian Mineral Processors*, 79–102. Ottawa: CIM.
49. Lishchuk, V., C. Lund, and Y. Ghorbani. 2019. Evaluation and comparison of different machine-learning methods to integrate sparse process data into a spatial model in geometallurgy. *Minerals Engineering* 134: 156–165.
50. Lai, K.C., S.K. Lim, C.P. Teh, and K.H. Yeap. 2016. Modeling electrostatic separation process using artificial neural network (ANN). *Procedia Computer Science* 91: 372–381.

Chapter 16
Advanced Analytics for Decreasing Greenhouse Gas Emissions in Surface Mines

Ali Soofastaei and Milad Fouladgar

Abstract This chapter demonstrates the practical application of artificial intelligence (AI) and machine learning (ML) to reduce greenhouse gas emissions and improve energy efficiency in surface mines. Mobile equipment in mine sites consumes a massive amount of energy, and the main part of this energy provides by diesel. The critical diesel consumers in surface mines are haul trucks, the huge machines that move mine materials in the mine sites. There are many effective parameters on haul trucks' fuel consumptions and gas emissions. AI and ML models can help mine managers to predict and minimize haul truck energy consumption and consequently reduce the greenhouse gas emission generated by these trucks. However, mine sites have different limitations to access the required data to train the developed models. Moreover, the main part of effective parameters on truck fuel consumption is not uncontrollable. This chapter presents a practical and validated AI approach to optimize three key parameters aimed to minimize haul truck fuel consumption in surface mines. This approach uses an artificial neural network to predict energy consumption and applies a genetic algorithm for optimization. The proposed integrated approach has been tested and validated in different mine sites, and the results of the developed AI model have been presented in this chapter.

Keywords Artificial intelligence · Energy efficiency · Fuel consumption · Haul trucks · Prediction · Optimization · Mining engineering

A. Soofastaei (✉)
Vale, Brisbane, Australia
e-mail: ali@soofastaei.net
URL: https://www.soofastaei.net

M. Fouladgar
Department of Mechanical Engineering, Najafabad Branch, Islamic Azad University, Najafabad, Iran
e-mail: milad_fouladgar@smc.iaun.ac.ir

A. Soofastaei (ed.), *Advanced Analytics in Mining Engineering*,
https://doi.org/10.1007/978-3-030-91589-6_16

Introduction

Climate change, energy security, water scarcity, land degradation, and dwindling biodiversity put pressure on communities, requiring more excellent environmental knowledge and resource-conscious economic practices. As a response to these genuine difficulties, both mining and industrial activities have adopted environmental plans.

The global accord, which 125 countries have signed, aims to reduce global GHG emissions by 80% by 2050 to achieve a low-carbon society. Thus far, the agreement has significantly impacted energy-related laws, such as carbon taxes and energy pricing.

However, following the Paris agreement, the energy costs in the mining industry have risen substantially in respect of overall operating costs. Six years ago, energy accounted for 10% of mining companies' operational costs; now, it is pushing close to 20%. This increases the cost base of companies significantly.

Mining is critical to our national security, economy, and the lives of individual citizens. Millions of tons of resources should be mined each year for each individual to maintain his or her quality of living [1]. In addition, the mining sector is a critical component of the world economy, supplying crucial raw materials such as coal, metals, minerals, sand, and gravel to manufacturers, utilities, and other enterprises [2]. To put it another way, mining will continue to be an essential part of the global economy for many years.

Mining necessitates much energy. Mining, for example, is one of the few non-manufacturing industrial sectors recognized as energy-intensive by the US Department of Energy [3]. It is also widely acknowledged that the mining industry could enhance its energy efficiency dramatically. Using the USA as an example, the US Department of Energy (DOE) estimates that the US mining sector consumes around 1315 PJ per year and that this annual energy consumption might be reduced to 610 PJ or about 46% of current annual energy usage [3]. According to the most recent data, energy consumption in Australia's mining sector was at 730 petajoules (PJ) in 2019–20, up 9% from the previous year [4]. This is slightly greater than the average rate of increase in energy use during the last decade. Mining consumes 175 PJ of energy per year in South Africa and is the largest consumer of electricity at 110.9 PJ per year, according to 2003 figures. The association between rising interest in energy efficiency and energy prices demonstrates increasing energy intensity on mining operating expenses [5, 6]. Given recent governmental moves by various governments to make industry pay for the expenses associated with carbon emissions, such high energy-intensive processes are not sustainable or cost-effective (carbon taxes and similar regulatory costs). As a result, all stakeholders have a vested interest in improving mine energy efficiency.

Since the rise in fuel prices in the 1970s, the importance of reducing energy usage has gradually grown. In addition, because the mining industry's primary energy sources are petroleum products such as electricity, coal, and natural gas, increasing margins through efficiency savings can also save millions of tons of gas emissions.

Mining companies are looking into reducing energy consumption and emissions to cut costs and emissions, especially considering any possible carbon emissions strategy. First, however, businesses must have a comprehensive understanding of their current energy usage, which involves using technology that allows employees to make decisions.

Mining businesses actively review their investment, capital expenditure, and operational plans to ensure that their operations are sustainable and ecologically beneficial. Sustainable practices and capital equipment investments must result in measurable cost savings. Mining businesses are looking to increase their energy efficiency to cut costs and lessen their environmental effect.

Sustainable investments were not thought to produce significant returns on investment in earlier years, but they are becoming more appealing with the quickly changing legislative and economic climate. When all the advantages of new technology and business practices are considered, including direct savings from increased efficiency as well as associated incentives such as carbon tax credits, investments become much more appealing. Furthermore, when considered over a longer time horizon, these same investments in energy savings, for example, become incredibly beneficial.

Data analytics represents a very appropriate approach to pulling together disparate data sources since it is the science of examining raw data to conclude that information. In addition, cost savings, faster and better decision making, and finally, new goods and services are some of the key benefits of data analytics [7].

Data analytics represents a very appropriate approach to pulling together these disparate data sources since it is the science of examining raw data to conclude that information. Cost savings, faster and better decision making, and finally, new goods and services are some of the most significant advantages of data analytics [7]. Data analytics is widely used and can be used in areas many might not have thought about before. One area that sees much potential in data analytics is the mining industry. Data analytics should be considered a necessity, not a luxury, for an industry that does trillions of dollars in business every year.

One of the advanced data analytic techniques discussed in this chapter aims to enhance the crucial issue of mining energy efficiency. The focus will be on open-pit mine haulage activities. This study aims to create a sophisticated data analytics model for assessing the complex connections that affect haul truck energy efficiency in surface mining. The application of artificial neural networks for predictive simulation and genetic algorithms (GAs) for optimization in the investigation of energy efficiency is the focus of this study.

Advanced Analytics for Mining Energy Efficiency

Global resource firms are currently struggling in challenging economic and regulatory environments. However, most companies in the mining business are now disclosing their performance in this area in response to growing social concern about the industry's numerous consequences and the birth of the idea of sustainable development. Many firm sustainability reports include total energy consumption and associated glasshouse gas (GHG) emissions in absolute and relative terms, indicating that energy consumption and its impact on climate change are priorities.

Mining companies are setting goals to improve these metrics, but there is also a global trend toward more complicated and lower-grade orebodies, which require more energy to process. As a result, mining businesses must be more innovative to improve their environmental sustainability and efficiency operations. In addition, companies must consider the specific energy usage of their processes to limit glasshouse gas emissions.

According to Australian government research, the most significant energy use industries in 2013–14 were transportation, metal manufacturing, oil and gas, and mining. Transportation consumes a quarter of Australia's annual energy. The manufacturing of metal products such as aluminum, steel, nickel, lead, iron, zinc, copper, silver, and gold accounted for over 16% of total energy consumption. The mining industry consumes 10% of all energy used by participants. Figure 16.1 shows the other industries that used the most energy in 2019–20.

Grinding (40%) and materials handling by diesel equipment are the most energy-intensive equipment types in the mining industry (17%) [8].

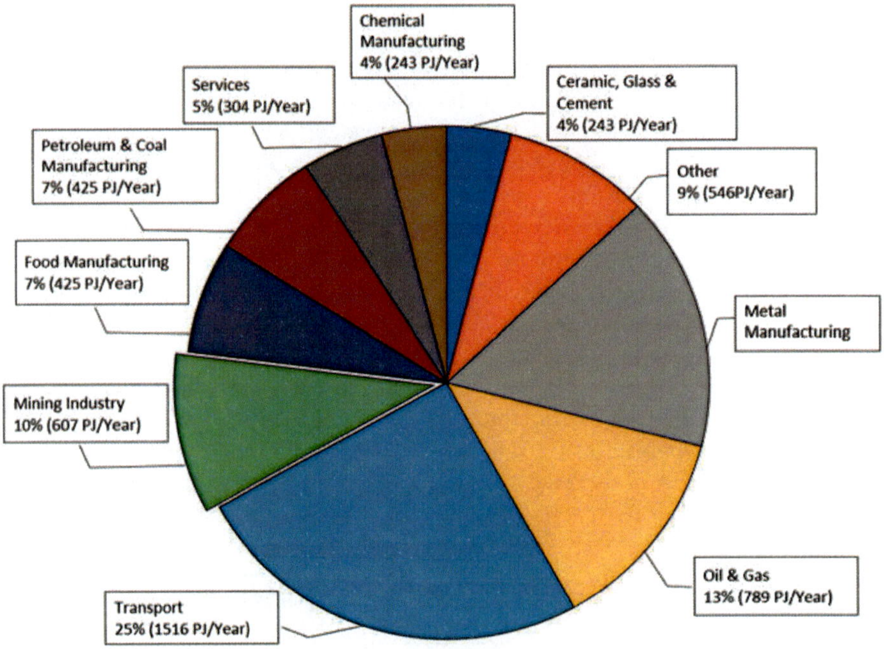

Fig. 16.1 Top energy users by industry sector 2019–20 (total 6069 PJ) [8]

According to the Australian Energy Statistics, Australian energy consumption has increased by an average of 0.6% a year for the past decade and reaching 6171 PJ in 2019–20.

Energy efficiency can significantly cut energy demand while also assisting in reducing glasshouse gas emissions at a low cost to industry and the larger economy. Therefore, it makes commercial and environmental sense to be aware of opportunities to maximize energy efficiency. The glasshouse gas emissions produced by mining companies were calculated using various fuels, including electricity, natural gas, and diesel. The mining companies' energy savings translated to a possible reduction in glasshouse gas emissions.

Data analytics is the science of examining raw data to discover useful information, reach conclusions about the meaning of the data, and support decision making. The foremost opportunity that data analytics presents for mining is its potential to identify, understand, and then guide the correction of complex root causes of high costs, poor process performance, and adverse maintenance practices. Therefore, data analytics can reduce costs and accelerate better decision making, which ultimately enables new products and services to be developed and delivered, creating added value for all [7].

Figure 16.2 illustrates the two dimensions of maturity: a time dimension (over which capability and insights are developed) and a competitive advantage dimension (the value of insights generated). At the lowest levels, analytics are routinely used to produce reports and alerts. These use simple, retrospective processing and reporting tools, such as pie graphs, top-ten histograms, and trend plots. They typically answer the fundamental question: "what happened and why?" Increasingly, sophisticated analytical tools, capable of working at or near real-time and providing rapid insights

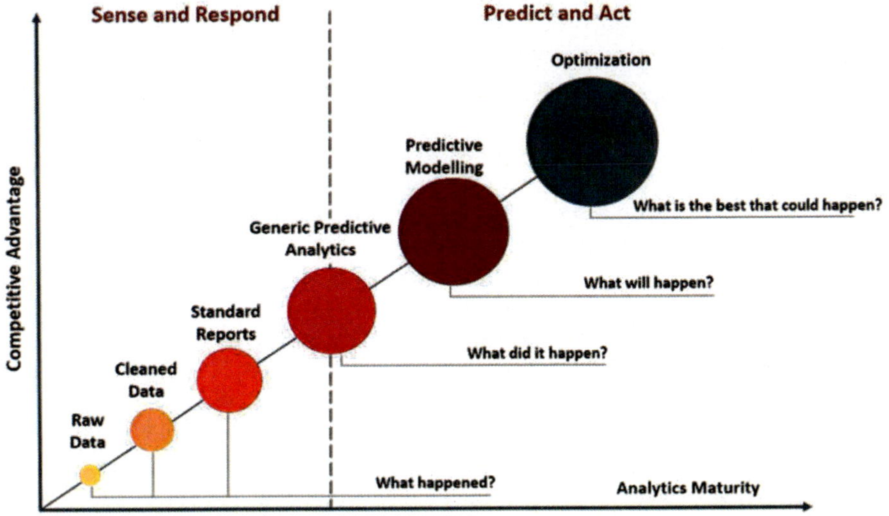

Fig. 16.2 Data analytics maturity levels [7]

for process improvement, can show the user "what just happened" and assist them in understanding "why" as well as the following best action to take. Toward the top end of the comparative advantage scale are predictive models and ultimately optimization tools, with the capability to evaluate "what will happen and the ability to identify the best available responses?"—"what is the best that could happen?".

The mining sector and governments have been pushed to perform research on energy consumption reduction due to the potential for energy (and financial) savings. As a result, a significant number of research studies and industrial projects have been conducted worldwide to achieve this in mining operations [8]. As a result, the mining industry might save roughly 37% of its current energy use by fully implementing state-of-the-art technology and installing new technology through research and development expenditure [9]. Furthermore, energy usage is significantly reduced when mining technologies and energy management systems improve. To put it another way, there are substantial further chances to minimize energy use in the mining business.

The four main phases of the mining process that data analytics can be used are (1) extraction of ore, (2) materials handling, (3) ore comminution and separation, and (4) mineral processing. The focus of many companies is efficiency improvements in the materials handling phase. For example, the hauling activity at an open-pit mine consumes a significant amount of energy and can be more energy-efficient [10]. The case study presented here—haulage equipment—is one of these potential areas for improving the mining energy efficiency as well as reducing greenhouse gas emissions.

Improve Haul Trucks Energy Efficiency

In a surface mining operation, truck haulage accounts for most costs. In surface mines, diesel fuel is used as an energy source for haul trucks, which is expensive and has a significant environmental effect. Energy efficiency is widely acknowledged as the easiest and most cost-effective strategy to manage rising energy bills and lower glasshouse gas emissions.

Depending on the production capacity and site layout, haul trucks are utilized in conjunction with other equipment such as excavators, shovels, and loaders. They collaborate to dig ore or waste material out of the pit and carry it to a disposal site, stockpile, or the next step in the mining operation [11].

The pace of energy consumption is determined by various factors that can be evaluated and tweaked to achieve optimal performance levels [12]. The energy efficiency of the mine fleet is affected by a variety of factors, including site production rate, vehicle age and maintenance, payload, speed, cycle time, mine layout, mine plan, idle time, tire wear, rolling resistance, dumpsite design, engine operating parameters, and transmission shift patterns. To improve energy efficiency, this knowledge can be incorporated into mining plan costing and design methods [8]. To assess the prospects for strengthening truck energy efficiency, a comprehensive analytical framework can be built.

We can not only save money each year by improving the energy efficiency of mine haulage systems, but we can also save considerable emissions of glasshouse gases and other air pollutants.

Data Analytics Models

A novel integrated model was proposed to improve haul truck energy usage's three most significant and critical effective characteristics. Payload (*P*), truck speed (*S*), and total resistance (*R*) are the three parameters (TR). However, the relationship between energy usage and these characteristics on an actual mining site is complicated. Therefore, to predict and reduce haul truck fuel consumption in surface mines, we apply two AI technologies to develop an advanced data analytic model (Fig. 16.3).

In the first step, an artificial neural network (ANN) model was developed to create a fuel consumption index (FC_{Index}) as a function of *P*, *S*, and TR. This index shows how many liters of diesel fuel are consumed to haul one ton of mined material in one hour. In this model, the main parameters used to control the algorithm were R^2 and MSE. After the first step, the optimum values of *P*, *S*, and TR will be determined using a novel multi-objective GA model. These improved parameters can be utilized to boost haul truck energy efficiency.

The proposed model's methods are all based on actual data obtained from surface mines. Below are the results of utilizing the developed model for two genuine major

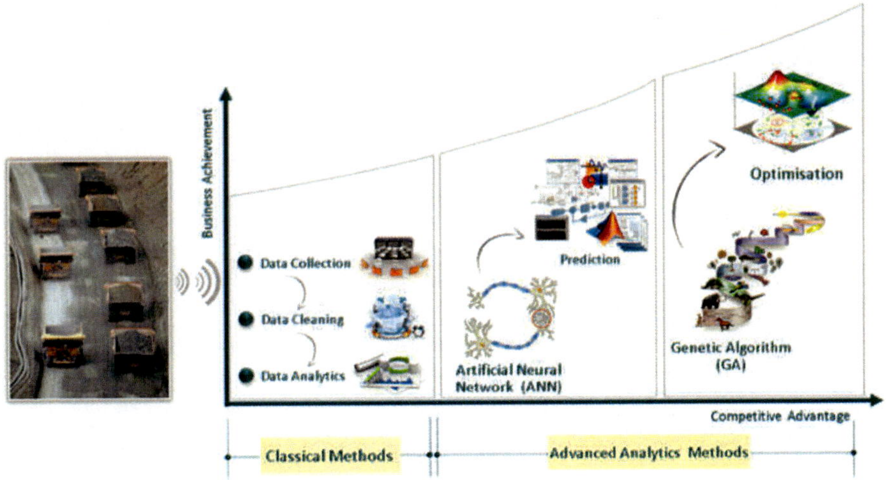

Fig. 16.3 A schematic of the developed model [8]

surface mines in Australia and Iran. On the other hand, the finished methods can be expanded for various mines by substituting the data.

Prediction Model—Artificial Neural Network

The artificial neural network (ANN) is a popular AI model and a robust computational tool based on the human brain's organizational structure [13]. ANNs are the representation of methods that the brain uses for learning which are known as neural networks (NNs), simulated neural networks (SNNs), or parallel distributed processing (PDP). ANN simulates the effect of multiple variables on one significant parameter by a fitness function. Thus, ANNs are excellent solutions for complex problems as they can signify the compound relationships between the various parameters involved in a problem.

ANN methods are established as powerful techniques to solve various real-world problems among the different machine intelligence procedures due to ANN's excellent learning capacity in recent decades. The approximate solution by ANN is found to be useful, but it depends upon the ANN model that one considers [14].

Layers are commonly used to organize neural networks. Layers are made from various interconnected "neurons/nodes," which include "activation functions." ANN processes information to solve problems through neurons/nodes in a parallel manner. First, ANN obtains knowledge through learning and is stored within interneuron connections' strength, expressed by numerical values called "weights." Then, these weights and biases are combined to calculating output signal values for a new testing input signal value. Next, patterns are provided to the network through the "input layer," which connects to one or more "hidden layers," where the actual processing is completed through a system of weighted "connections." The hidden layers then correlate to an "output layer," which generates the output through the activation functions (Eqs. 16.1, 16.2 and 16.3).

$$E_k = \sum_{j=1}^{q} (w_{ijk}x_j + b_{ik}) \quad k = 1,\ 2,\ \ldots,\ m \tag{16.1}$$

where i is the input, x is the normalized input variable, w is the weight of that variable, b is the bias, q is the number of input variables, and k is the counter of neural network nodes, and m is the number of neural network nodes in the hidden layer.

In general, the activation functions contain linear and nonlinear equations. The coefficients related to the hidden layer are grouped into matrices w_{ijk} and b_{ik}. Equation 16.2 is often used as the activation function between the hidden and output layers, where f is the transfer function.

$$F_k = f(E_k) \tag{16.2}$$

The output layer calculates the weighted sum of the signals provided by the hidden layer, and the related coefficients are grouped into matrices W_{ok} and b_o. Thus, the network output can be determined by Eq. 16.3.

$$\text{Out} = (\sum_{k=1}^{m} w_{ok} F_k) + b_o \tag{16.3}$$

The most significant component of neural network modeling is network training, which can be done in two ways: controlled and uncontrolled. Backpropagation is the most widely used training algorithm, which was established after examining several types of algorithms. A training algorithm modifies the coefficients (weight and bias) of a network to reduce the error between the estimated and actual network outputs.

The mean square error (MSE) and coefficient of determination (R^2) were used in this study to investigate the error and performance of the neural network output and determine the appropriate number of nodes in the hidden layer. Figure 16.4 depicts the created model's basic structure.

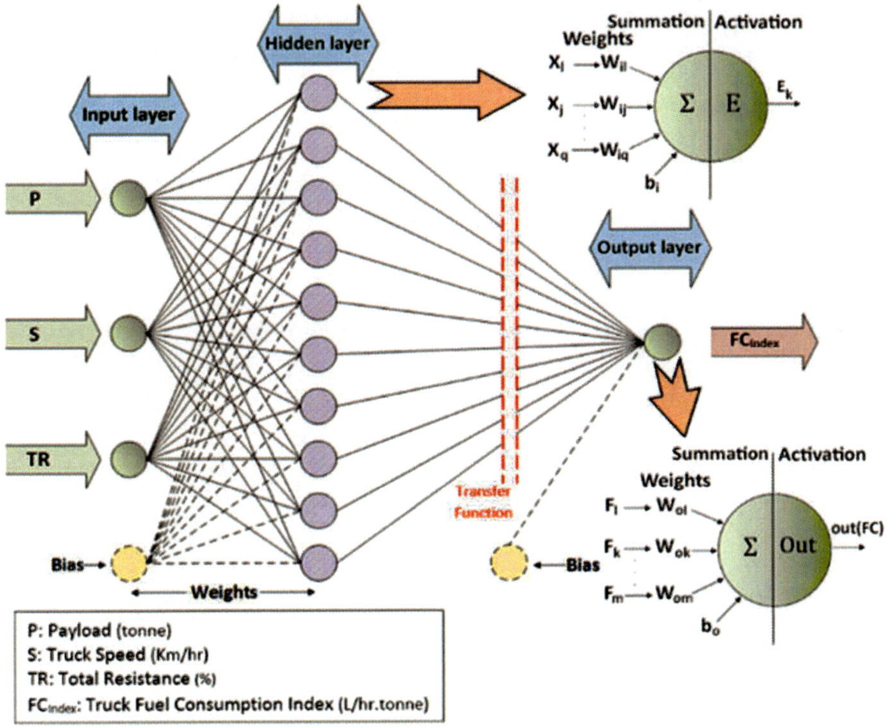

Fig. 16.4 Structure of artificial neural network [8]

The developed AI model was tested against actual data taken from standard trucks in two surface mines in Australia and Iran. Table 16.1 contains some information from these case studies.

For a standard range of loads, Figs. 16.5 and 16.6 show the correlation between *P*, S, TR, and FC_{Index} created by the constructed ANN model for two types of standard trucks employed in case studies.

The presented graphs show a nonlinear relationship between FC_{Index} and *P*. The fuel consumption rate increases dramatically with increasing TR. However, this rate does not change sharply with changing truck speed (*S*).

Table 16.1 Case studies information

Case study	Mine type	Mine details	Location	Investigated truck
Mine 1	Surface coal mine	The mine contains 877 million tons of coking coal reserves, making it one of Asia's and the world's most significant coal deposits. It can produce 13 million tons of coal per year	Queensland, Australia	CAT 793D
Mine 2	Surface iron mine	There are 36 million tons of iron deposits in the mine. It has a 15-million-ton ore and waste extraction capacity per year	Kerman, Iran	Komatsu HD785

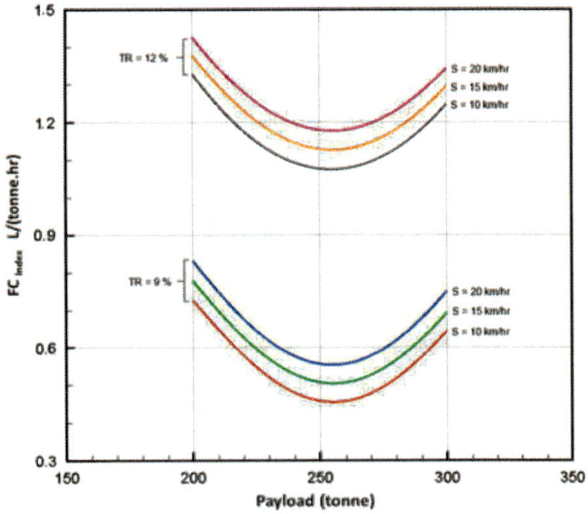

Fig. 16.5 Correlation between payload, *S*, TR, and FC_{Index} based on the developed ANN model for CAT 793D (mine 1)

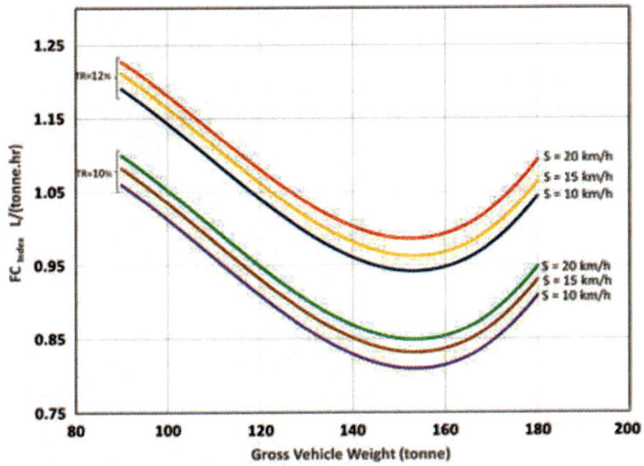

Figure 16.6 Correlation between GVW, *S*, T.R., and FCIndex based on the developed ANN model for HD785 (mine 2)

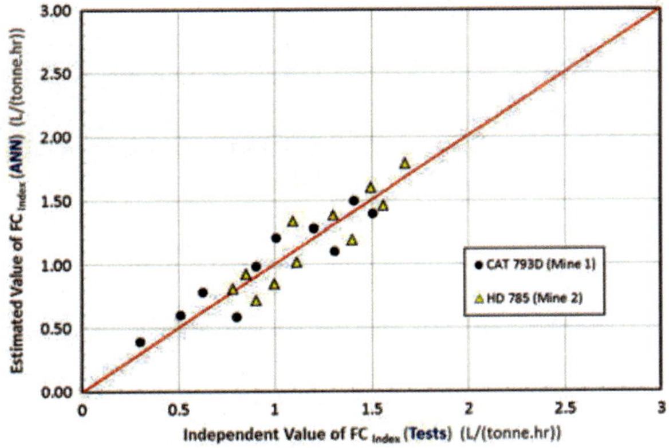

Fig. 16.7 Sample values for the estimated and the independent fuel consumption index

The results show good agreement between the estimated and actual values of fuel consumption. Figures 16.7 presents sample values for the independent (tested) and the estimated (using the ANN) fuel consumption to highlight the insignificance of the importance of the absolute errors in the analysis for studied mines.

Optimization Model—Genetic Algorithm

Optimization is a branch of computational science that shows how to find the best measurable solution to various issues. It is critical to consider the search area and goal function components when solving a specific problem. All the solution's possibilities are investigated in the search area. The objective function is a mathematical function that connects each point in the search area to an actual value that may be used to evaluate all search area members.

Traditional optimization methods are described by the stiffness of their mathematical models and limit their application in presenting dynamic and complex situations of "real life." Optimization techniques based on AI, underpinned by heuristic rulings, can reduce the problem of stiffness and are suitable to solve various kinds of engineering problems.

Some heuristic algorithms were developed in the 1950s to replicate biological processes in engineering. When computers were developed in the 1980s, it became possible to employ these algorithms to optimize functions and processes, whereas older methods failed.

During the 1990s, some new heuristic methods were developed by prior algorithms, such as swarm algorithms, simulated annealing, ant colony optimization, and GA. GA is one of the most widely used evolutionary optimization algorithms.

GAs were defined based on an abstraction of biological evolution using ideas from natural evolution and genetics to design and implement robust adaptive systems [15]. In optimization methods, using the new generation of GA is relatively novel. Moreover, they have good chances to escape from local minimums because of no need for any derivative information. As a result, their application in practical engineering problems can provide more satisfactory solutions than other traditional mathematical methods [16].

GAs are similar to the evolutionary aspects of natural genetics. The individuals are randomly selected from the search area. The fitness of the solutions is determined from the fitness function, subsequently. It is the result of the variable that is to be optimized. The individual that creates the best fitness in the population (a group of possible solutions) has the highest chance to return in the next generation with the opportunity of reproduction by the crossover with another individual, thus producing decedents with both characteristics. The possible solutions will converge to an optimal solution for the proposed problem by correctly developing a GA crossover, which contributes to the evolution based on selection, reproduction, and mutation.

Due to their potential as optimization techniques for complex functions, GAs have been used in various scientific, engineering, and economic problems [17–20]. There are four significant advantages of using GAs to optimize problems [21]:

- GAs do not have many mathematical requirements for optimization problems.
- They can use many objective functions and constraints (i.e., linear or nonlinear) defined in discrete, continuous, or mixed search spaces.
- They are very efficient at doing global searches due to the periodicity of evolution operators.
- They provide high flexibility to hybridize with domain-dependent heuristics to enable an efficient implementation for a problem.

It is crucial to investigate the impact of particular parameters on GA behavior and performance to determine their relevance to the problem requirements and available resources. Furthermore, the type of problem being addressed determines the impact of each parameter on the algorithm's performance. As a result, determining

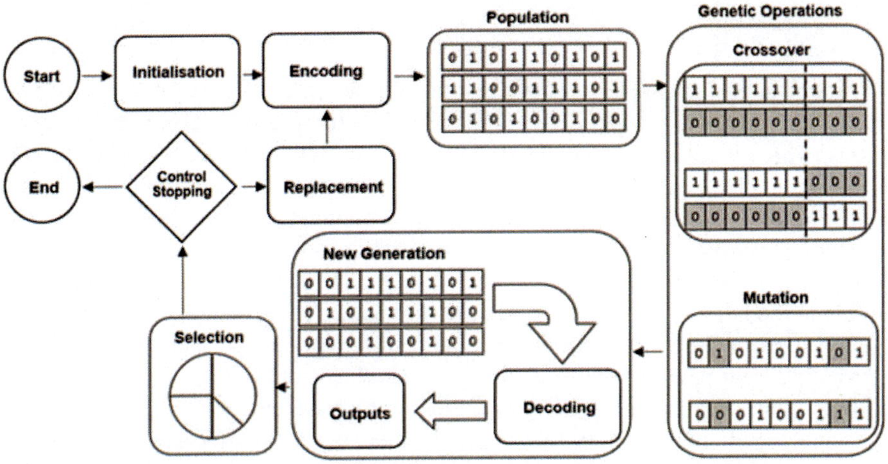

Fig. 16.8 GA processes (developed model) [8]

the best values for these characteristics will necessitate a significant amount of experimentation.

In the GA model, fitness function, individuals, populations and generations, fitness value, parents and children are the main parameters [17]. In addition, the population size impacts global performance and GA efficiency, and the mutation rate ensures that a given position does not remain fixed in value or the search becomes essentially random. Figure 16.8 depicts the basic framework of a GA model.

A GA model was created to optimize the significant, influential factors on the energy consumption of haul trucks. Tables 16.2 and 16.3 show the outcomes of utilizing the proposed model for actual case studies with an optimal range of variables.

Table 16.2 Result of the GA model for CAT 793D in mine (1)

Variables	Normal values		Optimized values	
	Minimum	Maximum	Minimum	Maximum
Gross vehicle weight (ton)	150	380	330	370
Total resistance (%)	8	20	8	9
Truck speed (km/h)	5	25	10	15

Table 16.3 Result of the GA model for HD 785 in mine (2)

Variables	Normal values		Optimized values	
	Minimum	Maximum	Minimum	Maximum
Gross vehicle weight (ton)	75	170	150	160
Total resistance (%)	8	15	8	10
Truck speed (km/h)	5	40	10	18

Conclusion

The purpose of this chapter was to demonstrate the value of modern data analytics models in improving energy efficiency in mining sectors, particularly in haulage operations, which are one of the most energy-intensive activities. However, improving haul truck fuel consumption for actual mining operations based on the link between influential factors, such as *P*, *S*, and TR, was difficult. Thus, two AI methods were utilized to construct a reliable model to assess the problem.

At first, an ANN model was utilized to simulate truck fuel consumption as a function of payload, truck speed, and total resistance. Then, the ANN was generated and tested using the collected accurate mine site datasets, and the results were showed good agreement between the actual and estimated values of FC_{Index}.

After that, to improve the energy efficiency in haulage operations, a GA method was developed to determine the optimal value of effective parameters on fuel consumption in haulage trucks. The developed model was used to analyze data for two surface mines in Australia and Iran. This model also can be applied to improve the haul truck fuel consumption for any dataset obtained from actual mine operations.

References

1. Golosinski, T.S. 2000. Mining education in Australia: A vision for the future.
2. Zheng, S., and H. Bloch. 2014. Australia's mining productivity decline: Implications for MFP measurement. *Journal of Productivity Analysis* 41 (2): 201–212.
3. Doe, U. 2007. *Mining Industry Energy Bandwidth Study*. Washington DC: BCS.
4. Allison, B., et al. 2016. *Australian Energy Update 2016*, 20–30. Office of the Chief Economist, Department of Industry Innovation and Science: Canberra, Australia.
5. Kecojevic, V., and D. Komljenovic. 2010. Haul truck fuel consumption and CO_2 emission under various engine load conditions. *Mining Engineering* 62 (12): 44–48.
6. Kecojevic, V., and D. Komljenovic. 2011. Impact of bulldozer's engine load factor on fuel consumption, CO_2 emission and cost. *American Journal of Environmental Sciences* 7 (2): 125–131.
7. Soofastaei, A., and J. Davis. 2016. Advanced data analytic: A new competitive advantage to increase energy efficiency in surface mines. *Australian Resources and Investment* 1 (1): 68–69.
8. Soofastaei, A. 2016. *Development of an Advanced Data Analytics Model to Improve the Energy Efficiency of Haul Trucks in Surface Mines*. Australia: The University of Queensland, School of Mechanical and Mining Engineering.
9. DOE. 2012. *Mining Industry Energy Bandwidth Study*, 26–33. Washington DC, USA: Department of Energy, USA Government.
10. Norgate, T., and N. Haque. 2010. Energy and greenhouse gas impacts of mining and mineral processing operations. *Journal of Cleaner Production* 18 (3): 266–274.
11. Darling, P. 2011. *SME Mining Engineering Handbook*, vol. 1. SME.
12. EEO. 2010. *Driving Energy Efficiency in the Mining Sector*, 18–22. Canberra, Australia: Australian Government, Department of Resources Energy and Tourism.
13. McCulloch, W.S., and W. Pitts. 1943. A logical calculus of the ideas immanent in nervous activity. *The Bulletin of Mathematical Biophysics* 5 (4): 115–133.
14. Chakraverty, S., and S. Mall. 2017. *Artificial Neural Networks for Engineers and Scientists: Solving Ordinary Differential Equations*. CRC Press.

15. Sivanandam, S.N., and S.N. Deepa. 2008. *Introduction to Genetic Algorithms*, vol. 3, 246–263. New Delhi, India: Springer.
16. Whitley, D., T. Starkweather, and C. Bogart. 1990. Genetic algorithms and neural networks: Optimizing connections and connectivity. *Parallel Computing* 14 (3): 347–361.
17. Velez-Langs, O. 2005. Genetic algorithms in the oil industry: An overview. *Journal of Petroleum Science and Engineering* 47 (1–2): 15–22.
18. Singh, A., and A. Rossi. 2013. A genetic algorithm based exact approach for lifetime maximization of directional sensor networks. *Ad Hoc Networks* 11 (3): 1006–1021.
19. Beigmoradi, S., H. Hajabdollahi, and A. Ramezani. 2014. Multi-objective aeroacoustic optimization of the rear end in a simplified car model by using hybrid robust parameter design, artificial neural networks, and genetic algorithm methods. *Computers & Fluids* 90: 123–132.
20. Sana, S.S., et al. 2019. Applying the genetic algorithm to job scheduling under ergonomic constraints in the manufacturing industry. *Journal of Ambient Intelligence and Humanized Computing* 10 (5): 2063–2090.
21. Yousefi, T., et al. 2013. Optimizing the free convection from a vertical array of isothermal horizontal elliptic cylinders via genetic algorithm. *Journal of Engineering Physics and Thermophysics* 86 (2): 424–430.

Chapter 17
Advanced Analytics for Haul Trucks Energy-Efficiency Improvement in Surface Mines

Ali Soofastaei and Milad Fouladgar

Abstract Rolling resistance as a part of total resistance plays a critical role in the productivity, fuel consumption, gas emissions, maintenance, and safety of haul truck operations in surface mines. This chapter aims to identify the most influential parameters on rolling resistance and complete an investigation about the effect of these parameters on the fuel consumption of haul trucks. This chapter has completed a comprehensive literature review to identify the influential parameters on rolling resistance. Through that process, 15 parameters have been identified. An online survey was conducted to determine the most influential of these parameters on rolling resistance, based on the knowledge and experience of many professionals within the mining and haul road industries. In this survey, 50 industry personnel have been contacted with a 76% response rate. The survey results have shown that road maintenance, tire pressure, and truck speed are the most important effective parameters on rolling resistance. In this study, based on the data collected from the literature review, the relationships between selected parameters and rolling resistance have been established. A correlation between the selected parameters and best performance fuel consumption for one type of common truck in Australian surface mines has been developed. As a case study, a computer model based on the nonlinear regression method has been created to find the correlation between fuel consumption and rolling resistance in a large coal surface mine in central Queensland, Australia. The relationships between the most influential parameters on rolling resistance and fuel consumption in this case study have also been developed. The case study results indicated that by decreasing the maintenance interval, increasing tire pressure, and decreasing truck speed, the fuel consumption of haul trucks could be decreased.

A. Soofastaei (✉)
Vale, Brisbane, Australia
e-mail: ali@soofastaei.net
URL: https://www.soofastaei.net

M. Fouladgar
Department of Mechanical Engineering, Najafabad Branch, Islamic Azad University, Najafabad, Iran
e-mail: milad_fouladgar@smc.iaun.ac.ir

A. Soofastaei (ed.), *Advanced Analytics in Mining Engineering*,
https://doi.org/10.1007/978-3-030-91589-6_17

Keywords Rolling resistance · Haul truck · Surface mine · Fuel consumption · Cost · Energy efficiency · Optimization · Advanced analytics

Introduction

Globally, the world marketed energy consumption was approximately 600,000 PJ in 2013 [1]. Australia consumed roughly 6000 PJ in the above period, with a forecasted growth of 1.1% over the next 10 years [2]. The mining industry annually consumes vast amounts of energy in operations such as exploration, extraction, transportation, and processing [3]. The Australian mining industry has observed a steady increase in energy consumption from 1976, consuming around 600 PJ of energy in 2013, or about 10% of the total energy consumption in Australia [2]. Many research studies and industrial projects have been carried out to reduce energy consumption in Australian mining operations [4, 5]. Haulage operations are one of the main energy consumers within the mining industry [3]. Of the total energy expenditure, loading, hauling, and dumping (LHD) operations represent approximately 60% of the total energy consumption in the Australian mining industry [6]. Service trucks, front-end loaders, bulldozers, hydraulic excavators, rear-dump trucks, and ancillary equipment, such as pick-up trucks and mobile maintenance equipment, are examples of the equipment powered by diesel engines used in mining operations [7]. Trucks in surface mines are used to haul ore and overburden from the pit to the stockpile, dumpsite, or the next stage of the mining process. According to the production capacity and the site layout, they are used in combination with other equipment such as excavators, diggers, and loaders [8]. The trucks used in the haulage operations of surface mines consume a large amount of energy, encouraging truck manufacturers and major mining corporations to carry out many research projects on the energy efficiency of haul trucks [5]. Understanding the energy efficiency of a haul truck is not limited to the analysis of vehicle-specific parameters. Mining companies can often benefit by expanding the analysis to include other factors that affect the energy use of trucks, such as effective parameters on haul road condition [9, 10]. There are a number of effective parameters on haul road condition that influence the energy used by trucks in a mine fleet, all of which need to be taken into account simultaneously for the optimization of fuel consumption. The consumption of fuel is dependent on many mine factors including the grade of the haul road, the rolling resistance, payload, speed, and truck engine characteristics [11]. By reducing the resistance a truck encounters during a hauling cycle, the overall fuel efficiency has the potential to be improved, without affecting cycle or productivity parameters [12, 13].

This chapter aims to use advanced analytics to investigate the effect of rolling resistance on haul truck fuel consumption and the key effective parameters affecting rolling resistance.

Haul Truck Fuel Consumption

Haul truck fuel consumption is a function of various parameters, the most significant of which have been identified and categorized into five main groups (see Fig. 17.1). The key parameters that affect the energy consumption of haul trucks include the truck characteristics, fleet management, haul road condition, mine plan, and environmental conditions, according to a study conducted by the Department of Resources, Energy, and Tourism [5]. In the present study, the effects of the Rolling Resistance (RR) on the fuel consumption of the haul trucks were examined. The RR is one of the main components of Total Resistance (TR). The TR is equal to the RR and the Grade Resistance (GR) sum when the truck moves against the haul road's grade [14].

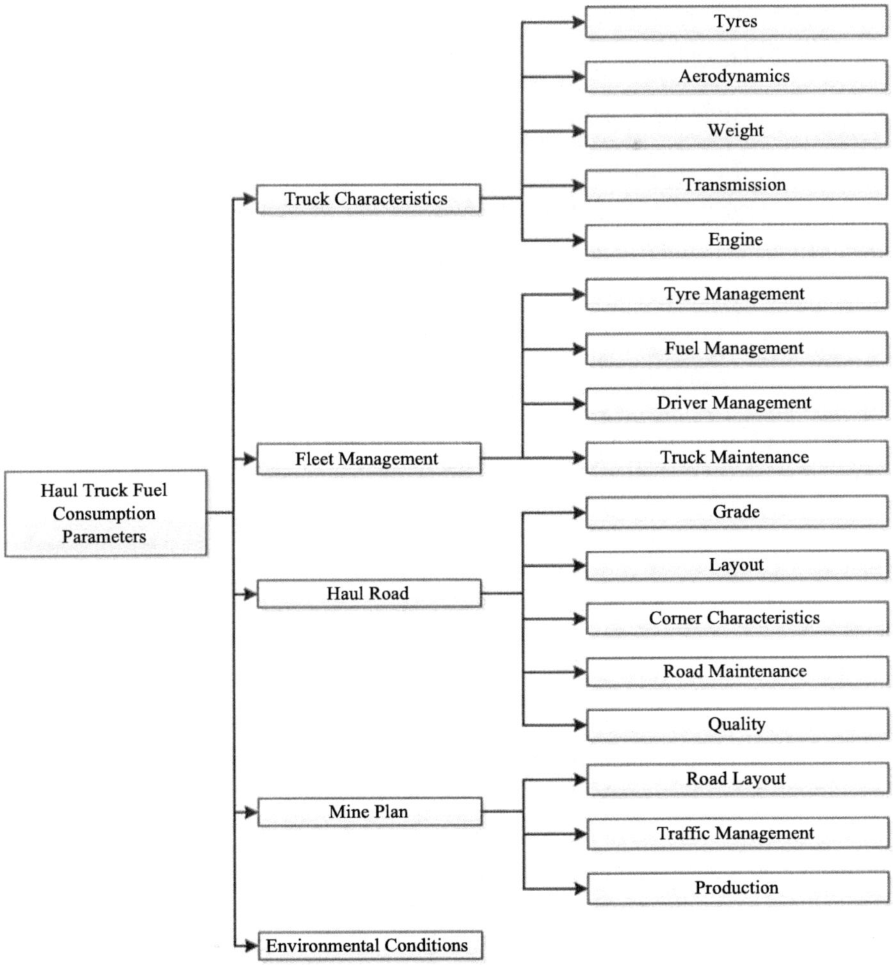

Fig. 17.1 Parameters affecting haul truck fuel consumption

$$TR = RR + GR \tag{17.1}$$

The RR depends on the tire and hauling road surface characteristics and is used to calculate the rolling friction force, which is the force that resists motion when the truck tire rolls on the haul road. The GR is the slope of the haul road. It is measured as a percentage and is calculated as the ratio between the rise of the road and the horizontal length (see Fig. 17.2). For example, a section of the haul road that rises 10 over 100 m has a GR of 10%. The GR is positive when the truck travels up the ramp and is negative when it travels down the ramp. The GR is positive for all the test conditions considered in this study, as the truck carrying the payload is traveling against the grade of the haul road.

Figure 17.3 presents a schematic diagram of a typical haul truck and the key factors that affect the performance of the truck, such as the Gross Vehicle Weight (GVW) (representing the sum of the empty truck weight and the payload), RR, Gradient (G), Rolling Friction Force (RFF), and Rimpull Force (RF). RF is the force available between the tire and the ground to propel the machine. It is related to the torque (T) that the machine is capable of exerting at the point of contact between its tires and the ground and the truck wheel radius (r) [15].

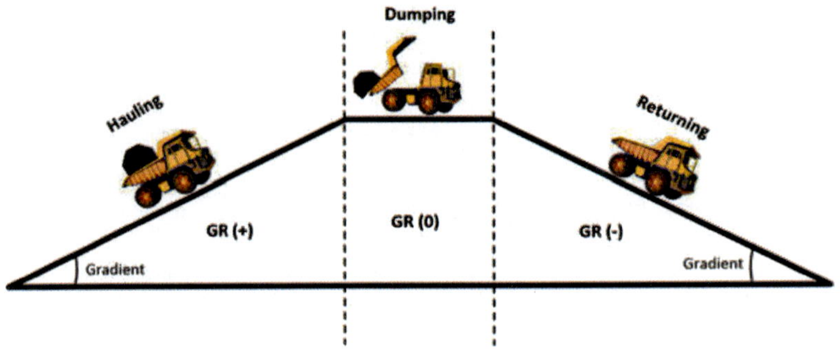

Fig. 17.2 Grade resistance (GR)

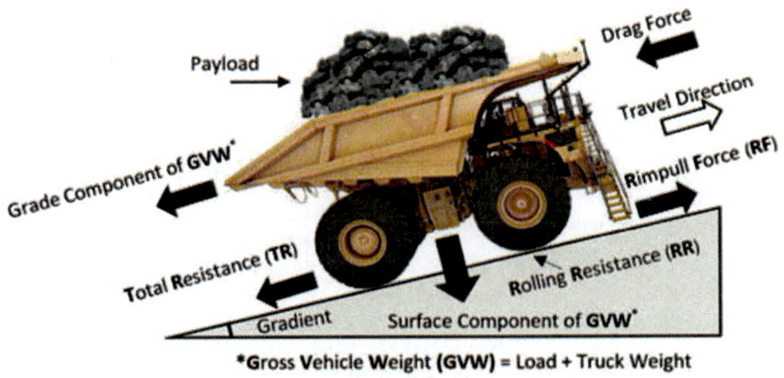

Fig. 17.3 A schematic diagram of a typical haul truck

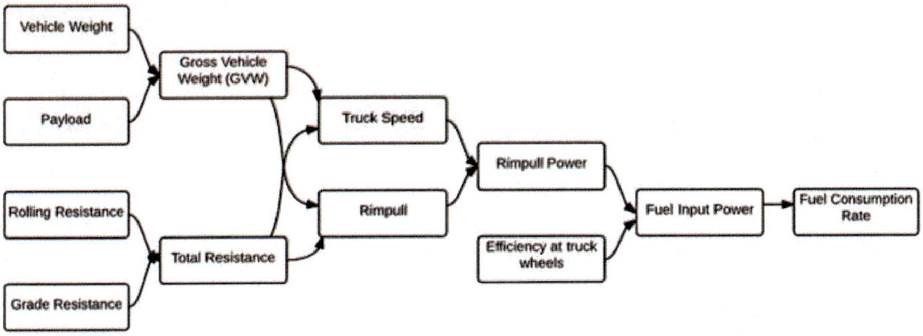

Fig. 17.4 Variable relationships required for truck fuel consumption estimation

$$\mathrm{RF} = \frac{T}{r} \tag{17.2}$$

Estimation of the fuel consumption rate requires many assumptions as well as calculations. Figure 17.4 illustrates the relationship between the haulage operation parameters and truck fuel consumption. It illustrates the variables to be initially defined and the values to be calculated to estimate energy consumption. Several input variables are initially required and include the vehicle weight, the weight of the unloaded truck and payload, or the weight of the material hauled by the truck. RR and GR are also required and are both measured as a percentage. RR can be estimated for the road or measured where possible.

In this study, a new parameter representing the fuel consumption by haul trucks has been defined. This parameter is the Fuel Consumption Index ($\mathrm{FC_{Index}}$). This index represents the quantity of fuel burnt by a haul truck to move one ton of mined material (Ore or Overburden) in an hour (L/h tons). The $\mathrm{FC_{Index}}$ can be estimated using the following Eqs. 17.3.

$$\mathrm{FC_{Index}} = \frac{F_i}{3600 \times P} \tag{17.3}$$

where F_i is Fuel Input Rate (L/s) and P is payload hauled by the truck (ton) [4];

$$F_i = \frac{\mathrm{PO_f}}{38{,}600} \tag{17.4}$$

where $\mathrm{PO_f}$ is Fuel Input Power (kW) [4];

$$\mathrm{PO_f} = \frac{\mathrm{PO_r}}{\mathrm{EEF}} \tag{17.5}$$

where $\mathrm{PO_r}$ is Rimpull Power and EEF is Energy Efficiency Factor [4];

$$PO_r = 0.28 \times S \times R \times g \quad (17.6)$$

where S is Truck Speed (km/h), R is Rimpull (tons), and g is the acceleration due to gravity (m/s^2).

Rolling Resistance

RR is defined as a measure of the force required to overcome the retarding effect between the tires and the road [16, 17]. This resistance is predominantly measured as a percentage of the GVW but can also be expressed as energy divided by a distance or a force [18, 19]. Tire RR can also be characterized by a Rolling Resistance Coefficient (RRC), a unit-less number [20, 21]. RR manifests itself predominantly in the form of hysteresis losses described as the energy lost, usually in the form of heat, when a section of vulcanized rubber is deformed regularly, such as during the operation of a haul truck [22].

RR can be measured by many different methods, detailed in the British Standard for Measuring RR [23]. Measurements can be made under laboratory conditions, generally on "test drum surfaces." This is a testing rig consisting of the tire to be tested and a drum with an irregular outer surface that can rotate, simulating the movement of the tire over a road surface. Many mathematical methods can then be applied to the values of Drum Torque, Power, and Tire Force measured during testing to determine the RR experienced by the tire [24].

Measurements of RR can also be obtained for specific mine haul roads using on-site testing. This method generally uses a specially designed trailer towed behind a truck to measure RR. A series of sensors attached to the trailer measure the force between the truck and trailer used to pull it across the road surface. They also measure the grade of the haul road and acceleration. This data is then used with the relevant mathematical expression to determine the RR of the haul road [25].

There are many effective parameters affecting RR, which can be categorized into four groups. These groups are Road, Tire, System, and Weather properties. Figure 17.5 illustrates the most influential parameters on RR. Haul road and tire properties are predictable, properties of the haul road and truck tires. System properties encompass operational parameters of the haul truck, and weather properties envelopes all parameters associated with weather conditions. These parameters are also categorized as a Design (D), Construction (C), Operational (O), or Maintenance (M) parameter. Table 17.1 illustrates the parameters affecting RR and the category to which it belongs.

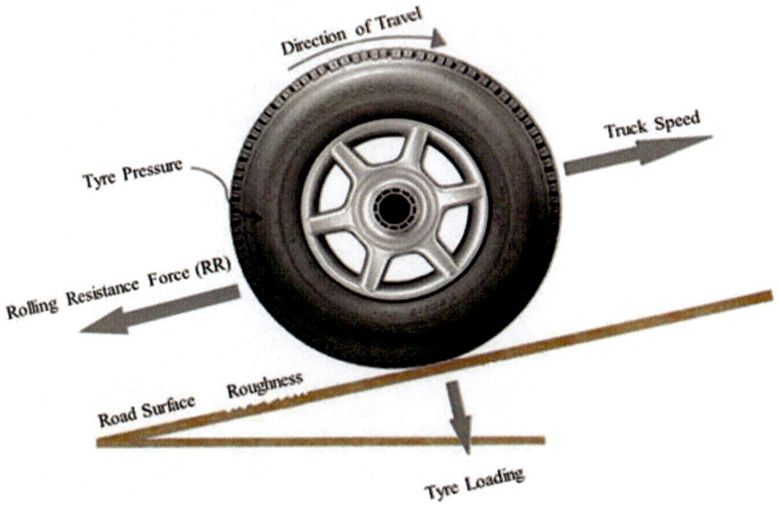

Fig. 17.5 Rolling Resistance and the most influential parameters

Table 17.1 Influential parameters on rolling resistance

Rolling resistance	Group	Category[a]				Parameter
		D	*C*	*O*	*M*	
	Road	✔			✔	Roughness
		✔	✔		✔	Defects
		✔	✔		✔	Material density
				✔		Moisture content
					✔	Road maintenance
	Tire	✔		✔	✔	Tire penetration
		✔				Tire diameter
				✔		Tire pressure
			✔	✔	✔	Tire condition
				✔		Tire loading
				✔		Tire temperature
	System			✔		Truck speed
				✔		Driver behavior
	Weather			✔		Humidity
				✔		Precipitation
				✔		Ambient temperature

[a]***D*** design, *C* construction, *O* operational, *M* maintenance

Table 17.2 Surface type and associated rolling resistance

Type of surface	Rolling resistance (%)
In-situ clay till	4–6.7
Compacted gravel	2–2.7
Compacted clay-gravel	3.9
Subsoil stockpile	4.4–8.3
Compacted clay till	4.1
The subsoil on mine spoil	7.3

Road Properties

Road properties are associated with the haul road itself, primarily the material forming the top layer or "wearing course" of the road [22–26]. Numerous studies have identified many road properties as affecting the RR experienced by trucks using the road.

The surface material of the haul road is identified as a major contributor to RR. Numerous studies have found that softer road surfaces with looser under-footing increased RR [25, 27–31]. Table 17.2 shows the results of one of these studies, displaying RR associated with surface type.

A study conducted by Sandberg [32] is concerned with effective parameters on RR. The focus of Sandberg's study is on-road roughness. Another study in this area completed by Mukherjee [33] shows that increasing the road roughness increased RR.

A study conducted by Thompson [9] is concerned with the impact of defects on RR. This study shows that the RR is also increased by increasing the Mean Profile Depth (MPD). Annand [34] presented an investigation about the estimation of RR. The presented results show that a higher degree of road compaction can decrease the RR. Thompson, in another study [18], investigated the effect of haul road maintenance on RR. The results of this study show that road maintenance plays a critical role in RR. The main objective of maintenance is to repair defects identified as significant contributors to RR and prevent them from occurring. The results obtained from Thompson's study showed that decreases in maintenance intervals or time between maintenance resulted in a decrease in RR.

Tire Properties

Tire properties are those associated with the tires of a haul truck and can be associated with the internal condition, tread properties, or operational characteristics. Many of these properties have been identified in many studies as influential on RR [32, 33, 35–40].

The Caterpillar research and development team completed a study about the effect of tire penetration as an influential parameter on RR [35]. In this study, a correlation between tire penetration and RR has been developed. The results show that by increasing the tire penetration, the RR increases. However, tire penetration is also affected by many parameters and varies depending on their influence. These parameters included tire pressure, where a lower pressure generally corresponds to increased tire penetration.

Tire diameter has been identified in several studies as a contributor to RR [23, 41]. The contact patch or area of the tire in contact with the road changes with tire diameter, and so this change in geometry results in a change in RR. A study by Sandberg [32] found that as tire diameter increased, RR decreased. The relationship was found to be constant among many different types of tires.

Tire pressure is a significant factor when assessing RR, with under or over-inflated tires displaying large changes in RR. A study found that increasing tire pressure resulted in decreasing RR [41]. Tire pressure is also affected by temperature, with a study by Paine [39] finding increasing tire temperature responsible for increased tire pressures.

Tire condition is also an important factor when considering RR, mainly manifesting in treadwear [42]. A study by Sandberg [32] found that tires in the worn condition with low tread height exhibited decreased RR. This relationship was observed among many different types of the tire.

Tire Loading is also considered in assessments and is the subject of several studies. Loading of a tire is affected by many other factors, including the vehicle and payload weights and operational parameters of trucks. Hall and Gai Ling [36, 38] found that increasing tire load resulted in increased RR. The results of Gai Ling's study [36] showed significant increases in RR with increasing tire load.

Tire temperature has been the subject of several studies, with its relationship with RR being the focus of experimental analysis [39, 41, 43, 44]. A study conducted by Janssen [44] found that increasing tire temperature resulted in decreased RR. The study used a unit-less factor, RRC, as the measurable representation of RR and decreased significantly with increasing tire temperature.

System Properties

System properties are those related to the operational parameters of the truck as well as environmental factors affecting truck operation. These are generally uncontrollable or dependent on the truck's operation and have been linked to RR by many studies.

Truck Speed (S) is a studied truck parameter, with experimental studies finding that increasing S resulted in increased RRC. Driver Behavior has not been the subject of extensive study concerning its direct effect on RR. It can, however, be linked to other parameters such as tire loading through basic physics. As a truck corner, it generates centripetal force, dependent on the mass of the truck, velocity, and radius of the turn [41]. Both turn radius and velocity are dependent on the driver's behavior. As a result, this behavior affects centripetal force, which affects tire loading [45].

Weather Properties

Whether properties are parameters associated with local conditions at a mine site, these are generally uncontrollable parameters and include temperature and the presence of rain or other environmental influences in surface mining.

Ambient or environmental temperature is an uncontrollable parameter in open surface mines and is considered due to its effect on tire pressure. Tire pressure has been identified as affecting RR. As a result, ambient temperature affects RR through its effects on tire pressure [44]. A study conducted by Michelin in France [37] found that an increase in ambient temperature resulted in decreased RR.

Rolling Resistance Parameters Selection

An online survey was conducted to determine the most influential parameters on RR, based on the knowledge and experience of many professionals within the mining and haul road industries. Fifty industry experts were contacted, and they answered with a 76% response rate. Of the experts surveyed, 12% worked in haul road planning, 48% in maintenance, 24% in design, and 16% in operations.

This survey allowed participants to estimate the influence of parameters identified as affecting RR. A score was assigned to each parameter between 0 and 100, representing the influence of a particular parameter on RR, where 0 is not influential, and 100 is highly influential. The survey results show that tire diameter has the lowest influence on RR, with a result of 40%. Defects, Tire Condition, Tire Temperature, Driver Behavior, and Ambient Temperature were all given rankings of approximately 50%. Maintenance, Tire Pressure, and Truck Speed were all identified as having the greatest influence on RR, with 80 and 90% scores. The remaining parameters all scored between 50 and 70% (see Fig. 17.6). Based on the collected data from the literature, the relationships between RR or RRC and Maintenance Interval (M) [18], Tire Pressure (TP) [32], and Truck Speed (S) [41] can be found in Figs. 17.7, 17.8, and 17.9, respectively.

Fuel Consumption Completed Correlations

This project has analyzed a real mine site dataset collected from a large surface mine in central Queensland, Australia. A sample of the collected mine site data is tabulated in Table 17.3. Caterpillar trucks are the most popular vehicles of the different brands

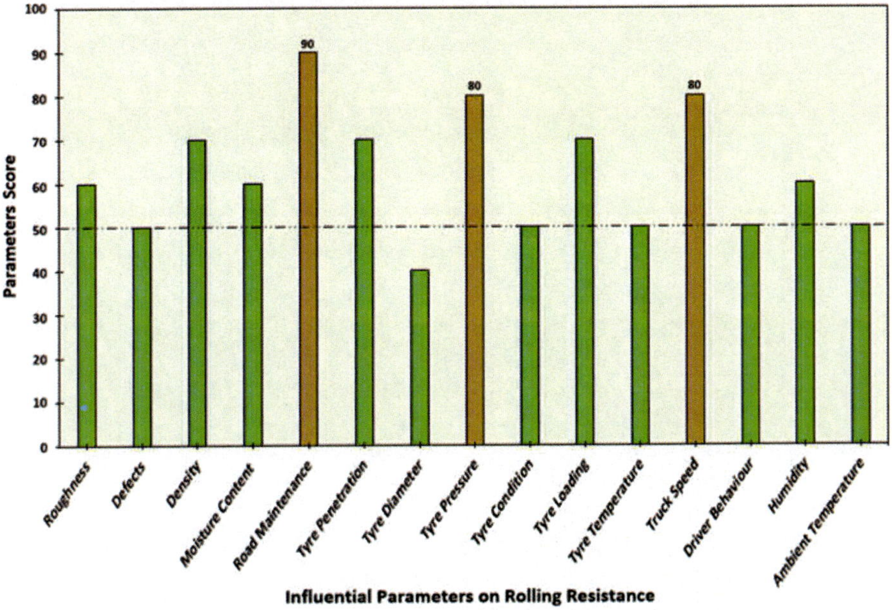

Fig. 17.6 Survey results

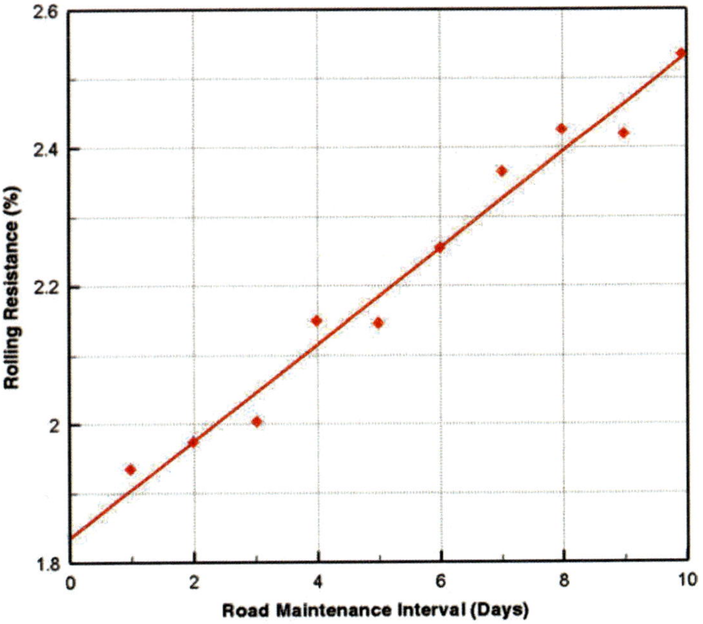

Fig. 17.7 Rolling resistance versus road maintenance interval [18]

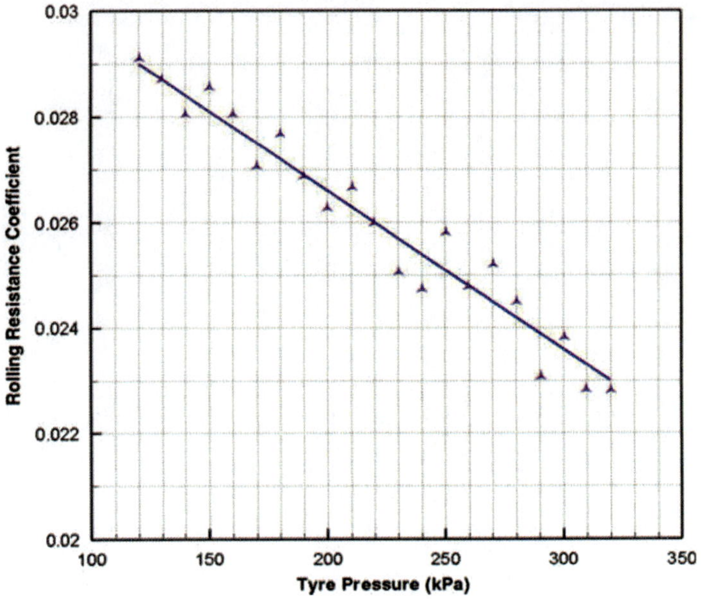

Fig. 17.8 Rolling resistance coefficient versus tire pressure [32]

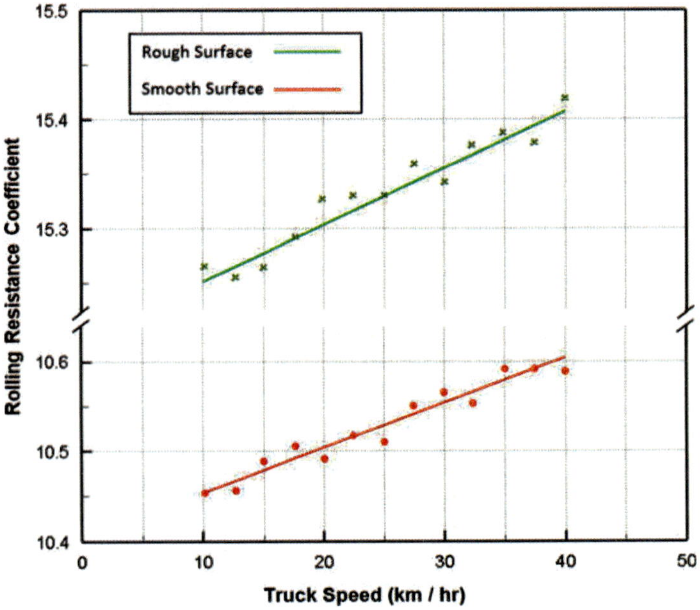

Fig. 17.9 Rolling resistance coefficient versus truck speed [41]

Table 17.3 A sample of dataset collected from a surface coal mine in central Queensland, Australia (CAT 793D)

Date	Payload (ton)	Truck speed (km/h)	Cycle time (h:m:s)	Cycle distance (km)	Rolling resistance (%)	Grade resistance (%)	Total resistance (%)	Fuel consumption (L/h)
23/01/2013	218.6	8.49	00:25:35	4.989	3.0	11.6	14.6	84.44
15/02/2013	219.4	11.39	00:16:17	5.150	3.0	8.7	11.7	90.26
13/03/2013	168.2	11.17	00:11:12	2.414	3.0	10.7	13.7	89.90
29/03/2013	158.9	14.04	00:17:42	5.150	3.0	9.1	12.1	93.78
22/04/2013	216.5	10.36	00:19:17	5.311	3.0	9.6	12.6	88.48
08/05/2013	202.1	12.06	00:18:45	5.311	3.0	9.4	12.4	91.28
25/06/2013	185.5	11.53	00:16:24	4.023	3.0	10.1	13.1	90.49
16/08/2013	175.9	11.94	00:18:48	4.667	3.0	10	13	91.10
07/10/2013	147.6	13.27	00:22:23	5.311	3.0	10.3	13.3	92.90
19/12/2013	214.3	11.58	00:17:55	5.150	3.0	8.9	11.9	90.56

used in the studied mine. Based on the power of the vehicle, mine productivity, haul truck capacity, and other key parameters, the Caterpillar CAT 793D (Table 17.4) was selected for the analysis presented in this study. Figure 17.10 presents the Rimpull-Speed-Grade curve extracted from the manufacturer's catalog [33]. This curve was used to determine the Rimpull (R) and the Maximum Truck Speed ($S_{\max}$) based on different values of TR for the different values of GVW.

Figure 17.11 demonstrates the relationship between Maintenance Interval and FCIndex. This relationship shows that the main solution for increasing the energy efficiency in haulage operation is decreasing maintenance. This Figure also shows that by reducing the maintenance interval from 10 to 5 days, FCIndex will be decreased from 0.65 to 0.4 L/h. ton. This amount of fuel savings can be a great opportunity for mine managers to reduce their operational costs.

Based on the completed online survey in this study, the second main effective parameter on RR is Tire pressure. Figure 17.12 illustrates the correlation between FCIndex and Tire Pressure for the CAT793D in different conditions. In Fig. 17.12, the relationship between haul truck fuel consumption and tire pressure has been completed for a normal range of tire pressures found for trucks in the studied surface mine. This relationship shows that by increasing the tire pressure, FCIndex will

Table 17.4 CAT 793D specifications

Feature	Value
Gross machine operating weight (GMW)	383,749 kg
Maximum payload capacity	218 tons
Top speed-loaded	54.3 km/h
Body capacity	129 m^3
Tires	40.00 R57

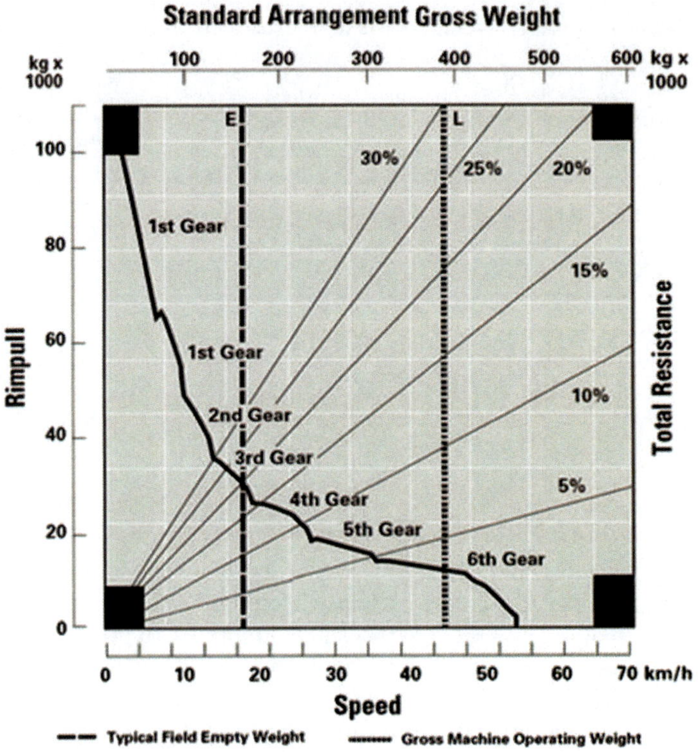

Fig. 17.10 Caterpillar 793D Rimpull curve [35]

be decreased sharply. Therefore, it is obvious that increasing the fuel efficiency in haulage operations will be possible by a regular pressure check.

The effect of *S* on FCIndex is illustrated in Fig. 17.13. The nonlinear correlation between FCIndex and *S* shows that fuel consumption by the truck increases by increasing the *S*. This Figure also shows that a correct approach for decreasing the fuel consumption is to decrease the total resistance, which could be achieved by reducing RR.

Conclusions

In this study, an investigation to identify the most influential parameters on rolling resistance was completed. After a comprehensive literature review, 15 parameters were identified, and an online survey was conducted to determine the most influential parameters on rolling resistance. In this survey, 50 industry personnel were contacted with a 76% response rate. Of the respondents, 12% worked in road planning, 48% in maintenance, 24% in design, and 16% in operations. The survey results revealed

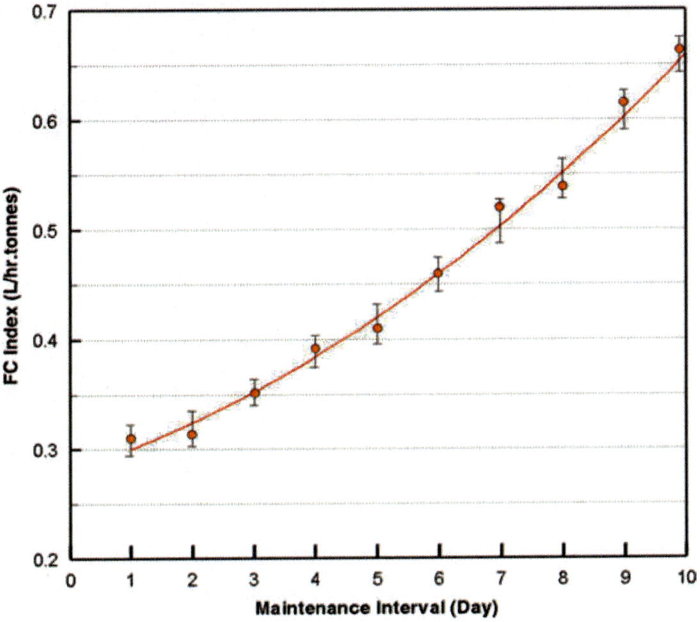

Fig. 17.11 Relationship between maintenance interval and FC_{Index}

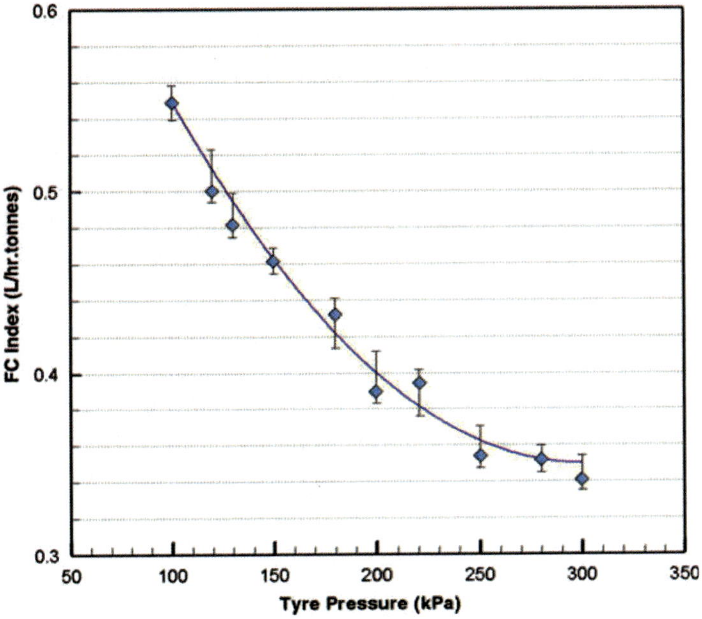

Fig. 17.12 Correlation between FC_{Index} and tire pressure

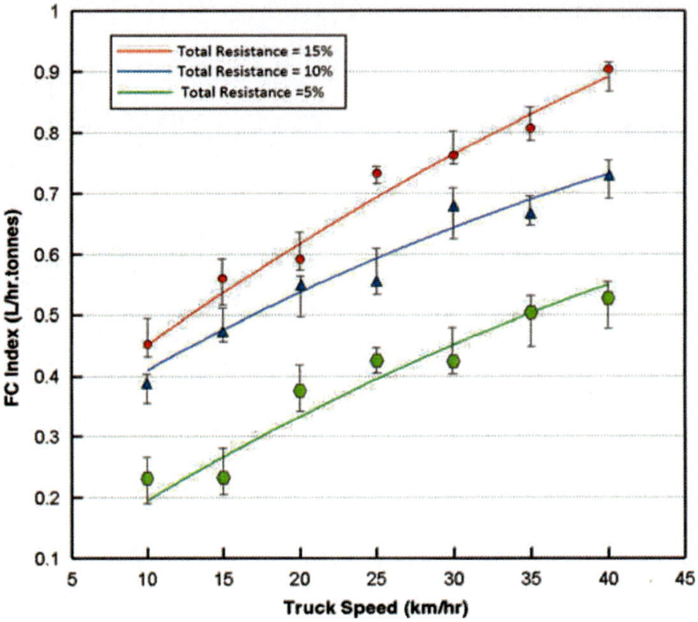

Fig. 17.13 The effect of truck speed on FC_{Index}

that road maintenance, tire pressure, and truck speed are the most important effective parameters on rolling resistance. The effect of these three main selected parameters was investigated on haul truck fuel consumption in a real mine site located in central Queensland, Australia. The nonlinear relationships between the selected parameters in the survey and fuel consumption in the real studied mine were developed. The results indicated that by decreasing the maintenance interval, increasing tire pressure, and decreasing truck speed, the fuel consumption of haul trucks could be decreased.

References

1. BP Energy Outlook 2035. 2014. 47–52.
2. BREE Australian Energy Update. 2014. *Bureau of Resources and Energy Economics*, 9–11.
3. DOE. 2012. US Mining industry energy bandwidth study.
4. EEO. 2010. *Energy-Mass Balance: Mining*, 21–28.
5. EEO. 2012. *Analyses of Diesel Use for Mine Haul and Transport Operations*, 2–12.
6. EEO. 2010. *Driving Energy Efficiency in the Mining Sector*, 18–22.
7. Stout, C.E., et al. 2013. Simulation of multiple large pit mining operations using GPSS/H. *International Journal of Mining and Mineral Engineering* 4 (4): 278–295.
8. Ghojel, J. 1993. Haul truck performance prediction in open mining operations. In *National Conference Publication-Institution of Engineers Australia NCP*. Australia: Institution of Engineers.
9. Thompson, V., and A. Visser. 2006. The impact of rolling resistance on fuel, speed and costs. *Continuous Improvement Case Study* 2 (1): 68–75.

10. Thompson, R., et al. 1998. Benchmarking haul road design standards to reduce transportation accidents. *International Journal of Surface Mining, Reclamation, and Environment* 12 (4): 157–162.
11. Franzese, O., and D. Davidson. 2011. *Effect of Weight and Roadway Grade on the Fuel Economy of Class-8 Freight Trucks*. Oak Ridge National Laboratory.
12. Goodyear Factors Affecting Truck Fuel Economy, 64–79 (2010).
13. Thompson, R. 1996. *The Design and Maintenance of Surface Mine Haul Roads*. University of Pretoria.
14. K. Burt, C.L.K. McShane, and O.T. Fong (Eds.). 2012. *Cost Estimation Handbook*, 2nd edn. Monograph 27. Australasian Institution of Mining and Metallurgy.
15. Assakkaf, I. 2003. *Machine Power. Construction Planning, Equipment, and Methods*, 123–142.
16. Descornet, G. 1990. Road-surface influence on tire rolling resistance. In *Surface Characteristics of Roadways: International Research and Technologies*. ASTM International.
17. Grover, P.S. 1998. Modeling of rolling resistance test data. *SAE Transactions*: 497–506.
18. Thompson, R., and A.T. Visser. 2003. Mine haul road maintenance management systems. *Journal of the Southern African Institute of Mining and Metallurgy* 103 (5): 303–312.
19. Iwashita, K., and M. Oda. 1998. Rolling resistance at contacts in the simulation of shear band development by DEM. *Journal of Engineering Mechanics* 124 (3): 285–292.
20. Thompson, R., and A. Visser. 2003. Mine haul road fugitive dust emission and exposure characterization. *WIT Transactions on Biomedicine and Health* 2003: 7.
21. Shida, Z., et al. 1999. A rolling resistance simulation of tires using static finite element analysis. *Tire science and Technology* 27 (2): 84–105.
22. Thompson, R., and A. Visser. 1997. A mechanistic, structural design procedure for surface mine haul roads. *International Journal o/Surface Mining, Reclamation, and Environment* 11 (3): 121–128.
23. Baafi, E. 1993. *Tyres, and wheels. Methods of Measuring Rolling Resistance*, 14–16. London: British Standards Institute.
24. McFarland, L.L., and B.D. Cargould. 1984. *Apparatus for Measuring the Rolling Resistance of Tires*. Google Patents.
25. Tannant, D., and B. Regensburg. 2001. *Guidelines for Mine Haul Road Design*.
26. Thompson, R., and A. Visser. 2000. The functional design of surface mine haul roads. *Journal of the Southern African Institute of Mining and Metallurgy* 100 (3): 169–180.
27. Holman, P., and I. St Charles. 2006. Caterpillar haul road design and management. St. Charles, IL: Big Iron University.
28. Lee, T.-Y. 2010. Development and validation of rolling resistance-based Haul road management.
29. Thompson, R., and A. Visser. 2007. Selection, performance and economic valuation of dust palliatives on surface mine haul roads. *Journal of the Southern African Institute of Mining and Metallurgy* 107 (7): 435–450.
30. DeRaad, L. 1978. The influence of road surface texture on tire rolling resistance. SAE Technical Paper.
31. Thompson, R.J., and A.T. Visser. 1999. Management of unpaved road networks on opencast mines. *Transportation Research Record* 1652 (1): 217–224.
32. Sandberg, U., et al. 2011. Rolling resistance: basic information and state-of-the-art measurement methods. Final version. 2011, Statens väg-och transportforskningsinstitut.
33. Mukherjee, D. 2014. Effect of pavement conditions on rolling resistance. *American Journal of Engineering Research* 3 (7): 141–148.
34. Anand, A. 2012. Scaled test estimation of Rolling Resistance.
35. Caterpillar. 2013. *Caterpillar Performance Handbook*, 10 edn, vol. 2. USA: US Caterpillar Co.
36. Ma, G.-L., H. Xu, and W.-Y. Cui. 2007. Computation of rolling resistance caused by rubber hysteresis of a truck radial tire. *Journal of Zhejiang University-Science A* 8 (5): 778–785.
37. Barrand, J., and J. Bokar. 2008. Reducing tire rolling resistance to save fuel and lower emissions. *SAE International Journal of Passenger Cars-Mechanical Systems* 1 (2008-01-0154): 9–17.

38. Hall, D.E., and J.C. Moreland. 2001. Fundamentals of rolling resistance. *Rubber Chemistry and Technology* 74 (3): 525–539.
39. Paine, M., M. Griffiths, and N. Magedara. 2007. The role of tire pressure in vehicle safety, injury, and environment. Road safety solutions.
40. Walczykova, M. 2001. Penetration resistance of the light soil influenced by tire pressure and the number of wheelings. In *Farm work science Facing the Challenges of the XXI Century. Proceedings Xxix Ciosta-Gigr V Congress*, Krakow, Poland, 25–27 June 2001. 2001. Wageningen Pers.
41. Popov, A.A., et al. 2003. Laboratory measurement of rolling resistance in truck tires under dynamic vertical load. *Proceedings of the Institution of Mechanical Engineers, Part D: Journal of Automobile Engineering* 217 (12): 1071–1079.
42. Grosch, K. 1996. The rolling resistance, wear and traction properties of tread compounds. *Rubber Chemistry and Technology* 69 (3): 495–568.
43. Ebbott, T., et al. 1999. Tire temperature and rolling resistance prediction with finite element analysis. *Tire Science and Technology* 27 (1): 2–21.
44. Janssen, M., and G. Hall. 1980. Effect of ambient temperature on radial tire rolling resistance. *SAE Transactions* 1980: 576–580.
45. Bellamy, D., and L. Pravica. 2011. Assessing the impact of driverless haul trucks in Australian surface mining. *Resources Policy* 36 (2): 149–158.

Chapter 18
Advanced Analytics for Mine Materials Handling

José Charango Munizaga-Rosas and Elmer Luque Percca

Abstract A material handling system in mining aims to move mineral materials/products timely, safely, and economically viable. Therefore, transporting material within the mine and outside requires a heavy dose of mechanization and machinery. Furthermore, there are different types of mineralization styles and geometries. Consequently, there is a great diversity of different machines/systems that serve the transport function in the mining industry. This chapter reviews some of those transportation options, and where appropriate, identifies opportunities for the application of advanced analytics within the domain. Mainly divided into surface and underground, there is a similar division provided in the chapter. However, other systems used outside the mine site are also discussed: pipelines, railways, and ships. These seem to have a significant impact, particularly in the relationship between a mine and the communities affected directly or indirectly by their presence. Finally, the chapter concludes with a small case study illustrating one specific technique used in mine transportation studies. Without going into great detail, the section attempts to give the reader an idea of how those analyses could contribute to their decision-making in material transport situations.

Keywords Materials handling · Advanced analytics · Machine learning · Discrete event simulation · Surface mining · Underground mining

Introduction

Transportation is an essential activity in every mining operation. The core of the mining process consists of moving material, first usually in the form of broken rock taken from the earth's crust and then potentially from intermediate stockpiles to the rest of the activities that transform the material both physically and chemically to

J. C. Munizaga-Rosas (✉)
Facultad de Ingeniería de Minas, Universidad Nacional del Altiplano, Puno, Peru
e-mail: jmunizaga@unap.edu.pe

E. L. Percca
Departamento de Ingeniería de Minas, Universidad de Chile, Santiago, Chile

A. Soofastaei (ed.), *Advanced Analytics in Mining Engineering*,
https://doi.org/10.1007/978-3-030-91589-6_18

produce concentrates or saleable product in final form. The number of steps that follow the initial breakage of the rock will depend on the mineralization type and style, the exploitation method, and the processing options chosen to transform the material into some form of saleable product. To illustrate the point, we need to highlight that in some cases, the product is sold as it is coming out of the mines, typical examples being bulk products such as coal, iron ore, or aggregates, which traditionally require little or no processing. In other cases, the final product is a highly refined metal in/or near-final form, typical examples being copper, nickel, gold, and silver. The one thing that all that mineral products share is that the movement of different types of materials is needed for their production. The aim of transportation systems in mining is to move in a timely, safe, and economical manner the materials required to produce mineral products and the movement of the finished product to buyers for them.

It is not difficult to quickly understand that moving large quantities of broken rock requires significant quantities of energy, particularly when they need to be moved over long distances. Early mining was a very labor-intensive activity when most if not all the transportation was performed using either human or animal energy with little mechanization. As technology developed, transportation systems in mining have readily adopted those advances to mechanize material handling in mining operations. Adopting technology has enabled mining companies to reduce the cost of transporting materials to produce mineral products, allowing them to exploit otherwise uneconomical mining projects.[1] In time, underground deposits have become deeper, open-pit mines larger, and production scale has grown by orders of magnitude from its humble origins.

Nowadays, mining transportation is a highly specialized activity with equipment specially developed to extract ore to be processed. These transportation systems also aim to promote a culture of safety, sustainable development, and respect for the environment. Transportation is a fundamental factor to ensure the continuity of operations of modern mining ventures. They should be carefully designed to consider the economic aspects and other constraints that make the decision-making process non-trivial.

It is worth mentioning that some critical goods such as fuel, explosives, giant tires, and drill rods also play an essential support role in the extraction process, and some readers may consider that they belong to the material transportation system. However, the provision of logistic services and the design of the support logistics has a complexity of its own and will not be considered in this chapter which will focus on the transportation of mineral products in different forms within modern mining operations. Nevertheless, for the sake of completeness, we can mention that

[1] More precisely, we need to mention that cut-off grade (COG) or the mineral grade that provides a break-even situation from a cost point of view, differentiates the mineral ore from the waste, i.e., the ore is material that after being processed and sold as metal can pay for its production costs at a profit. It is important to note here that this differentiation being established as a break-even condition depends on economic factors which include prices and production costs. Not only that, but they also include a whole series of other factors such as social, environmental, and government, that limit what can be extracted from the earth's crust. This subset of the whole mineralized body is known as "reserve," the mineralized body is known as "resource".

one possible way of dealing with some of the logistic complexity is to subcontract those to external entities that provide a guaranteed supply of those critical goods.

We also must bear in mind that the transportation system chosen by a mining company can expose society to some risks if not properly managed for some mineral products. For example, transportation of intermediate and final products could impact communities by exposing them to hazardous chemicals or minerals. Certainly, perception of a mining project could be negatively impacted if pollution, traffic, accidents, etc. are introduced into the lives of neighboring communities. Moreover, this contact with communities could become more significant as the geographic isolation of the mining operations increase, the reason being that isolated mining projects cannot afford to create their isolated transport infrastructure and must, out of necessity, share existing infrastructure (especially outbound logistics) with existing communities, thus irreversibly affecting them.

Depending on the type of product a mine produces, the outgoing logistics for finished/intermediate products could become a burden. This is particularly true for bulk materials, for which the transportation costs are a substantial component of the production cost, and for which the supply of materials in quantities and qualities agreed with the customer are specified in contracts that impose heavy penalties if the supply is interrupted or the quality is not met. Therefore, careful planning of extraction activities considering transport options and quality constraints for the finished product is essential to avoid those penalties.

Many operations have traditionally used rail to get the product to a port. However, in recent times, pipelines as a transport substitute have introduced more variety in outbound logistics. Nowadays, logistic networks enable multimodal transportation systems for which it is not strange to see interoperation of land and rail transport modes, usually enabled using multimodal transfer stations.

Regarding the future of transportation within the mining industry, it is believed that its importance remains at least equal, if not more significant, to the current levels. Increasing distances needing coverage for more isolated operations, deeper/larger deposits, and increased efficiencies in extraction methods will keep materials flowing through different transportation networks to supply the world with the metals and minerals needed for maintaining our lifestyle. In addition, further technological developments, which are currently disrupting the logistic function in other industries, are expected to make their way to mining, which implicitly requires constant technological vigilance and evaluation of newer alternatives.

Transport management has two main functions:

- appropriate choice and dimensioning of the transport system to be used, essentially the design of the transportation system; and
- planning of the material movements to be performed once a transport system has been chosen, essentially the optimal or near-optimal transportation system operation.

These two functions determine the economics of the transport system within a mining operation. Any decision-making process involving them must consider a variety of factors like as costs, limiting speeds, efficiencies, safety, reliability, etc., to name just a few. The number of interrelated decisions is overwhelming, and it is believed that it is impossible to go into detail for each of those within the constraints provided by a chapter. Hence, we try to illustrate the nature of the decisions, data requirements, constraints, etc., to the reader simply for some of the transportation methods/systems found in traditional mining operations.

As mentioned before, the chapter focuses on transporting intermediate/final mineral products within the mining industry. The rest of the chapter will describe the main characteristics of existing transportation alternatives in mining and discuss the opportunities for advanced analytical applications. The rest of the chapter is organized as follows: Sect. 18.2 discusses the archetypical transportation system in mining, i.e., shovel and truck systems traditionally used in surface mining operations. Section 18.3 puts the attention on the use of conveyor belts. Section 18.4 discusses the use of pipelines for the transportation of concentrates. Section 18.5 presents a discussion regarding the use of locomotives and railways transportation. Section 18.6 introduces maritime transportation options with a discussion of ships and shipments' role in outbound mining logistics. Section 18.7 goes to underground mining and some of the problems for those types of deposits. Section 18.8 focuses on a simple case study to illustrate the impact of using discrete-event simulation tools in a shovel and truck system for decision-making. Finally, Sect. 18.9 present some conclusions.

Surface Mining: Shovel and Truck Systems

Surface mining can informally be defined as the exploitation technique that extracts the mineral from the earth's crust advancing from the surface and going to deeper locations by creating a big crater-like manufactured structure, usually utilizing mechanized systems. More formally, open-pit mining consists of extracting material on the earth's surface to recover the mineral content or access the deeper mineral. This mining process causes the earth's surface to be continuously excavated, causing an increase in the pit's depth that will form until the end of the life of the mine [1].

It needs to be noted here that surface mining activities have the physical precedence of material within the reserve and need to consider the inclination of the walls created as the extraction progress (slope angle). If the extraction process does not respect this slope angle, it could cause the stability of the walls, potentially leading to the collapse of the wall with substantial economic and safety consequences for the operation.

The exploitation of an open-pit mine is usually operated in benches, which are not necessarily uniform in volume. The typical sequence of activities to move the material out of the pit for further processing or sending it to markets to be sold as final products are:

- drilling blast holes on each bench following a pre-designed pattern;
- fill the blast holes with explosives and primers;
- blast the bench using a sequence of firing within the available loaded blast holes;
- Using a shovel and trucks, move the broken material out of the pit. Each truck-load will have a destination based on the characteristics of the material.

Fig. 18.1 Open-pit at sunrise dam gold mine (Picture: Calistemon/Wikimedia Commons)[2]

- Before the current active benches (more than one active bench at any time) become exhausted, design the blasthole pattern for new benches and repeat the process.

The truck fleet and shovels are fixed in the previous process and must be allocated to perform tasks. Shovels are moved from bench to bench within the mine, and trucks are assigned tasks once they have delivered their load (either to the processing plant or the waste dump). Loading and transportation operations in open-pit mines have been the traditional subject of study due to the scale of the operations, its impact on costs, and the economic decision of allocating scarce resources to tasks to maximize the financial outcome of the operation.

Trucks and shovels systems are generally the bottlenecks in the surface mining production process. Not only that but the financial impact of the transportation system is also enormous. According to Alarie and Gamache [2], the transport of material is one of the essential components in an open-pit operation, representing, in the case of the shovel-truck system, around 50% of the operational costs and even 60% according to some authors. On the other hand, open-pit mining in the long term tends to lower productivity because of the deepening of extraction, so improving the efficiency and productivity of the equipment is essential (Fig. 18.1).

[2] Image file licensed under the Creative Commons Attribution-Share Alike 4.0 International license. Image date: 27 September 2009.

Fig. 18.2 Open-pit at sunrise dam gold mine (Picture: Calistemon/Wikimedia Commons)[3]

The shovel-truck system corresponds to the most widely used system for the material transport activity in open-pits worldwide because it is deemed one of the most flexible systems in terms of operation; additionally, it has a low initial investment cost other material handling systems. However, its operating cost is high due to the amount of equipment used [2]. Furthermore, the inherent uncertainties of mining operations coupled with regular updates of short-term plans, for example, after adding secondary information coming from blasthole drilling (named grade control), instill poor confidence in the plans [3], which in turn have an impact on the way the truck-shovel system is sized, deployed, and operated. Therefore, it is natural to conclude that one of the main problems in this system is the efficient use of the cargo and transport fleet to minimize costs and maximize efficiency (Fig. 18.2).

There are technical factors that need to be considered to decide the composition of the fleet. The work of Bozorgebrahimi et al. [4] focuses on how the size of the haulage trucks affects the performance of open-pit mines. Their method is centered around generating a set of long-range average cost curves for haulage trucks, basically by considering OEM data and further validated with empirical observations. One key conclusion stated in their work is that the haulage truck sizing decision does not consider the characteristics of the deposit.

Stated, the evidence suggests that bigger is not necessarily better. Larger equipment requires greater bench sizes which results in more insufficient dilution control and reduced selectivity, wider ramp requirements which result in a higher stripping ratio and shallower overall pit slope, coarser ore fragmentation is feasible with larger

[3] Image file licensed under the Creative Commons Attribution-Share Alike 4.0 International license. Image date: 12 October 2010.

Fig. 18.3 Bucyrus Erie 2570 dragline and CAT 797 haul truck at the North Antelope Rochelle opencut coal mine (Picture: Peabody Energy, Inc./Wikimedia Commons)[4]

equipment which impacts milling costs, and opportunity costs arising from machine failure are directly proportional to truck size.

Surface mining has been extensively studied, and we would venture to say that there are plenty of opportunities to achieve reasonable, advanced analytics solutions within this space. Different ideas are presented. However, the order in which they are presented does not indicate importance (Fig. 18.3).

One of the first essential problems that need to be addressed in the space is the fleet composition, both in numbers and characteristics of the equipment used. This problem has been around for some time, and there is no definitive solution to it. A good starting reference that the reader can access to see some current state-of-the-art survey is given by Burt and Caccetta [5]; their work presents a simplified mathematical model in the Appendix that illustrates the structure of the problem from an operations research point of view. For this problem, the standard data requirements related to the equipment that comes to mind should be (non-exhaustive list):

- equipment volume/weight capacity;
- equipment productivity/speed;
- availability of the equipment;
- maintenance needs;
- compatibility between different equipment types; and
- cost per unit of production.

[4] Image file licensed under the Creative Commons Attribution 3.0 Unported license. Image date: 18 November 2014.

However, other factors are equally important that affect the performance of the equipment. The most important of them is the mine environment. For example, the resistance of the ground to the torque of the tire affects the truck's forward motion. The design of the mine also plays a role as the inclination of the ramps and the location of the dump site and plant influence the equipment selection. The compatibility condition is an important one. To illustrate the point, consider the case where a haul truck can carry a nominal load of 315 tons and a shovel can load a nominal load of 60 tons. In this case, a simple calculation shows that with five passes of the shovel, we will load 300 tons (on average), which leaves 15 tons not utilized on average on every trip, which will definitively accumulate, not only that, but also the revenue referral associated to 15 tons for every trip will undoubtedly have an impact on the NPV for the project. The other possibility is to overload the truck and to use six passes which will give a nominal load of 360 tons. However, this action applied consistently will very likely increase the maintenance costs for the fleet substantially. The problem is further exacerbated if the swelling factor is considered in the calculations. The swelling factor indicates how the material changes its density when fragmented by blasting instead of intact in situ material. As the density changes, so do the volume it uses. In this case, the calculations need to be performed considering volume considerations, which will differ between mine sites. The fleet composition problem has another aspect of importance that requires consideration: the number of trucks and shovels that the project requires. This is a non-trivial problem due to the nature of the operations. As operations progress, they increase in depth, sustaining a production rate will require more trucks. Hence, the decision needs to be taken carefully.

Advanced Analytics Discussion and Opportunities

All the previous discussion calls for advanced analytics to characterize the data required to take those decisions properly. One of the most critical pieces of information required that has a bearing in the decisions relating to fleet composition is the appropriate characterization of load-and-haul cycle time and performance measurement. The whole system depends on having timely and curated information relating to the different times involved in the system's operation. Studies such as those presented in [6, 7] heavily depend on this type of data to produce answers. In particular, the link between equipment performance and its impact on mine plans is of importance. The usual technique to assess this impact is to use discrete-event simulation models to characterize the performance of a given fleet and to take design decisions regarding the fleet [8]. Also, a factor that influences the equipment's performance depends on the condition of the road surfaces. A poorly maintained road network will produce several problems such as excessive vibrations, dust emissions, which altogether accelerate the degradation of the trucks' reliability. There are specific options available to decrease the impact of this problem, such as watering the roads and scheduling regular road maintenance activities; however, the optimal point is difficult to reach as throughout the year, different weather conditions call for different measures. In any case, automated means to measure the quality of the haul roads within a mine seems to be an essential piece of information that could influence several decisions. However, analysis can become very complex. For example, the work presented in [9] uses artificial neural networks to identify road surface conditions by analyzing the vibrational signals obtained from equipment that must travel throughout those roads. In particular, the methodology performed very well in identifying discrete road faults such as bumps, depressions, or potholes. Unfortunately, the methodology could not be as effective in identifying the undulations present on a rough surface.

Another important application area of analytics is provided by equipment reliability studies [10], which identifies the aims of the research performed firstly to decrease trucks' sudden failures and breakdowns and also to improve the service life of the equipment; finally, another aim is to reduce maintenance costs. Reliability and availability are identified as two suitable metrics to measure the performance of a surface mining transport system. Reliability can be defined as the probability that the system operates without failure incidents within a given timeframe. On the other hand, availability is the probability that given equipment can operate at some timeframe. To measure reliability and availability, it is imperative to use historical data, as the operation of the transport system will be affected by the specific conditions of every mine site. It is necessary to fit those probabilities using historical data. If the mining project is new, then some form of analogous operation needs to estimate those probabilities. After the reliability and availability characterization is obtained, that information can be used to plan maintenance activities. There are two types of maintenance: corrective and preventive. Maintenance planning and reliability/availability characterization allow decisions related to spare parts procurement and, eventually, to schedule a repair plan for repairable parts. These are very traditional applications of complex data analytics in surface mining. Nevertheless, they can always benefit from improved techniques that give a fresh methodological view of the problem.

Other possibilities for advanced analytics in surface transport in mining are:

- Congestion modeling in road network: This can be used to inform dispatching to act based on traffic conditions (which vary throughout the day) instead of assuming average traffic conditions; and
- In-pit shift changes and further optimizations: Some mines perform shift changes at some predefined point, usually outside the mine. This introduces dead time in the regular operating hours of the mine as every activity needs to be stopped; the trucks need to travel to the reunion point where the change of drivers happens. One possibility to minimize the disruption of the shift change is to perform in-pit shift changes, which will require functional space in the pit to park the trucks. Along the same lines, progressive shift changes or other options can be analyzed to determine the less disruptive mechanism to perform such change; recall that the mine staff stays in specially located camps and that usually, transport to and from the mine site happens at certain predefined times during the day.

Dispatching algorithms: With the availability of GPS systems that inform the location of every truck and shovel at the mine site, the use of dispatch systems has become widespread in the industry. Most of the dispatching decisions are still taken manually. However, some attempts to automate the process using operations research (OR) models coupled with data analytics activities to feed the data to the OR models. This area represents an excellent opportunity to operate surface mining transport systems as it marries the operational characteristics of the fleet, availability, quality of the material being extracted, and mine plan. In particular, the development of operational models of the mine (so-called digital twins these days) enables experimentation with different potential future conditions to measure the resilience of the dispatch strategy. It is computationally involved; however, cloud computing and appropriate parallelization of algorithms help reduce the "solving" times, thus making it possible to approximate online systems. Some versions of similar ideas involve using video cameras and computer vision algorithms to allow the computer to understand equipment movements within the mine site and automatically extract operational characteristics fed into the operational model of the mine to assess the potential and robustness of the mine dispatching strategies.

Conveyor Belts

Conveyor belts are considered a potentially advantageous transport technology competitive and comparable to classical discontinuous haulage truck-shovel systems [11]. As opposed to the classical truck-shovel system, conveyor belts are a continuous material transport system. Therefore, despite being slightly more expensive in terms of CAPEX, mainly depending on the length of the transport system, the overall OPEX associated with conveyor belts should be smaller than traditional truck-shovel systems.

It may be thought that conveyor belts are a cheaper replacement for traditional haulage options. However, some factors influence the economics of the whole system that needs to be considered before deciding. On the one hand, conveyor belts are acknowledged as having less flexibility than the traditional discontinuous haulage system [12]. In this context, one salient problem is determining the belt conveyors distribution points (BCDP), which should be defined in terms of the mining sequence and location of the destination points for the material (ROM pad or waste dump). Although, on the other hand, the mine's geometric configuration will affect the design when using conveyor belts, and mines with steep slopes may represent some challenges.

Relative prices of energy and fuel are an important consideration too. One study places energy cost as being 40% of the operating cost of a belt conveyor system [13]. On the other hand, [11] indicates that about 40% of the total weight to move is the truck's weight, which, together with the fact that the return trip is empty, places the fuel consumption to move dead weight in the order of 60%. At this point of the discussion, it needs to be pointed out that conveyor belts allow the replacement of

Fig. 18.4 Conveyor belt at surface mine Schleenhain (Picture: Leonhard Lenz/Wikimedia Commons)[5]

large haul trucks, thus saving the associated fuel consumption, which is expected to play an even increasing role when meeting future carbon emission targets (Fig. 18.4)

A recent example of an open-pit mine transitioning to underground is Chuquicamata in Chile. The underground extraction will happen using the block caving technique. However, instead of using trucks to move the material from the current depth of the mine (about 850 m), a transport tunnel was built that covers 6.5 km which uses conveyor belts, overcoming an altitude difference of 1200 m. In total, when adding the distance to reach the treatment plant, Chuquicamata will use around a total of 13 km of conveyor belts.

The rock to be transported using conveyor belts needs to have specific maximum dimensions for the fragments (granulometry). This is usually ensured using mobile crushing equipment, which reduces the size of the material being handled to comply with the intended design specifications of the conveyor belt system. This extra element in the system allows the continuous transport of minerals and waste from the pit to its destination using the trucks-crusher-conveyor belt. The reader is encouraged to consult for a deeper review of in-pit crushing and conveying [14] (Fig. 18.5).

[5] Image file licensed under the Creative Commons Attribution-Share Alike 4.0 International license. Image date: 29 July 2018.

Fig. 18.5 Example of in-pit crusher (Picture: US Army Corps of Engineers Sacramento District)[6]

The design of conveyor belt systems requires determining the correct design and dimensions of the system components and, additionally, other parameters to guarantee smooth operation under loading and unloading conditions. Some components are the conveyor belt, motor, pulley and idlers, rollers, pneumatic cylinder, etc. According to [15], the design of a belt conveyor system should consider the following factors:

- dimension, capacity, and speed;
- roller diameter;
- belt power and tension.
- idler spacing;
- pulley diameter;
- motor;
- type of drive unit;
- location and arrangement of pulley;
- control mode;
- intended application; and
- maximum loading capacity.

[6] This image or file is a work of a US Army Corps of Engineers soldier or employee, taken or made as part of that person's official duties. As a work of the US federal government, the image is in the public domain. Image date: 22 November 2011.

Table 18.1 Comparison of advantages and disadvantages of using haul trucks or conveyors in mining (reproduced from [16])

Off-road truck	Conveyor
Advantages	
Lower investment cost	Long distances
More flexibility	High capacity
Short distances	High availability
No restriction for material size	Low cost of operation and maintenance
	Continuous production
	Easily extendable
	Correct environmentally
Disadvantages	
High operating costs and maintenance	High cost of investment
A large number of the equipment for high volumes	Less flexibility
High requirement of employee	Limitation on size of transported material
High maintenance of roads	Not safe for the operator

Cardozo et al. [16] present an excellent summary, in the form of a table, which we reproduce here comparing advantages and disadvantages between haul trucks and conveyors (see Table 18.1).

Advanced Analytics Discussion and Opportunities

The opportunities for advanced analytics in this transport system methodology are mainly centered around automation of specific tasks, which otherwise are work-intensive, manual, and subjective, to be performed by machines instead of human beings. One of such opportunities is the online inspection of conveyor belts using machine vision [17]. The conveyor belts are prone to failure in different modes: longitudinal rips, deviations, and surface damage. In any of those cases, the operation poses a severe threat that calls for corrective actions. The system integrates an image acquisition subsystem, algorithms for image processing, and finally, algorithms for target recognition, which is a very typical workflow when working with image processing applications. The main difference with other applications is that the system needs to be fast enough to monitor a rapidly moving belt, which imposes constraints at the algorithmic and hardware level for the proper acquisition of images. Once the system is in place, there are several options to identify features based on traditional algorithms, others based on more "modern" AI techniques. Another one consists of the use of non-destructive testing for automated damage detection [18].

Fig. 18.6 Conveyor belt at Boddington gold mine, WA (Picture: Calistemon/Wikimedia Commons)[7]

In the case discussed in this work, magnetic signals are acquired from the belt, using the known fact that the belt has an internal structure consisting of steel cords interlaced to provide strength to the rubber. Collecting the magnetic signals makes it possible to identify anomalies that damage the belt associated with those disturbances (Fig. 18.6).

Finally, automated mass flow measurement is another potential application in mining. The conveyor belt transports heterogeneous material. Particles of different sizes compose the material being transported by the belt. Using techniques based on image analysis and potentially some scanners to measure height profile in the belt, it is possible to estimate the volume and the granulometric composition of the material, which can be translated into mass flowing in the system. A similar setup can be used to identify alien objects in the conveyor belt to have a chance to remove them from the feed and avoid further problems in the plant. For example, some mines have remnants of old infrastructure with stopes built with wooden ties, which need to be removed before processing. Others have mesh and metal bars that also need to be removed before the material enters processing. For these cases, AI-based methodologies combining image processing can "see" the load on the conveyor belt

[7] Image file licensed under the Creative Commons Attribution-Share Alike 3.0 Unported license. Image date: 1 December 2009.

and decide when to activate an extraction mechanism to move part of the load away from the belt to isolate the offending elements.

Pipelines

Although financial performance is one of the critical drivers of decisions within the mining industry, other modifying factors need to be considered. One of such factors is the social environment in which mining projects are immersed. Therefore, just thinking and planning for excellent short-term profitability, not considering the social impact of the operations, could invariably lead to long-term harm. The reader may ask why this is being mentioned here in an advanced analytics and material transportation chapter. Because outgoing logistics play a significant role in the operation of mines, furthermore, with greater isolation of the newer project, many more kilometers need to be covered to have the mineral products reaching their destinations in the international markets.

One emblematical example, in connection specifically to pipelines, used in mining transportation, of how complex are social relations between the mine and the "host" community could become is given by the "Las Bambas" project, located in the Apurimac province in the Andes mountains in central Peru, which decided to change its original plan that contemplated building 206 km concentrates pipeline, changing it for an over the surface transport system consisting of transport by trucks on roads, to later connect this load with a railroad that would put the concentrates in the port of Matarani in the Pacific Ocean. This change in the initial planning, probably motivated by economic considerations, triggered a social conflict with the communities that were not close enough to the mine, which in turn received few benefits, but were exposed to potentially harmful chemicals due to this decision because the surface transport was meant to go through their communities. The conflict lasts to this day, in which successive road closures, strikes, and public disturbance organized by the local communities feeling insufficiently compensated and unnecessarily exposed, have created a considerable loss reported at tenths of millions of dollars for "Las Bambas."

Again, the reader may be asking why this previous discussion is essential concerning the use of pipelines as a mechanism for outgoing logistics from mines. The reason is simple and has a not-related cost: pipelines enclose the material to be transported, isolating everyone from potential contact with harmful chemicals or contamination of communities and ground water sources. A pipeline is a safe transport alternative, with definitive less pollution exposure, less dependence on the high-traffic public transportation corridors, and finally using less energy coming from fuels to move the concentrates. Also, it needs to be noted that pipeline transport is a good transport option for the case of mining projects that can benefit from a substantial height difference between the mine site and the port because the energy requirements are even less due to the positive impact the gravity will have on the transport of concentrates. It does not come as a surprise that emblematical projects in the

Fig. 18.7 Pipeline in Alaska (Picture: unknown author)[8]

Andes use pipelines to transport concentrate to the port: Antamina (Peru), Escondida (Chile), and Los Pelambres (Chile), to name just a few of them (Fig. 18.7).

The concentrate transport happens by the injection, at some pressure level, into the pipeline of slurry, which is a mixture of concentrate with water. The pipelines for slurry transportation are special in terms of their thickness. They need to have enough thickness to deal with the wear in the pipeline walls as the concentrate is particulate material, in some cases with the presence of some abrasive particles. For such cases, it is not uncommon to line the pipeline with high-density polyethylene (HDPE), or even in some cases, build the pipeline entirely of HDPE. Cristoffanini et al. [19] present the equations describing the movement of the concentrate slurry through the pipe. They are a transient form of the Navier–Stokes equations assuming that the slurry is slightly compressible. Therefore, we will not put the equations here. However, the reader must note that the Navier–Stokes system does not have an analytical solution and must be solved numerically. The fluid will behave in the pipeline, and the pressure profile of the slurry will depend on particle size distribution, viscosity, and pumping pressure.

Transport of concentrate slurry usually happens from the source (processing plant) to the sink of the system (port). In any case, the fact that the pipelines are safer does not mean that they are indestructible. For this reason, design specifications must usually comply with specific regulations dealing with the transport of hazardous materials. Furthermore, attention must be put into the flow within the pipeline as blockages

[8] This image is in the public domain under Creative Commons CCO license, "No Rights Re-served." Image date: 24 March 2017.

could happen (the same way that arteries in the human body accumulate plaque), affecting the flow of slurry and potentially leading to a collapse of the pipeline. For such a problem, the pipeline is washed to eliminate such blockages, thus imposing additional maintenance requirements to be scheduled within the system's operation. Also, it is important to note that ruptures within the pipeline could lead to leaks if remaining unnoticed, leading to the release of harmful chemicals to the environment with potentially catastrophic consequences. Hence, inspection is a significant cost driver in the operation of a pipeline system.

As we have seen, using pipelines brings many potential benefits. However, there are some disadvantages regarding their use that need consideration. The first one, and maybe very important to consider as it could be a deal-breaker relates to the high initial investment that a pipeline system requires. This initial investment will be justified if the economies of scale are such that the quantity transported pays this investment in time. Another disadvantage relates to the physical installed system. The volume of material that can be transported is bounded by the design specification defined by variables such as speed of slurry, pressure, specific gravity of the material, abrasiveness, and viscosity. It also must be noted that equipment must precondition the material to be mixed with water to be sent through the pipeline. This is usually the case with concentrates that already come out from the plant in a finely ground powder. However, other materials such as iron ore may not require too much comminution if using traditional transport systems requires additional equipment before mixing it with water (or some other liquid) to send it through the pipeline. Finally, and not for this reason least important, the consistent production of slurry requires liquid availability in appropriate quantities in the source of the pipeline and very likely a separation process in the loading port to separate the liquid from concentrate, which calls to environmentally friendly options to dispose or remediate the liquid, which depending on the nature of the mineral being transported could be a challenge of its own.

Advanced Analytics Discussion and Opportunities

There are some opportunities to build some exciting models around pipelines operation. One of such options is using AI models to model slurry hold-ups [20]. In any pipeline, a hold-up is a process by which different flow phases travel within the pipeline, causing some of them to be held by the quicker advance of some fluid phases. Such differences in phases could be due to differences in hydrostatic pressure, gravity, or chemical characteristics of different components in the heterogeneous liquid. This is a complex problem as the movement through the pipeline is affected by this behavior, and if too much pressure is put into the pipeline, waves and turbulence are formed, producing disturbances of the flow.

Another opportunity for advanced analytics is the use of experimentation and least squares models to characterize wear resistance [21]. Such models could be used to plan and apply proactive maintenance activities in the pipeline, thus avoiding the appearance of leaks in the pipeline. In particular, the selection of wear-resistant materials should be preferred to provide a cost-effective solution for the pipeline transport needs of a mining operation.

Another area of interest that represents an opportunity to apply decision models in transport systems based on pipelines is automated leak detection [22]. Leakage is classified into two main types, namely, rupture and background. Rupture leaks occur instantaneously due to increased flow over the maximum pipe capacity or excessive stress on the pipe wall. On the other hand, background leaks are more difficult to identify as they relate to the pipe's progressive deterioration, mainly due to corrosion. On the other hand, rupture leaks occur with a noticeable pressure reduction in the pipeline, so they are easy to identify by using sensors. The usual methods to identify the leaks range from basic visual inspections to more sophisticated techniques such as methods based on mass/volume balances or acoustic waves.

On a different tone, the location of pipelines is an interesting problem [23, 24]. On such problem class, a pipeline joining points A and B in a "map" needs to be designed, but considering the topography of the area, exclusion zones (like cities or national parks), land use, reliability and maintenance data, unit costs of procurement, probability of natural disasters (e.g., earthquakes), crossing with rivers, forests, etc. One objective of such a problem is to design a pipeline with minimum cost that joins A with B, satisfying all constraints, but there are versions of the problem where multiple objectives could be considered simultaneously.

Locomotive and Railways

Everyone knows locomotives, and probably, the reader has already experienced this method of transportation. In its simplest possible definition, a rail transport system has three main components: the locomotive, the wagons, and the railway. This transport system enabled the industrial revolution as it provided a transport option to move goods in greater distances, thus allowing goods exchange and specialization o labor. Since its inception, mining has taken advantage of the economies of scale that a rail system provides. Notably, the development of trains enabled the development of mines located far from ports or industrial centers, thus disrupting the financial performance of the mining ventures, allowing the exploitation of more isolated deposits.

The locomotive engine could be based on one of the following types of energy sources: battery, trolley, and diesel. The traction capability of the locomotives defines the efficiency of railway transport, which is affected by many factors, primarily the adhesion coefficients of the wheels to the lines, which depend on the operating conditions [25] (Fig. 18.8).

One of the essential operating conditions that affect the friction characteristics of the wheel–rail interaction is the weather (climate) conditions. Although theoretical and experimental studies have revealed that the lugs of the more solid surfaces penetrate the surface with the lower hardness, this mechanism for the interaction of the

Fig. 18.8 PeruRail's EMD GT42AC 812 and 808 haul a container train from Matarani toward the "Las Bambas" mine. The train is pictured just before arriving at Km 99, where a temporary facility is used to transfer the containers from rail to trucks for the remainder of the distance (Picture: David Gubler/Wikimedia Commons)[9]

rough surfaces points out to the study of the rail properties as they have a slightly higher hardness than the wheels [26] (Fig. 18.9).

The capital expenditure associated with constructing a railway transport system is usually significant; its construction takes time, and once installed, it tends to remain installed with little or no changes to it. Such significant investments in infrastructure are usually the result of careful planning made and promoted by governments to change the economic conditions of entire regions. Unfortunately, these decisions produce externalities which in the case of mining are particularly important. Usually, communities grow around train lines or in connection to train lines, and for that same reason, special consideration must be put into the transport of hazardous material. There are documented cases of contamination that can be attributed to mining related to the transport of the concentrates. The study of Heyworth and Mullan [27] investigates the lead and nickel contamination that the port town of Esperance in Western Australia has suffered in the past. They describe how a significant number of birds dying in the summer of 2006/2007 discovered elevated lead levels in the kidneys, livers, and brains of autopsied birds. This contamination was documented after the decision, in April 2005, by the Esperance Port Authority to begin shipping lead carbonate through the Esperance port. The lead was transported into the Esperance port via rail from a mine site near Willunga, around 900 km north of the town port. A

[9] Image file licensed under the Creative Commons Attribution-Share Alike 4.0 International license. Image date: 9 July 2017.

Fig. 18.9 Iron ore train leaving the Brockman 4 iron ore mine (Picture: Calistemon /Wikimedia Commons)[10]

similar problem has been observed in the Antofagasta port in Chile, where high lead levels have been detected in children, and inhabitants close to the lead stockpile are located where a train coming from Bolivia stocks the lead as part of an international treaty signed by the countries after the end of a war that involved them both [28] (Fig. 18.10).

As the infrastructure is usually fixed, the associated CAPEX to install the capacity is enormous, and another factor to consider is the location of the railway system concerning the deposits. As should be known, deposits are finite but fixed in location. This means that there is a continuous discovery of such deposits over time, with their location moving away from populated centers. This means that there is very likely a need to access the railway first from the mine. This rail load will then arrive at the port where the material is stocked and loaded into ships. This is an example of an intermodal transportation system that poses some challenges as they require excellent coordination between the different transportation modes that compose the system. Furthermore, from an economic point of view, having material sitting at an ore stockyard in a port facility exposes the company because depending on the way

[10] Image file licensed under the Creative Commons Attribution-Share Alike 4.0 International license. Image date: 26 July 2019.

Fig. 18.10 Iron ore stockyard (Picture: Nacho Manau/Wikimedia Commons)[11]

the ore is stored, it could be exposed to the elements. Otherwise, special stocking facilities need to be put in place, increasing the cost for the outbound logistics.

Advanced Analytics Discussion and Opportunities

A recent trend from different mining companies is to develop unmanned locomotive systems to reduce the reliance on human operators and to have better coordination and execution of tasks. When the locomotives are controlled by human drives, particularly in underground environments, the mine dispatching center has problems in establishing timely communication with the drivers, which leads to low efficiency in the operation of the system [29]. The use of autonomous systems gives the opportunity of collecting good volumes of information that need analysis for the optimal operation of the system. Here, there is a big opportunity for the use of advanced analytics within the transport systems in mining. Some examples of the potential opportunities and applications that come to mind are:

---

[11] Image file licensed under the Creative Commons Attribution-Share Alike 3.0 Unported license. Image date: 1 August 2003.

- predictive maintenance;
- characterization and prediction of operational characteristics under different weather conditions;
- operations research models for trip scheduling and operation of intermodal systems;
- network design;
- last-mile logistics; and
- estimation of effective network capacity.

Material Shipment

The minerals produced by mines are rarely further processed into final product form in the countries that own the mines. Usually, the consumption of those goods to produce final goods requires shipping minerals across vast distances. Archetypical examples of strategic commodities belonging to the domain of the non-renewable resource are iron ore, coal, and LNG. Then, of course, there is the transport of less bulky minerals, which are usually transported in the form of concentrates, for example, copper, nickel, aluminum, etc., but, if mining company or producing country/region has access to local smelters, then it is possible to ship final product in the form of ingots which is possibly the best way of transporting mineral products as the concentration of the commodity in the ingot is the highest possible, thus reducing the movement of deleterious elements and wastage and increasing the effectiveness of the transport function. It needs to be pointed out that smelters produce many negative environmental externalities, and current environmental legislation in most countries will make it very difficult to build a new smelter.

In some cases, the social opposition will also play an essential role in jeopardizing the smelter projects. Not every country has smelters for every mineral product, which means that they will need to be transported to overseas locations in some cases. In some other cases, it does not make economic sense for a company to build a smelter due to the limited life of their mine, and the cheapest transport option to a smelter is by ship. Whichever the case is, the increased demand for mineral products has created increased maritime traffic with all the implications that bring into the worldwide economics, port destinations, and communities (Fig. 18.11).

Apart from some specific types of vessels needed to transport certain types of commodities (iron ore, coal, or LNG tankers), there is not much difference between transportation in mining and other industries, e.g., grain and cereals, cattle, etc. However, some differences are due to the infrastructure required to warehouse some of the goods, or in the case of LNG products, ports are liquefied to be loaded in the tankers for transport to face a regasification process in the destination port, which calls for specific facilities in some ports to allow unloading of the goods.

Before going into the discussion of the possibilities to develop advanced analytics applications in connection with maritime transport, we need to point out that one of the main problems that can be associated with the transport of mining products to

Fig. 18.11 Fortescue's Herb Elliot Port, Port Hedland, WA (Picture: Fortescue Communications/Wikimedia Commons)[12]

different ports relates to the hazardous nature of some of the products. As it was pointed in the previous section of this chapter, the transportation and warehousing of concentrates have an impact on the surrounding environment, and there is the mobilization of particles to the environment, and in some cases, this produces serious health concerns, e.g., lead contamination of land and rainwater tanks in Esperance in Western Australia or lead exposure in Antofagasta, Chile. Material handling at port facilities should consider the harmful impact of the materials that require shipment and the route that material must cover before arriving at the port.

Advanced Analytics Discussion and Opportunities

There is consensus in the maritime industry that ships generate high volumes of information [30–32]. Furthermore, it is acknowledged that the environmental conditions under which the vessel must travel are dynamic and ever-changing [30]. Under these conditions, travel planning is a significant opportunity for data analytics. Travel planning relies on the analysis of historical information for past trips of vessels, which inform among others: automatic identification system (AIS), long range identification and tracking (LRIT), radar tracking, and earth observation [33]. This overload

[12] Image file licensed under the Creative Commons Attribution-Share Alike 4.0 International license. Image date: 21 June 2016.

of information makes manual processing to identify anomalous behavior an infeasible endeavor, and this calls for automation of the analysis of such information. For example, AIS ping signal information contains the vessel state vector and other kinematic information (e.g., latitude, longitude, speed, course), voyage related (e.g., destination, estimated time of arrival, etc.), and static information (vessel size, ship type, etc.) [33]. One great opportunity with this type of information is the mapping of maritime routes based on historical information. As the AIS information is received periodically with a specific frequency while the vessel is traveling and less frequently when it is stationary, this information can characterize a trip and correlate it with weather events or other disruptions of regular routes. Furthermore, analyzing the historical routes taken makes it possible to build estimates for trip duration and cost, which play a significant role in planning activities involving maritime transportation [34, 35]. This is a massive opportunity for the use of advanced analytics tools.

To illustrate the type of trade-off, consider maritime transport of a product such as LNG, which is a cryogenic cargo liquefied at temperatures of around − 160 °C. During the transportation of this type of cargo, there is a build-up of gas (boil-off gas), which is usually used to generate electricity in the LNG tanker transporting the gas. However, as the process is a continuous one, then at some point, the load will have evaporated so much LNG that it is no longer financially profitable to transport the gas. For this reason, in this case, the time to port estimates is essential because they heavily influence the financial performance of the shipment operation. Any perishable cargo faces a similar situation in nature, so it is a widespread problem in the maritime transportation industry. For such applications, the knowledge of routes and performance in terms of time and cost is invaluable for the planner, who could greatly benefit from using a mathematical model to make optimal freight transportation decisions. The reader interested in getting a broader perspective of the operations research approach's models into freight transportation is encouraged to consult [36] (Fig. 18.12)

As another topic of interest in this area that presents opportunities for applying advanced analytics, we need to consider supply chain integration. Usually, maritime transportation is the last link of the logistic chain that starts in the mine; however, it depends on coordination with all the previous links within the supply chain. Recall the case of "Las Bambas" in Peru, discussed in the previous section. The Matarani port receives copper concentrates unloaded from a train loaded at a transfer station in the mountains, supplied by road transportation from the mine. If either one of the previous links fail, be that the railway or the road transportation, then the material in the port stockyard will eventually run dry, and there will be nothing to continue the loading.

Furthermore, ports are usually managed by a respective port authority that establishes rules of operation within the port with some associated fees for concepts such as berthing, wharfage handling, insurance, extra handling charges for special cargo, charges for transshipment cargo, hire of cranes, and mooring and unmooring,. Not only that, but there is also a whole universe of paperwork, both legal and forms, that every cargo requires, for example, inward cargo manifests, outward cargo manifests, certificates, contracts, etc. [37]. Some of these charges are on a per-day basis

Fig. 18.12 Iron ore ship loader in action (Picture: Chad Hargrave/Wikimedia Commons)[13]

schedule, which means that keeping the vessel waiting to be loaded is probably not optimal as the cost quickly accumulates. This level of coordination certainly requires curated information available to all stakeholders involved to produce a better plan for this logistic function.

To finalize this section, we consider the work of Bag et al. [38], which concludes that industry 4.0 enablers bring positive operational changes in businesses' logistic networks. Industry 4.0 involves vertical and horizontal integration within specific companies and inter-firm integration with suppliers and customers. Furthermore, industry 4.0, also named the fourth industrial revolution, drives digitalization and intelligent systems, thus ensuring a seamless flow of information across the whole logistic network, being enabled by digital technologies such as RFID devices, intelligent objects, GIS, GPS systems, and wireless sensor networks, among others. These promising developments quickly find their way into mining, and impacts like those experienced in other industries are expected soon.

[13] Image file licensed under the Creative Commons Attribution 3.0 Unported license. Image date: 16 May 2011.

Underground Mining

Underground mining operations are not as standardized as surface operations. There are different extraction methods, each one better suited to specific types of mineralized bodies. Methods are selected based on characteristics of the mineralization such as depth, inclination, extension, and rock competency. This method-dependence characteristic of underground mining makes it a challenging problem, from a design and planning point of view, as every deposit is particularly unique with its characteristics which are highly tied to the nature of the mineralization but that have an impact on the way the ore is extracted and transported for further conversion into saleable mineral products (Fig. 18.13).

The haulage of materials in underground mining, which requires diverse equipment, consistent with the chosen exploitation method, and mainly due to the selective nature of underground operations, will represent a high percentage of the operating cost for the mining project. The haulage system is critical in every mining operation, be that an underground or surface one, but due to the specific constraints and nature of the underground operations, the criticality is perceived to be different in the underground context.

With selecting the mining method comes the selection of equipment, personnel requirements, maintenance planning, and an extensive list of other considerations that impact CAPEX and OPEX for a mining project. A classical reference for equipment selection covering both surface and underground mining is given by [39]. At a fundamental level, underground mining activities require a definition regarding different loading and carrying machines, considering size as a factor, constrained by the functional space available in the mine, which requires optimization in the context of the outbound transport system to maximize throughput. In the underground case, as in the surface case for mining, outbound material handling usually becomes a bottleneck, and the transportation capabilities are usually used to capacity.

As opposed to surface transportation systems in mines, the underground transport system consists of two main stages: primary and secondary. The primary stage involves transporting material from the production front to transfer points. The secondary stage usually consists of transportation from intermediate stocking points to the surface. Some underground processing activities are added in-between stages, mainly comminution processes but on occasion metallurgical processing. For the case of surface mining, transportation essentially moves the material out of the mine in one single stage, where it can potentially be stored in stockpiles but out of the mine in any case. In underground mining, it is very uncommon, save some exceptions, to directly transport the material out of the mining front to the surface outside the mine.

The material mined in the underground operation is usually transported vertically to the surface, whether upward or downward. However, depending on the location of the mine, in some cases, horizontal transportation out of the mine can become the main transportation mean, but not the only, as there is still some degree of vertical transportation needed to reach this horizontal transportation level. Vertical transport options include shafts, vertical conveyors, skips hoisting, inclined shafts or

Fig. 18.13 Hoisting tower of the Podsadzkowy shaft of the former KWK Bytom (Powstańców Śląskich) in Radzionków, Poland (Picture: Klaumich49 /Wikimedia Commons)[14]

inclined drifts, cages, and gravity flow transport. Horizontal transportation options include railroad, monorail, belt, rail conveyors, and trucks. In addition, other materials handling systems such as loaders are required for any transportation modes, mainly including load-haul-dump (LHD) machines.

Loaders are the main equipment used in underground ore handling systems. Capable of extracting ore from the extraction front and dumping it directly into an ore passageway, a loading dock, or into more significant transportation equipment (such as trucks) that takes the ore out of the mine. There are different loaders, such as rail rail-mounted, rubber rubber-mounted, track track-mounted, shuttle loaders, and tire-based loaders (LHD), mainly used in mines with hard rock characteristics, which can be diesel or electric powered.

LHD diesel equipment, which has been used to transport material in underground mining for decades, is considered a versatile and very mobile piece of machinery. It is claimed to have better traction, faster haulage speeds, increased productivity, and lower operating costs than its electric-powered counterparts. However, the same diesel LHD possesses high capital costs, more stringent maintenance requirements, noisier operation, emissions of toxic gases, and particulate material [40]. Electric LHDs have not been very common in underground mines [41]. However, they generate less heat which leads to fewer ventilation requirements and definitively fewer gas emissions. Furthermore, electric LHDs can be powered by batteries, power lines, or power cables, with each feature having a different impact on usage [41]. List the possible ways LHD equipment may receive electrical energy:

- from onboard battery pack;
- through umbilical power cable;
- from an overhead trolley line;
- using hybrid electric drive; and
- from fuel cells.

In underground mining, trucks have been widely used for long-distance transport of material from loading zones to transfer stations or directly to the surface [39]. Truck types can be diesel or electric and divided into rigid body rear dump trucks, articulated rear dump trucks, and tractor-trailer units. They are mainly used for development openings and primary levels, where they are commonly used to transport ore to pre-assigned locations.

The transport system joining underground with the surface (both ways) depends on the type of underground mine. Shallower underground mines with declines designed with reasonable slopes generally use a direct and simple transport system. Deeper mines with vertical and heavier inclined accesses will require specific transport systems connecting surface and underground (both ways).

Many mines use shafts with a hoisting system to transport personnel, material, and different work equipment required by the mine operations. One of the advantages is the quicker travel time. However, one of the system's defining characteristics is the load capacity that can be hoisted, depending on the installed equipment. Generally, a cage is hoisted up and down the shaft to move people and equipment. To transport ore in shafts, it is usual to use skips. It is important to note, though, that the density

Fig. 18.14 Load-Haul-Dump machine (Picture: Shahir Shundra/Wikimedia Commons)[15]

of the ore plays a significant role in this decision as higher densities could cause skip overload, risking the operation of the whole hoist system (Fig. 18.14).

As other options to consider, we have inclined shafts or drifts. For this system, transport by conveyor belt or strip conveyors is a sensible common choice. For this reason, inclinations are at around 16–18° with respect to the horizontal reference. In some mines, rail-less systems are used, such as low-profile trucks in excavations with slopes no steeper than 12°.

The monorail transport in underground mines is energy efficient, suitable for long-distance transport, has a high tonnage production per shift, can operate alone or in combination with trackless systems, and can be mounted on the floor or the ceiling. Ground-mounted trains are less flexible than trackless systems and can only be used in mines with small slopes. Although, in general, rail systems take a longer time to build, once operating, they can handle larger loads per trip than non-rail systems. Their main uses are to transport material, mechanical equipment, and personnel. The system can be designed to have multiple branches over long distances using different slopes. One big reason to use monorail systems is the deepening of the mining operations, which impose constraints on the size of the excavations (making them smaller) and favoring attempts to reduce the use of heat sources [42]. Simulation models have been used to estimate the performance of a monorail system for use in decline development [43]. This study shows that the cycle time for the decline development activities gets reduced by around 22%.

---

[14] Image file licensed under the Creative Commons Attribution-Share Alike 4.0 International license. Image date: 5 March 2014.

[15] Image file licensed under the Creative Commons Attribution-Share Alike 4.0 International license. Image date: 1 December 2016.

Fig. 18.15 Underground mining team (Picture: Dogad75/Wikimedia Commons)[16]

The rail conveyor system, which merges both belt conveyor and railway, produces a highly energy-efficient transportation system [44]. The system has steel track wheels running on steel rails that share a rolling resistance similar to railway systems with the relatively lightweight nature and continuous operational benefits of conveyor belts. The system's key innovation is that instead of having the belt supported by rods, it is supported by a series of linked carriages that rest on steel wheels on rail tracks. A rail conveyor has a low profile to allow small-scale operations, can travel on slopes of up to 20%, and has a wagon system. The cars are reversed for unloading, which maximizes the system's performance, considering having multiple loading and unloading points (Fig. 18.15).

Advanced Analytics Discussion and Opportunities

Using discrete-event simulation of underground operations represents a significant opportunity for advanced analytics in underground mine operations—Salama's Ph.D. Thesis [45] presents an interesting study that summarizes different underground haulage systems intending to determine how various material handling equipment may achieve desired production targets, ideally leading to lower costs. In this work, discrete-event simulation (DES) and mixed-integer programming (MIP) are combined to produce answers to this question. The methodology uses DES to estimate mine production for different haulage systems, and information is later used

[16] Image file licensed under the Creative Commons Attribution-Share Alike 4.0 International license. Image date: not specified.

to estimate production costs for each material handling system. MIP was used to generate the optimal production schedule, and a mine plan was later used to calculate NPV. This research avenue is an interesting one as it attempts to combine two usually separate worlds. Following this research, the opportunity is to characterize production uncertainty and solve a robust production schedule and mine plan for a given material handling system. Another example of the use of DES for analyzing underground transport systems is given in [46].

Several mines start as surface mines, and then they transition to underground. To determine the appropriate point when this transition happens is non-trivial. The work of Ben-Awuah et al. [47] takes an interesting viewpoint by comparing mining strategies for an orebody using mixed-integer linear programming (MILP) techniques to compare the NPV of open-pit mining, underground mining, and concurrent open-pit and underground mining. They conclude that open-pit mining performs better than underground mining. However, the concurrent option could provide some marginal benefits.

Overall, other research avenues have been explored by some researchers with applicability to any mining. One such avenue is using an integrated approach for the long-term planning of mines [48]; in this research, a global operation model comprises the different stages from extraction to final product production. This type of holistic modeling should involve transport definitions or decisions on its formulation. The research presents the following questions that such a model should at least answer and that we reproduce here [48]:

Extraction

- Which resources to extract and where to transport them?
- When to extract them? and
- Which technology to extract them?

Investment

- How much to invest?
- When to invest?
- Mine extraction capacity and expansion potential? and
- Plant processing capacity and expansion potential.

Processes

- How each plant should be operated? This includes defining the operating variables;
- How much to process in each plant; and
- Sale of byproducts and intermediate products.

More and more works are appearing that combine machine learning (ML) with decision-making in mining, and it seems to be a safe bet that the trend will only increase in the coming years. One such example is the work presented in [49]. This work uses a neural network model to compute material destination decisions that automatically respond to new information. For this purpose, a model of the flow of materials is created using a system dynamics approach. Additionally, the transportation aspects of the mining complex are modeled using DES, and in particular, such models are considered for determining response times of transport activities within the system. The holistic view of the problem is consistent with that presented by Caro et al. [48] a decade before, but a decade later, the tools and computational power are enablers for implementing some of those ideas.

Finally, a non-exhaustive list of other possibilities for the application of advanced analytics in underground mining is:

- characterization of load-and-haul cycle time and performance measurement;
- models to determine fleet composition based on fleet performance characterization;
- congestion modeling in underground access networks;
- equipment reliability and proactive maintenance models;
- underground infrastructure condition monitoring: ore passes, shafts, and tunneling condition; and
- dispatching algorithms and short-term scheduling.

Discrete-Event Simulation Model to Assess Production Uncertainty

As we have seen from the previous sections, plenty of possible tools are used in transportation within mining. However, providing detailed examples for every one of them could take too long, and it is probably out of the scope of a chapter for a book of this nature. Hence, to illustrate one of the possible techniques used in studying transport systems in mining, we will focus on one technique that seems to be very well used across the board in different contexts in the mining industry: discrete-event simulation. First, we will introduce some basic concepts. Afterward, we will describe the problem to implement the model finally and generate some runs. Then, we provide some details regarding the library used to model and run the discrete-event simulation. Furthermore, we provide details of the implementation of some of the objects used in the simulation.

Discrete-Event Simulation

This subsection attempts to introduce some of the concepts that relate to discrete event simulation (DES). This topic on its own is a large one, and complete treatment of it usually requires textbooks of its own. For this reason, this subsection is mainly intended to introduce the essential elements of the methodology conceptually and provide some pointers to sources so that the interested reader can deepen his/her knowledge in case this technique piques his/her interest. The starting point for a proper definition of the DES technique is the understanding of what simulation means. A dictionary definition states that simulation is "something that is made to look, feel, or behave like something else especially so that it can be studied or used to train people" [50]. An alternative definition provided by the same source says that simulation is "the imitative representation of the functioning of one system or process using the functioning of another." Regardless of the definition used, the defining characteristic of simulation seems to be the imitation of real-world processes or systems using some form of representation.

Imitating the actual system makes it possible to produce an artificial history of the system under study. This history can be usually produced at will as many times as needed to obtain a deep understanding of the system under scrutiny. The ability to produce artificial histories enables the researchers to conclude the behavior of the real-world process/system being modeled and use this ability to infer future behavior or establish causal relationships between factors that influence the outcome of the system. At his point of the discussion, it needs to be noted that simulation can be used as an aid for both the study of existing systems or as assistance in the design and forecasting of the behavior of new systems.

In the idealized world of engineering or mathematical modeling, systems are described using a set of mathematical expressions, and the resulting system of equations is thereby used to infer the behavioral characteristics of the system. Unfortunately, mathematical models usually introduce constraints/requirements that may lead to models that are not fully representative of the entire system being modeled. For example, it is common to invoke linearity even though the system may be nonlinear. Such simplifications are needed to guarantee some degree of solvability of the models, and a trade-off between realism/representativity and solvability of models is an issue that requires consideration. Simulation techniques are usually less constrained to mathematical constraints, allowing the researcher to obtain better representations of reality. For this reason, the simulation will be the preferred tool as it allows complex systems that are not well suited to a mathematical description to be described and analyzed. The typical use case is when simulation models enable numerical representations of real-world systems to perform controlled experiments to analyze system behavior. Simulation can have a variety of uses, particularly by considering the last typical use case. However, it is not the solution to all problems out there, and there are cases where simulation is not fit for purpose or inappropriate; simulation should not be used when [51]:

- the problem can be solved by common sense;
- the problem can be solved analytically;
- if it is easier to perform direct experiments;
- if the costs exceed the savings;
- if the resources or time to perform simulation studies are not available;
- if no data, not even estimates, is available;
- if there is not enough time or personnel to verify/validate the model;
- if managers have unreasonable expectations: overestimate the power of simulation; and
- if system behavior is too complex or cannot be defined.

Because simulation can be used to mimic a system, its use, once the simulation model has been positively validated and shown to reproduce reality, is well suited to forecasting future system behavior, the response of the system if changes are introduced, or to understand system variables relationships without the need of building/implementing the system or engaging in expensive experimentation. One particular use of simulation models is testing "what if" type questions regarding systems or processes; the responsiveness of the simulation model to test scenarios is by far a great help to test a relevant number of options before implementation of the actual system.

Despite the benefits commonly associated with using simulation models, not everything is shiny in their use. In some cases, there are some disadvantages related to the use of simulation models, and consequently, the reader needs awareness of those negative aspects. In particular, the construction of simulation models usually requires specialized knowledge, and it is a time-consuming endeavor. Furthermore, the fact that a model has been built for a particular system does not guarantee that it will be used appropriately in the decision-making process. At his point in the discussion comes to mind the usual aphorism "rubbish in, rubbish out": It does not matter how perfect the model is, if appropriate data is not supplied to it, it will produce rubbish outputs, and then any conclusion derived from its use is dubious/spurious. Similarly, even if the model is an excellent representation of reality and is fed appropriate data, it still needs to be used by a human decision-maker who can improperly utilize it to arrive at incorrect conclusions. Therefore, the analyst is responsible for deciding which information produced by the model will be used and how that information will be incorporated into the decision-making process.

The standard taxonomy of models classifies them according to their randomness, the role of time, and in some cases by how time is considered in the model. If the system does not have random components, then it is called deterministic. In such a case, the system's evolution can be precisely described using a set of equations and the future state of the system being perfectly predictable from any initial condition. On the contrary, if system randomness is present on some of its variables, then we call that system stochastic. For stochastic systems, future predictions based on equations are unreliable and prone to erratic behavior. Now, if the time variable does not play an essential role in the system's evolution, it is called static. Otherwise, we say that the system is dynamic. If a system can be described by using continuously evolving variables, then we say that the system is continuous. Otherwise, if some system variables change at specific instants in time, the system is called discrete. Thus, a discrete event system is characterized by being stochastic, dynamic, and discrete. A discrete event simulation model is a computational model used to represent a discrete-event system.

To build an appropriate computational model of the system, a definition of the system is required, which requires delineating what the system is and what is not. Informally, it can be said that a system is a collection of objects that interact together for which there is some interdependence. For example, an open-pit mine could be a system composed of mining entities such as mining equipment (loaders, trucks, etc.), the deposit, and the haul roads. Any other element that does not belong to the system is termed environment. Clear boundaries between a system and its environment are, in some cases challenging to determine. For example, for an open-pit mine: Do human miners belong to the system? Are the miners' families part of the system? etc. Without answering these questions, we acknowledge that their answers will define the system's boundaries. Hence, they could be subject to debate and interpretation, which is partially the modeler's responsibility. Other related terms are:

- ***entity***: an object of interest within the system;
- ***attribute***: is a property of an entity;
- ***activity***: represents a period of a specified length;
- ***state***: is defined for a system as a collection of variables that can be used to describe the system at any given time in relation to the entities of the system or the objectives of the study; and
- ***event***: is an instantaneous occurrence that changes the state of the system.

A typical simulation study is not a minor undertaking, and it usually consists of several steps, which can be broadly described as formulation, data collection, model implementation, validation, and runs. A more descriptive and precise list is presented below [51]:

- problem formulation;
- setting objectives and overall plan;
- model conceptualization (happens in parallel with data collection);
- data collection (happens in parallel with model conceptualization);
- model translation;
- verification (if the model is not verified, go back to model translation);
- validation (if the model is not validated, go back to data collection and model conceptualization);
- experimental design;
- production runs and analysis;
- additional runs (if more runs are needed, go back to experimental design and production runs and analysis);
- documentation and reporting; and
- implementation.

Problem formulation is an essential step, but surprisingly enough, not one performed appropriately in some cases. It suffices to say that if the problem is incorrectly identified/formulated, then you may end up with a correctly implemented simulation model that will not have been used in the real world, thus defeating the purpose of building the model in the first place. The other steps are undoubtedly necessary; however, they all follow the first and most crucial step of defining the problem. Another critical aspect of the previous list related to the fact that simulation models are "run" and not "solved," which implies that in order to check the correctness of the model, we need to check that the model can reproduce reality (or to be conversant with it); if the model is not able to be validated against known situations, then the full predictive capability is under question; this is then a fundamental step in every simulation study and one that frequently forces a re-examination of the problem definition, data availability, and implementation. Once the model has been validated, then there is a certain guarantee that it is used to conclude it, and this allows to proceed to run experiments with the model, previous experimental design. Once the bulk of the work is done, the remainder relates to documentation and communication of results to implement insight gathered using the simulation runs.

One key ingredient of simulation tools relates to the ability to generate random numbers. Most computer languages have tools included either within the language or provided as complementary libraries that generate some form of random number sequences, usually drawing from a $U[0, 1]$ distribution, and from there, other distributions can be generated. Complete textbooks describe different techniques used in the generation of these numbers and the statistical tests used to check that a particular generator behaves close to what would be expected from Uniform distribution. Essentially, any "generator" will attempt to create a sequence of numbers $R_1, R_2, \ldots$ that should be independent of each other and whose histogram ideally matches that of uniform histograms. As simple as the concept is, from a computational point of view, what is done is the generation of numbers that are not independent of one another, and for this reason, those numbers are called pseudo-random numbers.

One of the most common generators, called linear congruential, essentially produces a sequence of integers $X_1, X_2, \ldots$ between zero and $m - 1$ by using the following recursive relationship:

$$X_{i+1} = (aX_i + c)\,\mathrm{mode}\, m \forall i \in Z \tag{18.1}$$

in Eq. 18.1, the value X_0 is called the seed, a is called the multiplier, c is called the increment, and m is called the modulus. The selection of the values X_0, a, c, m severely affects the way the generator creates numbers. If the initial value X_0 (seed) is reached, then the whole sequence of numbers will start to repeat. The number of steps before this happens called the cycle length. Once a sequence of numbers $\{X_i\}_i$ has been successfully generated, it is further transformed to a number in the interval $[0, 1]$ by dividing by the modulus m. To summarize, random number generators create a sequence that starts with a seed, and then the generator is repeatedly applied to the result of the previous step. Several books cover the topic of random number generation with more detail, and the reader is encouraged to consult them [51–53].

To introduce a more formal definition, Heidergott et al. [54] define a DES model as one that generates a sequence $\overrightarrow{x}(k)$ of states, starting with an initial state $\overrightarrow{x}(0) = \overrightarrow{x}_0$. Under their paradigm, the state $\overrightarrow{x}(k+1)$ is obtained from $\overrightarrow{x}(k)$ by a state-transition mapping τ (measurable) that takes a vector of independent random variables $\overrightarrow{y}(k) = (y_1(k), .., y_m(k))$ and

$$X(k+1) = \tau(\vec{x}(k), \vec{y}(k)) \forall k \geq 0 \tag{18.2}$$

We observe the dynamic aspect of the system as modeled through the sequence of states $\overrightarrow{x}(k)$ (here k plays the role of time), and the vector gives the stochastic aspect of the model $\overrightarrow{y}(k)$ Moreover, the discrete event is given by the transitions that happen at discrete points in time $k \geq 0$.

The transitions described by Eq. 18.2 are modeled by adopting a simulation paradigm allowing the programmer to understand the computer objects involved in running the simulation. Several world views have been developed for DES programming, the most popular ones being [51]:

Activity-scanning: This world view focuses on activities and the conditions that allow an activity to begin. For this world view, time is broken down into small increments, and at each point in time, or called clock advance, the model checks each activity condition and starts those that satisfy them. Due to its nature, this approach could become extremely slow if time increments are too small. Unfortunately, larger time increments may not properly represent how the activities are to be executed in the system, as some activities may need to spend a considerable time waiting for the next clock advance to be started. On the other hand, if the time increments are too small, there will be multiple activity checks producing no change in the system. This is a significant trade-off to consider as it could severely undermine the quality of the simulation or the efficiency of computational resources. Given that today's increased computer power somehow eliminates this trade-off in favor of more frequent activity checks, this could nevertheless easily get out of control if too many activities requiring tracking or time increments are decided to be too small. This type of approach, while logical in its design, could lead to simulations taking several hours or even days to run.

Event-scheduling: This world view focuses on events and their impact on the system state. The time advance process is performed based on an ordered list called the future event list (FEL). When a new event is generated into the system, it needs to be scheduled and inserted in chronological order into the FEL. As opposed to the activity-scanning world view where the clock advances at regular steps, in the event-scheduling world view, the clock moves to the next impending event time corresponding to the first element of the FEL, processes the event, modifies the state of the system, and introduces all the changes needed into the FEL. The insertion operation into the list could become expensive if too many future events require scheduling. Nevertheless, this world view is definitively faster than the activity-scanning one as it avoids the unnecessary checking of activities at every clock advance.

Process-interaction: This world view focuses on processes instead of activities. A process is a collection of time-sequenced events, delays, and activities that define the life cycle of an entity as it moves through the system. In this world view, it is possible to have many processes active at any given time, and the interaction between processes could become quite complex. This approach produces a more modular code and is considered better from a programming point of view.

Case Study

The case study under consideration corresponds to a multi-pit mining complex producing a single mineral type that we will call dilithium, coming from three different surface mines. To not complicate the exercise unnecessarily, we will assume that the grades of dilithium are generally distributed within each mine, with each mine having a specific mean and standard deviation values, which can be fixed for the case study arbitrarily as we are extracting a fictitious mineral. We acknowledge that the grades have a distribution different from a normal one in a more realistic setting. However, the proper geostatistical simulation would unnecessarily complicate things and deviate attention from the central aspect of building the material transport system.

Each mine can deposit its material directly into the ROM pad or deliver it to one of four stockpiles differentiated by grade ranging from high grade to low grade. The material is transported by a fleet of trucks which are assigned to each mine. For the example, there are six trucks where two will be assigned to each surface mine, loading and unloading times are characterized as random variables following a normal distribution. For trucks unloading into the ROM pad, the unloading time also considers some allowance to account for unexpected traffic conditions.

There is also a crusher process that follows the ROM pad and is intended to break material. To observe the impact of hardness on the system's performance, the crusher element will act on the feed considering a processing time correlated with the hardness of the material fed.

Finally, a processing plant takes the material coming out of the crusher, transforming the material considering the recovery, which can be made a function of the head grade fed into the process. The simulation time will be in minutes, and the total simulation time will be 10,080 min which is equivalent to one week of operations. The simulation time can be adjusted to consider other time horizons, but it is considered that one week is sufficient for the purposes of the exercise. The structure of the simulation model is given in Fig. 18.16.

The main parameters used in the simulation are tabulated in Table 18.2.

Table 18.2 presents specific details for the different agents used in the case study (Tables 18.3, 18.4, 18.5 and 18.6).

The crusher has a processing capacity of 10 tons/min and is expected to break down every 500 min on average with a standard deviation of 40 min, a repair time estimated to be 10 min on average with a standard deviation of 1 min. Finally, the plant has a processing capacity of 10 tons/min and is expected to break down every 1500 min on average with a standard deviation of 50 min, a repair time estimated to be 5 min on average with a standard deviation of 1 min.

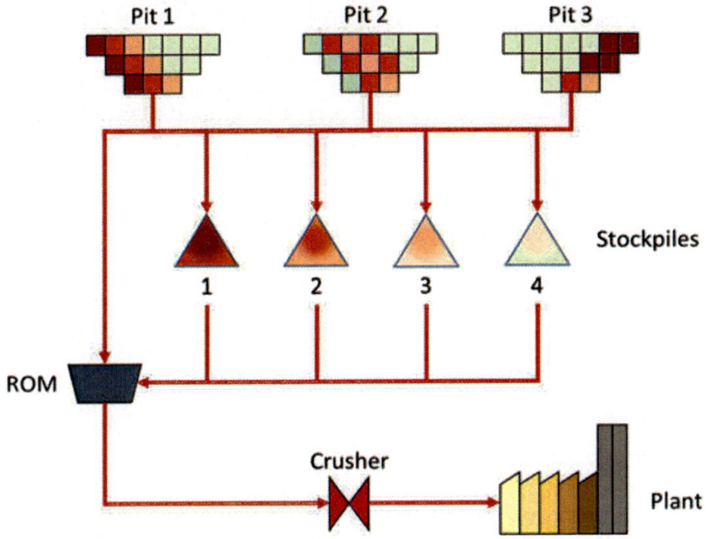

Fig. 18.16 Schematic representation of the case study

Model Implementation

Simpy

The problem described in the previous subsection requires tools that enable the implementation of a discrete simulation model. Simpy is a Python library for process-oriented discrete-event simulation. On the process-oriented discrete-event simulation paradigm, each simulation activity is modeled by a process. The other two paradigms are:

- Activity-oriented: The time is broken down into tiny increments. At each time point, the code will look at all activities and check for the possible occurrences of events,
- Event-oriented: This paradigm implements a future event list (FEL) where future activities are stored in an ordered manner. Then the clock is moved to the next event time in the FEL, thus reducing the time required by the activity-oriented paradigm.

Processes in simpy are defined by Python generator functions and can be used in many applications and the modeling of active components like customers, vehicles, or agents. Simpy also provides shared resources code facilities that can be used to model limited capacity congestion points, typically servers, checkout counters, front loaders, shovels, etc., to mention just a few, which are not being used in the context of the current work but are worth mention.

Table 18.2 Global parameters used in the simulation

Parameter	Unit	Value
Simulation time	min	10,080.0
Truck avg loading time	Minutes	3.0
Stdev truck loading time	min	0.2
Truck avg unloading time in ROM	min	20.0
Stdev truck unloading time in ROM	min	1.0
Truck avg unloading time in Stockpile	min	1.0
Stdev truck unloading time in Stockpile	min	0.1
Stockpile truck avg loading time	min	3.0
Stdev stockpile truck loading time	min	0.2
Stockpile truck avg unloading time in ROM	min	15.0
Stdev stockpile truck unloading time in ROM	min	1
Unit mine truck cost	$/ton	20
Unit crushing cost	$/ton	10
Unit plant cost	$/ton	5
Number of mines		3
Number of trucks		6
ROM pad		1
Number of stockpiles		4
Number of stockpile trucks		2
Number of crushers		1
Number of processing plants		1

Table 18.3 Physical characteristics of mines

Mine	Grade	Hardness	Distance	Truck capacity
1	$\mathcal{N}(1.5, 0.2)$	$\mathcal{N}(20.0, 1.0)$	350 mts	1
2	$\mathcal{N}(1.8, 0.1)$	$\mathcal{N}(21.0, 0.5)$	400 mts	1
3	$\mathcal{N}(3.5, 0.3)$	$\mathcal{N}(21.0, 1.0)$	410 mts	1

$\mathcal{N}$(a, b) is a normal distribution with mean a and standard deviation b, and distance represents the direct distance of the mine to the ROM pad, and truck capacity is essentially the number of shovels a mine has

Table 18.4 Time characteristics of mines

Mine	Breakdown	Repair	Norm. Extr. cost	Ext. Extr. cost
1	$\mathcal{N}(150, 2)$	$\mathcal{N}(10, 1)$	10.0	10.0
2	$\mathcal{N}(150, 2)$	$\mathcal{N}(10, 1)$	10.0	10.0
3	$\mathcal{N}(160, 3)$	$\mathcal{N}(8, 1)$	10.0	15.0

$\mathcal{N}$(a, b) is a normal distribution with mean a and standard deviation b, breakdown represents the frequency of failures, repair the time it takes to fix them, "norm. extra. cost" represents the normal cost of extraction, and "ext. extra. cost" represents the cost of extraction paid as extra hours

Table 18.5 Stockpiles characteristics

Stockpile	Min. grade	Max. grade	Capacity	Hardness
1	0.0	1.3	1,000,000	[19, 22]
2	1.3	1.6	1,000,000	[19, 22]
3	1.6	2.0	1,000,000	[19, 22]
4	2.0	20.0	1,000,000	[19, 22]

Table 18.6 Trucks characteristics

Mine trucks	Avg. tons	Stdev. tons	Speed
1	220	3	50
2	220	3	55
3	220	3	60
4	220	3	50
5	220	3	50
6	220	3	50
Stockpile trucks	Avg. tons	Stdev. tons	Speed
1	20	3	50
2	20	3	55

Simpy is based on the concept of a simulation environment. A simulation environment oversees running the simulation time and managing the scheduling and processing of events. A fundamental example of how an environment is initialized and used is presented in the following code listing (from simpy's documentation).

```
import simpy

def my_proc(env):
    yield env.timeout(1)
    return "Monty Python's Flying Circus."

env = simpy.Environment()
proc = env.process(my_proc(env))
env.run(until=proc)
```

`simpy` is declared using the directive import `simpy`. After that, the environment is declared using `env = simpy.Environment()` which later is run for a specified amount of time as exemplified by the line `env. run(until = proc)`. The rest of the lines shown in the example are used to create a process that is then executed according to the logic of the process. The better way that the `simpy` library is used is by integrating it into the classes, i.e., the processing logic becomes part of the class design. For example, the following listing adapted from the documentation illustrates the main structure of a code that runs the logic of processing inside a class in `simpy`.

```
import simpy
import random as rd

class Car(object):
    def __init__(self, env, id):
        self.env = env
        self.action = env.process(self.run())
        self.id = id

    def run(self):
        while True:
            print('Car', self.id, 'Start parking and charging at %f' %
self.env.now)
            charge_duration = rd.uniform(1, 3)
            yield self.env.process(self.charge(charge_duration))
            print('Car', self.id, 'Start driving at %f'
% self.env.now)
            trip_duration = rd.uniform(2, 4)
            yield self.env.timeout(trip_duration)

    def charge(self, duration):
        yield self.env.timeout(duration)


if __name__== '__main__':
    env = simpy.Environment()
    carList = []
    for i in range(3):
        carList.append(Car(env, i))
    env.run(until=12)
```

```
The output of this listing is given by:
```

```
Car 0 Start parking and charging at 0.000000
Car 1 Start parking and charging at 0.000000
Car 2 Start parking and charging at 0.000000
Car 1 Start driving at 1.194972
Car 2 Start driving at 1.996193
Car 0 Start driving at 2.279162
Car 0 Start parking and charging at 4.427830
Car 1 Start parking and charging at 4.825655
Car 2 Start parking and charging at 5.875815
Car 0 Start driving at 6.847299
Car 1 Start driving at 7.590702
Car 2 Start driving at 7.884991
Car 0 Start parking and charging at 9.352278
Car 2 Start parking and charging at 10.237889
Car 1 Start parking and charging at 10.471736
```

In general, we can see that the minimal structure that a Python class must have to support simpy is given by:

```
import simpy

class Class_Name(object):
    def __init__(self, env):
        self.env = env
        self.action = env.process(self.run())

    def run(self):
        while True:
            do_some_actions
            yield self.env.timeout(duration)
```

where the line `def __init__(self, env):` initializes the class and env is passed as the argument, the same one that later needs to be declared in the program using `env = simpy.Environment()`, that way, the object is linked to the environment's execution thread. The next line in terms of importance for the code is `self.action = env.process(self.run())` which declares the action that is attached to the environment which is defined on its logic by using `self.run()`.

Containers and Stores

In this subsection, we briefly mention a couple of particular objects defined in `simpy` that can be used in developing the implemented model: Containers and stores. Containers are `simpy` constructs that assist in modeling the production and consumption of homogeneous, undifferentiated bulk material. It may either be continuous (like water) or discrete (like apples). On the other hand, stores can model the production and consumption of concrete objects, in contrast to containers that store abstract amounts. Containers and stores both expose `capacity` as a property of the object, and in the case of containers, it is possible to query the level of the container at any point in time by using the `level` attribute.

Example of Implementation of Truck Class

As it is impossible to go into complete implementation details for every class in the simulation model, we will closely analyze the implementation of the class, which is perhaps the most important one in the whole code, as it performs most of the action. First, the complete code listing for the `Mine_Truck` class is presented in the Appendix, but it is briefly discussed here on the special aspects to DES or the library used.

`def __init__()` initializes the object and its properties. We observe that a `Mine_Truck` object is characterized by its identifier, capacity, speed, which mine is assigned to, etc. Other properties are needed for running the simulation, in particular, the `self.action = env.process(self.run())` is telling the program that the property action is, in fact, the outcome of using the simple library using the method run as an argument. `self.run()` contains the logic of a mining truck.

`self.run()` executes the following loop, which is valid during the simulation run:

- determines the characteristics of the material that is taking from the mine (hardness and grade), Request access to the shared resource (the shovel). Once the resource is free, the truck loads from the mine and takes a time as specified by a normal random variable.
- Then, depending on the capacity status of the ROM pad, it could decide to travel to it or go to a stockpile based on its grade.
- Then, it unloads the material on the corresponding destination and takes the corresponding time to do so.
- Finally, it travels back to the mine, where the whole process starts again.

The object performs other functions, such as updating running totals, monitoring costs, and other statistics. The object implements each action mentioned before, and for each one, there are conditions and logic that could be further discussed but probably would not add much more from a conceptual point of view. It needs to be noted that traveling times are considered as distance divided by speed; however, the distance used to calculate times is the distance between mines and stockpiles (as a group); from the mines or the stockpiles to the ROM pad, the travel time is modeled as a normally distributed variable that takes into account the travel time back and forth.

Model Validation

A trial run of 1500 min of simulation time was performed based on the configuration mentioned before. In the form of console console-printed statements, the output coming out of the program allows to follow the logic and check the consistency and flow of actions of the different agents of the simulation to show its logical correctness. The partial output for the run is shown below:

```
M Truck 3 Extracting from Mine the following (tons, grade, hardness)
[217.831,1.760,20.422] at time 2.697
M Truck 1 Extracting from Mine the following (tons, grade, hardness)
[220.567,1.482,20.116] at time 3.020
M Truck 5 Extracting from Mine the following (tons, grade, hardness)
[215.881,3.751,21.652] at time 3.051
M Truck 4 Extracting from Mine the following (tons, grade, hardness)
[214.566,1.736,20.534] at time 5.664
M Truck 2 Extracting from Mine the following (tons, grade, hardness)
[216.577,1.301,20.494] at time 5.789
M Truck 6 Extracting from Mine the following (tons, grade, hardness)
[221.616,3.579,19.483] at time 5.862
M Truck 3 Arriving to ROM with queue length 0 at time 9.363
M Truck 1 Arriving to ROM with queue length 0 at time 10.020
M Truck 5 Arriving to ROM with queue length 1 at time 11.251
M Truck 2 Arriving to ROM with queue length 2 at time 12.152
M Truck 4 Arriving to ROM with queue length 3 at time 13.664
M Truck 6 Arriving to ROM with queue length 4 at time 14.062
M Truck 3 Finishing Unloading in Tip Pocket at 29.462
M Truck 3 Back to Mine at 36.129
M Truck 3 Extracting from Mine the following (tons, grade, hardness)
[218.311,1.786,20.990] at time 38.990
Two trucks in the ROM Pad! Unloading in Stockpile...
We have found the right stockpile for you, and it is stockpiled S3 Min and
Max Grade: 1.6 2.0 Extracted Grade and Hardness: 1.7858610989046562
20.990338305494365
      ..... M Truck 3 Unloading into Stockpile S3 at 47.561
M Truck 1 Finishing Unloading in Tip Pocket at 49.299
SP Truck 1  Loading from Stockpile S3 at 52.597
SP Truck 1 Travelling to Tip Pocket at 53.597
M Truck 3 Back to Mine at 55.061
SP Truck 2  Loading from Stockpile S3 at 55.148
SP Truck 2 Travelling to Tip Pocket at 56.057
M Truck 1 Back to Mine at 56.299
M Truck 3 Extracting from Mine the following (tons, grade, hardness)
[223.365,1.942,21.157] at time 58.082
Two trucks in the ROM Pad! Unloading in Stockpile...
M Truck 1 Extracting from Mine the following (tons, grade, hardness)
[216.939,1.474,20.243] at time 59.164
Two trucks in the ROM Pad! Unloading in Stockpile...
We have found the right stockpile for you, and it is stockpiled S3 Min and
Max Grade: 1.6 2.0 Extracted Grade and Hardness: 1.94224811906223
21.156658324022434
      ..... M Truck 3 Unloading into Stockpile S3 at 66.540
We have found the right stockpile for you, and it is stockpiled S2 Min and
Max Grade: 1.3 1.6 Extracted Grade and Hardness: 1.473813320573364
20.243125248629667
      ..... M Truck 1 Unloading into Stockpile S2 at 68.119
M Truck 5 Finishing Unloading in Tip Pocket at 69.712
M Truck 3 Back to Mine at 74.040
M Truck 1 Back to Mine at 76.119
M Truck 3 Extracting from Mine the following (tons, grade, hardness)
[221.631,1.784,19.879] at time 77.107
Two trucks in the ROM Pad! Unloading in Stockpile...
M Truck 5 Back to Mine at 77.912
M Truck 1 Extracting from Mine the following (tons, grade, hardness)
[220.248,1.175,19.717] at time 79.435
```

```
Two trucks in the ROM Pad! Unloading in Stockpile...
M Truck 5 Extracting from Mine the following (tons, grade, hardness)
[226.610,3.533,20.515] at time 81.272
Two trucks in the ROM Pad! Unloading in Stockpile...
We have found the right stockpile for you, and it is stockpiled S3 Min and
Max Grade: 1.6 2.0 Extracted Grade and Hardness: 1.784294360858805
19.879248501868883
      ..... M Truck 3 Unloading into Stockpile S3 at 85.635
M Truck 2 Finishing Unloading in Tip Pocket at 87.400
We have found the right stockpile for you, and it is stockpiled S1 Min and
Max Grade: 0 1.3 Extracted Grade and Hardness: 1.1747266610549634
19.717296148058992
      ..... M Truck 1 Unloading into Stockpile S1 at 88.420
.
.
.
```

This step has been performed to check the logical integrity of the code. Usually, performing this process allows the modeler to discover logical, conversion, and implementation errors. The development process is not as straightforward as the industry would like. The simple example used in this case study, with all its simplifications, is already in the order of 1000 lines, mainly thanks to the simpy library, which significantly simplifies the development process. Without the library, it definitively would have taken many more lines. After satisfactory analysis of the simulation log and checking that things are well behaved, the xt step is to have a full run and gather statistics and then show the kind of information gathered.

Numerical Experiment

A complete simulation run for 10,080 min (one week) is performed. The code has been implemented with several monitors. A monitor is a function that takes care of gathering information regarding the status of a particular object. It has been decided that the ROM pad, the processing plant, stockpiles, mines, crusher, and trucks will be monitored. For the ROM pad, the tons and contained metal level is monitored. For the processing plant, we can monitor incoming grades and tons, cumulative tons, and metal. For stockpiles, we can monitor levels for tons and metal content. For the mines, accumulated running costs can be calculated. For the crusher, we can monitor incoming grades and tons, cumulative tons, and metal. Finally, for the trucks, we monitor accumulated distance traveled. Figures 18.17, 18.18 and 18.19 show cumulative distance traveled, histogram cumulative distance traveled, and cumulative tons received at the ROM pad.

We could be presenting pages and pages of graphs and summaries. However, for illustration purposes, we show few selected graphs so that the reader can get an idea of the type of output obtained from such a model. This has been done with the mine cumulative cost KPI, and the histogram is shown. From a financial point of view, once the distribution is established, it is possible to calculate a KPI analogous to

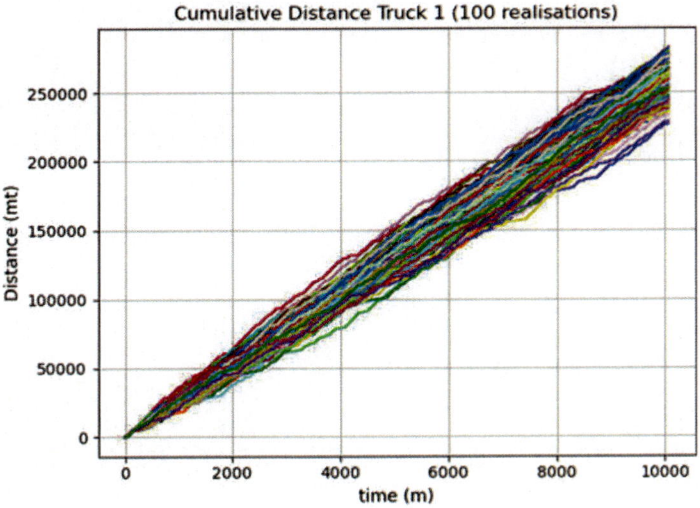

Fig. 18.17 Different trajectories for the cumulative distance traveled by one truck. Note that as time progress, the distribution becomes wider

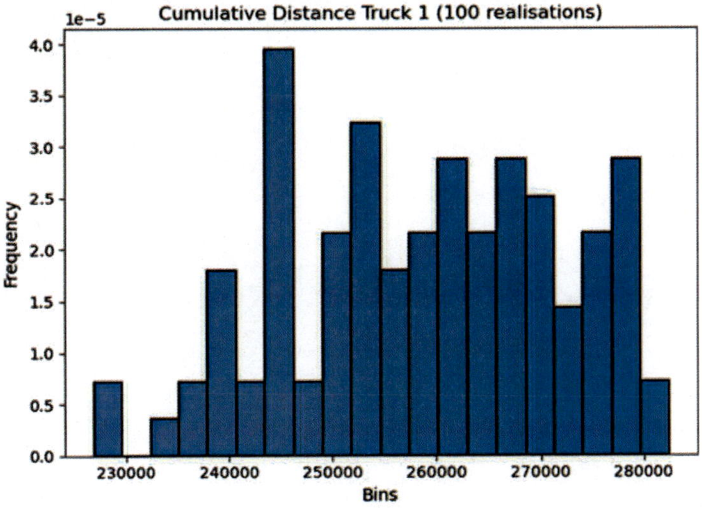

Fig. 18.18 Histogram for the final cumulative distance traveled by one truck

value at risk (VaR) that we can name cost at risk (CaR), which can be defined as analogous but opposite to the VaR (Fig. 18.20).

For the distribution of values at time 10,080 (one week), we have an average transport cost of \$1,290,209.9, and the percentile 95% for this distribution is \$2,639,250.4, which implies that the expected transport cost of \$1,290,209.9 could grow in as much

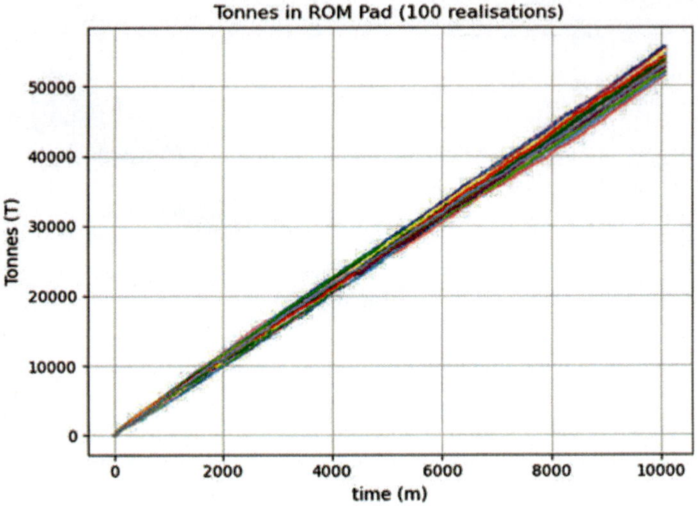

Fig. 18.19 Different trajectories for the cumulative tons received at the ROM pad

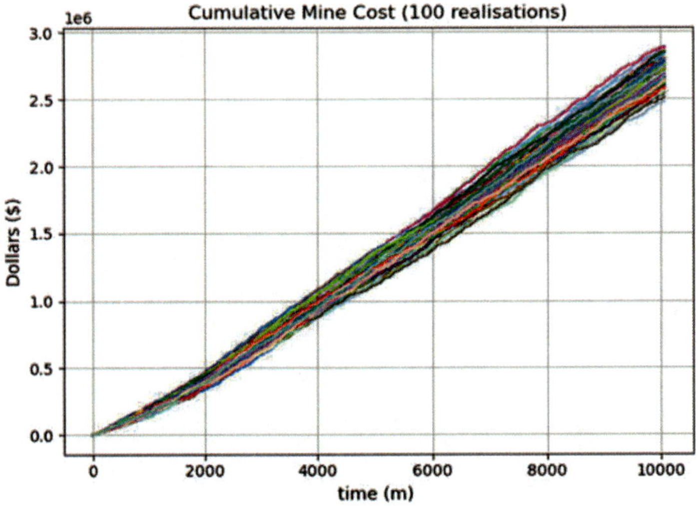

Fig. 18.20 Different trajectories for one mine cumulative transport cost

as $1,349,040.5, which in this case amounts to double the transport cost; all this due to the uncertainty in the transport function for this fictional example.

The same technique but looking at other factors can be applied. For instance, a model that focuses on the geographical location of equipment and traffic will not use the same code. However, the principle remains unaltered. This is the power of DES in the mining context. It allows the decision-maker to consider different options before committing to a final design specification, or at least, as the previous example has shown, it allows to understand the variables having a more significant impact on the potential damage to value for the mining project.

Conclusions

The chapter has reviewed different transportation options within the mining industry. First, there are the internal transportation systems that allow moving material from the mine to the processing facilities within it, and then there are the options that connect the mineral products of that mine to the external world. The role of transportation in mining is big. Depending on the extraction technique used to exploit a deposit, it could represent well in the order of 60% of the operational costs. Hence, any improvement has the potential to save millions of dollars every year.

If the reader does not have a mining background, it is expected that this chapter will introduce some of the terminology used in mining. However, a chapter of this nature in a book that has received contributions from different authors makes it difficult to try to condense an area that could be several books on its own. For this reason, a compromise has been made to introduce the main transportation systems used in mining briefly. However, the reader must be warned that there is much more that can be said about this.

Within each section of the chapter, some novel, interesting, and/or challenging analytics applications have been mentioned or discussed. Again, there are many uses, especially during recent times as the mining business is increasingly digital. Despite that, mining essentials remain the same: break material, transport to processing facilities, and transport the saleable products out of the mine. For obvious reasons, mining could not exist without transportation, but on the other hand, the mining environment is so challenging that it has helped develop and test technologies that later have found new life in other areas of human endeavor.

The chapter finished with an illustration of the type of modeling involved when using DES in mining. As best as possible, we have tried to keep the presentation of the case study without going into the unnecessary level of detail. However, both from a conceptual and practical point of view, the case study has tried to dig deeper into a well-used technique in mining transport. The case study mentioned in this chapter is far from finished. On the contrary, it could go on and on. However, the point was not to show highly technical modeling skills but to point out the process that is followed and the type of conclusions that can be obtained. That way, the practitioner can have a preliminary assessment of the potential benefits that such a technology can bring into their day-to-day activities before deciding to invest heavily in developing this type of model.

Appendix: Complete Code for the Mine Truck Class

```
class Mine_Truck(object):
    def __init__(self, avg_tonnes, stdev_tonnes, env,
             mine, speed, name, tip_pocket,
             stockpiles, mineset):
        self.avg_capacity = avg_tonnes
        self.stdev_capacity = stdev_tonnes
        self.tonnes = sps.norm(self.avg_capacity,
                       self.stdev_capacity)
        self.env = env
        self.mine = mine
        self.action = env.process(self.run())
        self.speed = speed

        self.extracted_grade = 0
        self.extracted_hardness = 0
        self.name = name
        self.tip_pocket = tip_pocket
        self.stockpile_used = False
        self.stockpiles = stockpiles
        self.last_k = 0
        self.proc = env.process(self.monitor_truck_cost())
        self.accumulated_distance = 0
        self.Accumulated_Distance_Travelled = []
        self.mineSet = mineset

    def run(self):
        while True:
            k = math.floor(self.env.now /
                       G_Simulation_Month_Time)
            if k > self.last_k:
                self.mine.hardness = sps.norm(
                    self.mine.avg_hardness[k],
                    self.mine.stdev_hardness[k])
                self.mine.grade = sps.norm(
                    self.mine.avg_grade[k],
                    self.mine.stdev_grade[k])

            # Loading from Mine
            with self.mine.res.request() as req:
                extraction_duration = sps.norm(
                    G_Mine_Truck_Loading_Time_Mean,
                    G_Mine_Truck_Loading_Time_StdDev).rvs()
                yield req
                yield self.env.process(
                     self.extract_from_mine(
                     extraction_duration, self.mine))

            # Conditional to ROM/Stockpile
            if len(self.tip_pocket.res.queue) < \
```

```
                self.tip_pocket.queue_cap:
            self.stockpile_used = False
            yield self.env.timeout(
                self.mine.distance / self.speed)
            self.accumulated_distance += \
                self.mine.distance
            print(self.name, 'Arriving to ROM with '
                             'queue length',
                  len(self.tip_pocket.res.queue),
                  'at time {0:.3f}'.format(
                      self.env.now))
            # Unload at ROM
            with self.tip_pocket.res.request() as req:
                unloading_duration = sps.norm(
                  G_Mine_Truck_Unloading_Tip_Pocket_Time_Mean,
                  G_Mine_Truck_Unloading_Tip_Pocket_Time_StdDev).rvs()
                yield req
                yield self.env.process(
                  self.unload_in_tip_pocket(
                     unloading_duration))
        else:
            self.stockpile_used = True
            print('Two trucks in the ROM Pad! '
                  'Unloading in Stockpile...')
            yield self.env.timeout((
              self.tip_pocket.distance_to_stockpile +\
              self.mine.distance) / self.speed)
            self.accumulated_distance += \
                self.tip_pocket.distance_to_stockpile\
                + self.mine.distance
            # Unload at Stockpile
            unloading_duration = sps.norm(
              G_Mine_Truck_Unloading_Stockpile_Time_Mean,\
              G_Mine_Truck_Unloading_Stockpile_Time_StdDev).rvs()
            yield self.env.process(
                self.unload_in_stockpile_set(
                 unloading_duration))

        self.mine = dispatch_truck(self, self.mineSet)

        # travel back
        if not self.stockpile_used:
            yield self.env.timeout(self.mine.distance
                                   / self.speed)
            self.accumulated_distance += \
                self.accumulated_distance
            print(self.name, 'Back to Mine at {'
                             '0:.3f}'.format(
                self.env.now))
        else:
            yield self.env.timeout((
              self.mine.distance +
               self.tip_pocket.distance_to_stockpile)
                                   / self.speed)
```

```
                self.accumulated_distance += \
                    self.tip_pocket.distance_to_stockpile\
                    + self.mine.distance
                print(self.name, 'Back to Mine at {'
                                 '0:.3f}'.format(
                    self.env.now))

    def extract_from_mine(self, duration, mine):
        yield self.env.timeout(duration)
        self.extracted_tonnes = self.tonnes.rvs()
        self.extracted_grade = mine.grade.rvs()
        self.extracted_hardness = mine.hardness.rvs()
        # Cost model
        if not self.mine.breakdown:
            self.mine.accumulated_cost += \
                self.extracted_tonnes*self.mine.normal_cost
            self.mine.accumulated_metal += \
                self.extracted_grade*self.extracted_tonnes
            self.mine.accumulated_tonnes += \
                self.extracted_tonnes
        else:
            self.mine.accumulated_cost += \
                self.extracted_tonnes*(
                  self.mine.normal_cost + self.mine.extra_cost)
            self.mine.accumulated_metal += \
                self.extracted_grade*self.extracted_tonnes
            self.mine.accumulated_tonnes += \
                self.extracted_tonnes
        # End of Cost Model
        print(self.name, "Extracting from Mine the "
                         "following (tons, grade, "
                         "hardness) "\
            "[{0:.3f},{1:.3f},{2:.3f}]".format(
            self.extracted_tonnes, self.extracted_grade,
            self.extracted_hardness),
              'at time {0:.3f}'.format(self.env.now))

    def unload_in_tip_pocket(self, duration):
        yield self.env.timeout(duration)
        if self.extracted_tonnes > 0:
            self.tip_pocket.grade = (self.extracted_grade
                * self.extracted_tonnes +
                    self.tip_pocket.metal.level) / (
                    self.extracted_tonnes +
                    self.tip_pocket.tonnes.level)
            self.tip_pocket.hardness = (
                self.extracted_hardness * self.extracted_tonnes +\
             self.tip_pocket.hardness * self.tip_pocket.tonnes.level) / (
                self.extracted_tonnes + self.tip_pocket.tonnes.level)
        yield self.tip_pocket.tonnes.put(
            self.extracted_tonnes) & self.tip_pocket.metal.put(
            self.extracted_grade * self.extracted_tonnes)
        print(self.name, 'Finishing Unloading in Tip '
                         'Pocket at {0:.3f}'.format(
            self.env.now))
```

```
def unload_in_stockpile_set(self, duration):
    # Identification of the Stockpile based on the
    # characteristics of the load
    selected_stockpile = self.stockpiles.set[
        G_Default_Stockpile_Key]
    dict = self.stockpiles.set
    Stockpile_Found = False
    for key, value in dict.items():
        if (self.extracted_grade >= value.min_grade) \
                and (self.extracted_grade <= value.max_grade) \
            and (self.extracted_hardness >=
                 value.min_hardness) and (
                self.extracted_hardness <=
                value.max_hardness):
            Stockpile_Found = True
            print('We have found the right stockpile '
                  'for you and it is stockpile', key,\
            'Min and Max Grade:', value.min_grade,
                  value.max_grade, 'Extracted Grade '
                                   'and Hardness:',\
                  self.extracted_grade, self.extracted_hardness)
            selected_stockpile = self.stockpiles.set[key]
            break
    if not Stockpile_Found:
        print('Sent to default stockpile... Extracted'
              'Grade:', self.extracted_grade,
              'Extracted Hardness: ',
              self.extracted_hardness)
    yield self.env.process(
        self.unload_in_particular_stockpile(
            selected_stockpile, duration))

def unload_in_particular_stockpile(self, stockpile,
                                   duration):
    with stockpile.res.request() as req:
        yield req
        yield self.env.timeout(duration)
        if self.extracted_tonnes > 0:
            stockpile.grade = (self.extracted_grade *
                self.extracted_tonnes +
                               stockpile.metal.level) / (
                self.extracted_tonnes +
                stockpile.tonnes.level)
            stockpile.hardness = (
              self.extracted_hardness *
              self.extracted_tonnes +
               stockpile.hardness *
               stockpile.tonnes.level) / (
                self.extracted_tonnes +
                    stockpile.tonnes.level)
        yield stockpile.tonnes.put(
            self.extracted_tonnes) & stockpile.metal.put(
            self.extracted_grade * self.extracted_tonnes)
        print('      .....', self.name, 'Unloading ' \
```

```
                    'into Stockpile', stockpile.name,
                'at {0:.3f}'.format(self.env.now))

    def monitor_truck_cost(self):
        while True:
            self.Accumulated_Distance_Travelled.append(
                self.accumulated_distance)
            yield self.env.timeout(1.0)
```

References

1. Newman, A.M., et al. 2010. A review of operations research in mine planning. *Interfaces* 40 (3): 222–245.
2. Alarie, S., and M. Gamache. 2002. Overview of solution strategies used in truck dispatching systems for open-pit mines. *International Journal of Surface Mining, Reclamation, and Environment* 16 (1): 59–76.
3. Upadhyay, S.P., and H. Askari-Nasab. 2018. Simulation and optimization approach for uncertainty-based short-term planning in open-pit mines. *International Journal of Mining Science and Technology* 28 (2): 153–166.
4. Bozorgebrahimi, A., R. Hall, and M. Morin. 2005. Equipment size effects on open pit mining performance. *International Journal of Surface Mining, Reclamation, and Environment* 19 (1): 41–56.
5. Burt, C.N., and L. Caccetta. 2014. Equipment selection for surface mining: A review. *Interfaces* 44 (2): 143–162.
6. Dembetembe, G.G., and V. Mutambo. 2018. *Optimisation of Materials Handling Fleet Performance at Nchanga Open Pit Mine, in First Zambia National Conference on Geology, Mining, Metallurgy, and Groundwater Resources: The Future Mining in Zambia. Held at Mulungushi International Conference.* Lusaka, Zambia.
7. Torkamani, E., and H. Askari-Nasab. 2015. A linkage of truck-and-shovel operations to short-term mine plans using discrete-event simulation. *International Journal of Mining and Mineral Engineering* 6 (2): 97–118.
8. Hashemi, A.S., and J. Sattarvand. 2015. Application of ARENA simulation software for evaluation of open-pit mining transportation systems—A case study. In *Proceedings of the 12th International Symposium Continuous Surface Mining-Aachen 2014*. Springer.
9. Ngwangwa, H.M., and P.S. Heyns. 2014. Application of an ANN-based methodology for road surface condition identification on mining vehicles and roads. *Journal of Terramechanics* 53: 59–74.
10. Morad, A.M., M. Pourgol-Mohammad, and J. Sattarvand. 2014. Application of reliability-centered maintenance for open pit mining equipment productivity improvement: A case study of Sungun Copper Mine. *Journal of Central South University* 21 (6): 2372–2382.
11. Lieberwirth, H. 1994. *Economic Advantages of Belt Conveying in Open-Pit Mining, Mining Latin America/Minería Latinoamericana*, 279–295. Springer.
12. Roumpos, C., et al. 2014. The optimal location of the distribution point of the belt conveyor system in continuous surface mining operations. *Simulation Modelling Practice and Theory* 47: 19–27.
13. Marx, D., and J. Calmeyer. 2004. A case study of an integrated conveyor belt model for the mining industry. In *2004 IEEE African. 7th Africon Conference in Africa (IEEE Cat. No. 04CH37590)*. IEEE.
14. Mohammadi, M., S. Hashemi, and F. Moosakazemi. 2011. Review of in-pit crushing and conveying (IPCC) system and its case study in copper industry. In *World Copper Conference.*

15. Daniyan, I., A. Adeodu, and O. Dada. 2014. Design of a material handling equipment: Belt conveyor system for crushed limestone using 3 roll idlers. *Journal of Advancement in Engineering and Technology* 1 (1): 2348–2931.
16. Ribeiro, B.G.C., W.T.D. Sousa, and J.A.M.D. Luz. 2016. Feasibility project for implementing conveyor belts in an iron ore mine. Study case: Fabrica Mine in Minas Gerais State, vol. 69, 79–83. Brazil. Rem: Revista Escola de Minas.
17. Yang, Y., et al. 2014. On-line conveyor belts inspection based on machine vision. *Optik* 125 (19): 5803–5807.
18. Blazej, R., L. Jurdziak, and R. Zimroz. 2013. Novel approaches for processing of multi-channels NDT signals for damage detection in conveyor belts with steel cords. In *Key Engineering Materials*. Trans Tech Publication.
19. Cristoffanini, C., M. Karkare, and M. Aceituno. 2014. *Transient Simulation of Long-Distance Tailings and Concentrate Pipelines for Operator Training*. Salt Lake City, UT, USA: SME.
20. Lahiri, S., and K. Ghanta. 2008. Development of an artificial neural network correlation for predicting hold-up of slurry transport in pipelines. *Chemical Engineering Science* 63 (6): 1497–1509.
21. Xie, Y., et al. 2015. Wear resistance of materials used for slurry transport. *Wear* 332: 1104–1110.
22. Liu, C., Y. Li, and M. Xu. 2019. An integrated detection and location model for leakages in liquid pipelines. *Journal of Petroleum Science and Engineering* 175: 852–867.
23. Marcoulaki, E.C., I.A. Papazoglou, and N. Pixopoulou. 2012. Integrated framework for designing pipeline systems using stochastic optimization and GIS tools. *Chemical Engineering Research and Design* 90 (12): 2209–2222.
24. Kang, J.Y., and B.S. Lee. 2017. Optimisation of pipeline route in the presence of obstacles based on a least-cost path algorithm and laplacian smoothing. *International Journal of Naval Architecture and Ocean Engineering* 9 (5): 492–498.
25. Gerasimova, A., A. Keropyan, and A. Girya. 2018. Study of the wheel-rail system of open-pit locomotives in traction mode. *Journal of Machinery Manufacture and Reliability* 47 (1): 35–38.
26. Keropyan, A., et al. 2019. Influence of roughness of working surfaces of the wheel-rail system of open-pit locomotives with an implementable adhesion coefficient. *Journal of Friction and Wear* 40 (1): 73–79.
27. Heyworth, J.S. 2009. Environmental lead and nickel contamination of tank rainwater in Esperance, Western Australia: An evaluation of the cleaning program. *Journal of Environmental Protection* 1 (01): 31.
28. Pérez-Bravo, F., et al. 2004. Association between aminolevulinate dehydrase genotypes and blood lead levels in children from a lead-contaminated area in Antofagasta, Chile. *Archives of Environmental Contamination and Toxicology* 47 (2): 276–280.
29. Jiang, Y., et al. 2017. Recent progress on smart mining in China: Unmanned electric locomotive. *Advances in Mechanical Engineering* 9 (3): 1687814017695045.
30. Schmitt, P., M.L. Bartosiak, and T. Rydberg. 2021. Spatiotemporal data analytics for the maritime industry. In *Maritime Informatics*, 335–353. Springer.
31. Mirović, M., M. Miličević, and I. Obradović. 2018. Big data in the maritime industry. *NAŠE MORE: znanstveni časopis za more i pomorstvo* 65 (1): 56–62.
32. Jović, M., et al. 2019. Big data management in maritime transport. *Pomorski zbornik* 57 (1): 123–141.
33. Alessandrini, A., et al. 2016. Mining vessel tracking data for maritime domain applications. In *2016 IEEE 16th International Conference on Data Mining Workshops (ICDMW)*. IEEE.
34. Coraddu, A., et al. 2015. Ship efficiency forecast based on sensors data collection: Improving numerical models through data analytics. In *OCEANS*. Genova: IEEE.
35. Coraddu, A., et al. 2018. Vessels fuel consumption: A data analytics perspective to sustainability. In *Soft Computing for Sustainability Science*, 11–48. Springer.
36. Gorman, M.F., et al. 2014. State of the practice: A review of the application of OR/MS in freight transportation. *Interfaces* 44 (6): 535–554.

37. Department of Justice, Government of Western Australia. Port Authorities Act Regulations 2001 [cited 2021 Feb 8]; Port Authorities Act 1999]. Available from https://www.legislation.wa.gov.au/legisla-tion/statutes.nsf/main_mrtitle_1932_homepage.html.
38. Bag, S., et al. 2020. Industry 4.0 and the circular economy: resource melioration in logistics. *Resources Policy* 68: 101776.
39. Atkinson, T. 1992. Selection, and sizing of excavating equipment. *SME Mining Engineering Handbook* 2: 1311–1333.
40. Novak, T., A. Gregg, and H. Hartman. 1987. Comparative performance study of diesel and electric face-haulage vehicles. *International Journal of Mining and Geological Engineering* 5 (4): 405–417.
41. Paraszczak, J., et al. 2014. Electrification of loaders and trucks–a step towards more sustainable underground mining. *In International Conference on Renewable Energies and Power Quality.*
42. Burke, P., and E. Chanda. 2007. Electro-monorails-an alternative operating system for deep mining. In *Proceedings of the Fourth International Seminar on Deep and High-Stress Mining.* Australian Centre for Geomechanics.
43. Chanda, E.K., and B. Besa. 2011. A computer simulation model of a monorail-based mining system for decline development. *International Journal of Mining, Reclamation, and Environment* 25 (1): 52–68.
44. Wheeler, C.A. 2019. Development of the rail conveyor technology. *International Journal of Mining, Reclamation, and Environment* 33 (2): 118–132.
45. Salama, A. 2014. *Haulage System Optimization for Underground Mines: A Discrete Event Simulation and Mixed-Integer Programming Approach.* Luleå tekniska universitet.
46. Greenberg, J., et al. 2016. Alternative process flow for underground mining operations: Analysis of conceptual transport methods using discrete event simulation. *Minerals* 6 (3): 65.
47. Ben-Awuah, E., et al. 2016. Strategic mining options optimization: Open-pit mining, underground mining or both. *International Journal of Mining Science and Technology* 26 (6): 1065–1071.
48. Caro, R., et al. 2007. An integrated approach to the long-term planning process in the copper mining industry. In *Handbook of Operations Research in Natural Resources*, 595–609. Springer.
49. Paduraru, C., and R. Dimitrakopoulos. 2019. *Responding to New Information in a Mining Complex: Fast Mechanisms Using Machine Learning.* Mining Technology.
50. Merriam-Webster. Merriam-Webster's Learner's Dictionary. 2020 [cited 2020 Nov 28]. Available from http://www.merriam-webster.com/dictionary/.
51. Jerry, B. 2001. *Discrete-Event System Simulation.* Series in Industrial and Systems Engineering.
52. Leemis, L.M., and S.K. Park. 2005. *Discrete-Event Simulation: A First Course.* Prentice-Hall, Inc.
53. Law, A.M., W.D. Kelton, and W.D. Kelton. 2000. *Simulation Modeling and Analysis*, vol. 3. New York: McGraw-Hill.
54. Heidergott, B., et al. 2010. Gradient estimation for discrete-event systems by measure-valued differentiation. *ACM Transactions on Modeling and Computer Simulation (TOMACS)* 20 (1): 1–28.
55. Chanchaichujit, J., and J.F. Saavedra-Rosas. 2018. *Using Simulation Tools to Model Renewable Resources.* Springer Books.

Chapter 19
Advanced Analytics for Mine Materials Transportation

Abhishek Kaul and Ali Soofastaei

Abstract Miners extract ore from underground or surface operations at the mine site. Once the ore is extracted, it needs to be transported to end customers. One of the most significant cost drivers in the mining industry is material transportation, and it can cost between 30 and 60% of the delivered price to the customer. Advanced analytics technologies offer mining companies the ability to make material transportation efficient, reduce costs, increase throughput, improve safety, and decarbonize the supply chain. This chapter describes eighteen advanced analytics use cases for most common equipment used in material transportation, including Haul Trucks, Railways, Conveyor belts, Stacker and Reclaimers, Car dumpers, Ships, and more. Further, for equipment, the use cases are defined according to the organization processes in the area of Supply Chain, Operations, Maintenance, Safety, Trading, and Chartering. Lastly, advanced analytics use cases for decarbonization in material transportation are discussed.

Keywords Advanced analytics · Mining · Material transportation · Prediction · Optimization

Introduction

Mining is a complex and fluctuating industry fraught with uncertainty around resource pricing, unpredictable resource fields, and significant projects that need to be managed right through their lifecycle.

Controlling costs for mineral exploration, construction, operation, and right through customer delivery is a monumental challenge. However, if the financial elements are managed well, it can help mining companies to be both competitive and profitable. The key to increasing profits is knowing the precise time

A. Kaul (✉)
IBM, Singapore, Singapore

A. Soofastaei
Vale, Brisbane, Australia
URL: https://www.soofastaei.net

A. Soofastaei (ed.), *Advanced Analytics in Mining Engineering*,
https://doi.org/10.1007/978-3-030-91589-6_19

Fig. 19.1 Open-pit mining

to increase production when there is strong demand using resource planning, improving machinery reliability with predictive and condition-based maintenance monitoring, fulfilling orders while controlling logistics costs by optimizing material transportation, and providing clarity-precise financial and operational reporting (Fig. 19.1).

Advanced analytics use cases have been applied across the complete mining industry value chain from exploration, mine management, extraction, processing, and transportation. There are considerable benefits of using an advanced analytics system to improve the quality of work at the mine site and reduce the human failures and hazards at the mine site. This chapter focuses on the mining supply chain, specifically on material transportation.

Material transportation is long and complex in mining. To illustrate, the ore extracted from a pit in Vale's Serra Sul mine site S11D in Serra dos Carajás, Brazil, may travel thousands of kilometers overhaul trucks, rail, conveyor belts, ships to eventually reach Baowu Group's steel plant in Wuhan central, China. Over many days, this journey will take place, with ore moving from one means of transport to the other. The ore is stored in a stockpile and retrieved based on the transport capacity for the following means of transport at each stage. The situation becomes more complex if blending is required i.e., mix two grades of ore to fulfill customer requirements (Fig. 19.2).

Material transportation is carried out by Haul Trucks, LHDs, Conveyer belts, Pipes, Railways, Stackers, Reclaimers, and Ships. Each mode of transportation is optimized for the specific transportation task based on the distance, throughput expectation, and cost of transportation per metric ton per km. For each mode of transportation, decisions need to be made for planning, operation, maintenance, and safety. For example, in Haul Trucks, few decisions are discussed below:

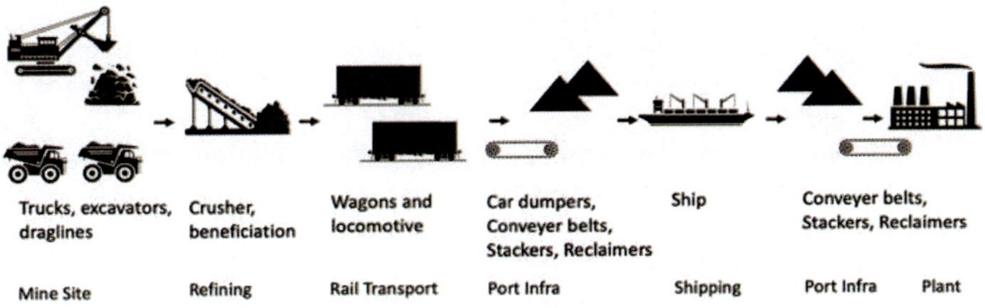

Fig. 19.2 Journey of ore from pit to customer

- Which pit or crusher should the truck move toward to reduce the idle time?
- What should be the speed on the incline for the truck to optimize fuel consumption?
- How is the health of the truck? How much more can I keep it running without breakdown or degradation in performance?
- When should the truck be sent for maintenance, balancing unplanned breakdown and throughput? And more.

Typically, these decisions are made using business rules based on operator experience, which is far from optimal. Advanced analytics can help miners to make better decisions based on data.

Advanced Analytics [1] refers to the autonomous or semi-autonomous examination of data using sophisticated techniques and tools, typically beyond those of traditional business intelligence (BI), to discover deeper insights, make predictions, or generate recommendations. Mining companies are increasingly using these techniques to improve revenue and reduce costs by making better decisions based on data.

We will discuss use cases by each segment of the material transportation. In order to move ore across a complex mining supply chain while controlling logistics costs requires insight, coordination, and reliable equipment. End-to-end planning for material transportation across multiple locations enables ore movement from pit to customer. The following use cases for the end-to-end supply chain are discussed in subsequent sections (Table 19.1).

Table 19.1 Supply chain use cases

Area	Topic	Use case description
1. Supply chain	Long-term planning (months)	Allocating demand as per customer profit, risk profile
2. Supply chain	Medium-term planning (weeks)	Optimizing mine to port logistics planning to ensure full berth capacity utilization
3. Supply chain	Short-term planning (days)	Predicting supply chain disruptions

The ore starts its journey from mine. Depending on underground or surface mining, typically Haul Trucks, LHDs, and Conveyer belts are used for material transportation at the mine site. The following use cases for Haul Trucks will be discussed in subsequent sections (Table 19.2).

Next, if the mine site is away from the port, the ore is transported through the railway to the port. Railway transportation can take few hours to days depending on the distance. The following use cases for the railway will be discussed in subsequent sections (Table 19.3).

Next, at the port, ore is unloaded manually or using automated Wagon Tipplers (also called Car Dumper). Then the ore is stored and blended at the port if required. Finally, stackers, reclaimers, conveyor belts are used to manage material transportation at the port. The following use cases for the port will be discussed in subsequent sections (Table 19.4).

Table 19.2 Haul truck use cases

Area	Topic	Use case description
4. Haul trucks	Operations	Predict queuing time at crushers or excavators for trucks to reduce overall idle time
5. Haul trucks	Operations	Reduce the fuel consumption for Haul Trucks by optimizing fuel burn index
6. Haul trucks	Maintenance	Predict health score and failure for trucks to take proactive maintenance actions
7. Haul trucks	Maintenance	Optimize the maintenance schedule based on failure prediction, mine schedule, and workshop load
8. Haul trucks	Safety	Ensure safety for autonomous trucks

Table 19.3 Railway use cases

Area	Topic	Use case description
9. Railway	Operations	Reduce diesel consumption for locomotives by optimizing the speed
10. Railway	Maintenance	Predict failure of systems
11. Railway	Safety	Inspecting tracks using AI-based visual recognition technologies

Table 19.4 Port use cases

Area	Topic	Use case description
12. Port	Operations [*Conveyor belts*]	Conveyor belt wear and deviation prediction model
13. Port	Maintenance [*Stackers and reclaimers*]	Predicting failure of stackers and reclaimers
14. Port	Maintenance [*Wagon tippler/Car dumper*]	Predict failure of Wagon tippler based on Remaining useful life
15. Port	Safety [*Worker safety*]	Video Analytics to detect PPE violations

Table 19.5 Trading, chartering, operations use cases

Area	Topic	Use case description
16. Trading and chartering	Market insights	Insights from commodity flow
17. Operations	Demurrage costs and Safe operations	Predicting ship ETA and deviations from the route

Table 19.6 Decarbonization use cases

Area	Topic	Use case description
18. Decarbonization	Shipping	Improving the performance of vessels

In parallel, the trading and sales team engages in selling the ore through long-term contracts or short-term spot buy. The chartering team also fixes the vessels for transportation as per trading, sales team requirements. Then the operations team monitors the vessels till it reaches the customer. The following use cases for Trading, Chartering, and Operations will be discussed in subsequent sections (Table 19.5).

Finally, once ore reaches the destination port, similar equipment is responsible for transportation to the customer. In this journey, emissions are emitted by equipment responsible for transportation. Therefore, decarbonization in material transportation is an essential topic for meeting the sustainability commitments for miners. The following use cases for decarbonization will be discussed in subsequent sections (Table 19.6).

Before each use case is described, the fundamentals of advanced analytics are discussed briefly in the next section.

Benefits of Using Advanced Analytics

Advanced analytics use cases are poised for exponential growth in the industry as the technology to analyze data becomes easily accessible, storage becomes cheaper, and there is ubiquitous connectivity with 5G.

In this chapter, advanced analytics is used as the umbrella term to reference Artificial Intelligence, Machine Learning, Deep Learning, Supervised learning, Unsupervised learning, Natural language processing, Computer vision, Neural networks, Algorithms, and more.

Advanced analytics is the evolution of traditional analytics, where capabilities were classified as:

- descriptive analytics—processing data for visualization and analyzing business performance;
- predictive analytics [2]—using mathematical, statistical techniques to understand patterns in data and predict future; and

- prescriptive analytics [3]—using optimization techniques to identify the best business outcome given a set of constraints.

Technology and learning disciplines were the two critical drivers for the evolution of traditional analytics to advanced analytics. First, technology helped with instrumentation, real-time device connectivity, low cost of data storage, an exponential increase in processing power. This helped the mining companies acquire, ingest, and store granular operations and maintenance data from multiple source systems.

Secondly, learning disciplines from Computer science, Data technology, Visualization, Mathematics, Operations research, and Statistics merged to into what is now called the Data Science discipline. This helped mining companies develop teams to analyze this vast amount of data and provide insights, prediction, optimization, and automation.

Miners have piloted and deployed many advanced analytics use cases on a large scale. At its core, the business can develop intelligent behavior systems by analyzing their environment and taking actions with some degree of autonomy to achieve specific goals. These systems can be purely software-based (predicting events, video analytics, and more) or embedded in hardware devices (robots, autonomous trucks, digital twin, and more).

Many of the miners have been able to reduce cost (lower energy consumption and reduce unplanned maintenance), increase productivity (increase fleet and equipment efficiency), automate (optimized planning and reduce human resources) and make the area of material transportation more sustainable. Further, the use cases are discussed from a department/process and material transportation equipment perspective. Essential functions covered are Planning, Operations, Maintenance, and Safety across the use cases. In the next section, end-to-end planning is covered.

End-to-End Planning

Material transportation starts with planning the ore movement from the mine site to the customer.

Balancing the customer demand with a long lead time supply within customer order fulfillment expectations at the optimal cost is essential for mining supply chains.

Miners have developed many advanced analytics and optimizations models to plan these complex supply chains. However, with the advancement of technology capability in terms of artificial intelligence, machine learning, and the ability to process a large amount of accurate time information from internal and external sources, there is a significant opportunity to improve the static optimization models.

Below, three use cases from advanced analytics are described across each of the planning segments in terms of long-term planning (months), mid-term planning (weeks), and short-term planning (days) (Table 19.7).

Table 19.7 Supply chain use case details

Area	Topic	Use case description
Supply chain	Long-term planning (months)	Allocating demand as per customer profit, risk profile
Supply chain	Medium-term planning (weeks)	Optimizing mine to port logistics planning to ensure full berth capacity utilization
Supply chain	Short-term planning (days)	Predicting supply chain disruptions

Use Case 1: Supply Chain—Long-Term Planning: Allocating Demand as Per Customer Profit, Risk Profile

Many papers have been published in mine schedule optimization, which maximizes the profit from a mine site keeping existing mine design, people, and equipment. For example, a paper on "Stochastic optimization in mine planning scheduling" by Giovanni et al. discusses solving the mine scheduling problem by considering the geological, technical, and market uncertainties. The paper on "Optimizing underground mining strategies" by A M Ebbels discusses how to schedule optimization techniques to take into account not only production requirements, equipment productivity but also scenarios on grade, metal content, capital, and operating costs that can help to improve the net present value (NPV) of the project.

However, the focus of this use case is on sales allocation to customers, which maximizes the profitability and reduces the long-term risk for the miner. Miners typically enter into long-term contracts with customers to commit their output. The contracts stipulate an agreed volume over a period of time at an agreed price with adjustment based on the quantity, quality, or any significant changes in the underlying assumptions. These contract characteristics are different for each customer, and so is the risk associated with each contract. In a study by McKinsey & Company [4], they discuss the topic of risk and the importance of factoring that into the deal to ensure profitability.

The traditional analysis focuses only on historical data points. However, as more and more data is available both from internal and external sources, advanced analytics techniques are applied to the data to develop sophisticated risk models to estimate the actual profitability of contracts based on projected underlying conditions. This information helps the miners make informed decisions on the profit, risk profile, and allocate the demand to the limited supply of ore. Multiple scenario plans can be executed based on projections (Fig. 19.3).

Advanced analytics techniques like Artificial Neural Networks (ANNs) can be used in this scenario to solve the problem of allocating supply to customers.

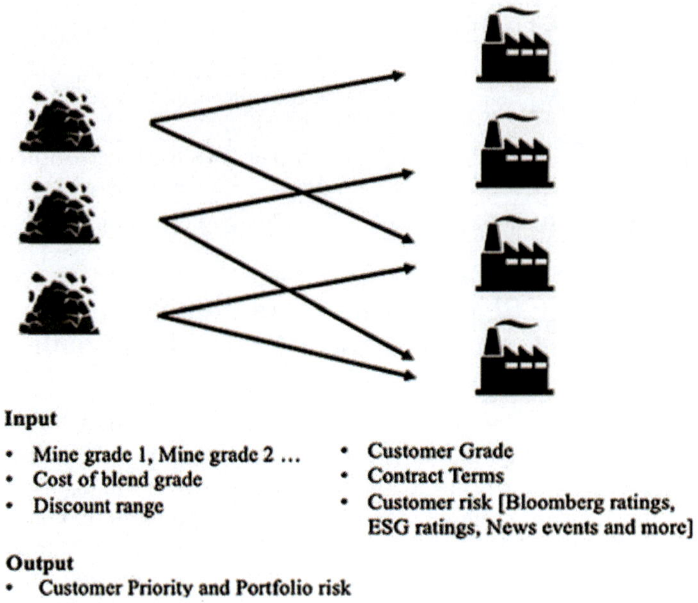

Fig. 19.3 Customer prioritization model

Use Case 2: Supply Chain—Medium-Term Planning: Optimizing Mine to Port Logistics Planning to Ensure Full Berth Capacity Utilization

Optimizing techniques are used to manage scarce resources efficiently. For example, most miners use optimization engines to synchronize the long and short-term production plans across the value chain. This helps them maintain a delicate balance between push production and pull demand.

Traditional optimization techniques are good. However, they do not take advantage of the highly instrumented equipments which provides real-time information on their status. The scope of optimization is also focused on solving one leg of the problem. Advanced analytics enables optimization from mine to port with dynamic rescheduling based on equipment conditions. The dynamic rescheduling capability enables faster response to disruptions and adjustments to plans to meet goals.

In Fig. 19.4, there are three groups of constraints:

- Ship/berth allocation—optimize the allocation of the ship to a berth, determining the stockpiles from the stockyard to load the ship, and deciding the ship sail time-based on high tide timings and percentage of shipload;
- PortSide scheduling—optimize the handling of the ore from car dumpers and transporting them to Stockyard, including managing process requirements (like screening and blending); and

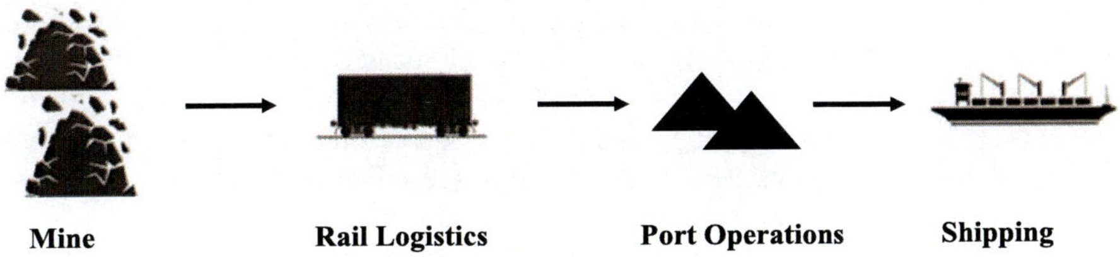

Fig. 19.4 Mine to port optimization

- Rail scheduling—optimize the loading of ores in the mines, transport them to the Port, and make them available at the car dumpers.

Note, the mine schedule optimization is not in the scope of this chapter.

In the paper "Mathematical models for the berth allocation problem in dry bulk terminals [5]," the authors define an optimization model specifically for the first part—i.e., Ship to Berth allocation. Optimizing these constraints simultaneously in near real-time using the data coming from these equipments helps to reschedule for any disruption dynamically. The KPIs optimized are minimize demurrage, maximize the fulfillment of the ship loading plan, maximize the fulfillment of the stockyard fulfillment plan, and minimize inventory deficit at the stockyard.

Most machine learning algorithms build an optimization model and learn objective function parameters from the given data. Optimization methods like Adam, Newton's Method, Generalized Gauss–Newton Matrix can be implemented using Neural networks and Deep neural networks [6].

The other side of the optimization model when a loaded ship arrives at a berth for unloading is very well described in the paper "Rule-based optimization for a bulk handling port operations [7]" In this paper, the authors define a decision support model which optimizes the unloading time of ship at port, congestion in the stockyard, and loading time of rakes.

By optimizing the end-to-end supply chain, significant benefits can be achieved to meet customer deliveries and keep the material transportation costs low.

Use Case 3: Supply Chain—Short-Term Planning: Predicting Supply Chain Disruptions

Short-term disruptions can occur due to unforeseen events, for example, a typhoon near the mine site or an unstable geopolitical situation. If the mining company can get early notifications of the events, it can help them better anticipate and plan for the disruption.

Many papers focus on the forecasting problem by suggesting novel ways to increase the demand forecast accuracy and running simulation scenarios like the one written by the Mosaic [8] team.

This section focuses on providing early information on disruptive events by monitoring the news media feeds. Nowadays, there is a massive amount of information available in the news domain, whether news websites or social media feeds. Using natural language processing techniques (NLP), this external data can be monitored to screen for specific events, and alerts can be triggered. NLP techniques like topic modeling and BERT are seen as very effective in this domain.

In the article "Leveraging on NLP to gain insights in Social Media, News & Broadcasting [9]," the authors define data analysis techniques for the organization (Fig. 19.5).

Similarly, DHL Resilience360 Supply Watch [10] analyzes millions of online sources in real-time to detect early indicators of potential supply distresses before they occur.

The insights for miners regarding supply chain risk can help them predict disruptions and take mitigation actions.

Truck Advanced Analytics Use Case

The truck is the primary means to haul the ore over a short distance in addition, and trucks are used to haul ore and waste out of the mine to a crusher or conveyor belt. There has been a gradual increase in the size of trucks over time. The tonnage carried by mining trucks can range from 25 to 450 tons in one load. There has also been a trend of moving from direct-current (DC) wheel drives to mechanical transmissions or alternate-current (AC) wheel drives [11].

Trucks move in a circular path from loading to unloading. This is defined as the haulage cycle. The truck's haulage cycle begins from the excavator where the ore is loaded to the crusher where ore is dumped.

Cycle time for haulage will depend on loading time, travel time, dumping time, waiting time, and breakdown time [12].

This journey is made over few kilometers in harsh conditions. As a result, there is often a loss of productivity and high costs incurred due to breakdowns, high waiting time, or high fuel consumption.

Trucks gather many data and typically have an online or offline data sharing capability. With this data, many advanced analytics use cases can be developed (Fig. 19.6).

Advanced analytics use cases have been applied over the Operations, Maintenance, and Safety for Trucks.

Primary use cases for trucks are (Table 19.8).

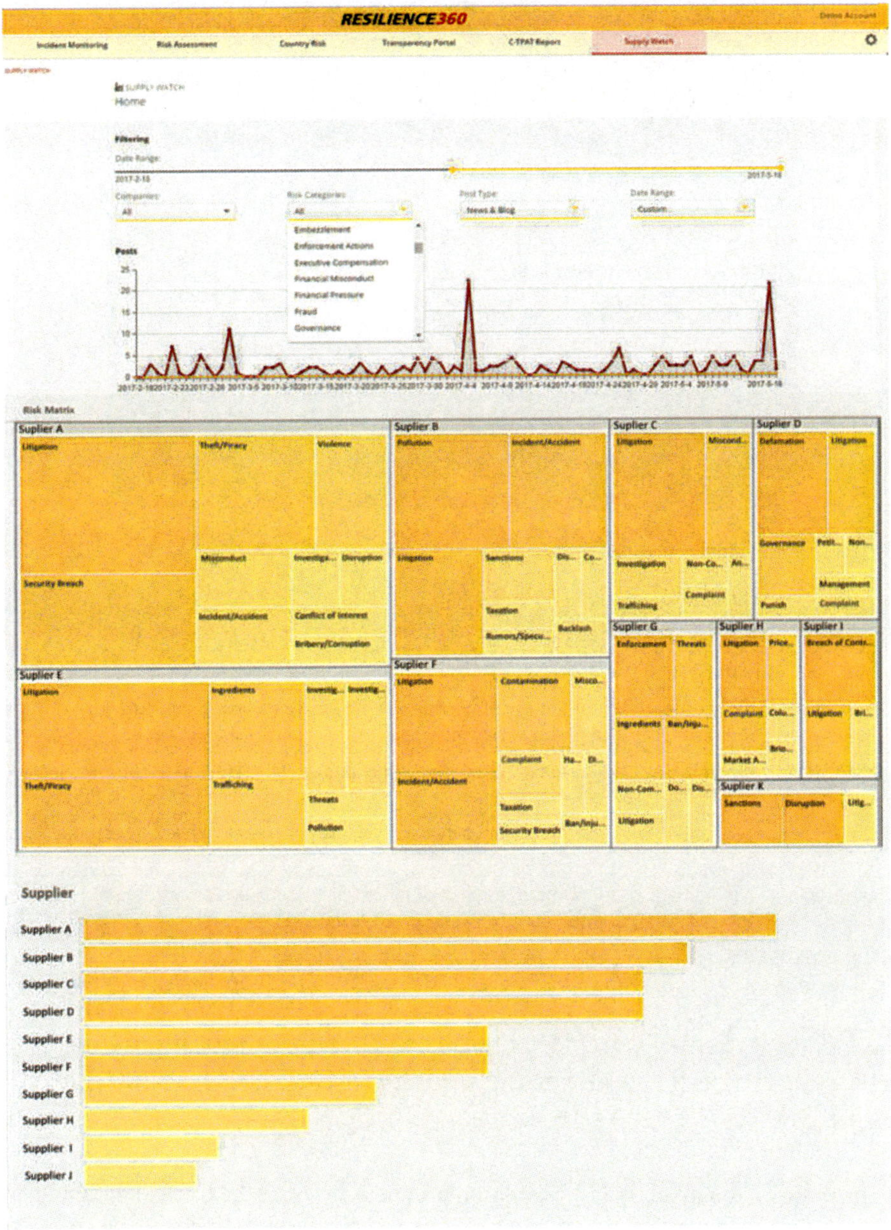

Fig. 19.5 DHL supply Reslience360. *Source* https://mms.businesswire.com/media/20170524005934/en/588816/5/R360_Supply_Watch.jpg

Fig. 19.6 Excavator loading truck

Table 19.8 Haul truck use case details

Area	Topic	Use case description
Haul trucks	Operations	Predict queuing time at crushers or excavators for trucks to reduce overall idle time
Haul trucks	Operations	Reduce the fuel consumption for Haul Trucks by optimizing fuel burn index
Haul trucks	Maintenance	Predict health score and failure for trucks to take proactive maintenance actions
Haul trucks	Maintenance	Optimize the maintenance schedule based on failure prediction, mine schedule, and workshop load
Haul trucks	Safety	Ensure safety for autonomous trucks

Use Case 4: Haul Truck Operations—Predict Queuing Time at Crushers or Excavators for Trucks to Reduce Overall Idle Time

In operations, the objective is to get high productivity from the trucks i.e., transport the maximum amount of ore and reduce the cost of transportation.

Waiting time or Idle time for trucks affects productivity and is caused due to queuing time at crusher or excavator. Many commercially off-the-shelf fleet management solutions [13] are available, which help to manage the truck haulage cycle.

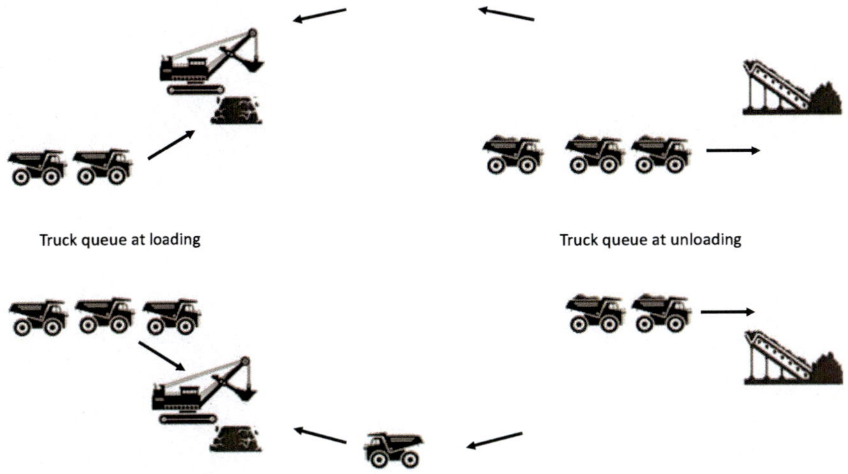

Fig. 19.7 Haul truck dispatch cycle

These solutions help communicate between the central dispatcher and truck, crusher, excavator operators, thereby channeling the trucks to operational crushers or excavators with shorter queues. Further, these solutions do have some ability to identify where the current shortest queues are present (Fig. 19.7).

When we use advanced analytics use cases, the problem changes to predict when there will be high queuing time for excavators, crushers, and proactive actions.

In order to predict the queuing time—say at excavator, multiple features like truckload factor, lithography, time are more considered to provide the dispatcher forecast into how much idle time will the truck have in the queue. Based on this forecast, dispatcher can take a proactive decision to update the plan and reduce the idle queuing time. Advanced analytics Algorithms like reinforcement learning [14], XGBoost have been quite successful at forecasting the queuing time.

Use Case 5: Haul Truck Operations—Reduce the Fuel Consumption for Haul Trucks by Optimizing Fuel Burn Index

In operations, it is also essential to reduce fuel consumption. Multiple studies and papers have been published on the topic of fuel consumption and how to reduce them. In a joint study [15] by the Department of Resources, Energy, and Tourism, they analyzed diesel use in mining operations of Fortescue Metals Group Ltd, Downer EDI Mining Pty Ltd, and Leighton Contractors Pty Limited. Results of the study identified and quantified the energy costs associated with stopping Haul Trucks unnecessarily, developed performance indicators that use an "equivalent flat haul" calculation to

	Factors	Influence
Short term Controllable factors	• Truck Speed • Transmission shift patterns • Truck Payload • Total resistance • Idle time • Maintenance of the Trucks	• Driver behaviour • Payload variance • Tyre pressure, Weather, road condition • Queuing time • Preventative and predictive maintenance
Long term design factors	• Mine plan and mine layout • Dump site design • Cycle time • Tyre wear • Age of Trucks • Engine operating parameters	• Mine, road design, elevation gain • Geographic location and weather • Type and age oftrucks

Fig. 19.8 Factors influencing fuel consumption

account for elevation changes on a specific mine route, and developed a Best Truck Ratio model to evaluate and benchmark the efficiency of fleet operations across a single site and multiple operations, where the nature of the work undertaken varied greatly.

Many factors affect fuel consumption like road conditions, operator driving behavior (speed), payload, elevation gain, travel time, rain, temperature, and more. By analyzing these factors, significant efficiency gains can be made in the fuel consumption index (Fig. 19.8).

In the paper "Reducing Fuel Consumption of Haul Trucks in Surface Mines Using Artificial Intelligence Model [16]" by Soofastaei et al., he provides recommendations on speed per elevation gain for optimal fuel consumption by first identifying the important controllable factors affecting fuel consumptions—namely payload, truck speed, and total resistance and then suggesting techniques like reducing payload variance, road maintenance, tire pressure, recommendation on driver behavior to reduce the fuel consumption index.

Advanced analytics techniques like artificial neural networks help establish the correlation between the controllable factors and fuel consumption. Once that is established, the factors that influence the controllable factors can be used to develop an objective function for optimization to reduce the fuel consumption in Haul Trucks. Optimization techniques like generic algorithms can be applied to say provide the recommended speed at each elevation gain for reduced fuel consumption and more.

Use Case 6: Haul Truck Maintenance—Predict Health Score and Failure for Trucks to Take Proactive Maintenance Actions

Maintenance refers to the set of activities that an organization takes up to manage their equipment performance, risks, and operating expenses over the entire life cycle. There are many maintenance types for Haul Trucks, ranging from Routine inspection Maintenance, Preventive Maintenance, Breakdown or Corrective Maintenance, Overhaul and Shutdown Maintenance, and Predictive Maintenance.

The cost to avoid the failure of trucks (predictive maintenance) is much lower than the cost of truck fails (breakdown maintenance). Predictive Maintenance (PdM) relies on the reliable prediction that a truck or any of its critical equipment will fail. Hence if the action is taken to do the maintenance activities, it pre-empts the high cost of breakdown maintenance. Today, trucks are highly instrumented. The vehicle health monitoring systems (VHMS) data (running hours, load, engine temperate) combined with road conditions data, weather data, and historical maintenance provide the patterns required by advanced analytics models to predict failure. Many mining companies [17, 18] are already enjoying the benefits of these techniques (Fig. 19.9).

Advanced analytics techniques like survival analysis predict the impending failure with a certain confidence level before its occurrence for Haul Trucks. Survival analysis is done using COX regression which uses the inputs like age, type of equipment, working hours, march hours, idle hours to the breakdown event and develops the probability cure to breakdown event. In industry terms, this is called as Remaining useful life (RUL) for the truck and major equipments. Multiple RULs from major

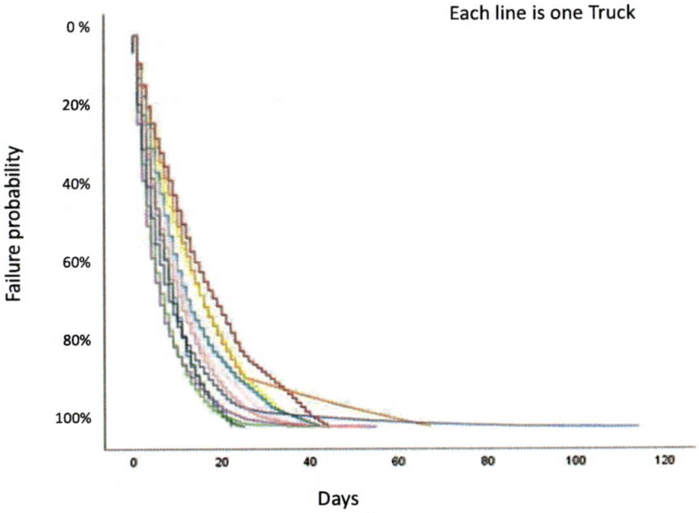

Fig. 19.9 Probability of truck failure using COX regression

truck components are aggregated to provide a comprehensive health score. If the score falls below a certain threshold, it is recommended to send the truck to the maintenance workshop or provide slack time so that maintenance engineers can inspect the truck.

Use Case 7: Haul Truck Maintenance—Optimize the Maintenance Schedule Based on Failure Prediction, Mine Schedule, and Workshop Load

The maintenance schedule depends on many parameters. If the mining company can optimize the maintenance schedule, then they can achieve the reduction in maintenance time, reduce the transit of fleet to the workshop, reduce the impact for idle hours, and increase equipment availability.

Optimizing the maintenance schedule requires merging and optimizing the preventative and predictive maintenance plans with the workshop shift calendars and projected weather conditions to reduce maintenance costs (Fig. 19.10).

The maintenance schedule can be optimized if we reduce the total loss by maintaining the asset at the optimal day, not before nor after [19]. Thus, the objective function is defined as the sum of production loss, maintenance loss (sum of cost curves for breakdown, preventative and predictive maintenance cost), and the delta operation cost (deteriorated health, thus lower efficiency). Optimization engines like CIPLX or Gurobi can perform the best in these scenarios.

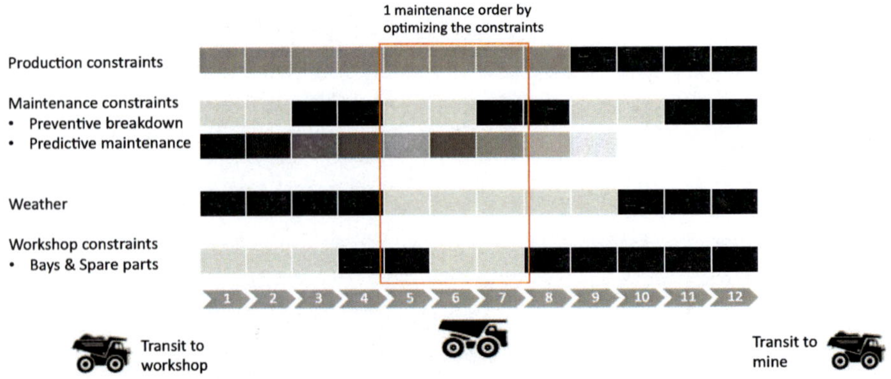

Fig. 19.10 Optimization in the maintenance schedule

Use Case 8: Haul Truck Safety—Ensure Safety for Autonomous Trucks

Autonomous Haulage Systems (AHS) technology has been deployed in many mine sites and provides significant business benefits. For example, in Rio Tinto's autonomous fleet, each AHS-operated truck was estimated to have operated for 700 h more than conventional trucks and 15% lower load and haul unit costs [20]. In addition, although these trucks are safe to operate, they have had few incidents in the past [21].

With current technology, the AHS-operated truck follows a preprogrammed path from one point to the next point and is monitored through remote control rooms with remote operator manual fallback. However, as automation technology evolves, the trucks will become truly autonomous to adjust their paths to accommodate obstruction, other variations which will lead to moral dilemma as described for automatic cars [22] (Fig. 19.11).

With advancements in Computer vision technologies, AI models are trained with the video feed to perform automated detection for object detection, object classification, object tracking, and raise proactive alerts in real-time to detect and mitigate any safety incidents. Advanced analytics deep learnings models like convolution neural networks (CNN) are best suited. Many pre-trained classifiers are available from an open-source domain like VGG, Inception, ResNet, Yolo, which can speed up the implementation of these models.

Locomotive, Railways Advanced Analytics Use Cases

Material transport from the crusher to the port over a long distance is done through rail haulage (trains). The train consists of one or more locomotives pulling the ore-filled wagons (or cars) on railway tracks. It is common to have 25,000 tons of ore transported by one train with 200 wagons (or cars) being pulled by four locomotives over 200 km. These wagons are discharged using a rotary car dumper (or wagon tippler) at the port side (Fig. 19.12).

In this section, the Operations, Maintenance, and Safety use cases are covered from rail transport (Table 19.9).

Use Case 9: Railway Operations—Reduce Diesel Consumption for Locomotives by Optimizing the Speed

Mine locomotives generally consume diesel fuel to transport the ore from the mine site to port. The speed of the locomotive is a significant factor that influences diesel

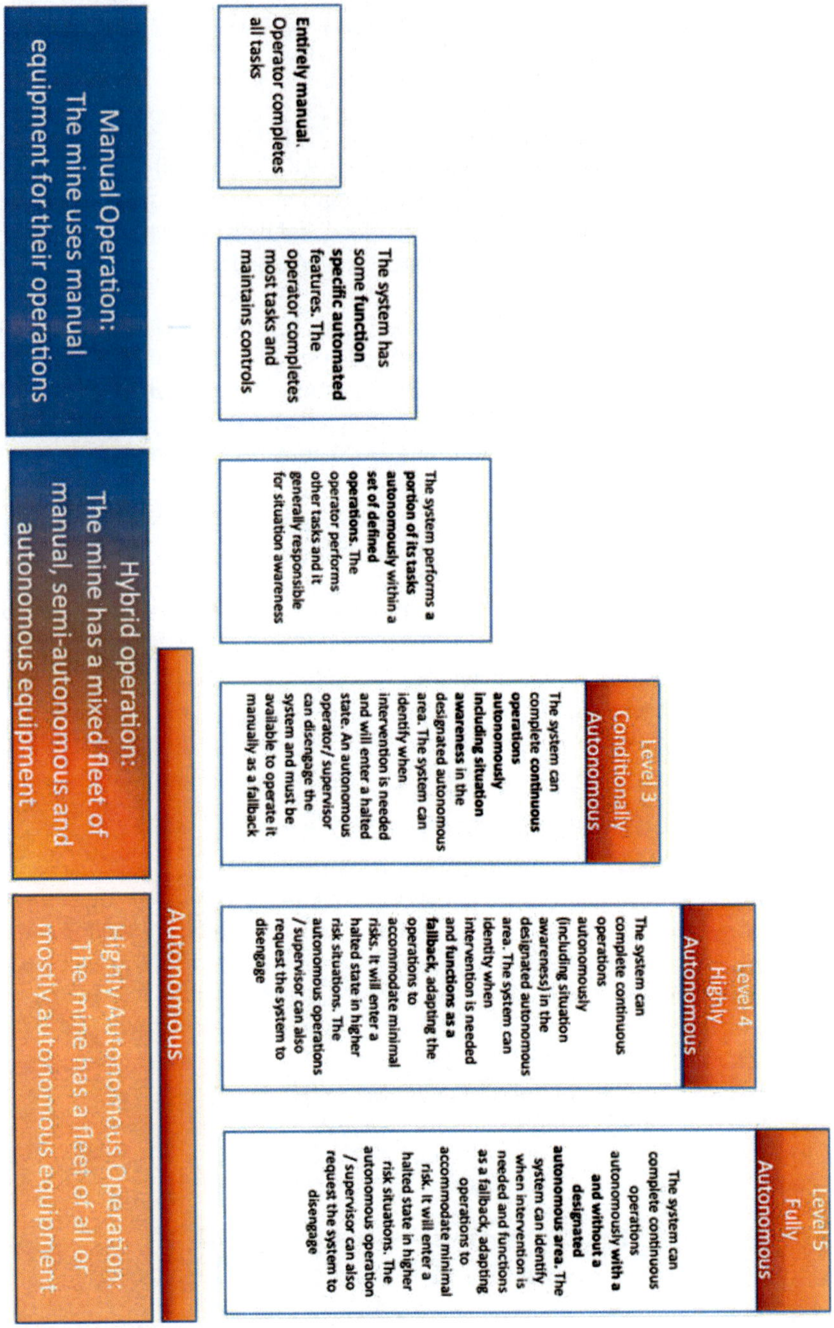

Fig. 19.11 Evolution of autonomous mining operations

Fig. 19.12 Locomotive on tracks

Table 19.9 Railway use case details

Area	Topic	Use case description
Railway	Operations	Reduce diesel consumption for locomotives by optimizing the speed
Railway	Maintenance	Predict failure of systems
Railway	Safety	Inspecting tracks using AI-based visual recognition technologies

consumption. Along with speed, travel time also has to be considered for performance improvement.

Typically, the locomotive has a diesel engine, generator, electric motor, and drive system. For example, in the paper "Application of critical velocities to the minimization of fuel consumption in the control of trains [23]," the authors conclude that for a given sequence of throttle settings, the fuel consumption is minimized if the settings are changed only when the velocity reaches a certain level. Further, in another study, "Optimal strategies for the control of a train [24]," the authors describe a method for control strategies to reduce fuel consumption by changing the throttle position after fulfilling certain conditions (Fig. 19.13).

The fuel burn of a locomotive is a function of torque and speed. Assuming that fuel burn remains constant at a given throttle position, Advanced analytics techniques can be used to simulate and reduce the fuel burn index for diesel engines. Leveraging the work done for electric vehicles on look-forward controls by Kouzani and Ganji [25], a similar approach can be used in diesel locomotives to develop an AI model to control the throttle position future inclination and speed limits to optimize fuel

Fig. 19.13 Locomotive fuel burn to influence factors

consumption. Artificial neural network algorithms and Genetic Algorithms to optimize the parameters for reducing the fuel burn index are seen to work best in these use cases.

Use Case 10: Railway Maintenance—Predict Failure of Systems

Railway Maintenance systems include regular, preventive, and predictive maintenance for wagons, tracks, and signaling equipment.

As per the Association of American Railroad [26], railroads leverage massive data to predict equipment and track defects. Many sensors pick up data, and advanced analytics models identify patterns and predict what network elements may soon need to be repaired or replaced. Typical sensors include acoustic, infrared, ultrasonic, and laser to detect bearing faults, wheel cracks, brake, and axle faults. In one case, using advanced analytics Norfolk Southern engineers predicted locomotive's water coolant would break a week before it happens (Fig. 19.14).

In the paper "Prediction of Railcar Remaining Useful Life by Multiple Data Source Fusion [27]" by Zhiguo Li and Qing He, they describe the methodology to predict the Remaining useful life (RUL) of both wheels and trucks (wagons) by merging data from three types of detectors, including wheel impact load detector, machine vision systems, and optical geometry detectors.

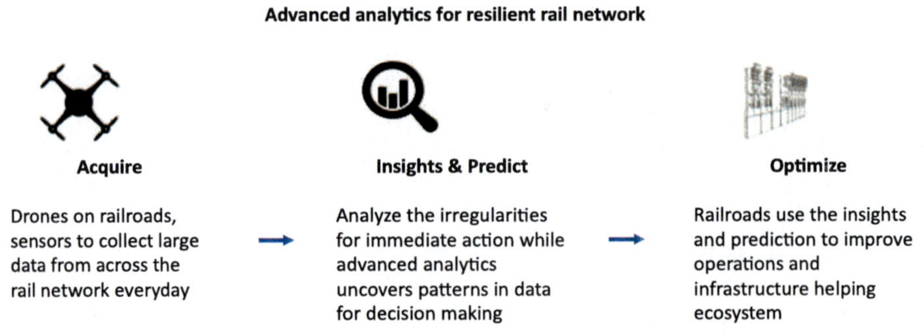

Fig. 19.14 Data analytics for the resilient rail network

Similarly, mining companies can leverage the work done in railways to develop their maintenance plans, integrating data from multiple sensors to predict wagons, tracks, and signaling equipment requirements.

A recent study surveyed the extensive data analysis [28] (advanced analytics) work done in railway transportation, found that for predictive analytics, classification, pattern recognition models, time series analysis, and stochastic models were prominent in research. ANN and SVM are two of the most popular techniques.

Use Case 11: Railway Safety—Inspecting Tracks Using AI-Based Visual Recognition Technologies

Track inspection is done for track geometry and track structure. Track geometry degradation means the poor condition of geometry parameters like profile, alignment, and gage. Track structure condition means rail, ballast, sleepers, sub-grade, and drainage system. This use case focuses on geometric degradation.

The paper on "Modeling Track Geometry Degradation Using Support Vector Machine Technique [29]" by Can Hu and X. Liu discusses data analysis processes for modeling track geometry degradation for—surface, cross-level, and dip by using the Support vector machine (SVM) model.

In the paper "Using multiple adaptive regression to address the impact of track geometry on development of rail defects [30]," Allan et al. use multivariate regression splines for understanding the railroad track defect behavior.

According to the Association for American railroad [31], BNSF uses machine learning to analyze the data collected by its track geometry cars, which travel along train tracks, identifying anomalies. Data analytics software identifies patterns in this data, enabling rail engineers to predict track problems 30 days out.

With the advancement in Computer vision technologies, many industries are taking advantage of automated inspections. Below, an approach is presented in the railway context to monitor the wheels' lateral movement relative to the rail in regular operations (Fig. 19.15).

The paper "Deep Learning-Based Virtual Point Tracking for Real-Time Targetless Dynamic Displacement Measurement in Railway Applications [32]" by Dachuan et al. developed a novel approach using virtual point tracking. This enables them to do automatic calibration, virtual point detection for each frame, and then apply rules to measure the lateral displacement of the wheel on the rail during operation. This approach has been proved to be feasible and provides real-time tracking of geometric degradation. Furthermore, the authors use a lightweight CNN architecture that provides accurate results even with a noisy background.

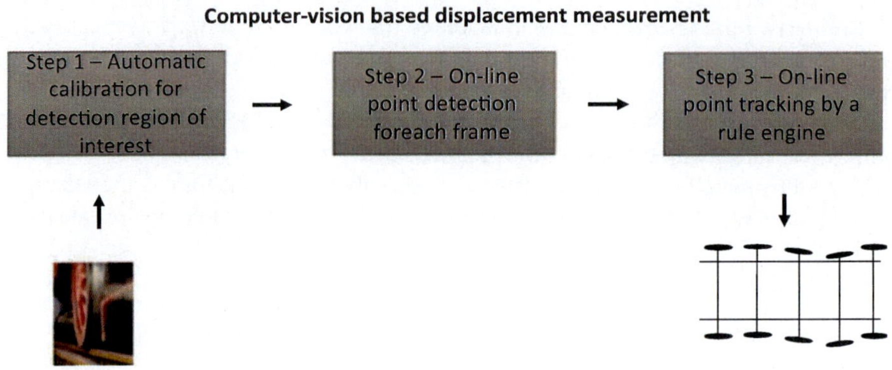

Fig. 19.15 Approach to track virtual points on track and wheel

Port Operations

Material transportation between resource-rich countries like Brazil, South Africa, and Australia to the manufacturing-heavy countries like China and India is done using sea transport. Ports enable the anchorage, berthing, loading, and unloading of ships. They also act as stockyards for storing the ore and blending the ore to meet customer grade requirements.

Major equipment in the port is conveyor belts, stackers and reclaimers, car dumper, and cranes. The car dumper empties the loaded wagons by turning them upside down. Conveyor belts transport the material between car dumper, stockyards, loading, and unloading. Stackers and reclaimers deposit, retrieve the ore to and from stockyards (Fig. 19.16).

Fig. 19.16 Tugboats mooring a large vessel at the port

Table 19.10 Port use case detailed

Area	Topic	Use case description
Port	Operations [*Conveyor belts*]	Conveyor belt wear and deviation prediction model
Port	Maintenance [*Stackers and reclaimers*]	Predicting failure of stackers and reclaimers
Port	Maintenance [*Wagon tippler/car dumper*]	Predict failure of Wagon tippler based on Remaining useful life
Port	Safety [*Worker safety*]	Video Analytics to detect PPE violations

The focus of this section is on the advanced analytics use cases for the critical equipment (Table 19.10).

Use Case 12: Port Operations [Conveyor Belts]—Conveyor Belt Wear and Deviation Prediction Model

Conveyor belts have two or more pulleys rotating and moving the belt, which carries the ore from source to destination. Belt width is dependent on the lump size and is supported by 3 or 5 idlers across the cross-section with a general angle of 35 degrees on both sides. Conveyor belt throughput is dependent on the speed, width, density and can vary from a few hundred tons per hour to a few thousand tons per hour. Any downtime in conveyor belts causes considerable losses to the miners.

In traditional root cause analysis for belt failure, key factors [33] identified are bearing failure, motor overload, and belt wear. One area to mention here is the access to real-time data. Models are greatly benefited if the data on vibration, temperature, tension, the deviation is available at higher frequency through instrumentation [34] rather than through periodic measurements.

In the paper "Integrated decision making for predictive maintenance of belt conveyor systems [35]" by Xiangwei et al., they suggest a framework for maintenance decision-making based on thresholds (Fig. 19.17).

Advanced analytics techniques can help miners deploy predictive maintenance models to identify patterns in data based on ore grade, tonnage, duration of use, temperature-vibration-deviation measurements, current draw, and historical maintenance activities. For example, conveyor belt sub-system models can predict belt wear rate, predict belt tension deviation [36], predict belt splicing, predict roller failure (bearing), and predict motor failure. Typical techniques used are regression analysis on sparse matrix, fourier transforms, and cox regression. This can help move unplanned breakdowns to planned maintenance activities by triggering early maintenance notifications, thereby saving costs and costly downtime.

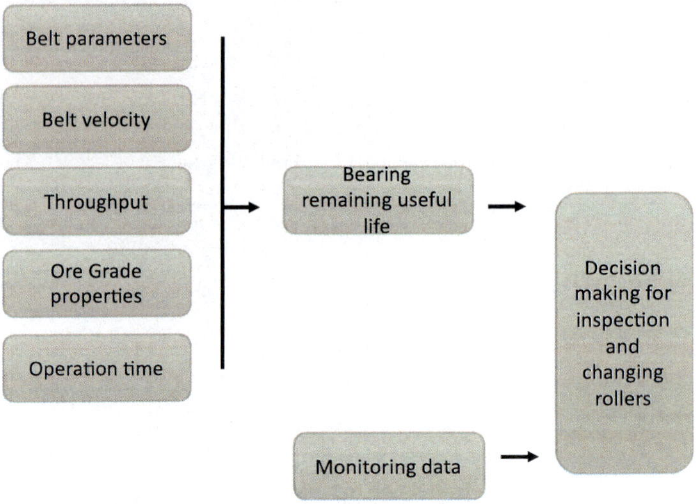

Fig. 19.17 Predicting belt failure

Use Case 13: Port Maintenance [Stacker–Reclaimer]—Predicting Failure of Stackers and Reclaimers

Stacker is used to stockpiling the ore, and reclaimer is used to recover the ore from the stockpile at the stockyards [37]. They generally travel on rails between stockpiles and are connected with conveyor belts. Many times the blending is done by depositing the ore from different grades to the same stockpile. In recent years, much work has gone into automating these instruments for remote operations [38], collision avoidance, stockpile mapping, and performance improvements.

Traditionally, failure mode and effect analysis (FEMA), fault tree analysis, and Pareto are done to identify the root cause of failure. Depending on the level of instrumentation, most stackers and reclaimers have historians where operation data is stored. This operation data, including sensor data, is used in advanced analytics to predict the maintenance requirements (Fig. 19.18).

Advanced analytics like Remaining useful life (RUL) are popular for predicting subcomponent failure. Key subcomponents for stackers and reclaimers are bucket-wheel assembly, undercarriage, travel wheels, slew arms, motors, and gears. Advanced analytics models use the vibration, current, and acoustic sensor data [39] to identify deviations from normal operation and pro-actively raise maintenance notifications for slew bearing wear and abnormal motor current draw.

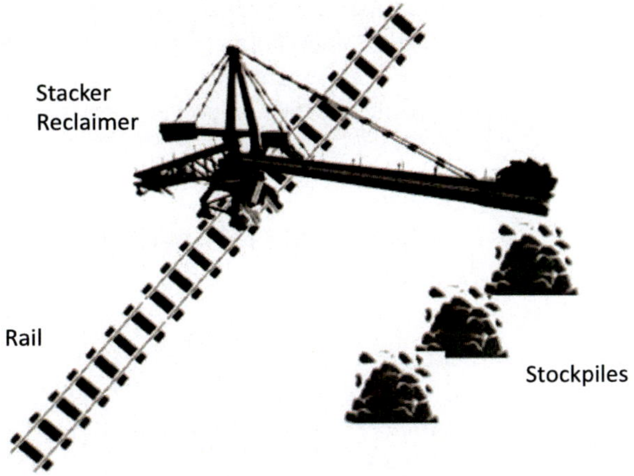

Fig. 19.18 Stacker-reclaimer at the port stockyard

Use Case 14: Port Maintenance [Wagon Tippler/Car Dumper]—Predict Failure of Car Dumper Based on Remaining Useful Life

A wagon tippler or car dumper is used to unload ore from the railway wagon at the ports. In the rotary type wagon tippler, wagons are placed in the barrel and unloaded by rotating them upside down.

Advanced analytics models in wagon tippler have been used to combine the vibration sensor data with the current draw from the variable speed drives helps to predict equipment failure [40, 41]. In the past, some analysis focused on identifying the patterns in the sequence of alarms to predict significant equipment failure (Fig. 19.19).

Recently, in one case, it was found that the difference in power consumed between the two motors of wagon tipper was a significant contributor to impending failure.

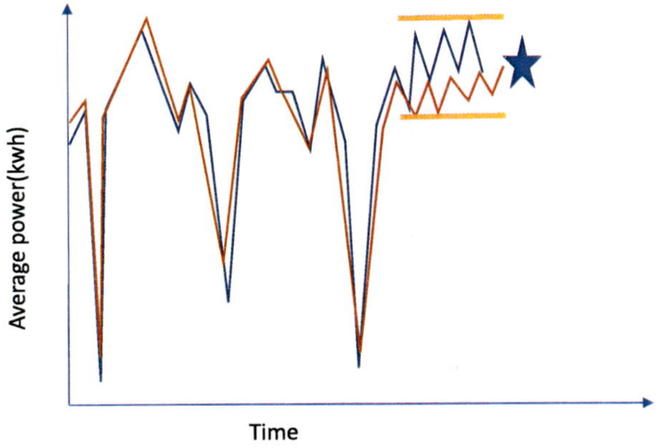

Fig. 19.19 Power consumption deviation between two variable speed drives before failure

Similarly, predicting the power consumption based on operation, weight, and speed and analyzing the difference between predicted and actual power was also a solid contributor to failure. This difference was attributed to bearing failure and used to trigger maintenance notification to prevent costly downtime.

Use Case 15: Port Safety [Worker Safety]—Video Analytics to Detect PPE Violations

Video surveillance data (CCTV cameras) is used in many work areas to ensure the security and safety of the site. Typically, hundreds of cameras feed data to the security/safety office. Since it is impossible to review the feed from all cameras in real-time by the human operator, the use case for video surveillance tends toward post-facto video retrieval and analysis for historical incidents and disputes. With advancements in Computer vision technologies, advanced analytics models are trained with the surveillance video feed to perform automated analysis for object detection, object classification, object tracking and raise proactive alerts in real-time to detect and mitigate any safety or security violations (Fig. 19.20).

Advanced analytics Computer vision technologies are being used for PPE violation detection to improve the safety of workers. Alerts are generated automatically by the model if it detects a violation of the safety rules, like not wearing a helmet or safety vest. Convolutional neural network (CNN) models are seen as most effective

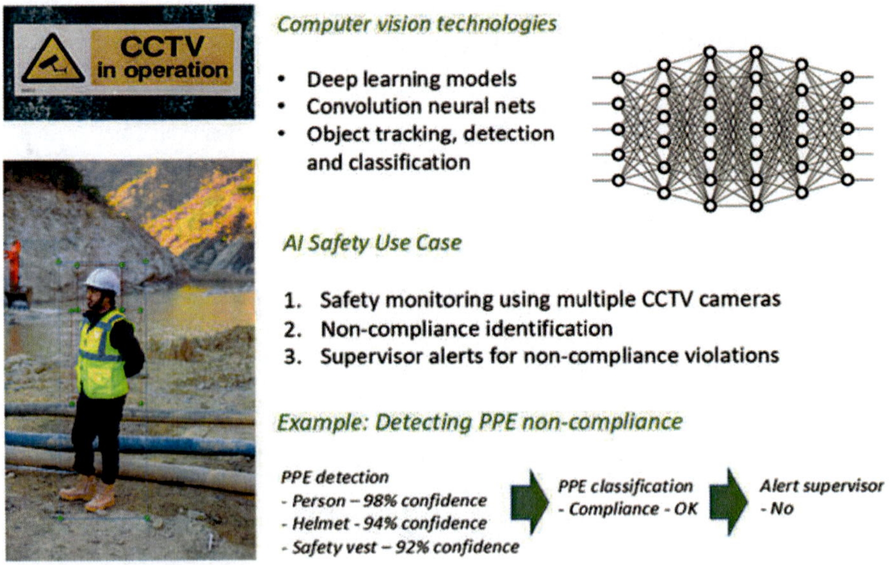

Fig. 19.20 Computer vision models to detect PPE non-compliance

in these use cases. Typical challenges include providing the labeled or annotated images for training the deep learning CNN models.

Advanced analytics technologies in computer vision are leapfrogging every few months, like understanding the Spatio-temporal relationships of objects, which can monitor people's behavior based on the change of posture, time of day, relationship with equipment, movement between areas, and more.

Trading, Chartering, and Operations Advanced Analytics Use Cases

Much of the long-distance transportation of bulk minerals is done using marine transportation. Advanced analytics is being applied in marine transportation by sales, chartering, and operation teams to make more informed decisions by negotiating with the customer on price, fixing time-charter vessels, or monitoring safety.

Most commercial ships need to report AIS (automatic identification system) data. Typically, this data contains status, speed, heading, the ship's name, port of origin, size, draft, ETA, and more. The data gathered from AIS is used for advanced analytics use cases and becomes more powerful if the data is combined with port line-ups and other internal data sets (Fig. 19.21).

In this section, the following use case will be discussed (Table 19.11).

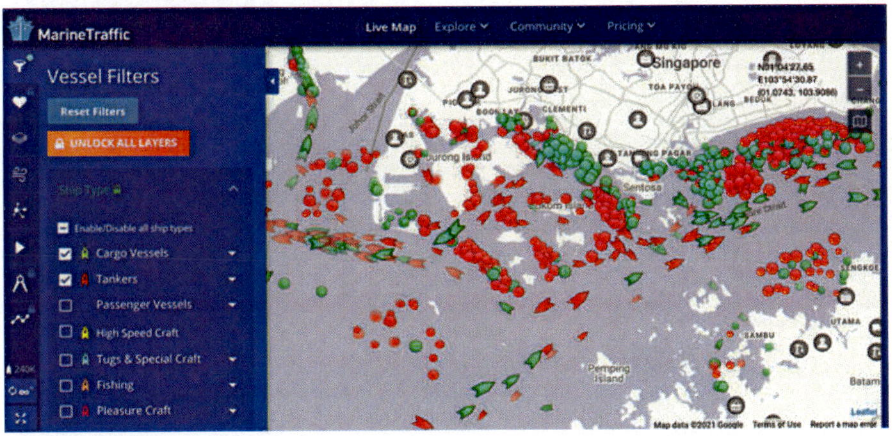

Fig. 19.21 Maritime traffic for vessels. *Source* https://www.marinetraffic.com/

Table 19.11 Trading, chartering, operations use case details

Area	Topic	Use case description
Trading and chartering	Market insights	Insights and prediction on commodity flows
Operations	Demurrage costs and safe operations	Predicting ETA and deviations from the route

Use Case 16: Trading and Chartering [Market Insights]—Insights and Prediction on Commodity Flows

Supply and demand affects exchange-traded goods, including the ore price and ships' time charter price. Since ships are the major means of transport for bulk commodities, analyzing the ship voyages provides powerful insights.

AXS Marine [42] and IHS Markit [43] both provide trade flow data. Miners' unique differentiation is based on fusing this external trade flow data with the internal voyage, port line-up, and other data sets (Fig. 19.22).

In the paper "Can big maritime data be applied to shipping industry analysis? Focussing on commodities and vessel sizes of dry bulk carriers [44]," the authors estimate the global trade flow pattern of dry bulk cargo by commodity by using AIS data.

Leveraging the base commodity flow and AIS data, advanced analytics models can provide insights into, say, which mill has consumed what grade of ore and how

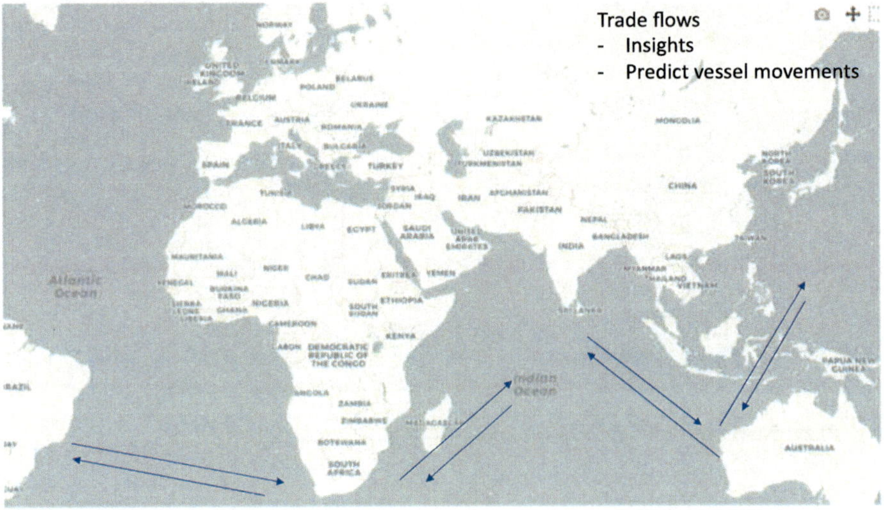

Fig. 19.22 Predicting vessel movements

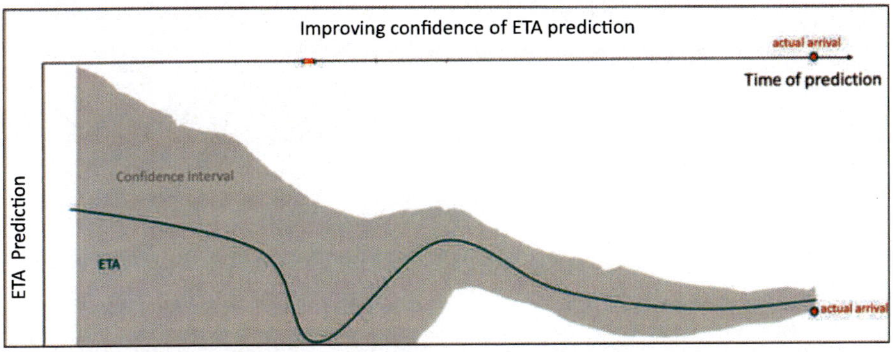

Fig. 19.23 Predicting vessel ETA

much more ore does it require. Further, the models can also predict the future 1–3 months of supply and demand of ships based on the voyage patterns to help charters make informed decisions on when to charter the vessel.

Use Case 17: Operations [Demurrage Costs and Safe Operations]—Predicting ETA and Deviations from the Route

Estimating time to arrival (ETA) is an essential parameter for understating when the ship will arrive at a particular destination. By monitoring the AIS data, analyzing ship trajectory, and accurate prediction can be made on ship ETA. In the paper "Vessel estimated time of arrival prediction system based on a path-finding algorithm [45]," the authors use reinforcement learning to predict trajectory and Markov Chain property and Bayesian Sampling to estimate speed. Further in the paper "ETA Prediction for Vessels using Machine Learning [46]," the authors explore multiple machine learning models to predict the ETA of ships (Fig. 19.23).

Decarbonization in Mining Industry

In many studies, it has been shown that ESG policy adoption and commitments improve firms' performance in the long term [47] compared to the competition. However, to meet the "E" in ESG commitments, mining companies have to reduce their environmental impact [48]. One of the key contributors to the environmental and climate impact is emission of greenhouse gases. In order to reduce emissions, miners are exploring alternative fuels in material transportation that have lower emissions, making equipment more efficient and developing trade models with responsible suppliers and customers.

A recent study by UNDP [49] has mapped the mining industry issue areas to the UN sustainable development goals as below (Fig. 19.24).

In this mapping, specifically for SDG13 Climate Action, the critical discussion is around:

- Reduce emissions—by improving energy efficiency, using renewable energy, use low-emission fuels, align with INDCs, measure and report direct, indirect, and product-related emissions;
- Build climate change resilience—plan for climate change impacts on mines and communities, strengthen emergency response plans, model climate-related environmental impact; and
- Recognize climate change in planning and investment—use scenario planning to inform views on climate and energy risks, use climate projects in design and placement of operations and infrastructure, adopt corporate climate change carbon

Fig. 19.24 Mapping UN SDGs to Mining Industry by UNDP [49]

management and disclosure policies, use shadow carbon prices to inform portfolio evaluation investment decisions and include climate change in the board agenda.

Specifically, to reduce emissions, the recommendation is to start with measuring real emissions. This means deploying a robust data platform that can ingest data from multiple mine sites, and equipment is to provide the actual figure on emissions. Second, by using advanced analytics to focus on efficiency improvement, miners can unlock capital which can then fund the investments required in moving to electrification or low-emission energy sources like electric trucks.

Use Case 18: Decarbonization in Shipping—Improving the Performance of Vessels

Many physical and digital technologies are being experimented with by shipping majors [50], start-up incubators [51], and governments [52] to improve the efficiency of ships and move toward low carbon emission fuels.

Some work in the area of physical technologies include:

- air lubrication [53]—improve the efficiency of vessels by pumping air down the hull to create bubbles which reduce the draft;
- waste heat recovery [54]—recover the waste heat to power the generator for auxiliaries;
- fuel additives [55]—to detect and remove fuel contamination;
- special Coating [56]—to reduce friction;
- CCS capture [57]—dropping CO_2 ice on the sea bed; and
- low carbon fuels [58] explore alternative fuels like ammonia, methane, green hydrogen, and nuclear.

In addition to the physical technologies, digital technologies like enterprise data analytics platforms are also being used to gain efficiencies (Fig. 19.25).

The enterprise data analytics platform houses the data and analytics capability. Specifically for vessels, data from multiple sources like Vessel reference data, AIS, wind, weather route is used to analyze ship performance and reduce fuel consumption.

Conclusion

There is a huge opportunity to reduce the cost of material transportation and gain efficiencies by using advanced analytics for decision-making.

The eighteen use cases described in the chapter provide practical examples of how miners can gain cost benefits by taking advantage of the data from each piece of equipment, process, and analyze patterns in data to improve efficiency. Efficiency

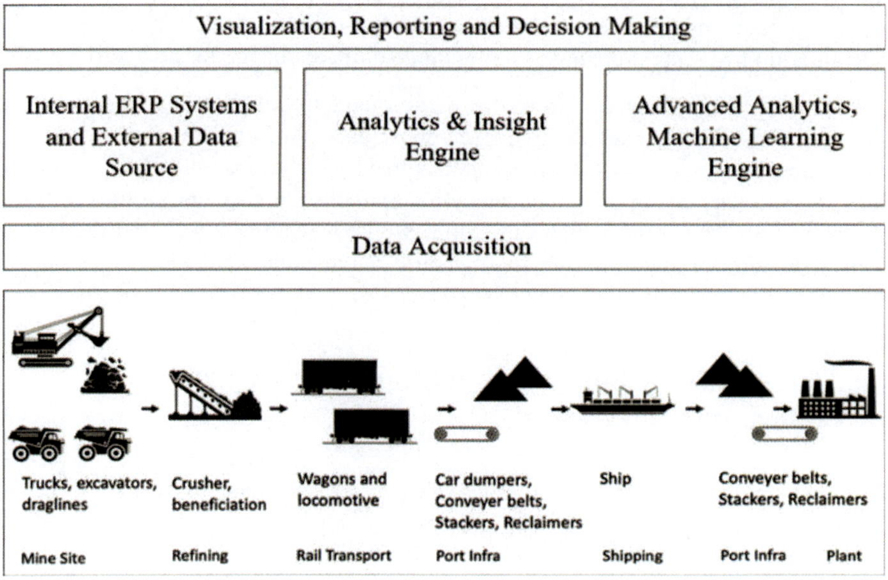

Fig. 19.25 Enterprise data lake for reporting emissions and increasing efficiencies to reduce impact

improvements also result in reducing emissions and align with the long-term goals of decarbonization in material transportation.

References

1. Gartner Glossary Advanced Analytics [cited 2021 Jun 1]. Available from: https://www.gartner.com/en/information-technology/glossary/advanced-analytics.
2. Siegel, E. 2013. *Predictive Analytics: The Power to Predict Who Will Click, Buy, Lie, or Die*. Wiley.
3. Dantzig, G.B. 2014. *The Nature of Mathematical Programming* [cited 2021 Jun 1]. Available from: https://glossary.informs.org/second.php?page=nature.html#:~:text=A%20mathematical%20program%20is%20an%20optimization%20problem%20of%20the%20form%3A&text=The%20relations%2C%20x%20in%20X,find%20an%20equivalent%20standard%20form.
4. Kiewell, D., K. McLellan, and S. Saatchi. 2014. *How to Price Risk to Win and Profit* [cited 2021 Jun 1]. Available from: https://www.mckinsey.com/business-functions/marketing-and-sales/our-insights/how-to-price-risk-to-win-and-profit.
5. Ernst, A.T., et al. 2017. Mathematical models for the berth allocation problem in dry bulk terminals. *Journal of Scheduling* 20 (5): 459–473.
6. Sun, S., et al. 2019. A survey of optimization methods from a machine learning perspective. *IEEE Transactions on Cybernetics* 50 (8): 3668–3681.
7. Pratap, S., et al. 2018. Rule based optimization for a bulk handling port operations. *Journal of Intelligent Manufacturing* 29 (2): 287–311.

8. Combat Supply Chain Disruptions with Data Science [cited 2021 Jun 1]. Available from: https://www.mosaicdatascience.com/2020/03/17/combat-supply-chain-disruptions-with-data-science/.
9. Regkas, G. 2020. *Leveraging on NLP to Gain Insights in Social Media, News & Broadcasting* [cited 2021 Jun 1]. Available from: https://towardsdatascience.com/leveraging-on-nlp-to-gain-insights-in-social-media-news-broadcasting-ca89752ef638.
10. DHL Supply Watch: Machine Learning to Mitigate Supplier Risks [cited 2021 Jun 1]. Available from: https://www.businesswire.com/news/home/20170524005934/en/DHL-Supply-Watch-Machine-Learning-to-Mitigate-Supplier-Risks.
11. Different types of Mining Transportation. 2017. [cited 2021 Jun 1]. Available from: https://steemit.com/science/@jonelq/different-types-of-mining-transportation.
12. Daemen, J.J. 2003. *Mining Engineering*.
13. DISPATCH Fleet Management System (FMS) Helps Mine Optimize Its Haulage Cycle and Dramatically Reduce Truck Idle Times [cited 2021 Jun 1]. Available from: https://www.modularmining.com/case-studies/dispatch-fms-helps-mine-optimize-haulage-cycle/.
14. Mining Logistics Optimization with AI Re-routes Haul Trucks and Increases Ore Production [cited 2021 Jun 1]. Available from: https://pathmind.com/mining-logistics-optimization-with-ai-re-routes-haul-trucks-and-increases-ore-production/.
15. Analyses of Diesel Use for Mine Haul and Transport Operations [cited 2021 Jun 1]; Available from: https://www.energy.gov.au/sites/default/files/analyses_of_diesel_use_for_mine_haul_and_transport_operations.pdf.
16. Soofastaei, A., et al. 2016. Reducing Fuel Consumption of Haul Trucks in Surface Mines Using Artificial Intelligence Models.
17. Artificial Intelligence Predicts When Haul Trucks Need Maintenance. 2018 [cited 2021 Jun 1]. Available from: https://nma.org/2018/12/14/artificial-intelligence-predicts-when-haul-trucks-need-maintenance/.
18. Fiscor, S. 2015. Applying Advanced Analytics to Optimize Open-pit Operations and Maintenance [cited 2021 Jun 1]. Available from: https://www.e-mj.com/features/applying-advanced-analytics-to-optimize-open-pit-operations-and-maintenance/.
19. Ageeva, Y. 2018. *Predictive Maintenance Scheduling with AI and Decision Optimization* [cited 2021 Jun 1]. Available from: https://towardsdatascience.com/predictive-maintenance-scheduling-with-ibm-data-science-experience-and-decision-optimization-25bc5f1b1b99.
20. Rio Tinto to Expand Autonomous Truck Operations to Fifth Pilbara Mine Site. 2018. [cited 2021 Jun 1]. Available from: https://www.riotinto.com/en/news/releases/Automated-truck-expansion-Pilbara.
21. Latimer, C. 2015. *Autonomous Truck and Water Cart Collide on Site UPDATE* [cited 2021 Jun 1]. Available from: https://www.australianmining.com.au/news/autonomous-truck-and-water-cart-collide-on-site-update/.
22. Moral Machine. [cited 2021 Jun 1]. Available from: https://www.moralmachine.net/.
23. Jiaxin, C., and P. Howlett. 1992. Application of critical velocities to the minimisation of fuel consumption in the control of trains. *Automatica* 28 (1): 165–169.
24. Howlett, P. 1996. Optimal strategies for the control of a train. *Automatica* 32 (4): 519–532.
25. Ganji, B., and A.Z. Kouzani. 2010. A study on look-ahead control and energy management strategies in hybrid electric vehicles. In *IEEE ICCA 2010*. IEEE.
26. Today's Railroads Can Predict the Future. [cited 2021 Jun 1]. Available from: https://www.aar.org/article/todays-railroads-can-predict-the-future-heres-how/.
27. Li, Z., and Q. He. 2015. Prediction of railcar remaining useful life by multiple data source fusion. *IEEE Transactions on Intelligent Transportation Systems* 16 (4): 2226–2235.
28. Ghofrani, F., et al. 2018. Recent applications of big data analytics in railway transportation systems: A survey. *Transportation Research Part C: Emerging Technologies* 90: 226–246.
29. Hu, C., and X. Liu. 2016. Modeling track geometry degradation using support vector machine technique. In *ASME/IEEE Joint Rail Conference*. American Society of Mechanical Engineers.
30. Zarembski, A.M., D. Einbinder, and N. Attoh-Okine. 2016. Using multiple adaptive regression to address the impact of track geometry on development of rail defects. *Construction and Building Materials* 127: 546–555.

31. Freight Rail Technology. [cited 2021 Jun 1]. Available from: https://www.aar.org/topic/freight-rail-tech.
32. Shi, D., et al. 2021. *Deep Learning based Virtual Point Tracking for Real-Time Target-less Dynamic Displacement Measurement in Railway Applications*. arXiv preprint arXiv:2101.06702.
33. Artificial Intelligence Applied to Mining: Monitoring System for Predictive Maintenance of Conveyor Belts. 2020. [cited 2021 Jun 1]. Available from: https://coreal.cl/en/artificial-intelligence-applied-to-mining-monitoring-system-for-predictive-maintenance-of-conveyor-belts/.
34. Predictive Maintenance of Machine Parts at Port-Based Coal Conveying Facility. [cited 2021 Jun 1]. Available from: https://www.bannerengineering.com/us/en/solutions/iiot-data-driven-factory/predictive-maintenance-of-rotating-parts-on-conveyor.html.
35. Liu, X., et al. 2019. Integrated decision making for predictive maintenance of belt conveyor systems. *Reliability Engineering & System Safety* 188: 347–351.
36. Predict Conveyor Belt Tensioning Date. [cited 2021 Jun 1]. Available from: https://www.seeq.com/resources/use-cases/predict-conveyor-belt-tensioning-date.
37. Stackers and Reclaimers Information. [cited 2021 Jun 1]. Available from: https://www.globalspec.com/learnmore/material_handling_packaging_equipment/material_handling_equipment/stackers_reclaimers#:~:text=Stackers%20and%20reclaimers%20are%20used,large%20piles%20or%20circular%20stacks.
38. AMECO: The Power of Know-How. 2017. [cited 2021 Jun 1]. Available from: https://www.ameco-group.com/news/ameco-the-power-of-know-how/
39. Condition Monitoring. [cited 2021 Jun 1]. Available from: https://www.rockwellautomation.com/en-us/products/hardware/allen-bradley/condition-monitoring.html.
40. Pal, A. 2013. Failure analysis of gear shaft of rotary wagon tippler in a thermal power plant-a theoretical review. *International Journal of Mechanical and Production Engineering Research and Development* 3 (2): 99–104.
41. Farias, O.S., et al. 2008. A real time expert system for faults identification in rotary railcar dumpers. In *ICINCO-ICSO*.
42. AXS Marine Tradeflows [cited 2021 Jun 1]. Available from: https://public.axsmarine.com/tradeflows.
43. Commodities at Sea. [cited 2021 Jun 1]. Available from: https://ihsmarkit.com/products/commodities-at-sea.html.
44. Kanamoto, K., et al. 2021. Can maritime big data be applied to shipping industry analysis? Focussing on commodities and vessel sizes of dry bulk carriers. *Maritime Economics & Logistics* 23 (2): 211–236.
45. Park, K., S. Sim, and H. Bae. 2021. Vessel estimated time of arrival prediction system based on a path-finding algorithm. *Maritime Transport Research* 2: 100012.
46. Flapper, E. 2020. *ETA Prediction for Vessels Using Machine Learning*. University of Twente.
47. Eccles, R., L. Ioannou, and G. Serafeim. 2012. *Is Sustainability Now the Key to Corporate Success?* [cited 2021 Jun 1]. Available from: https://www.theguardian.com/sustainable-business/sustainability-key-corporate-success.
48. O'Brien, J., and H. Stoch. 2021. *Trend 3: ESG: Getting Serious About Decarbonization* [cited 2021 Jun 1]. Available from: https://www2.deloitte.com/us/en/insights/industry/mining-and-metals/tracking-the-trends/2021/decarbonization-mining-and-climate-change.html.
49. Atlas, A. 2016. *Mapping Mining to the Sustainable.*
50. Decarbonising Shipping. [cited 2021 Jun 1]. Available from: https://www.shell.com/energy-and-innovation/the-energy-future/decarbonising-shipping.html.
51. Decarbonizing Shipping. [cited 2021 Jun 1]. Available from: https://discover.rainmaking.io/decarbonizing-shipping.
52. Singapore Looks to Tech to Bolster Its Role in Shipping. [cited 2021 Jun 1]. Available from: https://www.wsj.com/articles/singapore-looks-to-tech-to-bolster-its-role-in-shipping-11617909788.
53. WÄRTSILÄ Encyclopedia of Marine Technology. [cited 2021 Jun 1]. Available from: https://www.wartsila.com/encyclopedia/term/air-lubrication.

54. Waste Heat Recovery Technology from Orcan Energy on Energy Efficient Tanker for the First Time. 2020. [cited 2021 Jun 1]. Available from: https://www.orcan-energy.com/en/details/waste-heat-recovery-technology-from-orcan-energy-on-energy-efficient-tanker-for-the-first-time.html.
55. Comprehensive Monitoring of Fluid Quality [cited 2021 Jun 1]. Available from: https://dravam.com/.
56. SLIPS® for Marine Applications. [cited 2021 Jun 1]. Available from: https://adaptivesurface.tech/marine/.
57. Eason, C. 2019. *Dropping Shipping's CO_2 Straight into the Seabed May Work or Maybe Not* [cited 2021 Jun 1]. Available from: https://fathom.world/co2-spears-decarbonise-ccs/.
58. Konrad, J. 2020. *Bill Gates Launches A Nuclear Ship Battery Partnership* [cited 2021 Jun 1]. Available from: https://gcaptain.com/bill-gates-nuclear-ship-battery/.

Chapter 20
Advanced Analytics for Energy-Efficiency Improvement in Mine-Railway Operation

Ali Soofastaei and Milad Fouladgar

Abstract The mining industry consumes an enormous amount of energy globally, the main part of which is conservable. Diesel is a key source of energy in mining operations, and mine locomotives have significant diesel consumption. Trian speed has been recognized as the primary parameter affecting locomotive fuel consumption. In this study, an artificial intelligence (AI) look-forward control is developed as an online method for energy-efficiency improvement in mine-railway operation. An AI controller will modify the desired train-speed profile by accounting for the route's grade resistance and speed limits. The travel-time increment is applied as an improvement constraint. Recent models for mine-train-movement simulation have estimated locomotive fuel burn using an indirect index. An AI-developed algorithm for mine-train-movement simulation can correctly predict locomotive diesel consumption based on this paper's considered values of the transfer parameters. This algorithm finds the mine-locomotive subsystems and satisfies the practical diesel-consumption data specified in the locomotive's manufacturer catalog. The model developed in this study has two main sections designed to estimate locomotive fuel consumption in different situations using an artificial neural network (ANN). An optimization section applies a genetic algorithm (GA) to optimize train speed to minimize locomotive diesel consumption. The AI model proposed in this paper is learned and validated using real datasets collected from a mine-railway route in Western Australia. The simulation of a mine train with a commonly used locomotive in Australia General Motors SD40-2 (GM SD40-2) on a local railway track illustrates a significant reduction in diesel consumption along with a satisfactory travel-time increment. The simulation results also demonstrate that the AI look-forward controller has faster calculations than dynamic programming control systems.

A. Soofastaei (✉)
Vale, Brisbane, Australia
e-mail: ali@soofastaei.net
URL: https://www.soofastaei.net

M. Fouladgar
Department of Mechanical Engineering, Najafabad Branch, Islamic Azad University, Najafabad, Iran
e-mail: milad_fouladgar@smc.iaun.ac.ir

A. Soofastaei (ed.), *Advanced Analytics in Mining Engineering*,
https://doi.org/10.1007/978-3-030-91589-6_20

Keywords Fuel consumption · Energy efficiency · Locomotive · Mining · Railway · Simulation · Optimization · Artificial intelligence · Neural network · Genetic algorithm · Look-forward control

Introduction

Locomotive and railway-transportation systems are generally considered an energy-efficient methods for transferring mining material [1]. However, this type of transportation system is one of the principal fuel consumers in the mining industry. Therefore, decreasing locomotive diesel consumption and improving energy efficiency in the mine-railway-transportation system would have substantial economic and environmental benefits. Research in the field of energy-efficiency improvement in railway transportation has evolved during the past 16 years [1–3]. Using the maximum principle and practical analysis, and optimized train-speed profile on a route lacking a gradient [3]. The effect of train speed limits on locomotive fuel consumption was also investigated by Howlett and Pudney [2]. Hansen and Pachl explained improved train-speed profile concerning train movement between two stations on a flat route [4]. This improved train-speed profile has four stages: (1) accelerating to reach maximum velocity; (2) maintaining constant velocity; (3) coasting exclusively of traction forces; (4) decelerating.

There is an opportunity to reduce fuel consumption by up to 15% using advanced data-analytics models to optimize energy efficiency in locomotives and railways [5]. Howlett and Cheng applied an innovative model to solve a typical optimization problem for a route with a constant gradient [1]. In 2009, Howlett and Pudney created a set of specific equations based on the Kuhn equations. In their project, locomotive fuel-burn optimization was resolved for each given route segment by a local minimization of fuel burned by the locomotive [6]. The developed optimization model attained the best point at which to increase train speed before an upward hill. Khmelnytskyi applied the maximum-value method to decrease the locomotive diesel consumption in underground routes [7]. In 2011, Kang proposed a GA optimization model to improve train speed [6]. In this research, the primary parameter was the coasting point, and a fitness function was introduced for a situation in which the locomotive could pass through the distance between two positions in fixed travel time. In 2013, Li and colleagues proposed a stochastic operation algorithm for locomotive fuel efficiency; in their suggested model, the coasting stage was replaced by a quasi-coasting stage [8]. In other research, train transportation in a subway network was examined [7, 9]. This research developed the comprehensive evaluation index (CEI) to investigate railway transportation combined optimization models by estimating energy costs and valuable process time.

In all the studies discussed, the researchers used a simple model for locomotive traction force, and train energy consumption was calculated indirectly. Locomotive fuel burn is a function of an engine's operation torque and speed. Therefore, considering the working points and function limitations of locomotive subsystems can provide more correct simulation outcomes.

The numerical models reviewed in the present study deliver reputable results. However, these models use offline calculations [2–4, 6–8, 10] because they consider the static condition of the problem. Thus, if any changes occur in the railway route ahead, the models must be run again. This disadvantage highlights the importance of developing online optimization models that consider dynamic conditions in the locomotive cabin or the train control room.

The look-forward control method proposed by Kouzani and Ganji in 2010 [8] is an innovative method for optimizing vehicles' fuel consumption on normal roads. This method was developed based on applying future road information to make deceleration and acceleration control signals. The look-forward control method has been used as an effective instrument for reducing fuel burn by applying dynamic programming to tackle the optimization control problem [10]. In research projects completed in 2010, the results illustrated significant proficiency in road transportation [11, 12]. In research conducted in 2011, the look-forward control method was used as an effective approach for optimizing fuel consumption in hybrid electric cars [13]. Khayyam also applied this approach in 2011 to optimize ventilation systems [14].

The research reviewed in this study demonstrates that look-forward control algorithms do not consider vehicle speed limits in segments of the route ahead. In some cases, the developed algorithms suggested increasing vehicle speed before reaching an uphill segment, but technically, it is not correct to increase velocity when approaching a curved route or other speed limits in a segment. Moreover, the look-forward control method has not been applied practically in the field of mine-railway transportation. The present study aims to use the look-forward control approach to increase energy efficiency in mine-railway transportation. The proposed AI prediction and optimization algorithms can be applied to reduce mine-locomotive fuel consumption.

An integrated AI look-forward algorithm as a dynamic method with online calculations is applied in this study by considering the route grade and velocity limits of the route ahead. The developed model is a combination of a prediction and optimization of AI-based algorithms. To estimate locomotive fuel consumption, an ANN is developed and explained in Sect. 20.5. The optimization problem in mine-railway operation is defined in Sect. 20.3. To reduce locomotive fuel consumption, a GA is developed and combined with the previously developed prediction model. This algorithm will optimize train speed to decrease the fuel burn of the locomotive engine. This optimization algorithm is explained in Sect. 20.5. The cost function is defined as travel time, and locomotive fuel burn is considered a limitation. The designed AI look-forward controller is presented in Sect. 20.4. Section 20.6 explains the efficiency of the proposed AI model compared with the standard speed controller based on dynamic programming using a simulation of a mine train with the GM SD40-2 locomotive on a local railway track in Western Australia. Finally, Sect. 20.7 presents the conclusion.

Mine-Train-Movement Simulation

Examining the algorithms developed for mine-train-movement simulation reveals that indirect indexes estimate fuel burn and that no practical model has been developed to test locomotive and railway-transportation fleets in the mining industry. Therefore, the present study develops an innovative AI algorithm for mine-train-movement simulation to estimate locomotive fuel burn accurately. This algorithm considers all the mine-locomotive subsystems that play critical roles in generating and transmitting power. The main block diagram of the proposed model is presented in Fig. 20.1.

The components of the mine-train system, including the wheels, final drive, electric motor, generator, and diesel engine, are illustrated in Fig. 20.1. In the developed AI model, mine-locomotive fuel burn is estimated based on the specific fuel-burn graphs completed by locomotive manufactures as a function of the diesel engine's torque and speed. Thus, the developed model can estimate locomotive fuel burn straight and more accurately than current models of mine-train-movement simulation.

Mine locomotives use an internal control loop with a diesel-engine governor, companion equipment, and a load regulator as the principal components [15]. The real engine speed and train driver's throttle position control some external inputs of the governor. The governor controls the fuel-injector setting that regulates the load-regulator position and engine fuel rate. The load regulator is primarily a potentiometer that controls the output power of the locomotive generator by changing the loading applied to the main engine. As the load on the main engine changes, its rotational speed will also change. The engine governor senses this through a change in the engine-speed feedback signal. The effect of this change is to adjust both the load-regulator position and the fuel consumption rate. As a result, the diesel engine's torque and speed will remain constant for any given throttle position, regardless of actual mine-train speed on the route.

A mine-train driver can change the locomotive speed by using the brake handle or setting the throttle position. Therefore, the mine-locomotive traction force F_{traction} can be calculated by Eq. 20.1.

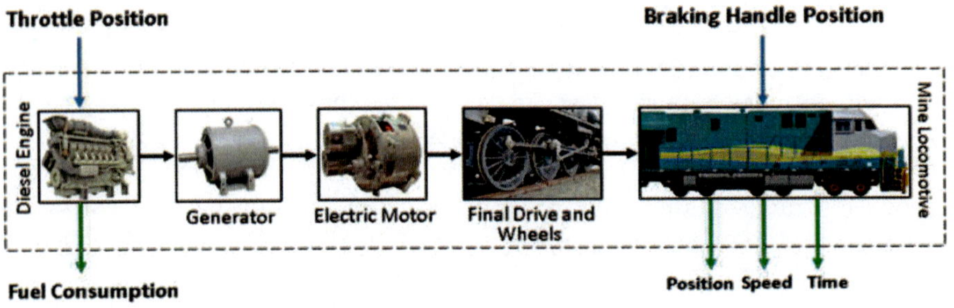

Fig. 20.1 Mine-train model

$$hF_{\text{traction}} = \min(F_{\text{motor traction}}, F_{\text{adhesion}}) \tag{20.1}$$

where $F_{\text{motor traction}}$ is the force generated by electrical motors and F_{adhesion} is the maximum adhesion force between the rail and train wheels. This constriction is measured in the axle and wheel block. The maximum adhesion force between the rail and train wheels is calculated by Eq. 20.2.

$$F_{\text{adhesion}} = 9.8\alpha \times W \times N \tag{20.2}$$

where α is the adhesion coefficient, N is the number of locomotive axles, and W is the locomotive axial load (kg). The adhesion coefficient is represented by Eqs. 20.3 and 20.4 for different weather conditions [4].

$$\alpha = 0.161 + \frac{7.5}{44 + 3.6S} \quad \text{Dry Condition} \tag{20.3}$$

$$\alpha = \frac{3.78}{23.6 + S} \quad \text{Wet Condition} \tag{20.4}$$

where S is the train speed (km/h).

The experimental equations of locomotive movement are applied in the mine-train longitudinal dynamics block. In this block model, resistance forces are calculated by the Davis equation, where the resistance force is presented as a parabolic function of the mine-train speed [16] (see Fig. 20.2).

Figure 20.2 shows some effective forces in mine-locomotive operation. These forces are aerodynamic resistance, Davis resistance, route resistance, curvature resistance, and traction force provided by the locomotive engine.

Davis resistance force $F_{\text{resistance-Davis}}$ is estimated separately for the locomotive and the cars; their summation is subsequently considered. The Davis equation is written as follows:

$$F_{\text{resistance-Davis}} = 9.8\left(A + \frac{B}{0.001W} + C \times S + \frac{E \times D \times S^2}{0001W \times N}\right) \tag{20.5}$$

Fig. 20.2 Mine-locomotive and Resistance Forces

where N is the number of axles; W is the locomotive axial load (kg); and A, B, C, D, and E are the Davis coefficients.

In this case, the additional resistance force is the route geometry resistance due to route curvature and grade. Therefore, route geometry resistance $F_{\text{resistance-route}}$ (N/kg) is calculated by Eq. 20.6:

$$F_{\text{resistance-route}} = 0.01\left(F_{\text{resistance-grade}} + 0.04F_{\text{resistance-curvature}}\right) \tag{20.6}$$

where $F_{\text{resistance-grade}}$ (N/kg) depends on elevation variations (m) per 1000 m and $F_{\text{resistance-curvature}}$ (N/kg) is the curved-route resistance force represented by:

$$F_{\text{resistance-curvature}} = \begin{cases} \frac{650}{r-55}\ r \geq 500 \\ \frac{650}{r-30}\ r < 500 \end{cases} \tag{20.7}$$

where r is the curve radius (m) [4].

The traction force becomes zero by braking, and the braking force is replaced with the traction force. Therefore, the braking force F_{brake} (N/kg) is estimated by Eq. 20.8.

$$F_{\text{brake}} = 5\phi_k \times \theta_b \tag{20.8}$$

where θ_b is the brake weight fraction, a function of locomotive speed and route grade [4]; and φ_k is the friction coefficient, expressed by Eq. 20.9.

$$\phi_k = 0.32 \times \frac{S + 100}{5S \times 100} \tag{20.9}$$

where S is the mine-train speed (km/h) [4]. Therefore, the locomotive speed, acceleration, train position, and travel time can be estimated by applying Newton's second law of motion.

Mine-Train Optimization

The reviewed literature demonstrates that one of the best approaches to optimizing effective parameters for mine-train-movement performance is improving locomotive fuel consumption and travel time. However, these two parameters perform oppositely. This means that locomotive fuel-burn reduction may lead to an increase in train travel time. Thus, finding an accurate parameter that can change mine-train travel time to its corresponding locomotive fuel burn is complex. Defining the objective function as locomotive-engine diesel consumption and mine-train travel time as a constraint is potentially one of the most effective proposed approaches to the optimization problem of mine-train operation [6]. The diesel burned by a locomotive engine must be minimized when the train travel-time increment is less than a predefined limit. In the present study, mine-train travel-time restriction is considered at most a 5% increase compared with the travel time achieved by following the initially desired speed profile. This profile is calculated by the mine-train speed-limit signs on the route. Another limitation in the optimization problem is the speed deviation from the initially desired speed profile. This study considers the speed-deviation limitation of 20 km/h below and above the initial desired speed.

It should be noted that the initially desired speed profile is generally below the maximum permissible speed on the railway track. Thus, 20 km/h deviation from the desired speed profile will not cause safety problems. Therefore, the mine-train optimization problem is defined as follows.

Minimum locomotive fuel consumption is subject to:

$$\begin{aligned} &\Delta t < 0.05 t_{\mathrm{IDS}} \\ &S_{\mathrm{IDS}} - 20\,(\mathrm{km/h}) < S_d < S + 20\,(\mathrm{km/h}) \end{aligned} \tag{20.10}$$

where Δt is the mine-train travel-time increment, and t_{IDS} is the mine-train travel time reached by the initially desired speed profile estimated by the train speed-limit signs on the route. S_{IDS} is the initially desired speed profile, and S_{d} is the desired speed.

AI Controller Design

An AI controller is designed to minimize mine-train fuel burn. This controller has an intelligent system and changes the initially desired speed profile by using AI algorithms according to well-prepared road gradients and train speed limits of the route ahead. For example, if there is an uphill segment in the route ahead, the desired speed should be increased before the uphill segment. As a result, the train can pass through the uphill segment with less effort. A simple structure of the proposed AI controller is presented in Fig. 20.3.

The speed controller is the main regulator that controls the braking and throttle position based on the received speed error from the desired locomotive speed profile. The proposed AI control system is divided into two components: the AI-designed unit and the AI controller. An AI controller block diagram is illustrated in Fig. 20.4.

As presented in Fig. 20.4, the route gradient (grade resistance), initial desired speed, and speed limits are required to use the AI control system. Therefore, the locomotive speed limit, curve radius, road gradient, and start and endpoints should be available in provided datasets for each route in the data lake. The AI-designed unit calculates locomotive speed limit and the average gradient for a selected segment of the route ahead, known as the "look-forward window." This window will be considered in the route segment ahead with a specified distance from the prompt position of the train (see Fig. 20.5).

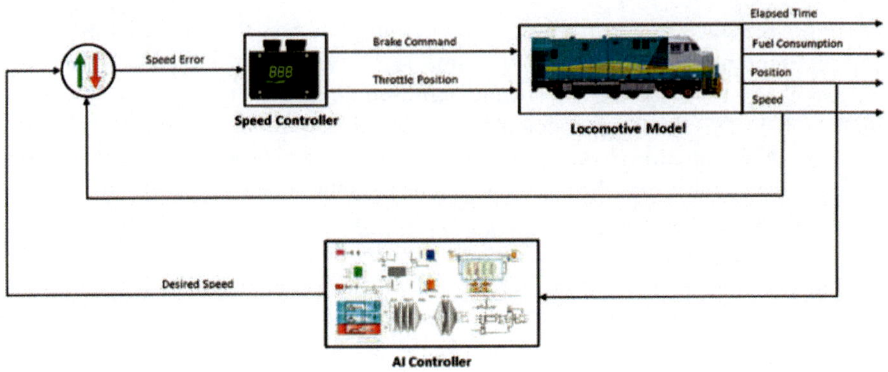

Fig. 20.3 Mine-train-speed AI controller

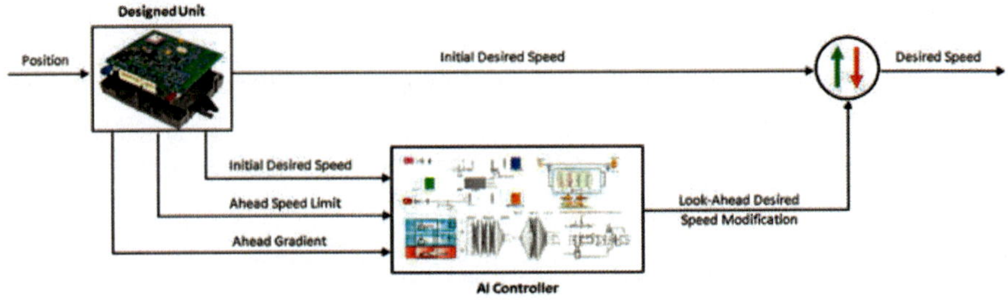

Fig. 20.4 AI control system's internal components

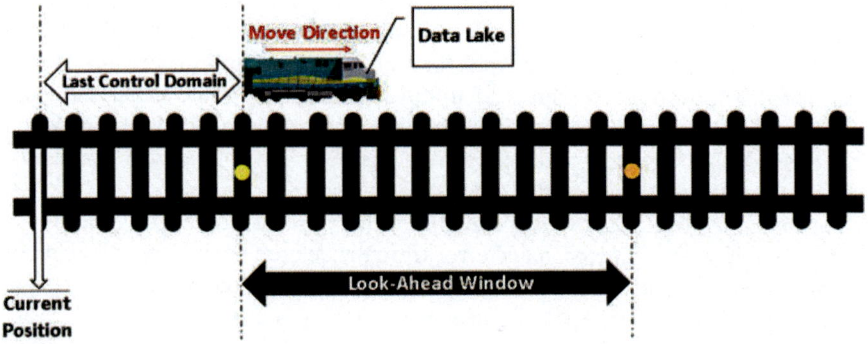

Fig. 20.5 Look-forward window

In this window, previous control commands affect train speed. The look-forward window is moving at the same speed in front of the mine train. The look-forward window immediately scans the route segments ahead. Therefore, the fluctuations in the route segments ahead are sensed by the opened window. The effect of each control command is anticipated to appear in the start point of its corresponding look-forward window.

The route is discretized as a set of similarly spaced repeated points where the requested information is obtainable in the data lake to estimate the average upcoming route gradients and locomotive speed limits. If the previous control command's area and the look-forward window include m_1 and m_2 points, respectively, then the forward position (locomotive speed limit and route gradient) will be defined by Eq. 20.11.

$$\text{Forward Situation} = \frac{\sum_{m_1+1}^{m_1+m_2} \text{Points Situation}}{m_2} \tag{20.11}$$

Therefore, the "forward position" is the average position of the points inside the look-forward window.

The second part of the AI control algorithm is the AI controller. This unit performs as a feed-forward controller and changes the initially desired speed profile (S_{IDS}) based on the train-speed limits and route gradient of the route ahead. The model of train movement is a complex model based on lookup tables and working limitations of the electric motors, generators, and diesel engines. Therefore, developing an AI controller may be more appropriate than the other models reviewed in the present study. Furthermore, AI models can be trained based on the heuristically applied experiences of mine-train drivers. For example, if there is no locomotive speed limit in the route ahead and the route gradient increases, the locomotive speed should increase. However, if there is a locomotive speed limit in the route ahead and the gradient increases, there is no need to increase the locomotive speed. As a result, the locomotive's energy is missed during the braking phase. As a result, using an AI controller can help mine-train drivers reduce braking force by planning the optimal speed to use in the different route segments.

AI Models

AI has existed since the 1950s [17]. Technologies pioneered by AI researchers that many organizations use extensively include deep-learning or neural networks, natural language processes (used in conversational user interfaces), and image processing (used in products such as self-driving cars). In recent years, cheap computer processing and storage have transformed old AI techniques into practical technologies. The application of AI in the mining industry can be categorized into three main groups: development of prediction models, optimization applications, and decision-making algorithms [18]. This study has developed two popular AI models (ANN and GA) to predict and minimize the mining industry's locomotive fuel burn (respectively).

ANN (Prediction)

ANNs represent the approaches the brain uses for training. They are a series of mathematical algorithms that reproduce several available features of natural nerve systems and sketch analogies of adaptive natural training. The critical element of an ANN model is the infrequent structure of the data-processing system. A typical neuronal model thus comprises weighted connectors and an activation function (see Fig. 20.6).

ANNs are designed for and used in numerous computer applications to solve multipart problems. They are straightforward and fault-tolerant algorithms that do not need the information to recognize the associated parameters and do not need mathematical explanations of the phenomena involved in the procedure.

The main part of an ANN structure is a "node". Nodes commonly sum the signals received from many sources in different ways and then perform a nonlinear act on the outcomes to generate the productions. Thus, neural networks naturally have an input layer, one or more hidden layers, and an output layer. Each input is multiplied by its connected weight, and in the purest state, these quantities and biases are joint; they then pass through the activation functions to create the output (see Fig. 20.7 and Eqs. 20.12, 20.13, 20.14).

$$E_k = \sum_{j=1}^{q} (w_{i,j,k}x_j + b_{i,k})\; k = 1,\ 2,\ ...,\ m \tag{20.12}$$

where x is the normalized input variable; w is the weight of that variable; i is the input; b is the bias; q is the number of input variables; and k and m are the counter and number of neural network nodes, respectively, in the hidden layer.

Overall, the activation functions contain both linear and nonlinear equations. The coefficients related to the hidden layer are congregated into matrices $W_{i,j,k}$ and $b_{i,k}$.

Equation 20.13 can be used as the activation function between the hidden and the output layers.

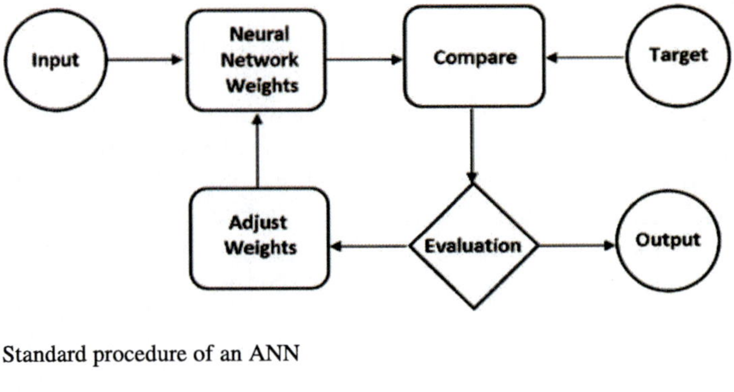

Fig. 20.6 Standard procedure of an ANN

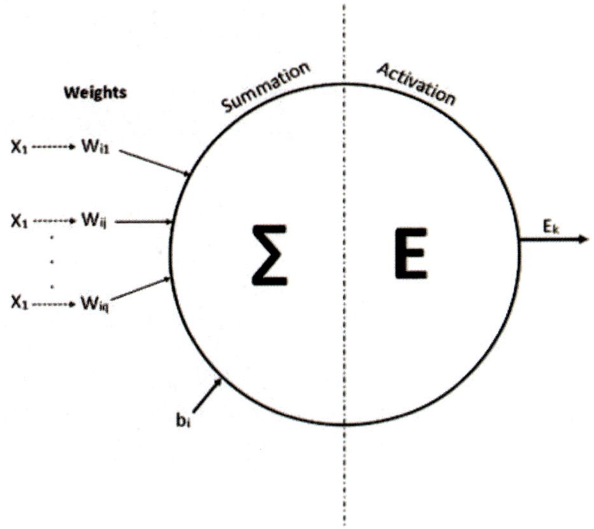

Fig. 20.7 Data processing in a neural network cell [19]

$$F_k = f(E_k) \tag{20.13}$$

where f is the transfer function.

The output layer calculates the weighted sum of the signals delivered by the hidden layer, and the corresponding coefficients are gathered into matrices $W_{o,k,}$ and b_o. Using the matrix notation, the network output can be specified by Eq. 20.14.

$$\text{Out} = \left(\sum_{k=1}^{m} w_{o,k} F_k\right) + b_o \tag{20.14}$$

Learning the network is the most significant part of the neural network demonstration and is performed using two approaches: controllable and uncontrollable learning. The most common learning algorithm is back-propagation [20]. A learning algorithm is well defined as a technique that consists of adjusting the coefficients (weights and biases) of a network to minimize the error function between the predicted network outputs and the actual outputs.

This study presents a different type of algorithm that has been examined to determine the best back-propagation learning algorithm. Unlike other back-propagation algorithms, the Levenberg–Marquardt (LM) back-propagation learning algorithm has the minimum root mean square error (RMSE), mean square error (MSE), and correlation coefficient (R^2).

Further, network learning with the LM algorithm can run efficiently with the minimum expanded memory specification (EMS) and a fast learning process. RMSE, MSE, and R^2 are the statistical criteria utilized to assess the accuracy of the results according to the following equations:

$$\mathrm{MSE} = \frac{1}{p}\sum_{r=1}^{p}(y_r - z_r)^2 \quad (20.15)$$

$$\mathrm{RMSE} = \left(\frac{1}{p}\sum_{r=1}^{p}(y_r - z_r)^2\right)^{\frac{1}{2}} \quad (20.16)$$

$$R^2 = 1 - \frac{\sum_{r=1}^{p}(y_r - z_r)^2}{\sum_{r=1}^{p}(y_r - \overline{y})^2} \quad (20.17)$$

where y is the target (real); z is the output (estimated) of the model; $\overline{y}$ is the average value of the targets, and p is the number of the network outputs.

In this study, the MSE and R^2 approaches are applied to examine the performance of the neural network output, and the LM optimization algorithm is utilized to find the optimum weights of the network.

GA (Optimization)

The GA was first proposed by Holland (1975) to represent the concept of biological development and to illustrate ideas from natural development and genetics for enterprise and the implementation of strong adaptive structures. The new generation of GAs represents reasonably recent optimization approaches. These GAs do not apply any information on derivate, which means they have an excellent opportunity for trapping local minimums. Their application in related engineering problems usually carries to the global optimal, or at least to more acceptable answers than those gained by traditional mathematical approaches. These GAs apply a straight analogy of the phenomena of development in nature. The individuals are randomly nominated from the research area. The fitness of the answers results from the variable to be optimized determined afterward from the fitness function. The individual that produces the best fitness within the population has the maximum chance of returning in the following generation. Thus, the opportunity to regenerate by crossover with another individual means creating decedents with the characteristics of both individuals. The population will converge to an optimal answer for the projected problem if a GA is adequately established. The processes that have a greater influence on the development are crossover, selection, reproduction, and mutation. Figure 20.8 presents the data-processing phases in a simple GA model.

GAs have been used in various engineering, scientific, and economic problems [20] due to their potential as optimization methods for multifaceted functions. There are four significant benefits in using GAs for optimization problems. First, GAs do not have a lot of mathematical necessities in optimization problems. Second, GAs can handle objective functions and limitations defined in continuous, discrete, or mixed search spaces. Third, the periodicity of development operators makes GAs operative during global accomplishment searches. Fourth, GAs provide extreme flexibility for

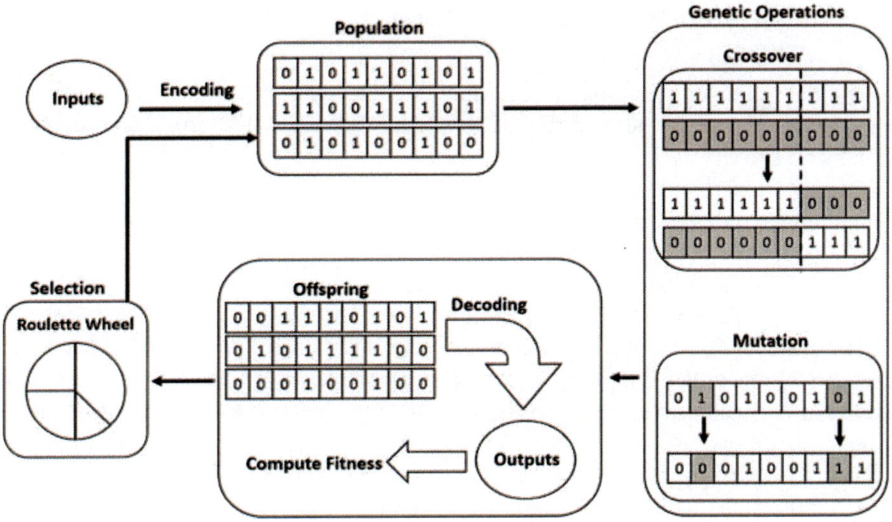

Fig. 20.8 Data processing in a GA model [18]

hybridizing with domain-dependent heuristics to allow a well-agonized application for a specific problem.

Moreover, it is significant to analyze the effect of some parameters in the behavior and the performance of GAs to create them conferring to the problem requirements and the existing resources. The effect of each parameter in the algorithm's performance depends on the class of problems that is being treated. Therefore, determining an optimized collection of values for these parameters depends on an excessive number of experiments and examinations. There are several key parameters applied in the GA technique. Details of GA main processes and parameters are presented in Table 20.1.

The primary genetic parameters are the dimension of the population, which affects the global performance and the effectiveness of the algorithm, and the mutation rate, which avoids a specified position remaining stationary in value or the search becoming fundamentally random.

Table 20.1 GA parameters [18]

GA parameter	GA parameter
Fitness function	The main function for optimization
Individuals	Any parameter to affect the fitness function; the value of the fitness function for an individual is its score
Populations and generations	An array of individuals; at each iteration, the algorithm makes a series of calculations on the existing population to create a new population; each succeeding population is referred to as a "new generation"

Implementation of Proposed Method

To test the proposed AI application, a mine locomotive was tested on a real railway track in a 100 km segment of the Goldsworthy railway in Western Australia. The Goldsworthy railway is owned and operated by a large mining company and is a private rail network in the Pilbara district built to transport iron ore. The Goldsworthy heavy railway is 208 km long, joining the Yarrie mine to Finucane Island near Port Hedland (see the red line in Fig. 20.9).

The mine trains on the Goldsworthy railway track have 90 wagons per train. Each wagon transports up to 126 tons of iron ore. The railroad grade is in the variety of -10 to $+\ 10$ per 1000 m. The minimum road curve radius is 300 m. The grade profile for the 100 km of the railway (i.e., tested segment) is presented in Fig. 20.10. The way is discretized into a set of points with an equal distance of d$x = 10$ m between them.

The mine-locomotive model used by the study is the GM SD40-2, which is a commonly used model in the Australian railway transportation system. The functional specifications of the trained model and the locomotive are presented in Table 20.2.

Fig. 20.9 Western Australia railway map

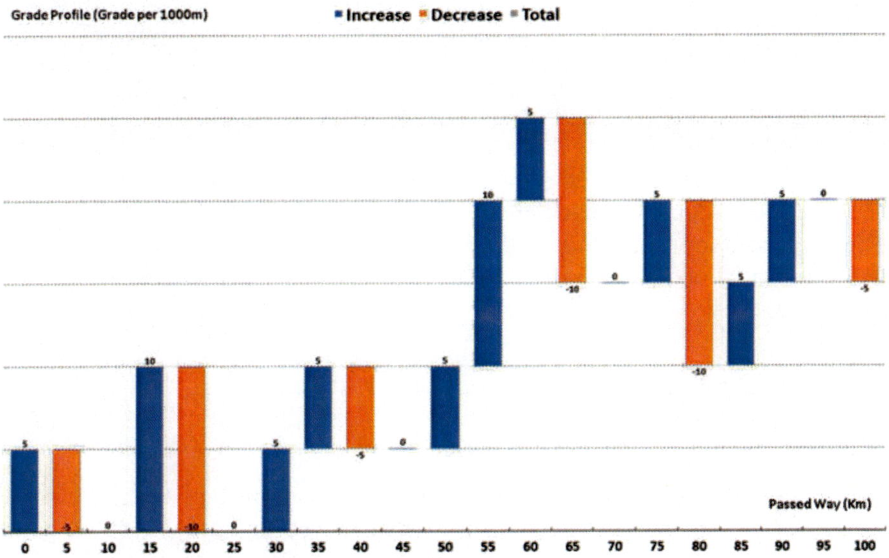

Fig. 20.10 Road grade profile of path for tested segment

Table 20.2 Tested locomotive and train parameters

Parameter	Value	Parameter	Value
Locomotive model	GM SD40-2	Wheel mass	7000 kg
Initial friction coefficient	2.7	Wheel radius	0.96 m
Secondary friction coefficient	0.03 s/m	Wheel-friction coefficient	0.3
Drag coefficient	0.0024	Generator model	AR1OJBA
Locomotive frontal area	11.148 m^2	Generator mass	7200 kg
Center of mass height	2.6 m	Generator maximum current	4200 A
Weight ratio on the front bogie	0.5	Generator minimum voltage	1250 V
Distance between two bogies	13.8 m	Generator maximum output power	5250 kW
Locomotive mass	167,000 kg	Electric-motor model	D77
Cars' mass	300,000 kg	Electric-motor power	360 kW
Diesel-engine model	645 E3C	Electric-motor mass	2722 kg
Diesel-engine speed range	300–1100 rpm	Electric-motor maximum current	1120 A
Diesel-engine maximum power	2250 kW	Bus voltage	1250 V
Diesel-engine mass	14,742 kg	Final-drive mass	1200 kg
Final-drive ratio	4	Final-drive loss	0.2 input torque

Table 20.3 Locomotive fuel consumption

Throttle position	N	1	2	3	4	5	6	7	8
Average fuel burn (l/h)	6.1	9.8	23.7	45.3	62.5	78.5	104.8	135.5	167.4

The present study analyzed real data collected over six months to estimate the amount of diesel burned by a locomotive in the selected segment of railway track in different conditions. The results of these analyses are presented in Table 20.3.

The fuel burn of the mine diesel locomotive is at a constant rate in each throttle position.

Prediction Model

The construction of the planned ANN algorithm for function calculation is a feed-forward multilayer-perceptron neural network. The activation functions in the hidden layers (f) are the continuous, differentiable nonlinear tangents sigmoid (see Eq. 20.18).

$$f = \tan sig(E) = \frac{2}{1 + \exp(-2E)} - 1 \tag{20.18}$$

where E can be determined by Eq. 20.12.

Where x is the normalized input variable; w is the weight of that variable; i is the input; b is the bias; q is the number of input variables; and k and m are the counter and number of neural network nodes, respectively, in the hidden layer.

Equation 20.13 can be used as the activation function between the hidden and output layers. In this equation, F is the transfer function.

The output layer computes the weighted sum of the signals provided by the hidden layer and the related coefficients. The network output is specified by Eq. 20.14.

MSE and the coefficient of determination (R^2) were considered for different nodes in the hidden layer to find the optimal number of nodes in the hidden layer. As a result, the minimum MSE and the maximum R^2 (best performance) were found for ten nodes in the hidden layers. The schematic structure of the designed neural network is presented in Fig. 20.11.

To learn the proposed ANN model, 85,000 pairing data were randomly selected from the 175,000 values of the collected real data. To examine network accuracy and validate the model, 90,000 independent samples were used. The results show good agreement between the actual and estimated values of locomotive fuel consumption.

The validation results of the synthesized network are presented in Fig. 20.12, where the vertical and horizontal axes demonstrate the real fuel-consumption values and the estimated fuel consumption values by the developed model, respectively. The locomotive throttle position is illustrated on the right side of Fig. 20.12. Figure 20.12

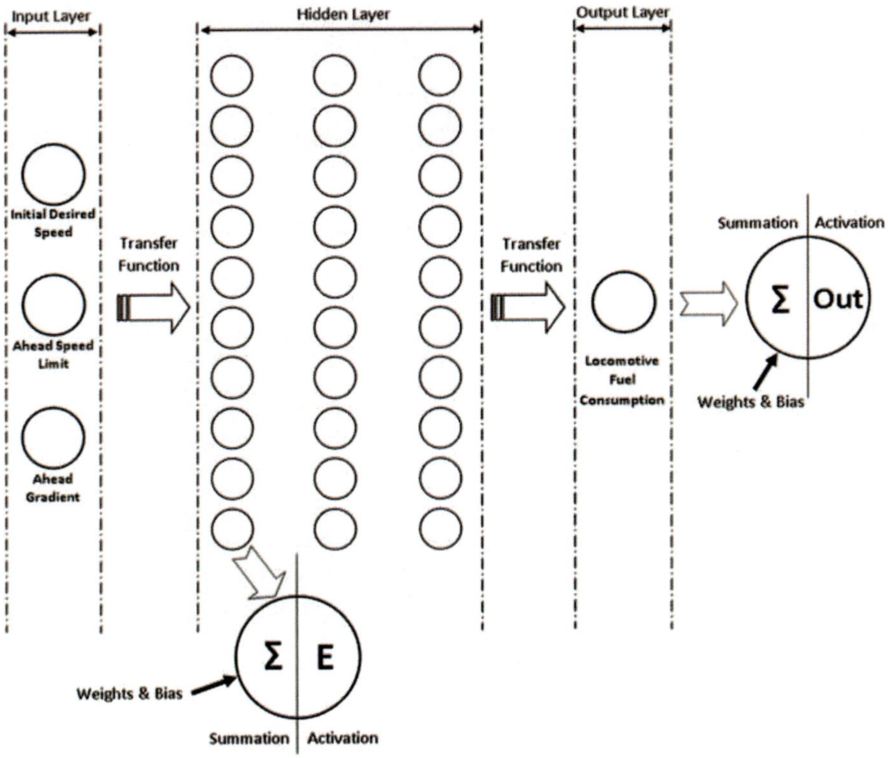

Fig. 20.11 Data processing in a neural network model

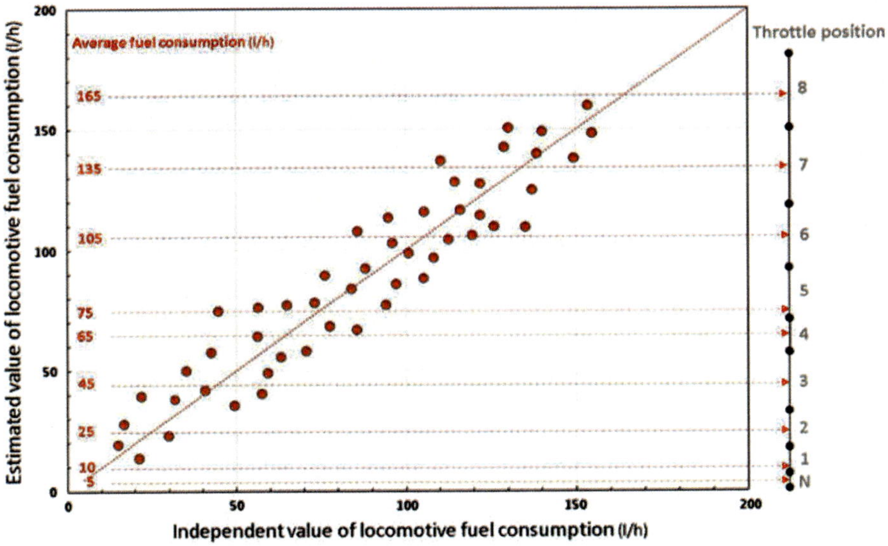

Fig. 20.12 Prediction-model validation

Table 20.4 GA processes [18]

Process	Details
Initialization	Generate initial population of candidate solutions
Encoding	Digitalize initial population value
Crossover	Combine parts of two or more parental solutions to create new
Mutation	Divergence operation; this operation is intended to occasionally break one or more members of a population out of minimum local space and potentially discover a better answer
Decoding	Change digitalized format of the new generation to the original one
Selection	Select better solutions (individuals) out of worse ones
Replacement	Replace the individuals with better fitness values as parents

also presents the average estimated fuel consumption for each throttle position (shown in red on the left).

The achieved results demonstrate that the developed ANN model can estimate locomotive fuel burn with an acceptable error. Furthermore, the fitness function produced by ANN is fed to the developed AI optimization model aimed to optimize the train speed in different railway track segments to minimize the fuel burned by the locomotive.

Optimization Model

This study has developed a GA algorithm to optimize train speed to minimize locomotive diesel consumption. In this model, the fitness function is created by the developed ANN to feed to the GA algorithm. In the developed GA algorithm, the following seven main processes were defined: initialization, encoding, crossover, mutation, decoding, selection, and replacement. The details of these seven processes are presented in Table 20.4. In the completed model, the main factors used to control the algorithm are R^2 and MSE.

Results

The first step of running the developed optimization model is defining the minimum and maximum value of the variable (train speed). The range of possible values for variables in the established model is based on the collected real dataset. The parameters used to control the established models are R^2 and MSE. The value of MSE was very close to 0, and the value of R^2 was approximately 0.96 after the fifty-seventh generation. These values did not change until the GA model was stopped in the sixty-third generation. In addition, the values of the control parameters were constant after the fifty-seventh generation, but the model continued all processes until the sixty-third. This is because a confidence interval was defined for the model to find reliable results. Figure 20.13 presents the rate of locomotive fuel consumption before and after optimization for a selected railway path. The rate of locomotive fuel consumption changes based on the road profile and grade resistance. The large range of presented fuel consumption rates in Fig. 20.13 returns to the illustrated road profile in Fig. 20.10.

The result shows that increasing the road grade increases locomotive fuel consumption and reduces train speed considerably. The lowest level of fuel consumption for the locomotive is predicted for the flat segments where the grade resistance is equal to zero and in some segments of railway path that have a negative grade.

Table 20.5 presents the range of locomotive fuel consumption in real and optimized conditions (before and after using the developed model) in the investigated case study in Western Australia.

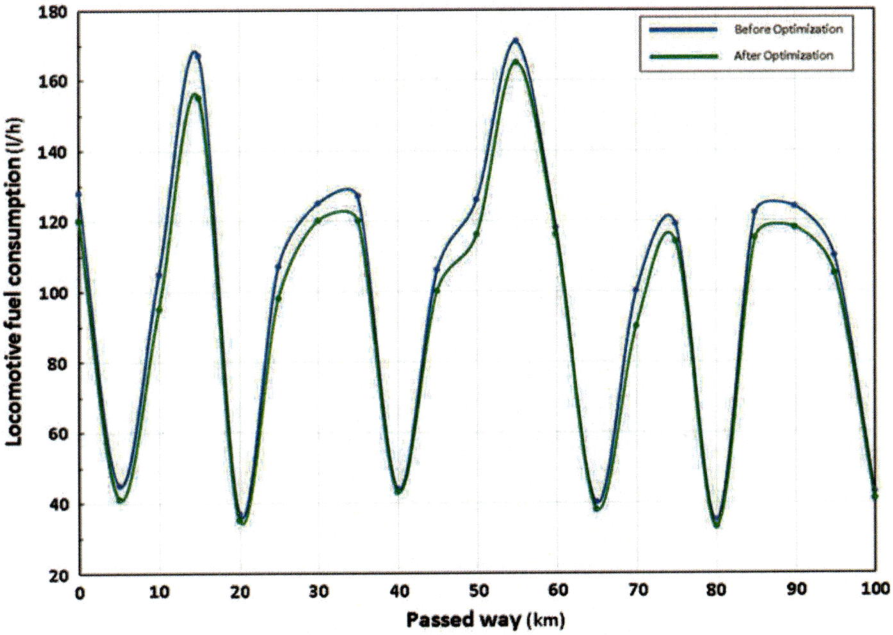

Fig. 20.13 Optimization result for selected railway path

Table 20.5 Locomotive fuel consumption—Optimization results

Real fuel consumption (l/h)		Minimized fuel consumption (l/h)		Fuel-consumption improvement (%)	
Min	Max	Min	Max	Min	Max
35	171	33	165	1.69	9.85

The results presented in Table 20.5 confirm that using the proposed and validated AI model can provide practical help that will allow the operation team to reach the minimum 1.69% and maximum 9.85% energy-efficiency improvement in the studied transportation system of the mine railway.

Conclusion

This study developed an AI look-forward control as an online approach for energy-efficiency improvement. The AI controller modifies the desired train-speed profile by accounting for the route's grade resistance and speed limits. The travel-time increment was applied as an improvement constraint. The AI model developed for train-movement simulation accurately predicted locomotive fuel consumption based on the values of the transfer parameters considered in this study. The developed model considered the locomotive subsystems and satisfied the experimental fuel-consumption data specified in the locomotive's catalog. The developed model in this study had two main sections for estimating locomotive fuel consumption in different situations: one section applies ANN, and the other section (the optimization section) applies GA to optimize the train speed that will minimize locomotive diesel consumption. The proposed AI model in this study was trained and tested using real data collected from a mine-railway route in Western Australia. The simulation of a train with a GM SD40-2 locomotive on a local railway track presented a significant reduction in locomotive fuel burn along with a satisfactory travel-time increment. The model's achievements were that the AI look-forward controller has faster computations than the controller based on the dynamic-encoding method. The results achieved in this study illustrate that using the developed AI model means that reaching an average of 5.77% energy-efficiency improvement is practically possible. Developing an AI look-forward controller with a dynamic look-forward window can be suggested as an avenue for future research to achieve even further locomotive fuel consumption reduction.

References

1. Jiaxin, C., and P. Howlett. 1992. Application of critical velocities to minimize fuel consumption in the control of trains. *Automatica* 28 (1): 165–169.
2. Howlett, P., I. Milroy, and P. Pudney. 1994. Energy-efficient train control. *Control Engineering Practice* 2 (2): 193–200.
3. Howlett, P. 1996. Optimal strategies for the control of a train. *Automatica* 32 (4): 519–532.
4. Hansen, I., J. Pachl, and A. Radtke. 2008. Infrastructure modelling. In Railway Timetable & Traffic–Analysis, Modelling, Simulation. Germany: Eurail Press Hamburg.
5. Howlett, P.G., P.J. Pudney, and X. Vu. 2009. Local energy minimization in optimal train control. *Automatica* 45 (11): 2692–2698.
6. Kang, M.-H. 2011. A GA-based algorithm for creating an energy-optimum train speed trajectory. *Journal of International Council on Electrical Engineering* 1 (2): 123–128.

7. Feng, X., et al. 2013. *A Review Study on Rail Transport Traction Energy Saving*. Discrete Dynamics in Nature and Society.
8. Ganji, B., and A.Z. Kozani. 2010. A study on look-ahead control and energy management strategies in hybrid electric vehicles. In *IEEE ICCA 2010*. IEEE.
9. Li, X., et al. 2013. Train energy-efficient operation with stochastic resistance coefficient. *International Journal of Innovative Computing, Information and Control* 9 (8): 3471–3483.
10. Hellström, E., et al. 2009. Look-ahead control for heavy trucks to minimize trip time and fuel consumption. *Control Engineering Practice* 17 (2): 245–254.
11. Hellström, E., J. Åslund, and L. Nielsen. 2010. Design of an efficient algorithm for fuel-optimal look-ahead control. *Control Engineering Practice* 18 (11): 1318–1327.
12. Sahlholm, P., and K.H. Johansson. 2010. Road grade estimation for look-ahead vehicle control using multiple measurement runs. *Control Engineering Practice* 18 (11): 1328–1341.
13. Ganji, B., A.Z. Kozani, and M.-A. Hessami. 2011. Backward modeling and look-ahead fuzzy energy management controller for a parallel hybrid vehicle. *Control and Intelligent Systems* 39 (3): 179.
14. Khayyam, H., et al. 2011. Intelligent energy management control of vehicle air conditioning via the look-ahead system. *Applied Thermal Engineering* 31 (16): 3147–3160.
15. Motors, G. 1972. *SD-40 Locomotive Service Manual*. Revision E, Electro-Motive Division of General Motors.
16. Davis, W.J. 1926. *The Tractive Resistance of Electric Locomotives and Cars*. General Electric.
17. Soofastaei, A. 2017. Intelligence predictive analysis to reduce production cost. *Australian Resources and Investment* 11 (1): 24–25.
18. Soofastaei, A. 2016. *Development of an Advanced Data Analytics Model to Improve the Energy Efficiency of Haul Trucks in Surface Mines*.
19. Soofastaei, A., et al. 2016. Development of a multi-layer perceptron artificial neural network model to determine to haul trucks energy consumption. *International Journal of Mining Science and Technology* 26 (2): 285–293.
20. Fu, L.-M. 2003. *Neural Networks in Computer Intelligence*. Tata McGraw-Hill Education.

Chapter 21
Advanced Analytics for Hard Rock Violent Failure in Underground Excavations

Amin Manouchehrian and Mostafa Sharifzadeh

Abstract In recent decades, mining projects have required deeper excavations to access resources due to increased excavating depth, the risks of mining operations, and rockburst increase. As an important line of defense, ground control measures and burst-resistant rock support are used to prevent or minimize excavation damage and thus enhance workplace safety. In this chapter, the first previous research of stable and unstable failures will be discussed. Then different approaches have been used to study different aspects of unstable rock failure. Then numerical modeling of the rock failure is proposed to predict the potential occurrence of rockburst in underground openings, and two examples were presented to show the interpretation of the failure type in numerical models. Then Loading System Reaction Intensity and kinematic energy were selected as indicators of unstable failure, and it was shown that a decrease in Loading Screen System causes a more violent failure mode and the release of more kinetic energy.

Keywords Deep mining · Advanced analytics · Violent failure · Rockburst

Introduction

Earth crust is the main source of human needs for life, such as water, minerals, fuel, etc. Continuous extraction of natural recourses in the past has evacuated many reserves at the surface and low depths. Therefore, it is needed to excavate deeper inside the earth to explore resources. However, as we go deeper inside the earth, the stress state transits from low stress to high-stress conditions, making the excavating activities riskier and more challenging. A hazardous challenge of excavating at the deep ground is rockburst. Rockburst is a sudden rock failure accompanied by rapid ejection of broken pieces of rock, which threatens the safety of mine crew, machinery, and, consequently, the mining economy.

A. Manouchehrian
School of Resources and Environmental Engineering, Jiangxi University, Nanchang, China

M. Sharifzadeh (✉)
MEC Mining / Deep Mining Geomechanics, Perth, WA 6000, Australia

A. Soofastaei (ed.), *Advanced Analytics in Mining Engineering*,
https://doi.org/10.1007/978-3-030-91589-6_21

Advances in excavating technologies allowed mines and tunnels to be constructed on deeper grounds. This has resulted in occasional rockburst occurrences in different underground spaces around the world. The case histories of rockburst have documented many tragic events, including loss of lives, heavy damages to the equipment, and total collapse of the underground construction [1, 2]. Consequently, rockburst in underground openings became a threatening problem for deep mining and civil constructions. Rockburst is an unstable rock failure known to be one of the most dangerous problems in deep excavations. A key condition for rockburst to occur is high stress. Thus, rockburst is more frequent in the deep hard-rock mines and tunnels where the stresses are high. As a general rule, rockburst occurrences increase as the excavating activities progress to deeper grounds. In the last several decades, extensive research has been conducted to study unstable rock failure and rockburst. Despite the extensive research, rockburst control remains a challenging engineering problem, and further studies are needed to control rockburst.

Background of Stable and Unstable Failures

Rocks are a solid material composed of minerals or crystals made from atoms, molecules, and lattices on a smaller scale. From a micro-scale point of view, the particles of rocks are attached with chemical bonds. In essence, the bonds are the matter of energy between microstructures that governs the mechanical rock properties, specifically strength, brittleness, or ductility. In rocks, different forms of microstructural defects may exist. These defects include:

- atomic disorder and dislocations in pure and homogenous rocks;
- crystal lattice boundaries in crystalline and foliated rocks due to two-dimensional covalent networks;
- heterogeneity (adjacency of weak and strong rock particles);
- pore spaces during generation mainly due to gas escape in volcanic rocks;
- cleavages due to overtime deformation and residual stresses; and
- micro-crack or structural defects due to stresses.

These microstructural defects are the main reasons behind crack initiation, propagation, and failure on larger scales [3]. When a rock specimen undergoes loading, stresses concentrate around micro-scale deficiencies, and when stresses exceed the strength of the materials around deficiencies, new micro-cracks are nucleated, or existing micro-cracks propagate. An increase of loading on the specimen leads to micro-crack initiation, coalescence, and propagation in the whole specimen and results in the final failure. In rocks, the failure may happen in either quasi-static or dynamic forms. In a quasi-static failure, the rock shows a stable behavior in the post-failure stage with a gradual drop of stress as the loading continues (stable failure). However, the failure may be dynamic with an unstable and uncontrolled failure process in the post-failure stage. In an unstable rock failure, the stress drops quickly, and broken pieces of rock are ejected around. The rock failure is unstable if the stored energy in the loading system transfers to the rock. In a system, when the energy amount that the rock can dissipate is less than the released energy from the loading system, then the system is in an unstable equilibrium. An unstable equilibrium exists upon the failure of a brittle rock specimen if the loads are applied to the rock by a soft loading system. First, Cook [4] described the failure instability in rocks using the stress–strain curve and the associated components of energy. Later, Salamon [5] clarified the stored energy contribution to the rock's failure instability. In this section, the concept of failure instability is explained theoretically.

Figure 1a shows a spring-rock system composed of a linear elastic spring (loading system) and a rock specimen. The system is in equilibrium. A compressive load Ps is applied to the system at point O1 until the rock fails. It results in a downward displacement in the upper end (O1) and lower end (O2) of the spring, indicated in Fig. 1a by γ and s, respectively. In Fig. 1b, the red line indicates the linear elastic force–displacement behavior of the spring. The force (F_s) in the spring is a function of the spring stiffness (k) and the change in its length ($\gamma - s$). Therefore, the compressive force P_s on the spring can be calculated by Eq. 21.1.

$$P_s = k.(\gamma - s) \tag{21.1}$$

The relation between force and displacement in the rock is expressed by

$$P_r = F(s) \tag{21.2}$$

For the system in equilibrium, $P_s = P_r$, that is,

$$k(\gamma - s) = f(s) \tag{21.3}$$

As demonstrated in Fig. 1b.

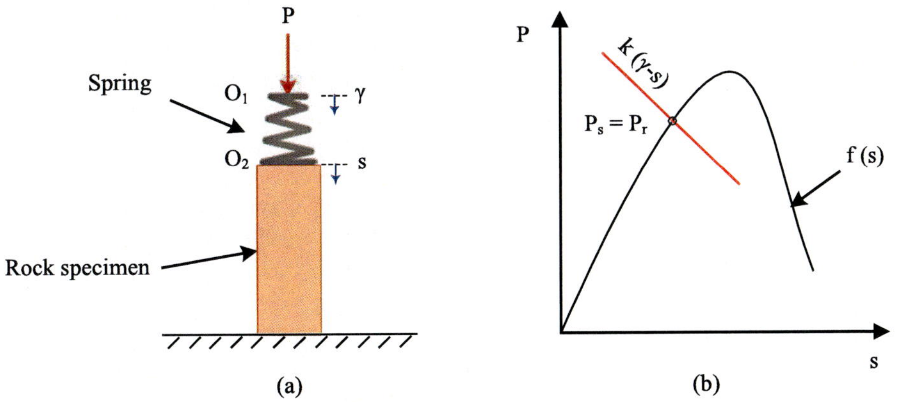

Fig. 21.1 **a** A simple loading system, and **b** the illustration of the equilibrium between load and rock resistance (reproduced and modified from [5])

If no further displacement is applied to the specimen's end by the spring, the equilibrium state remains stable Eq. 21.2. In this condition, no more external energy is transferred to the rock from the spring. In other words, if γ does not change ($\Delta\gamma = 0$), no external energy goes to the spring-rock system. Thus, the equilibrium state is stable if the work done by the spring (ΔW_s) during a virtual displacement (Δs) of point O_2 is smaller than the work required to induce the same displacement in the rock (ΔW_r), which is,

$$\Delta W_r - \Delta W_s > 0 \quad \text{(stable)} \tag{21.4}$$

The change in the work done by the spring and the work required to deform the rock are expressed by Eqs. 21.5 and 21.6, respectively.

$$\Delta W_s = \left(P + \frac{1}{2}\Delta P_s\right).\Delta s \tag{21.5}$$

$$\Delta W_r = \left(P + \frac{1}{2}\Delta P_r\right).\Delta s \tag{21.6}$$

The changes in forces on the spring and specimen can be calculated from Eqs. 21.1 and 21.2, respectively ($\Delta\gamma = 0$),

$$\Delta P_s = -k.\Delta s \tag{21.7}$$

$$\Delta P_r = f'(s)\Delta s = \lambda\Delta s \tag{21.8}$$

where the slope of the force–displacement curve of the specimen is defined by λ. Substituting Eqs. 21.7 and 21.8 into Eqs. 21.5 and 21.6 and then into Eq. 21.4 returns,

$$\frac{1}{2}.(k + \lambda).(\Delta s)^2 < 0 \tag{21.9}$$

It indicates that the condition for the failure to be stable is

$$k + \lambda > 0 \quad \text{(stable)} \tag{21.10}$$

When a brittle rock fails in such a system, a negative slope would be observed. By definition, the stiffness of the spring k is always positive. Thus the failure is unstable if the negative slope of the rock specimen after failure gets greater than the spring's stiffness. To sum up the above, the failure instability may be assessed either from the capacity of the loading system to apply more forces than will be resisted by the falling rock or through the amount of the excess energy which is resulted from the unstable equilibrium.

In deep grounds where stresses are naturally high, rocks show higher strength than their unconfined strengths. As depicted in Fig. 2a, the specimen shows low

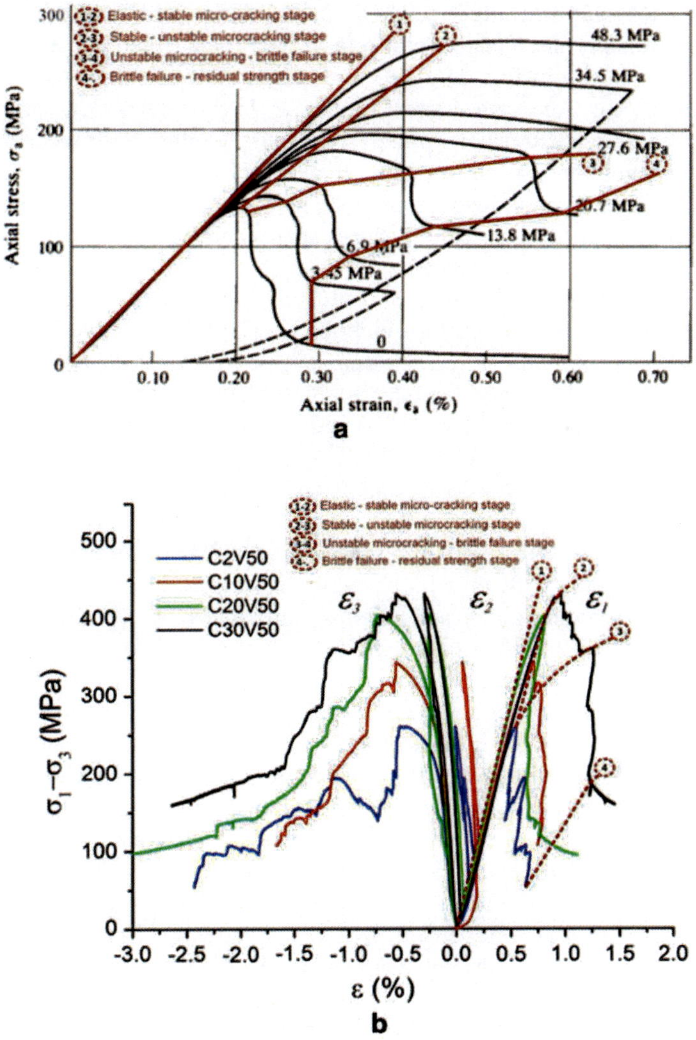

Fig. 21.2 Intact rock behavior modes and their postulated transition limit test results from **a** conventional triaxial on marble specimens [6] and **b** true-triaxial test results on granitic rock specimens

strength and high brittleness with very low or no residual strength in the uniaxial testing condition. With the gradual increase in confining stresses, rocks' peak and residual strength increase, and rock brittleness reduce. The stress–strain curves from conventional and true-triaxial tests are shown in Figs. 2a, b, respectively [3]. Based on Fig. 21.2, the whole stress–strain behavior can be categorized into four distinct groups. It clearly shows the rock behavior evolution and transition under different confining stresses, which is briefly explained below:

I. Elastic to stable cracking zone (1–2): Based on Fig. 21.2, the area between line 1 (elastic) and line 2 (stable cracking or ductile) is called the stable cracking zone. As shown in Fig. 21.2, increases in confining stresses increase elastic behavior range or delay in crack initiation. In this zone, first, rock behavior deviates from linear elastic behavior, and existing pores and cracks in the rock are activated, and possibly minor tensile cracks may develop, where cracks will be closed with unloading and rock will recover the pre-test condition. At this stage, rock mechanical properties such as cohesion (C), friction (φ), and poisons ratio (υ) are constant.

II. Stable to unstable cracking (2–3): In Fig. 21.2, the area between line 2 (stable cracking) and line 3 (unstable cracking) is called the unstable cracking zone. In this zone, new cracks nucleate and develop from existing pores and cracks. The length of crack development is highly dependent on the stress level. At this stage, an increase in loads leads to crack propagation and coalescence, where the loading energy stored in the specimen causes an uncontrolled crack development. The majority of the cracks are created at this stage. The crack development rate and the unstable crack development length highly depend on the confining stress levels. In other words, at low confining stresses, stable cracking rapidly ends up with unstable cracking and brittle failure, but with increasing confining stresses length of unstable cracking increases significantly. The mechanical properties of rock may react differently during unstable cracking. Cohesion reduces because of crack opening, while friction increases due to shear cracking. Poison's ratio and Young's modulus reduce at this stage.

III. Unstable cracking to brittle failure (3–4): At this stage, the accumulated energy in the specimen is released and creates unstable cracks, which leads to the major failure plane in the specimen. The increase in the confining stresses has a remarkable impact on the failure, brittleness mode (energy releasing–energy absorbing), and magnitude of the strength loss. Line 3 in Fig. 21.2 indicates that the stress drops rapidly after failure, which can be read as a brittle failure. By increasing the confinement, firstly, rock strength increases. Figure 21.2 shows that the specimens fail at higher strains. Also, the magnitude of strength drop reduces, and specimens tend to behave more ductile than brittle mode.

IV. Brittle failure to residual strength (4 onward): At this stage, fractured pieces of a rock slide on each other, and only frictional stress or apparent cohesion resists against the applied loads. Under low confinements, the residual strength is zero or negligible; the residual strength increases by increasing the confining stress.

In the triaxial compression test, the rock mechanical behavior strongly depends on the stresses. Therefore, in practical engineering, site-specific conditions should be considered for the assessment of mechanical rock properties. Laboratory results show that a small increase in confining stresses can significantly improve rock properties such as peak and residual strength, ductility, and reduced brittleness. This concept is used in underground excavation by providing support reinforcement to absorb ground energy and limit the possibility of failure.

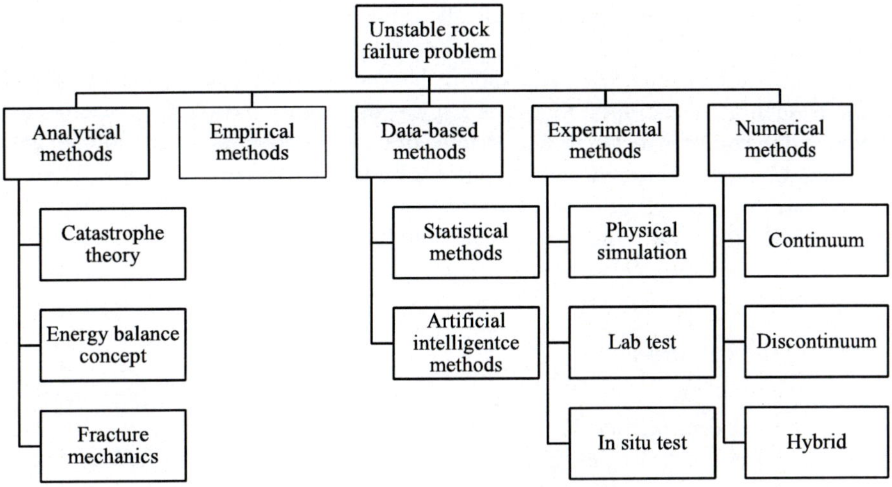

Fig. 21.3 Different methods for assessing unstable rock failure [8]

Techniques for Analysis of Unstable Rock Failure

The unstable rock failure problem in underground excavations first arose in South African and Indian mines at the beginning of the twentieth century, and it remains a threatening problem in many mines and tunnels around the world [1, 7]. In the past several decades, unstable rock failure (rockburst) has been a challenging and hot research topic among engineers and scientists to find ways to predict and control it. In the rock mechanics and rock engineering literature, many studies have been devoted to the study of rockburst and unstable failure of rock [8]. Different approaches have been used to study different aspects of unstable rock failure. The methods used for studying unstable rock failure problems are summarized in Fig. 21.3.

Analytical Methods

Generally, analytical methods provide solutions for problems with less effort and can highlight the main variables that influence the solution of a problem. Therefore, analytical methods are very helpful in geomechanics. In geomechanics, the analytical methods give solutions with good accuracy for calculating problems with simple geometry and boundary conditions. However, these methods fail to find acceptable solutions for many real-world engineering problems. Therefore, in analytical methods, some simplifications are made for solving problems. For instance, in most analytical methods, the rock is assumed to be an elastic, homogeneous, isotropic, and continuous material, while in reality, rock is a non-elastic, heterogeneous, anisotropic, and discontinuous material. Due to these simplifications, analytical methods have a limited application in geomechanics for solving practical problems.

In some studies, unstable rock failure has been investigated using analytical methods such as catastrophe theory, energy equilibrium, and fracture mechanics. However, the results from analytical studies of unstable rock failure are too general and cannot be applied to real-world engineering problems. Furthermore, analytical methods are not appropriate to assess a complex problem such as unstable rock failure but can better understand the mechanisms that drive an unstable rock failure.

Empirical Methods

Advances in geomechanics have provided appropriate tools for designing and analyzing different structures successfully. However, sometimes, some decisions are made based on the personal experience of experts. In practical geomechanics, empirical criteria are used to evaluate different engineering parameters and conditions, quantitatively or qualitatively. Empirical criteria are mathematical equations that are developed to be applied for site-specific conditions. The complex conditions of the field are not considered, and only a few parameters are taken into account. However, in many cases, empirical methods have been used efficiently to evaluate different engineering parameters in an acceptable level of confidence.

In many geotechnical projects, especially those in shallow depths, the rockburst potential is only assessed using empirical criteria. The most applied empirical criteria are the elastic strain energy criterion, brittleness criterion, and the maximum tangential stress criterion. Because the empirical criteria are developed based on experience and engineering judgments, the rockburst potential assessed with different criteria may be different. Studies demonstrate that for a particular case, the rockburst evaluation results from different criteria may be different. For example, a criterion may evaluate a light rockburst while another may anticipate a violent rockburst [9]. Thus, in the structures excavated in deeper grounds, more detailed investigations must be conducted to evaluate the rockburst potential.

Data-Based Methods

The collected data from different systems are valuable sources of information that can be utilized to develop relations for studying complex systems. Generally, the collected data can be used for finding trends and patterns of different parameters and phenomena. Different sciences such as medicine, economics, sociology, and engineering have benefited from data-based analyses. In geomechanics, uncertainties always exist. Thus, data-based analysis can provide valuable information about the ground to be used for solving problems. In recent years, the data collected from different projects have been used to analyze rockburst using statistical and artificial intelligence methods. In many cases, the data-based studies helped engineers find rockburst trends and patterns in the construction sites. However, the interaction between different influencing factors on the rockburst occurrence is complicated, and large datasets are required to develop accurate predictive models.

Experimental Methods

In geomechanics, physical testing is an inseparable part of any engineering project. Laboratory and in situ experiments are needed to evaluate different engineering parameters of rock. Moreover, the rock behaviors under different loading conditions are investigated through physical experiments. Despite the high costs and the difficulty, laboratory and in situ tests are the most important methods for obtaining information from the real conditions of the ground. Extensive research has been done to study the unstable failure of rock in mines and tunnels using physical experiments [10]. Physical experiments can reproduce the rock failure process and help to study the characteristics of rockburst. However, it is difficult to apply the in situ conditions to simulate rockburst. In addition, direct measurement of some parameters such as energy components and stress fields is still difficult in physical testing. Physical testing along with numerical simulations can give the best results for studying unstable rock failure.

Numerical Methods

In modern engineering, the application of analytical solutions is limited to a few problems with many simplifying assumptions. However, advances in computing technology have empowered scientists and engineers to develop detailed models for the simulation of complex systems. Nowadays, the engineering world is very dependant on numerical simulations. In the geomechanical literature, the successful application of numerical models for the conduction of safer projects has been well documented [11].

Continuum and discontinued models have been utilized to study the unstable rock failure (rockburst) in excavations [12, 13]. With the aid of appropriate numerical simulations, different aspects of the rockburst can be investigated. The mechanisms that drive rockburst are complex because many factors may influence the rockburst damage [14]. The rockburst case history clearly shows that tragic disasters may happen if the influencing factors on rockburst are not considered in different construction stages. Because of the rockburst complexity, computer models must consider different influencing factors and their interactions to assess the rockburst potential in underground excavations.

It should be noted that numerical simulations only show rockburst potential locations, but it does not necessarily mean that the rockburst will occur at those locations. Thus, field observations, geological surveys, and seismic data should support the numerical simulation results to make reliable assessments. In practice, computer simulation and microseismic monitoring are very useful for assessing the rockburst potential and damage locations in tunnels and mines.

Numerical Modeling of Unstable Rock Failure

In geomechanical engineering, numerical simulation is an important task for engineering planning and design. Many simple and sophisticated geomechanical problems have been solved using numerical models. For example, extensive studies have been carried out to investigate the unstable failure of rock using numerical models [15, 16]. In computer simulations, recognition of stable and unstable rock failures is not straightforward. Usually, one or more indicators should be used to interpret the modeling outputs. Researchers have used various indicators to judge the mode of failure in the numerical models. In this section, two examples are presented to show the detection of the failure mode in the modeling results.

Example 1: Unstable Rock Failure in UCS Test

Laboratory experiment results show that the rock failure mode depends on the relative stiffness of the rock in the post-peak region and the loading system [6]. For an elastic column with a uniform cross-section area and material property, the stiffness can be calculated by

$$k = \frac{AE}{L} \tag{21.11}$$

where E is the elastic modulus, A is the cross-section area, and L is the length of the column. In Fig. 21.4, the concept of equilibrium state related to the stiffness of the loading system and material is illustrated. In Fig. 21.4, the unloading stiffnesses of soft and stiff loading systems are indicated by the slopes of the red, and the green dashed lines, respectively. Also, the post-peak stiffness of the material is illustrated by the slope of the solid line. When the loading system stiffness is smaller than the material's characteristic post-peak stiffness, failure happens violently and is unstable. In contrast, when the loading system stiffness is larger than the material's characteristic post-peak stiffness, the energy released from the loading system is absorbed by the material, which results in a stable failure of the material. Therefore, by comparing the stiffnesses of the material in the post-peak region and the loading system, the failure mechanisms in the stable and unstable modes can be investigated.

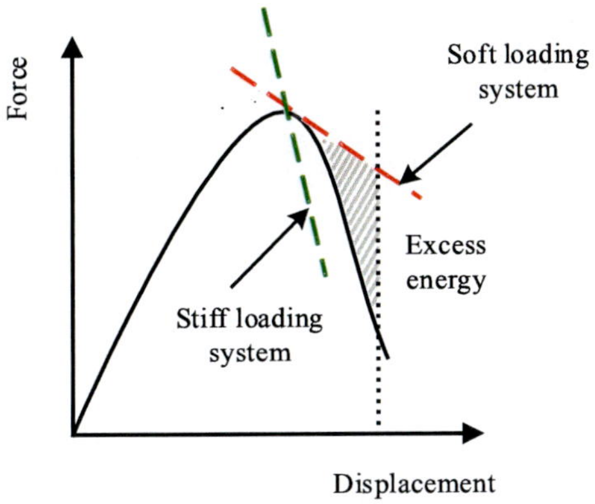

Fig. 21.4 Equilibrium stability concept

Because the unstable rock failure is a nonlinear dynamic phenomenon, and usually implicit numerical models cannot simulate it. This study used an Abaqus explicit code to simulate the unstable rock failure in UCS tests. For the current study, Tianhu granite was considered. The physical and mechanical properties of Tianhu granite are listed in Table 21.1 [17]. In the numerical modeling, the platen length was changed to account for different loading system stiffness, and its elastic modulus was kept constant as the steel's elastic modulus (i.e., $E = 200$ GPa). The loading screen system (LSS) was calculated using Eq. 21.11.

The UCS test shown in Fig. 21.5 was simulated. One end was fixed with a roller constraint, and the load was applied to the platen with a constant velocity of 0.015 m/s. A frictionless contact was assigned between platens and the specimen. Figure 21.6 presents the stress–strain curves resulted from UCS test simulations with different LSS values. To obtain the characteristic behavior of the rock, a case with an infinitely rigid loading condition was simulated (labeled as "Material" and represented by a dashed line in Fig. 21.6). By applying the loads directly to the specimen's ends, a rigid loading condition was simulated. The material's post-peak stiffness at the steepest part in the stress–strain curve was 4.23 GN/m. The modeling results indicate that when the LSS is smaller than the rock post-peak stiffness, the specimen's post-peak curve deviates significantly from the rock's characteristic post-peak curve, which indicates unstable rock failure. This is the result of the platen rebound upon the rock failure. In the models with the LSS values larger than the post-peak stiffness of the

Table 21.1 The physical and mechanical properties of Tianhu granite [18]

Parameter	Value
Density (kg/m^3)	2650
Young's modulus (GPa)	51
Poisson's ratio	0.27
Uniaxial compressive strength (UCS) (MPa)	160
Brazilian tensile strength (MPa)	7.1

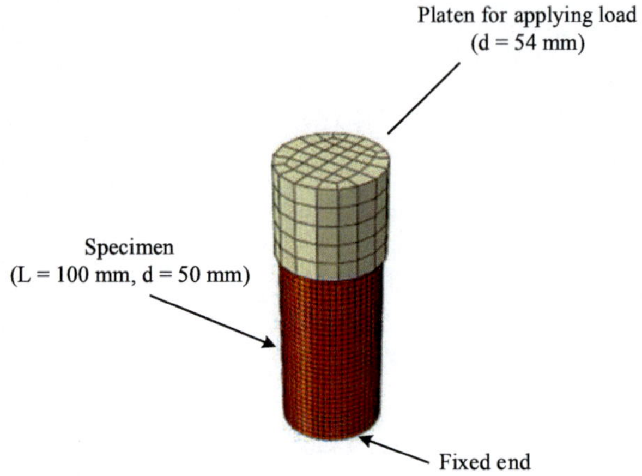

Fig. 21.5 The geometry of the UCS model

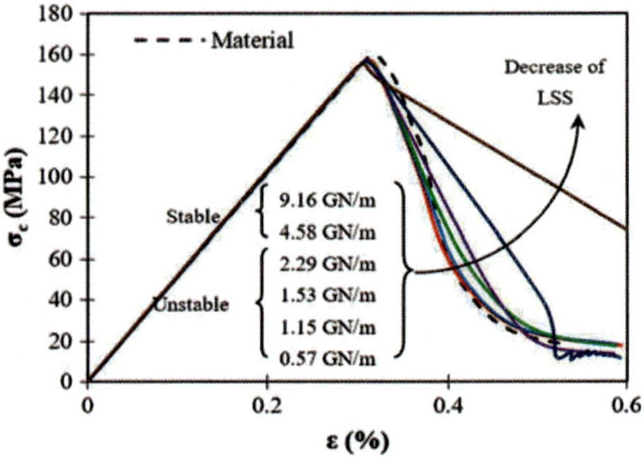

Fig. 21.6 Stable and unstable rock failures simulated using Abaqus explicit

rock, the deviation of the specimen's post-peak curve from the rock's characteristic post-peak curve is negligible. The results confirm that the developed models can be used to simulate unstable rock failure.

When rock failure occurs, the stored energy in the loading system is released and transferred to the rock. If the released energy from the loading system is more than the absorbable energy by the falling rock, then the excess energy will be released in the form of kinetic energy. The larger the released energy is, the reaction of the loading system is faster and stronger. The system reaction can be observed in laboratory tests. When the failure is violent, the reaction is very obvious and can be noticed from recorded videos. When the failure is non-violent, the reaction becomes less obvious and may require sensitive monitoring equipment to capture it. The reaction phenomenon was studied in numerical modeling. The velocities in all nodes in the model can be tracked so that the loading system reaction can be used as an indicator of unstable failure. In this study, the Loading System Reaction Intensity (LSRI) is used

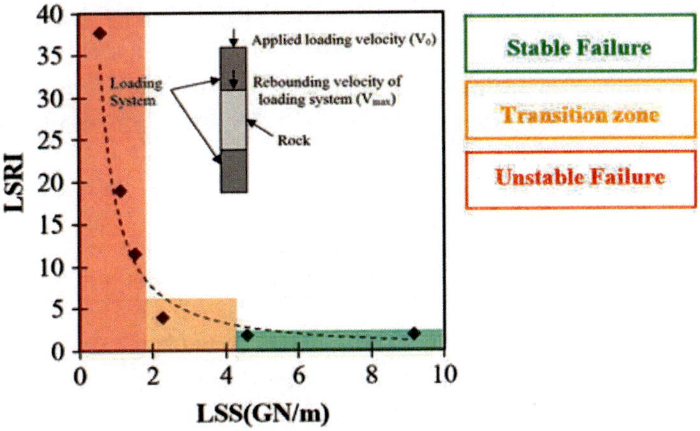

Fig. 21.7 Relationship between LSRI and LSS

as an indicator to detect the failure modes in the numerical models (Eq. 21.12). The developed indicator LSRI is calculated by dividing the platen's maximum velocity (V_{max}), which occurs at the contact of the platens with the rock specimen, to the applied loading velocity (V_0 at the platen's top end).

$$\mathrm{LSRI} = \frac{V_{max}}{V_0} \tag{21.12}$$

The relation between LSRI and LSS is plotted in Fig. 21.7. The figure shows that when the LSS is much smaller than the material's post-peak stiffness, the LSRI value is high ($\gg 1$). Moreover, in the models with larger LSS values comparing to the material's post-peak stiffness, failure is stable, and the LSRI value is small ($<$ 2). However, the failure is not necessarily violent in the models in which the LSS is slightly smaller than the material's post-peak stiffness. This transition zone is shown in Fig. 21.7 (orange zone). In this condition, the failure might be in a transition mode between stable and unstable failures. In laboratory experiments, this condition may happen in the form of sudden spalling of rock specimen flakes. Sudden spalling of rock slabs can be an example of this condition at the excavation boundaries. The presented modeling result indicates that there is no clear boundary between stable and unstable failures. For the simulated rock specimens, the LSRI values between 2 and 6 most likely show a less violent transitional failure of the rock under uniaxial compression.

Example 2: Unstable Rock Failure in True-Triaxial Test

In rock masses, when an opening is excavated, the stresses are redistributed around the opening (Fig. 21.8). Upon the excavation, the radial stress decreases at the opening boundaries, and the tangential stress increases. If the tangential stress is high enough, then a rockburst may occur. Hence, it is vital to assess the rockburst potential under unloading conditions. He et al. [19] developed a true-triaxial rock test system with an unloading face to simulate the rockburst under unloading conditions. The apparatus was designed to reproduce the loading and unloading processes in three principal stress directions independently.

In this section, a numerical study is carried out to simulate a rock test in a true-triaxial rockburst test machine (Fig. 21.9). In this study, true-triaxial unloading tests

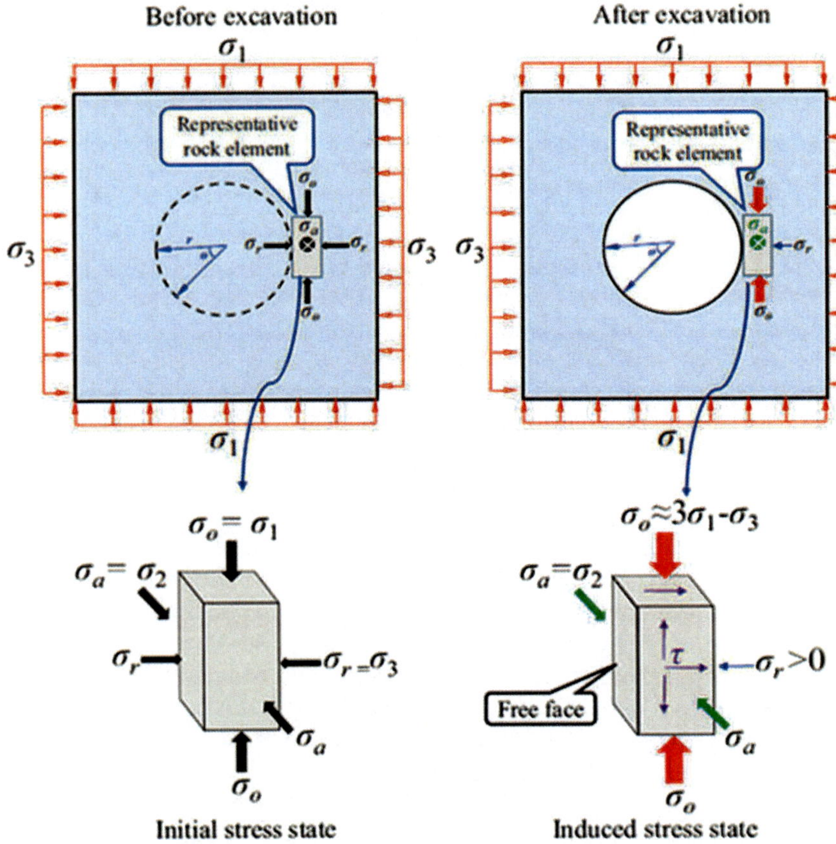

Fig. 21.8 Representation of stress state near underground opening boundaries

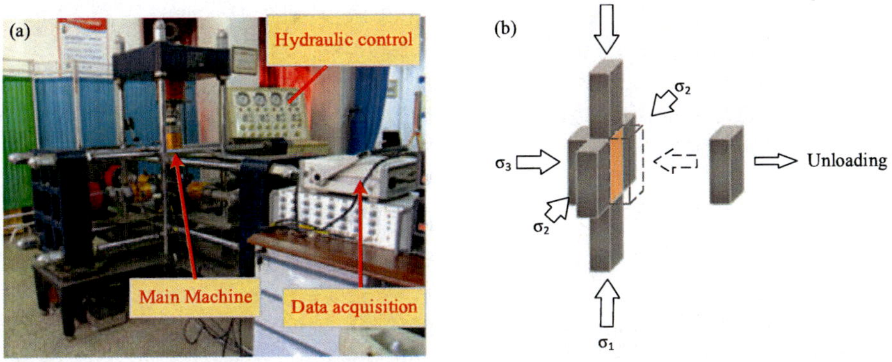

Fig. 21.9 **a** The true-triaxial strain burst test machine; **b** a schematic diagram showing unloading in the $\sigma 3$ direction (reproduced from [19])

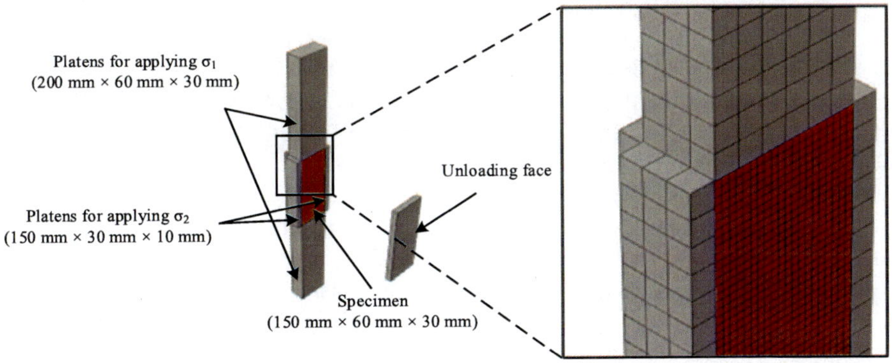

Fig. 21.10 Model geometry and mesh size [8]

are simulated using a simplified model shown in Fig. 21.10. The prism-shaped specimen with properties listed in Table 21.1 was loaded in three directions. The dimensions of the rock specimen were 150 (height) × 60 (width) × 30 (thickness) mm^3. The lateral stresses of $\sigma_1 = 20$ MPa, $\sigma_2 = 13$ MPa, and $\sigma_3 = 12$ MPa were applied to the specimen. Then the specimen was unloaded in the direction of σ_3 while σ_2 was kept constant. Subsequently, σ_1 was increased until the specimen failed. The loading system stiffness was simplified and represented by three pairs of the platen. To define the tangential behavior of the interface between the rock specimen and the platens, a friction coefficient of $\mu = 0.2$ was assigned to the model.

To simulate unstable failure with different LSS, the platens (L) length was changed, and the elastic modulus of the platens was kept constant at $E = 200$ GPa. By varying the length of the platens from $L = 200$ to 1600 mm, the LSS changes from 1.8 to 0.225 GN/m. In addition, four different heights (H) to width (W) ratios of the specimen were considered to investigate how the H/W ratio affects failure mode. The FEM model shown in Fig. 21.10 is for $H = 150$ mm, $W = 60$ mm, and

$t = 30$ mm, where t is the thickness of the rock specimen. Hexahedral eight-node linear elements are used for all FEM models.

The strain energy inside an elastic column under uniaxial stress can be estimated using

$$U = \frac{\sigma^2 AL}{2E} \tag{21.13}$$

where σ is the uniaxial stress, A is the cross-sectional area, L is the length, and E is the elastic modulus. Equation 21.13 implies that when a loading system is softer (e.g., with smaller elastic modulus), larger or longer (high A and L), or under higher stress, it stores more strain energy. Therefore, if longer platens are used in the simulation while keeping A and E constant, more energy will be released when the rock failure happens. In this situation, the failure will be more violent if the unstable failure conditions are satisfied.

In this study, models with different LSS were built to investigate the influence of loading system stiffness on the rock failure and the released energy. The test machine system has approximately equal loading system stiffness in all three principal stress directions. Because the platens applying load in the σ_1 direction results in the rock failure, only LSS in this direction was varied. The LSS in the σ_2 and σ_3 directions were not varied. For this purpose, the platens in the σ_1 direction were sized to 60 mm in width and 30 mm in thickness. Also, the height of the platens varied from 200 to 1600 mm, which resulted in LSS values variation from 0.225 to 1.8 GN/m. The typical elastic properties of steel (i.e., $E = 200$ GPa, $\upsilon = 0.3$, and $\rho = 7700$ kg/m^3) were assigned to the platens. To apply the compressive load in the σ_1 direction, a velocity of 0.015 m/s was applied to the top platen. Figures 21.11 and 21.12 indicate the influence of LSS on the released kinetic energy from the specimen and the LSRI, respectively. Figure 21.11 indicates that in the models with longer platens, more time was needed to move the platens for rock breakage. Also, Fig. 21.11 shows that when a softer loading system was used to break the specimen, more kinetic energy

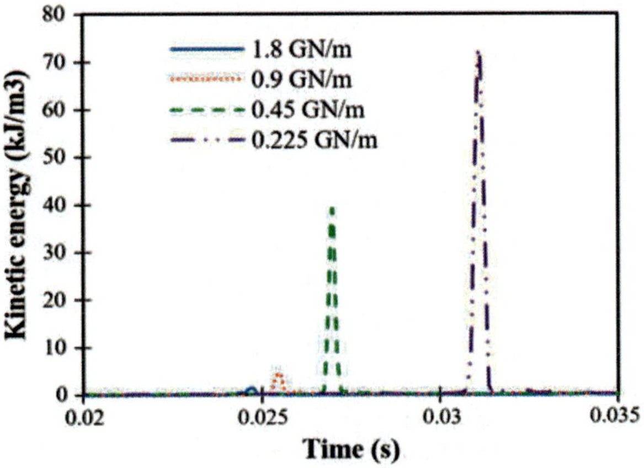

Fig. 21.11 Influence of LSS on the released energy in the specimen

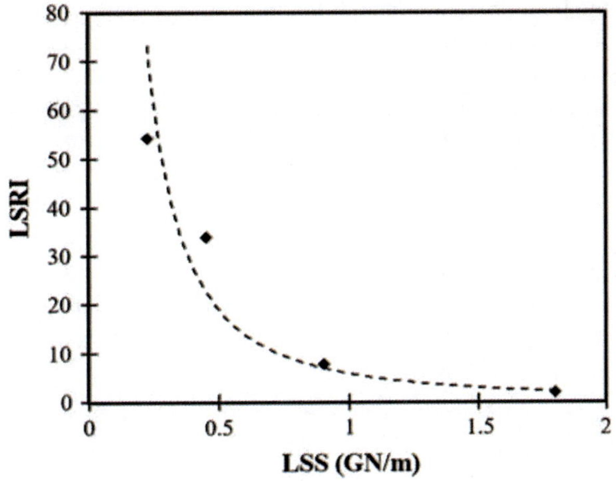

Fig. 21.12 Influence of LSS on LSRI from true-triaxial unloading simulation

was released from the specimen. Figure 21.12 shows that as the LSS increased, the LSRI values decreased. The UCS test simulation calculated a post-peak stiffness of 4.23 GN/m for the rock specimen. In a UCS test, the failure is unstable if the LSS value is smaller than the post-peak stiffness of the rock. In the model with LSS = 1.8 GN/m, the LSRI value was 1.9, indicating that the failure was stable. With the decrease of the LSS to 0.225 GN/m, the LSRI value was increased to 54.4, which indicates an unstable failure.

Conclusions

The case histories of rockburst in deep excavations have demonstrated the relentless face of unstable rock failure. In this research, analysis of the unstable rock failure with an emphasis on numerical methods was discussed. Firstly, the concept of unstable rock failure was described theoretically. Moreover, the brittle rock failure in laboratory compressive tests was explained. Also, different approaches for studying unstable rock failure were discussed—this helps better understand violent, sudden failure at deep underground excavations.

In the later part of this research, the numerical modeling of unstable rock failure was discussed. A large number of studies have proved the suitability of numerical methods for studying unstable rock failure. However, in numerical modeling, recognition of stable and unstable rock failures is not straightforward. Usually, one or more indicators are needed to interpret the simulation results. In this study, two examples were presented to show the interpretation of the failure type in numerical models. First, several UCS tests with different LSS values were simulated, and the stress–strain curves were plotted. In models with different LSS, the deviation of the stress–strain curve from the rock's characteristic post-peak curve was interpreted as the sign of failure instability. Also, LSRI was suggested as an indicator for recognition of the failure mode. Next, the rock failure in the true-triaxial compression tests with an unloading face and different LSS was simulated. In this simulation, LSRI and the released kinetic energy were chosen as unstable failure indicators. Results showed that with a decrease in the LSS, the failure mode tended to be more violent and released more kinetic energy.

The unstable rock failure in underground excavations is a complex phenomenon in which the influence of many factors has been proven. However, the contribution of influencing factors on rockburst occurrence and the mechanisms that drive rockburst are still not fully understood. Therefore, to mitigate the rockburst damages, the factors that influence rockburst should be given sufficient attention. Therefore, appropriate analyses and investigations such as numerical simulation, field observations, geologic survey, and seismic data monitoring are suggested to evaluate the rockburst potential and damage locations in underground excavations.

References

1. Blake, W., and D.G. Hedley. 2003. *Rockbursts: Case Studies from North American Hard-Rock Mines*. SME.
2. Hedley, D. 1988. *Rockbursts in Ontario mines during 1985*, vol. 87. Canada Centre for Mineral and Energy Technology.
3. Sharifzadeh, M.Z., et al. 2017. Challenges in multi-scale hard rock behavior evaluation at deep underground excavations. In *12th Iranian and 3rd Regional Tunneling Conference: Tunnelling and Climate Change*.
4. Cook, N. 1965. A note on rockburst considered as a problem of stability. *Journal of the Southern African Institute of Mining and Metallurgy* 65 (8): 437–446.
5. Salamon, M. 1970. Stability, instability and design of pillar workings. *International Journal of Rock Mechanics and Mining Sciences & Geomechanics Abstracts*.
6. Wawersik, W.R., and C. Fairhurst. 1970. A study of brittle rock fracture in laboratory compression experiments. *International Journal of Rock Mechanics and Mining Sciences & Geomechanics Abstracts* 7 (5): 561–575.
7. Aga, I., P. Shettigar, and R. Krishnamurthy. 1990. Rockburst hazard and its alleviation in Kolar gold mines-a review. In *Rockbursts: Global experiences. Plenary Scientific Session of Working Group on Rockbursts of International Bureau of Strata Mechanics*, p. 5.
8. Manouchehrian, S.M.A. 2016. *Numerical Modeling of Unstable Rock Failure*. Laurentian University of Sudbury.
9. Miao, S.-J., et al. 2016. Rock burst prediction based on in-situ stress and energy accumulation theory. *International Journal of Rock Mechanics and Mining Sciences* 100 (83): 86–94.
10. Milev, A., et al. 2002. *Meaningful Use of Peak Particle Velocities at Excavation Surfaces to Optimize the Rockburst Criteria for Tunnels and Stopes*.
11. Bobet, A. 2010. Numerical methods in geomechanics. *The Arabian Journal for Science and Engineering* 35 (1B): 27–48.
12. Manouchehrian, A., and M. Cai. 2017. Analysis of rockburst in tunnels subjected to static and dynamic loads. *Journal of Rock Mechanics and Geotechnical Engineering* 9 (6): 1031–1040.
13. Manouchehrian, A., and M. Cai. 2018. Numerical modeling of rockburst near fault zones in deep tunnels. *Tunneling and Underground Space Technology* 80: 164–180.
14. Kaiser, P.K., and M. Cai. 2012. Design of rock support system under rockburst condition. *Journal of Rock Mechanics and Geotechnical Engineering* 4 (3): 215–227.

15. Manouchehrian, A., and M. Cai. 2016. Influence of material heterogeneity on failure intensity in unstable rock failure. *Computers and Geotechnics* 71: 237–246.
16. Manouchehrian, A., and M. Cai. 2015. Simulation of unstable rock failure under unloading conditions. *Canadian Geotechnical Journal* 53 (1): 22–34.
17. Zhao, X., and M. Cai. 2015. Influence of specimen height-to-width ratio on the starburst characteristics of Tianhu granite under true-triaxial unloading conditions. *Canadian Geotechnical Journal* 52 (7): 890–902.
18. Zhao, XG., Wang, J., Cai, M., Cheng, C., Ma, L.K., Su, R., Zhao, F., Li, D.J. 2014. Influence of unloading rate on the strainburst characteristics of Beishan granite under true-triaxial unloading conditions, Rock Mechanics and Rock Engineering 47 (2), 467–483.
19. He, M., et al. 2012. Experimental investigation of bedding plane orientation on the rockburst behavior of sandstone. *Rock Mechanics and Rock Engineering* 45 (3): 311–326.

Chapter 22
Advanced Analytics for Heat Stress Management in Underground Mines

Ali Soofastaei and Milad Fouladgar

Abstract This chapter presents the results of an experimental investigation on the effects of coal dust contamination on the accuracy of wet-bulb temperature measurements and heat stress analysis. A coal dust-contaminating unit was designed and built to contaminate the wick of a wet-bulb temperature sensor uniformly. Contaminated sensors were obtained by applying different volumes of coal dust to the wick of the sensors. To examine the accuracy of wet-bulb temperature measurements, both contaminated and clean temperature sensors were used inside a standard air conditioning unit. Experiments were conducted for two different heating and cooling processes to study the temperature measurement accuracy in increasing and decreasing temperature processes. A direct correlation was found between the accuracy and wick contamination in both heating and cooling processes. It was also found that the interpretation of heat stress indices and heat management decisions strongly depends on the accuracy of wet-bulb temperature measurements.

Keywords Advanced analytics · Coal dust · Wet-bulb temperature · Heat stress index · Underground coal mine

Introduction

The temperature and heat management problem is becoming more prevalent as underground mines become deeper [1]. A hot work condition in underground mines reduces the safety and ability of both miners and machines to operate at their maximum efficiency [2]. The wet-bulb temperature (WBT) measurement plays an essential role

A. Soofastaei (✉)
Vale, Brisbane, Australia
e-mail: ali@soofastaei.net
URL: https://www.soofastaei.net

M. Fouladgar
Department of Mechanical Engineering, Islamic Azad University, Najafabad Branch, Najafabad, Iran
e-mail: milad_fouladgar@smc.iaun.ac.ir

A. Soofastaei (ed.), *Advanced Analytics in Mining Engineering*,
https://doi.org/10.1007/978-3-030-91589-6_22

in the design of mine ventilation and heat management strategies [3]. Temperature sensors used in underground mines need to accurately measure WBT to ensure miners' health, safety, comfort, and productivity. However, the wetted wick, which covers the common WBT sensors, may be exposed to different types of airborne contaminants such as coal dust in underground coal mines due to the underground environmental conditions. Any measurement errors caused by these contaminants result in inaccurate heat management decisions [4].

Literature includes many research studies on WBT measurements and heat management analysis [5–8]. However, the studies on inaccuracies of WBT measurements are very limited. Moulsley and Fryer [9] and Cotton [10] are among the first researchers who reaffirmed that contaminants transferred to the wetted wick increase the inaccuracy of readings. The contamination on the wick affects the absorption and evaporation processes. WBT measured by a contaminated WBT sensor is expected to be higher than measured by a clean sensor. Ramsey et al. [11] evaluated the effect of contamination build-up on the wetted wick of thermal sensors on their accuracy measurements as a part of experimental research in an underground coal mine. They compared two wet-bulb thermometers: one with its wick changed daily and the other one with its wick unchanged for three weeks. The results showed minor effects of wetted wick contamination on the accuracy of measurements. This was because the WBT sensor was exposed to a low level of airborne contaminants. Lee [12] conducted a series of experiments to investigate the effects of wick contamination and thermal component variation on thermal indices such as WBT. He found that the wick contamination caused considerable errors in the measurement of natural WBTs.

Pain [13] experimentally examined the effects of different contaminants such as coal dust, salt, engine oil, and diesel particulate matter on the accuracy of WBT readings in various processes such as heating, cooling, and moisturizing. He concluded that contaminations play an essential role in the accuracy of WBT measurements. Ustymczuk and Giner [14] recently investigated the relative humidity errors when measuring dry-bulb temperature (DBT) and WBT. Other researchers examined the uncertainty of humidity measurements [15–17] and the effect of WBT instrument uncertainty on the calculation of absolute humidity [18]. Finally, Davies-Jones [19] presented an efficient and accurate method for computing the WBT, a long pseudo-adiabatic.

None of those mentioned above studies has quantitatively examined the sensitivity of the WBT senses to different degrees of contaminations. This chapter is the first to experimentally investigate the effects of various degrees of contaminations on WBT measurement difference and hence on heat stress analysis in underground mines.

Heat Stress Indices for Underground Mines

Despite the significant impact of hot and humid environments on the safety and productivity of operations and comfort of mineworkers, less attention has been paid to the thermal comfort and heat management strategies in underground coal mines. The thermal condition of any underground coal mine depends on the location and depth of the mine, mining method, ambient temperature, thermal characteristics of strata, and mining equipment. In underground coal mines, where extreme climate conditions become critical and face equipment, heat loads are significant; high workplace temperatures affect comfort, safety, and production. Therefore, heat management and refrigeration must be introduced in the mine ventilation system when workplace temperatures exceed 27 °C.

The most applicable heat stress indices for underground mines are generally classified into three types:

1. Empirical index: Wet-bulb globe temperature (WBGT),
2. Rational index: Air-cooling power (ACP), and
3. Direct measurement index: WBT.

Wet-bulb globe temperature (WBGT) is the most commonly used heat stress index, being endorsed by the International Standards Organization (ISO), the American Conference of Governmental Industrial Hygienists, and the National Institute of Occupational Safety and Health (NIOSH). The WBGT index was chosen to specify the environment because it employs relatively simple measurements. An advantage of the WBGT index is that air velocity does not need to be measured directly. Occupational Health and Safety Administration (OHSA) heat stress standards were developed to establish work conditions that would ensure that workers' body temperatures do not exceed 38 °C. This limit was based on recommendations by a panel of experts from the World Health Organization (WHO), who considered the WBGT index the most suitable means of specifying the work environment. Table 22.1 presents the recommended threshold limit values for WBGT for three different workload conditions and two different air velocities ranges [20].

Air-cooling power (ACP) is another heat stress index that can address the satisfactory working criteria for underground mines. ACP combines all the pertinent environmental parameters of the cooling capacity of air, such as WBT and air velocity, and relates directly to engineering design parameters. Figure 22.1 presents the values of ACP at different values of WBT and air velocity and for three different types of clothing [21]. When using rational indices such as ACP, the effect of clothing (unclothed, heavy, or light clothing) becomes important due to its insulating effect on heat transfer and the resultant body temperature. The prescribed actions for various levels of ACP in Australia are presented in Table 22.2 [20].

Table 22.1 Wet-bulb globe temperature limit for workloads in hot environments [20]

Workload	WBGT (°C)	
	Air velocity < 1.5 m/s	Air velocity > 1.5 m/s
Light	30.00	32.20
Moderate	27.80	30.60
Heavy	26.10	28.90

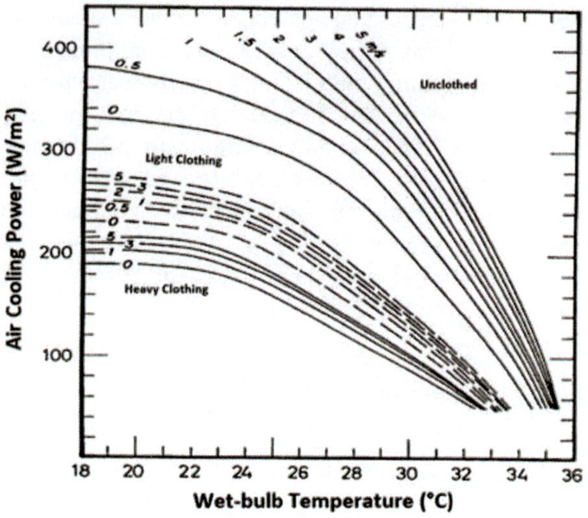

Fig. 22.1 Air-cooling power for different wet-bulb temperatures, air velocities, and different clothing [21]

Table 22.2 Prescribed actions for various levels of ACP in Australian underground mines [20]

ACP (W/m^2)	The prescribed course of action
<115	Remove workers from the area
115–140	Monitor conditions
140–220	Acclimatize exposed workers
>220	Acceptable conditions

Wet-bulb temperature (WBT) refers to the temperature of moisture evaporation and hence the relative moisture content of the air [22]. WBT is a principal determinant of the WGBT index and the ACP index. It is also used as a direct absorption index for monitoring and assessing heat stress in underground mines. The maximum acceptable WBT for the design purposes in underground mines should be between 27 and 28 °C. It has been recommended that routine work should not be permitted when WBT exceeds 32 °C or DBT exceeds 37 °C. This condition minimizes the risk of heat illnesses and enhances labor force productivity [20]. Therefore, the literature recommends that a heat stress index, preferably ACP, should be used to design an underground mine ventilation system. However, WBT should be used once the monitoring and control systems are implemented.

WBT is measured by moving air through a moist wick that covers the temperature sensor head. The instrumentation required for measuring WBT is inexpensive and easy to use by non-specialists. The WBT sensors that are common in underground coal mines are non-electrical and explosion-free sensors. Examples of these sensors are handheld psychrometers or whirling hygrometers (Fig. 22.2). When used in contaminated environments such as underground coal mines, the hygrometer wick becomes dirty over time, possibly reducing the accuracy of measurements.

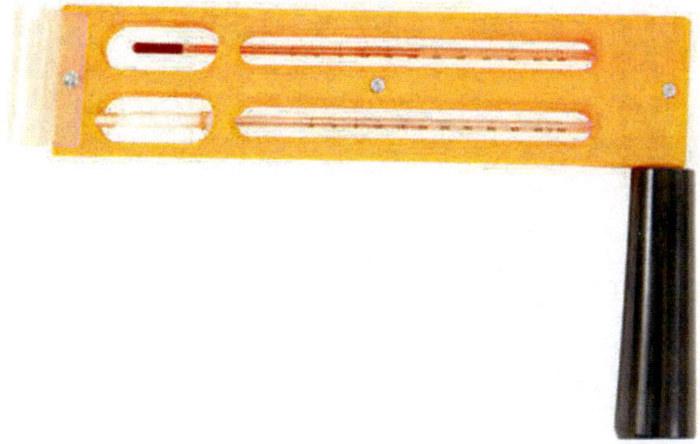

Fig. 22.2 Photograph of a whirling hygrometer

Nevertheless, accurate WBT measurements are essential to ensure safe underground mines and actual heat management plans. In underground mines, accurate WBT is one of the main parameters that help shift supervisors decide whether to continue or stop the operation. Accurate WBT measurements also allow precise calibration of ventilation models to meet the requirements of the mine. This study examines the effects of different levels of coal dust accumulation on the wick of WBT sensors on the accuracy of measurements and the heat stress analysis.

Experimental Setup

A series of experiments were designed and conducted in the Newcrest Heating, Ventilating, and Air Conditioning (HVAC) laboratory at the University of Queensland, Australia. These experiments made use of the P.A. Hilton air conditioning unit (Fig. 22.3). The unit is equipped with digital temperature sensors and computerized data acquisition. This allows accurate temperature readings to three decimal places to be recorded in real time by the attached data logger and supplied computer software. There are two heating elements in the ducting: a steam injector and an evaporative cooling unit. These components allow the unit to perform three processes on humid air, including dry heating, moisturizing and cooling processes (Fig. 22.3). In addition, the unit incorporates four separate temperature stations consisting of one wet-bulb and one dry-bulb thermocouple (Fig. 22.4). Thus, in this unit, there is four main stations to measure the temperature. Temperature station "A" measures the ambient air conditions of the environment in which the unit is operating, and as such, it is not affected by any of the heating or cooling processes. Station "B" occurs immediately downstream of the re-heaters and the steam injector and, as such, responds to the changes induced by these processes. Temperature station "C" measures conditions immediately after the cooling unit. Finally, temperature station "D" measures the air requirements after subjected to all heating, cooling, and humidification processes before exiting the unit.

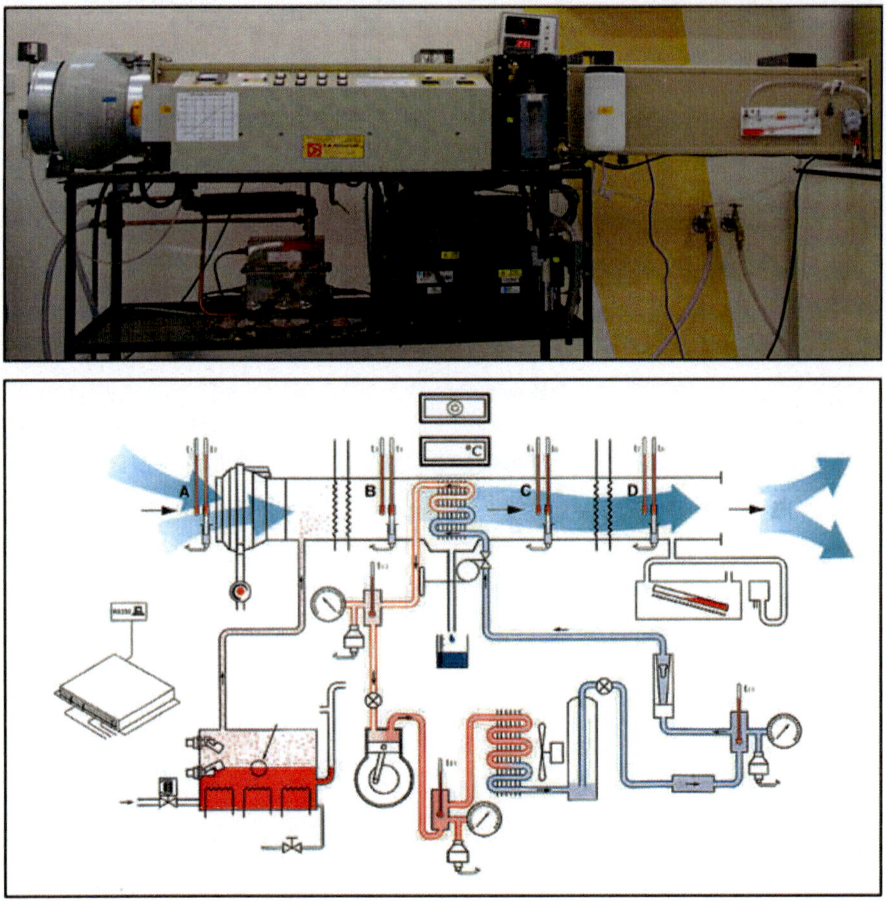

Fig. 22.3 P.A. Hilton air conditioning laboratory unit

Experimental Procedure

Contaminating Wick Samples

A coal dust contaminator unit was designed, built, and used to uniformly contaminate seven cloth samples at a different rate of contamination used to make the contaminated wicks (Fig. 22.5). Figure 22.5a shows the vacuum pump with controllable vacuum rate, elastic vacuum hose, and stainless steel vacuum duct. The vacuum pump power can change from 100 to 2000 W. Figure 22.5b presents a close-up of the coal dust distributor that consists of four polypropylene ducts with a diameter of 3 cm. The mesh screens (Fig. 22.5c) and cloth samples (Fig. 22.5d) are installed in

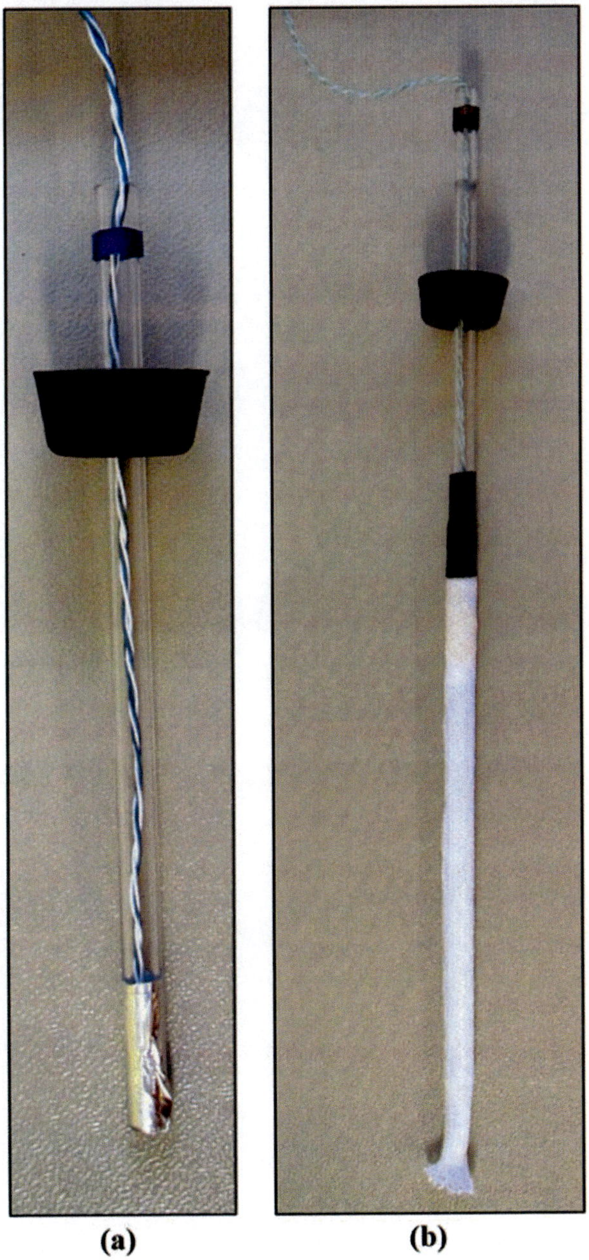

Fig. 22.4 Dry-bulb (**a**) and wet-bulb (**b**) temperature sensors used in P.A. Hilton air conditioning laboratory unit

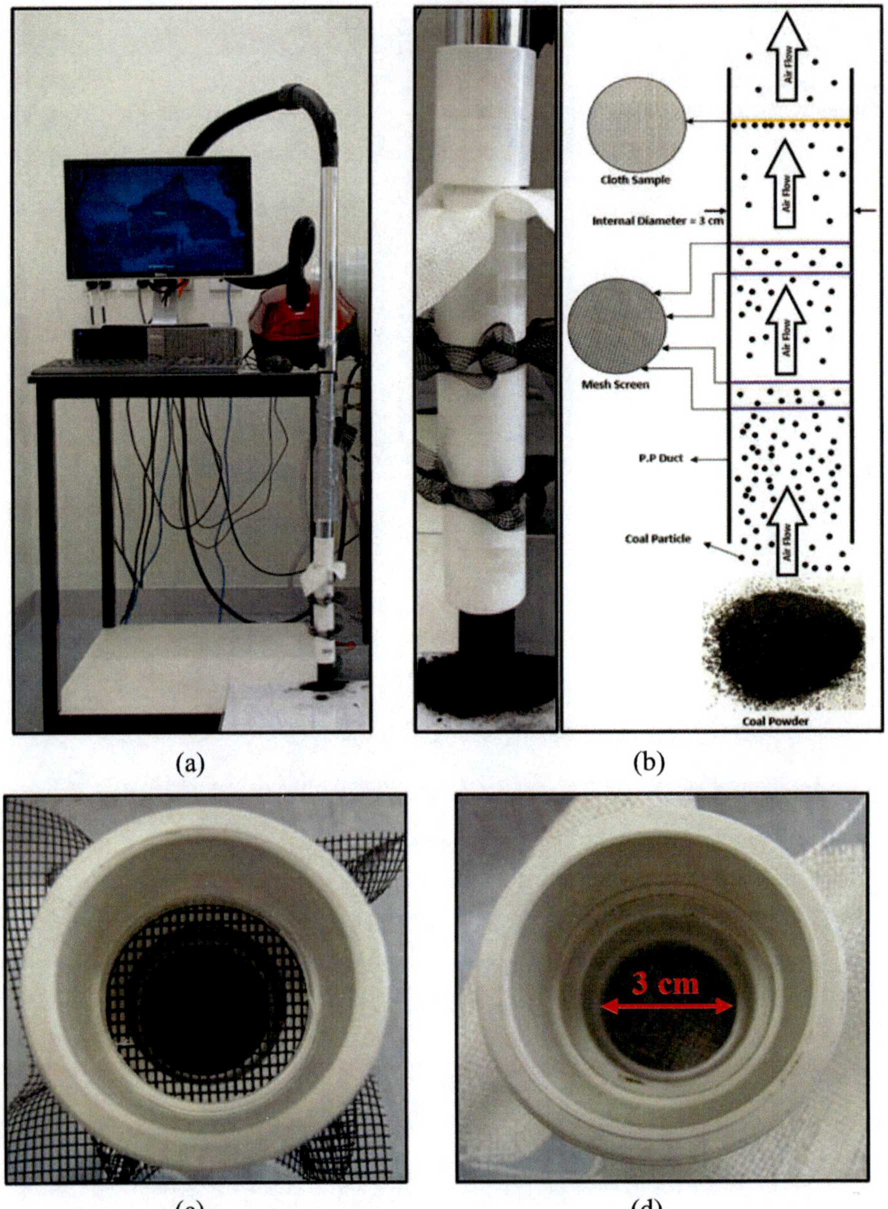

Fig. 22.5 Coal dust contaminator unit, **a** vacuum pump, **b** coal dust distributor, **c** mesh screen, **d** cloth sample

Table 22.3 Experimental characteristics of contaminated cloth samples

Sample number	Exposure time (s)	Contaminant mass (mg)	Contaminant concentration (mg/cm^2)
1	40	39.9	5.7
2	60	59.5	8.5
3	70	69.3	9.9
4	90	88.9	12.7
5	120	118.3	16.9
6	200	198.1	28.3
7	250	247.8	35.4

the coal dust distributor. This section provides uniform distribution of coal dust on the cloth samples.

The vacuum pump creates an airflow that sucks the coal dust into the distributor. The coal dust first travels through four stages of mesh screens to provide uniform distribution of coal dust. The power of the vacuum pump was 500 W, and the travel velocity of coal dust was 1.6 m/s. The exposure time was varied from 40 to 250 s to make seven different contaminated cloth samples. The contaminant mass for each cloth sample was determined based on the difference between the mass of contaminated and clean cloth samples at the end of contamination exposure time. Table 22.3 presents the exposure time (s), the contaminant mass (mg), and the contaminant concentration (mg/cm^2) for each cloth sample. Figure 22.6 shows the macro and micro-images of seven contaminated cloth samples with the same area of approximately 7.0 cm^2. The contamination process for each cloth sample was carried out until the required length for the contaminated wick (3 cm) covering the WBT sensor (shown in Fig. 22.7) was obtained.

Wet-Bulb Temperature Measurements

The experiments were conducted using the P.A. Hilton air conditioning unit. Station "D" of the unit was modified to accommodate two WBT sensors. One of the sensors was kept clean and used to measure the reference WBT. The other sensor was used to measure WBT under the influence of contaminated wicks with different volumes of coal dust (Fig. 22.8). WBT sensors with no contamination were first cross-calibrated to ensure consistent temperature readings. Then, the repeatability of the measurements was examined by conducting many tests at the same flow conditions and for the WBT sensor with the same amount of contamination. Finally, a series of tests for different contaminated wick samples were conducted to examine the effects of wick contamination on the accuracy of measurements. Two different heating and cooling processes on the humid air were considered, and both clean and contaminated sensors measured WBT.

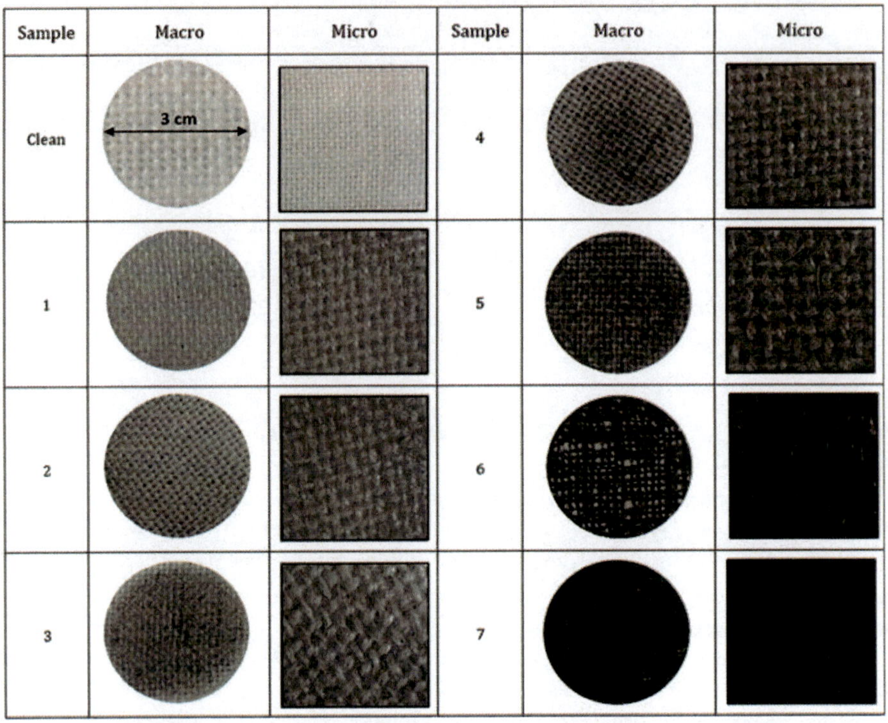

Fig. 22.6 Macro and micro-images of contaminated cloth samples

Results and Discussion

Temperature Sensor Calibration

Two WBT sensors used in this study were first cross-calibrated to ensure consistent temperature readings. Next, the calibration process was carried out for the sensors with clean wicks that measured the variation of WBT in a heating process and at a volumetric flow rate of 0.126 m^3/s. The temperature profiles for the two sensors before and during the heating process are presented in Fig. 22.9. The results show that both temperatures increase sharply and reach a reasonably established condition with minor fluctuation once the electrical heater was turned on. These fluctuations were due to the error of instrumentations. One of the sensors showed a WBT of 20.50 ± 0.05 °C, and the other one showed 20.60 ± 0.05 °C. The calibration data were obtained for a period of 800 s during the stabilized condition. The results show an apparent offset between the temperatures measured by the two sensors. This offset was then corrected by applying a linear calibration curve that is presented in Fig. 22.10. Finally, the calibration was applied to the sensor results that will be

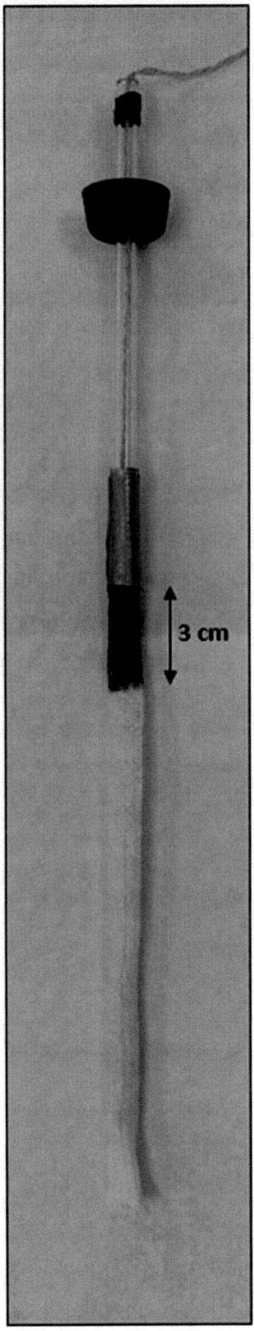

Fig. 22.7 Sample of contaminated wick on the wet-bulb temperature sensor (scale 1:2)

Fig. 22.8 Clean and contaminated wet-bulb temperature sensors inside air conditioning unit

Fig. 22.9 Wet-bulb temperature profiles measured by two clean sensors

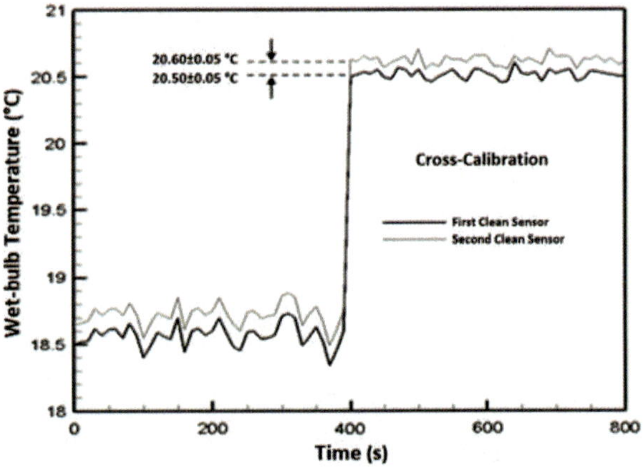

contaminated during the experiments. The bias error of the calibration was found to be 0.025 °C.

Comparison analysis between the experimental and theoretical values of WBT in the heating process was carried out based on the psychrometry analysis. It was

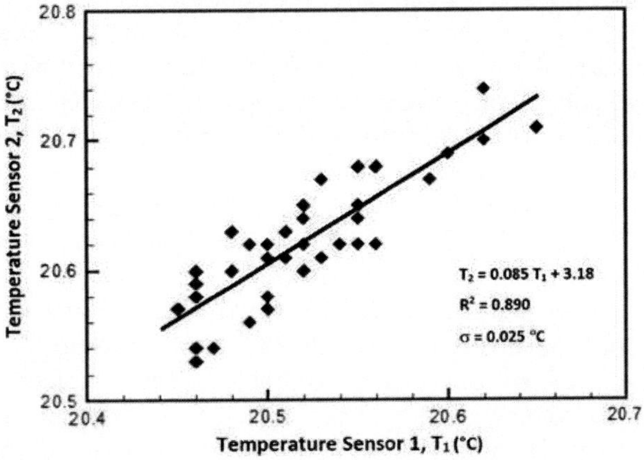

Fig. 22.10 Cross-calibration curve for two wet-bulb temperature sensors

considered that the initial average of WBT was 18.50 °C, the average DBT was 28.00 °C, the ambient pressure was 101 kPa, the volumetric flow rate of air through the system was 0.126 m^3/s, and the heat added to the system was 1 kW. The mass flow rate of dry air was calculated by:

$$\dot{m}_a = Q/\text{ASV} \tag{22.1}$$

where $\dot{m}_a$ is the mass flow rate of dry air (kg/s), Q is the volumetric flow rate of air (m^3/s), and ASV is the apparent specific volume (m^3/kg). The sigma heat after the heating process was calculated by:

$$S_2 = \frac{\text{HEAT}}{\dot{m}_a} + S_1 \tag{22.2}$$

S_1 and S_2 are the sigma heat before and after the heating process, respectively (kJ/kg), and HEAT is the heat added to the system (kW). The theoretical WBT was then determined using the psychrometry analysis. A difference of 0.04 °C was obtained between the theoretical result (20.54 °C) and the experimental measurement (20.50 °C). This is an indication of a good agreement between experiments and theory.

Effects of Wick Contaminant Concentration on Wet-Bulb Temperature Measurements

To determine the effects of coal dust contamination on the accuracy of measurements, WBTs were measured using seven contaminated sensors against the clean sensor in four heating and four cooling stages for a period of 3400 s. In the heating process, the electrical heaters, each having 1 kW power, were turned on one by one forming the four heating stages. In the cooling process, they were turned off one by one, forming the four cooling stages. Figure 22.11 illustrates these stages for the most contaminated wick sample. Figure 22.11 also shows the differences in WBT and response time between the contaminated and clean sensors. It is shown that there is a clear difference between WBTs measured by the contaminated and clean sensors for both heating and cooling processes. It is also evident that the response time of the contaminated sensor is lower than the clean sensor for both heating and cooling processes. The reason is that the coal dust contamination causes a delay in water evaporation from the sensor wick.

Figures 22.12 and 22.13 show WBT and response time differences for the seven contaminated sensors against the clean sensor in the heating and cooling process, respectively, for a total period of 800 s. The volumetric flow rate of air through the air conditioning unit was 0.126 m^3/s. The heating process (Fig. 22.12) was carried out in the first stage of heating by adding 1 kW heat to the air, and the cooling process (Fig. 22.13) was carried out in the last stage of cooling by removing 1 kW heat from the air. The clean sensors show that the WBT increases from 18.50 to 20.50 °C in the first stage of heating and decreases from 20.50 to 18.50 °C in the last stage of cooling. The clean sensor measurements were compared with seven contaminated sensor measurements for both heating and cooling processes. The results generally show that the WBT measurement and response time differences increase as the wick contamination concentration increases for both heating and cooling processes. This is because the coal dust contaminants fill and cover the pores of the wick, causing a reduction in water evaporation.

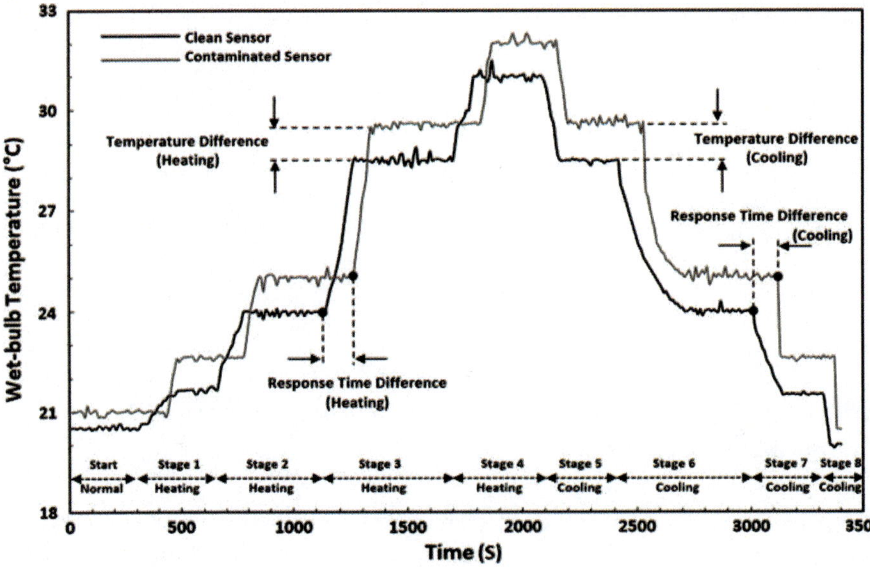

Fig. 22.11 General trend of contamination affects the accuracy of wet-bulb temperature measurements

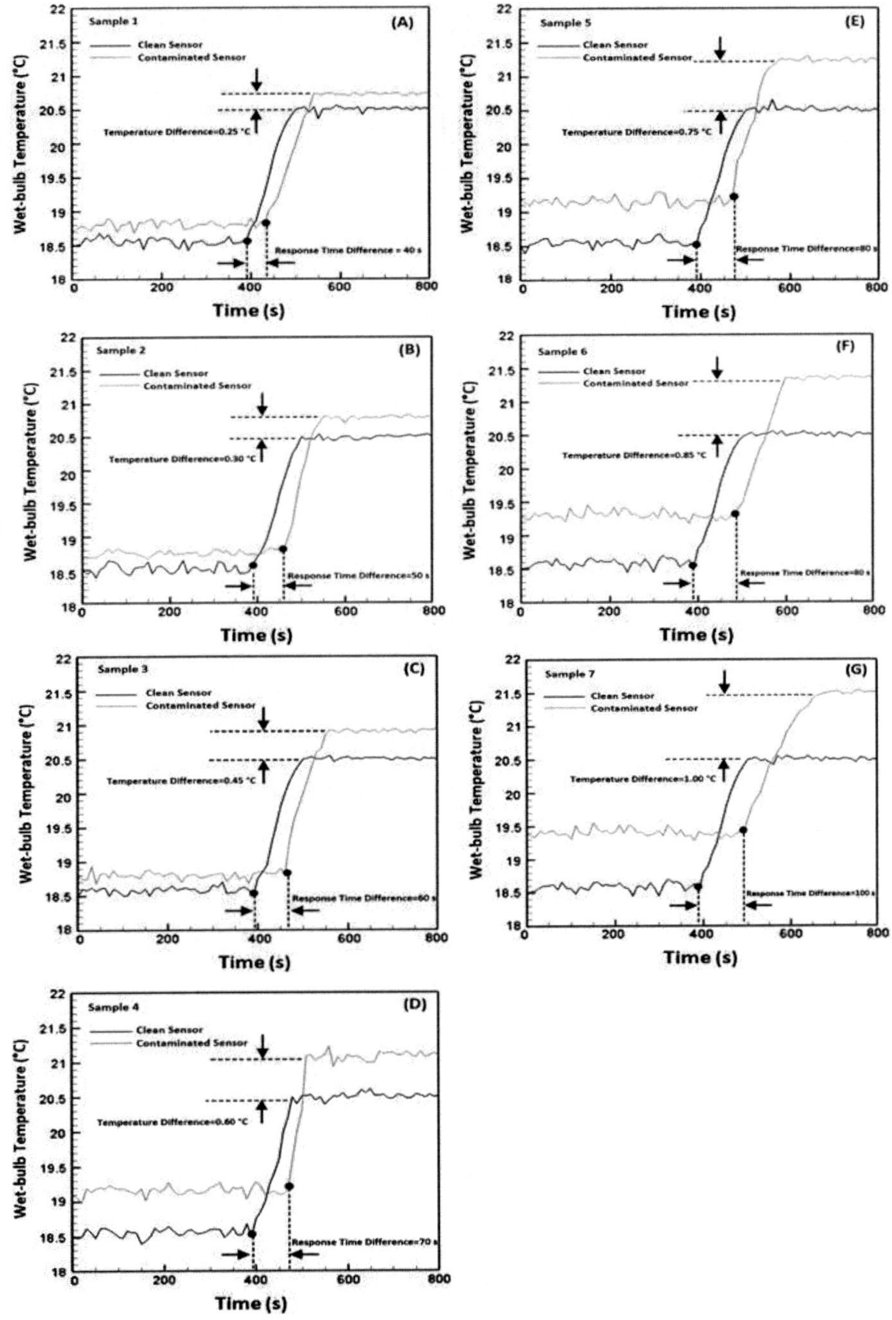

Fig. 22.12 Wet-bulb temperature measurement and response time differences for different wick contaminant concentrations in a heating process

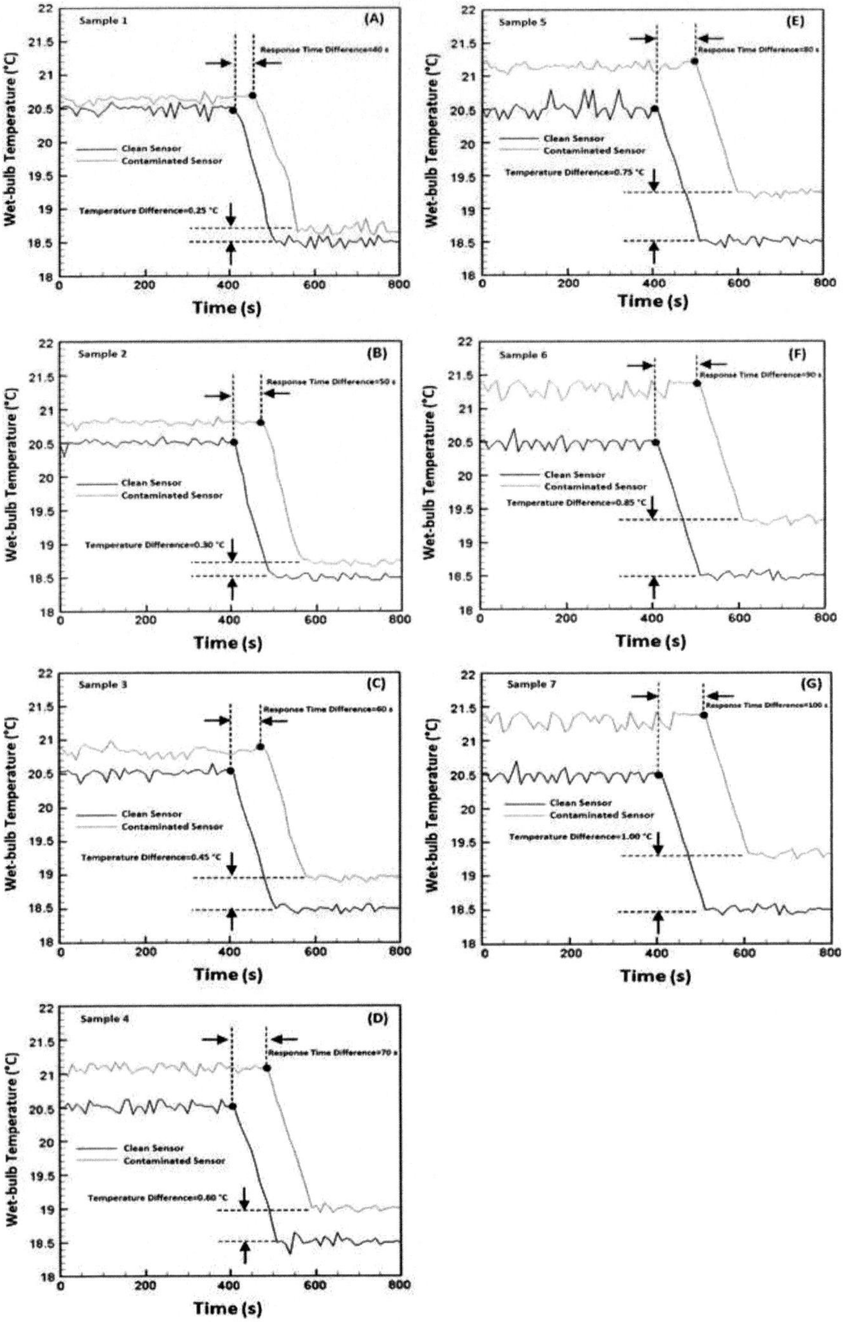

Fig. 22.13 Wet-bulb temperature measurement and response time differences for different wick contaminant concentrations in a cooling process

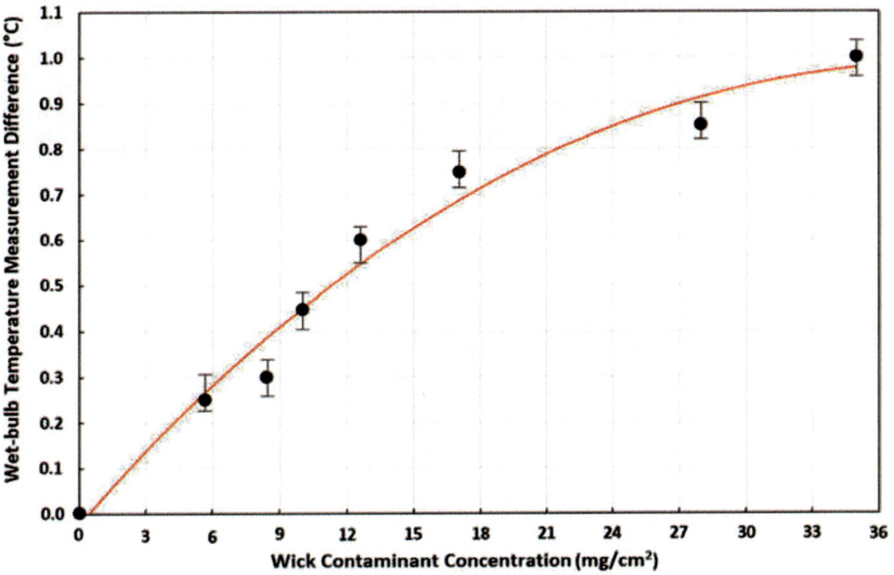

Fig. 22.14 Variation of wet-bulb temperature measurement difference with wick contaminant concentration

Figure 22.14 clearly shows the increasing WBT measurement difference concerning the wick contaminant concentration for heating and cooling processes. The WBT difference increases from 0.23 to 1.01 °C as the wick contaminant concentration increases from 5.7 to 35.4 mg/cm^2 in heating and cooling processes. Figure 22.15 shows that the response time difference of the contaminated and clean sensors increases as the wick contaminant concentration increases—the difference increases from 40 to 100 s from contaminated wick samples 1 to 7. The reason is that the thermal resistance of the wick increases due to the addition of coal dust contaminants, causing a reduction of the heat transfer rate.

Conclusions

Measurements of WBT were made using clean and contaminated temperature sensors in heating and cooling processes using a standard P.A. Hilton air conditioning unit. Seven wick samples with different volumes of coal dust were prepared using a coal dust-contaminating unit designed and built for the present investigation. Two WBT sensors used in the entire experiment were first cross-calibrated. This was to ensure consistency in the WBT readings by the sensors with clean wicks. A general comparison between two WBT sensors with clean and contaminated wicks showed a constant misreading of the WBT by the contaminated sensor. This temperature measuring difference was observed in both heating and cooling processes. It was also found

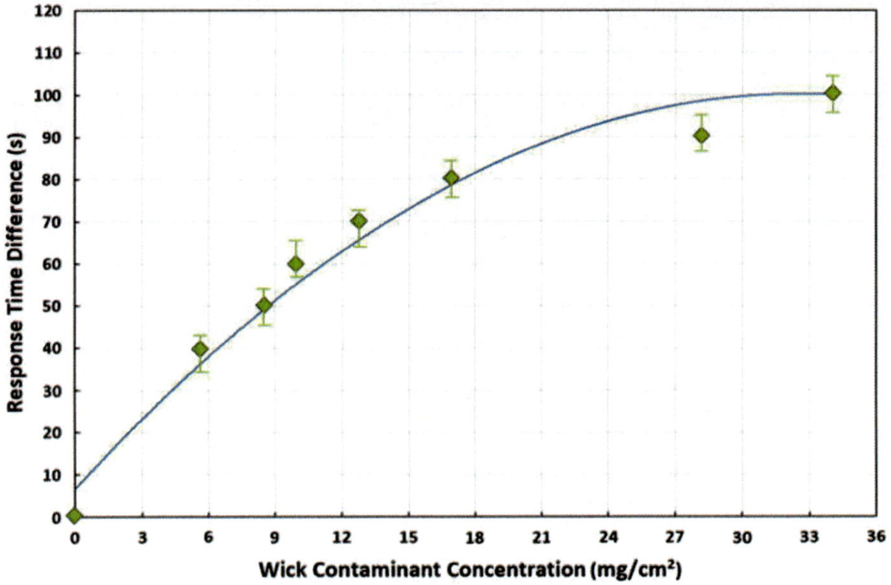

Fig. 22.15 Variation of response time difference with wick contaminant concentration

that there was a direct relationship between the amount of coal dust contamination and the WBT measurement difference. This relationship was found to be slightly different for heating and cooling processes. Finally, the effect of the WBT measurement difference on the heat stress index, air-cooling power, was examined. It was found that any error in WBT measurements causes a corresponding error in the heat stress index. There was a dramatic change of the prescribed action for some ranges of the errors, such as "monitor conditions" to "remove workers from the area." It is recommended that the WBT sensors should be maintained clean and free of airborne particles when used in contaminated environments such as dusty underground coal mines.

References

1. Gillies, A. 1991. *The problem of heat stress in the Australian mining industry: Report*. Department of Mining & Metallurgical Engineering, The University of Queensland.
2. Lavenne, F., and J. Brouwers. 1983. *Heat acclimatization in The Encyclopaedia of Occupational Health and Safety*. U.S.A.: International Labour Organisation.
3. Hardcastle, S., and K. Butler. 2008. A comparison of the globe, wet and dry temperature and humidity measuring devices available for heat stress assessment. Presented at the 12th North American Mine Ventilation Symposium, Ontario, Canada.
4. Blunden B. 2012. *Air quality impact assessment in underground mines ventilation*. Professional report, Queensland (10 (2)).

5. Aich, V., et al. 2011. Development of wet-bulb-temperatures in Germany regarding conventional thermal power plants using wet cooling towers. *Meteorologische Zeitschrift* 601–614.
6. Alfano, F.R.D.A., B.I. Palella, and G. Riccio. 2012. On the problems related to natural wet bulb temperature indirect evaluation for assessing hot thermal environments using WBGT. *Annals of Occupational Hygiene* 56 (9): 1063–1079.
7. Stull, R. 2011. Wet-bulb temperature from relative humidity and air temperature. *Journal of Applied Meteorology and Climatology* 50 (11): 2267–2269.
8. Tomas, S., et al. 2012. Internal stress measurement during drying of rubberwood lumber: Effects of wet-bulb temperature in various drying strategies. *Holzforschung* 66 (5): 645–654.
9. Moulsley, L., and J. Fryer. 1976. A resistance thermometer psychrometer. *Journal of Agricultural Engineering Research* 21 (1): 101–102.
10. Cotton, R.F. 1969. A resistance thermometer psychrometer for use in high solar radiation conditions. *Journal of Agricultural Engineering Research* 14 (3).
11. Ramsey, J.D., C.L. Burford, F.N. Dukes-Dobos, F. Tayyari, and C.H. Lee. 1986. Thermal environment of an underground mine and its effect upon miners. In *Annals of the American Conference of Governmental Industrial Hygienists*.
12. Lee, C.H. 1986. *Effects of wick contamination and thermal component variation on thermal indices*. Texas Tech University.
13. Pain, D. 2011. *Investigating the effects of contamination on wet-bulb temperature readings*. Australia: School of Mechanical and Mining Engineering, The University of Queensland.
14. Ustymczuk, A., and S. Giner. 2011. Relative humidity errors when measuring dry and wet bulb temperatures. *Biosystems Engineering* 110 (2): 106–111.
15. Hubbard, K.G., et al. 2005. Sources of uncertainty in calculating design weather conditions. *ASHRAE Transactions* 111: 317.
16. Lovell-Smith, J. 2009. The propagation of uncertainty for humidity calculations. *Metrologia* 46 (6): 607.
17. Mathioulakis, E., G. Pancras, and V. Belessiotis. 2011. Estimation of uncertainties in indirect humidity measurements. *Energy and Buildings* 43 (10): 2806–2812.
18. Slayzak, S.J., and J.P. Ryan. 1998. *Instrument uncertainty effect on calculating absolute humidity using dewpoint, wet-bulb, and relative humidity sensors*. Golden, CO: National Renewable Energy Lab.
19. Davies-Jones, R. 2008. An efficient and accurate method for computing the wet-bulb temperature along pseudoadiabats. *Monthly Weather Review* 136 (7): 2764–2785.
20. Webber, R., R.M. Franz, W.M. Marx, and P. Schutte. 2003. A review of local and international heat stress indices, standards, and limits regarding ultra-deep mining. *Journal of the Southern African Institute of Mining and Metallurgy* 103 (5): 313–323.
21. McPherson, M.J. 2012. *Subsurface ventilation, and environmental engineering*. Springer Science & Business Media.
22. Karmoush, M. 2008. *Air quality monitoring plan for underground mines, in Donaldson Coal Pty Limited professional document*. New South Wales.

Chapter 23
Advanced Analytics for Autonomous Underground Mining

Mohammad Ali Moridi and Mostafa Sharifzadeh

Abstract In the challenging environment and dynamic topology of an underground mine, reliable and effective communication is a high-stakes issue. Automation by remote and automated systems has resulted in enhanced real-time response to events and improvements in employees' workplace health and safety, operational management, energy efficiency, and cost-efficiency. An integrated wireless ad hoc network (WANET) and GIS system are introduced to ensure continuous underground mine communication, monitoring, and control. The proposed system improves health and safety and decreases capital expenditures (CAPEX) and operating expenditures (OPEX). Considering the ZigBee network and ArcGIS applications, the functions for real-time underground monitoring (temperature, humidity, and gas concentration), ventilation system control, and communication in emergency conditions by the surface user would be practicable. The system is fortified with automated or/and remote triggers action plans for measured environmental elements. At normal (green) conditions, the received attributes are within safe limits. In this state, readings are recorded at 30-min intervals, and mining operations continue. At the transient (yellow) condition, the attributes are between average and threshold value limits. In this situation, trigger actions are set up to automatically switch on the auxiliary fans and texting messages to authorized personnel and reduce the reading's intervals to 15-min. At unsafe (red) conditions, the progress of measurements is more significant than threshold values. In this state, the system sends text messages to all underground personnel to immediately evacuate hazardous levels, and 5-min reading intervals are monitored. Additionally, the WANET-incorporated GIS runs multi-user operations and 3D monitoring to understand the environmental attributes and miners' conditions in underground mines. It plays a critical role in growing Artificial Intelligence (AI) in the mining industry by mitigating risks and reducing costs.

Keywords Advanced analytics · Mining · Mine safety · Real-time monitoring · Autonomous underground mine · Communication

M. A. Moridi
Evolution Mining Limited, Mungari Operations, PO Box 10398, Kalgoorlie, WA 6433, Australia

M. Sharifzadeh (✉)
MEC Mining / Deep Mining Geomechanics, Perth, WA 6000, Australia

A. Soofastaei (ed.), *Advanced Analytics in Mining Engineering*,
https://doi.org/10.1007/978-3-030-91589-6_23

Introduction

Underground mine safety and health remain challenging issues in the mining industry. Human errors were concluded from reports as the most significant reasons for mining fatalities. Therefore, safety and health issues are always a significant concern in mining operations and deserve priority in management and engineering designs to provide and maintain a safe and healthy workplace. In response to these challenges, mine automation by new technologies such as wireless sensor network (WSN) assisted with geographic information system (GIS) has been widely utilized toward autonomous underground mine communication, monitoring, and control to enhance safety and health, productivity, and reduce operational costs.

To this end, an integration system is developed to mitigate underground safety and health concerns. Based on developing a wireless ad hoc network (WANET), this system is introduced to sense the underground mine environment, regulate the ventilation system, and communicate between surface offices and miners. Thus, reduced power consumption, near real-time environment monitoring, and bilateral communication between the surface and underground personnel are achieved. In addition, experimental tests were carried out to verify network reliability and security of the packet delivery in underground mines. The architecture of underground monitoring and communication for the system integration is illustrated in Fig. 23.1. Temporal WANET data, including messages, operation orders, and environmental attribute readings such as temperature, humidity, and gas concentration, are transferred to the GIS management server in the surface control center. The transmitted data are received and stored by the WANET program then provided for manipulation in the control center. Thus, risk situations are immediately identified and responded through a logical data analysis process in the GIS management server before reaching dangerous (unsafe) levels and accidents. The ventilation system management is also used for workplace health and safety compliance and optimizing mine site power usage.

The remainder of the chapter is organized as follows. The fundamental knowledge considering ZigBee technology as one of the wireless ad hoc networks and GIS are first described. Then, the implementation and structure of system integration are demonstrated. Finally, the strategic process of combining WANET data and map information through the GIS management server is modeled for monitoring, communicating, and controlling the environmental attributes in an underground mine. In this chapter, the applications and functions of the underground mine monitoring and communication systems are also considered based on the capability of developed WANET nodes.

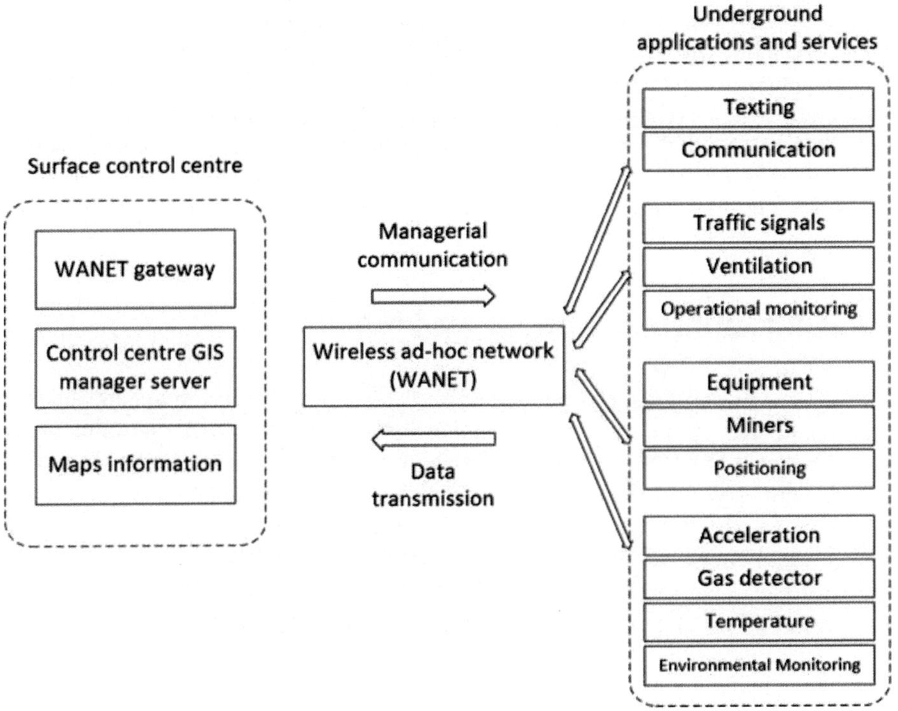

Fig. 23.1 Architecture of monitoring and communication system in underground mines based on the wireless ad hoc network

WANET Trends

The underground WANETs consist of a few to several hundred nodes between a surface gateway and specified sensor nodes at underground levels [1]. In the former study, the implementation and verification of a Wi-Fi ad hoc communication system in an underground mine environment were proven [2]. In this study, ZigBee technology, another wireless ad hoc network, is selected to develop a system integration to investigate underground mine applications and services. Based on IEEE 802.15.4 protocol, ZigBee is a new wireless sensor technology with more benefits than other WSNs for underground monitoring and communication systems [3]. Even though ZigBee technology provides only a low data rate, its benefits are low power consumption, very cost-effective nodes, network installation, and maintenance [4]. It can also provide networking applications for data transmission between nodes (node to node relays) with high performance based on many wireless hops. Moreover, it does not require any access point or central node to transmit data between clusters. The significance of ZigBee in underground mines compared to other WSNs was evaluated in the recent publication of authors [5].

GIS is a new technology used for spatial data analysis to capture, store, analyze, manage, and present data linked to locations [6]. GIS allows users to view, understand, question, interpret, and visualize data in many ways that reveal relationships, patterns, and trends in the form of maps, globes, reports, and charts. Web-GIS is an inevitable trend that helps solve the problems of spatial information integration and sharing in the technical aspect of web media [7, 8]. In addition, researchers have recently focused on the GIS supports to manage emergency and unsafe conditions [9–11].

Data Management in Underground Mines

GIS is based on computer programs used to store, model, retrieve, map, and analyze geographic data. In this system, the spatial features of a specified environment are stored and manipulated in a coordinate system, which refers to a specific place. GIS merges multi-layers of required geographic and spatial data for the user evaluation and helps determine the locations and times of possible incidents in advance. GIS servers can manage and process data for a substantial number of attributes coming from different sources. It also can distribute and share data between users based on the internet or intranet, and data could be saved, manipulated, or informed by other users. Therefore, GIS can decrease the time and cost of sharing geographic data and its attributes [12].

Underground Mine System Integration

In the challenging environment and changing topology of a mine, reliable and simplified communication is a high-stake issue with safe and efficient mining operations objectives. Automation of remote and automatic systems has improved workplace safety and health for miners, yielded cost-effectiveness, management improvement of technical problems, energy savings, and real-time response to incidents. In response to these challenges, the integration of technologies has a significant role in underground mining automation. According to WANETs' specific features of high reliability and multi-hop networking, WANET can create an integrated wireless network between nodes in the underground mine tunnels and the surface gateway. In this study, ZigBee's capability of monitoring underground environmental attributes is combined with geographic information to provide potential applications in communication, operational, and environmental monitoring systems of underground mining.

In order to achieve such an autonomous underground mine system, integrating maps information and Spatio-temporal data from WANET nodes into a database at a control center is required. Figure 23.2 illustrates data processing and results from management in the surface control center.

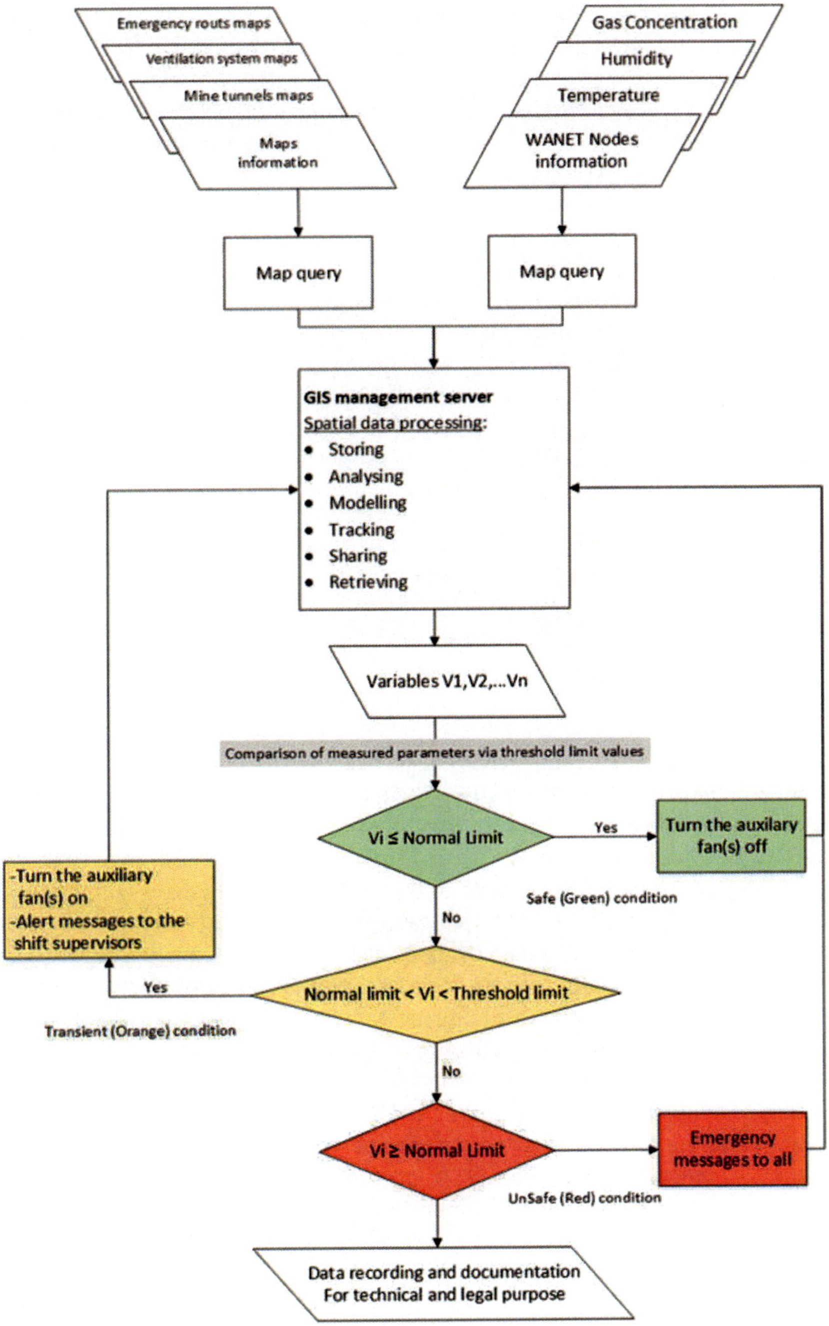

Fig. 23.2 Flow chart of data processing and result management

The network demanded in an underground mine must provide bilateral communications between the surface control center and all underground wireless nodes interactively. According to the threshold limit values for the different variable parameters ($V1$, $V2$, ..., Vn) of the underground mine environment, safe, transient, and unsafe conditions were set. Thus, the remote or automatic countermeasures in a GIS management server were arranged to control ventilation fans and send alerts or alarm messages to relevant authorities. Additionally, immediate texting messages are bilaterally communicated between underground personnel and the surface operator in emergency conditions.

This system has achieved the required safety and health outcomes and improved underground mining operations, near real-time monitoring data, remote and automatic controls, and communication by texting messages. Furthermore, such achievements are more efficient for emergency management when system configuration enables control, monitoring, and communication between users in various places connected by internet medium access.

Communication System Structure

Wireless Network Set Up

The entire system of the tested underground WSN is composed of different WANET nodes such as coordinator, routers, and end devices. These products were developed in collaboration with Tokyo Cosmos Electric Co., Ltd. The JN5148-EK010 kit (Jennic) stacks were employed to create a ZigBee network. The wireless network is initially created by the coordinator (gateway) to join other nodes. Then, a ZigBee coordinator is connected to the laptop (PC) used in the experiments. Bilateral communication between the coordinator and end devices was provided to send and receive messages and readings instantaneously taken by their sensors. Routers with the ability to sense the environment were employed to relay communication through the network. In addition, they are sending and receiving messages, and remote control of ventilation fans is enabled by the surface coordinator based on the designed software.

To set up WANETs, power consumption and high reliability of packet delivery are the most concerns. For the former case, ZigBee nodes are configured to transmit data in more extended periods when the mine is in safe and transient conditions caused to extend the life of batteries. In the latter case, different time intervals are considered for data delivery of environment sensing to avoid network congestion and the possibility of packets loss. The power usage of direct and alternating currents (DC/AC) for the WANET nodes (except the coordinator) was designed to operate under battery and mine site power supply, respectively. Thus, alternating currents power usage results in battery life extension and ZigBee nodes can continue long-time data telemetry during power outages at any accident. The ZigBee nodes can last a few days to several months depending on their data rate and applications.

Sensing Environment

The safety and health of coal and metal/non-metal mining operations were raised considerably due to the wireless monitoring of environmental attributes. Digital temperature-humidity compound sensor on-board of each JN5148 with advanced sensitivity and long-term stability for mine sites is utilized in the system. Methane, Oxygen, CO_2, CO, NOX, and SO_2 concentration sensors (readers) are easily connected to ZigBee nodes to sense the environment. The sensors were configured the single-line communication to transmit real-time data to the nodes. The measurement of CO_2 concentration in this study was considered to manage safety and health risks nearby coal strata in coal mines or fumes-filled spaces in metal/non-metal mines.

Text Messaging Operators

Developed WANET nodes, ZigBee nodes in this experiment, can connect with laptops and mobile phones for sending and receiving text messages. Portable radio stations to connect laptop (Tablet) designed to be placed in an underground refuge chamber. The radio station plays a significant role in wireless communication with the surface operator during accidents, particularly when cable damage or power outage occurs. Even though its primary role is the remote control of ventilation fans. ZigBee nodes were placed in the boxes to minimize environmental effects on their operation [12].

Ventilation Control

Air ventilation deficiency in underground mines is a critical issue to mine personnel's occupational safety and health. Moreover, optimizing the fan's power consumption to supply underground fresh air is considered on ventilation system design. Therefore, adding auxiliary fans to the ventilation system is economically required to improve air quality during hot seasons, blasting, gas leakages, and exhaust fumes. In the proposed system, remote and automatic controls of auxiliary fans were programmed with the software installed on PCs located at the surface office and refuge chamber.

Data Management Server Using GIS

The prototype model developed in this study relies on WANET data and geo-processing data of GIS. Data management server was developed on ESRI's established ArcMap 3D software, part of the ArcGIS software package.

Input Data

The first step of our designed management server is to communicate with the outside world to receive the required information. Input datasets in the database comprise map information, WANET nodes data, WANET text messages, WANET node positions, threshold limit values, and contact details. Map query is the primary process of map information to merge and display required features in the GIS server to geographically represent the fundamental layers of underground tunnels. These layers are revised according to the progress of underground mining activities. Then, other input data are analyzed and located on the layers for further manipulation.

The quality of the input dataset is considered for processing and analyzing any particular database. Consequently, the quality of input data in our designed GIS management server is divided between long-term and short-term datasets. Maps, WANET node positions, threshold limit values, and contact details are determined to be long-term input data into the database, which may be periodically updated. These data are stored in attribute tables that are associated with ArcGIS geo-processing models. On the other hand, WANET node data that measured mine tunnels' environmental properties are derived as short-term (temporal) data. In this case, the datasets of environmental phenomena such as temperature, humidity, and gas concentrations change from time to time or remain relatively continuous. Therefore, Spatio-temporal data models, which show both spatial and temporal characteristics of the environment, are considered input data in the GIS management server. The Spatio-temporal data are stored and manipulated in the ArcGIS geo-processing based on the related or joined table command to digital data collection tables by WANET gateway software.

Process Strategy

Real-time process strategy for safe working environments involves combining data models and programs in GIS management server to monitor and communicate underground mine automatically and remotely. A pattern of decision-making in managing Spatio-temporal data was modeled as a procedure to monitor the environmental attributes of underground mine tunnels (Fig. 23.2). To this end, a near real-time and flexible scheduling strategy was planned to apply the performance of WANET in an emergency status. In this model, a gateway was located in the surface control office to receive and transmit data through the underground network. The network is extended by ZigBee routers between the surface gateway and underground end devices based on optimized communication ranges. In addition, WANET sensor nodes are mounted in the working area, which sense environment attributes such as temperature, humidity, and gas concentration.

The transmitted data are firstly stored in the GIS management server located at the control center. The ability of the map visualization on GIS (ArcGIS) allows the position and component of the attributes in the underground mine environment to

Table 23.1 Threshold limit values for working environments in underground mine

Variables (V_i)	Event procedure conditions		
	Safe (Green)	Transient (Yellow)	Unsafe (Red)
Temperature (T_1, T_2, …, T_n), °C	$Ti \leq 28$	$28 < Ti < 40$	$Ti \geq 40$
Humidity (H_1, H_2, …, H_n), %	$Hi \leq 75$	$75 < Hi < 85$	$Hi \geq 85$
Gas concentration for CO_2 (G_1, G_2, …, G_n), ppm	$Gi \leq 2000$	$2000 < Gi < 5000$	$Gi \geq 5000$

be visually displayed on the screen. Then, the Spatio-temporal data tables stored by WANET software in the database were joined or related to the attribute tables of geographic node positions in the geo-processing services of the GIS management server. A joining table of Spatio-temporal data and geographic node position is created in ArcGIS (ArcMap). Each node position is connected to the related and measured variable temperature, humidity, and gas concentration parameters.

Following this, the Spatio-temporal data were analyzed, modeled, and retrieved in the GIS management server. Finally, a geo-processing model based on Python (ArcPy) was designed to track and control the environmental attributes in different conditions. Typical and threshold limit values to assess environmental attributes according to underground mining standards were then derived.

Typical and threshold limit values for the discrete conditions of safe and unsafe statues are presented in Table 23.1. According to the normal and threshold limit values, the status of the working environment in the underground mine was assessed in three conditions of safe (green), transient (yellow), and unsafe (red).

Finally, a loop of conditional procedures and trigger actions were set. The measured parameters (Spatio-temporal data) were stored while these data are less than or equal to normal limit values (safe condition). The loop was periodically retrieved every 30 min to consume less power, extend the battery life of WANET nodes, and reduce congestion through the network. Otherwise, a trigger plan was set for the values mounted in the range between normal and threshold limits (transient condition) or greater than the threshold limit (unsafe condition). The trigger action plan applied in the GIS management server to respond to the deviation of values from normality is presented in Table 23.2.

In the transient (yellow) condition, the auxiliary fans designed for the emergency ventilation system would be automatically or remotely turned on. In this state, the model was also set up to send alert messages to shift supervisors. The periodic time of data reading in the yellow state is reduced to 15 min to ensure the safety and health conditions of the underground environment in the shorter time possible. Emergency (alarm) messages in the event of unsafe (red) conditions would be texted to surface

Table 23.2 Trigger action response plan

Variables (V_i)	Countermeasure implements Safe (Green)	 Transient (Yellow)	 Unsafe (Red)
Reading time interval (min)	30	15	5
Temperature (T_1, T_2, …, T_n) Humidity (H_1, H_2, …, H_n) Gas concentration (G_1, G_2, …, G_n)	• Next reading	• Turn the auxiliary fan(s) on • Text message to the shift supervisors • Next reading	• Text message to all to evacuate from an unsafe place(s) • Next reading

authorize and to underground personnel for immediate evacuee from the hazardous places. The cycle time of data acquisition is minimized to 5 min in this situation.

Output

Mine safety and health were improved by intelligent maps supporting Spatio-temporal data and coordinate WANET nodes in this experiment. The final outputs of the GIS management server are comprised of 3D visualization monitoring of underground mine tunnels and messages texting for alert and alarm conditions. The web-GIS is another application supporting the GIS management server to promote the underground monitoring and communication system.

Data Storage

Data storage and management in the central data repository of the server is an essential part of the integrated system. All geographic and spatial data are stored and managed in ArcMap's geodatabase, which accesses the database at any time over the long term. The organizational structure for storing datasets and creating relationships between datasets was also provided for further analysis and interpretation in the geodatabase. Besides, multi-user access is enabled to work and command orders from different mine site offices.

An integrated data management and documentation to generate geospatial metadata was another approach of geodatabase automation. Metadata can create geospatial data documents to investigate any genuine or non-genuine claims.

Conclusion

An integrated system based on the wireless ad hoc network (WANET) and GIS was introduced to automate underground mine communication, monitoring, and control. The proposed system enhances safety and health, operational management, and reduces capital costs. Considering the WANET and ArcGIS capability, the applications of real-time underground monitoring (temperature, humidity, and gas concentration), ventilation system control, and communication in emergency conditions by the surface user would be achievable. The system is equipped with automatic or remote triggers action plans for measured environmental attributes. The measured data were classified into three categories consisting of normal (green), transient (yellow), and unsafe (red) conditions based on their values compared to normal and threshold limit values. At normal (green) conditions, the measured attributes are below the normal value limits. The mining operation is continuing as it was, and readings are recorded at 30-min intervals. At the transient (yellow) condition, the measurements are between normal and threshold value limits. In this state, trigger actions are become automatically active to switch the auxiliary fan on and texting messages to shift supervisors. In addition, reading intervals are reduced to 15-min in this situation. At unsafe (red) conditions, the measurements are getting greater than threshold value limits, and the system texts messages to all underground personnel for immediate evacuee from hazardous places. Reading intervals are reduced to 5-min. Furthermore, the system provides multi-user surface operation and 3D visualization for a realistic understanding of the underground environment and miners' conditions, and it could be a useful approach for high-tech underground mining.

References

1. Karl, H., and A. Willig. 2007. *Protocols and architectures for wireless sensor networks*. Wiley.
2. Ikeda, H., et al. 2019. Implementation and verification of a Wi-Fi ad hoc communication system in an underground mine environment. *Journal of Mining Science* 55 (3): 505–514.
3. Chen, S., J. Yao, and Y. Wu. 2012. Analysis of the power consumption for wireless sensor network node based on Zigbee. *Procedia Engineering* 29: 1994–1998.
4. Shu-Guang, M. 2011. Construction of wireless fire alarm system based on ZigBee technology. *Procedia Engineering* 11: 308–313.
5. Moridi, M.A., et al. 2014. An investigation of underground monitoring and communication system based on radio waves attenuation using ZigBee. *Tunnelling and Underground Space Technology* 43: 362–369.
6. Esri. 2012. *ArcGIS for emergency management*. USA: Esri.

7. Ghorbani, M., et al. 2012. Geotechnical, structural and geodetic measurements for conventional tunnelling hazards in urban areas—The case of Niayesh road tunnel project. *Tunnelling and Underground Space Technology* 31: 1–8.
8. Huang, X., W. Zhu, and D. Lu. 2010. Underground miners localization system based on ZigBee and WebGIS. In *2010 18th International Conference on Geoinformatics*. IEEE.
9. Kawamura, Y., et al. 2014. Using GIS to develop a mobile communications network for disaster-damaged areas. *International Journal of Digital Earth* 7 (4): 279–293.
10. Şalap, S., M.O. Karslıoğlu, and N. Demirel. 2009. Development of a GIS-based monitoring and management system for underground coal mining safety. *International Journal of Coal Geology* 80 (2): 105–112.
11. Sharifzadeh, M., Y. Mitani, and T. Esaki. 2008. Rock joint surfaces measurement and analysis of aperture distribution under different normal and shear loading using GIS. *Rock Mechanics and Rock Engineering* 41 (2): 299–323.
12. Moridi, M.A., et al. 2015. Development of underground mine monitoring and communication system integrated ZigBee and GIS. *International Journal of Mining Science and Technology* 25 (5): 811–818.

Chapter 24
Advanced Analytics for Spatial Variability of Rock Mass Properties in Underground Mines

Luana Cláudia Pereira, Eduardo Antonio Gomes Marques, Gérson Rodrigues dos Santos, Marcio Fernandes Leão, Lucas Bianchi, and Jandresson Dias Pires

Abstract One of the critical features required for rock engineering is achieving a reliable estimate of rock masses properties. However, representing the geomechanical properties of rock mass remains a challenge in rock mechanics, and determining these parameters directly is time-consuming, expensive, and the reliability of the results of these tests is sometimes questionable. Therefore, this chapter aims to predict the rock mass properties via Random Forests by a Case Study. Random Forest is an algorithm that can act as a classifier and regressor, using a collection of decision trees. To do this, a simplified Rock Mass Rating (RMR) model was developed using the information obtained from the drill hole. Also, a data treatment method was used to prevent information quality harm by conflicting information, and a qualitative analysis was performed. Finally, the proposed method results are satisfactory and showed that by validating and calibrating the database, using this method for geomechanical modeling can be successful.

Keywords Underground mining · Artificial intelligence · Rock mass properties · Random Forest

Introduction

The comprehension of a rock mass behavior requires the knowledge of its mechanical parameters. These data are typically obtained in situ or through laboratory tests, in samples collected on drill holes, which information is only punctual. Due to rock genesis and discontinuity distribution, rock masses can be highly complex, which representativeness can be simplified as a function of the applied geomechanical characterization method. For rocks that present dissolution susceptibility, geomechanical behavior is even more complex, as this dissolution easiness is a function of rock mass discontinuities distribution.

L. C. Pereira (✉) · E. A. G. Marques · G. R. dos Santos · M. F. Leão · L. Bianchi · J. D. Pires
Federal University of Viçosa, Viçosa, Brazil

A. Soofastaei (ed.), *Advanced Analytics in Mining Engineering*,
https://doi.org/10.1007/978-3-030-91589-6_24

On ideal elastic conditions, the acquisition of material parameters, be they soils, rocks, or intermediate materials (weathered rocks), on statistic (stress–strain relationships), or dynamic (elastic wave velocity), should arrive at the same result. Nevertheless, rock masses are not usually elastic, mainly composed of heterogeneous and anisotropic, such as phyllites. This factor can be intensified considering the mother rock genesis, resulting in differences in elasticity moduli values, estimated by the methods mentioned earlier, influenced by constitutive properties of rock materials, such as porosity.

In this context, understanding and representing the spatial variability of rock mass geomechanical parameters are still a challenge for rock mechanics, and, in this process, the construction of geomechanical models is of great interest.

A model consists of a simplified representation of a natural environment. According to Landim [1], the quantitative management of geological data requires simplification. Kaufmann and Martin [2] sustain that the first step for constructing a 3D geological model is collecting, organizing, and selecting the data to be used. According to these authors, the initial effort is considerable but essential to guarantee a reliable model. In the sequence, data need to be processed and stored in a consistent database. Still, according to these same authors, selected data need to be referred in conformity to a geographical system and need to have a working scale defined. Finally, due to natural variability and guaranteeing the quality of data, a thorough validation is necessary.

Kaufmann and Martin [2] report that, in many cases, the primary source of geological data is geological mapping and that punctual data should be obtained from drill holes. These authors also describe a method for building a geological model composed of several steps, as shown in Fig. 24.1.

Among these steps, one consists of automation so that agility of data processing is provided, allowing short time updating. Another step is the reinterpretation and validation of data.

Hadjigeorgiou [3] reports that data must be analyzed, summarized, and transformed into useful information for many geotechnical projects. Otherwise, its collection will have been the last activity itself.

The construction of geomechanical models is still a challenge, as there is a necessity to encompass several parameters to guarantee the representativeness of field reality, which is not a simple task. Furthermore, selecting one variable (or some variables) is necessary to construct geomechanical models in some cases. This choice needs to be trustworthy to field observations, and also, redundant or irrelevant variables should be left out of the model to reduce computational cost. So, to reduce subjectivity, the error on the selection process and improving the prevision precision of the built model, one can use some methods to rank the variables available in the data set, and multivariate and Bayesian statistical and regression analyses and machine learning, using Random Forest, are the most used techniques.

The regression analyses use methods to investigate the association between some observable quantities and, in the affirmative case, the nature of such associations. Multivariate statistics can be applied to analyses in which the data set encompasses simultaneous measurements of several variables, whose purpose is to measure,

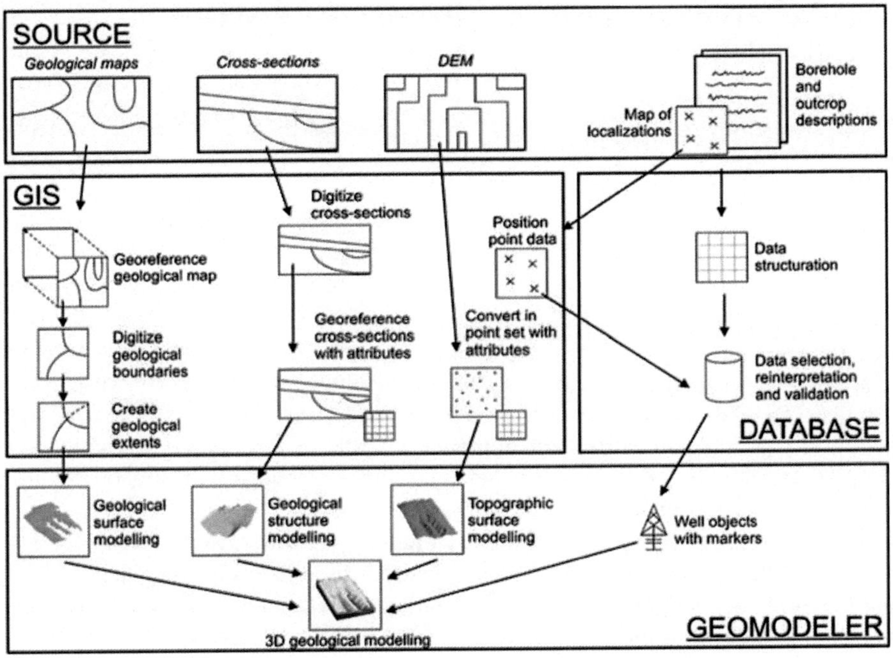

Fig. 24.1 Organogram of the method proposed by Kaufmann and Martin [2] for building a 3D geological model

explain, and predict the interrelationship degree between these variables. Bayesian statistics can also be used to quantify the uncertainties of a selected variable based on probability. There is necessary to test theoretical assumptions for each of these methods, which may not be determined in practice. So, in this context, the option for using machine learning techniques, such as Random Forest, can be used, as in the present study.

The purpose of this chapter is to present the results of a method developed in R language [4] to perform drill holes data treatment, classical exploratory data analyses, spatial exploratory data analyses, univariate interpolation analyses through kriging, multivariate kriging through machine learning (Random Forest), and kriging, and, finally, to rank variables used for simplified classification of rock masses (4D interactive maps).

Application of Multivariate Analyses of Rock Masses Classification

The construction of a geomechanical model necessarily uses interpolation techniques, as there will always be places where no field (primary) information is available, so it is necessary to know how a specific variable behaves in these areas. So, interpolation techniques are essential to understand the behavior of a specific parameter from information collected in the study area.

However, when there are many variables to construct a model, working with all these variables can be challenging. Besides, one can have variables that will not influence the results in a significant manner, and that should not be, in theory, analyzed, or considered. With this in mind, identifying the variables with a bigger (or smaller) influence on a determined area using multivariate analyses is a significant activity for building a geomechanical model. In order to define which variable has a significant influence on a rock mass behavior and, consequently, which one will be most important for the construction of the geomechanical model, it is necessary to provide joint analyses of those variables, so defining an order of importance. One way to perform this sensitivity analysis is to use a machine learning technique, such as the Random Forest algorithm. With this method, selecting the explanatory variables essential to describe a problem sets aside the less relevant ones [5].

Once known and defined as the variable (or more than one variable) that controls the rock mass behavior, mathematic, and statistic methods, such as interpolation, can be used to enable the construction of a geomechanical model. The production of secondary information is based on a determined model governed by assumptions and conditions. One basic assumption is that the information on the surface places of the phenomenon in which sampling was performed presents a satisfactory level of spatial dependency (autocorrelation). The mathematic function used is approaching the continuous phenomenon that is being analyzed. So, each interpolation method can result in different representations of the same set of data. The use of a determined interpolator depends on the input data set from the interpolator's intrinsic characteristics. Each interpolator has a particularity, so it must be carefully observed before its application [6–9].

Random Forest (RF)

Random Forest is a deterministic interpolation method that integrates what is known as machine learning, which basic principle states that systems can learn from analyzed data, identify patterns, and, finally, making decisions with or without a minimum of human intervention. Trees are classification or regression models that allow the use of continuum and/or discrete variables. These models are adjusted by dividing successively a data set into more and more homogeneous groups. The classification can be translated as a categorical or discrete value prediction and aims to construct models and define rules, from a set of correctly pre-classified examples, for further classification of new and unknown examples. On the other hand, regression seeks to find a function that can map a data item for a variable prediction with a continuous numerical value [10, 11].

Random Forest (RF) is an algorithm that can act as a classifier and regressor, using a collection of decision trees initially developed by Breiman [12]. It is a random decision tree model for non-linear prediction between statistical variables whose main characteristics include calculating essential measures of the variable (variable ranking), automatic calculation of errors, automatic treatment of missing data, and the possibility of verification of variable dependency. They also avoid overfitting and are less sensitive to noise. It can and has been used in several fields of knowledge [13–15].

As they are also known, binary decision trees use binary recursive partitioning, in which data from a primary variable are successfully divided along the gradient of the explanatory variables into two subsets, or nodes, descending. These divisions occur in a way that, for any node, the division is selected to maximize the difference between two sets or ramifications. The mean value of the primary variable in each node can then be used to map the variable through the region of interest [10].

The RF aims to perform the conception of several decision trees using a subset of randomly selected attributes with reposition around 2/3 of the original set and reserving the other 1/3, named out-of-bag, for validation. So, there is the use of two main approaches.

The first approach uses initial aggregation known as bagging (bootstrap sample), ensuring that each new selected set could have some registers included more than once and others not included. A sample $L(\theta)$ of "n" size is selected from the test set (L) of N size as a training set modified for each new tree. Each predictor $T_{L(\theta)}$ is dependent on a random vector θ that indicates that the samples selected belong to the entire set (L). In the end, the predictor (f) is the majority vote—in other words, the one with the highest number of votes on the tree—or the mean of the trees (with y'_η as a predicted answer for samples x_η and with K equal to the size of the sample).

The second approach is related to the restriction of variables, randomly selected, in each node. Thus, the vector θ indicates both the primary selection and the randomization [11, 13, 15–18].

The CART algorithm was proposed for the learning of classification and regression trees. The algorithm stated that given a training sample L with N samples, formed by M predictor variables x_i ($i = 1, 2, \ldots, M$) as the entrance space X and one response variable y, the CART recursively divide the entrance space to obtain a predictor for the T_L tree (with y' as the response, variable), according to Eq. (24.1). Having all the entrance space, the algorithm tries to find a binary partition to maximize the purity of the response on the subspace formed by the suggested partition. The algorithm depends on the homogeneity of the response classes; a commonly used measure to classify it is the Gini's impurity measure, while the mean error is used for regression. The binary partitioning is repeated on each new subspace up to homogeneity of sub-spatial response is achieved. The subspace estimate for a particular point is the majoritarian value, classification, or the mean, for regression, of responses of training on the subspace [13, 19].

$$Y' = T_L(X) \tag{24.1}$$

The decision trees are algorithms wherein the main idea is founded in principle "*divide – and – conquer*," in which data are divided in a training dataset in a recursively way. They present characteristics such as: are non-parametric, work both with homogeneous or heterogeneous data (or both), are easy to interpret, and are relatively resistant to the presence of outliers [20, 21].

As already mentioned, Random Forest is an algorithm that uses the majority voting (for classification) or the mean (for regression) to make predictions. For a set to be more precise than its members, two conditions must be met: the set member must have better individual predictions than the random ones and the members of the set must be diverse, meaning that the errors in prevision are not correlated [13].

Equations (24.2) and (24.3) present the calculations for the use of RF for regression and classification, respectively.

$$y' = f\left(x_\eta\right) = \frac{1}{K}\sum_{K=1}^{K} T_{L(\theta_K)}(x_\eta)_1^K \tag{24.2}$$

$$y' = f\left(x_\eta\right) = \text{voto majoritario}\left\{T_{L(\theta_K)}(x_\eta)\right\}_1^K \tag{24.3}$$

In Fig. 24.2, the structures of decision trees, showing the "*if–then*" for each branch, with the partitioning of the subspace associated with a hypothetic two-dimensional space. The individual predictions of all trees are collected and combined as a unique prevision by voting or by mean [13]. The key parameters of the RF model are *mtry*,

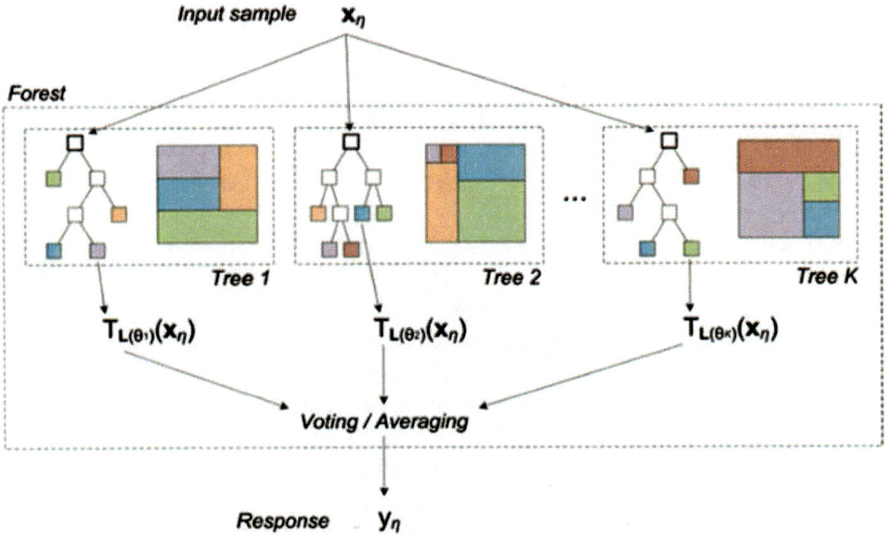

Fig. 24.2 Scheme of the Random Forest proposed by Auret and Aldrich [7]

and *ntree*, in which *mtry* represents the square root of the number of factors and *ntree* represents the number of trees in the forest.

For the growing of trees, one has a tree that can present high growth, causing *overfitting*, which can be explained as the decomposition of the generalization error in terms of bias and variance [21]. Oshiro [22] argue that to avoid this problem, it is necessary to use a technique known as *pruning*, in which there is the generation of a more generic hypothesis from the training dataset. There are two types of *pruning*:

- a pre-pruning, in which the tree construction is interrupted during its growth due to a pre-established criterion; and
- the post-pruning, in which a complete tree is built and, later, sub-trees are removed, based on the estimate error calculation of a determined node. The first type can induce more errors.

Other relevant characteristics in using RF are the no need to perform cross-validation or a separated test to obtain an impartial test error estimate. This fact is caused by the fact that each tree is built using one bootstrap sample, different from the original data, and the third part of the omitted cases is not used in tree construction, is set under the tree. Instead, a classification test is performed and, later, the out-of-bag error is estimated. In this way, it is possible to note that the error rate is related to the correlation between any two trees of the forest and each tree's strength on it [16–18, 20–23].

Girolamo Neto [11] endorses that several evaluation and performance measures can be applied to analyze the behavior of an RF, such as performance measures derived from a matrix that represents matchings and errors of the model (named confusion matrix), followed by a graphical analysis of ROC (Receiver Operating Characteristics) type and the agreement index, Kappa. However, it is not advisable to use the index.

Finally, Matin et al. [24] quote that there was an increasing interest in RF models in the last decade, but they have not yet appeared expressively in rock mechanics literature, for example.

Use of R Language to Rock Mass Geomechanical Classification—Case Study

The origin of R statistical, computational environment, and language [4] is linked to the implementation of S computational language, developed by ATT&T Bell Laboratories. It integrates the GNU Project (GNU's Not Unix), a world project that has developed the GNU computational language, aiming to create and maintain a free complete software system (available as a Free Software as a font code). It is compatible with several platforms, such as Linux, Windows, MacOs. R is a statistical calculation system and graphic construction that uses its programing language. It calculates mathematic, statistic, and data management functions, and its functions operate on data structure, including lists, vectors, lettering, factors, and matrices. R language is a functional programming environment in which the users can use internal functions or write their functions [25] in a free and applied way in several fields of knowledge.

From the following R packages: gstat [26], sp [27], spacetime [28], raster [29], rgeos [30] (2020), rgdal [30], lattice [31], moments [32], plotKML [33], GSIF [34], ranger [35], geoR [36], plotly [37], DescTools [38], readxl [39], psych [40], ggplot2 [41], dplyr [42] caret [43], corrplot [44], spatstat [45], maptools [30], scatterplot3d [46], tcltk2 [47], doParallel [48], GGally [49], e1071 [50], rpart [51], mlbench [52], randomForest [53], party [54], MASS [55], nycflights13 [56], gapminder [56], Lahman [57], and htmlwidgets [58] it was possible to, through R language, to develop the geomechanical model based on drillhole data, defining a script that better represented the geomechanical characteristics of rock masses in a underground mine. On Fig. 24.3, it is presented the area in which the proposed methodology was applied.

The study area is located in the middle of a shear zone where mineralized hydrothermal fluids have percolated, originating zinc, and other associated minerals. The host rocks of the zinc deposit are composed of political carbonates, very susceptible to dissolution and weathering, generating cavities into the rock mass, which can be filled with many different weathering and soil-like materials. According to the previous studies and in situ information, the rock mass is fractured, with intense water percolation, that originates thick dissolution and weathered zones. This process is responsible for cavity formations, which, usually, are filled with clay plastic material with rock blocks. A thick shear zone where several anastomosed surfaces cross one to another composes the mineralized body, isolating lenticular bodies with metric decametric dimensions. In addition, to mass failures, it can be

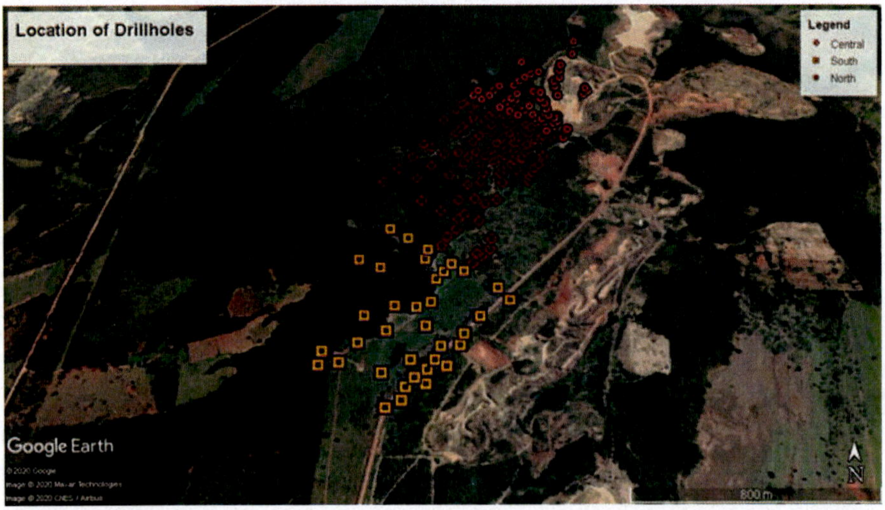

Fig. 24.3 Location of drill holes on north sector of the underground mine from which the database was collect for the study

noted that those are commonly controlled by shear zones, originating discontinuities into the rock masses. These discontinuities have an irregular spacing, form metric to a decametric, low strength (plan to low undulated), and are generally filled by clay material with millimetric thickness.

RMR (Rock Mass Rating) and *Q* (Tunneling Quality Index) were insufficient to identify those subtle differences in rock mass quality, and a specific and local geomechanical classification was developed for the mine and is presented in Table 24.1.

For the design of this local classification, the empirical knowledge of the mine rock mass and drill hole and laboratory tests data were used. Based on this information, it was concluded that the most relevant parameters for local rock mass classification should be: fracturing and weathering degree, structural pattern, crack presence, drill hole recovery, and RQD. A rock bolting design was defined for each geotechnical rock mass class possessing the structural pattern and back analyses of observed failures. The support type indicated for each class was defined by a trial and error procedure based on a theoretical and practical foundation.

Data Bank: Treatment and Analyses

In practical mining activities, the estimate of mineral resources uses drill hole data, in which information such as extension and geometry of mineralized ore bodies and physical and chemical data [59]. The drilling of drill holes is a routine activity in which, once drilled, the drill holes are made a lithological and geomechanical description of collected samples. This description is made for segments of similar behavior, from the borehole mouth to its end. Usually, a description of different lithotypes, geological structures, and geomechanical characteristics is performed. All this information is stored in a databank and later used for different purposes and designs, such as geological modeling, block modeling, or to construct geological-geotechnical cross-sections. Jakubec and Esterhuizen [60] alert that misconceptions can occur during description, such as difficulties in differentiating natural and mechanical fractures, recognizing filling materials, and others. So, this activity must be performed in a criterion manner by specialized technicians.

Schepers et al. [61] state that drill holes supply the need for precise information on physical parameters of the rock formations, but this information is limited to a limited distance from the drill hole. Data obtained from drill holes are fundamental for decision-making in mining. Definitions on which direction the ore body extends; the ore content; the best technique to extract this ore; and which are the geomechanical characteristics of the sampled rock masses are possible due to drill hole data. So, the most reliable is drill hole data. The best will be the quality of geological and geotechnical models based on such information. This quality is related to drilling the drill hole itself and mainly to the quality of the description of the information obtained from samples collected in the drill hole.

Table 24.1 Adapted geomechanical classification for the rock mass under study

Geomechanical class									
Rock mass class	Weathering degree	Fracturing degree	Rock quality	Structural pattern	Lithology	Crack presence	Block fragmentation	Recovery	RQD (%)
II-A	A2	F2-F3	Very good	Floor slab	Dolomites	Do not occur	m^3	>95	>60
II-B	A2	F2-F3	Very good	Wedge	Breccia	Do not occur	m^3	>95	>60
III-A	A2-A3	F3	Very good	Floor slab	Dolomite	Do not occur	dm^3 a m^3	90 a 95	50 a 75
III-B	A2-A3	F3	Very good	Wedge	Breccia	Do not occur	dm^3 a m^3	90 a 95	50 a 75
IV-A	A3	F3-F4	Good/average			Thick cm to dm	cm^3 a dm^3	>90	25 a 50
IV-B	A3-A4	F3-F4	Average/poor			Thick cm to dm	cm^3 a dm^3	75 a 95	25 a 50
V	A3-A4	F4	Very poor			Thick cm to dm	cm^3 a dm^3	50 a 75	<25
VI	A2	F4-F5	Very poor	Floor slab	Phyllite	Do not occur	cm^3 a dm^3	>95	<25
VII	A4	F4-F5	Very poor			Metric thickness		<50	<25
Pillar	Regardless of the above parameters, the effect of induced stresses must be considered								

Weathering degree
A2—Little weathering, some minerals are discolored. Rock matrix loses very little strength to hammer blow, and it is not friable
A3—Very discolored rock; mineral alteration is very penetrative, assuming colors predominately orange and reddish. Material is not friable—moderate strength to hammer blow
A4—Completely weathered rock with well-developed clay portions. Friable fragments are present and show the original rock structure
A5—Soil. The original rock texture is wholly destroyed, and the soil generally presents reddish color
Fracturing degree
F1—Low fractured: average spacing of fractures is bigger than 2 m, forming rock blocks with a volume equal to cubic meter
F2—Low fractured: average spacing of fractures from 0.6 to 2 m, forming rock blocks with a volume equal to cubic meter
F3—Moderately fractured: average spacing of fractures from 0.06 to 0.2 m, forming rock blocks with volume varying from cubic centimeters to cubic decimeters
F4—Very fractured: average spacing of fractures from 0.06 to 0.2 m, forming blocks with volume from cubic centimeters to cubic decimeters
F5—Very fractured: average spacing of fractures lower than 0.06 m, forming rock blocks with volumes lower than or up to some cubic centimeters

Bieniawski [62] advocates that geomechanical need to be collected in sufficient quantity and quality and that the interpretation of this data should be specific for each project.

In this context, to support the geomechanical model of the study area, a total of 513 drill holes, several laboratory tests, and local (adapted) geomechanical classification data were used. Nevertheless, no lithological, geomechanical, and structural maps/models were provided. The mining sector used for this study is subdivided into three subsectors: South, Central, and North. In Fig. 24.3, it is presented the drill holes available for each of these subsectors—73 for South, 170 for Central, and 270 for North.

For each one of the subsectors, the following data was provided: (i) drill hole coordinates, both for drill hole "mouth" and in-depth, allowing the definition of the actual position of each drill hole into space; (ii) geomechanical parameters, for each segment of the drill hole, such as rock mass local classification, weathering, fracturing, recovery, and RQD; (iii) lithotypes for each segment of the drill hole; and (iv) the inclination of each drill hole and the geological-geomechanical cross-section that each one intercept.

The method used for the databank construction was partially based on the proposal by Kaufmann and Martin [2] and has been considered input data for the drill holes. All data were used from the information collected from drill holes, except the variable recovery, because inconsistencies were observed, such as high values of recovery associated with high weathering and fracturing levels and high RQD values.

Data treatment has begun with the adequacy of the coordinate system that was all modified from the local coordinate system to UTM, Datum Córrego Alegre 23 S.

A script was developed in R [4] to compile drill hole position information and all its geomechanical classification parameters—weathering, fracturing, recovery and RQD, rock mass classification, and lithology. The script has furnished, as an answer, a final datasheet with the following columns: one column with drill hole id and its name; three columns with x, y, and z coordinates, respectively; two columns with the geomechanical classification of the rock mass, one with the local classification and the second with a simplified RMR classification; one column with the lithotypes; and, finally, three columns with weathering, fracturing, and RQD. In Table 24.2, it is presented an example of a final spreadsheet, without the id and coordinates data, for confidentiality reasons.

The joint evaluation of variables recovery and RQD would aim to identify the occurrence of gaps (cracks—FD) associated with the regions of the rock mass with high RQD values. However, on the provided databank, it was possible to observe that for some depth segments, there was a low value of recovery associated with high values of RQD and low fracturing and weathering values, which does not make sense. So, because of such inconsistencies, the variable recovery was excluded from the analyses.

The quality and quantity of collected data are defined by the geotechnical project's purposes to be developed. However, on the case study presented here, the type and the quantity of data needed a thorough analysis to guarantee that they would fit the necessary quality for the proposed use.

Table 24.2 Example of a final spreadsheet with final databank

ID	X	Y	Z	Class	Class local	Lithology	Weathering	Fracturing	RQD
Drillhole 1	X_0	Y_0	0	5	0	S/R	0	0	0
Drillhole 1	$X_0 +$ 0.01	$Y_0 +$ 0.01	44.95	5	0	S/R	0	0	0
Drillhole 1	X_1	Y_1	44.95	3	2.5	DORO	2.5	2	84.72
Drillhole 1	$X_1 +$ 0.01	$Y_1 +$ 0.01	51.1	3	2.5	DORO	2.5	2	84.72
Drillhole 1	X_2	Y_2	51.1	4	6	DORO	3.5	2	39.02
Drillhole 1	$X_2 +$ 0.01	$Y_2 +$ 0.01	55.2	4	6	DORO	3.5	2	39.02
Drillhole 1	X_3	Y_3	55.2	5	9	DORO	3.5	1	13.6
Drillhole 1	$X_3 +$ 0.01	$Y_3 +$ 0.01	87.7	5	9	DORO	3.5	1	13.6
Drillhole 1	X_4	Y_4	87.7	4	2.5	DORO	3	2	69.23

So, in possession of the compiled spreadsheet, data were analyzed and treated. Initially, the quality of data was sought to allow its use for multidimensional geomechanical modeling. So, some criterion was used to grant its quality and also to allow its reading and analyses.

The first criterion was compatible with the lithological and geomechanical information segments to obtain equal intervals that contained both information. Then, each drill hole segment's geographical coordinates (x and y) were obtained based on the drill hole plunge azimuth and orientation; i.e., a correction of drill hole inclination was performed. Six drill holes were excluded from the databank because they do not have any information on its geomechanical parameters. Some others were excluded, as they do not have any information of its azimuth and plunge.

The initial and final value of each segment was considered. The final value of each segment was coincident with the initial one from the following segment. So, to avoid duplicate information at one only depth, a value equal to 0.01 was added for x and y variables to the first point of each segment, starting with the second segment up to the end of the drill hole, as shown in Table 24.3.

The last criterion is RQD, as the value provided on the drill holes description was used on the databank. All other geomechanical parameters—weathering, fracturing, and local geomechanical classification were reclassified according to a numerical scale, as presented in Tables 24.4, 24.5 and 24.6. For the cases in which, for example, the weathering grade was classified as A2/A3, a value equal to 2.5 was considered on

Table 24.3 Example of the artifice used for definition of drillhole depth

ID	X	Y
Drillhole 1	X_0	Y_0
Drillhole 1	$X_0 + 0.01$	$Y_0 + 0.01$
Drillhole 1	X_1	Y_1
Drillhole 1	$X_1 + 0.01$	$Y_1 + 0.01$
Drillhole 1	X_2	Y_2
Drillhole 1	$X_2 + 0.01$	$Y_2 + 0.01$
Drillhole 1	X_3	Y_3
Drillhole 1	$X_3 + 0.01$	$Y_3 + 0.01$
Drillhole 1	X_4	Y_4

Table 24.4 Numerical scale adopted for weathering

Weathering	
Drillhole description	Numerical scale
A1	1
A2	2
A3	3
A4	4
A5	5

Table 24.5 Numerical scale adopted for fracturing

Fracturing	
Drillhole description	Numerical scale
F1	1
F2	2
F3	3
F4	4
F5	5

Table 24.6 Numerical scale adopted for local geomechanical classification

Local geomechanical classification	
Drillhole description	Numerical scale
II-A	1
II-B	2
III-A	3
III-B	4
IV-A	5
IV-B	6
V	7
VI	8
VII	9

Table 24.7 Legend of lithotypes present on the databank

Legend	Lithotypes
ARD	Slate
BXD	Hydrothermal breccia
BXD/Dcz	Dolomitic breccia/gray dolomite
BXD/Doro	Dolomitic breccia/pink dolomite
BXH	Hematitic breccia
BXW	Willemitic breccia
DCZ	Gray dolomite
DCZ/bx	Gray dolomite
DO	Dolomite
DORO	Pink dolomite
DORObd	Pink dolomite
DORObx	Pink dolomite
DOROfr	Pink dolomite
DRA	Dolarenite
DRU	Dolorudite
FD	Crack
FIL	Phyllite
FO	Shale
HC	Compact hematite
MB	Metabasic rock
MGA	Marl
MT	Terrain material
MT/CAL	Terrain material/calamine/calamine
MTARG	Clay material
R/S	Destroyed
S/R	Not recovered
SOLO	Soil

the numerical scale. The same reasoning was used for fracturing and local geomechanical classes. Due to the absence of drill hole bulletins, RQD was determined for equal fracturing segments, not maneuver segments. Lithotypes present on the data bank are listed in Table 24.7, and no modification of such data was done.

After finishing data treatment, a quality analysis was performed, as for some drill hole segments, no geomechanical data (weathering, fracturing, and/or RQD) was available. The lack of information for some segments can be related to not recovering rock material during drilling. In this case, an S/R or R/S symbol was used. In addition, geological factors such as unconsolidated material, discontinuities intersection, secondary porosity, fault zones, and others can affect the material recovering on drill holes. Another possible cause can be inability during drill hole description, as trained technical staff must be used for this procedure.

Fe and FD symbols on some drill hole segments indicate voids/cracks/gaps into rock masses. Therefore, for these segments, geomechanical parameters were also considered to be equal to zero.

For all segments in which the lithotypes were classified as Soil, R/S (destroyed), and S/R (Without Recovery), geomechanical values equal to zero were given, as there is no sense in giving rock mechanical parameters for soil-like materials.

Pells et al. [63] argue that there are problems in the use and determination of RQD around the globe and that the use of such an index must be carefully performed. The problems mentioned above go from errors during samples drill hole samples description to misunderstanding the index itself and comprise the lack of ability of the professional responsible for its quantification. However, despite all these problems, RQD is still a used parameter used in rock mechanics worldwide, and its use was kept for the present study.

In order to classify the rock mass understudy, a modified RMR was used, as there was not all the information needed to the traditional RMR—no information on discontinuities characteristics and uniaxial compressive strength. Therefore, on this modified RMR classification, the weight of rock matrix strength was not considered, and the weight related to discontinuities, only the weight for weathering was considered. Also, the same weight for water presence was used for all drill holes (equal to 4, based on field information). So, the modified RMR classification was based on the following parameters: RQD, fracturing equivalent to spacing, weathering, and water presence.

As presented in Table 24.8, the weight for RQD, weathering, fracturing, and water presence was kept equal to the original RMR classification, but the final sum of weights was reorganized for each RMR class.

By comparing the simplified RMR (Table 24.7) to the local classification (Table 24.1), one can note that the differences between the classifications are mainly because the local classification considers, besides the RMR criterion, the rock quality, the structural pattern, and the structural pattern the block fragmentation. These data are not available on the available data bank and had to be discarded.

The modifications made to the data bank, previously described, allowed an improvement in data quality and reliability. At the end of the analysis, a CLASS variable corresponding to the simplified RMR value was included in the data bank. All this data treatment process was automatically performed by using R language and manually revised lately. So, after this process, the data bank for the South subsector comprises a total of 70 drill holes; the Central subsector a total of 255 drill holes; and the North subsector a total of 149 drill holes, with a drill hole reduction of 4.1%, 5.6%, and 12.4%, respectively.

Table 24.8 Simplified RMR classification

Parameters	Coefficients					
1	Rock matrix strength	Not considered				
2	RQD	91–100	76–90	51–75	26–50	<25
	Weight	20	17	13	8	3
3	Fracturing (interval—from drillholes, Table 24.3)	>2 m (0–1.5)	0.6–2 m (1.6–2.5)	200–600 mm (2.6–3.5)	60–200 mm (3.6–4.5)	<60 mm (4.6–5)
	Weight	20	15	10	8	5
4	Weathering (interval from drillholes, Table 24.3)	Not weathered (0–1.5)	Slightly weathered (1.6–2.5)	Moderately weathered (2.6–3.5)	Very weathered (3.6–4.5)	Decomposed (4.6–5)
	Weight	6	5	3	1	0
5	Presence of water	Completely dry	Interstitial water	Humid	Low flow	Water inflow
	Weight	15	10	7	4	0
Global weight		49–61	38–48	30–37	22–29	<21
Simplified RMR classification		I	II	III	IV	V

Application of Random Forest to the Data Bank

Random Forest was used to listing which variable (or variables) would be the most important to explain the mechanical behavior of the rock mass understudy on the underground mining in the North sector. Kaufmann and Martin [2] argue that the primary goal of selecting variables is: to improve interpretation capacity, eliminating irrelevant input to create a model with a small number of variables; to reduce noise, avoiding overfitting risk; to accelerate modeling time, improving model precision; and, finally, to furnish a better definition of the process of data generation. Still, according to those authors, the use of RF to measure the importance of variables has several advantages in comparison to other similar methods, such as it can determine the impact of each predictor variable individually, as well as on multivariate interactions with other predictors; it grants a measure of impartial importance and can be applied to several fields of knowledge.

RF provides two measures of importance—the Mean Decrease Accuracy (*IncMSE*) and the Mean Decrease Gini (*IncNodePurity*). These two parameters can be used to classify and select variables. The IncMSE quantifies the importance of a variable by measuring the change into prediction precision when this variable's values are randomly commuted compared to the original observations. It is a measure of how much the precision reduces if a variable is excluded from the model. The IncNodePurity is the sum of all the reductions on Gini impurity due to a determined variable (whenever this variable is used to form a division into RF), normalized by the number of trees. The IncNodePurity measures the contribution of each predictor variable to the impurity of the trees resulting from the RF model and reflects how much each variable contributes to the homogeneity of the nodes and leaves [64–67].

The results obtained for all subsectors are presented in Fig. 24.4 and Table 24.9. For all three subsectors South, Central, and North, a regression of variables was performed throughout RF. As a result, the number of used trees in all subsectors was equal to 1000 (*ntree* = 1000), and the number of nodes of the analyzed trees was equal to 6 (*mtry* = 6).

Table 24.9 presents a resume of the determination coefficients, obtained errors for predicted and actual values, as well as the percentage of each variable explained by RF. As observed in Table 24.9, it is found that more than 99% (*% var explained*) of the rock mass behavior of the North Edge of the underground mine can be explained by the results obtained throughout RF. Concomitantly, the errors (RMSE and MAE) were very low, and the determination coefficient (*R-squared*) was very close to 1. So, it can be concluded that the ranking of the importance of the variables by using RF is reliable to explain the behavior of the studied rock mass.

The IncMSE of the South sector presented the variables RQD, weathering, and fracturing as the ones that most influence rock mass strength and lithology is the less important. For the Central subsector, the variables that better explain the strength behavior of the rock mass are RQD, fracturing, and weathering, and again, the less critical variable was lithology. Finally, the most important variables were observed in the South subsector—RQD, weathering and fracturing, and lithology was the less important one for the North subsector. Based on these results, it became clear that RQD is the variable that can better explain the behavior of the rock masses on all analyzed subsectors, followed by weathering and fracturing. Based on this result, RQD should be considered the independent variable to construct geomechanical data of the underground mine. So, RQD and RMR simplified maps were generated for each subsector.

The relevance of RQD was already expected and has proved to be coherent, as this parameter compiles weathering and fracturing characteristics in only one index and is a measure of drill hole recovery.

By comparing the results of RF with data from exploratory analyses, it can be noted that the mean RQD value is higher for the Central subsector. This is because this region has a minor structural control compared to the other two subsectors, as this subsector is out of the fault zone. On the other hand, the North subsector presents the lower mean value for RQD, and this result can be linked to several faults and structural lineaments in this area.

Among the results of RF, it was expected that, with the increase of depth, there would be an increase in strength and, consequently, a better rock mass behavior. However, as the region has an expressive presence of carbonatic rocks and underground water, the association of these two geological characteristics explains the fact that the increase of depth does not present a direct and transparent relationship with the strength increase, as even for very high depths (up to 600 m) there is the presence of dissolution voids/gaps/cracks, filed or not with soil-like materials.

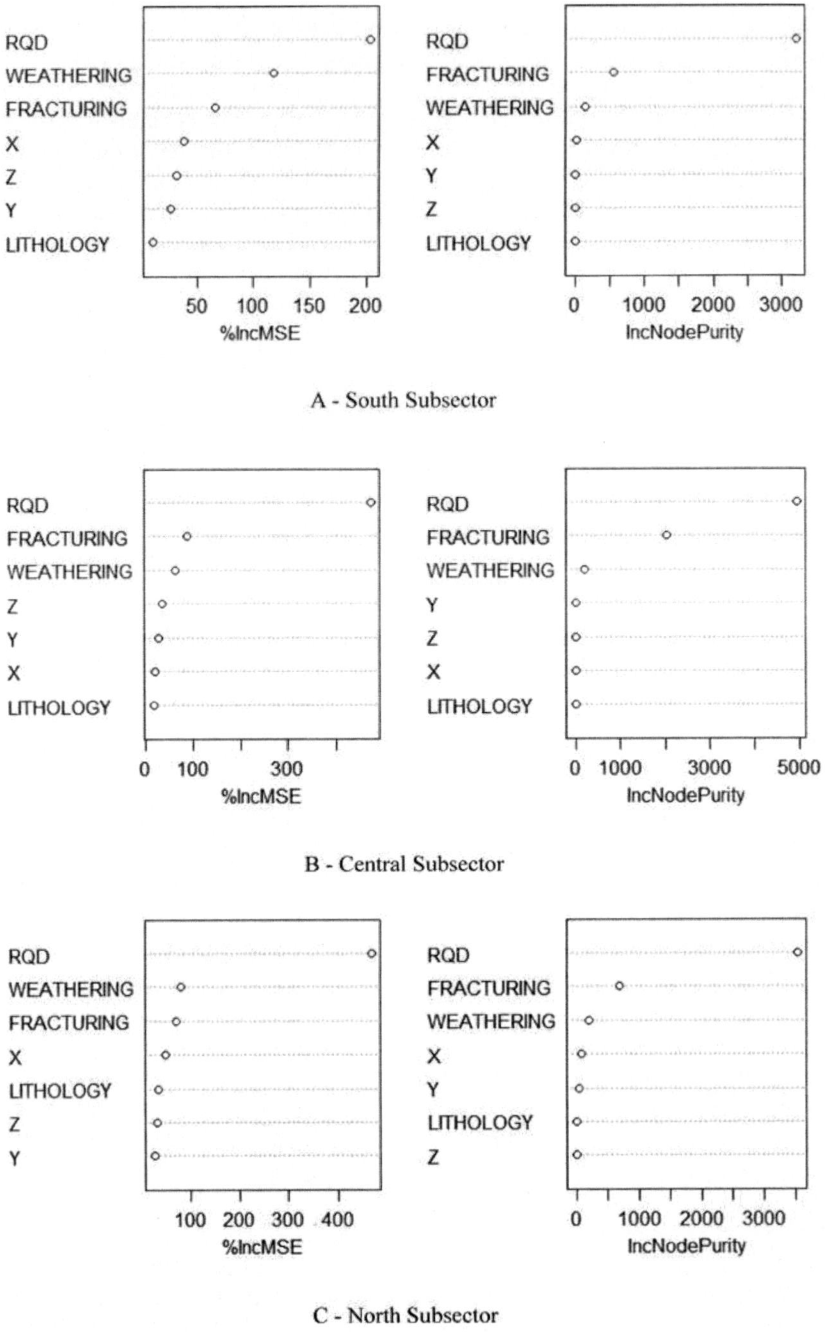

Fig. 24.4 Ranking of the importance of each variable of the rock mass. On the abscissae axis, the importance of each variable; and on the ordinate axis, the variables, with information about the mean decrease precision (InsMSE) and of Gini mean decrease (IncNodePurity)

Table 24.9 Resume of determination coefficients and errors obtained through RF

Subsector	RMSE (average square error)	R-squared (determination coefficient)	MAE (absolute square error)	% of explained variable
South	0.0157	0.9999	0.0018	99.96
Central	0.0454	0.9987	0.0041	98.86
North	0.0286	0.9997	0.0021	99.81

The technical literature related to geomechanical behavior consulted by the authors do not present works in which the RF was used to predict the importance of variables and its later use for geomechanical modeling. However, some studies use RF to predict one variable's behavior based on a set of variables for both rock matrix and rock masses. Among the relevant contributions consulted, Tiryaki [68] has used neural networks, regression trees techniques, principal components, and factor analyses to predict the uniaxial rock strength and the elasticity module of three different rock types. Dong et al. [69] have used RF to classify if the rock-burst phenomenon would occur and its intensity in underground mining. The authors have analyzed factors that can start the phenomena, such as: in situ stresses, uniaxial compressive strength, tension strength, and others. Kaufmann and Martin [2] has used RF to predict the elasticity module and uniaxial compressive strength based on some selected variables: porosity, Point Load Index (Is_{50}), wave propagation velocity (V_p), and Schmidt accelerometry. Finally, Xie and Peng [70] have used regression RF to predict damage zones on excavations in an underground coal mine, considering that these zones were affected by disturbs caused by explosives, stress relief due to excavation process, and by adjustments of in situ stresses.

Joint Analyses

5D iterative graphics and animations were built and allowed joint analyses of different variables of the databank. Among the interactive graphics, a graphic is presented in Fig. 24.5, in which it is possible to visualize the simplified RMR classification for each lithotype. On this graphic, the variable CLASS was interpolated using Ordinary Kriging, and each colored dot indicates which lithotype was observed in the field. So, there is an association of predicted values for the CLASS variable with the sampled lithology points obtained from drill holes. So, in Fig. 24.5, only the visualization of voids/gaps/cracks can be seen, and it allows observing that some low RMR regions (Class 4) can be associated with these structures' presence.

For the CLASS × RQD graph (Fig. 24.6), the same behavior of Fig. 24.5 can be observed, as the CLASS variable is interpolated with RQD, and it can be seen that high RQD regions are related to good RMR (Class 3) materials for South subsector.

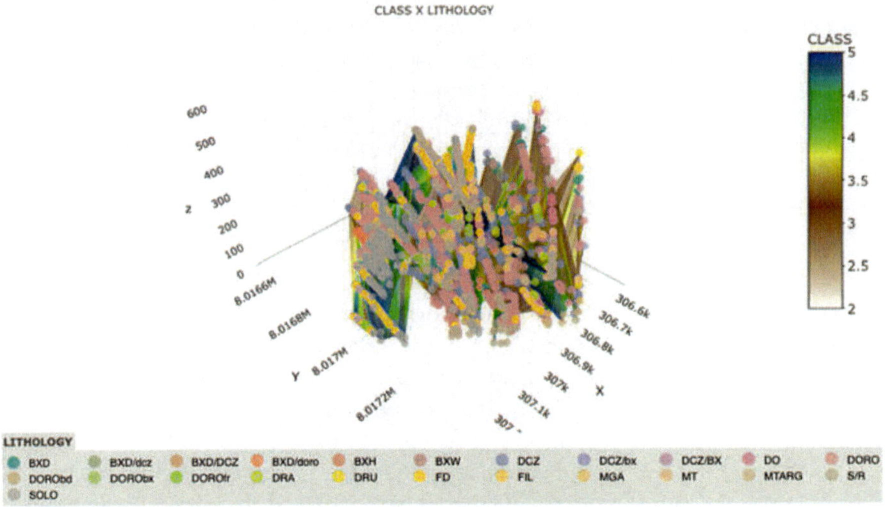

Fig. 24.5 Simplified RMR classification and its interaction with lithotypes

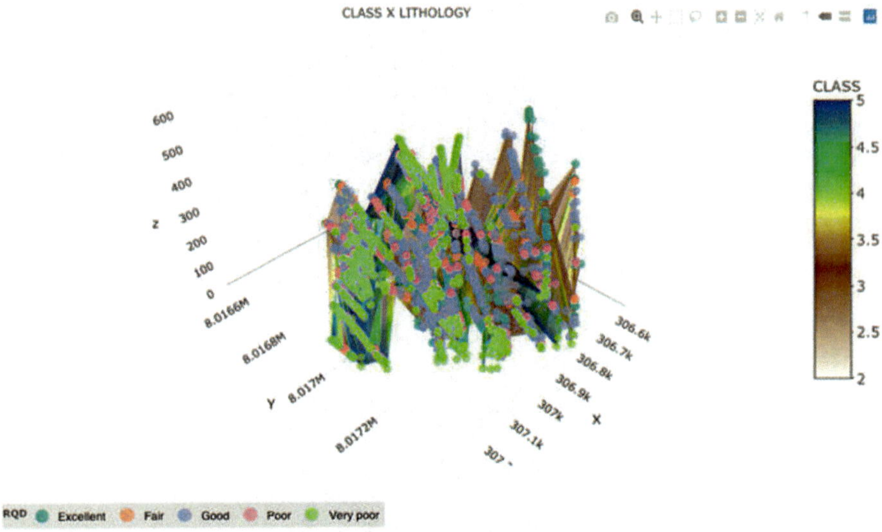

Fig. 24.6 Simplified RMR classification and its interaction with RQD values

In Fig. 24.7, the relationship between CLASS and weathering can be seen, and one can note that weathering materials classified as equal to 2 are related to several RMR classes for the North subsector.

Iterative graphics were also generated for joint analyses of the CLASS variable with depth. The same iterative graphs mentioned for the rock mass classification were also generated for RQD, as, for example, RQD × Weathering.

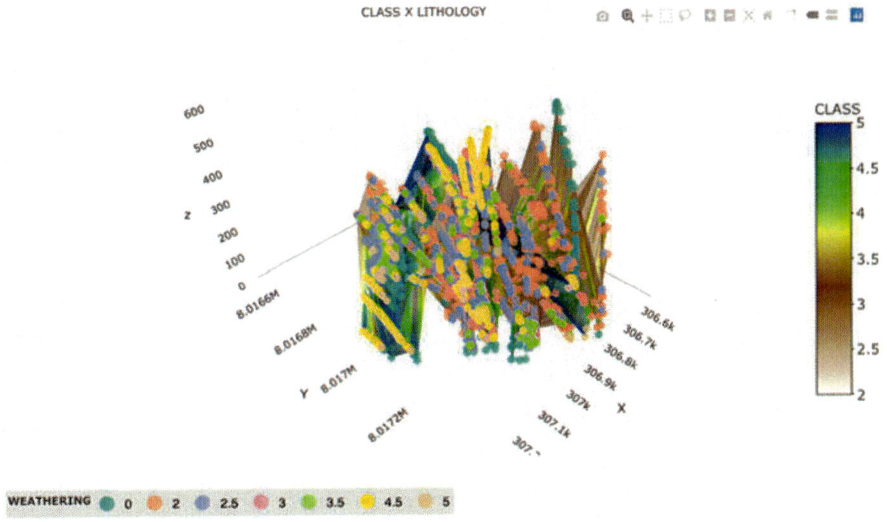

Fig. 24.7 Simplified RMR classification and its interaction with weathering

Still, as an analyses tool, animations in which three or more variables were jointly analyzed. For example, lithology × fracturing × CLASS (Fig. 24.8) or lithology × fracturing × RQD (Fig. 24.9). It must be highlighted that to present/explain this kind of toll only in 2D harms its understanding, as only one frozen image can be shown and not the role animation itself.

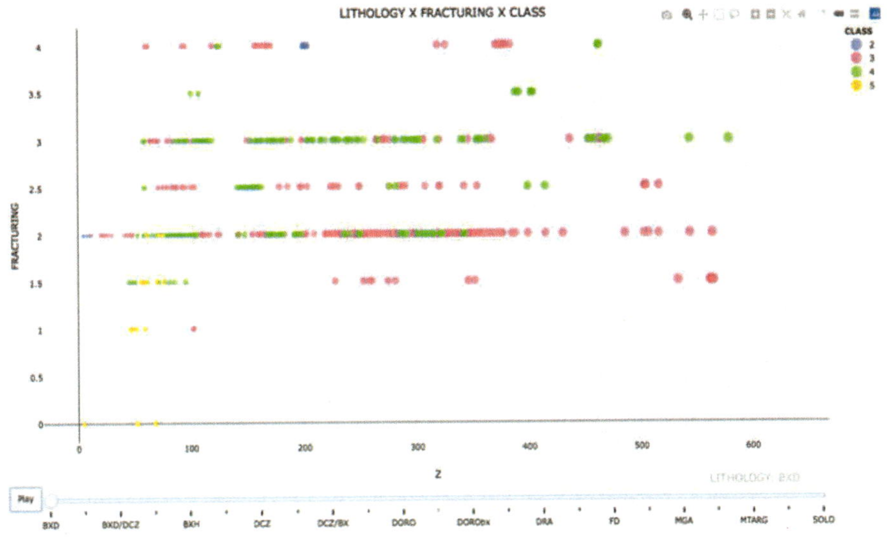

Fig. 24.8 Image of an animation correlating three variables: lithology × fracturing × CLASS

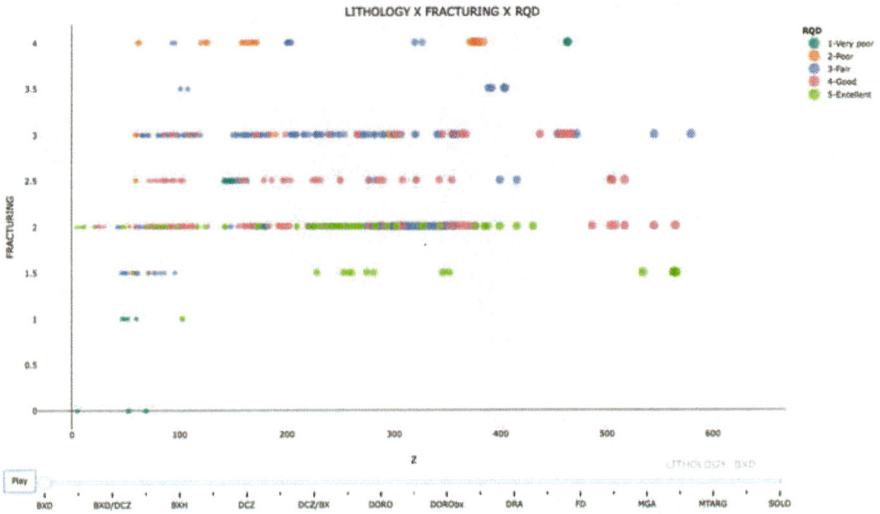

Fig. 24.9 Image of an animation correlating three variables: lithology × fracturing × RQD

The iterative graphs have two significant advantages on geological-geotechnical modeling. The first is to analyze the essential variables that govern the geomechanical behavior in an easy, rapid, and iterative way. On the graphs that correlate RQD × Weathering, as an example, it is possible to visualize points/regions in the interior of rock masses that have low RQD values. So, it is possible to verify how Weathering behavior in these same places. The second advantage is to allow a better comprehension by professionals of correlate areas. Herwanger [71] advocate that 3D and 4D geomechanical models integrate data from several disciplines and allow that scientist and engineers collaborate and communicate with each other on the way for more successful decision-making.

Conclusions

The information collected on drill holes was used for developing a simplified RMR classification named CLASS. On this classification, those segments described as voids/cracks/gaps (named FD) without recovering (S/R or R/S) or soil were classified as having a value equal to zero. In addition, previous treatment of the used databank was performed to withdraw any conflicting information that could harm the quality of the geomechanical modeling.

Using a machine learning technique (Random Forest), it was possible to identify, among all considered variables for rock mass classification, the one that better explains the geomechanical behavior of the rock mass of all subsectors of the North of an underground mine is the RQD. Despite being already expected, this result did not result in disregarding the other variables for the analysis of the observed mechanical behavior.

The proposed method for developing geomechanical modeling using drill hole data and free programming software (R software) was satisfactory, both for the visualization of the results and the quality of the information itself. However, it is still necessary for the specific study area to calibrate and validate the results with field data or in a new drill hole database. Meanwhile, it became clear that the potential of the developed tool for geomechanical modeling is encouraging and should be better explored in future.

References

1. Landim, P. 2003. *Statistical analysis of geological data*. Sao Paulo: UNESP.
2. Kaufmann, O., and T. Martin. 2009. Reprint of "3D geological modelling from boreholes, cross-sections and geological maps, application over former natural gas storages in coal mines" [Comput. Geosci. 34 (2008) 278–290]. *Computers & Geosciences* 35 (1): 70–82.
3. Hadjigeorgiou, J. 2012. Where do the data come from? *Mining Technology* 121 (4): 236–247.
4. Team, R.C. 2013. *R: A language and environment for statistical computing*.
5. Bhattacharya, S., et al. 2019. Application of predictive data analytics to model daily hydrocarbon production using petrophysical, geomechanical, fiber-optic, completions, and surface data: A case study from the Marcellus Shale, North America. *Journal of Petroleum Science and Engineering* 176: 702–715.
6. Silva, C.D.R.L., and J.A.S. Centeno. 2007. Study of the interpolation of the inverse distance to a power method. In *Simpósio Brasileiro de Geomática*, 57–62. São Paulo: Presidente Prudente.
7. Auret, L., and C. Aldrich. 2012. Interpretation of non-linear relationships between process variables by use of random forests. *Minerals Engineering* 35: 27–42.
8. Sajid, A., R. Rudra, and G. Parkin. 2013. Systematic evaluation of kriging and inverse distance weighting methods for spatial analysis of soil bulk density. *Canadian Biosystems Engineering* 55.
9. Setianto, A., and T. Triandini. 2013. Comparison of kriging and inverse distance weighted (IDW) interpolation methods in lineament extraction and analysis. *Journal of Applied Geology* 5 (1).
10. Li, J., and A.D. Heap. 2008. A review of spatial interpolation methods for environmental scientists.
11. Girolamo Neto, C.D. 2014. Potential of data mining for mapping of coffee plantation areas. Monography for SER-300 course. Cited 2021 June 4. Available from: http://wiki.dpi.inpe.br/lib/exe/fetch.php?media=ser300:alunos2014:cesare_monografia.pdf.
12. Breiman, L. 2001. Random forests. *Machine Learning* 45 (1): 5–32.
13. Strobl, C., et al. 2007. Bias in random forest variable importance measures: Illustrations, sources and a solution. *BMC Bioinformatics* 8 (1): 1–21.
14. Harris, J.R., et al. 2014. A comparison of different remotely sensed data for classifying bedrock types in Canada's Arctic: Application of the robust classification method and random forests. *Geoscience Canada: Journal of the Geological Association of Canada/Geoscience Canada: Journal de l'Association Géologique du Canada* 41 (4): 557–584.
15. Lorenzett, C.D.C.T.A. 2016. Comparative study between algorithms of data mining, random forest and J48 for decision making. *Simpósio de Pesquisa e Desenvolvimento em Computação* 2 (1).
16. Couteiro, T.A.F. 2010. Relational statistics methods for prevision of medical results. In *Integrated master in engineering, IT and computing*, 67. Porto: Faculdade de Engenharia, Universidade do Porto.

17. Ou, C.-M., W.-J. Hwang, and S.-M. Yang. 2013. FPGA-based online learning hardware architecture for kernel fuzzy c-means algorithm.
18. Harris, J., and E.C. Grunsky. 2015. Predictive lithological mapping of Canada's North using Random Forest classification applied to geophysical and geochemical data. *Computers & Geosciences* 80: 9–25.
19. Breiman, L. 1993. *Classification and regression trees*. New York: Chapman & Hall. ISBN 0412048418.
20. Carvalho, H.M. 2014. Machine learning applied to data mining: Decision trees. Undergraduation conclusion thesis, in software engineering, Universidade de Brasília, Brasília, 90.
21. Louppe, G. 2014. Understanding random forests: From theory to practice. arXiv preprint arXiv: 1407.7502.
22. Oshiro, T.M. 2013. An approach for building a single tree from a random forest for classifying the basis of genetic expression. In *Biocomputing*. São Paulo: Universidade de São Paulo.
23. Soofastaei, A. 2020. *Data analytics applied to the mining industry*. CRC Press.
24. Matin, S., et al. 2018. Variable selection and prediction of uniaxial compressive strength and modulus of elasticity by random forest. *Applied Soft Computing* 70: 980–987.
25. Grunsky, E. 2002. R: A data analysis and statistical programming environment—An emerging tool for the geosciences. *Computers & Geosciences* 28 (10): 1219–1222.
26. Pebesma, E.J. 2004. Multivariable geostatistics in S: The gstat package. *Computers & Geosciences* 30 (7): 683–691.
27. Pebesma, E. 2012. spacetime: Spatio-temporal data in R. *Journal of Statistical Software* 51 (7): 1–30.
28. Pebesma, E., and R.S. Bivand. 2005. S classes and methods for spatial data: The sp package. *R News* 5 (2): 9–13.
29. Hijmans Robert, J. 2020. *raster: Geographic data analysis and modeling*. R package version 3.1-5.
30. Bivand, R., and C. Rundel. 2017. *RGeos: Interface to geometry engine-open source ('GEOS')*. R package version 0.3–26.
31. Sarkar, D. 2008. *Lattice: Multivariate data visualization with R*. Springer Science & Business Media.
32. Komsta, L., and F. Novomestky. 2015. *Moments, cumulants, skewness, kurtosis and related tests*. R package version 0.14.
33. Hengl, T., et al. 2015. plotKML: Scientific visualization of spatio-temporal data. *Journal of Statistical Software* 63 (5): 1–25.
34. Hengl, T., et al. 2017. *GSIF: Global soil information facilities*. R package version 0.5–4 [Software].
35. Wright, M.N., and A. Ziegler. 2015. ranger: A fast implementation of random forests for high dimensional data in C++ and R. arXiv preprint arXiv:1508.04409.
36. Ribeiro Jr., P.J., et al. 2020. *Package 'geoR'*.
37. Sievert, C. 2020. *Interactive web-based data visualization with R, plotly, and shiny*. CRC Press.
38. Signorell, A., et al. 2020. *DescTools: Tools for descriptive statistics*. R package version 0.99.36.
39. Wickham, H. 2018. *nycflights13: Flights that departed NYC in 2013*. R package version 1(0).
40. Revelle, W. 2019. *psych: Procedures for personality and psychological research*. Northwestern University. Version 1: 12.
41. Hadley, W. 2016. *Ggplot2: Elegant graphics for data analysis*. Springer.
42. Wickham, H., R. François, L. Henry, and K. Müller. 2020. *dplyr: A grammar of data manipulation*. R package version 0.8.4.
43. Kuhn, M., et al. 2020. *caret: Classification and regression training*. R package version 6.0-86. Available at: https://cran.r-project.org/web/packages/caret/caret.pdf. Accessed 20 Mar 2020.
44. Wei, T., and V. Simko. 2017. *R package "corrplot": Visualization of a correlation matrix* (version 0.84).
45. Baddeley, A., E. Rubak, and R. Turner. 2015. *Spatial point patterns: Methodology and applications with R*. CRC Press.

46. Ligges, U., and M. Mächler. 2002. Scatterplot3d—An R package for visualizing multivariate data. Technical report.
47. Grosjean, P., E. Lecoutre, and J.C. Faria. 2019. *SciViews-R*. Citeseer.
48. Calaway, R., et al. 2019. *doParallel: Foreach parallel adaptor for the 'parallel' package*. R package version 1.0. 14. Vienna: Comprehensive R Archive Network.
49. Schloerke, B., et al. 2020. *GGally: extension to 'ggplot2'*. R package version 1.5.0. R Foundation for Statistical Computing.
50. Meyer, D., et al. 2019. *Misc functions of the statistics department, probability theory group*. R package version 1.
51. Therneau, T., et al. 2015. Package 'rpart'. Available online: http://cran.ma.ic.ac.uk/web/packages/rpart/rpart.pdf. Accessed 20 Apr 2016.
52. Leisch, F., and E. Dimitriadou. 2010. *mlbench: Machine learning benchmark problems*. R package version 2.0-2010. URL http://CRAN.R-project.org/package=mlbench.
53. Liaw, A., and M. Wiener. 2002. Classification and regression by randomForest. *R News* 2 (3): 18–22.
54. Hothorn, T., K. Hornik, and A. Zeileis. 2006. Unbiased recursive partitioning: A conditional inference framework. *Journal of Computational and Graphical Statistics* 15 (3): 651–674.
55. Venables, W., and B. Ripley. 2002. *Modern applied statistics with S*, 4th ed. World.
56. Wickham, H., et al. 2019. *Package 'readxl'*.
57. Friendly, M., et al. 2014. *Lahman: Sean Lahman's baseball database*. R package version 3.0-1.
58. Vaidyanathan, R., et al. 2019. *htmlwidgets: HTML widgets for R*. R package version 1.5.1.
59. Webber, T., J.F.C.L. Costa, and P. Salvadoretti. 2013. Using borehole geophysical data as soft information in indicator kriging for coal quality estimation. *International Journal of Coal Geology* 112: 67–75.
60. Jakubec, J., and G.S. Esterhuizen. 2007. Use of the mining rock mass rating (MRMR) classification: Industry experience. In *Proceedings of the International Workshop on Rock Mass Classification in Underground Mining*.
61. Schepers, R., et al. 2001. Application of borehole logging, core imaging and tomography to geotechnical exploration. *International Journal of Rock Mechanics and Mining Sciences* 38 (6): 867–876.
62. Bieniawski, Z. 1988. Towards a creative design process in mining. *Mining Engineering* 40 (11).
63. Pells, P., et al. 2017. Rock quality designation (RQD): Time to rest in peace. *Canadian Geotechnical Journal* 54 (6): 825–834.
64. Calle, M.L., and V. Urrea. 2011. Letter to the editor: Stability of random forest importance measures. *Briefings in Bioinformatics* 12 (1): 86–89.
65. Verikas, A., A. Gelzinis, and M. Bacauskiene. 2011. Mining data with random forests: A survey and results of new tests. *Pattern Recognition* 44 (2): 330–349.
66. Mohammady, M., H.R. Pourghasemi, and M. Amiri. 2019. Land subsidence susceptibility assessment using the random forest machine-learning algorithm. *Environmental Earth Sciences* 78 (16): 1–12.
67. Zaimes, G.N., D. Gounaridis, and E. Symenonakis. 2019. Assessing the impact of dams on riparian and deltaic vegetation using remotely-sensed vegetation indices and Random Forests model. *Ecological Indicators* 103: 630–641.
68. Tiryaki, B. 2008. Predicting intact rock strength for mechanical excavation using multivariate statistics, artificial neural networks, and regression trees. *Engineering Geology* 99 (1–2): 51–60.
69. Dong, L.-J., X.-B. Li, and K. Peng. 2013. Prediction of rockburst classification using Random Forest. *Transactions of Nonferrous Metals Society of China* 23 (2): 472–477.
70. Xie, Q., and K. Peng. 2019. Space-time distribution laws of tunnel excavation damaged zones (EDZs) in deep mines and EDZ prediction modeling by random forest regression. *Advances in Civil Engineering* 2019.
71. Herwanger, J. 2019. 4D geomechanical simulations for field development planning. *Oil Geomechanic* 3 (1): 34–44.